From drawing by Thomas Sully

Old Swedes' (Wicaco or Gloria Dei) Church, Philadelphia

Peter Kalm's Travels In North America

The English Version of 1770

Revised from the original
Swedish and edited by

ADOLPH B. BENSON

COMPLETE IN ONE VOLUME

DOVER PUBLICATIONS, INC.
New York

Published in Canada by General Publishing Company, Ltd., 30
Lesmill Road, Don Mills, Toronto, Ontario.
Published in the United Kingdom by Constable and Company, Ltd.,
10 Orange Street, London WC2H 7EG.

This Dover edition, first published in 1987, is a republication in a
single volume of the two-volume 1966 Dover edition, which was an
unabridged republication of the work originally published by Wilson-
Erickson, Inc., in 1937.
The present edition contains two additional illustrations from
Forster's English edition of 1770–71 (see plates facing pages 110 and
253). The map from the same English version is here reproduced
approximately two and one-half times larger than in the 1937 edition.

Manufactured in the United States of America
Dover Publications, Inc., 31 East 2nd Street, Mineola, N.Y. 11501

Library of Congress Cataloging-in-Publication Data

Kalm, Pehr, 1716–1779.
Peter Kalm's travels in North America.

Includes only those portions of the original work dealing with the
United States and Canada.
Reprint. Originally published: New York: Wilson-Erickson, 1937.
"A bibliography of Peter Kalm's writings on America" p.
Includes index.
1. United States — Description and travel — To 1763. 2. Natural
history — United States. 3. Canada — Description and travel — To
1763. 4. Natural history — Canada. 5. Kalm, Pehr, 1716–1779 —
Diaries. I. Benson, Adolph B. (Adolph Burnett), 1881–1962.
II. Title. III. Title: Travels in North America. IV. Title: America
of 1750.
E162.K173 1987 917.4'041 87-541
ISBN 0-486-25423-2 (pbk.)

Title Page of Dutch Translation

TABLE OF CONTENTS

LIST OF ILLUSTRATIONS

INTRODUCTION

With the death of Charles XII in 1718 Sweden turned gladly from her disastrous wars to constructive industry and agriculture. The number of her scientific investigators increased and her peace-time pursuits flourished. To stimulate and direct scientific study the Swedish Academy of Sciences was founded, 1739, and soon its leaders became conscious of the excellent opportunities for original, practical research in America.

The primary function of the Academy, in the last analysis, was public service—to further, by suggestions and study, the economic welfare of the country and increase the wealth and prosperity of its people. It corresponded roughly to a modern agricultural experiment station. One of the main problems considered by the Academy was how to enlarge the number of varieties of useful plants and trees in Sweden by the importation and planting of foreign seeds. To this end it was decided to send a naturalist either to Siberia, Iceland, or to some other northern country, to obtain seed-material of new herbs and trees hardy enough to thrive on Swedish soil. More food and fodder were needed for folk and cattle, and industrial promoters were searching for extraneous dye-stuffs. A specific object of the proposed expedition—looking forward to an independent silk industry—was to find a mulberry tree that would endure the severity of the Swedish climate. So, at the suggestion of none other than Linné (Linnæus), well-informed and far-sighted that he was, it was finally agreed that the Academy should send a representative to North America on this important mission. Who would be the best man for the enterprise?

One hundred and ninety years ago the mere trip itself to the western continent was a matter of some moment and anxiety, how much more so an official scientific journey undertaken with great economic responsibilities and attended by innumerable uncertainties. It was not an easy task and required a man of courage, knowledge, industry, discrimination, travelling experience, and strong physique. Successful pioneer work among the Colonists demanded a kindly and diplomatic personality, a scientist with an alert, open, and reasonably suspicious mind, one who could separate the chaff from the wheat in the narratives of the natives and settlers, and who possessed an objective, independent judgment, a keen observation and a faculty for scrupulous accuracy in the recording of his findings. A candidate with these qualifications was found, recommended and endorsed, the unanimous choice for the expedition being Pehr Kalm (generally Englished into Peter K.), a member of the Swedish Academy of Sciences and a pupil of Linné.[1]

Pehr Kalm, the posthumous son of a clergyman from Österbotten, Finland, was born 1716 in the province of Ångermanland, Sweden, whither the family had fled during the Carolinian wars. In 1735 he matriculated as a graduate student at the Åbo Academy, Finland—which then belonged to Sweden,—and displayed early a leaning toward theology. Bishop Johan Brovallius, however, a divine and naturalist of broad learning, saw greater possibilities for the young man in the natural sciences and encouraged his budding interest in them. Through the bishop's influence Kalm was befriended by Baron Sten Carl Bjelke, who defrayed the poor student's expenses during his scientific travels in Finland and Sweden. He soon became, at Uppsala, a pupil of the renowned Carl von Linné, who helped the promising naturalist in every way. Kalm's travels were extended and an account of his results published. In 1744 he accompanied his teacher on a tour through Russia and Ukraine, and was publicly acknowledged by Linné in an academic thesis the same year. Kalm appeared to be particularly interested in medicinal and dye-yielding plants, and in all his scientific work evinced a distinct utilitarian tendency. His science had to show promise of something useful. His motto was *Nisi utile est quod facium, stulta est gloria.* He collected seeds for the Academic Garden at Uppsala and studied the harm or benefit of the various plants. Special attention was paid to toxic herbs. In his reports he proved to be unusually severe in his judgment of careless farmers, a fact which we should remember in reading his description of American tillers of the soil.

---

[1] Kalm's qualifications are enumerated in a letter by Linné of January 11, 1746. See *Bref och Skrifvelser af och till Carl von Linné*, edited by Th. M. Fries, 1908, I, II, 59.

His election to the Swedish Academy of Sciences came in
1745; the following year he was granted, without the usual academic examination, the title of docent in "natural history and
economy," which indicated practical husbandry or agriculture;
and in 1747 he was made professor in that subject at Åbo, the
first one appointed to that position. But Kalm was given leave
of absence almost immediately to accept the flattering offer of
going to America.

While in New Jersey, incidentally, the pastor of the Swedish
congregation at Raccoon, Johan Sandin, having died, Kalm, who
still retained a penchant for things religious and ecclesiastical,
often substituted in the pulpit when no regular clergyman could
be obtained. And his interest proved also personal, for in 1750
he married the pastor's widow, b. Anna Magaretha Sjöman.

After his return to Sweden and Finland, in 1751, Kalm, besides
attending to his academic lectures, labored diligently for many
years in taking care of the American plants which he had brought
home—for there was no public garden in Åbo—and in preparing the diary notes on his investigations for the press. In 1757,
while still retaining his college professorship in the natural
sciences, he was ordained a Lutheran clergyman—which in his day
brought with it certain practical academic benefits,—was honored
by the University of Lund with a doctor's degree in theology,
1768, and in 1775 was spoken of in some circles as a suitable
candidate for the office of Bishop of Åbo. In the interim he had
as professor presided at 146 academic disputations, including six
on American topics, such as Esquimaux and dye plants. An
offer to become professor of botany at St. Petersburg at a salary
of a thousand rubles per annum was rejected, and he was elected
to membership in some native and foreign societies. A royal
Swedish decoration also fell to his lot, one of the first clergymen,
it is said, to have been so honored. Kalm died in 1779.

"Kalm's journey to the New World," with which we are most
concerned, "was not without its exciting adventures and annoying
features. First of all, it was no small matter to collect the necessary funds for such a costly expedition, making due allowances
for delays, accidents, and unfavorable rates of exchange. Much
was contributed by stipends and other academic gifts, for the
scientist was to travel at 'public expense', but Kalm was eventually

obliged to draw on his own savings as well, and frankly confessed upon his return that the amount he had left for buying furniture and settling down 'could easily be counted.' At all events, animated by a hopeful, enterprising spirit and accompanied by an expert gardener, Lars Jungström, Kalm started from Uppsala, 'in the name of the Lord right after dinner,' on the fifth of October, 1747 (Oct. 16, new style) for England via Gothenburg. Embarking on the 11th of December, a storm drove his vessel against the coast of Norway, where Kalm, while waiting for another boat, made unintentional but opportune investigations until February 8, 1748. On February 17 we find our travellers in London, now facing the cold prospects of waiting another six months, because of the lack of ships, before securing passage for America. But as always, Kalm made good use of his time, studying English conditions, making the acquaintance of eminent Englishmen, improving his own knowledge of the English language, and obtaining valuable letters of introduction to prominent Colonial families. Finally, on August 5, Kalm and his companion were duly installed on the *Mary Gally,* Captain Lawson, bound for the new continent. This part of the voyage proved unusually pleasant, and the naturalist had good opportunity to study the seaweed, the fish, fowl, porpoises, and other phenomena of the sea, and to take regular meteorological observations using the newly invented Swedish centigrade thermometer. On September 13 the *Mary Gally* ran onto a sandbar off the coast of Maryland, but managed to get afloat again, and landed in Philadelphia two days later."

"Naturally Kalm had no definite itinerary with fixed dates mapped out for his work in America; but, in brief, his explorations extended to the States of Pennsylvania, New York and New Jersey, and to southern Canada. The remainder of the first year, 1748, was spent in the States attending to the more specific duties of his mission, the collecting and dispatching of seeds to Sweden. The following year he continued his wanderings to Lake Champlain and Canada, returning to 'New Sweden' about Christmas time. In 1750 he explored western Pennsylvania and penetrated northward to Niagara Falls. In October the botanist came back to Philadelphia, which he left on February 13, 1751, for Europe, regretting that his work was yet unfinished. He saw America

TRAVELS

INTO

NORTH AMERICA;

CONTAINING

Its NATURAL HISTORY, AND

A circumſtantial Account of its Plantations
and Agriculture in general,

WITH THE

CIVIL, ECCLESIASTICAL AND COMMERCIAL
STATE OF THE COUNTRY,

The MANNERS of the INHABITANTS, and ſeveral curious
and IMPORTANT REMARKS on various Subjects.

By PETER KALM,

Profeſſor of Oeconomy in the Univerſity of *Aɩbo* in Swediſh
Finland, and Member of the *Swediſh* Royal Academy of
Sciences.

TRANSLATED INTO ENGLISH

By JOHN REINHOLD FORSTER, F.A.S.

Enriched with a Map, ſeveral Cuts for the Illuſtration of
Natural Hiſtory, and ſome additional Notes.

VOL. I.

WARRINGTON:

PRINTED by WILLIAM EYRES.

MDCCLXX.

Title Page of English Translation

En
Resa
Til
Norra AMERICA,
På
Kongl. Swenska Wetenskaps
Academiens befallning,
Och
Publici kostnad,
Förrättad
Af
PEHR KALM,
Oeconomiæ Professor i Åbo, samt Ledamot af
Kongl. Swenska Wetenskaps-Academien.

Tom. I.

Med Kongl. Maj:ts Allernådigste *Privilegio.*

STOCKHOLM,
Tryckt på LARS SALVII kostnad 1753.

Title Page of Swedish Original
(Yale Copy)

for the last time on the 18th; came in sight of England on March 23; was the victim of an accident on the Thames soon after so that he was forced to proceed to London by land; but eventually arrived back in Stockholm June 3 [May 23, old style], having been absent almost four years." [1] The printed record of his travels began to appear two years later and in the meantime Linné had honored his pupil by giving to the American mountain laurel the scientific name of *Kalmia latifolia*.

Kalm while in the Colonies had become a welcome guest in the best society and private homes, and public buildings were opened to him under conditions which involved more than ordinary privileges. Kalm tells us, for instance, that he was allowed to borrow books without charge from the recently established Philadelphia library, a no mean privilege for that time. His unquestionable thirst for knowledge extended actively far beyond his primary interest or profession, hence the value of his records. He was especially interested in public institutions like churches, hospitals and convents. Kalm had, too, the advantage of frequent personal contacts with Benjamin Franklin, the intellectual leader, of course, among the Colonists, at whose house the young Swede found a congenial host and ready counsellor. He had also the opportunity of interviewing the descendants of the earliest Delaware Swedes, one of whom in 1749 was 91 years old, and of sound mind, thus obtaining valuable historical information from the middle of the seventeenth century.

"Large indeed is the scope of subjects that attract Kalm's attention,[2] and striking the simplicity, straightforwardness, poise, conscientiousness", and the almost humorous naiveté "with which

---

[1] Quotations in this Introduction are from my article on "Pehr Kalm's Journey to North America" in *The American-Scandinavian Review* for June, 1922.

Kalm soon began to dream of a second trip to America, but it never materialized. In a letter to Linné of February 11, 1753, Kalm suggests returning to North America to finish his work, and says he is willing to be ordained, so that he can at the same time—in part, for financial reasons—serve as pastor in New Sweden.

[2] The English translator of Kalm's *Travels* was much impressed by the scope of information in them, and in his version (in the custom of the time) expanded the relatively simple title of the original into the following elaborate but significant summary: "Travels into North America; containing its natural history, and a circumstantial account of its plantations and agriculture in general, with the civil, ecclesiastical and commercial state of the country, the manners of the inhabitants, and several curious and important remarks on various subjects." Of course, Forster was anxious, also, to sell his translation; hence the potent need for an impressive title.

he makes the heterogeneous entries of his observations in his diary. The construction and operation of a cider-press or a new type of fence gets the same relative space as the description and classification of a flower, rare shrub, lumber tree, cereal plant, or medicinal herb. Kalm is as much interested in the preserving of mushrooms and in the preparation of delicate dishes of food as in the character and distribution of diseases in America. All receive proper attention. Mineral and ore deposits are perhaps more valuable to him, but hardly so fascinating as birch canoes. Geography, topography, American history and antiquity are treated by our diarist; and domesticated animals are not forgotten. Architecture and building materials; servants' wages, the medium of exchange, and the monetary system; windmills, fortresses, and beaver dams; word formation in the Algonquin Indian dialect; Roman Catholicism in Canada; the probable reason for the prevalence of poor teeth among Americans—these and dozens of other topics, connected chronologically, are thrown together, as it were, and yet discussed with lucidity, forming an exceptionally readable report. Kalm sees the thoughtlessness in agriculture, and listens to the frequent complaints about the disappearance of fish and game because of ruthless deforestation. He deprecates the number of destructive insects that abound on our continent, and soon becomes aware of the changeableness of the climate in eastern America and its dangers to public health. Though coming from the frigid North, Kalm himself suffered not a little during the rather severe winters spent here."

While pure science and practical usefulness go ever hand in hand with Kalm, he is perhaps "most all interested in some of our distinctly American animals. The bullfrog's vocal organs and the intensity of his croaking; the twilight call of the whip-poor-will; the singular development of the seventeen-year locust; the poisonous sting of the New Jersey mosquito compared with that of the European species; and the habits of black snakes— all are treated with solicitous fullness." Descriptions of the latter, incidentally, demanded special caution and tact because of the numerous conflicting reports about their strength, size, ferocity and alleged attacks upon women and children. Virtually bombarded with hair-raising tales—some of them shockingly realistic— about the ravages of these swarthy reptiles, by respectable citizens whose veracity might not be openly questioned at least, the Swedish observer was considerably puzzled, particularly when the stories apparently contradicted his first-hand information. In such a case, however, Kalm made a compromise in his final report: he gave the native's tale *verbatim,* throwing the responsibility for it on the narrator by appending his name, and then by way of personal reservation, or opinion, stated modestly that according to his own experience every black snake that he had ever seen was quite willing to run away from one of human kind, if he had a chance, though he might show fight in the mating season. The rest was left, as often it had to be, without direct personal comment, to the judgment of the reader.

The Åbo professor—to give another specific illustration— makes a telling comparison between the women of Montreal and those of Quebec. The latter, especially the unmarried ones, were vain and lazy, says Kalm; and although the fair sex in Montreal also thought a little too much of themselves, they were in general more modest and handsome, domestic and intelligent. Besides, they would never hesitate to go out and purchase a watermelon or a pumpkin and carry it home themselves, a fact which Kalm is inclined to believe shows a commendable quality. His references to the brevity of women's skirts in certain parts of Canada bear the stamp of an interested but dispassionate neutrality.

It is obvious that Pehr Kalm's activity in Colonial America should command the attention of a larger circle of readers, instead of it being limited, as it is so often, to a comparatively small group of specializing scholars. He has much more illuminating material to offer us about our early history than is generally supposed. It is true that the interest in him has recently increased, but he is still neglected, and many people have never heard of him. Moreover, the fact that he was apparently the first efficient naturalist to conduct comprehensive and methodical studies in the Colonial settlements *and make his findings known in a large way* is in itself an achievement worthy of more attention. North America was tolerably well known of course before Kalm's time, but extensive scientific reports about it had been scarce.

The official Swedish account of Kalm's American trip, *En Resa til Norra America,* appeared in three volumes in Stockholm,

1753-1761. The manuscript of a fourth volume was finished but never published and was finally destroyed in a fire at Åbo, 1827. A few years ago, however, the diary notes for this unpublished part were discovered in the university library at Helsingfors, Finland, by George Schauman and were edited and published in their original form by Fred. Elfving, 1929. This part is now offered in English for the first time. In the interim the first three parts had been republished also, in Swedish, by Elfving and Schauman, 1904-1915.

The original travelogue was translated into German almost immediately, by three different men—Carl Ernst Klein and the brothers J. P. and J. A. Murray; the English translation by John Reinhold Forster, which omitted the part about England, appeared in 1770-1771, at Warrington; and a second, abridged edition of this was printed in London the following year. Another printing made its appearance in 1812, but the part dealing with England, and translated by Joseph Lucas, was not published until 1892. A remarkable Dutch version based on Forster's English and the Murrays' German translation, with copper etchings, appeared in Utrecht, 1772, in two volumes. A French work by J. P. Rousselot de Surgy was to a large extent a translation of Kalm's *Travels*, and in 1880 L. W. Marchand published in the memoirs of the Montreal historical society his French version of the same material, but with the part treating of the United States much condensed and the emphasis placed, as we might expect, upon Kalm's sketches of Canada.—Kalm's work was more or less extensively reviewed or quoted in all countries that became acquainted with his *Travels*, and for the most part favorably.

Kalm was the first man to describe the Niagara Falls in English from first-hand information, and many versions of his account were published between September 20, 1750 (in the *Pennsylvania Gazette*) and 1921, when Kalm's description reappeared in C. M. Dow's *Anthology and Bibliography of Niagara Falls*. It is maintained that this account was, also, translated into French and German, but the present writer has never found such a translation.

It should be mentioned here, too, that in addition to the above, Kalm published in Swedish between the years 1749 and 1778 seventeen scientific articles in the transactions of the Swedish Academy of Sciences (*Kungliga Vetenskaps-Academiens Handlingar*) on topics that were either wholly or in part based on his American studies. They dealt with such subjects as maple sugar, Indian corn, American ticks, grasshoppers, wild doves, spruce beer, walnut trees, and rattlesnakes. Besides, among the academic theses that he directed at Åbo there were, as noted above, six items of *Americana*. (For a more complete bibliography see "A Bibliography of Peter Kalm's Writings on America" at end of this work, pp. 770 ff.).

The present version is the first exclusively American edition of Kalm's *Travels,* in English—for the Canadian one of 1880 is in French—and the first one dealing with the part on United States and Canada to appear in this country.[1] It should go far to make this notable source accessible both to the historian and to the general reader. The part on Norway and England has been omitted for obvious reasons: we are interested here only in America. The various translations have long been out of print and are now available only in the largest libraries. Even the relatively recent reprint of the original, in Finland, is getting scarce. So a new edition in our own tongue is highly desirable. The hitherto untranslated portion of Kalm's work has been done into English by Miss Edith M. L. Carlborg of the Brown University Library and the present editor. The remainder, and by far the larger part, is based on Forster's translation, which has been carefully revised, however, and compared with the Swedish original, since Forster (and his precocious son, who did most of the work) used the German version. Except for desirable or necessary changes in diction and vocabulary no attempt has been made to alter seriously the translator's style or language where his facts and renderings are correct and intelligible. In fact, the editor feels that a preservation of the eighteenth century idiom, flavor, quaintness and general mode of expression may, in this case, be an added attraction. But an effort has been made to remove glaring instances of obsolete words and to simplify and smoothen highly involved, awkward and ponderous sentences.

---

[1] Lucas's translation, published in London and New York, dealt only with England, as indicated above.

Most of the parts left out by Forster have been reintroduced, to make the translation as complete as possible, the spelling of some words has been modernized, and the emasculated parts restored. The English translator reveals a tendency to make the work seem favorable to the English; the present editor has therefore tried to correct any prejudiced alterations or omissions of the original text. Unless otherwise stated the notes and corrections are by the latter. Where Forster's notes have been retained they are indicated by an -F. A serious effort has been made to make the index of important references complete. Notes, either by Kalm himself or by any one of his translators or editors—which we shall reproduce, with proper acknowledgment, wholly or in part, as we proceed— will clarify most obscure points of the text and explain such attendant circumstances as the reader must know; but one other fact should be emphasized again before reading this travelogue. Kalm was a scientist rather than a writer or stylist, and he was extremely anxious to please his academic benefactors by giving a practical account of himself in his reports. As a result literary embellishments are wanting, as Kalm himself admits in his own Introduction. The sober, all-absorbing statement of fact, as he sees it or as it is reported to him, is the determining requisite, and style is wholly secondary. But the very logical simplicity, the childish honesty, and occasional credulity, which are evidenced in his descriptions, coupled with an obvious and expressed desire to be of some use to his fatherland, produce an effect and atmosphere which eliminate in the reader any yearning for flowery language or strict sequence of scenes. In the last 200 pages especially, which reproduce essentially unaltered the unpolished diary notes of a weary traveller, we should not expect to find any pretensions to humor, strict logical arrangement or literary style.

The more scientifically-minded readers who are interested in concrete facts about Kalm's contributions to the knowledge of American plants will find such information summarized in an article by H. O. Juel and John W. Harshberger entitled "New Light on the Collection of North American Plants made by Peter Kalm," which appeared in the *Proceedings of the Philadelphia Academy* (The Academy of Natural Sciences of Philadelphia) for the year 1929.[1] We shall quote below a few sentences from this résumé.

In 1751 L. J. Chenon published a dissertation *Nova Plantarum Genera* in which "we are told that Kalm had recently returned from America bringing with him a huge treasure of conchylia, insects, and amphibia, in addition to dried and living plants and also seeds. In *Species Plantarum*, 1753, Linnæus described 700 North American plants, and in 90 cases we find Kalm mentioned as the collector of the species. . . . Not all of the 90 species were new, but about 60 species undoubtedly were. . . . In the later works of Linnæus the total number of North American plants described by him increased until the grand total reached about 780 species, of which 60 species were founded upon specimens collected by Kalm. Professor H. O. Juel in a paper entitled 'Early Investigations of North American Flora with Special Reference to Linnæus and Kalm'[2] lists all of the North American plants first described by Linnæus, and in this lists the plants collected by Kalm."

Three sets of Kalm's North American plants were preserved originally. One is in the possession of the Linnæan Society of London; one was probably destroyed in the Åbo fire of 1827; and the third is in the Museum of Natural History at Uppsala. The collection mounted in Åbo by Kalm contains about 380 specimens. The list of Kalm's North American plants in the Botanical Museum in Uppsala is printed in the above article and amounts to over 300 specimens. Kalm was the author of the three genera: Gaultheria, Lechea, and Polymnia. His "Flora of Canada," mentioned by Linné in his *Species Plantarum* was probably lost in manuscript in the Åbo fire.

As editor I acknowledge a debt of gratitude to *Svenska Litteratursällskapet i Finland* and to Professor Fred. Elfving for permission to translate the part edited by him; to Dr. Emil Lindell of Växjö, Sweden, for invaluable bibliographical information; to Mr. Nils Gösta Sahlin for help in matters of form; to Dr. Emil L. Jordan of the New Jersey College for Women for gift of rare illus-

---

[1] LXXXI, 297-303.
[2] In *Svenska Linnésällskapets Årsskrift*, III, 1920.

tration material; to Mr. Henry D. Paxson, Jr., of Philadelphia, for permission to reproduce cuts from *Where Pennsylvania History Began,* a work by his father, the late Col. Henry D. Paxson; and to my wife for preparing the index and aiding in the revision of the manuscript.

New Haven, Connecticut,
October, 1936.

ADOLPH B. BENSON

PETER KALM'S TRAVELS
IN NORTH AMERICA

PETER KALM'S TRAVELS

AUGUST THE 5TH, 1748

TOGETHER with my servant Lars Jungström (who joined to his abilities as gardener a tolerable skill in mechanics and drawing) I went on board the *Mary Gally,* Captain Lawson, at Gravesend, bound for Philadelphia, and though it was as late as six o'clock in the afternoon, we weighed anchor and sailed a good way down the Thames before we stopped again.[1]

AUGUST THE 6TH

Very early in the morning we resumed our voyage, and after a few hours sailing we came to the mouth of the Thames, where we turned into the channel and sailed along the Kentish coast, which consists of steep and almost perpendicular chalk hills, covered at the top with some soil and a fine verdure, and including strata of flint, such as are frequently found in this type of chalk hills in the rest of England. And we were delighted in seeing on them excellent grain fields, covered for the greatest part with ripening wheat.

At six o'clock at night we arrived at Deal, a small well-known town, situated at the entrance of a bay exposed to the southern and easterly winds. Here commonly the outward bound ships provide themselves with greens, fresh victuals, brandy, and many more articles. This trade, a fishery, and in the last war the equipping of privateers, has enriched the inhabitants.

AUGUST THE 7TH

When the tide was out I saw numbers of fishermen resorting to the sandy shallow places, where they find small round eminences caused by the excrements of the log worms, or sea worms,

---

[1] The account of Kalm's voyage over the ocean has, as in Forster's translation, been condensed and rearranged.

(*Lumbrici marini* L) who live in the holes leading to these hillocks, sometimes eighteen inches deep, and they are then dug out with a small three-tined iron fork and used as bait.

AUGUST THE 8TH

At three o'clock we tided down the channel, passed Dover, and saw plainly the opinion confirmed of the celebrated Camden [1] in his *Britannia,* that here England had formerly been joined to France and Flanders by an isthmus. Both shores form here two opposite points; and both are formed by the same chalk hills, with the same configuration, so that a person acquainted with the English coasts and approaching those of Picardy afterwards, without knowing them to be such, would certainly take them to be the English ones.[2]

AUGUST THE 9TH—12TH

We tided and alternately sailed down the channel and passed Dungness, Fairlight, the Isle of Wight, Portsmouth, the Peninsula of Portland and Bolthead, a point behind which Plymouth lies, and during all this time we had very little wind.

AUGUST THE 13TH AND 14TH

Towards night we sailed out of the English channel into the Bay of Biscay. The following day we had contrary wind, and this increased the rolling of the ship, for it is generally remarked that the Bay of Biscay has the greatest and broadest waves, which are of equal size with those between America and Europe. They are commonly half an English mile in length, and have a height proportional to it. The Baltic and the North Sea have on the contrary short and broken waves.

Whenever an animal is killed on board the ship, the sailors commonly hang some fresh pieces of meat for a while in the sea, for it is said it then keeps better.

---

[1] William Camden (1551-1623), English antiquary and historian. His *Britannia* had appeared in 1586.

[2] The same opinion has been confirmed by Mr. Buffon in his *Hist. Naturelle.* tom. I, art. XIX.—F.

August the 15th

The same swell of the sea continued, but the waves began to smooth, and a foam swimming on them was said to forebode, in calm weather, a continuance of the same for a few days.

About noon a northeasterly breeze came up, in the afternoon it blew more, and this gave us a fine spectacle; for the great waves rolled the water in great sheets in one direction, and the northeasterly wind curled the surface of these waves in a different one. By the beating and dashing of the waves against one another, with a more than ordinary violence, we could see that we passed a current whose direction the captain could not determine.

August the 16th—21st

The same favorable breeze continued to our great comfort and amazement, for the captain observed that it was very uncommon to meet with an easterly or northeasterly wind between Europe and the Azores (which the sailors call the Western Islands) for more than two days in succession; for the more common wind is here a westerly one: but beyond the Azores they find a great variety of winds, especially about this time of the year; nor do the westerly winds continue long beyond these isles. To this it is attributed that when navigators have passed the Azores, they think they have performed one half of the voyage, although in reality it is but a third. These isles come seldom in sight, for the navigators keep off them on account of the dangerous rocks under the water surrounding them. Upon observation and comparison of the journal, we found that we were in forty-three degrees twenty-four minutes north latitude and thirty and a half degrees west longitude from London.

August the 22nd—26th

About noon the captain assured us, that in twenty-four hours we should have a southwest wind: and upon my inquiring into the reasons of his foretelling this with certainty, he pointed at some clouds in the southwest, whose points turned towards northeast, and said they were occasioned by a wind from the opposite quarter. At this time I was told we were about half way to Pennsylvania. The next day about seven o'clock in the morning the expected southwest wind arrived and soon accelerated our course so much that we went at the rate of eight knots an hour. On the following day the wind shifted and was in our face. We were told by some of the crew to expect a little storm, the higher clouds being very thin and striped and scattered about the sky like parcels of combed wool, or so many skeins of yarn, which they said foreboded a storm. These striped clouds ran northwest and southeast in the direction of the wind we then had. Towards night the wind abated and we had a perfect calm, which is a sign of a change of wind. On the twenty-fifth a west wind came up and grew stronger and stronger, so that at last the waves washed our deck. This continued throughout the following day.

August the 27th

In the morning we got a better wind, which passed through various points of the compass and towards night brought on a storm from the northeast.

Our captain related an observation founded on long experience, viz. that though the winds changed frequently in the Atlantic Ocean, especially in summer time, the most frequent however was the western, and this accounts for the passage from America to Europe commonly being shorter than that from Europe to America. Besides this, the winds in the Atlantic during summer are frequently partial, so that a storm may rage on one part of it, and within a few miles of the place little or no storm at all may be felt. In winter the winds are more constant, extensive and violent, so that the same wind reigns on the greater part of the ocean for a good while and causes greater waves than in summer.

August the 30th

As I had observed the night before some strong flashes of lightening without any subsequent clap of thunder, I inquired of our captain, whether he could assign any reasons for it. He told me these phenomena were pretty common, and the consequence of a preceding heat in the atmosphere; but that when lightenings were observed in winter, prudent navigators were accustomed to reef their sails, as they were by this sign certain of an impending storm. So likewise in that season, a cloud rising from the northwest is an infallible forerunner of a great tempest.

September the 7th

We had contrary wind on the first day of the month; on the second it shifted to the north; was again contrary the third, and fair the fourth and following days. The fifth we were in forty degrees three minutes north latitude and between fifty-three and fifty-four degrees west longitude from London.

Besides the common waves rolling with the wind, we met on the fourth and fifth inst. with waves coming from the southwest, which the captain gave as a mark of a former storm from that quarter in this neighborhood.

September the 8th

We crossed by a moderate wind a sea with the highest waves we met on the whole passage, attributed by the captain to the division between the great ocean and the inner American waters,[1] and soon after we met with waves greatly inferior to those we had observed before.

September the 9th

In the afternoon we remarked that in some places the color of the sea (which hitherto had been of a deep blue) was changed into a paler hue. Some of these spots were narrow strips of twelve or fourteen fathoms breadth, of a pale green color, which is supposed to be caused by the sand, or, as some say, by the weeds under the water.

September the 12th

We were becalmed that day, and as we in this situation observed a ship which we suspected to be a Spanish privateer, our fear was

---

[1] The Gulf Stream.

very great; but a few days after our arrival at Philadelphia we saw the same ship arrive there, and heard that the crew seeing us had been under the same apprehension as ourselves.

SEPTEMBER THE 13TH

Captain Lawson, who kept his bed for the greater part of the voyage, on account of an indisposition, assured us yesterday that we were to all appearances very near America: but as the mate was of a different opinion, and as the sailors could see no land from the head of the mast, nor find ground by the lead, we steered on directly towards the land. About three o'clock in the morning the captain gave orders to heave the lead, and we found but ten fathoms: the second mate himself took the lead and called out ten and fourteen fathoms, but a moment afterwards the ship struck on the sand, and this shock was followed by four other very violent ones. The consternation was incredible; and very justly might it be so; for there were above eighty persons on board, and the ship had but one [life] boat: but happily our vessel got off again, after having been turned. At daybreak, which followed soon after (for the accident happened half an hour past four), we saw the continent of America within a Swedish mile[1] before us: the coast was whitish, low, and higher up covered with pines. We found out that the sandbar we struck lay opposite Arcadia in Maryland, in thirty-seven degrees fifty minutes north latitude.

We coasted by the shores of Maryland all day, but not being able to reach Cape Henlopen, where we intended to take a pilot on board, we cruised all night before the Bay of Delaware. The darkness of the night made us expect rain, but we encountered only a heavy fall of dew which made our coats wet, and the pages of a book, accidently left open on the deck, were in half an hour's time after sunset likewise wet. We were told by the captain and the sailors that both in England and in America a heavy dew was commonly followed by a hot and sultry day.

---

[1] About six American miles.

September the 14th

We saw land on our larboard in the west, which appeared to be low, white, sandy, and higher up the country covered with pines. Cape Henlopen is a headland running into the sea from the western shore, and has a village on it. The eastern shore belongs to New Jersey, and the western to Pennsylvania. The Bay of Delaware has many sandbars and from four to eleven fathoms of water.

The fine woods of oak, hickory and pine covering both shores made a good appearance, and were used in part for shipbuilding at Philadelphia. For this purpose some English captains take passage every autumn to the town and during winter superintend the building of new ships in which they go to sea the following spring: and at this time it was more common, as the French and Spanish privateers had seized many English merchant ships.

A little after noon we reached the mouth of the Delaware River, which is here about three English miles broad but decreases gradually so much that it is scarcely a mile wide at Philadelphia.

Here we were delighted in seeing now and then between the woods some farm houses surrounded with grain fields, pastures well stocked with cattle, and meadows covered with fine hay; and more than one sense was agreeably affected when the wind brought to us the finest effluvia of odoriferous plants and flowers or that of the fresh-mown hay: these agreeable sensations and the fine scenery of nature on this continent, so new to us, continued till it grew dark.

Here I shall return to the sea and give the reader a short account of the various occurences and phenomena belonging to natural history, during our crossing the Ocean.

Of sea weeds (*Fucus* Linn.) we saw on August the sixteenth and seventeenth a kind which had a similarity to a bunch of onions tied together. These bunches were of the size of a fist, and of a white color. On September the eleventh, near the coast of America and within the American water, we met likewise with several forms of sea weeds, one species of which was called by the sailors rockweed; another kind looked like a string of

pearls; and another was white, about a foot long, narrow, everywhere equally wide, and quite straight. From August the twenty-
fourth to September the eleventh we saw no other weeds but
those commonly going under the name of gulfweed, because
they are supposed to come from the Gulf of Florida (i. e. Mexico).
Others call it sargasso, and Dr. Linné, *Fucus natans*. Its stalk
is very slender, rotundate-angulated, and of a dark green; it has
many branches and each of them has numerous leaves arranged
in a row. They are extremely thin, are ferrated, and are a line[1]
or a line and a half wide, so that they bear a great resemblance
to the leaves of Iceland moss. Their color is a yellowish green.
Its fruit in a great measure resembles unripe juniper berries, is
round, greenish yellow, almost smooth on the outside, and grows
under the leaves on short footstalks, of two or three lines length.
Under each leaf are from one to three berries, but I never have
seen them exceed that number. Some berries are small, and when
cut are hollow and consist of a thin peel only, which is calculated
to communicate their buoyancy to the whole plant. The leaves
grow narrower in proportion as they approach the extremities
of the branches: their upper sides are smooth, the ribs are on the
under sides, and there likewise appear small roots of two, three
or four lines length. I was told by our mate that gulfweed, dried
and pounded, was given in America to women in childbed, and
besides this it is also used there in fevers. The whole ocean
seems covered with this weed, and it must also exist in immense
quantities in the Gulf of Florida, from whence all this drifting on
the ocean is said to come. Several little shells pointed like horns,
and *Escharæ* or horn wracks are frequently found on it: and
there is seldom one bundle of this plant to be met with which does
not contain either a minute shrimp, or a small crab, the latter of
which is the *Cancer minutus* of Dr. Linné. Of these I collected
eight, and of the former three, all of which I put in a glass with
water. The little shrimp moved as swift as an arrow around the
glass, but sometimes its motion was slow, and sometimes it stood
still on one side or at the bottom of the glass. If one of the
little crabs approached, it was seized by its forepaws, killed
and sucked, for which reason they were careful to avoid their

---

[1] A measure of length, usually one-twelfth of an inch.

fate. The crab had the shape of a shrimp. When swimming it was always sideways or backwards, the sides and the tail moving alternately. It was capable of putting its forepaws entirely into its mouth: its antennæ were in continual motion. Having left these little shrimps together with the crabs during the night, I found in the morning all the crabs killed and eaten by the shrimps. The former moved when alive with incredible swiftness in the water. Sometimes when they were at the bottom of the glass, with a motion something like to that of a *Puceron* or *Podura* of Linné they came in a moment to the surface of the water. In swimming they moved all their feet very close, sometimes they held them down as other crabs do, sometimes they lay on their backs, but as soon as the motion of their feet ceased, they sank to the bottom. The remaining shrimps I preserved in spirits, and the loss of my little crabs was soon repaired by other specimens which are so plentiful in each of the floating bundles of gulf-weed, and for a more minute description of which I must refer the reader to another work I intend to publish.[1] In some places we saw a crab the size of a fist, swimming by the continual motion of its feet, which being at rest, the animal began immediately to sink. And one time I met with a great red crawfish, or lobster, floating on the surface of the sea.

Blubbers, or *Medusæ* L. we found of three kinds: the first is the *Medusa aurita* L.; it is round, purple-colored, opens like a bag, and in it are, as it were, four white rings. Their size varies from one inch diameter to six inches. They have not that nettling and burning quality which other blubbers have, as for instance those found near the coast of Norway and in the ocean. These we met chiefly in the Channel and in the Bay of Biscay.

After having crossed more than half of the ocean between Europe and America, we met with a kind of blubber which is known to sailors by the name of the Spanish or Portuguese man-of-war, it looks like a great bladder, or the lungs of a quadruped, compressed on both sides, about six inches in diameter, of a fine

---

[1] Kalm refers often in his *Travels* to more detailed, scientific articles, on highly specialistic subjects which he intended to prepare for the learned journals of the time. Some of these were later published in the transactions of the Swedish Academy of Sciences (*Svenska Vetenskapsakademiens handlingar*), but some never appeared for lack of time or funds.

purple-red color. When touched by the naked skin of the human body, it causes a greater burning than any other kind of blubber. They are often overturned by the rolling of the waves, but they are upright again in an instant, and keep the sharp or narrow side uppermost.

Within the American gulf [Delaware Bay] we saw not only these Spanish men-of-war, but another kind too, for which the sailors had no other name but that of blubber. It was of the size of a pewter plate, brown in the middle, with a pale margin, which was in continual motion.

I saw on the thirtieth of August a log of wood floating on the ocean quite covered with *Lepas anatifera* L. or goose barnacles. Of insects I saw in the channel, when we were in sight of the Isle of Wight, were several white butterflies, very similar to the *Papilio brassicæ* L. They never settled, and by their venturing at so great a distance from land they caused us real astonishment.

Some common flies were in our cabin during the whole voyage, and it cannot therefore be determined whether they were originally in America or whether they came over with the Europeans.

Of cetaceous fish, we met with porpoises, or as some sailors call them sea-dogs (*Delphinus phocæna* L), first in the channel and then everywhere on this side of the Azores, where they are the only fish [1] navigators found. Beyond these isles they are seldom seen; but in the neighborhood of America we saw them equally frequent up to the very mouth of the Delaware River. They always appeared in schools, some of which consisted of almost a hundred individuals. Their swimming was very swift, and though they often swam along side of our ship, being attracted as it were by the noise caused by the cutting of the waves, they soon passed it when they were tired of staring at it. They are from four to eight feet long, have a bill like that of a goose, a white belly, and frequently leap four feet up into the air, and a distance of from four to eight feet in length. Their snorting indicates the effort which a leap of that nature costs them. Our sailors made many vain attempts from the forecastle to strike

---

[1] To-day, of course, it would not be strictly scientific to call the porpoise a "fish," since it is a mammal. "Cetaceous fish" is strictly a whale-like fish (from *cetus*, whale).

one of them with the harpoon when they came within reach, but their velocity always eluded the sailors' skill.

Another cetaceous fish, of the dolphin kind, which we met is called by the sailors bottle-nose.[1] It swims in great schools, has a head like a bottle, is killed with a harpoon, and is sometimes eaten. These fish are very large, some fully twelve feet long. Their shape and manner of tumbling and swimming make them nearly related to porpoises. They are to be met with everywhere in the ocean from the channel to the very neighborhood of America.

One whale we saw at a distance, and knew it by the water which it spouted up.

A dog-fish of a considerable size followed the ship for a little while, but it was soon out of sight, without our being able to determine to which species it belonged. This was the only cartilaginous fish we saw on the whole passage.

Of the bony fish we saw several beyond the Azores, but never one on this side of those isles. One of them was of a large size, and we saw it at a distance. The sailors called it an albecor, and it is Dr. Linné's *Scomber thynnus*.

The *dolphin* of the English is the *dorado* of the Portuguese and Dr. Linné calls it *Coryphaena hippurus*.[2] It is about two feet and a half long, near the head six inches deep, and three inches broad; from the head the dolphin decreases in size on all sides towards the tail, where its perpendicular depth is one inch and a half, and its breadth hardly one inch. The color of the back near the head is a fine green on a silver ground, but near the tail a deep blue. The belly is white, and sometimes mixed with a deep yellow. On the sides it has some round pale brown spots. It has six and not seven fins as was imagined; two of them are on the breast, two on the belly, one at the tail extending to the anus, and one along the whole back, which is of a fine blue. When the fish is just taken the extremities of the most outward rays in the tail are eight inches apart. Their motion when they swim behind or alongside of the ship is very slow and gives a fair opportunity to strike them with the harpoon, though some are taken with a hook and line and a bait of chicken bowels, small fish,

---

[1] Bottle-nosed dolphin (*Tursiops tursio*), commonly called porpoise by sailors.
[2] This species will attain a length of six feet or more.

pieces of his own species, or the flying fish, the latter being their chief food. It is by their chasing them that the flying fish leave their element to find shelter in one to which they are strangers. The dolphins sometimes leap a fathom out of the water, and love to swim about casks and logs of wood that sometimes drift in the sea. They are eaten with butter, when boiled, and are sometimes fried, and afford a palatable but rather dry food. In the bellies of the fish of this species which we caught, several animals were found, *viz.* an ostracion; a little fish with blue eyes, which was yet alive, having been swallowed just the moment before, and measuring two inches in length. Another little fish, a curious marine insect, and a flying fish, none being damaged by digestion, I preserved in spirits.

The flying fish (*Exocoetus volitans* L.) are always seen in great schools, sometimes a hundred or more coming out of the water at once, being pursued by greater fish, and chiefly by dolphins. They rise about a yard and even a fathom above the water in their flight, but they attain this height only when they take their flight from the top of a wave; and sometimes it is said they fall on the deck of ships. The greatest distance they fly is a good musket-shot, and this they perform in less than half a minute's time; their motion is somewhat like that of the yellow-hammer (*Emberiza citrinella* L.). It is very remarkable that I found the course they took always to be against the wind, though I was contradicted by the sailors, who affirmed that they flew in any direction. I nevertheless was confirmed in my opinion by a careful observation during the whole voyage, according to which they fly constantly either directly against the wind, or in a somewhat oblique direction.

We saw likewise the fish called bonitos (*Scomber pelamys* L.). They too were in schools, hunting some smaller fish, which chase caused a noise like that of a cascade because they were all swimming close in a body; but they always kept out of the reach of our harpoons.

Of amphibious animals or reptiles, we met twice with a turtle, one of which was sleeping; the other swam without taking notice of our ship; none was two feet in diameter.

Birds are pretty frequently seen on the ocean, though aquatic birds are of course more common than land birds.

The petrel (*Procellaria pelagica* L.) was our companion from the Channel to the shores of America. Flocks of this bird were always about our ship, chiefly in that part of the sea which, being cut by the ship, forms a smooth surface, where they frequently seem to settle, though always on the wing. They pick up or examine everything that falls accidentally from the ship or is thrown overboard. Little fish seem to be their chief food. In the daytime they are silent, in the dark noisy; they are reputed to forebode a storm, for which reason the sailors, disliking their company, complimented them with the name of *witches*; but they are as frequent in fair weather without a storm following their appearance. To me it appeared as if they stayed sometimes half an hour and longer under the waves, and the sailors assured me they did. They look like swallows and like them they sometimes skim on the water.

The shearwater (*Procellaria puffinus* L.) is another sea-bird which we saw everywhere on our voyage from the Channel to the American coasts. It has much the appearance and size of the dark gray sea-gull or of a duck. It has a brown back and commonly a white ring around its neck, and has a peculiar slow way of flying. We plainly saw some of these birds feed on fish.

The Tropic bird (*Phaëton æthereus* L.) has nearly the shape of a gull, but two very long feathers in its tail distinguish it from any other bird; its flight is often exceedingly high. The first of this kind that we met was at about forty degrees north latitude and forty-nine or fifty degrees west longitude from London.

Common gulls (*Larus canus* L.) we saw when we were opposite Land's End, the most westerly cape of England, and when, according to our reckoning, we were opposite Ireland.

Terns (*Sterna hirundo* L.), though of a somewhat darker color than the common ones, we found to be very plentiful after the forty-first degree of north latitude and forty-seventh degree west longitude from London, and sometimes in flocks of several hundred; occasionally they settled, as if tired, on our ship.

Within the American gulf [Delaware Bay] we discovered a sea-bird at a little distance from the ship, which the sailors called a sea-hen.

Land-birds are now and then seen at sea and sometimes at a

good distance from any land, so that it is often difficult to account for their appearance in so uncommon a place. August the eighteenth we saw a bird which settled on our ship and was just like the great titmouse (*Parus major* L.). Upon an attempt to catch it, it got behind the sails and could never be caught.

On September the first we observed some land-birds flying about our ship which we took for sand martins (*Hirundo riparia* L.). Sometimes they settled on our ship or on the sails. They were of a grayish white and the tail somewhat furcated; a heavy shower of rain drove them away afterwards. On September the second a swallow fluttered about the ship and sometimes it settled on the mast; it seemed to be very tired; several times it approached our cabin windows as if it were willing to take shelter there. These cases happened at about forty degrees north latitude and between forty-seven and forty-nine degrees west longitude from London, and also at about twenty degrees longitude, i. e. at more than nine hundred and twenty sea miles from any land whatsoever.

September the tenth in Delaware Bay a large bird, which we took for an owl, and likewise a little bird, settled on our sails.

A woodpecker settled on our rigging on September the twelfth: its back was of a speckled gray and it seemed extremely fatigued. And another land-bird of the passerine class endeavored to take shelter and rest on our ship.

Before I entirely take leave of the sea, I will communicate my observation on two curious phenomena.

In the Channel and on the Ocean we saw at night-time sparks of fire, as if floating on the water, especially where it was agitated, sometimes one single spark swam for the space of more than one minute on the ocean before it vanished. The sailors observed that they appeared commonly during and after a storm from the north, and that often the sea was as if full of fire, and that some such shining sparks would likewise stick to the masts and sails. Sometimes this light had not the appearance of sparks, but looked rather like a phosphorescence of putrid wood.[1]

The Thames, which furnished our provision of fresh water, is reputed to have the best of any river. And yet it not only

---

[1] What Kalm saw were undoubtedly tiny living organisms, myriads of marine "flagellate protozoans" of the genus Noctiluca, which are remarkable for their phosphorescence.

settled in the oak casks in which it was kept, but in a short time became stinking when stopped up; however it soon lost this nauseous smell, after being poured into large stone jugs and exposed to the fresh air for two or three hours. Often the vapors arising from the cask which had been kept closed for a great while took fire, if a candle were held near them when the cask was opened, and the Thames water was thought to have more of this quality than any other, though I was told that this happened with any other water under the same circumstances.

Now I can resume my narrative and observe that we afterwards sailed on the river with a fair wind pretty late at night. In the early evening we passed by Newcastle, a little town on the western shore of the Delaware River. It was already so dark that we hardly knew it except for the light which appeared through some of the windows. The Dutch are said to have been the first founders of this place which is therefore reckoned the oldest in the country, even older than Philadelphia. But its trade can by no means be compared with that of Philadelphia, though its location has more advantages in several respects, one of which is that the river seldom freezes there and consequently ships can come in and go out at any time. Near Philadelphia it is covered with ice almost every winter so that navigation is interrupted for a few weeks. But since the country about this city and farther up is highly cultivated, and the people bring all their goods to that place, Newcastle must always be inferior to it.

I mentioned that the Dutch laid the foundations of this town. This happened at a time when the territory was still subject to Sweden. Later the Dutch crept in and intended by degrees to dispossess the Swedes as a people who had taken possession of their property. They succeeded in their attempt, for the Swedes, not being able to bear this encroachment, came to a war in which the Dutch won.[1] But they did not enjoy the fruits of their victory long: for a few years afterwards [1664] the English came and deprived them of their acquisition and have ever since continued in the undisturbed possession of the country.—Somewhat later at night we cast anchor, the pilot not venturing to guide the ship up the river in the dark, several sandbars being in the way.

---

[1] The Dutch captured New Sweden in 1655. Fortunately the "war" which brought it about was a bloodless one.

September the 15th

At dawn we weighed anchor and continued our voyage up the river. The country was inhabited almost everywhere, on both sides. The farmhouses were, however, rather far apart. About eight o'clock in the morning we sailed by the little town of Chester on the western side of the river. In this town our mate, who was born in Philadelphia, showed me the places which the Swedes still inhabit.

At last we arrived in Philadelphia about ten o'clock in the morning. We had not been more than six weeks, or (to speak more accurately) not quite forty-one days on our voyage from Gravesend to this place, including the time we spent at Deal in supplying ourselves with the necessary fresh provisions, etc. Our voyage was therefore reckoned one of the shortest. It was common in winter time to be fourteen, nineteen or more weeks in coming from Gravesend to Philadelphia. Hardly anybody ever had a more pleasant voyage over this great ocean than we. Captain Lawson affirmed this several times. Nay, he assured us he had never seen such calm weather on this ocean, though he had crossed it often. The wind was generally so favorable that a boat of a middling size might have sailed in perfect safety. The sea never went over our cabin and but once over the deck, and that was only in a swell. The weather was, indeed, so clear that a great number of the Germans on board slept on the deck. The cabin windows did not need the shutters. All these are circumstances which show the uncommon quality of the weather.

Captain Lawson's civility increased the pleasure of the voyage, for he showed me all the friendship that he could have shown to any of his relations.

As soon as we had come to town and cast anchor many of the inhabitants came on board to inquire for letters. They took all those which they could carry, either for themselves or for their friends. Those which remained the captain ordered to be carried on shore and to be brought into a coffee-house, where everybody could make inquiry for them, and by this means he was rid of the trouble of delivering them himself. I afterwards went on shore with him. But before he went he strictly charged the second mate to let no one of the German refugees out of the ship unless he paid for his passage or somebody else paid for him or bought him.

On my leaving London I received letters of recommendation from Mr. Abraham Spalding [1], Mr. Peter Collinson [2], Dr. Mitchell [3], and others to their friends here. It was easy for me therefore to get acquainted. Mr. Benjamin Franklin, to whom Pennsylvania is indebted for its welfare and the learned world for many new discoveries in electricity, was the first who took notice of me and introduced me to many of his friends. He gave me all necessary instruction and showed me kindness on many occasions.

I went to-day accompanied by Mr. Jacob Bengtson, a member of the Swedish consistory, and the painter, Gustavus Hesselius [4], to see the town and the fields which lay before it. (The latter is brother of the Rev. Messrs. Andrew and Samuel Hesselius, both ministers at Christina in New Sweden, and of the late Dr. John Hesselius in the provinces of Närke and Värmland). My new friend had followed his brother Andrew in 1711 to this country and had since lived in it. I found that I had now come into a new world. Whenever I looked on the ground, I found everywhere such plants as I had never seen before. When I saw a tree I was forced to stop and ask its name of my companions. The first plant which struck my eyes was an *Andropogon,* or a kind of grass, and grass is a part of botany I always delighted in. I was seized with a great uneasiness at the thought of learning so many new and unknown parts of natural history. At first I only considered the plants, without venturing a more accurate examination.

At night I took up my lodging with a grocer who was a Quaker, and I met with very good honest people in this house, such as most members of this faith appeared to be. Jungström, the companion of my voyage, and I had a room, candles, beds, attendance and three meals a day, if we chose to have so many,

[1] Abraham Spalding (1712-1782), Swedish merchant in London.

[2] Peter Collinson (1694-1768), English merchant and antiquary who took a great interest in the American Colonies. See p. 105.

[3] Dr. John Mitchell (d. 1768), an English botanist who had spent most of his life in America. He discovered several new plants and wrote many scientific treatises.

[4] Gustavus Hesselius (1682-1755), a native of Dalarna, Sweden, and pioneer of American painting and organ building. In 1721 he received the "first public art commission in the colonies of which we have any record." This was an altar-piece, "The Last Supper." He was a relative of Swedenborg and subsequently joined the Moravian Brethren.

for twenty shillings per week in Pennsylvania currency. But wood, washing and wine, if required, were extra.

SEPTEMBER THE 16TH

Before I proceed I must give a short description of Philadelphia, which I shall mention frequently in the course of my travels. I here put down several particulars which I marked during my stay there as a help to my memory.

Philadelphia, the capital of Pennsylvania, a province which forms a part of what formerly was called New Sweden, is one of the principal towns in North America and next to Boston the greatest. It is situated almost in the center of the English colonies and its latitude is thirty-nine degrees and fifty minutes, while its west longitude from London is near seventy-five degrees.

This town was built in the year 1683, or as others say in 1682, by the well-known Quaker William Penn, who got this whole province by a grant from Charles the Second, King of England, after Sweden had given up its claims to it. According to Penn's plan the town was to have been built upon a piece of land which is formed by the union of the rivers Delaware and Schuylkill, in a quadrangular form, two English miles long and one broad. The eastern side would therefore have been bounded by the Delaware, and the western by the Schuylkill. They had actually begun to build houses on both these rivers; for eight capital streets, each two English miles long, and sixteen lesser streets (or lanes) across them, each one mile in length, were marked out with a considerable width, and in straight lines. The place was at that time almost entirely a wilderness covered with thick forests, and belonged to three Swedish brothers called Svenssöner (Svensons, sons of Sven or Swen) who had settled on it.[1] They reluctantly left the place, the location of which was very advantageous. But at last they were persuaded to leave it by Penn, who gave them, a few English miles from there, twice the space of the land they inhabited. However, Penn himself and his descendants after him have, by repeated mensurations, considerably lessened the ground belonging to the Swedes, under pretence that they had taken more than they should.[2]

---

[1] Sven, Olof and Anders, sons of Sven Gunnarson. See pp. 33-34.
[2] See pages 645 and 726.

But settlers could not be induced to come in sufficient numbers to fill a place of such size. The plan therefore about the river Schuylkill was laid aside till more favorable circumstances should occur, and the houses were built only along the Delaware. This river flows along the eastern side of the town, is of great advantage to its trade, and gives a fine prospect. The houses which had already been built upon the Schuylkill were moved hither by degrees. This town accordingly lies in a very pleasant country, from north to south along the river. It measures somewhat more than an English mile in length, and its breadth in some places is half a mile or more. The ground is flat and consists of sand mixed with a little clay. Experience has shown that the climate of this place is very healthy.

The streets are regular, pretty, and most of them fifty feet, English measure, broad. Arch Street measures sixty-six feet in breadth, and Market Street, the principal thoroughfare, where the market is kept, near a hundred. The streets which run longitudinally or from north to south are seven in number, exclusive of a small one which runs along the river to the south of the market and is called Water Street. The lanes which go across and were intended to reach from the Delaware to the Schuylkill, number eight. They do not run exactly from east to west, but deviate a little from that direction. All the streets, except two which are nearest to the river, run in a straight line and make right angle: at the intersections. Some are paved, others are not, and it seems less necessary since the ground is sandy and therefore soon absorbs the wet. But in most of the streets is a pavement of flags, a fathom or more broad, laid before the houses, and four-foot posts put on the outside three or four fathoms apart. Those who walk on foot use the flat stones, but riders and teams use the middle of the street. The above-mentioned posts prevent horses and wagons from injuring the pedestrians inside the posts, and are there secure from careless teamsters and the dirt which is thrown up by horses and carts. Under the roofs are gutters which are carefully connected with pipes, and by this means, those who walk under them when it rains or when the snow melts need not fear being wetted by the water from the roofs.

The houses make a good appearance, are frequently several stories high and built either of bricks or of stone; but the former are more commonly used, since they are made near the town and are of good quality. The stone which has been employed in the building of houses is a mixture of a loose and quite small-grained limestone and of a black or grey glimmer, running in undulated veins which run scattered between the bendings of other veins and are of a gray color excepting here and there some single grains of sand of a paler hue. The glimmer forms the greatest part of the stone, but the mixture is sometimes of another kind, as I shall relate below under the date of the eleventh of October. This stone is now obtained in great quantities in the country, is easily cut, and has the good quality of not attracting moisture in a wet season. Very good lime is burnt everywhere hereabouts for masonry.

The houses are covered with shingles. The wood for this purpose is taken from the *Cupressus thyoides* L. or a tree which Swedes here call the "white juniper tree", and the English "the white cedar". Swamps and morasses formerly were full of them, but at present these trees are for the greatest part cut down and no attempt has as yet been made to plant new ones. The wood is very light, rots less than any other in this country, and for that reason is exceedingly good for roofs, for it is not too heavy for the walls and will last forty or fifty years. But many people already begin to fear that these roofs will in time be looked upon as having been very detrimental to the city. For being so very light, most people who have built their houses of stone or bricks have been led to make their walls extremely thin. At present this kind of wood is almost entirely gone. Whenever, therefore, in process of time these roofs decay, the people will be obliged to have recourse to the heavier materials of tiles or the like, which the walls will not be strong enough to bear. The roof will therefore require more support or the people be obliged to pull down the walls and build new ones, or to take other steps for securing them. Several people have already in late years begun to make roofs of tiles.

Among the public buildings I shall first mention churches, of which there are several, for God is served in various ways in this country.

1. The *English established church* stands in the northern part of the town, at some distance from the market, and is the finest

of all. It has a small, insignificant steeple, in which a bell is rung when it is time to go to church, and at burials. It has likewise a clock which strikes the hours. This building, which is called Christ Church, was founded towards the end of the last century, but has lately been rebuilt and more adorned. It has two ministers who get the greatest part of their salary from England. In the beginning of this century the Swedish minister, the Rev. Mr. Rudman[1], performed the functions of a clergyman in the English congregation for nearly two years during the absence of their own clergyman.

2. The *Swedish church*, which is otherwise called the Church of Wicaco, is in the southern part of the town, almost outside of it on the riverside, and its location is therefore more agreeable than that of any other. I shall have an opportunity of describing it more exactly when I speak of the Swedes who live in this place.

3. The *German Lutheran church* is on the northwest side of the town. On my arrival in America it had a little steeple, but having been put up by an ignorant builder before the walls of the church had become quite dry, the latter were forced out by its weight and the steeple had to be pulled down again in the autumn of the year 1750. About that time the congregation received a fine organ from Germany. They have only one minister, who also preaches at another Lutheran church in Germantown. He preaches alternately one Sunday in that church, and the other in this. The first clergyman which the Lutherans had in this town was the Rev. Mr. Mühlenberg[2] who laid the foundations of the church in 1743, and being called to another place afterwards, the Rev. Mr. Brunnholtz[3], a Dane from Schleswig, was his successor, and he is still there. Both these gentlemen were sent here from

---

[1] Andreas Johannes Rudman (1668-1708) arrived in America, 1697, and became the founder of the present church of Gloria Dei at Philadelphia, then Wicaco. He also preached for the Dutch in New York, and "officiated at the Oxford Church, near Frankford." He was associated with Christ Church, Philadelphia, when he died, at the early age of forty.

[2] Rev. Heinrich Melchior Mühlenberg (1711-1781) landed in Philadelphia, November, 1742, to take charge of the German and Swedish Lutherans. His name and work are of course well known in the history of Pennsylvania.

[3] Rev. P. Brunnholtz was at first Mühlenberg's assistant, assigned to Philadelphia and Germantown. Since he came from Schleswig Professor A. B. Faust naturally calls him a German. See his *The German Element in the United States*, second ed., p. 119.

Halle in Saxony, and have been of great value to the church by their peculiar talent of preaching in an edifying manner. A little while before this church was built the Lutheran Germans had no clergyman of their own so that the much-beloved Swedish minister at Wicaco, Mr. Dylander[1], preached to them also. He therefore gave three sermons every Sunday; the first early in the morning to the Germans: the second to the Swedes; and the third in the afternoon to the English. Besides this he went all week into the country and instructed the scattered Germans who lived there. He frequently preached sixteen sermons a week. And only after his death, which happened in November, 1741, did the Germans write to Germany for a clergyman. This congregation is at present very large, so that every Sunday the church is crowded. It has two galleries but no vestry. They do not sing the collects but read them before the altar. The sermon is given from the pulpit.

4. The *Old Presbyterian church* is not far from the market and on the south side of Market Street. It is of a middling size and built in the year 1704, as the inscription on the northern pediment shows. The roof is built almost hemispherical, or at least forms a half-hexagon. The building stands north and south, for the Presbyterians are not so particular as other people whether their churches face a certain point of the heavens or not.

5. The *New Presbyterian church* was built in the year 1750[2] by the "New-lights" in the northwestern part of the town. By the name of New-lights are understood the people who have, from different religions, become proselytes of the well-known Whitefield[3], who in the years 1739, 1740, and likewise in 1744 and 1745, travelled through almost all the English colonies in North America. His delivery, his extraordinary zeal, and other talents so well adapted to the intellects of his hearers, made him so popular that he frequently, especially in the two first years got an audience of from eight to twenty thousand people. His intention in these travels, was to collect money for an orphans' hos-

---

[1] John Dylander (c. 1709-1741), Swedish Lutheran clergyman who came to America in 1737 to take charge of the Gloria Dei Church at Wicaco (now Southwark, Philadelphia). He married a daughter of the merchant Peter Kock.

[2] This information was supplied from notes taken later of course than 1748.

[3] George Whitefield (1714-1770), evangelist and leader of Calvinistic Methodists. See long article on him in *Dictionary of National Biography*.

pital which had been erected in Georgia. Here he frequently collected seventy pounds sterling at one sermon; nay, at two sermons which he preached in the year 1740, both on one Sunday, at Philadelphia, he received a hundred and fifty pounds. The proselytes of this man, or the above-mentioned "New-lights", are at present merely a sect of Presbyterians. For though Whitefield was originally a clergyman of the English church, he deviated little by little from her doctrines, and on arriving in the year 1744 at Boston in New England the Presbyterians argued with him about their teachings so much that he embraced them almost entirely. For Whitefield was no great disputant and could therefore easily be led by these cunning people whithersoever they would have him. This also, during his latter stay in America, caused his audience to be less numerous than during the first. The New-lights first built in the year 1741 a large house in the western part of the town in which to hold divine service. But a division arising amongst them after the departure of Whitefield, and also for other reasons, the building was sold to the town in the beginning of the year 1750 and destined for a school. The New-lights then built a church which I call the new Presbyterian one. On its eastern pediment is the following inscription, in golden letters: *Templum Presbyterianum, annuente numine, erectum, Anno Dom. MDCCL.*

6. The *Old German Reformed* (Calvinistic) *church* is built in the west-northwest part of the town and looks like the church in Ladugårdsgärdet near Stockholm. It is not yet finished, though for several years the congregation has kept up divine service in it. These Germans attended the German service at the Swedish church whilst the Swedish minister Mr. Dylander lived.—But as the Lutherans procured a clergyman of their own on the death of the last, those of the Reformed church likewise made preparations to obtain one from Dordrecht, Holland, and the first who was sent to them was the Rev. Mr. Slaughter whom I found on my arrival. But in the year 1750, another clergyman of the Reformed church arrived from Holland and by his artful behavior so insinuated himself into the favor of the Rev. Mr. Slaughter's congregation that the latter lost almost half of his audience. The two clergymen then disputed for several Sundays about the pulpit; nay, people relate that the newcomer mounted the pulpit

on a Saturday and stayed in it all night, the other being thus excluded. The two parties in the audience made themselves the subject both of the laughter and of the scorn of the whole town by beating and bruising each other and committing other excesses. The affair was inquired into by the magistrates and decided in favor of the Rev. Mr. Slaughter, the person who had been abused.

7. The *New Reformed church* was built at a little distance from the old one by the party of the clergyman who had lost his cause. The newcomer, however, had influence enough to bring over to his party almost the whole audience of his antagonist at the end of the year 1750, and therefore this new church will soon be useless.

8. & 9. The *Quakers* have two meeting-houses, one in the market and the other in the northern part of the town. Among them, according to their custom, there are neither altars nor pulpits nor any other ornament usual in churches, but only seats and some sconces. They meet thrice every Sunday in them, and besides that at certain times every week or every month. I shall mention more about them hereafter.

10. The *Anabaptists* have their service in the northern part of the town.

11. The *Roman Catholics* have in the southwest part of the town a large building which is well adorned within and has an organ.

12. The *Moravian or Zinzendorfian Brethren* have hired a large house in the northern part of the town, in which they perform service both in German and English, not only twice or three times every Sunday but every night after it has grown dark. In the winter of the year 1750 they were obliged to drop their evening meeting, some wanton young fellows having several times disturbed the congregation by an instrument sounding like the note of a cuckoo; for this noise they made in a dark corner not only at the end of every stanza but likewise at that of every line whilst they were singing a hymn.

Those of the English church, the New-lights, the Quakers, and the Germans of the Reformed religion have their burying places out of town and not near their churches, though the first of these sometimes makes an exception. All the others bury their dead in their church-yards, and the Moravian Brethren bury where they

can. The negroes are buried in a separate place out of town.
I now proceed to mention the other public buildings in Philadelphia.

The *Town Hall,* or the place where the assemblies are held, is situated in the western part of the town. It is a fine, large building having a tower with a bell, and is the greatest ornament in the town. The deputies of each province commonly meet in it every October, or even more frequently if circumstances require it, in order to consider the welfare of the country and to hold their parliament or diets in miniature. There they revise the old laws and make new ones.

On one side of this building stands the *Library* which was first begun in the year 1742 on a public spirited plan formed and put into execution by the learned Mr. Franklin. For he persuaded first the most substantial people in town to pay forty shillings at the outset, and afterwards annually ten shillings, in Pennsylvania currency, towards purchasing all kinds of useful books. The subscribers are entitled to make use of them. Other people are likewise at liberty to borrow them for a certain time, but must leave a pledge and pay eight-pence a week for a folio volume, sixpence for a quarto, and fourpence for all of a smaller size. As soon as the time allowed a person for the perusal of a volume has elapsed, it must be returned or he is fined. The money arising in this manner is employed for the salary of the librarian and for purchasing new books. There is already a fine collection of excellent works, most of them English; many French and Latin, but few in any other language. The subscribers were kind enough to order the librarian, during my stay here, to lend me every book which I should want without any payment. The library is open every Saturday from four to eight o'clock in the afternoon. Besides the books, several mathematical and physical instruments and a large collection of natural curiosities are to be seen in it. Several little libraries were founded in the town on the same principle or nearly so.

The *Court House* stands in the middle of Market Street to the west of the market. It is a fine building with a small tower and a bell. Below and around this building the market is properly held every week.

The building of the *Academy* is in the western part of the city. It was formerly, as I have mentioned before, a meeting-house of the followers of Whitefield, but they sold it in the year 1750, and it was destined to become the seat of higher learning, or to express myself in more exact terms, to be a college. It was therefore fitted up for this purpose. The young men here are taught only those things which they learn in our common schools and gymnasia; but in time such lectures are intended to be given as are usual in real universities.

At the close of the last war a *redoubt* was erected here, on the south side of the town near the river, to prevent the French and Spanish privateers from landing. But this was done after a very strong debate. For the Quakers opposed all fortifications as contrary to the tenets of their religion, which do not allow Christians to make war either offensive or defensive, but direct them to place their trust in the Almighty alone. Several papers were then handed around for and against the opinion. But the enemy's privateers having taken several vessels belonging to the town in the river, many of the Quakers, if not all of them, found it reasonable to further the building of the fortification as much as possible, at least by a supply of money.

Of all the natural advantages of the town, its temperate climate is the most considerable, the winter not being over severe, and its duration but short and the summer not too hot; the country round about bringing forth those fruits in the greatest plenty which are raised by husbandry. Their September and October are like the beginning of the Swedish August. And the first days in their February are frequently as pleasant as the end of April and the beginning of May in Sweden. Even their coldest days in some winters have been no severer than the days at the end of spring are in the middlemost parts of Sweden and the southern ones of Finland.

The good and clear water in Philadelphia is likewise one of its advantages. For though there are no fountains in the town, there is a well in every house and several in the streets, all of which furnish excellent water for boiling, drinking, washing and other uses. The water is commonly found at the depth of forty feet. The water of the River Delaware is likewise good. But in making the wells a fault is frequently committed which in several places of the town spoils the water which is naturally good. I shall later take an opportunity of speaking further about it.

Trade. The Delaware is exceedingly convenient for trade. It is one of the largest rivers: it is three English miles broad at its mouth, two miles at the town of Wilmington, and three quarters of a mile at Philadelphia. This city lies within ninety or a hundred English miles from the sea, or from the place where the river Delaware discharges itself into the bay of that name. Yet its depth is hardly ever less than five or six fathoms. The largest ships therefore can sail right up to the town and anchor in good ground in five fathoms of water on the side of the bridge. The water here has no longer a saltish taste, and therefore all destructive worms which have fastened themselves to the ships in the sea and have pierced holes into them either die or drop off, after the ship has been here for a while.

The only disadvantage which commerce has here is the freezing of the river almost every winter for a month or more. For during that time navigation is entirely stopped. This does not happen at Boston, New York and other towns which are nearer the sea.

The tide comes up to Philadelphia and even goes thirty miles higher to Trenton. The difference between high and low water is eight feet at Philadelphia.

The cataracts of the Delaware near Trenton and of the Schuylkill at some distance from Philadelphia make these rivers useless further up the country in regard to the conveyance of goods either from or to Philadelphia. They must therefore be carried on wagons or carts. It has therefore already been thought of making these two rivers navigable [for greater distances and] for larger vessels.

Several ships are built annually of American oak, in the docks which are found in several parts of and near the town, yet they can by no means be compared with those built of European oak in point of goodness and durability.

The town carries on a great trade both with the inhabitants of the country and with other parts of the world, especially the West Indies, South America and the Antilles, England, Ireland, Portugal and the various English colonies in North America. Yet none but English ships are allowed to come into this port.

Philadelphia reaps the greatest profits from its trade with the West Indies. For thither the inhabitants ship almost every day a quantity of flour, butter, meat and other victuals, timber, planks and the like. In return they receive either sugar, molasses, rum, indigo, mahogany and other goods or ready money. The true mahogany which grows in Jamaica is at present almost all cut down. Philadelphians send both West India goods and their own productions to England; the latter comprise all sorts of woods, especially black walnut and oak planks for ships, ships ready built, iron, hides and tar. Yet this latter is properly bought in New Jersey, the forests of which province are consequently more ruined than any others. Ready money is likewise sent over to England from whence in return they get all sorts of goods manufactured, *viz.* fine and coarse cloth, linen, iron ware and other wrought metals, and East India goods. For it is to be observed that England supplies Philadelphia with almost all stuffs and articles which are wanted here.

A great quantity of linseed goes annually to Ireland, together with many of the ships which are built here. Portugal gets wheat, corn, flour and grain which is not ground. Spain sometimes takes some grain. But all the money which is gotten in these several countries, must immediately be sent to England in payment for goods from thence, and yet those sums are not sufficient to pay all the debts.

But to show more exactly what the town and province have imported from England in different years I shall here insert an extract from the English customhouse books, which I obtained from the engineer, Lewis Evans,[1] at Philadelphia and which will sufficiently answer the purpose. This gentleman had desired one of his friends in London to send him a complete account of all the goods shipped from England to Pennsylvania during a certain number of years. He got this account, and though the goods are not enumerated in it yet their value in money is calculated. Such extracts from the customhouse books have been made for every North American province, in order to convince the English parliament that those provinces have taken greater quantities of goods in that kingdom after they set up their own paper currency.

I have taken the copy from the original itself and it is to be observed that it begins with the Christmas of the year 1722, and ends about the same time of the year 1747. In the first column is the value of the foreign goods, the duty for which has already been paid in England. The second column shows the value of the goods manufactured in England and exported to Pennsylvania. And in the last column these two sums are added together, and at the bottom

---

[1] Mr. Lewis Evans (c. 1700-1756), engineer and geographer. He published in 1749 "A Map of Pennsylvania, New Jersey, New York and the three Delaware Countries."

each column is added up. But this table does not include the goods
which are annually shipped in great quantities to Pennsylvania from
Scotland and Ireland, among which is a great quantity of linen.

THE VALUE OF THE GOODS ANNUALLY SHIPPED FROM ENGLAND TO PENNSYLVANIA

The year, from one Christmas to another	Foreign Goods for which the duty has already been paid, & which therefore only req. receipts.			English manufactured Goods.			The Sums of these two preceding columns added together.		
	£.	s.	d.	£.	s.	d.	£.	s.	d.
1723	5199	13	5	10793	5	1	15992	19	4
1724	9373	15	8	20951	0	5	30324	16	1
1725	10301	12	6	31508	1	8	42209	14	2
1726	9371	11	6	28263	6	2	37634	17	8
1727	10243	0	7	21736	10	0	31979	10	7
1728	14073	13	3	23405	6	2	37478	19	11
1729	12948	8	5	16851	2	5	29799	10	10
1730	15660	10	11	32931	16	6	48592	7	5
1731	11838	17	4	32421	18	9	44260	16	1
1732	15240	14	4	26457	19	3	41698	13	7
1733	13187	0	8	27378	7	5	40585	8	1
1734	19648	15	9	34743	12	1	54392	7	10
1735	18078	4	3	30726	7	1	48804	11	4
1736	23456	15	11	38057	2	5	61513	18	4
1737	14517	4	3	42173	2	4	56690	6	7
1738	20320	19	3	41129	5	0	61450	4	3
1739	9041	4	5	45411	7	6	54452	11	11
1740	10280	2	0	46471	12	9	56751	14	9
1741	12977	18	10	78032	13	1	91010	11	11
1742	14458	6	3	60836	17	1	75295	3	4
1743	19220	1	6	60120	4	10	79340	6	4
1744	14681	8	4	47595	18	2	62214	6	6
1745	13043	8	8	41237	2	3	54280	10	11
1746	18013	12	7	55595	19	7	73699	12	2
1747	8585	14	11	73819	2	8	82404	17	7
Total	343,789	16	0	969,049	1	6	1,312,838	17	6

The whole extent of the Philadelphia trade may be comprehended from the number of ships, which annually arrive at and sail from this town. I intend to insert here a table of a few years which I have taken from the gazettes of the town. The ships coming and going in one year are to be reckoned from the twenty-fifth of March of that year to the twenty-fifth of March of the next.

The Year.	Ships arrived.	Ships sailed.
1735	199	212
1740	307	208
1741	292	309
1744	229	271
1745	280	301
1746	273	293

But it is much to be feared that the trade of Philadelphia and of all the English colonies will rather decrease than increase in case no provision is made to prevent it. I shall hereafter plainly show upon what foundation this decrease of trade is likely to take place. The town not only furnishes most of the inhabitants of Pennsylvania with the goods which they want, but several inhabitants of New Jersey come every day to trade.

The town has two great fairs every year, one on May sixteenth and the other on November sixteenth. But besides these fairs there are every week two market days, *viz.* Wednesday and Saturday. On those days the country people in Pennsylvania and New Jersey bring to town a quantity of food and other products of the country, and this is a great advantage to the town. It is therefore to be wished that a similar regulation be made in our Swedish towns. You are sure to find on market days every produce of the season which the country affords. But on other days they are sought for in vain.

Provisions are always to be got fresh here, and for that reason most of the inhabitants never buy more at a time than what will be sufficient till the next market day. In summer there is a market almost every day, for the victuals do not keep well in the great heat. There are two places in town where these markets are kept, but that near the court-house is the principal one. It begins about four or five o'clock in the morning and ends about nine in the forenoon.

The town is not enclosed and has no other customhouse than the large one for the ships.

The governor of the whole province lives here and though he is nominated by the heirs of Penn he cannot take that office without being confirmed by the king of England.

The Quakers of almost all parts of North America have their great assembly here once a year.

In the year 1743 a society for the advancement of the sciences was started here. Its activity was to have embraced the curiosities of the three kingdoms of nature, mathematics, physics, chemistry, economics and manufactures. But the war, which ensued immediately, stopped all designs of this nature and since that time nothing has been done to revive it.

The declination of the magnetic needle was here observed on the thirtieth of October, 1750, old style, to be five degrees and forty-five minutes west. It was examined in relation to the new meridian which was drawn at Philadelphia in the autumn of the same year and extended a mile in length. By experience it appears that this declination lessens about a degree in twenty years time.

The greatest difference in the rising and falling of the barometer is, according to the observations made for several years by Mr. James Logan,[1] that recorded at 28.59″ and 30.78″.

There are three printers here and every week two English newspapers and one German are printed.

In the year 1732 on the fifth of September, old style, a little earthquake was felt here about noon and at the same time at Boston in New England and at Montreal in Canada, which places are above sixty Swedish miles apart.

In the month of November of the year 1737 the well known prince from Mount Lebanon, Sheich Sidi, came to Philadelphia on his travels through most of the English American colonies.—In the same year a second earthquake was felt about eleven o'clock at night, on the seventh of December. But it did not continue above half a minute and yet it was felt according to the accounts of the gazettes at the same hour in Newcastle, New York, New London,

---

[1] Mr. James Logan (1674-1751) well known jurist, scientist and educationalist; mayor of Philadelphia 1723; chief justice of the supreme court of the Province; one of the founders of the University of Pennsylvania. He was the author of *Experimenta de Plantarum Generatione;* He also made a translation of Cicero's *De Senectute.*

Boston and other towns of New England. It had, therefore, an areal influence of several miles.

Count Zinzendorf[1] arrived here in December of the year 1741 and stayed until the following spring. His uncommon behavior persuaded many Englishmen of rank that he was disordered in his head.

Population. I have not been able to find the exact number of inhabitants in Philadelphia. In the year 1746 they were reckoned to be above ten thousand, and since that time their number has incredibly increased. Neither can it be ascertained from the lists of mortality since they are not kept regularly in all the churches. I shall, however, mention some of those which appeared either in the gazettes or in printed lists separately.

Year.	Dead.	Year.	Dead.	Year.	Dead.
1730	227	1741	345	1745	420
1738	250	1742	409	1748	672
1739	350	1743	425	1749	758
1740	290	1744	410	1750	716

From these mortality lists it also appears that the diseases which are most fatal are consumption, fevers, convulsions, pleurisy, hemorrhages and dropsy.

The number of births cannot be determined since in many churches no order is observed with regard to this affair. The Quakers, who are the most numerous in this town, never baptize their children, though they keep a pretty exact account of all who are born among them. It is likewise impossible to guess at the number of inhabitants from the dead because the town gets such great supplies of immigrants annually from other countries. In the summer of the year 1749 nearly twelve thousand Germans came over to Philadelphia, many of whom stayed in that town. In the spring of the same year the houses in Philadelphia were counted and found to be two thousand and seventy-six in number.

The town is now well filled with inhabitants of many nations, who in regard to their country, religion and trade are very different from each other. You meet with excellent masters in all trades and many things are made here fully as well as in England. Yet no

---

[1] Count N. L. Zinzendorf, a well-known head of the Moravian Brethren.

manufactures, especially for making fine wool cloth, are established. Perhaps the reason is that it can be got with so little difficulty from England and that the breed of sheep which is brought over degenerates in process of time and affords but a coarse wool.

There is a great abundance of provisions here and their prices are very moderate. There are no examples of an extraordinary dearth.

Freedom. Everyone who acknowledges God to be the Creator, preserver and ruler of all things, and teaches or undertakes nothing against the state or against the common peace, is at liberty to settle, stay and carry on his trade here, be his religious principles ever so strange. No one is here molested because of misleading principles of doctrine which he may follow, if he does not exceed the above-mentioned bounds. And he is so well secured by the laws, both as to person and property, and enjoys such liberties that a citizen here may, in a manner, be said to live in his house like a king. It would be difficult to find anyone who could wish for and obtain greater freedom.

Rapidity of Urban Growth. On careful consideration of what I have already said it will be easy to conceive why this city should rise so suddenly from nothing into such grandeur and perfection without any powerful monarch contributing to it, either by punishing the wicked or by giving great supplies of money. And yet its fine appearance, good regulations, agreeable location, natural advantages, trade, riches and power are by no means inferior to those of any, even of the most ancient, towns in Europe. It has not been necessary to force people to come and settle here; on the contrary foreigners of different languages have left their country, houses, property and relations and ventured over wide and stormy seas in order to come hither. Other countries, which have been peopled for a long space of time, complain of the small number of their inhabitants. But Pennsylvania which was no better than a wilderness in the year 1681, and contained hardly fifteen hundred people, now vies with several kingdoms in Europe in the number of inhabitants. It has received hosts of people which other countries, to their infinite loss, have either neglected, belittled or expelled.

Oldest Building. A wretched old wooden building on a hill near the river, located a little north of the Wicaco is preserved on purpose as a memorial to the poor condition of the place before the town was built on it. It belonged formerly to one of the Svensons, from whom, as before-mentioned,[1] the ground was bought upon which to build Philadelphia. Its antiquity gives it a kind of superiority over all the other buildings in town, though in itself it is the worst of all. This hut was inhabited whilst yet stags, deer, elks and beavers at broad daylight lived in the future streets, church yards and marketplaces of Philadelphia. The noise of a spinning wheel was heard in this house before the manufactures now established were thought of or Philadelphia was built. But with all these advantages the house is ready to fall down, and after a few years it will be as difficult to find the place where it stood as it was unlikely at the time of its erection that one of the greatest towns in America should in a short time stand close to it.

A Custom. It was a custom in Philadelphia, upon meeting a woman on the street, to let her go on the side nearest the houses. To make her walk on the outside was considered boorish and unrefined. I have mentioned before that the streets here are like those of London, with pavements for pedestrians and wooden posts that prevent driving upon the people. Similarly, when walking with a lady, she must be allowed the side next to the houses. The same practice obtains when promenading with a gentleman of higher social station than oneself. I have seen men so vain in this effort to give honor to another that they have constantly shifted from the right to the left side of a person, depending upon the number of times they crossed a street together. The custom is supposed to have arisen from an attempt to protect the walking companion from the filth of the street, hence, the side next to it is held to be less honorable.

SEPTEMBER THE 17TH

Birds and Snakes. Mr. Peter Cock,[2] a merchant of this town, assured me that he had last week himself been a spectator of a snake's swallowing a little bird. This bird, which from its cry has the name of catbird (*Muscicapa carolinensis* L.) flew from one branch of a tree to another and was making a doleful tune.[3] At the bottom of

---

[1] See page 18.

[2] Mr. Peter Cock, a noted settler and native of Karlskrona, Sweden.

[3] It is not certain that this bird was a catbird. *Muscicapa* means flycatcher. See L. W. Marchand, *Voyage de Kalm en Amérique*, I, p. 18, note 1. The modern scientific name of the catbird is *Galeoscoptes carolinensis*.

the tree, but at a fathom's distance from the stem lay one of the great black snakes, with its head continually upright, pointing towards the bird, which was always fluttering about and now and then settling on the branches. At first it kept only in the topmost branches, but by degrees it came lower down and even flew upon the ground and hopped to the place where the snake lay, which immediately opened its mouth, caught the bird and swallowed it; but it had scarce finished its repast before Mr. Cock came up and killed it. I was afterwards told that this kind of snake was frequently observed to pursue little birds in this manner. It is already well known that the rattlesnake does the same.

Trees. I walked out today into the fields in order to get more acquainted with the plants hereabouts; I found several European and even Swedish plants among them. But those which are peculiar to America are much more numerous.

The *Virginian maple* grows in plenty on the shores of the Delaware. The English in this country call it either buttonwood or water beech, which latter name is most usual. The Swedes call it *vattenbok* or *vassbok*. It is Linné's *Plantanus occidentalis*. It grows for the greatest part in low places but especially on the edge of rivers and brooks. But these trees are easily transplanted to drier places, if they be only filled with good soil, and as their leaves are large and their foliage thick, they are planted about the houses and in gardens to afford a pleasant shade in the hot season, to the enjoyment of which some seats were placed under them. Some of the Swedes had boxes, pails and the like made of the bark of this tree by the native Americans. They say that those people whilst they were yet settled here, made little dishes of this bark for gathering whortleberries. The bark was a line in thickness. This tree likewise grows in marshes or in swampy fields where ash and red maple commonly grow. They are frequently as tall and thick as the best of our fir trees. The seed stays on them till spring, but in the middle of April the pods open and shed the seeds. Are they not ripe before that time and consequently sooner fit for planting? This American maple is remarkable for its quick growth, in which it exceeds all other trees. There are such numbers of them on the low meadows between Philadelphia and the ferry at Gloucester, on both sides of the road, that in summertime you go as it were through a shady walk. In that part of Philadelphia which is near the Swedish church some large trees of this kind stand on the shore of the river. In the year 1750, on the fifteenth of March, I saw the buds still on them and in the year 1749 they began to flower on the eighth of May. Several trees of this sort are planted at Chelsea near London and in point of height they now vie with the tallest oak.

Dysentery. The following prescription was regarded as an infallible cure for dysentery: Boil some cinnamon bark in water; take a quantity of this water and pour it with some brandy in a bowl; over it place a couple of splinter (or pipe-stems) close beside each other, with a piece of sugar on top of them; set fire to the brandy and let it burn until the sugar is quite burnt. Let the patient then eat this sugar and drink some of the brandy and the cinnamon water. It is claimed that one dose was often enough for an immediate cure. Another excellent remedy was said to be fried red English cheese eaten on a sandwich.

SEPTEMBER THE 18TH

In the morning I went with the Swedish painter, Mr. Hesselius, to the country seat of Mr. Bartram [1] which is about four English miles to the south of Philadelphia, at some distance from the highway to Maryland, Virginia and Carolina. I had therefore the first opportunity here of getting an exact knowledge of the state of the country, which was a plain covered with all kinds of deciduous trees. The ground was sandy, mixed with clay. But the sand seemed to be in greater quantity. In some parts the wood was cut down and we saw the habitations of some country people, whose corn fields and plantations were round their farm houses. The wood was full of mulberry trees, walnut trees of several kinds, chestnut trees, sassafras and the like. Several sorts of wild vines clasped their tendrils round, and climbed up to the summits of the highest trees, and in other places they twined round the fences so thick that the latter almost sunk down under their weight. The persimmon, or *Diospyros Virginiana* L., grew in the marshy fields and about pools. Its little apples looked very well, but are not fit for eating before the frost has affected them and then they have a fine taste. Hesselius

---

[1] Mr. John Bartram (1699-1777). See page 61. Kalm often conferred with him about his scientific findings. He was the most prominent American naturalist of the time. He was in 1777 elected a member of the Swedish Academy of Sciences.

Photograph by Philip B. Wallace   Courtesy of Henry D. Paxson, Jr.
Birthplace of John Bartram, Darby, Pa.

gathered some of them and desired my servant to taste of this fruit of the land, but the poor credulous fellow had hardly bit into them when he felt the qualities they have before the frost has touched them, for they contracted his mouth so that he could hardly speak and got a very disagreeable taste. This disgusted him so much that he was with difficulty persuaded to taste of it during the whole of our stay in America, notwithstanding it loses all its acidity and acquires an agreeable flavor in autumn and towards the beginning of winter. For the fellow always imagined that though he should eat them ever so late in the year they would still retain the same obnoxious taste.

To satisfy the curiosity of those who are willing to know how the woods look in this country and whether or no the trees in them are the same as those found in our forests, I here inserted a small catalogue of those which grow wild in the woods nearest to Philadelphia. I exclude such shrubs as do not attain any considerable height. I shall put that tree first in order which is most plentiful and so on with the rest, and therefore trees which I have found but single, though near the town, will be last.

1. *Quercus alba,* the white oak in good ground.
2. *Quercus nigra,* or the black oak.
3. *Quercus Hispanica,* the Spanish oak, a variety of the preceding.
4. *Juglans alba,* hickory, a kind of walnut tree, of which three or four varieties are to be met with.
5. *Rubus occidentalis,* or American blackberry shrub.
6. *Acer rubrum,* the maple tree with red flowers, in swamps.
7. *Rhus glabra,* the smooth leaved sumach, in the woods, on high glades and old corn fields.
8. *Vitis labrusca* and *vulpina,* grape vines of several kinds.
9. *Sambucus Canadensis,* American elder tree, along the hedges and on glades.
10. *Quercus phellos,* the swamp oak, in morasses.
11. *Azalea lutea,* the American upright honey-suckle, in the woods in dry places.
12. *Cratægus Crus galli,* cockspur thorn (the Virginian azarole), in woods.
13. *Vaccinium* ...., a species of whortleberry shrub.
14. *Quercus prinus,* the chestnut oak in good ground.

15. *Cornus florida*, the cornelian cherry, in all kinds of soil.
16. *Liriodendron tulipifera*, the tulip tree, in every kind of soil.
17. *Prunus Virginiana*, the wild cherry tree.
18. *Vaccinium* . . . ., a frutex swamp whortleberry, in good ground.
19. *Prinos verticillatus*, the winterberry tree in swamps.
20. *Platanus occidentalis*, the water-beech.
21. *Nyssa aquatica,* the tupelo tree; on fields and mountains.[1]
22. *Liquidambar styraciflua*, sweet gum tree, near springs.
23. *Betula alnus*, alder, a variety of the Swedish; it was here but a shrub.
24. *Fagus castanea*, the chestnut tree, on corn fields, pastures and on wooded hills.
25. *Juglans nigra*, the black walnut tree, in the same place with the preceding tree.
26. *Rhus radicans*, the twining sumach, climbed up the trees.
27. *Acer negundo*, the ash-leaved maple, in morasses and swampy places.
28. *Prunus domestica*, the wild plum tree.
29. *Ulmus Americana*, the white elm.
30. *Prunus spinosa*, sloe shrub, in low places.
31. *Laurus sassafras*, the sassafras tree, in a loose soil mixed with sand.
32. *Ribes nigrum*, the currant tree, grew in low places and in marshes.
33. *Fraximus excelsior*, the ash tree in low places.
34. *Smilax laurifolia*, the rough bindweed with the bay leaf, in woods and near fences.
35. *Kalmia latifolia*, the American dwarf laurel, on the northern side of hills.
36. *Morus rubra*, the mulberry tree on fields, hills and near the houses.
37. *Rhus vernix*, the poisonous sumach, in wet places.
38. *Quercus rubra*, the red oak, but a peculiar variety.
39. *Hamamelis Virginica*, the witch hazel.

---

[1] Dr. Linné mentions only one species of Nyssa, namely *Nyssa aquatica;* Mr. Kalm does not mention the name of the species; but if his is not a different species it must at least be a variety, since he says it grows on hills, whereas the aquatica grows in the water.—F.

40. *Diospyros Virginiana,* the persimmon.
41. *Pyrus coronaria,* the anchor tree.
42. *Juniperus Virginiana,* the red juniper, in a dry poor soil.
43. *Laurus æstvalis,* spice-wood in a wet soil.
44. *Carpinus ostrya,* a species of hornbeam in good soil.
45. *Carpinus betulus,* a hornbeam, in the same kind of soil with the former.
46. *Fagus sylvatica,* the beech, likewise in good soil.
47. *Juglans cinerea,* a species of walnut tree on hills near rivers, called by the Swedes "butternutsträ."
48. *Pinus Americana,* Pennsylvania fir tree; on the north side of mountains and in valleys.[1]
49. *Betula lenta,* a species of birch on the banks of rivers.
50. *Cephalantus occidentalis,* button wood, in wet places.
51. *Pinus tæda,* the New Jersey fir tree, on dry sandy heaths.
52. *Cercis Canadensis,* the sallad tree, in good soil.
53. *Robinia pseudacacia,* the locust tree, on the corn fields.
54. *Magnolia glauca,* the laurel-leaved tulip tree, in marshy soil.
55. *Tilia Americana,* the lime tree, in a good soil.
56. *Gleditsia triacanthos,* the honey locust tree, or three thorned acacia, in the same soil.
57. *Celtis occidentalis,* the nettle tree, in the fields.
58. *Annona muricata,* the custard apple, in a fertile soil.

Visiting the Swedes. We visited several Swedes who were settled here and were at present in very good circumstances. One of them was called Anders Rambo;[2] he had a fine house built of stone, two stories high, and a great orchard near it. We were everywhere well received, and stayed over night with the above-mentioned country-man. We saw no other marks of autumn than that several fruits of this season were already ripe. For, besides this, all the trees were yet as green and the ground still as much covered with flowers as in our summer. Thousands of frogs croaked all night long in the marshes and brooks. The locusts and grasshoppers made likewise such a great noise that it was hardly possible for one person when speaking

---

[1] This species is not found in Linné, *Spec. Plant.*—F.
[2] Anders Rambo, a prominent member of the early Delaware settlers. There were several families by that surname in New Sweden, and many descendants have retained their prominence to this day. Osmond Rambo, Jr. was until recently treasurer of the Swedish Colonial Society.

to understand another. The trees, too, were full of all sorts of birds, which by the variety of their fine plumage, delighted the eye, while the infinite variety of their tunes was continually re-echoed.

The orchards along which we passed to-day were only enclosed by hurdles. But they contained all kinds of fine fruit. We wondered at first very much when our leader leaped over the hedge into the orchards and gathered some agreeable fruit for us. And our astonishment was still greater when we saw that the owners in the garden were so little concerned at it that they did not even look at us. Our companion told us that the people here were not so particular in regard to a few specimens of fruit as they are in other countries where the soil is not so fruitful. We afterwards found very frequently that the country people in Sweden and Finland guarded their turnips more carefully than the people here do the most exquisite fruits.

SEPTEMBER THE 19TH

Dew. As I walked this morning into the fields I observed that a heavy dew had fallen, for the grass was as wet as if it had rained. The leaves of the plants and trees had contracted so much moisture that the drops ran down. I found on this occasion that the dew was not only on the superior but likewise on the inferior side of the leaves. I therefore carefully inspected many leaves both of trees and of other plants, not only those which are situated high up but those which are nearer the ground. I found in all of them that both sides of the leaves were equally bedewed, except those of the *Verbascum thapsus,* or great mullein, which though their superior side was pretty well covered with the dew, yet their inferior had but little.

Fruit trees. Every countryman, even a common peasant has commonly an orchard near his house in which all sorts of fruit, such as peaches were now almost ripe. They are rare in Europe, particularly in Sweden, for in that country hardly any people besides the rich can eat them. But here every countryman had an orchard full of peach trees which were covered with such quantities of fruit that we could scarcely walk in the orchard without treading upon the peaches that had fallen off, many of which were left on the ground. Only a part of them was sold in town, and the rest was consumed by the family and strangers, for everyone that passed by was at liberty to go into the orchard and gather as many of them as he wanted. Nay, this fine fruit was frequently given to the swine.

This fruit is, however, sometimes kept for winter use, and for this purpose is prepared in the following manner: it is cut into four parts, the stone thrown away and the fruit put upon a thread, on which they are exposed to the sunshine in the open air till they are sufficiently dry. They are then put into a vessel for winter. But this manner of drying them is not very good, because the rain of this season very easily spoils and putrifies them, while they hang in the open air. For this reason a different method is followed by others which is by far the most successful. The peaches are as before cut into four parts, are then either put upon a thread, or laid upon a board, and so hung up in the air when the sun shines. Being dried in some measure, or having lost their juice by this means they are put into an oven, out of which the bread has just been taken, and are left in it for a while. They are soon taken out and brought into the fresh air; and after that they are again put into the oven and this is repeated several times till they are as dry as they ought to be. For if they were dried all at once in the oven, they would shrivel up too much and lose part of their flavor. They are then put up and kept for the winter. They are either baked into tarts and pies, or boiled and prepared as dried apples and pears are in Sweden. Several people here dry and preserve their apples in the same manner as their peaches.

The peach trees have, as I am told, been first planted here by the Europeans. But at present they succeed very well, and require even less care than our apple and pear trees.

The orchards have seldom other fruit than apples and peaches. Pear trees are scarce in this province, and those that have any of them have planted them in their orchards. They sometimes have cherry trees, but commonly by the sides of the roads leading to the house, or along the fences. Mulberry trees are planted on some hillocks near the house and sometimes even in the courtyards of the house. The black walnut trees, or *Juglans nigra*, grow partly on hills and in fields near the farmhouses and partly along the fences; but most commonly in the forests. No other trees of this kind are made use of here. The chestnuts are left in the fields; here and there is one in a dry field or in a wood.

Plants. The *Hibiscus esculentus*, or okra,[1] is a plant which grows wild in the West Indies but is planted in the gardens here. The fruit, which is a long pod, is cut while it is green and boiled in soups, which thereby become as thick as porridge. This dish is reckoned a dainty by some people and especially by the negroes.

Capsicum annuum or Guinea pepper is likewise planted in gardens. When the fruit is ripe it is almost red; it is added to a roasted or boiled piece of meat, a little of it being strewed upon it or mixed with the broth. Besides this, cucumbers are pickled with it, or the pods are pounded while they are yet tender, and being mixed with salt are preserved in a bottle. This spice is served on roasted or boiled meat or fried fish and gives them a fine taste. But the fruit itself is as sharp as common pepper.

The Sumach. This country contains many species of the plant which Dr. Linné calls *Rhus*, and the most common is the *Rhus foliis pinnatis ferratis lanceolates retrinque nudis*, or the *Rhus glabra*. The English call this plant sumach. But the Swedes here have no particular name for it and therefore make use of the English name. Its fruit or berries are red. They are made use of for dyeing and afford a color like their own. This tree is like a weed in this country, for if a cornfield is left uncultivated for a few years it appears on it in great abundance, since the berries are spread everywhere by the birds. And when the ground is to be plowed the roots interfere with the operation. The fruits stays on the shrub during the whole winter. But the leaves drop very early in autumn, after they are turned reddish, like those of our Swedish mountain ash. The branches boiled with the berries yield a black ink-like tincture. The boys eat the berries, there being no danger of falling ill after the repast; but they are very sour. The sumach seldom grows to a height above three yards. On cutting the stem, it appears that it contains nothing but pith. I have cut several in this manner and found that some were ten years old, and that most of them were above one year old. When the cut is made a yellow juice comes out between the bark and the wood. One or two of the most outward circles are white, but the innermost are of a yellowish green. It is easy to distinguish one from the other. They contain a large propor-

---

[1] In Miller's *Garden Dictionary*, it is called *Ketmia Indica folio ficus, fructu pentagono recurvo esculento, graciliori et longiori*. The modern scientific name is *Abelmoschus esculentus*.

tion of pith, the diameter of which is frequently half an inch and sometimes more. It is brown and so loose that it is easily pushed out by a little stick in much the same manner as in blackberry bushes. This sumach grows near fences and round the cornfields, but especially on fallow ground. The wood seems to burn well and makes no great crackling while on fire.

SEPTEMBER THE 20TH

In the morning we walked in the fields and woods near the town, partly for gathering seeds and partly for gathering plants for my herbal, which was our principal occupation; and in the autumn of this year we sent part of our collection to England and Sweden.

Poisonous trees. A species of *Rhus*, which was frequent in the marshes here was called the "poison tree" by both English and Swedes. Some of the former gave it the name of "swamp sumach," and my countrymen gave it the same name. Dr. Linné in his botanical works calls it *Rhus Vernix*.[1] An incision being made into the tree, a whitish yellow juice, which has a nauseous smell, comes out between the bark and the wood. This tree is not known for its good qualities, but greatly so for the effect of its poison, which though it is noxious to some people, yet does not in the least affect others. One person can handle the tree as he pleases, cut it, peel off its bark, rub it or the wood upon his hands, smell it, spread the juice upon his skin, and make more experiments, with no inconvenience to himself. Another person on the contrary dares not meddle with the tree while its wood is fresh, nor can he venture to touch a hand which has handled it, nor even expose himself to the smoke of a fire from this wood without soon feeling its bad effects; for the face, the hands, and frequently the whole body swells excessively and is affected with a very acute pain. Sometimes bladders or blisters arise in great numbers and make the sick person look as if he were infected by leprosy. In some people the external thin skin, or cuticle peels off in a few days, as is the case when a person has scalded or burnt any part of his body. Nay, the nature of some persons will not even allow them to approach the place where the tree grows, or to expose themselves to the wind when it carries the effluvia or exhalations of this tree with it, without letting them feel

---

[1] *Species plantarum, I,* 380.—F.

the inconvenience of the swelling which I have just now described. Their eyes are sometimes shut up for two or more days by the swelling. I know two brothers, one of whom could without danger handle this tree in whatever manner he pleased, whereas the other could not come near it without trouble. A person sometimes does not know that he has touched this poisonous plant or that he has been near it before his face and hands show it by their swelling. I have known old people who were more afraid of this tree than of a viper, and I was acquainted with a person who merely by the noxious exhalations of it was swelled to such a degree that he was as stiff as a log of wood and could only be turned about in sheets.

On relating in the winter of the year 1750 the poisonous qualities of the swamp sumach to Jungström, who attended me on my travels, he only laughed and looked upon the whole idea as a fable, in which opinion he was confirmed by his having often handled the tree in the autumn before, cut many branches of it, which he had carried for a good while in his hands in order to preserve its seeds, and put many into the herbariums, all without feeling the least inconvenience. He would, therefore, being a kind of philosopher in his own way, take nothing for granted of which he had no sufficient proofs, especially since he had his own experience in the summer of the year 1749 to support the contrary opinion. But in the next summer his opinion was altered, for his hands swelled and he felt a violent pain and itching in his eyes as soon as he touched the tree, and this inconvenience not only attended him when he meddled with this kind of sumach but even when he touched the *Rhus radicans* (poison ivy), or that species of sumach which climbs on the trees and is not by far so poisonous as the former. By this adventure he was so convinced of the power of the poison tree that I could not easily persuade him to gather more seeds of it for me. He not only felt the noxious effects of it in summer when he was very warm, but even in winter when both he and the woods were cold. Hence it appears that though a person be secure against the power of this poison for a while, yet in course of time he may be affected by it just as much as people of a weaker constitution.

I have likewise tried experiments of every kind with the poison tree on myself. I have spread its juice upon my hands, cut and broken its branches, peeled off its bark and rubbed my hand with it, smelt of it, carried pieces of it in my bare hands, and repeated all this frequently without feeling the baneful effects so commonly attributed to it; yet once I experienced that the poison of the sumach was not entirely without effect upon me. On a hot day in summer, as I was in some degree of perspiration, I cut a branch of the tree and carried it in my hand for about half an hour and smelled of it now and then. I felt no effects from it till in the evening. But the next morning I awoke with a violent itching of my eyelids and the parts thereabouts, and this was so painful that I could hardly keep my hands from it. It ceased after I had washed my eyes for a while with very cold water. But my eyelids were very stiff all that day. At night the itching returned and in the morning as I awoke I felt it as much as the morning before and I used the same remedy against it. However it continued for almost a whole week, and my eyes were very red and my eyelids scarcely movable during that period. My pain ceased entirely afterwards. About the same time I had spread the juice of the tree very thick upon my hands. Three days afterward they occasioned blisters which soon went off without affecting me much. I have not experienced anything more of the effects of this plant, nor had I any desire to do so. However I found that it could not exert its power upon me when I was not perspiring.

I have never heard that the poison of this sumach has been fatal, and the pain ceases after a few days' duration. The natives formerly made their flutes of this tree because it has a great deal of pith. Some people assured me that a person suffering from its unhealthy exhalations would easily recover by spreading a mixture of the wood, burnt to charcoal, and hog's lard, upon the swelled parts. Some asserted that they had really tried this remedy. In some places this tree is rooted out on purpose so that its poison may not affect the workmen.

Minerals. I received today several curiosities belonging to the mineral kingdom which were collected in the country. The following were worth the most attention. The first was a white and quite transparent crystal.[1] Many of this kind are found in Pennsylvania in several kinds of stone, especially in a pale gray limestone. The pieces are of the thickness and length of the little finger and commonly as transparent as possible. But I have likewise gotten crystals here of the length of a foot and of the thickness of a middle-sized man's leg. They are not so transparent as the former.

[1] *Nitrum Crystallus montana* L., *Systema naturæ*, 3. p. 84.

The cubic pyrites [1] of Bishop Brovallius [2] was of a very regular texture. But its cubes were different in size, for in some of them the planes of the sides only amounted to a quarter of an inch, while in the biggest they were fully two inches. Some were exceedingly brilliant so that it was very easy to perceive that they consisted of sulphureous pyrites. But in some only one or two sides glittered so much, and the others were dark brown. Yet most of these marcasites had this same color on all its sides. On breaking them they showed the pure pyrites. They are found near Lancaster in this province and sometimes lie quite above the ground; but commonly they are found at the depth of eight feet or more from the surface, on digging wells and the like. Mr. Hesselius [3] had several pieces of this kind of stone which he used in his painting. He first burnt them, then pounded or ground them to a powder and at last rubbed them still finer in the usual way and this gave him a fine reddish brown color.

Few black pebbles are found in this province which on the other hand yields many kinds of marble, especially a white one, with pale gray, bluish spots, that is found in a quarry at the distance of a few English miles from Philadelphia and is very good for working, though it is not one of the finest kinds of marble. People make tombstones and tables, enchase chimneys and doors and lay floors and flags in front of fireplaces, of this kind of marble. A quantity of this commodity is shipped to different parts of America.

Muscovy glass [4] is found in many places hereabouts and some pieces of it are pretty large and as fine as those which are brought from Russia. I have seen some of them which were a foot and more in length. I have several in my collection that are nearly nine inches square. The Swedes on their first arrival here made their windows of this native glass.

A pale gray fine limestone [5] of a compact texture lies in many places hereabouts and produces a fine lime. Some pieces of it are so full of fine transparent crystals that almost half of the stone consists of nothing else. But besides this limestone they make lime near the seashore from oyster shells and bring it to town in winter, which is

---

[1] *Pyrites crystallinus* L., *Ibid*, p. 113.
[2] Bishop Brovallius, see Introduction. It was he who encouraged Kalm to study the natural sciences.
[3] Gustavus Hesselius, the painter.
[4] *Mica membranacea* L., *Syst. nat.* 3, p. 58.
[5] *Marmor rude* L., *Ibid*, p. 41.

said to be poorer for masonry but better for whitewashing than that
which is gotten from the limestone.

Coal has not yet been found in Pennsylvania, but people claim to
have seen it higher up in the country among the natives. Many
agree that it is found in a great quantity farther north, near Cape
Breton.[1]

The ladies make wine from some of the fruits of the land. They
principally take white and red currants for that purpose, since the
shrubs of this kind are very plentiful in the gardens and succeed
very well. An old sailor who had frequently been in Newfoundland
told me that red currants grew wild in that country in great quantity. They likewise make a wine of strawberries, which grow in
great plenty in the woods but are sourer than the Swedish ones. The
American blackberries, or *Rubus occidentalis*, are also used for this
purpose, for they grow everywhere about the fields almost as abundantly as thistles in Sweden and have a very agreeable taste. In
Maryland a wine is made of the wild grapes which grow in the
woods of that province. Raspberries and cherries which are cultivated and well taken care of also give a fine wine. It is unnecessary
to give an account of the manner of making the currant wine, for in
Sweden this art is of a higher perfection than in North America.

SEPTEMBER THE 21ST

Description of country. The common privet, or *Ligustrum vulgare*
L., grows among the bushes in thickets and woods. But I cannot
determine whether it belongs to the indigenous plants or to those
which the English have introduced, the fruits of which the birds
may have dispersed everywhere. The fences and pales are generally
made here of wooden planks and posts. But a few good economists,
having already thought of sparing the woods for future times, have
begun to plant quick hedges round their fields; and for this purpose
they take the above-mentioned privet, which they plant in a little
bank that is thrown up for it. The soil everywhere hereabouts is
a clay mixed with sand and of course very loose. The privet hedges,

---

[1] This has been confirmed since Cape Breton is in the hands of the English, and it
is reported that the strata of coals run through the whole isle and some basset out
today near the seashore, so that this isle will afford immense treasures of coals when
the government will find it convenient to have them dug for the benefit of the
Nation.—F.

however, are only suitable for keeping out domestic cattle and other such animals here, for the hogs all have a triangular yoke about their necks, and the other cattle are not very unruly. But in places where the latter seek to break through the fences, hedges of this kind make but a poor defence. The people who live in the neighborhood of Philadelphia are obliged to keep their hogs enclosed.

In the afternoon I rode with Mr. Peter Cock, who was a merchant, born in Karlskrona, Sweden,[1] to his countryseat about nine miles from the town to the northwest.

The country on both sides of the road was covered with a large forest. The trees were all deciduous and I did not see a single fir or pine. Most of the trees were different sorts of oak. But we likewise saw chestnut, walnut, locust, apple and hickory trees, with blackberry bushes and the like. The ground ceased to be so even as it had been before and began to look more like the English soil, diversified with hills and valleys. We found neither mountains nor great stones, and the wood was so much thinned and the ground so uniformly even that we could see a great way between the trees under which we rode without any inconvenience, for there were no bushes to stop us. In some places where the soil was thrown up we saw some little stones of a kind of which the houses here are so generally built. I intend to describe them later.

As we went on in the forest we continually saw at moderate distances little fields which had been cleared of the wood. Each one of these was a farm. These farms were commonly very pretty, and a walk of trees frequently led from them to the highroad. The houses were all built of brick or of the stone which is found here everywhere. Every countryman, even the poorest peasant, had an orchard with apples, peaches, chestnuts, walnuts, cherries, quinces and such fruits, and sometimes we saw vines climbing in them. The valleys were frequently blessed with little brooks of crystal-clear water. The fields by the sides of the road were almost all mown and of grain crops only corn and buckwheat were still standing. The former was to be met with near each farm in greater or lesser quantities; it grew very well and to a great length, the stalks being from six to ten feet high and covered with fine green leaves. Buckwheat likewise was quite common, and in some places

---

[1] See page 34.

the people were beginning to reap it. I intend further on to be more specific about the qualities and use of these kinds of grain.

Germantown. After a ride of six English miles we came to Germantown; this settlement has only one street but is nearly two English miles long. It is for the greatest part inhabited by Germans, who come from their country to North America and settle here because they enjoy such privileges as they are not possessed of anywhere else. Most of the inhabitants are tradesmen and make almost everything in such quantity and perfection that in a short time this province will want very little from England, its mother country. Most of the houses are built of stone, which is mixed with glimmer and found everywhere around Philadelphia, but is scarcer further on. Several houses, however, are made of brick. They are commonly two stories high and sometimes higher. The roofs consist of shingles of white cedar wood. Their shape resembles that of the roofs in Sweden, but the angles they form at the top may be either obtuse, right-angled, or acute, according as the slopes are steep or gentle. They sometimes form either half an octagon or half a dodecagon.

Many of the roofs were made in such a manner that they could be walked upon, having a balustrade round them. Many of the upper stories had balconies before them from whence the people had a prospect into the street. The windows, even those in the third story, had shutters. Each house had a fine garden. The town had three churches, one for the Lutherans, another for the Reformed Protestants, and a third for the Quakers. The inhabitants were so numerous that the street was always full. The Mennonites have likewise a meetinghouse.

SEPTEMBER THE 22ND

After I had been at church, I employed the remainder of the day in conversing with several gentlemen in town who had lived here for a long while and I inquired into the curiosities hereabouts.

Mr. Cock had a fine spring near his countryseat; it came from a sandy hill and afforded water enough constantly to fill a little brook. Just over this spring Mr. Cock had erected a building from those above-mentioned glittering stones, into which were put

many jugs and other earthen vessels full of milk, for it kept very
well in cold water during the great heat of the summer.

I afterwards found many houses which were located like this
one over springs, and therefore were intended to keep the meat
and milk fresh.

Fences. Almost all the fences round the grain fields and meadows
hereabouts were made of rails fastened in a horizontal direction.
I only perceived a hedge of privet in one single place. The fences
were not made like ours, for the people here take posts from four
to six feet in height and make two or three holes in them, so that
there was a distance of a foot or more between them. Such a post
does the same service as two [in the Swedish type of fence] and
sometimes three poles are scarcely sufficient. The posts were set
in the ground at two or three fathoms [1] distance from each other,
and the holes in them kept up the rails which were nine inches
and sometimes a foot broad and were inserted one above the other
in the posts. Such a fence therefore looked at a distance like the
hurdles in which we enclose the sheep at night in Sweden. They
were really no closer than these, being only destined to keep out
the bigger animals such as cows, sheep and horses. The hogs are
kept near the farmhouses everywhere about Philadelphia, and
therefore this fence does not need to be made tighter on their
account. Chestnut trees were commonly used for this purpose,
because this wood kept longest against rotting, and a fence made
of it could stand for thirty years. But where no chestnut wood was
to be gotten, the white and black oaks were taken. Of all kinds of
wood that of the red cedar lasts the longest. A very large quantity
of it is brought here, for near Philadelphia it is not plentiful
enough, and many fences near the town are made of this wood.

Fuel. The best wood for fuel in everybody's opinion is hickory,
or a species of walnut, for it heats well; but it is not good for
fences since it cannot well withstand rotting after it is cut. The
white and black oaks are next in goodness for fuel. The woods
with which Philadelphia is surrounded would lead one to conclude
that fuel must be cheap there. But it is far from being so, because
the large and high forest near the town is the property of some
people of quality and fortune who do not regard the money which
they could make from it. They do not fell so much as they require

---

[1] A *fathom* was formerly sometimes only five feet instead of six.

for their own use and much less would they fell it for others. But they leave the trees for times to come, expecting that wood will become much more expensive. However, they fell it for joiners, coach-makers and other artisans, who pay exorbitantly for it. For a quantity of hickory, eight feet in length and four in depth, and the pieces being likewise four feet long, [i. e. a cord] they paid at present eighteen shillings of Pennsylvania currency. But the same quantity of oak only came to twelve shillings. The people who came at present to sell wood in the market were farmers who lived at a great distance from the town. Everybody complained that fuel in the space of a few years, had risen to a price many times as much as it had been formerly, and to account for this the following reasons were given: the town had increased to such a degree that it was four or six times bigger and more populous than some old people knew it to be when they were young. Many brick kilns had been made hereabouts which require a great quantity of wood. The country is likewise more cultivated than it used to be, and consequently large forests have been cut down for that purpose; and the farms built in those places also consume a quantity of wood. Lastly, people melt iron out of ore in several places about the town, and this work goes on without interruption. For these reasons it is concluded that in future times Philadelphia will be obliged to pay a high price for wood.

Wines. The wine of blackberries, which has a very fine taste, is made in the following manner: the juice of the blackberries is pressed out and put into a vessel; with half a gallon of this juice, an equal quantity of water is well mixed. Three pounds of brown sugar are added to this mixture, which must then stand for a while, and after that it is fit for use. Cherry wine is made in the same manner, but care must be taken that when the juice is pressed out, the stones be not crushed, for they give the wine a bad taste.

They make brandy from peaches here after the following method: the fruit is cut asunder and the stones are taken out. The pieces of fruit are then put into a vessel, where they are left for three weeks or a month till they are quite putrid. They are then put into the distilling vessel and the brandy is made and afterwards distilled over again. This product is not good for people who have a more refined taste, but it is only for the common people such as workmen and the like.

Apples yield a brandy when prepared in the same manner as the peaches. But for this purpose those apples are chiefly taken which fall from the tree before they are ripe.

The American nightshade, or *Phytolacca decandra* L., grows abundantly near the farms, on the highroad in hedges and bushes and in several places in the fields. Whenever I came to any of these places I was sure of finding this plant in great abundance. Most plants had already ripe berries, which grew in bunches and looked very tempting, though they were not at all fit for eating. Some of these were yet in flower. In some places such as in hedges and near the houses they sometimes grow two fathoms high. In the fields they were always low, yet I could nowhere perceive that the cattle had eaten of them. A German of this place, by the name of Sleidorn, who was a confectioner, told me that the dyers gathered the roots of this plant and made a fine red dye of them.

Animals. Here are several species of squirrels. The ground squirrels, or *Sciurus striatus* L., are commonly kept in cages because they are very pretty: but they cannot be entirely tamed. The larger squirrels, or *Sciurus cinereus* L., frequently do a great deal of mischief in the plantations, but particularly destroy the corn, for they climb up the stalks, cut the ears in pieces and eat only the loose and sweet kernels inside. They sometimes come by hundreds upon a cornfield and then destroy the whole crop of a farmer in one night. In Maryland, therefore, everyone is obliged annually to kill four squirrels, and their heads are given to a local officer to prevent deceit. In other provinces everybody who kills squirrels receives twopence apiece for them from the public on delivering the heads. Their flesh is eaten and reckoned a dainty. The skins are sold, but are not much esteemed. Squirrels are the chief food of the rattlesnake and other snakes, and it is a common fancy with the people hereabouts that when the rattlesnake lies on the ground and fixes its eyes upon a squirrel, the latter will be as if charmed, and that though it be on the uppermost branches of a tree it will come down by degrees till it leaps into the snake's mouth. The snake then licks the little animal several times and makes it wet all over with his spittle so that it may go down the throat easier. It then swallows the whole squirrel at once. When the snake has made such a good meal it lies down to rest without any concern.

The quadruped, which Dr. Linné in the memoirs of the Royal

FLYING SQUIRREL.

GROUND SQUIRREL

From English edition

Academy of Sciences has described by the name of *Ursus cauda elongata*, and which he calls *Ursus Lotor*, in his *Systema Naturæ*, is here called a raccoon. It is found very frequently and destroys many chickens. It is hunted by dogs, and when it runs up a tree to save itself a man climbs up after it and shakes it down to the ground, where the dogs kill it. The flesh is eaten and is reputed to taste well. The bone of its male parts is used for a pipe cleaner. The hatters purchase their skins and make hats of them, which are next in quality to those of beavers. The tail is worn round the neck in winter and therefore is likewise valuable. The raccoon is frequently the food of snakes.[1]

Oyster Shells. Some Englishmen asserted that near the river Potomac in Virginia a great quantity of oyster shells were to be found and that they themselves had seen whole mountains of them. The place where they are found is said to be about two English miles from the seashore. The proprietor of that ground burns lime out of them. This stratum of oyster shells is two fathoms or more deep. Such quantities of shells have likewise been found in other places, especially in New York, on digging in the ground, and in one place at a distance of a few English miles from the sea a vast quantity of oyster shells and other shells was found. Some people conjectured that the natives had formerly lived in that place and had left the shells of the oysters, which they had consumed in such great numbers. Others could not conceive how it had happened that they were thrown in such immense quantities into one place.

Indians. Everyone was of the opinion that the American savages were a very good-natured people, if they were not attacked. Nobody was so strict in keeping his word as a savage. If any one of their allies came to visit them they showed him more kindness and greater endeavors to serve him than he could have expected from his own countrymen. Mr. Cock gave me the following account as a proof of their integrity. About two years ago an English merchant travelling among the savages in order to sell them necessaries and to buy other goods was secretly killed without the murderers being found out. But about a year afterward the savages found out the guilty person amongst themselves. They immediately seized him, bound his hands on his back, and thus sent him with a guard to

---

[1] We doubt that this rather large animal, if full-grown, can be a food for snakes in the United States.

the governor at Philadelphia, and sent him word that they could no longer acknowledge this wretch (who had been so wicked toward an Englishman) as their countryman, and therefore would have nothing more to do with him, and that they delivered him up to the governor to be punished for his villainy as the laws of England directed. This Indian was afterwards hanged at Philadelphia.

Their good natural parts are proved by the following account which many people have given me as a true one. When their ambassadors are sent to the English colonies in order to settle things of consequence with the governor, they sit down on the ground as soon as they come near him and listen with great attention to the governor's demands which they are to answer. His demands are sometimes many. Yet they have only a stick in their hand and make their marks on it with a knife, without writing anything else down. When they return the next day to give their decisions they answer all the governor's articles in the same order in which he delivered them, without leaving one out, and give such accurate answers as if they had a full account of them in writing.

Mr. Sleidorn related another story which gave me great pleasure. He said he had been at New York and had found a venerable old American savage amongst several others in an inn. This old man began to talk with Sleidorn as soon as the liquor was getting the better of his head and boasted that he could write and read in English. Sleidorn therefore desired leave to ask a question, which the old man readily granted. Sleidorn then asked him whether he knew who was first circumcised, and the old man immediately answered "Father Abraham," but at the same time asked leave to propose a question in his turn which Sleidorn granted. The old man then said, "Who is the first Quaker?" Sleidorn said it was uncertain: that some took one person for it and some another. But the cunning old fellow told him, "You are mistaken, sir; Mordecai was the first Quaker, for he would not take off his hat to Haman."[1] Many of the savages who are yet heathens are said to have some obscure notion of the Deluge. But I am convinced from my own experience that not all are acquainted with it.

Ancient Giants. I met with people here who maintained that giants had formerly lived in these parts, and the following particu-

---

[1] Mordecai-Haman. See *Book of Esther.* Mordecai was a cousin of Esther, and through her influence frustrated Haman's attempt to destroy the Jews.

lars confirmed them in this opinion. A few years ago some people digging in the ground met with a grave which contained human bones of an astonishing size. The *tibia* is said to have been fourteen feet long and the *os femoris* to have measured as much. The teeth are likewise said to have been of a size proportional to the rest. But more bones of this kind have not yet been found. Persons skilled in anatomy, who have seen them, have declared that they were human bones. One of the teeth has been sent to Hamburg to a person who collects natural curiosities. Among the savages in the neighborhood of the place where the bones were found is an account handed down through many generations from father to son, that in this neighborhood on the banks of the river there lived a very tall and strong man, in ancient times, who carried the people over the river on his back and waded in the water, though it was very deep. Everybody to whom he did this service gave him some corn, some skins of animals, or the like. In fine he got his livelihood by this means and was, as it were, the ferryman of those who wanted to cross the river.

The *soil* here consists for the greatest part of sand which is more or less mixed with clay. Both the sand and the clay are of the color of pale bricks. To judge by appearances the ground is none of the best; and this conjecture is verified by the inhabitants of the country. When a grain field has been obliged to bear the same kind of product for three years in succession it does not after that produce anything at all if it be not well manured, or allowed to lie fallow for a few years. Manure is very difficult to obtain and therefore people rather leave the field uncultivated. In that interval it is covered with all sorts of plants and trees, and the countryman in the meanwhile cultivates a piece of ground which has till then been fallow, or he chooses a part of the ground which has never been plowed before and he can in both cases be pretty sure of a plentiful crop. This method can here be used with great convenience. For the soil is loose so that it can easily be plowed and every farmer has commonly a great deal of land for his property. The cattle here are neither housed in winter nor tended in the fields, and for this reason they cannot gather a sufficient quantity of dung.

All the *cattle* have originally been brought over from Europe. The natives have never had them and at present few care to get any. But the cattle degenerate here and gradually become smaller.

The cows, horses, sheep and hogs are all larger in England, though those which are brought over here are of the same breed. But the first generation decreases a little in bulk and the third and fourth is of the same size as the cattle already common here. The climate, the soil and the food together contribute towards producing this change.

Aging of Americans. It is remarkable that the inhabitants of this country commonly acquire understanding sooner but likewise grow old sooner than the people in Europe. It is nothing uncommon to see little children giving sprightly and ready answers to questions that are proposed to them, so that they seem to have as much under-standing as old men. But they do not attain to such an age as the Europeans, and it is an almost unheard-of thing that a person born in this country lives to be eighty or ninety years of age.[1] But I only speak of the Europeans that settle here; for the savages, or first inhabitants, frequently attain a great age, though at present such instances are uncommon, which is chiefly attributed to the great use of brandy which the Indians have learned of the Europeans. Those who are born in Europe attain a greater age than those who are born here of European parents. In the last war it plainly ap-peared that these new Americans were by far less hardy than the Europeans in expeditions, sieges and long sea voyages, and died in large numbers. It is very difficult for them to accustom them-selves to a climate different from their own. The women cease bearing children sooner than in Europe. They seldom or never have children after they are forty or forty-five years old, and some leave off in their thirties. I inquired into the causes of this, but none could give me a good answer. Some said it was owing to the affluence in which the people live here. Some ascribed it to the inconstancy and changeableness of the weather and believed that there was hardly a country on earth in which the weather changed so often in a day as it does here. For if it were ever so hot, one could not be certain whether in twenty-four hours there would not be a piercing cold. Nay, sometimes the weather would change five or six times a day.

Quality of wood. The trees in this country seem to have the same qualities as its inhabitants. The ships which are built of American wood, are by no means equal in point of strength to

---

[1] But see *infra*, pp. 212, 647 and 648.

those which are constructed in Europe. This is what nobody attempts to contradict. When a ship built here has served eight or twelve years it is worth little, and if one is seen which has been in use longer and is yet serviceable it is reckoned very astonishing. It is difficult to find out the reason for this condition. Some lay the fault to the badness of the wood; others condemn the method of building the ships, which is to make them of timber that is green and which has had no time to dry. I believe both causes to be right, for I found oak which at the utmost had been cut down about twelve years and was covered by a hard bark, and upon taking off this bark the wood below it was almost entirely rotten and like flour so that I could rub it into powder between my fingers. How much longer will not our European oak last before it crumbles!

At night we returned to Philadelphia.

September the 23rd

There are no hares in this country, but some animals which are a medium between our hares and rabbits and make a great devastation whenever they get into fields of cabbage and turnips.

Plants. Many people have not been able to find out why the North American plants which are carried to Europe and planted there for the greatest part flower so late and do not produce ripe fruit before the frost overtakes them, although it appears from several accounts of travels that the winters in Pennsylvania, and more so those in New York, New England and Canada, are fully as severe as our Swedish winters and therefore are much severer than those which are felt in England. Several men of judgment charged me for this reason to examine and inquire into this phenomenon with all possible care. I shall for an answer reproduce a few remarks which I have made upon the climate and plants of North America, and leave my readers at liberty to draw their own conclusions.

1. It is true that the winters in Pennsylvania, and much more so those in the more northern provinces, are frequently as severe as our Swedish winters and much colder than the English ones, or those of the southern parts of Europe. For I found at Philadelphia, which is above twenty degrees more southerly than several prov-

inces in Sweden, that the thermometer of Professor Celsius,[1] fell twenty-four degrees below the freezing point in winter. Yet I was assured that the winters I spent here were none of the coldest, but only common ones, which I could likewise conclude from the fact that the Delaware was not frozen strong enough to bear a carriage at Philadelphia during my stay, though this often happens. On considering the breadth of the river, which I have already mentioned in my description of Philadelphia, and the difference between high and low water, which is eight English feet, it will plainly appear that an intense frost is required to cover the Delaware with such thick ice.

2. But it is likewise true that though the winters are severe here they are commonly of no long duration, and I can justly say that they do not continue above two months and sometimes even less, at Philadelphia; and it is something so uncommon when they continue for three months that it is noted in the newspaper. Nearer the pole the winters are somewhat longer of course, and in the parts far north they are as long as the Swedish winters. The daily meteorological observations which I have made during my stay in America, and which I intend to annex at the end of this work, will give more light in this matter.[2]

3. The heat in summer is excessive and without intermission. I own I have seen the thermometer rise to nearly the same degree at Åbo in Finland. But the difference is that when the thermometer rose to thirty degrees C. above the freezing point once in two or three summers in Åbo, the same thermometer in America, did not only for three months stand at the same degree, but even sometimes rose higher, not only in Pennsylvania, but likewise in New York, Albany, and a great part of Canada. During the summers which I spent at Philadelphia, the thermometer rose two or three times to thirty-six degrees above the freezing point.[3] It may therefore with great certainty be said that in Pennsylvania the greatest part of April, the whole of May, and all the following months till October are like our Swedish months of June and July. So excessive and continued a heat must certainly have very great effects.

---

[1] Anders Celsius (1701-1744), Swedish inventor of the Centigrade thermometer.

[2] In this edition the meteorological observations are placed at the end of the second volume.

[3] About 97° Fahrenheit.

I here again refer to my meterorological observations. It must likewise be ascribed to the effects of this heat that the common melons,
the watermelons, and the pumpkins of different sorts are sown
in the fields without any bells or the like put over them, and yet
are ripe as early as July; further, that cherries are ripe at Philadelphia about the twenty-fifth of May, and that in Pennsylvania
the wheat is frequently reaped in the middle of June.

4. The whole of September and half if not the whole of October,
are the finest months in Pennsylvania, for the preceding ones are
too hot. But these represent our July and half of August. The
majority of the plants are in flower in September, and many do
not begin to bloom before the latter end of the month. I have no
doubt that the character of the season which is enlivened by a clear
sky and a tolerably hot sunshine, greatly contributes towards this
last effort of the flora. Yet, though these plants come out so late,
they are quite ripe before the middle of October. But I am not
able to account for their blooming so late in autumn, and I rather
ask, why do not the *Centaurea Pacea*, the *Gentiana, Amarella Virga aurea* flower before the end of summer? or why do the common
noble liverwort, or *Anemone hepatica*, the wild violets (*Viola
martia* L.), the mezereon (*Daphne Mezereum* L.) and other plants
put forth their flowers so early in spring? It has pleased the Almighty Creator to give them this disposition.[1] The weather at
Philadelphia during these months is shown by my meteorological
tables. I have taken the greatest care in my observations, and have
always avoided putting the thermometer into any place where
the sun could shine upon it, or where it had before heated the wall
by its beams; for in those cases my observations would certainly
not have been exact. The weather during our September and October is too well known to need explanation.

5. However, there are some wild plants in Pennsylvania which
do not every year bring their seeds to maturity before the cold
begins. To these belong some species of *Gentiana*, of asters, and
others. But in these too the wisdom of the Creator has wisely
ordered everything in its turn. For almost all the plants which
have the quality of flowering so late in autumn, are perennial, or
such as, though they have no seed to propagate themselves, can

---

[1] Kalm was always very religious, and to-day we should call him orthodox. He
reverently regarded all plants and animals as the direct creation of a personal God.

revive by shooting new branches and stalks from the same roots every year. But perhaps a natural cause may be given for the late growth of these plants. Before the Europeans came into this country it was inhabited by savage nations who practised agriculture but little or not at all, and lived chiefly upon hunting and fishing. The woods therefore had never been meddled with, except that sometimes a small part had been destroyed by fire. The accounts which we have of the first landing of the Europeans here show that they found the country all over covered with thick forests.[1] Hence it follows that, excepting the higher trees and the plants which grow in the water or near the shore, the rest must for the greatest part have been obliged to grow perhaps for a thousand years in shade, either below or between the trees, and they therefore naturally took on characteristics which are only peculiar to woody and shady places. The trees in this country drop their leaves in such quantities in autumn that the ground is covered with them to the depth of four or five inches. These leaves lie a good while in the next summer before they moulder, and this must of course hinder the growth of the plants which are under the trees, at the same time depriving them of the few rays of the sun which come down to them through the thick leaves at the top of the trees. These causes joined together make such plants flower much later than they would otherwise do. May it not therefore be said, that in so many centuries these plants had at last contracted a "habit" of coming up very late, and that it would now require a great space of time to make them lose this habit and quicken their growth?

September the 24th

We employed this whole day in gathering the seeds of plants of all kinds, and in putting rare plants into the herbarium.

September the 25th

Petrified wood. Mr. Hesselius made me a present of a little piece of petrified wood, which was found in the ground here in New Sweden. It was four inches long, one inch broad, and one-fourth

---

[1] *Vide* Hackluyt's *Collect. Voy.*, III, 246.—F.

Photograph by Philip B. Wallace

Courtesy of Henry D. Paxson, Jr.

Home of John Bartram

of an inch thick. It might plainly be seen that it had formerly been wood, for in the places where it had been polished all the longitudinal fibres were easily distinguishable, and it might have been taken for a piece of ash which had been cut smooth. My piece was the part of a still larger one. It was here thought to be petrified hickory. I afterwards got more of it from other people. Mr. Lewis Evans told me that on the boundaries of Virginia, a great petrified block of hickory had been found in the ground, with the bark on it, which was likewise petrified.

Mr. *John Bartram* is an Englishman, who lives in the country about four miles from Philadelphia.[1] He has acquired a great knowledge of natural philosophy and history, and seems to be born with a peculiar genius for these sciences. In his youth he had no opportunity of going to school, but by his own diligence and indefatigable application he got, without instruction, so far in Latin as to understand all books in that language and even those which were filled with botanical terms. He has in several successive years made frequent excursions into different distant parts of North America with an intention of gathering all sorts of plants which are scarce and little known. Those which he found he has planted in his own botanical garden and likewise sent over their seeds or fresh roots to England. We owe to him the knowledge of many rare plants which he first found and which were never known before. He has shown great judgment and an attention which lets nothing escape unnoticed. Yet with all these qualities he is to be blamed for his negligence, for he did not care to write down his numerous and useful observations. His friends in London once induced him to send them a short account of one of his travels, and they were ready, with a good intention though not with sufficient judgment, to get this account printed. But the book did Mr. Bartram more harm than good, for, as he is rather backward in writing down what he knows, this publication was found to contain but few new observations. It would not however be doing justice to Mr. Bartram's merit if it were to be judged by this performance. He has not filled it with a thousandth part of the great knowledge which he has acquired in natural philosophy and history, especially in regard to North America. I have often

---

[1] Cf. page 36. Cf., also, Henry D. Paxson, *Where Pennsylvania History Began* (1926), pp. 147 ff.

been at a loss to think of the sources whence he obtained many things which came to his knowledge. I, also, owe him much, for he possessed that great quality of communicating everything he knew. I shall therefore in this work frequently mention this gentleman. I should never forgive myself if I were to omit the name of a discoverer and claim that as my contribution which I had learned from another person.

In this locality the term *West Indies* is applied to the whole of Southern America, and to those places that lie in the torrid zone, especially its islands.

Many *mussel shells*, or *Mytili anatini*, are to be seen on the northwest side of the city in the clay pits, which were at present filled with water from a little brook in the neighborhood.

These mussels seem to have been washed into the place by the tide when the water in the brook was high. For these clay pits are not old; they were made recently. Poor boys sometimes go out of town, wade in the water, and gather great quantities of these mussels, which they sell very easily, they being reckoned a dainty.

The Virginian azarole with a red fruit, or Linné's *Cratægus Crus galli*, is a species of hawthorn, and is planted in hedges instead of the variety which is commonly used for this purpose in Europe. Its berries are red and of the same size, shape and taste as those of our hawthorn. Yet this tree does not seem to make a good hedge, for its leaves had already fallen, while other trees still preserved theirs. Its spines are very long and sharp, their length being two or three inches, and have some little practical value as pins, pipe cleaners, etc. Each berry contains two stones.

Mr. Bartram assured me that the *North American oak* cannot resist decay nearly as long as the European. For this reason the boats (which carry all sorts of goods down from the upper parts of the country) upon the river Hudson, which is one of the largest in these parts, are made of two kinds of wood. That part which must always be under water is made of black oak; but the upper part, which is now above and now under water and is therefore more exposed to rotting, is made of red cedar or *Juniperus Virginiana*, which is reckoned the hardiest wood in the country. The bottom is made of black oak, because that wood is very tough. The river being full of stones and the boats frequently running against them, the black oak gives way and therefore does not easily crack, but the cedar would not do for this purpose because it is hard and brittle. The oak, also, does not decay so rapidly when it is kept under water.

In autumn I could always get good pears here, but everybody acknowledged that the pear tree does not grow very fast in this country.

All my observations and remarks on the qualities of the rattle-snake, are inserted in the *Memoirs of the Swedish Academy of Sciences* for the years 1752, 1753, and thither I refer the reader.[1]

Bears are very numerous higher up in the country and do much mischief. Mr. Bartram told me that when a bear catches a cow he kills her in the following manner: he bites a hole into the hide and blows with all his power into it till the animal swells excessively and dies, for the air expands greatly between the flesh and the hide.[2] An old Swede called Nils Gustafson, who was ninety-one years of age, said that in his youth the bears had been very frequent hereabouts, but that they had seldom attacked the cattle: that whenever a bear was killed its flesh was prepared like pork, and that it had a very good taste. The flesh of bears is still prepared like ham, on the river Morris. The environs of Philadelphia and even the whole province of New Sweden in general contain very few bears, they having been extirpated by degrees. In Virginia they are killed in several ways. Their flesh is eaten by both rich and poor, since it is reckoned equal in goodness to pork. In some parts of this province, where no hogs can be kept on account of the great numbers of bears, the people try to catch and kill them, and to use them instead of hogs. The American bears, however, are said to be less fierce and dangerous than the European ones.

---

[1] *Vide "Medical etc., cases and experiments,"* translated from the Swedish, London 1758, p. 282.—F.

[2] This has all the appearance of a vulgar error: neither does the succeeding account of the American bears being carnivorous agree with the observations of the most judicious travellers, who deny the fact.

However it might be easy to reconcile both opinions. For Europe has two or three kinds of bears, one species of which is carnivorous, the other lives only on vegetables: the large brown species, with its small variety, are reputed to be carnivorous, the black species is merely phytivorous. In case therefore both species are found in North America, it would be very easy to account for their being both carnivorous and not.—F.

Was Mr. Bartram trying to test Kalm's credulity?—Ed.

Plants. The broad plantain, or *Plantago major,* grows on the high-roads, foot paths, meadows, and in gardens in great quantity. Mr. Bartram had found this plant in many places on his travels, but he did not know whether it was an original American plant or whether the Europeans had brought it over. This doubt had its rise from the savages (who always had an extensive knowledge of the plants of the country) pretending that this plant never grew here before the arrival of the White Men. They therefore gave it a name which signified the (Englishman's) foot, for they say that wherever a European had walked, this plant grew in his footsteps.

The *Chenopodium album,* or goosefoot with sinuated leaves, grows in plenty in the gardens. But it is more scarce near the houses, in the streets, on dunghills and grain fields. This seems to show that it is not a native of America but has been brought over with other seeds from Europe. In the same manner it is thought that the tansy (*Tanacetum vulgare* L.), which grows here and there in the hedges, on the roads, and near houses, was produced from European seeds.

The common vervain, with blue flowers, or *Verbena officinalis,* was shown to me by Mr. Bartram, not far from his house in a little plain near Philadelphia. It was the only place where he had found it in America. And for this reason I suppose it likewise was sown here amongst other European seeds.

Soil Strata. Mr. Bartram was at this time building a house in Philadelphia and had sunk a cellar to a considerable depth, the soil of which was thrown out. I here observed the following strata. The upper loose soil was only half a foot deep, and of a dark brown color. Under it was a stratum of clay so much blended with sand that it was in greater quantity than the clay itself, and this stratum was eight feet deep. These were both brick-colored. The next stratum consisted either of a clear or of a dark quartz;[1] they were quite smooth and roundish on the outside, and lay in a stratum

---

[1] *Quartzum hyalinum* L., Syst. *nat.* 3, p. 65; *Quartzum solidum pellucidum,* Wallerii, *Miner.* 91.; The common Quartz, Forster's *Mineralogy,* p. 16; and *Quartzum coloratum* L., *Syst. nat.* 3, p. 65; *Quartzum solidum opacum coloratum* Wall., *Min.* 99. The impure Quartz, Forst. *Min.* p. 16.—F.

which was a foot deep. The brick-colored clay mixed with sand appeared again. But the depth of this stratum could not be determined. Query: could the river formerly have reached to this place and formed these strata?

Mr. Bartram has not only frequently found oyster shells in the ground, but has likewise met with such shells and snails as undoubtedly belong to the sea, at a distance of a hundred or more English miles from the shore. He has even found them on the ridges of mountains which separate the English plantations from the habitations of the savages. These mountains which the English call the Blue Mountains, are of considerable height and extend in one continuous chain from north to south, or from Canada to Carolina. Yet in some places they have gaps which exist, as it were, to afford a passage for the great rivers that flow down into the lower country.

The *Cassia chamaecrista* L. grew on the roads through the woods and sometimes on uncultivated fields, especially when shrubs grew in them. Its leaves are like those of the sensitive plant or *Mimosa*, and have likewise the quality in common with the leaves of the latter, of contracting when touched.

The crows in this country are little different from our common crows in Sweden.[1] Their size is the same as that of our crows, and they are as black as jet in every part of their body. I saw them flying together today in great numbers. Their voices are not quite like those of our crows, but have rather the cry of the rook, or Linné's *Corvus frugilegus*.

Potholes. Mr. Bartram related that on his journeys to the northern English colonies he had discovered great holes in the mountains on the banks of rivers which according to his description must exactly have been such "giants' pots,"[2] as are to be met with in Sweden, and which I have described in a particular dissertation read in the Royal Swedish Academy of Sciences. Mr. Bartram has likewise addressed some letters to the Royal Society at London upon this subject. Some people insisted that these holes were made by

---

[1] The crows in Sweden have gray wings, not black as the American variety.

[2] In Sweden and in the north of Germany the round holes in rivers, with a stoney or rocky bed, which the whirling of the water has made are called giants' pots; these holes are likewise mentioned in Mr. Grossey's *New observations on Italy.* Vol. I, p. 8.—F.

the savages, that they might in time of war hide their corn and other valuable effects in them. But he wrote against this opinion and accounted for the origin of these cavities in the following manner: when the ice settles, many pebbles stick in it. In spring when the snow melts, the water in the rivers swells so high that it reaches above the place where these holes are now found in the mountains. The ice therefore will of course float as high. And then it often happens that the pebbles which had been imbedded in it ever since autumn, when it first settled on the banks of the river, fall out of the ice upon the rocky bank, and are from thence carried into a cleft or crack by the water. These pebbles are then continually turned about by the water which comes in upon them and by this means they gradually form the hole. The water at the same time polishes the stone by its circular motion round it and helps to make the hole or cavity round. It is certain that by this turning and tossing the stone is at last unfit for its purpose; but every spring the river throws other stones in place of it into the cavity and they are turned round in the same manner. By this whirling both the mountain and the stone produce either a fine or a coarse sand which is washed away by the water when in spring or at other times it is high enough to direct its waves into the cavity. This was the opinion of Mr. Bartram about the origin of these holes. The Royal Society of Sciences at London has given a favorable reception to and approved of them.[1] The remarks which I made in the summer of the year 1743, during my stay in the country in Sweden will prove that I was at that time of the same opinion in regard to these holes. I have further explained this opinion in a letter to the Royal Academy of Sciences, and this letter is still preserved in the Academy's Memoirs, which have not yet been published. But there is great reason to doubt whether or not all cavities of this kind in mountains have the same origin.

Mulberry Trees. Here are different species of mulberry trees which grow wild in the forests of North and South America. In these parts the red mulberry trees are more plentiful than any other. However, Mr. Bartram assured me that he had likewise seen the white mulberry trees growing wild, but that they were scarcer. I

---

[1] How far this approbation of the Royal Society ought to be credited, is to be understood from the advertisements published at the head of each new volume of the *Philosophical Transactions.*—F.

asked him and several other people of this country why they did not set up silk manufactures, having such a quantity of mulberries which succeed so well. For it has been observed that when the berries fall upon the ground where it is not compact but loose, they soon sprout and put out several fine delicate shoots. But they replied that it would not be worth while to erect any silk mills here because labor was so expensive.[1] A man gets from two and a half to three shillings and upwards, for one day's work, and the women are paid in proportion. They were therefore of the opinion that the cultivation of all sorts of grain, of hemp, and of flax would be of greater advantage, and that at the same time it did not require nearly so much care as the feeding of silk worms. Besides, the Colonial cereal products found a ready market in South America. But from the trials of a governor in Connecticut, whose estate lay far north of New York City, it is evident that silk worms thrive very well here and that this kind of mulberry tree is very good for them. The governor brought up a great quantity of silk worms in his courtyard; and they succeeded so well and spun so much silk that they gave him a quantity sufficient for clothing himself and all his family.

Grapevines. Several sorts of vines grow wild hereabouts. Whenever I made a little excursion out of town, I saw them in numerous places climbing up trees and hedges. They clasp and cover them sometimes entirely and even hang down on the sides. This has the same appearance at a distance as the tendrils of hops grasping the trees. I inquired of Mr. Bartram why they did not plant vineyards or press wine from the grapes of the wild vine. He answered that the chief objection to that was the same as to the erection of a silk factory— the necessary labor was too scarce and it was therefore more rational to make agriculture their chief employment. But the true reason undoubtedly is that the wine which is pressed out of most of the North American wild grapes is sour and sharp and has not nearly such an agreeable taste as that which is made from European grapes.

The Virginian wake-robin, or *Arum Virginicum*, grows in wet places. Mr. Bartram told me that the savages boiled the spadix and the berries of this flower and devoured them as a great delicacy. When the berries are fresh they have a harsh, pungent taste, which they lose in great measure upon boiling.

---

[1] We wonder what the people of 1750 would say if they were still living in 1936.

The *Sarothra gentianoïdes* grows abundantly in the fields and under the bushes in a dry sandy ground near Philadelphia. It looks much like our whortleberry bushes when they first begin to grow green and when the points of the leaves are still red. Mr. Bartram sent the plant to Dr. Dillenius,[1] but this gentleman did not know how to classify it. It is reckoned a very good traumatic, and this quality Mr. Bartram himself experienced, for once being thrown and kicked by a vicious horse in such a manner as to have both his thighs greatly hurt, he boiled the Sarothra and applied it to his wounds. Thereupon it not only immediately appeased his pain, which before had been violent, but by its assistance he recovered in a short time.

The Larch Tree. Having read in Mr. Miller's *Botanical Dictionary*[2] that Mr. Peter Collinson had a particular larch tree from America in his garden I asked Mr. Bartram whether he was acquainted with it. He answered that he himself had sent it to Mr. Collinson, that it grew only in the eastern parts of New Jersey, and that he had found it in no other English plantation. It differs from the other species of larch, its cones being much smaller. I afterwards saw it in great numbers in Canada.

Native Fruit. Mr. Bartram was of the opinion that the apple tree was brought into America by the Europeans, and that it never was there before their arrival. But he looked upon peaches as an original American fruit, and as growing wild in the greatest part of America. Others again were of the opinion that they were first brought over by the Europeans. But all the French in Canada agreed that on the banks of the river Mississippi and in the country thereabouts peaches were found growing wild in great quantity.[3]

---

[1] Dr. John James Dillenius, M. D., (1687-1747), professor of botany at Oxford was born and educated in Germany, but came to England in 1721. He published many notable works in his field. Linné spent a month with him at Oxford in 1736 and dedicated his *Critica Botanica* to him.

[2] *The Gardener's Dictionary* by Philip Miller (1691-1771), a famous gardener, had appeared in 1731-1739. The seventh edition of 1759, adopted the Linnean system of nomenclature. A sketch of this, his chief work, had been published in two volumes in 1724 under the name of *The Gardener's and Florist's Dictionary or a Complete System of Horticulture*. The dictionary was very important and was translated into French and German.

[3] Thomas Herriot, servant to Sir Walter Raleigh, who was employed by him to examine into the productions of North America, makes no mention of the peach among the other fruits he describes and M. du Pratz, who has given a very good account of Louisiana and the Mississippi, says, that the natives got their peaches from the English colony of Carolina, before the French settled there.—F.

SEPTEMBER THE 27TH

The tree which the English here call persimmon, is the *Diospyros Virginiana* of Linné. It grows for the greatest part in wet places, round the water pits. I have already mentioned that the fruit of this tree is extremely bitter and sharp before it is ripe and that being eaten in that state it contracts one's mouth and has a very disagreeable taste. But as soon as the persimmons are ripe, which does not happen till they have been quite softened by the frost, they are a very agreeable fruit. They are here eaten raw and seldom any other way. But in a certain book, which contains a description of Virginia, you find different ways of preparing the persimmon under the article of that name.[1] Mr. Bartram, related that they were commonly put upon the table amongst the sweetmeats and that some people made a tolerably good wine of them. Some of these persimmons had dropped on the ground in his garden and were almost ripe, having been exposed to the great heat of the sun. We picked up a few and tasted them, and I must own that those who praised this fruit as an agreeable one have but done it justice. It really deserves a place among the most palatable fruits of this land, when the frost has entirely removed its bitterness.

The *Verbascum thapsus*, or the great white mullein, grows in great quantity on roads, in hedges, on dry fields and high meadows in a soil mixed with sand. The Swedes here call it wild tobacco, but confessed that they did not know whether or not the Indians really used this plant instead of tobacco. The Swedes tie the leaves round their feet and arms, when they have the ague. Some of them prepared a tea from the leaves for dysentery. A Swede likewise told me that a decoction of the roots was injected into the wounds of cattle when afflicted with worms, and that this killed them and made them fall out.[2]

SEPTEMBER THE 28TH

The hay fields which are surrounded by woods, and are at present

---

[1] Kalm does not give the name of the book on Virginia.

[2] These worms are the Larvæ of the *Oestrus* or Gadfly, which deposits its eggs on the back of cattle and the larvæ being hatched from these eggs, cause great sores, wherein they live till they are ready for their change. In the south of Russia they use for the same purpose the decoction of *Veratrum*, or the white *Hellebore.*—F.

mown, have a fine fresh verdure. But those on hills, in the open, or on some elevated location, especially those exposed to the sun, looked brown and dry. Several people from Virginia told me that on account of the great heat and drought the meadows and pastures almost always had a brown color and looked as if they were burnt. The inhabitants of those parts do not therefore enjoy the pleasure which a European feels at the sight of our verdant meadows.

The American nightshade, or the *Phytolacca decandra*, grows abundantly in the fields and under the trees on little hills. Its black berries are now ripe. We observed to-day a few small birds with blue plumage, the size of our hortulans and yellow hammers (*Emberiza citrinella* and *Emberiza hortulanus*) flying down from the trees in order to settle upon the nightshade and eat its leaves.

Towards night I went to Mr. Bartram's country seat.

September the 29th

The *Gnaphalium margaritaceum* grows in astonishing quantities upon all uncultivated fields, glades, hills and the like. Its height varies with the soil and location. Sometimes it is very ramose and sometimes very small. It has a strong but agreeable smell. The English call it "life everlasting," for its flowers, which consist chiefly of dry, shining, silvery leaves (*Folia calycina*) do not change when dried. This plant is now everywhere in full blossom. But some have already lost their flowers and are beginning to drop the seeds. The English ladies are accustomed to gather great quantities of this life everlasting and to pick them with the stalks. For they put them into pots, with or without water, amongst other fine flowers which they gather in the gardens and in the fields, and place them as an ornament in the rooms. English ladies in general are much inclined to keep flowers all summer long about or upon the chimneys, upon a table or before the windows, either on account of their beauty or because of their sweet scent. The above-mentioned *Gnaphalium* was one of those which they kept in their rooms during the winter, because its flowers never altered from what they were when they grew in the ground. Mr. Bartram told me another use of this plant: a decoction of the flowers and stalks is used to bathe pained or bruised parts of the body, or they may be rubbed with the plant itself tied up in a thin cloth or bag.

Instead of flax several people made use of a kind of dog's bane, or Linné's *Apocynum cannabinum*. The people prepared the stalks of this plant in the same manner as we prepare those of hemp or flax. It was spun and several kinds of stuffs were woven from it. The savages are said to have had the art of making bags, fishing nets and the like from it for many centuries before the arrival of the Europeans.

Geological Changes. I asked Mr. Bartram whether he had observed in his travels that the water had receded and that the sea had formerly covered places which were now land. He told me that from what he had experienced he was convinced that the greatest part of this country, even for several miles, had formerly been under water. The reasons which led him to credit this opinion were the following:

1. On digging in the Blue Mountains, which are over three hundred English miles from the sea, you find loose oyster and other sorts of shells, and they are also found in the valleys formed by these mountains.

2. A vast quantity of petrified shells are found in limestone, flint, and sandstone on the same mountains. Mr. Bartram assured me at the same time that it was incredible what quantities of them there were in the different kinds of stones of which the mountains consist.

3. The same shells are likewise dug in great quantities, in quite a non-decayed state, in the provinces of Virginia and Maryland, as also in Pennsylvania and in New York.

4. On digging wells (not only in Philadelphia, but likewise in other places) the people have struck trees, roots and leaves of oak, which were for the greatest part not yet rotten, at the depth of eighteen feet.

5. The best soil and the richest mould is to be found in the valleys hereabouts. These are commonly crossed by a rivulet or brook, and on their borders a mountain commonly rises, which in those places where the brook passes close to it, looks as if it were cut on purpose. Mr. Bartram believed that all these valleys formerly were lakes; that the water had by degrees hollowed out the mountain and opened a passage for itself through it; and that the great quantity of mud which is contained in the water and which had

subsided to the bottom of the lake was the rich soil that exists at present in the valleys and is the cause of their great fertility. But such valleys and cloven mountains are very frequent in the country, and of this kind is the peculiar gap between two mountains through which a river takes its course on the boundaries of New York and Pennsylvania. The people in jest say that this opening was made by the Devil, who wanted to get out of Pennsylvania into New York.

6. The whole appearance of the Blue Mountains plainly shows that water formerly covered a part of them. For many are broken in a peculiar manner, though the highest are plain.

7. When the savages are told that shells are found on these high mountains, and that from thence there is reason to believe that the sea must formerly have extended to them and even in part flowed over them, they answer that this is not new to them, they having a tradition from their ancestors among them that the sea formerly covered these mountains.

8. The water in rivers and brooks likewise decreases in volume. Mills which sixty years ago were built on rivers and at that time had a sufficient supply of water almost all the year long, have at present so little that they cannot run except after a heavy rain or when the snow melts in the spring. This diminution of water in part arises from the great quantity of land which is now cultivated, and from the extirpation of great forests for that purpose.

9. The seashore increases also in time. This comes from the quantity of sand which is continually being thrown on shore from the bottom of the sea by the waves.

Mr. Bartram thought that some special attention should be paid to another factor about these observations. The shells which are to be found petrified on the northern mountains are of a kind not to be gotten in the sea in the same latitude, and they are not found on the shore till you come to South Carolina. Mr. Bartram therefore took an occasion to defend Dr. Thomas Burnet's [1] opinion

---

[1] Dr. Thomas Burnet (d. 1715), English schoolmaster, who took a great interest in the geological history of the earth, and wrote a number of highly original Latin treatises on the subject. "Burnet maintained that the earth resembled a gigantic egg; the shell was crushed at the Deluge, the internal waters burst out, while the fragments of the shell formed the mountains, and at the same catastrophe the equator was diverted from its original coincidence with the ecliptic." [*Dictionary of National Biography*] I cannot find that he had a doctor's degree.

that the earth before the Deluge was in a different position towards the sun. He likewise asked whether the great bones which are sometimes found in the ground in Siberia and which are supposed to be elephant's bones and tusks did not confirm this opinion. For the present those animals cannot live in such cold countries; but if according to Dr. Burnet the sun once formed different zones about our earth from those it now makes, the elephant may easily be supposed to have lived in Siberia. However, it seems that all which we have hitherto mentioned may have been the effect of different causes. To those belong the universal Deluge, the increase of land which is merely the work of time, and the changes of the course of rivers, which when the snow melts so that they become great floods leave their first beds and form new ones.

More Potholes. At some distance from Mr. Bartram's country house, a little brook flowed through the wood and likewise ran over a rock. The attentive Mr. Bartram here showed me several little cavities in the rock and we plainly saw that they must have been generated in the manner I before described, that is, by supposing a pebble to have remained in a cleft of the rock and to have been turned round by the violence of the water till it had formed such a cavity in the mountain. For on putting our hands into one of these holes, we found that it contained numerous small pebbles, whose surfaces were quite smooth and round. And these stones we found in each of the holes.

Mr. Bartram's Herbarium. Mr. Bartram showed me a number of plants which he had collected into a herbarium on his travels. Among these were the following, which likewise grow in the northern parts of Europe, and of which he had either gotten the whole plants or only broken branches.

1. *Betula alba.* The common birch tree, which he had found on the Catskills.
2. *Betula nana (foliis orbiculatis serratis).* This species of birch grows in several low places near the hills.
3. *Comarum palustre,* in the meadows, between the hills in New Jersey.
4. *Gentiana lutea,* the great gentian, from the fields near the mountains. It was much like our variety, but had not so many flowers under each leaf.

5. *Linnæa borealis,* from the mountains in Canada. It creeps along
    the ground.
6. *Myrica Gale,* from the neighborhood of the river Susquehanna,
    where it grows in a wet soil.
7. *Potentilla fruticosa,* from the swampy fields and low meadows,
    between the river Delaware and New York.
8. *Trientalis Europæa,* from the Catskills.
9. *Triglochin maritimum,* from the salt springs near the country
    of the Five Nations.

An Indian Grave. Mr. Bartram showed me a letter from East
Jersey in which he received the following account of the discovery
of an Indian grave. In the April of the year 1744, as some people
were digging a cellar, they came upon a great stone, like a tomb-
stone, which was at last removed with great difficulty, and about
four feet under it they discovered a large quantity of human bones
and a cake of corn. The latter was yet quite untouched, and several
of the people present tasted it out of curiosity. From these circumstances it was concluded that this was a grave of a person of
note among the savages. For it is their custom to bury along with
the deceased meat and other things which he liked best. The stone
was eight feet long, four feet broad, and even some inches more
where it was broadest, and fifteen inches thick at one end, but only
twelve inches at the other. It consisted of the same coarse kind of
material that is to be found everywhere in this locality. There
were no letters nor other characters visible on the stone.

Indian Planting. The grain which the Indians cultivate chiefly
is corn, or *Zea mays* L., and they have little fields for that purpose.
But besides this they plant a great quantity of squashes, a species
of pumpkin or melon, which they have always cultivated, even in
the remotest ages. The Europeans settled in America got the seeds
of this plant from the Indians, and at present their gardens are full
of it. The fruit has an agreeable taste when it is well prepared.
The squashes are commonly boiled, then crushed (as we are used
to do with turnips when we make a porridge of them) and seasoned
with some pepper or other spice, whereupon the dish is ready.
The Indians likewise sow several kinds of beans, which for the
greatest part they have secured from the Europeans. But peas,
which they sow also, they have always had amongst them before
any foreigners came into the country. The squashes of the Indians,

which now are cultivated by the Europeans too, belong to those kinds of gourds (*cucurbita*) which ripen before any other. They are a very delicious fruit, but will not keep. I have, however, seen them kept till pretty late in winter.

SEPTEMBER THE 30TH

Planting Time. Wheat and rye are sown in autumn about this time and commonly reaped towards the end of June or the beginning of July. These kinds of grain, however, are sometimes ready to be reaped in the middle of June, and there are even instances when they have been cut in the beginning of that month. Barley and oats are sown in April and they commonly begin to ripen at the end of July. Buckwheat is sown in the middle or at the end of July and is about this time, or somewhat later, ready to be harvested. If it is sown before the above-mentioned time, as in May or in June, it only produces flowers and little or no grain.

Origin of Pennsylvania Cows. Mr. Bartram and other people assured me, that most of the cows which the English have here are the offspring of those which they bought of the Swedes when they were masters of the country. The English themselves are said to have brought over but few. The Swedes either brought their cattle from home, or bought them of the Dutch, who were then settled here.

Near the town, I saw an ivy or *Hedera helix,* planted against the wall of a stone building which was so covered by the fine green leaves of this plant as almost to conceal the whole. It was doubtless brought over from Europe for I have never perceived it anywhere else on my travels through North America. But in its stead I have often seen wild vines made to run up the walls of the houses.

Trees Versus Environment. I asked Mr. Bartram whether he had observed that the trees and plants decreased in size in proportion as they were brought further to the north, as Catesby [1] pretends. He answered that the question should be more limited and then his opinion would prove more worthwhile. There are some trees which

---

[1] Mark Catesby (c. 1679-1749) naturalist and traveler and member of the Royal Society, was born in England. Beginning in 1722 he spent three years "studying and collecting the fauna and flora of South Carolina, Georgia and Florida." His *Natural History of Carolina, Florida, and the Bahama Islands* appeared in 1731-1743. It was illustrated with etchings made by the author. Catesby died in London.

grow better in southern countries, and become smaller as you advance to the north. Their seeds or berries are sometimes brought into colder climates by birds and by other accidents. They gradually decrease in growth, till at last they will not grow at all. On the other hand, there are other trees and herbs which the wise Creator destined for the northern countries, and they grow there to an amazing size. But the further they are transplanted to the south the smaller they grow, till at last they degenerate so much as not to be able to grow at all. Other plants love a temperate climate, and if they be carried either south or north, they will not succeed well, but always grow smaller. Thus, for example, Pennsylvania contains some trees which grow exceedingly well, but always decrease in proportion as they are carried further off either to the north or to the south.

Afterwards on my travels I had frequent proofs of this truth. The sassafras, which grows in Pennsylvania under forty degrees of latitude and becomes a pretty tall and thick tree, was so small at Oswego and Fort Nicholson, between forty-three and forty-four degrees of latitude, that it hardly reached the height of two or four feet, and was seldom as thick as the little finger or a half-grown person. This was likewise the case with the tulip tree. For in Pennsylvania it grows as high as our tallest oaks and firs, and its thickness is proportional to its height. But about Oswego it was not above twelve feet high, and no thicker than a man's arm. The sugar maple, or *Acer saccharinum*, is one of the most common trees in the woods of Canada and grows very tall. But in the southern provinces, as New Jersey and Pennsylvania, it grows only on the northern side of the Blue Mountains and on the steep hills which are on the banks of the river, and which face the north. Yet there it does not attain to a third or fourth part of the height which it has in Canada. It is needless to mention more examples.

OCTOBER THE 1ST

Mosquitoes. The gnats which are very troublesome at night here, are called mosquitoes. They are exactly like the gnats in Sweden, only somewhat smaller, and the description which is to be seen in Dr. Linné's *Systema Naturæ and Fauna Suecica* fully agrees with that, and they are called by him *Culex pipiens.* In daytime or at night they come into the houses, and when the people have gone to bed they begin their disagreeable humming, approach nearer and nearer to the bed, and at last suck up so much blood that they can hardly fly away. When the weather has been cool for some days the mosquitoes disappear; but when it changes again, and especially after a rain, they gather frequently in such quantities about the homes that their numbers are legion. The chimneys of the English, which have no dampers for shutting them up, afford the gnats a free entrance into the houses. On sultry evenings they accompany the cattle in great swarms from the woods to the houses or to town, and when they are driven before the houses, the gnats fly in wherever they can. In the greatest heat of summer, they are so numerous in some places that the air seems to be full of them, especially near swamps and stagnant waters, such as the river Morris in New Jersey. The inhabitants therefore make a fire before their homes to expel these disagreeable guests by smoke. The old Swedes here said that gnats had formerly been much more numerous; that even at present they swarmed in vast quantities on the seashore near the salt water, and that those which troubled us this autumn in Philadelphia were of a more venomous kind than they commonly used to be. This last quality appeared from the blisters, which were formed on the spots where the gnats had inserted their sting. In Sweden I never felt any other inconvenience from their sting than a little itching, while they sucked. But when they stung me here at night, my face was so disfigured by little red spots and blisters that I was almost ashamed to show myself.

Fences. I have already mentioned something about the fences that are usual here. I now add that most of the rails which are put horizontally, and of which the fences in the environs of Philadelphia chiefly consist, are of red cedar, which is here reckoned more durable than any other. But where this could not be obtained either white or black oak supplied its place. The people were very glad if they could get cedar for posts, but otherwise they took white oak or chestnut, as I was told by Mr. Bartram. But it seems that these kinds of wood in general do not keep well in the earth for any length of time. I saw some chestnut posts that had been put into the ground only the year before, which were already for the greatest part rotten below.[1]

---

[1] Generally, chestnut posts, when they can be obtained, last much longer than a year.—Ed.

The sassafras tree, or *Laurus sassafras* L., grows in abundance in the country, and stands scattered up and down the woods and near bushes and fences. On old land which is left uncultivated it is one of the first that comes up and is as plentiful as young birches are on those Swedish fields which are formed by burning the trees on them.[1] The sassafras grows in a dry loose earth of a pale brick color, which consists for the greatest part of sand mixed with some clay. It seems to be but a poor soil. The hills round Gothenburg in Sweden would afford many places rich enough for the sassafras to grow on, and I even fear they would be too rich. I here saw it both in the woods amidst other trees and more frequently by itself along fences. In both places it thrives equally well. I have never seen it on wet or in low places. The people here gather its flowers and use them instead of tea. But the wood itself is of no use in husbandry; for when it is set on fire it causes a continual crackling and gives no good account of itself. The tree spreads its roots very much, and new shoots come up from them in some places; but these shoots are not good for transplanting, because they have so few fibres besides the root which connects them to the main stem, that they cannot well strike into the ground. If therefore anyone wishes to plant sassafras trees he must endeavor to get their berries, which however is difficult, since the birds eat them before they are half ripe. The cows are very greedy for the tender new shoots and look for them everywhere.

The bark of this tree is used by the women here in dyeing worsted a fine lasting orange color, which does not fade in the sun. They use urine instead of alum in dyeing, and boil the dye in a brass boiler, because in an iron vessel it does not yield so fine a color. Mr. Bartram told me that a woman in Virginia had successfully employed the berries of the sassafras to cure a severe pain in one of her feet, which she had had for three years in such a degree that it almost hindered her from walking. She was advised to boil the berries of sassafras, and to rub the painful parts of her foot with the oil which by this means would be gotten from the berries.

---

[1] In Mr. Osbeck's *Voyage to China,* vol. I, p. 50, in a note, an account is given of this kind of land, which the Swedes call *Swedjeland,* where it is observed, that the trees being burnt, their ashes afford fertilizer sufficient for three years, after which they are left uncultivated again, till after twenty or more years, a new generation of trees being produced on them, the country people burn them and cultivate the country for three years again.—F.

She did so, but at the same time it made her vomit; yet this was not sufficient to keep her from following the prescription three times more, though as often as she made use thereof it always had the same effect. However she was entirely freed from the pain and recovered completely.

OCTOBER THE 2ND

Woodpeckers. A black woodpecker with a red head, or the *Picus pileatus* L., is seen frequently in the Pennsylvania forests and stays all winter, as I know from my own experience. It is counted among those birds which destroy corn, because it settles on the ripe ears and destroys them with its bill. The Swedes call it *tillkråka;* but all other woodpeckers, those with golden yellow wings excepted, are called *hackspettar* in the Swedish language. I intend to describe them all more exactly in a special work. I only observe here that almost all the different species of woodpeckers are very injurious to the corn when it begins to ripen: for by picking holes in the membrane round the ear, the rain gets into it and causes the ear with all the corn it contains to rot.

OCTOBER THE 3RD

Journey to Wilmington, Delaware. In the morning I set out for Wilmington, which was formerly called Christina by the Swedes and is thirty English miles to the southwest of Philadelphia. Three miles south of this city I crossed the river Schuylkill in a ferry, beyond which the country appears almost a continual chain of mountains and valleys. The mountains have an easy slope on all sides and the valleys are commonly traversed by brooks with crystal water. The greater part of the country is covered with several kinds of deciduous trees; for I scarcely saw a single tree of the evergreen variety except a few red cedars. The trees of the forest were tall but branchless below, so that it left a free view to the eye, and no underwood obstructed the passage between them. It would have been easy in some places to have gone under the branches with a carriage for a quarter of a mile, the trees standing at great distances from each other, and the ground being very level. In some spots little glades opened which were either meadows, pastures or grain fields; of the latter some were cultivated and others not. In a few places several houses were built close to one another, but for the most part they were single. In some parts of the fields the wheat had already been sown, in the English manner without trenches but with shallow furrows pretty close together. I sometimes saw the country people very busy sowing their rye. Near every farmhouse was a little field with corn. The inhabitants hereabouts were commonly either Englishmen or Swedes.

All day long I saw a continual variety of trees; walnut trees of different sorts which were full of nuts; chestnut trees quite covered with fine chestnuts; mulberries, sassafras, liquidamber, tulip trees and many others.

Grapevines. Several species of vines grow wild hereabouts. They run up to the summits of the trees, their clusters of grapes and their leaves covering the stems. I even saw some young oaks five or six fathoms high whose tops were crowned with grape vines. The ground is like that so common hereabouts, which I have already described, *viz.* a clay mixed with a great quantity of sand and covered with a rich soil or vegetable earth. The vines are principally seen on trees which stand single near grain fields or at the borders of wooded areas where the meadows, pastures and fields begin, and likewise along fences, where they cling with their tendrils round the trees that stand there. The lower parts of the plant were now full of grapes that hung below the leaves, were almost ripe, and had a pleasant sourish taste. The country people gather them in great quantities and sell them in town. They are eaten without further preparation, and are commonly offered to guests when they come to pay a visit.

The *soil* does not seem to be deep in this section, for the upper black stratum is hardly two inches. This I had an occasion to see, both in the places where the ground is dug up and in such where the water during heavy showers of rain has made cuts, which are pretty numerous here. The upper soil has a dark color and the next a pale hue like bricks. I have observed everywhere in America that the depth of the upper soil does not by far agree with the computation of some people, though we can almost be sure that in some places it has never been stirred since the Deluge.[1] More about this later.

---

[1] Though a scientist, Kalm believed implicitly in the literal account of the Deluge as related in the Bible.

Weeds. The *Datura stramonium*, or thorn apple, grows in great quantities near all the villages. Its height is different according to the soil it is in. For in a rich earth it grows eight or ten feet high, but in a hard and poor ground it will seldom come up to six inches. The *Datura*, together with the *Phytolacca*, or American nightshade, grow here in those places near the gardens, houses and roads which in Sweden are covered with nettles and goosefoot. These European plants are very scarce in America. But the *Datura* and *Phytolacca* are the worst weeds here, nobody knowing any particular use for them.

Turnip fields are sometimes to be seen.

In the middle of the highroad I perceived a dead black snake, which was four feet six inches long and an inch and a half in thickness. It belonged to the viper group.[1]

Late at night a great halo appeared round the moon. The people said that it prognosticated either a storm or rain, or both together. The smaller the ring is, or the nearer it comes to the moon, the sooner this weather sets in. But this time neither of these changes happened, and the ring had indicated a coldness in the air.

Chermes on the alder (*Chermes alni*) were today found in great abundance on the branches of that tree, which for that reason looked almost white and at a distance appeared as if it were covered with mould.

October the 4th

Description of Country. I continued my journey early in the morning, and the country had the same appearance wherever I went. It was a continual chain of rather high hills with an easy ascent on all sides and of valleys between them. The soil consisted of a brick-colored mould, mixed with clay and a few pebbles. I rode now through woods of several sorts of trees and now over pieces of land which had been cleared of the wood and which at present were grain fields, meadows and pastures. The farmhouses stood

---

[1] This is apparently a mistake. Black snakes are not poisonous and cannot therefore belong to the vipers in the usual present sense of that term, although "viper" is sometimes applied loosely to any snake supposedly venomous. The Swedish original, *huggorm,* implies a poisonous reptile, though not one necessarily or even commonly fatal to man.

single, sometimes near the roads, and sometimes at a little distance from them, so that the space between the road and the houses was taken up with small cultivated tracts and meadows. Some of the houses were built of stone, two stories high, and covered with shingles of white cedar. But most of them were wooden and the crevices stopped up with clay instead of moss, which we make use of for that purpose. No dampers were to be found in the chimneys and the people did not even know what I meant by them. The ovens were commonly built at some distance from the houses, and were either under a roof or without any covering against the weather. The fields bore partly buckwheat, which was cut, partly corn, and partly wheat, which had been but lately sown; but sometimes they lay fallow. The grape vines climbed to the top of several trees and hung down on both sides. Other trees again were surrounded by the ivy (*Hedera quinquefolia* L.) which with the same flexibility ascended to a great height. The *Smilax laurifolia* always joined with the ivy, and together with it twisted itself round the trees. The leaves of the ivy were at this time commonly reddish, but those of the vine were still quite green. The trees which were surrounded with ivy leaves looked at a distance like those which are covered with hops in our country (and on seeing the combination from afar, one might expect to find wild hops climbing upon them). Walnut and chestnut trees were common near the fences, in woods and on hills, and at present were loaded with their fruit. The persimmon was likewise plentiful near the roads and in the forests. It had a great quantity of fruit, but it was not yet fit for eating, since the frost had not softened it. At some distance from Wilmington I passed a bridge over a little river which flows north into the Delaware. The rider here pays twopence toll for himself and his horse.

Towards noon I arrived at Wilmington.

Wilmington is a little town about thirty English miles southwest of Philadelphia. It was founded in the year 1733. Part of it stands upon the grounds belonging to the Swedish church which annually receives certain rents, out of which they pay the minister's salary, and employ the rest for other uses. The houses are built of stone and look very pretty; yet they are not built close together, but large open places are left between them. The Quakers have a meeting-house in this town. The Swedish church, which I intend to men-

tion later, is half a mile east of the town. The parsonage is in the
city. A little river called Christina-kill passes by the town and then
empties into the Delaware. The river is said to be sufficiently deep
so that the greatest vessel may come right up to the town, for at
its mouth or juncture with the Delaware it is shallowest, and yet
its depth even there when the water is lowest is from two to two
and a half fathoms. But as you go higher its depth increases to
three, three and a half, and even four fathoms. The largest ships
therefore may safely and with their full cargoes come to and from
the town with the tide. From Wilmington you have a fine view of
a great part of the river Delaware and the ships sailing on it. On
both sides of the river Christina-kill, almost from the place where
the redoubt is built to its juncture with the Delaware, are low
meadows which yield a great quantity of hay. The town carries
on a considerable trade, and would have been larger if Philadelphia
and Newcastle, which are towns of a more ancient date, were not
so near on both sides of it.

The redoubt upon the river Christina-kill was erected this summer when it was known that the French and Spanish privateers
intended to sail up the river and to attempt a landing. It stands,
according to the late Rev. Mr. Tranberg,[1] on the same spot where
the Swedes had built theirs. It is remarkable that on working in
the ground this summer to make this redoubt, an old Swedish silver
coin of Queen Christina's reign, not quite so big as a shilling,[2] was
found among some other things at the depth of a yard. Mr. Tranberg afterwards presented me with it. On one side were the arms
of the house of Vasa with the inscription: CHRISTINA. D. G.
DE. RE. SVE., that is, Christina, by the grace of God, elected
Queen of Sweden; and near this the year of our Lord 1633. On
the reverse were these words: MONETA NOVA REGNI SVEC.,
or, a new coin of the kingdom of Sweden. At the same time a
number of old iron tools, such as axes, shovels, and the like, were
discovered. The redoubt that is now erected consists of bulwarks
of planks with a rampart on the outside. Near it is the powder
magazine in a vault built of bricks. At the erection of this little

---

[1] Rev. Petrus Tranberg had arrived in America, 1726, to take charge of the Swedish
Church at Raccoon and Penn's Neck. He had moved to Christina, now Wilmington,
Delaware, in 1742, and died in 1748.
[2] The Swedish, *rundstycke,* indicates a copper coin a little smaller than our quarter.

fortification it was remarkable that the Quakers whose tenets reject even defensive war, were as busy as the other people in building it. For the fear of being every moment suddenly attacked by privateers conquered all other thoughts. Many of them scrupled to put their own hands to the work, but promoted it by supplies of money and by getting ready everything which was necessary.

OCTOBER THE 5TH

It was my design to cross the Delaware and to get into New Jersey with a view to get acquainted with the country; but as there was no ferry to bring my horse over I set out on my return to Philadelphia. I went partly along the highway and partly deviated on one or the other side of it in order to make more exact observations of the land and of its natural history.

Corn was seen in several places. In some its stalks had been cut somewhat below the ear, dried and put up in narrow high stacks, in order to keep them as food for the cattle in winter. The lower part of the stalk had likewise leaves, but as they commonly dry of themselves the people do not like to feed the cattle with them, all their flavor being lost. But the upper ones are cut while they are yet green.

The valleys between the hills ordinarily have brooks; but they are not very broad and require no bridges, so that carriages and horses can easily pass through them, for the water is seldom above six inches deep.

The leaves of most trees were still quite green, such as those of oaks, chestnuts, black walnut trees, hickory, tulip trees and sassafras. The two latter species are found in plenty on hillsides, on the fallow fields, near hedges, and on the road. The persimmon likewise still had its leaves; however some trees of this kind had dropped them. The leaves of the American bramble were at present almost entirely red, though some of these bushes yet retained a vivid green in the leaves. The Cornelian cherry also had already a mixture of brown and pale leaves. Those of the red maple were also red.

I continued my journey to *Chichester*, a borough upon the Delaware, where travellers cross the river in a ferry. They build here every year a number of small ships for sale. From iron works which lie higher up in the country they carry iron bars to this place for shipment.

Canoes [1] are boats made of one piece of wood and are much in use among the farmers and other people upon the Delaware and some little rivers. For that purpose a very thick trunk of a tree is hollowed out; the red juniper or cedar (*Juniperus Virginiana*), the white cedar, the chestnut, the white oak and the tulip tree are commonly used. Canoes of red and white cedar are reckoned the best because they float very lightly upon the water and last twenty years. But of these the red cedar canoes are most preferred. Those made of chestnut will likewise last for a good while, while those of white oak are hardly serviceable more than six years and also float deep because they are so heavy. The liquidambar tree, or *Liquidambar styraciflua* L., is large enough, but unfit for making canoes because it imbibes the water. The size of the canoes varies with the purposes for which they are destined. They can carry six persons, who, however must in no way be unruly, but sit at the bottom of the canoe in the quietest manner possible, lest the boat capsize. The Swedes in Pennsylvania and New Jersey, near the rivers, seldom have any other boats in which to go to Philadelphia, which they commonly do twice a week on market days, though they be several miles distant from the town, and meet sometimes with severe storms. Yet misfortunes from the overturning etc. of these canoes are seldom heard of, though they might well be expected on account of the small size of the boats. Still, a great deal of attention and care is necessary in managing the canoes when the wind is a little violent; for they are narrow, round below, have no keel and therefore may easily be upset. Accordingly when the wind is more brisk than ordinary the people make for the land.

The common garden cresses grow in several places on the roads about Chichester, and undoubtedly come from the seeds which were by chance carried out of the many gardens about that town.

---

[1] Kalm was much interested in canoes, and when he returned to Finland, he had a pupil write a thesis on the subject, presumably from material furnished by Kalm. Cf. editor's article on "Pehr Kalm's Writings on America," *Scandinavian Studies and Notes* for May, 1933. See, also, p. 776 of this work, item 32.

Johan Lorentz Odhelius in his Swedish memorial address on Kalm (*Åminnelse-Tal över Herr Pehr Kalm*), 1781, tells us (p. 20, note) that during a visit to Åbo in 1752 King Adolf Fredrik of Sweden was entertained by a boat ride in a birch canoe, "in a manner which Professor Kalm had observed among the American Savages."

The *American brambles* (*Rubus occidentalis* L.) are here in great abundance. When a field is left uncultivated they are the first plants to appear on it, and I frequently observed them in such fields as are annually plowed and have grain sown on them. For when these bushes are once rooted they are not easily extirpated. Such a bush runs out tendrils that are sometimes four fathoms from its root, and then grows a new one, so that on pulling it up you meet with roots on both ends. On some old land which had long been uncultivated there were so many bushes of this kind that it was very troublesome and dangerous walking among them. A wine is made of the berries, as I have already mentioned. The berries are likewise eaten when they are ripe, and taste well. No other use is made of them.

OCTOBER THE 6TH

The *Chenopodium anthelminticum* is very plentiful on the road and on the banks of the river, but chiefly in dry places in a loose sandy soil. The English who are settled here call it wormseed and Jerusalem oak. It has a disagreeable scent. In Pennsylvania and New Jersey its seeds are given to children as a cure for worms, and for that purpose they are excellent. The plant itself is common in both provinces.

The *environs of Chichester* contain many gardens which are full of apple trees, sinking under the weight of innumerable apples. Most of them are winter fruit, and therefore were yet quite sour. Each farm has a garden, and so has each house of the better sort. The extent of these gardens is also not inconsiderable, and therefore furnishes the possessor all the year long with abundant supplies for his housekeeping, both for eating and drinking. I was frequently surprised at the prudence of the inhabitants of this country. As soon as one has bought a piece of ground which is neither built upon nor sown, his first care is to get young apple trees and make a garden. He next proceeds to build his house, and lastly prepares the uncultivated land for grain. It is well known that the trees require many years before they arrive at the bearing age, and this makes it necessary to plant them first. I now perceived near the farms mills, wheels and other instruments which are made use of in crushing the apples for cider.

From Chichester I went on towards Philadelphia. The oaks were the most plentiful trees in the wood. But there were several species of them, all different from the European ones. The swine now went about in great herds in the oak woods, where they fed upon the acorns which fell in great abundance from the trees. Each hog had a wooden triangular yoke about its neck by which it was hindered from squeezing through the holes in the fences, and for this reason they are made very slender and easy to put up, and do not require much wood. No other fences are in use but those which are like sheep hurdles. A number of squirrels were in the oak woods, some running on the ground and some leaping from one branch to another; and at this time they fed chiefly upon acorns.

I seldom saw beech trees, but I found them quite the same as the European ones. Their wood is considered very good for joiners' planes.

Ants. I do not remember seeing any other than the black ants, or *Formica nigra*, in Pennsylvania. They are as black as coal and of two sorts, some very little, like the smallest of our ants, and others of the size of our common reddish ants. I have not yet observed any hills of theirs, but only seen some running about singly. In other parts of America I have found other species of ants, which I intend to describe later.

The *common privet*, or *Ligustrum vulgare*, is made use of in many places as a hedge round grain fields and gardens, and on my whole voyage I did not see any other shrubs or trees used for this purpose, though the Englishmen here well know that the hawthorn makes a much better hedge. The privet hedges grow very thick and close, but having no spines the hogs and even other animals easily break through them, and when they have once made a hole it requires a long while before the opening is closed by further growth. But when the hedges consist of spinose bushes, the cattle will hardly attempt to get through them.

About noon I came through *Chester*, a little market town which lies on the Delaware. A rivulet running down from the country passes through this place and empties into the Delaware. There is a bridge over it. The houses stand dispersed. Most of them are built of stone and are two or three stories high; some are however made of wood. In the town is a church and a marketplace.

Wheat is now sown everywhere. In some places it was already

green, having been sown four weeks before. The wheat fields were prepared in the English manner, having no ditches in them but numerous furrows for draining the water, at a distance of four or six feet from one another. Great stumps of trees which had been cut down are everywhere seen on the fields, and this shows that the country has been but lately cultivated. The roots of the trees do not go deep into the ground but spread horizontally. I had opportunities of observing this in several places where the trees were dug up, for I seldom saw one whose roots went above a foot deep into the ground, though it was a loose soil.

A Forge. About two English miles beyond Chester I passed by an iron forge, which was on the right by the roadside. It belonged to two brothers, as I was told. The ore however is not dug here but thirty or forty miles away, where it is first melted in a furnace and then carried to this place. The bellows were made of leather, and both they and the hammers, and even the hearth, were but small in proportion to ours. All the machines were worked by water. The iron was wrought into bars.

To-day I noticed what I have since frequently seen on my travels in this country that horses are very fond of apples. When they are let into an orchard to feed upon the grass and there are any apples on the ground, they frequently leave the fresh green grass and eat the apples, which however are not reckoned a good food for them; and besides that it is too expensive.

The red maple, or *Acer rubrum*, is plentiful in these places. Its native location is chiefly swampy, wet places, in which the alder commonly is its companion. Out of its wood they make plates, spinning wheels, spools, feet for chairs and beds, and many other kinds of turnery. With the bark they dye both worsted and linen, giving it a dark blue color. For this purpose it is first boiled in water, and some copperas,[1] such as the hat makers and the shoemakers commonly use, is added, before the stuff (which is to be dyed) is put into the boiler. This bark likewise yields a good black ink. When the tree is felled early in spring a sweet juice runs out of it, like that which comes out of our birches. This sap is not made use of here, but in Canada they make both syrup and sugar of it. There is a variety of this tree which they call the curled

---

[1] Formerly, either green, blue or white vitriol, i. e. the sulphates of iron, copper or zinc.

maple, the wood being as it were marbled within; it is much used in all kinds of joiners' work, and the utensils made of this wood are preferable to those of any other kind in this country, and are much dearer than those made of the wild cherry (*Prunus Virginiana*) or of black walnut. But the most valuable utensils are those made of curled black walnut, for that is an excessively scarce kind of wood. The curled maple is likewise very uncommon, and you frequently find trees whose outsides are marbled but their inside not. The tree is therefore cut very deep before it is felled, to see whether it has veins in every part.

In the evening I reached Philadelphia.

October the 7th

New Jersey. In the morning we crossed the Delaware in a boat to the other side which belongs to New Jersey, each person paying fourpence for his passage. The land here is very different from that in Pennsylvania, for here the ground is almost entirely sand, while in the other province it is mixed with a good deal of clay, and this makes the soil very rich. The descriptions which I made to-day of insects and plants I intend to mention in another work.

Corn. One might be led to think that a soil like this in New Jersey could produce nothing, because it is so dry and poor. Yet the corn which is planted on it grows extremely well, and we saw many fields covered with it. The earth is the kind in which tobacco commonly thrives, but it is not nearly so rich. The stalks of corn are usually eight feet high, more or less, and are full of leaves. The corn is planted as usual in squares, in little hills, so that there is a space of five feet and six inches between each hill, in both directions.[1] From each of these little hills [that we saw] three or four stalks had come up, which were not yet cut for the cattle; each stalk again had from one to four ears, which are large and full of corn. A sandy ground could never have been better employed. In some places the ground between the corn is plowed and rye sown in it, so that when the corn is cut the rye remains upon the field.

We frequently saw *asparagus* growing near the fences in a loose soil on uncultivated sandy fields. It was also plentiful between the

---

[1] This distance between the corn hills seems a little large to the present-day farmer.

stands of corn, and was at present full of berries; but I cannot tell whether or not the seeds are carried by the wind to the places where I saw them. It is however certain that I have likewise seen it growing wild in other parts of America.

The *wormseed* is also plentiful on the roads in a sandy soil such as that near the ferry opposite Philadelphia. I have already mentioned that it is given to children as a remedy for worms. It is then put into brandy, and when it has been there for a few hours it is taken out again, dried and given to the children, either in beer sweetened with syrup or in some other liquor. Its effects are a matter of dispute. Some people say it kills the worms; others again pretend that it furthers their increase. But I know from my own experience that this wormseed has had very good results upon children.

The *portulaca* which we cultivate in our gardens grows wild in great abundance in the loose soil amongst the corn. It was there creeping on the ground, and its stems were pretty thick and succulent, which circumstances very justly gave reason to wonder whence it could get juice sufficient to supply it in such dry ground. It is plentiful in such soil in other places of this country.

The *Bidens bipinnata* is here called Spanish needles. It grows single about farmhouses, near roads, pales and along hedges. It was yet partly in flower, but for the greatest part it had already finished blooming. When its seeds are ripe it is very disagreeable walking where it grows, for they stick to the clothes and make them black, and it is difficult to remove the spots which they occasion. Each seed has three spines at its extremity, and each of these again is full of numerous little hooks, by which the seed fastens itself to the clothes.

In the woods and along the hedges in this neighborhood some small red ants (*Formica rubra*) crept about, and their antennæ or feelers were as long as their bodies.

Towards night we returned to Philadelphia.

OCTOBER THE 8TH

Oysters. The shore of Pennsylvania has a great quantity of the finest oysters. About this time the people begin to bring them to Philadelphia for sale. They come from that part of the shore which is near the mouth of the Delaware River. They are considered as good as the New York oysters, of which I shall make special mention later. However, I believe that the latter kind is generally larger, fatter and more palatable. It is remarkable that they commonly become edible about the time when the agues have spent their fury (i. e. in October). Some men are seen with whole carts full of oysters crying them about the streets. This is unusual here when anything is to be sold, but in London it is very common. The usual way of preparing oysters here is to fry them on live coals until they begin to open. They are then eaten with a sandwich of soft wheat bread and butter. Since they are sooty outside from the fire it is customary to hold them in the left hand with a rag or napkin while eating. The oyster shells are thrown away, though formerly a lime was burnt from them, which has been found unnecessary, there being better material from which to make lime in this neighborhood. The people showed me some houses in this town which were built of stone, and in the mason work of which the lime of oyster shells had been employed. The walls of these houses are always so wet two or three days before a rain, that great drops of water can plainly be perceived on them, and thus they serve as hygrometers.[1] Several people who had lived in them complained of these inconveniences.

OCTOBER THE 9TH

A Pea Pest. Peas are not much cultivated in Pennsylvania at present, though formerly, according to the accounts of some old Swedes, every farmer had a small field of them. In New Jersey and the southern parts of New York also, peas are no longer planted in such quantities as they used to be. But in the northern parts of New York, or about Albany, and in all the parts of Canada which are inhabited by the French, the people sow great quantities of them, and have a large crop. In the former colonies a little despicable insect has compelled the people to give up this useful product of agriculture. The little insect was formerly little known, but a few

---

[1] As the shells of oysters are a marine animal production, and their cavities are full of particles of sea water, the moisture evaporates, leaving behind its salt. When the shells are burnt, and the lime is slacked, the salt mixes with the lime; and though the mortar of such a lime grows ever so dry, the particles of salt immediately attract the moisture of the air, and cause that dampness complained of here.—F.

years ago it multiplied excessively. It couples in summer about the time when the peas are in blossom, and then deposits an egg into almost every one of the little peas. When these are ripe their outward appearance does not expose the worm, which however is found within when they are cut. If undisturbed this worm lies in the pea all winter and part of the spring, and in that space of time consumes the greatest part of the inside of the pea. In spring therefore little more than the mere thin outward skin is left. This larva at last changes into an adult insect, of the Coleoptera class, and in that state creeps through a hole of its own making in the husk, and flies off to look for new fields of peas in which it may couple with its cogeneric insects and provide food sufficient for its posterity.

This noxious insect has spread from Pennsylvania to the north. The country about New York, where it is common at present, had not been plagued with it until some twelve or fifteen years ago: before that time the people had sown peas every year without any inconvenience and had excellent crops. But by degrees these little enemies came in such numbers that the inhabitants were forced to leave off sowing peas. The people complained of this in several places. The farmers about Albany have still the pleasure to see their pea fields left untouched by these beetles, but are constantly afraid of their approach, as it has been observed that they come nearer to that territory every year.

I know not whether this insect could live in Europe, for I should think our Swedish winters would kill the larva, were it ever so deeply imbedded in the pea. But it is often as cold in New York (where this insect is so abundant) as in our country, and yet it continues to multiply there every year and to proceed farther north. I was very near bringing some of these vermin into Europe without knowing it. At my departure from America I took some sweet peas with me in a paper, and they were at that time quite fresh and green. But on opening the paper after my arrival at Stockholm, on August the first, 1751, I found all the peas hollow, and the head of an insect peeping out of each. Some of these insects even crept out in order to try the weather of the new climate; but I made haste, to shut the paper again, in order to prevent the spreading of this noxious insect.[1] I own that when I first perceived them I was

---

[1] Though Mr. Kalm has so carefully avoided haunting Europe with this insect, Dr. Linné assures us in his *Systema Naturæ*, that the southern countries of Europe are al-

more frightened than I would have been at the sight of a viper. For I understood at once the damage which my dear country would have suffered if only two or three of these terrible insects had escaped me. The posterity of many families, and even the inhabitants of whole provinces, would have had sufficient reason to detest me as the cause of so great a calamity. I afterwards sent some of them, though well secured, to Count Tessin[1] and to Dr. Linné, together with an account of their destructive qualities. Dr. Linné has already inserted a description of them in an academical dissertation, which has been drawn up under his presidency, and treats of the damages made by insects.[2] He there calls this insect the *Bruchus* of North America.[3] It was very peculiar that every pea in the paper was eaten without exception.

When the inhabitants of Pennsylvania sow peas procured from abroad, they are not commonly attacked by these insects for the first year; but in the next they take possession of the pea. It is greatly to be wished that none of the ships which annually depart from New York or Pennsylvania will bring them into the European countries. From the above the power of a single despicable insect will plainly appear; and also, that the study of entomology is not to be looked upon as a mere pastime and useless employment.[4]

Poison Ivy. The *Rhus radicans* is a shrub or tree[5] which grows abundantly in this country and has in common with the ivy, called *Hedera arborea*, the quality of not growing [to any height] without the support either of a tree, a wall or a hedge. I have seen it

---

ready infested with it; Scopoli mentions it among his *Insecta Carniolica* p. 63, and Geoffroy among his *Parisian Insects,* Vol. I, p. 267, t. 4. f. 9, has given a fine drawing of it.—F.

[1] Carl Gustaf Tessin (1741-1770), President of the Royal Swedish Academy of Sciences and one of the gentlemen who was most instrumental in sending Kalm to North America.

[2] Diss. *de Noxa Insectorum, Amœn. Acad.* vol. 3, p. 347.—F.

[3] In his *Systema Naturæ,* he calls it *Bruchus Pisi,* or the peas beetle; and says that the *Gracula Quiscula,* or purple daw of Catesby, in the greatest destroyer of them, and though this bird has been proscribed by the legislature of Pennsylvania, New Jersey and New England, as a corn thief, they feel however the imprudence of extirpating this bird; for a quantity of worms which formerly were eaten by these birds destroy their meadows at present.—F.

[4] If the peas were steeped before they are sown in a lye of lime water, and some dissolved arsenic, the pupa or aurelia of the insect would be killed.—F.

[5] It is a little difficult for the non-scientific person to think of poison ivy as a shrub or tree. Why not call it simply a vine or a climber?

climbing to the very top of high trees in the woods, and its branches shoot out little roots everywhere, which fasten upon the tree and, as it were, enter into it. When the stem is cut, it emits a pale brown sap of a disagreeable scent. This sap is so strong that the letters and characters made upon linen with it cannot be removed, but grow blacker the more the cloth is washed. Boys commonly marked their names on their linen with this juice. If you write with it on paper the letters never go out, but grow blacker as time goes on.

This species of *sumach* has the same noxious qualities as the poisonous sumach or poison tree, which I have above described, being injurious to some people, though not to every one. Therefore all that has been said of the poison tree is likewise applicable to this, excepting that the former has the stronger venom. However, I have seen people who have been as swollen from the noxious exhalations of the latter, as they could have been from those of the former. I also know that of two sisters one could handle the tree without being affected by its poison, and the other immediately felt it as soon as the exhalations of the tree came near her, or whenever she came a yard too close to it, and even when she stood in the way of the wind which blew directly from this shrub.[1] But upon me this species of sumach has never exerted its power, though I made above a hundred experiments upon myself with the largest stems, and the juice once squirted into my eye without doing me any harm. On another person's hand which I had covered very thick with the sap, the skin a few hours afterward became as hard as a piece of tanned leather, and peeled off during the following days as if little scales had fallen from it.

OCTOBER THE 10TH

In the morning I accompanied Mr. Cock to his countryseat, which is about nine miles north of Philadelphia.

Ship Building. Though the woods of Pennsylvania have many oaks, and more species of them than are found further north, people do not build so many ships in this province as they do in the northern part, and especially in New England. But experience has taught

---

[1] Recently some Wesleyan University scientists found that the virulent essence of the poison ivy does not vaporize; hence the poisoning from it can come only through contact, direct or indirect.

them that the same kind of tree is more durable the further north it grows, and that this advantage decreases the more it grows in warm climates. It is likewise plain that the trees in the south grow more every year, and form thicker ringlets than those in the north. The former have likewise much greater tubes for the circulation of the sap than the latter. And for this reason they do not build so many ships in Pennsylvania, as they do in New England, though more than in Virginia and Maryland. Carolina builds very few, and its merchants get all their ships from New England. Those which are here made of the best oak are hardly serviceable above ten, or at the most twelve years; for then they are so rotten that nobody ventures to go to sea in them. Many captains of ships come over from England to North America in order to get ships built here. But most of them choose New England, that being the most northerly province; and even if they come over in ships which are bound for Philadelphia, they frequently on their arrival set out from Pennsylvania for New England. The Spaniards in the West Indies are said to build their ships of a peculiar sort of cedar, which holds out against decay and moisture, but it is not obtainable on the continent in the English provinces. Here are more than nine different sorts of oak, but with regard to quality not one of them is comparable to the single species we have in Sweden, and therefore a ship of European oak costs a great deal more than one made of the American variety.

Red Beets. Many people whose chief occupation was gardening had found in a succession of years, that the red beet which had been grown from New York seed became very sweet and had a very fine taste, but that every year it lost part of its goodness if it was cultivated from seeds which were procured here. The people were therefore obliged to get as many seeds of red beets every year from New York as were wanted in their gardens. It has also been generally observed, that the plants which are produced from English seeds are much better than those which come from native ones.

In the garden of Mr. Cock was a radish which in the loose soil had grown so big as to be seven inches in diameter. Everybody that saw it owned it was uncommon to see one of such size.

Sweet Potatoes. That species of *Convolvulus* which is commonly called Batatas, has here the name of "Bermudian potatoes." Common people and gentry without distinction planted them in their gardens. This is done in the same manner as with the common potatoes. Some people made little hillocks, into which they put these potatoes; but others only planted them in flat beds. The soil must be a mixture of sand and earth, and neither too rich nor too poor. When farmers are about to plant these they cut them, as common potatoes, taking care however that a bud or two is left upon each piece. Their color is commonly red without and yellow within. They are bigger than the common sort, and have a sweet and very agreeable taste, which I cannot find in the other potatoes, in artichokes, or in any other root, and they almost melt in the mouth. It is not long since they were first planted here. They are prepared in the same manner as common potatoes and are either mixed and served with them or eaten separately. They grow very fast and very well here; but the greatest difficulty consists in keeping them over winter, for they stand neither cold, great heat, nor wet. They must therefore be kept during winter in a box with sand in a warm room. In Pennsylvania where they have no dampers in their chimneys, they are put in such a sand box, at some distance from the fire, and there they are secure both against frost and heat. It will not answer the purpose to put them into dry sand in a cellar, as is usually done with the common sort of potatoes, for the moisture which is always in cellars penetrates the sand and makes them rot. It would probably be very easy to keep them in Sweden in warm rooms during the cold season. But the difficulty lies wholly in bringing them over to Sweden. I carried a considerable number of them with me on leaving America and took all possible care in preserving them. But we had a very violent storm at sea, through which the ship was so greatly damaged that the water came in everywhere and wet our clothes, beds and other goods so much that we could wring the water out of them. It is therefore no wonder that my Bermuda potatoes became rotten; but as they are now cultivated in Portugal and Spain, nay even in England, it will be easy to bring them into Sweden. The drink which the Spaniards prepare from these potatoes in their American possessions is not usual in Pennsylvania.[1]

Paper Mill. Mr. Cock had a paper mill on a little brook, and all

[1] Mr. Miller describes this liquor in his *Gardener's Dictionary* under the article of *Convolvulus,* species the 17th, and 18th.—F.

the coarser sorts of paper were manufactured in it. It is now annually rented for fifty pounds Pennsylvania currency.

OCTOBER THE 11TH

The Apple Crop. I have already mentioned, that every countryman has a greater or lesser number of apple trees planted round
his farmhouse from which he gets large quantities of fruit, a part
of which he sells. From another part he makes cider, and some are
used in his own family for pies, tarts and the like. However, he
cannot expect an equal amount of fruit every year, and I was told
that this season had not by far yielded such a quantity of apples as
the preceding, the cause being the continual severe drought in the
month of May which had hurt all the blossoms of the apple trees
and made them wither. The heat had been so great that it dried
up all the plants and the grass in the fields.

The *Polytrichum commune,* a species of moss, grew plentifully
on wet and low meadows between the woods, and in several places
quite covered them, as our mosses cover the meadows in Sweden.
It was likewise very plentiful on hills.

Agriculture was in a very bad state hereabouts. Formerly when
a person had bought a piece of land, which perhaps had never been
plowed since Creation, he cut down a part of the wood, tore up
the roots, tilled the ground, sowed seed on it, and the first time he
got an excellent crop.—But the same land after being cultivated
for several years in succession, without being manured, finally loses
its fertility of course. Its possessor then leaves it fallow and proceeds to another part of his land, which he treats in the same manner. Thus he goes on till he has changed a great part of his possessions into grain fields, and by that means deprived the ground of
its fertility. He then returns to the first field, which now has pretty
well recovered. This he tills again as long as it will afford him a
good crop; but when its fertility is exhausted he leaves it fallow
again and proceeds to the rest as before.

Careless Farming. It being customary here to let the cattle go
about the fields and in the woods both day and night, the people
cannot collect much dung for manure. But by leaving the land
fallow for several years a great quantity of weeds spring up in it,

and get such strength that it requires a considerable time to extirpate them. This is the reason why the grain is always so mixed
with the seed of weeds. The great richness of the soil which the
first European colonists found here, and which had never been
plowed before, has given rise to this neglect of agriculture, which
is still observed by many of the inhabitants. But they do not consider that when the earth is quite exhausted a great space of time
and an infinite deal of labor are necessary to bring it again into
good condition, especially in these countries which are almost every
summer scorched by the excessive heat and drought. The soil of
the grain fields consists of a thin mould, greatly mixed with a
brick-colored sand and clay and a quantity of small particles of
glimmer (mica). The latter comes from stones which are found
here almost everywhere at the depth of a foot or so. These little
pieces of mica make the ground sparkle when the sun shines
upon it.

Building Stone. Almost all the houses hereabouts are built either
of stone or brick, but those of stone are more numerous. Germantown, which is about two English miles long, has no other houses,
and the country homes in that locality are all built of stone. But
there are several varieties of that material which is commonly made
use of in building. Sometimes it consists of a black or gray mica,
running in undulated veins, the spaces between them being filled
up with a gray, loose, fine-grained limestone which is easily crushed.
Some transparent particles of quartz are scattered in the mass, of
which mica forms the greatest part. It is very easily cut, and with
proper tools can readily be shaped into any form. Sometimes however the pieces consist of a black fine-grained mica, a white similarly-
grained sandstone, and some particles of quartz, the several constituent parts being well mixed together. Sometimes the stone has
broad stripes of white limestone without any addition of mica, but
more commonly they are blended together and of a gray color.
Again, this stone is found to consist of quite fine black pieces of
mica, and a gray, soft and very fine-grained limestone. This is likewise very easily cut, being incompact.

These varieties of stone are commonly found close together. They
are abundant everywhere at a small depth, but not in equal quantity or quality, and cannot always be readily broken. When therefore a person intends to build a house, he inquires where the best

stone can be obtained. It is to be found on grain fields and meadows, at a depth which varies from two to six feet. The pieces are different in size. Some are eight or ten feet long, two broad, and one thick. Others are still bigger, but frequently much less. Hereabouts they lie in strata one above another, the thickness of each stratum being about a foot. The length and breadth are different, but commonly such as I have before mentioned. Ordinarily the builders have to dig three or four feet before they reach the first stratum. The loose ground above that stratum is full of little pieces of this stone. This ground is the common brick-colored soil, which is universal here, and consists of sand and clay, though the former is more plentiful. The loose pieces of mica which shine so much in it seem to have been broken off from the great strata of stone.

It must be observed that when people build with this stone they take care to turn the flat side of it outwards. But as that cannot always be done, the stone being frequently rough on all sides, it is easily cut smooth with tools, since it is soft, and not very difficult to be worked. The stones however are unequal in thickness, and therefore by putting them together they cannot be kept in such straight lines as bricks. It sometimes happens also that pieces break off when they are cut and leave holes on the outside of the wall. But in order to fill up these holes, the little pieces of stone which cannot be made use of are pounded, mixed with mortar, and put into the holes. The places thus filled up are afterwards smoothed, and when dry they are hardly distinguishable from the rest, at a distance. Finally, on the outside of the wall, lines in the mortar are made, which cross each other horizontally and perpendicularly, so that it looks as if the wall consisted wholly of equal, square stones, and as if the white strokes were the places where they were joined with mortar. The inside of the wall is made smooth, covered with mortar and whitewashed. It has not been observed that this kind of stone attracts the moisture in a rainy or wet season. In Philadelphia and its environs you find several houses built of this kind of material.

Houses. The houses here are commonly built in the English manner. Upon investigating I found that the lowest room was entirely underground, and its walls made of the above-mentioned stone, although the house above it was built of brick. It was used for a cellar, pantry, wood-shed, or sometimes a kitchen, and merchants occasionally kept their goods in it. There was no real, useless garret as with us in Sweden, for the building was so constructed that rooms for dwelling purposes extended up to the roof, generally with a fireplace in them, and sometimes with dormer windows, so that the servants at least could in the summer live in them comfortably. Besides, clothes and other household goods could be stored there. If a fireplace was built in the middle of a gable-wall there were cupboards set up on each side. [Then follows a description by Kalm of the type of window which prevailed and prevails to this day in America, showing the advantages of a window that opens by moving it up and down on cords instead of outward on hinges.] Most of the houses in Philadelphia were built of brick, and in the country, also, the same material was coming into use, because the clay hereabout was of excellent quality. It was not a custom to whitewash the houses. They looked like some old stone structures or churches, and had the color of brick with mortar between. The walls did not seem to have been injured by the weather. Similarly, there was no chimney in Philadelphia covered with lime or white-wash.

The Badger. One of Mr. Cock's negroes showed me the skin of a badger which he had killed a few days ago, and which convinced me that the American badger is the same as the Swedish one. It was here called ground hog.[1]

Towards night I returned to Philadelphia.

OCTOBER THE 12TH

In the morning we went to the river Schuylkill, partly to gather seeds, partly to collect plants for the herbarium, and to make all sorts of observations. The Schuylkill is a narrow river, which flows into the Delaware, about four miles south of Philadelphia; but narrow as it is, it rises on the west side of those high mountains, commonly called the Blue Mountains, and flows two hundred English miles, and perhaps more. It is a great disadvantage to this section that there are several cataracts in the river as far down as Philadelphia, for which reason there can be no navigation on it. To-day

---

[1] To-day this designation is generally given to the woodchuck (*Marmota monax*). The badger is the *Taxidea taxus;* a North European species, the *Meles taxus.* Did Kalm see a woodchuck?

I made some descriptions and remarks on such plants as the cattle liked, or such as they never touched.

Moles. I observed several little subterranean passages in the fields, running under ground in various directions, the opening of which was big enough for a mole: the earth, which formed as it were a vault above it, and lay elevated like a little bank, was nearly two inches high, fully as broad as a man's hand, and about two inches thick. On uncultivated land, I frequently saw these subterraneous passages, which could be noticed from the ground thrown up above them, and which when trod upon gave way and made it inconvenient to walk.

These passages are inhabited by a kind of mole[1] which I intend to describe more accurately in another work. Its food is commonly roots, and I observed the following qualities in one which was caught. It had greater rigidity and strength in its legs than I ever saw in any other animal, in proportion to its size. Whenever it intended to dig, it held its legs obliquely, like oars. I laid my handkerchief before it, and it began to stir in it with the snout, and taking away the handkerchief to see what it had done to it I found that in the space of a minute it had made it full of holes, and it looked as if it had been pierced by an awl. I was obliged to put some books on the cover of the box in which I kept this animal, otherwise it would have been flung off immediately. It was very irascible, and would bite great holes into anything that was put in its way. I held a steel pen-case close up to it: at first it bit at the case with great violence, but having felt its hardness, it would not venture again to bite at anything. These moles do not make such hills as the European ones, but only such passages as I have already mentioned.

OCTOBER THE 13TH

Bayberries. There is a plant here from whose berries the settlers make a kind of wax or tallow, and for which reason the Swedes call it the "Tallow Shrub." The English call the same tree the

---

[1] This animal is probably the *Sorex cristatus* of Dr. Linné, who says it is like the mole and lives in Pennsylvania.—F.

The species common in the eastern United States are the *Scalopus aquaticus,* which has partially webbed feet, and the *Condylura cristata.*—Ed.

candleberry-tree, or bayberry-bush; and Dr. Linné gives it the name
of *Myrica cerifera*. It grows abundantly in wet soil, and it seems
to thrive particularly well in the neighborhood of the sea; in fact
I have never found it high up in the country far from the water.
The berries grow abundantly on the female shrub, and look as if
flour had been strewed upon them. They are gathered late in
autumn, being ripe about that time, and are then thrown into a
kettle or pot full of boiling water. By this means their fat melts
out, floats at the top of the water, and is skimmed off into a
vessel. The skimming is continued until there is no tallow left.
The latter, as soon as it is congealed, looks like common tallow or
wax, but has a dirty green color. It is for that reason melted over
again and refined, by which means it acquires a fine and rather
transparent green color: this tallow is dearer than common tallow,
but cheaper than wax. In Philadelphia they paid a shilling Pennsylvania currency for a pound of this tallow, while a pound of the
common kind only came to half that money, and wax costs as much
again.

Candles. From this tallow people make candles in many parts of
the province, but they usually mix some common tallow with it.
Candles of this kind do not easily bend, and do not melt in summer as common candles do; they burn better and slower; nor do
they cause any smoke, but rather yield an agreeable smell, when
they are extinguished. An old Swede ninety-one years of age [1] told
me that these candles had formerly been much in use with his
countrymen. At present, however, they do not make so many of
this kind, if they can get the tallow of animals, it being too troublesome to gather the berries. However, these candles are made use
of by poor people, who live in the neighborhood of a place where
the bushes grow, and have not cattle enough to kill to supply them
with a sufficient quantity of ordinary tallow. From the wax of the
candleberry tree they likewise make a soap here, which has an
agreeable scent, and is the best for shaving. This wax is also used
by doctors and surgeons, who reckon it exceedingly good for plas-
ters upon wounds. A merchant of this town once sent a quantity
of these [bayberry] candles to those American provinces which had
Roman Catholic inhabitants, thinking he would be well paid, since
wax candles are used in the Roman Catholic churches; but the

---

[1] Nils Gustafsson, see p. 63.

clergy would not take them. An old Swede mentioned that the root of the candleberry tree was formerly utilized by the Indians as a remedy against the toothache, and that he himself having had the affliction very violently had cut the root in pieces and applied it round his tooth. The pain had been lessened by it. Another Swede assured me that he had been cured of the toothache by applying the peel of the root to it. In Carolina they not only make candles out of the wax of the berries but also sealing-wax.

OCTOBER THE 14TH

Pennyroyal is a plant which has a peculiar strong scent, and grows abundantly on dry places in the country. Botanists call it *Melissa pulegioides*.[1] An extract from it is reckoned very wholesome to drink as a tea when a person has a cold, as it promotes perspiration. I was likewise told that on feeling a pain in any limb, this plant, if applied to it, would give immediate relief.

Exports. The goods which are shipped to London from New England are the following: all sorts of fish caught near Newfoundland and elsewhere; train-oil of several sorts; whale-bone; tar, pitch, and masts; new ships, of which a great number is annually built; a few hides; and sometimes certain kinds of wood. The English islands in America, as Jamaica and Barbadoes, get from New England, fish, flesh, butter, cheese, tallow, horses, cattle; all sorts of lumber, such as pails, buckets, and hogsheads; and have returns made in rum, sugar, molasses, and other produce of the country, or in cash, the greatest part of which is sent to London (the money especially) in payment for the goods received from there. Yet all this is insufficient to pay off the debt.

OCTOBER THE 15TH

The *alders* grew here in considerable abundance on wet and low places, and even sometimes on pretty high ones, but they never reached the height of the European varieties, and commonly stood like a bush about a fathom or two high. Mr. Bartram and other

---

[1] Forster and Marchand, in the English and French translations, respectively, call it *Cunila p.* The modern term is *Hedeoma pulegioides*.

gentlemen who had frequently travelled in these provinces told me
that the further south you went, the smaller were the alders, but
that they were higher and thicker the more you advanced to the
north. I found afterwards, myself, that the alders in some places
of Canada were but little inferior to the Swedish ones. Their bark
is employed here in dyeing red and brown. A Swedish inhabitant of
America told me that he had once cut his leg to the very bone,
and that some blood had already congested within; that he had been
advised to boil the alder bark, and to wash the wound often with
the water; that he had followed this advice, and had soon got his
leg healed, though it had been very dangerous at first.

The Pokeweed. The *Phytolacca Americana* is called poke by the
English. The Swedes have no particular name for it, but make use
of the English term, with some slight variation into *påk*. When the
juice of its berries is put upon paper or the like, it dyes it a deep
purple, which is as fine as any in the world, and it is a pity that no
method has as yet been discovered of making this color last on
woolen and linen cloth, for it fades very soon. Mr. Bartram mentioned that having hit his foot against a stone, he had gotten a violent pain in it; that he had then bethought himself of putting a leaf
of the *Phytolacca* on his foot, by which he had lost the pain in a
short time and got his foot well soon after. The berries are eaten by
the birds about this time. The English and several Swedes make
use of the leaves in spring, when they just come out and are still
tender and soft, and eat them, partly as green kale and partly in
the manner we eat spinach. Sometimes they prepare them in the first
of these ways, when the stalks have already grown a little longer,
breaking off none but the upper sprouts which are yet tender and
not woody. But in the latter case, great care has to be taken, for if
you eat the plant when it is large and its leaves are no longer soft,
you may expect death as a consequence, a calamity which seldom
fails to follow, for the plant has then got a power of purging the
body to excess. I have known people who by eating large full-
grown leaves of this plant have been afflicted by such a strong dysentery that they came near dying from it. Its berries, however, are
eaten in autumn by children without any ill consequences.

Tanning Bark. Woolen and linen cloth is dyed yellow with the
bark of hickory. This likewise is done with the bark of the black
oak, or Linné's *Quercus nigra*, [and that variety of it which Catesby

in his *Natural History of Carolina*[1] calls *Quercus marilandica*] The flowers and leaves of the *Impatiens Noli tangere* or balsamine, likewise dye all woolen stuffs with a fine yellow color.

The Horse Balm. The *Collinsonia Canadensis*[2] was frequently found in little woods and bushes, in a good rich soil. Mr. Bartram, who knew the country well, was sure that Pennsylvania and all parts of America in the same latitude constituted the true and original habitat of this plant. For farther to the south, neither he nor Messrs. Clayton and Mitchell[3] had ever found it, though the latter gentlemen had made accurate observations in Virginia and parts of Maryland. From his own experience he knew that it did not grow in the northerly parts. I have never found it more than fifteen minutes north of forty-three degrees. The time of the year when it comes up in Pennsylvania is so late that its seed has just sufficient time in which to ripen, and therefore it seems unlikely that it can succeed further north. Mr. Bartram was the first to discover it and send it to Europe. Mr. Jussieu,[4] during his stay at London, and Dr. Linné afterwards, called it *Collinsonia*, from the celebrated Mr. Peter Collinson, a merchant in London, and fellow of the English and Swedish Royal Societies. He well deserved the honor of having a plant called after his name, for there are few people that have promoted natural history and all useful sciences with a zeal like his, or that have done as much as he towards collecting, cultivating and making known all sorts of plants. The *Collinsonia* has a peculiar scent, which is agreeable, but very strong. It always gave me a pretty violent headache whenever I passed by a place where it stood in plenty, and especially when it was in flower. Mr. Bartram was acquainted with a better characteristic of this plant, namely that of

---

[1] Vol. I, tab. 19.

[2] Also called citronella.

[3] Messrs. Clayton and Mitchell: John Clayton (1685-1773), botanist. The chief results of his work are embodied in *Flora Virginiana*, by J. F. Gronovius. He corresponded with Linné and probably with Kalm. His work was of special importance to the former.

John Mitchell (d. 1769), English botanist who spent much time in America. He discovered several new species of plants. Linné named one *Mitchella repens* after him. One of his botanical works was dedicated to Sir Hans Sloane and another, *Nova Plantarum genera,* to Peter Collinson.

[4] Bernard de Jussieu (1699-1758), noted French botanist, friend and sometime collaborator of Linné. He preferred, however, a natural system in the classification of plants rather than the somewhat arbitrary but more scientific method advocated and introduced by Linné.

being an excellent remedy against all sorts of pain in the limbs, and for a cold, when the parts affected were rubbed with it. And Mr. Conrad Weiser,[1] interpreter of the language of the Indians in Pennsylvania, had told him of a more wonderful cure with this plant. He had once been among a company of Indians, one of which had been stung by a rattlesnake. The savages gave him up, but Mr. Weiser boiled the Collinsonia and made the poor wretch drink the water, from which he happily recovered. Farther north and in New York they call this plant horseweed, because the horses eat it in spring, before any other plant comes up.

OCTOBER THE 16TH

Land Formerly Covered with Water. I asked Mr. Benjamin Franklin and other gentlemen who were well acquainted with this country, whether they had come upon any evidence that places which were now a part of the continent had formerly been covered with water; and I received the following answer.

1. On travelling from here to the south, you meet with a place where the highway is very low in the valley between two mountains. On both sides you see nothing but oyster and mussel shells in immense quantities. Yet the place is many miles from the sea.

2. Whenever colonists dig wells or build houses in town, they find the earth lying in several strata above each other. At a depth of fourteen feet or more they find globular stones, which are as smooth on the outside as those which lie on the seashore and have been made round by the rolling of the waves. After having dug through the sand and reached a depth of eighteen feet or more they discovered in some places a mud like that which the sea throws up on the shore, and which commonly lies at its bottom and in rivers; this mud is full of stumps, leaves, branches, reed, charcoal, etc.

3. It has sometimes happened that new houses have sunk on one side in a short time, and have obliged the people to pull them down again. On digging deeper for hard ground to build upon they have found a quantity of the above mud, wood, roots, etc.

Are not these reasons sufficient to make one suppose that those

---

[1] Conrad Weiser, son of Johann Conrad Weiser, had as a boy lived with the Mohawk Indians and learned their language and customs. See A. B. Faust, *The German Element in the United States,* second ed. pp. 94 ff.

places in Philadelphia which are at present fourteen feet and more under ground were formerly the bottom of the sea, and that by several violent changes, sand, earth, and other things were carried upon them? or, that the Delaware formerly was broader than it is at present? or that it has changed its course? This last still happens at present, the river tearing off material from the bank on one side, and depositing it on the other. Both the Swedes and English often showed me such places.

OCTOBER THE 18TH

Plants in Blossom. At present I did not find above ten different kinds of plants in blossom: they were a *Gentiana,* two species of aster, the common golden rod, or *Solidago Virga aurea,* a species of *Hieracium,* the yellow wood sorrel, or *Oxalis corniculata,* the foxglove, or *Digitalis purpurea,* the *Hamamelis Virginiana,* or witch-hazel, our common milfoil, or *Achillæa millefolium,* and our dandelion, or *Leontodon taraxacum.* All other plants had for this year laid aside their gay colors. Several trees, especially those which were to flower early in spring, had already formed such large buds that on opening them all the parts of reproduction, such as calyx, corolla, stamen and pistil were plainly distinguishable. It was therefore easy to determine the genus to which such trees belonged. Such were the red maple, or *Acer rubrum,* and the *Laurus æstivalis,* a species of bay. Thus nature prepared to bring forth flowers with the first mild weather of the next year. The buds were at present quite hard and their parts pressed so close together that all cold might be excluded.

The *black walnut trees* had for the greatest part dropped their leaves, and many of them were entirely without them. The walnuts themselves had already fallen off. The green peel which enclosed them, if frequently handled, would yield a black color, which could not be got off the fingers in two or three weeks' time, though the hands were washed ever so much. It was therefore not advisable to carry these nuts with their green shells in any cloth, for it would then be spoiled.

Dogwood. The *Cornus florida* was called dogwood by the English, and grew abundantly in the woods. It looks beautiful when it is adorned with its numerous great white flowers in spring. The wood is very hard, and is therefore made use of for weavers' spools, joiners' planes, wedges, etc. When the cattle fall down in spring for want of strength, the people tie a branch of this tree on their neck, thinking it will help them.

OCTOBER THE 19TH

The *tulip tree* grows everywhere in the woods of this country. The botanist calls it *Liriodendron tulipifera*, because its flowers, both with respect to their size and exterior form and even in some measure with regard to their color, resemble tulips. The Swedes call it "canoe tree," for both the Indians and the Europeans often make their canoes from its trunk. The Englishmen in Pennsylvania give it the name of poplar. It is considered to grow to the greatest height and thickness of any in North America, and vies in that point with our greatest European trees. The white oak and the fir in North America, however, are almost as tall. It must therefore be very agreeable in spring, at the end of May (when it is in blossom) to see one of the largest trees covered for a fortnight with flowers, which with regard to their shape, size, and color are like [some] tulips. The leaves have likewise something peculiar; the English therefore in some places call the tree the "old woman's smock," because the leaves resemble one.

Its wood is here used for canoes, boards, planks, bowls, dishes, spoons, doorposts, and all sorts of joiners' work. I have seen a barn of considerable size whose walls and roof were made of a single tree of this kind, split into boards. Some joiners reckoned this wood better than oak, because the latter frequently is warped, while the former is not, and can easily be worked. Others again valued it very little. It is certain that it contracts enough in hot weather to occasion great cracks in the boards, and in wet weather it swells so as to be near bursting. The people hardly know of a wood in these parts which varies so much in contracting and expanding. The carpenters, however, use it a great deal in their work. They say there are two species of it; but there are merely two varieties, one of which in time turns yellow within, and the other is white. The former is said to have a looser texture. The bark (like Russia glass) is divisible into very thin layers, which are very tough like bast, though I have never seen it employed as such. The leaves when crushed and applied to the forehead are said to be a remedy for headache. When horses are plagued with worms, the bark is pounded and given to them in dry form. Many people believe its roots to be as efficacious against the fever as the Jesuits' bark.[1] The trees grew in all sorts of dry soil, both on high and low grounds, but too wet a soil will not agree with them.

October the 20th

Sweet Bay Tree. The beaver tree grows in several parts of Pennsylvania and New Jersey, in a poor swampy soil, or on wet meadows. Dr. Linné calls it *Magnolia glauca;*[2] both the Swedes and English call it the "beaver tree," because beavers are caught by using its root as bait. However the Swedes sometimes give it a different name, and the English as improperly called it "swamp sassafras" and "white laurel." The trees of this kind dropped their leaves early in autumn, though some of the young trees kept them all winter. I have seldom found the beaver tree north of Pennsylvania, where it begins to flower about the end of May. The scent of its blossoms is agreeably strong, for by it you can discover within three quarters of an English mile whether these little trees stand in the neighborhood, provided the wind be not against it. The whole air is filled with their sweet and pleasant scent. It is agreeable beyond description to travel in the woods about that time, especially towards night. They retain their flowers for three weeks and even longer, according to the quality of the soil on which the trees stand; and during their whole blossom time they spread their odoriferous exhalations. The berries likewise look very fine when they are ripe, for they have a rich red color, and hang in bunches on slender stalks. The cough and other pectoral diseases are cured by putting the berries into rum or brandy, of which a draught may be taken every morning. The virtues of this remedy are universally extolled, and even praised for their salutary effects in consumption. The bark being put into brandy, or boiled in any other liquor, is said not only to ease pectoral diseases, but likewise to be of some service against all internal pains and fever, and it was thought that a decoction of it could stop dysentery.

---

[1] A Peruvian bark of the genus *Cinchona*. It has much the same properties as quinine.

[2] Now *Magnolia Virginiana* or sweet bay tree.

Persons who had caught cold, boiled the branches of the beaver tree in water, and drank it to their great relief. A Swede, called Lars Låck, gave the following account of a cure effected by this tree. One of his relations, an old man, had an open sore in his leg, which would not heal up again, though he had much advice and used many remedies. An Indian at last effected the cure. He burnt some of this wood to charcoal, which he reduced to powder, mixed it with the fresh fat of pork and rubbed the open places several times with it. This dried up the holes, which before were continually open, and the legs of the old man remained sound to his death. The wood is likewise made use of for joiners' planes.

OCTOBER THE 22ND

Domesticated Animals and Birds. Upon trial it has been found that the following animals and birds, which are wild in the woods of North America, can be made nearly as tractable as domestic animals.

The *Wild Cows* and *Oxen.* Several people of distinction have taken young calves from these wild cows which exist in Carolina, and other provinces to the south of Pennsylvania, and brought them up among the domesticated cattle. When grown up they were perfectly tame, but at the same time very unruly, so that there was no fence strong enough to hold them if they had a mind to break through it; for as they possess a great strength in their neck, it was easy for them to knock down the posts with their horns, and to get into the grain fields; and as soon as they had made a hole, all the tame cattle followed them. They likewise copulated with the latter, and by that means generated as it were a new breed. This American species of oxen is Linné's *Bos Bison.*

American deer can likewise be tamed, and I have seen such myself in several places. A farmer in New Jersey had one in his possession which he had caught when it was very young; and at present it was so tame that in the daytime it ran into the wood for its food, and towards night it returned home, and frequently brought a wild deer out of the wood, giving its master an opportunity to shoot it. Several people have therefore tamed young deer and made use of them for hunting wild ones, or for decoying them home, especially at the time of their rutting.

Beavers have been made so tame that they have gone out fish-

MOCKING BIRD.

REDBREASTED THRUSH.

RACOON.

AMERICAN POLE-CAT.

From English edition

ing and brought home what they had caught to their masters. This is often the case with otters, of which I have seen some which were as tame as dogs and followed their master wherever they went. If he went out in a boat, the otter went with him, jumped into the water, and after a while came up with a fish. The opossum can likewise be tamed, so as to follow people like a dog.

The *raccoon* which we (Swedes) call "Siupp," [1] can in time be made so tame as to run about the streets like a domestic animal; but it is impossible to make it give up its habit of stealing. In the dark it creeps to the poultry and kills a whole flock in one night. Sugar and other sweet things must be carefully hidden from it, for if chests and boxes are not locked up, it gets into them, eats the sugar, and after plunging into molasses licks it off its paws. The women therefore have every day some complaint against it, and for this reason many people would rather forbear the diversion which this ape-like animal affords.

The *gray and flying squirrels* are so tamed by the boys that they sit on their shoulders and follow them everywhere.

The *turkey* cocks and hens run about in the forests of this country, and differ in no particular from our domestic ones, except in their superior size, and redder, though more palatable flesh. When their eggs are found in the woods and put under tame turkey hens, the young ones become tame also. However, when they grow up it sometimes happens that they fly away. Their wings are therefore usually clipped, especially when young. But the tamed turkeys are as a rule much more irascible than those which are naturally domestic. The Indians likewise tame and keep them near their huts.

Wild geese have likewise been tamed in the following manner. When the wild geese first come hither in spring and stop a little while (for they do not breed in Pennsylvania) the people try to shoot them in the wing, which however is generally mere chance. They then row to the place where the wild geese fall, catch and keep them for a time at home; and by this means many of them have been made so tame, that when they were let out in the morning they returned in the evening. Yet to be doubly sure of them their wings were commonly clipped. I have seen wild geese of this kind, which the owner assured me that he had kept for more than twelve years;

---

[1] *Siupp,* also called *tvättbjörn* in modern Swedish, i. e. *wash-bear* (Ger. *Waschbär*), because he is supposed to wash his food before eating it.

but though he kept eight of them, in all this time they had never mated or laid eggs.

Partridges, which are here in abundance, may likewise be so tamed as to run about all day with the poultry, and to come along with them to be fed when they are called. In the same manner I have seen wild pigeons which were made so tame that they flew away and returned again. In some winters there are immense numbers of wild pigeons in Pennsylvania.

OCTOBER THE 24TH

The *humming bird* is the most admirable of all the rare birds of North America, or at least most worthy of special attention. Several reasons induce me to believe that few parts of the world can produce its equal. Dr. Linné calls it *Trochilus colubris*. The Swedes and some Englishmen call it the kingsbird, but the name is not as common as that of humming bird. Catesby in his *Natural History of Carolina*,[1] has drawn it, in its natural size, with the proper colors, and added a description of it. In size it is not much bigger than a large bumble bee, and is therefore the smallest of all birds.[2] It is doubtful if there is a lesser species in the world. Its plumage is most beautifully colored, most of its feathers being green, some gray, and others forming a shining red ring round its neck. The tail glows with fine feathers, changing from green into a copper color. These birds come here in spring about the time when it begins to grow very warm, and make their nests in summer, but towards autumn they retreat again into the more southern countries of America. They subsist only upon the nectar of sweet juice of flowers contained in that part which botanists call the nectarium, and which they suck up with their long bills. Of all the flowers, they like those most which have a long tube, and I have observed that they have fluttered chiefly about the *Impatiens Noli tangere*, and the *Monarda* with crimson flowers. An inhabitant of the country is sure to have a number of these beautiful and agreeable little birds before his window all the summer long, if he takes care to plant a bed with all sorts of fine flowers under them. It is indeed a diverting spectacle to

---

[1] Vol. I, p. 65.

[2] There is a much lesser species of humming-bird, by Linné called *Trochilus minimus,* being the smallest bird known; Sir Hans Sloane's living one, weighed only twenty grains and Mr. Edward's dry one forty-five.—F.

see these little active creatures flying about the flowers like bees, and sucking the juices with their long and narrow bills. The flowers of the above-mentioned *Monarda* grow verticillated, that is, at different distances they surround the stalk, as the flowers of our mint (*Mentha*), bastard hemp (*Galeopsis*), mother-wort (*Leonurus*), and dead nettle (*Lamium*). It is therefore diverting to see them putting their bills into every flower in the circle. As soon as they have sucked the honey of one flower, they flutter to the next. One that has not seen them would hardly believe in how short a space of time they have had their tongues in all the flowers of a plant. When the flowers are large, and have long tubes, the little bird by putting its head into them looks as if it crept in with half its body.

While they are sucking the juice of the flowers they never settle on them, but flutter continually like bees, bend their feet backwards, and move their wings so quick that they are hardly visible. During this fluttering they hum like bees, or make a sound like the turning of a little spinning wheel. After they have thus, without resting, fluttered for a while, they fly to a neighboring tree or post and rest awhile. They then return to their humming and sucking. They are not very shy, and I in company with several other people have been not fully two yards from the place where they fluttered about and sucked the flowers; and though we spoke and moved they were in no way disturbed, but on going towards them they would fly off with the swiftness of an arrow. When several of them were on the same bed, there was always a violent combat between them: in meeting each other at the same flower (for envy was likewise predominant amongst these little creatures) they attacked with such impetuosity, that it would seem as if the strongest would pierce its antagonist through and through with its long bill. During the fight they seem to stand in the air, keeping themselves up by the incredibly swift motion of their wings. When the windows towards the garden are open, they pursue each other into the rooms, fight a little, and flutter away again. Sometimes they come to a flower which is withering, and has no more juice in it; they then in a fit of anger pluck it off and throw it on the ground, that it may not mislead them in the future. If a garden contains a great number of these little birds, they are seen to pluck off the flowers in such quantities that the ground is quite covered with them, and it seems as if this proceeded from envy.

Commonly you hear no other sound than their humming, but when they fly against each other in the air, they make a chirping noise like a sparrow or chicken. I have sometimes walked with several other people in small gardens, and these birds have on all sides fluttered about us, without appearing very shy. They are so small that one could easily mistake them for large humming bees or butterflies, and their flight resembles that of the former, and is incredibly swift. They have never been observed to feed on insects or fruit; the nectar of flowers seems therefore to be their only food. Several people have caught humming birds on account of their singular beauty, and have put them into cages, where they died for want of proper food. However, Mr. Bartram kept a couple of them for several weeks by feeding them with water in which sugar had been dissolved, and I am of the opinion that it would not be difficult to keep them all winter in a hot-house.

The humming bird always builds its nest in the middle of a branch, and it is so small that it cannot be seen from the ground, but he who intends to see it must get up to it. For this reason it is looked upon as a great rarity if a nest is accidentally found, especially as the trees in summer have so thick a foliage. The nest is likewise one of the smallest in nature. The one in my possession is quite round, and consists on the inside of a brownish and quite soft down, which seems to have been collected from the leaves of the great mullein or *Verbascum thapsus*, which are often found covered with a soft wool of this color, and the plant is plentiful here. The outside of the nest has a coating of green moss, such as is common on old pales or fences and on trees; the inner diameter of the nest is hardly an inch at the top, and its depth half an inch. It is however known that the humming birds make their nest likewise of flax, hemp, moss, hair and other such materials: they are said to lay two eggs, each of the size of a pea.

OCTOBER THE 25TH

I employed this day and the next in packing up all the seeds gathered this autumn, for I had an opportunity of sending them to England by the ships which sailed about this time. From England they were to be forwarded to Sweden.

October the 27th

In the morning I set out on a little journey to New York in company with Mr. Peter Cock, with a view of seeing the country and inquiring into the safest road which I could take in going to Canada, through the wild or uninhabited country between it and the English provinces.

That part where we travelled then was pretty well inhabited on both sides of the road, by Englishmen, Germans and other Europeans. Plains and hills of different dimensions were seen alternately, mountains and stones I never saw, excepting a few pebbles. Near almost every farm was a great orchard with peach and apple trees, some of which were still loaded with fruit.

Yokes and Hobbles. The fences were in some parts low enough for the cattle to leap over with ease; to prevent this the hogs had a triangular wooden yoke and this custom was, as I have already observed, common over all the English plantations. To the horse's neck was fastened a piece of wood, which at the lower end had a tooth or hook which would catch in the fence and stop the horse just when it lifted its fore feet to leap over; but I know not whether this is a safe invention with regard to horses. They were likewise kept in bounds by a piece of wood, one end of which was fastened to one of the fore feet, and the other to one of the hind feet [a hobble]. It forced them to walk pretty slowly, and at the same time made it impossible for them to leap over the fence. To me it appeared that the horses were subject to all sorts of dangerous accidents from this contrivance.

Near New Frankfurt we rode over a little stone bridge, and somewhat further, eight or nine English miles from Philadelphia, we passed over another, which was likewise of stone. There are no milestones put up in the country, and the inhabitants only compute the distances by guess. We were afterwards brought over a river in a ferry, where we paid threepence a person for ourselves and our horses.

The Mocking-bird. At one of the places where we stopped to have our horses fed, the people had a mocking-bird in a cage; and it is here considered the best singing bird, though its plumage is very simple and not showy at all. At this time of the year it does not sing. Linné calls it *Turdus polyglottos*, and Catesby in his *Natural History of Carolina*,[1] has likewise described and drawn this bird. The people said that it built its nests in the bushes and trees, but it is so shy that if anybody comes and looks at its eggs it leaves the nest and never comes back. Its young require great care in their bringing up. If they are taken from their mother and put into a cage, she feeds them for three or four days but seeing no hope of setting them at liberty she flies away. It then often happens that the young ones die soon after, doubtless because they cannot accustom themselves to eat what the people give them. But it is generally imagined that the last time the mother feeds them, she finds means to poison them in order the sooner to deliver them from slavery and wretchedness. These birds stay all summer in the colonies, but retire in autumn to the south and stay away all winter. They have got the name of mocking-birds on account of their skill in imitating the note of almost every bird they hear. The song peculiar to them is excellent, and varied by an infinite change of notes and melody; several people are therefore of the opinion that they are the best singing birds in the world. So much is certain that few birds come up to them; this is what makes them precious; the Swedes call them by the same name as the English.

About noon we came to *New Bristol*, a small town in Pennsylvania, on the banks of the Delaware, about fifteen miles from Philadelphia. Most of the houses are built of stone, and stand separated. The inhabitants carry on a small trade, though most of them get their goods from Philadelphia. On the other side of the river, almost directly opposite to New Bristol, lies the town of Burlington, in which the governor of New Jersey resides.

We now saw country estates on both sides of the road. We came into a lane bordered with pales on both sides and enclosing rather large cultivated fields. Next followed a wood, and we perceived for the space of four English miles nothing else, except a very poor soil on which the *Lupinus perennis* grew plentifully and succeeded well. I was overjoyed to see a plant thrive so well in these poor dry places, since it served to make such places useful. But I afterwards had the mortification to find that the horses and cows eat almost all other plants, save the lupine, which was however very green, looked very

---

[1] Vol. I, p. 27.

luxuriant, and was extremely soft to the touch. Perhaps means may be found of making this plant palatable to cattle. In the evening we arrived at Trenton, after having previously passed the Delaware on a ferry.

OCTOBER THE 28TH

Trenton is a long narrow town, situated at some distance from the Delaware River, on a sandy plain; it belongs to New Jersey, and they reckon it thirty miles from Philadelphia. It has two small churches, one for the people belonging to the Church of England, the other for the Presbyterians. The houses are built partly of stone, though most of them are made of wood or planks, commonly two stories high, together with a cellar below the building, and a kitchen under ground close to the cellar. The houses stand at a moderate distance from one another. They are commonly built so that the street passes along on one side of the houses, while gardens of different dimensions bound the other side; in each garden is a well; the place is reckoned very healthy. Our landlord told us that twenty-two years ago, when he first settled there, there was hardly more than one house; but from that time on Trenton has increased so much that there are at present nearly a hundred houses. They were divided into several rooms by thin partitions of boards. The inhabitants of the place carried on a small trade with the goods which they got from Philadelphia, but their chief income consisted in attending to the numerous travellers between that city and New York, which are usually brought by the Trenton yachts between Philadelphia and Trenton. But from Trenton to New Brunswick, the travellers go in wagons which set out every day for that place. Several of the inhabitants however subsist on the transportation of all sorts of goods, which are sent every day in great quantities, either from Philadelphia to New York, or from there to the former place. Between Philadelphia and Trenton all goods are transported by water, but between Trenton and New Brunswick they are carried by land, and both these means of transportation belong to people of this town.

On the boats which ply between this place and the capital of Pennsylvania, people usually pay a shilling and sixpence of Pennsylvania currency per person, and everyone pays besides for his baggage. Every passenger must provide meat and drink for himself, or pay some settled fare: between Trenton and New Brunswick a person pays two shillings and sixpence, and the baggage is paid for separately.

We continued our journey in the morning; the country through which we passed was for the greatest part level, though sometimes there were some long hills; some parts were covered with trees, but by far the greater part of the country was without woods; on the other hand I never saw any place in America, the city excepted, so well peopled. An old man, who lived in the neighborhood and accompanied us a short distance, assured me however that he could well remember the time when between Trenton and New Brunswick there were not above three farms, and he reckoned it was about fifty and some odd years ago. During the greater part of the day we saw very extensive cultivated fields on both sides of the road, and we observed that the country generally had a noticeable declivity towards the south. Near almost every farm was a spacious orchard full of peaches and apple trees, and in some of them the fruit had fallen from the trees in such quantities as to cover nearly the whole surface of the ground. Part of it they left to rot, since they could not take care of it all or consume it. Wherever we passed by we were welcome to go into the fine orchards and gather our hats and pockets full of the choicest fruit, without the owner so much as looking at us. Cherry trees were planted near the farms, on the roads, etc.

The *barns* had a peculiar kind of construction in this locality, of which I shall give a concise description. The main building was very large almost the size of a small church; the roof was high, covered with wooden shingles, sloping on both sides, but not steep. The walls which supported it were not much higher than a full grown man; but on the other hand the breadth of the building was all the greater. In the middle was the threshing floor and above it, or in the loft or garret, they put the unthrashed grain, the straw, or anything else, according to the season. On one side were stables for the horses, and on the other for the cows. The young stock had also their particular stables or stalls, and in both ends of the building were large doors, so that one could drive in with a cart and horses through one of them, and go out at the other. Here under one roof therefore were the thrashing floor, the barn, the stables, the hay loft, the coach house, etc. This kind of building is used chiefly by the Dutch and Germans, for it is to be observed that the country between Trenton and New York is not inhabited by many Englishmen, but mostly by Germans or Dutch,[1] the latter of which are especially numerous.

Indians. Before I proceed I must mention one thing about the Indians or old Americans; for this account may find readers, who, like many people of my acquaintance, have the opinion that North America is almost wholly inhabited by savage or heathen nations; and they may be astonished that I do not mention them more frequently in my account. Others may perhaps imagine that when I state in my journal that the country is widely cultivated, that in several places houses of stone or wood are built, round which are grain fields, gardens and orchards, that I am speaking of the property of the Indians. To undeceive them I shall here give the following explanation. The country, especially that along the coasts in the English colonies, is inhabited by Europeans, who in some places are already so numerous that few parts of Europe are more populous. The Indians have sold the land to the Europeans, and have retired further inland. In most parts you may travel twenty Swedish miles, or about a hundred and twenty English miles, from the coast, before you reach the first habitation of the Indians. And it is very possible for a person to have been at Philadelphia and other towns on the seashore for half a year without so much as seeing an Indian. I intend further on to give a more circumstantial account of them, their religion, manners, economic conditions, and other particulars. At present I return to the continuation of my journal.

The Soil. About nine English miles from Trenton, the ground began to change its color. Hitherto it had consisted of a considerable quantity of hazel-colored clay, but at present the earth was a reddish brown, so that it sometimes had a purple color, and sometimes looked like logwood. This color came from a red limestone which closely resembled the kind on the mountain Kinnekulle in Västergötland [Sweden] and made a particular stratum in the rock. The American red limestone therefore seemed to be merely a variety of what I saw in Sweden. It lay in strata of two or three fingers thick-

---

[1] This kind of building is frequent in the north of Germany, Holland, and Prussia, and therefore it is no wonder that it is employed by people who were used to them in their own country.—F.

ness, but was divisible into many thinner layers, whose surface was seldom flat and smooth, but rough. The strata themselves were frequently cut off by perpendicular cracks. When this stone was exposed to the air, it weathered by degrees and at last turned into dust. The people of this neighborhood did not know what use to make of it, and the soil above was sometimes rich and sometimes poor. In places where the people had lately dug new wells, I perceived that most of the rubbish which was thrown up consisted of such species of stone. This reddish brown earth was seen almost all the way up to New Brunswick, where it is particularly plentiful. The banks of the river showed in many places nothing but strata of limestone, which did not run horizontally, but dipped very much.

About ten o'clock in the morning we came to Prince-town, (*Princeton*), which is situated on a plain. Most of the houses were built of wood, and were not contiguous, so that there were gardens and pastures between them. As these parts had been inhabited by Europeans earlier than Pennsylvania, more woods had been cut down and the country more cultivated, so that one might easily have imagined oneself in Europe.

We thought of continuing our journey, but as it began to rain very heavily, and kept on doing so all day and part of the night, we were forced to stay until the next morning.

OCTOBER THE 29TH

This morning we proceeded on our journey. The country was pretty well peopled; yet there were great wooded areas in many places, all of which were covered with deciduous trees, and I did not perceive a single tree of the evergreen kind till I came to New Brunswick. The ground in general was level, and did not seem to be everywhere of the richest quality. In some places, however, there were hills losing themselves almost imperceptibly in the plains, the latter commonly crossed by a rivulet. Near almost every farmhouse were large orchards. The houses were as a rule built of timber, and at some distance by themselves stood the ovens for baking, constructed usually of clay.

On a hill covered with trees and called Rockhill I saw several pieces of stone or rock so big that they would have required three men to move them. But besides these there were few large stones

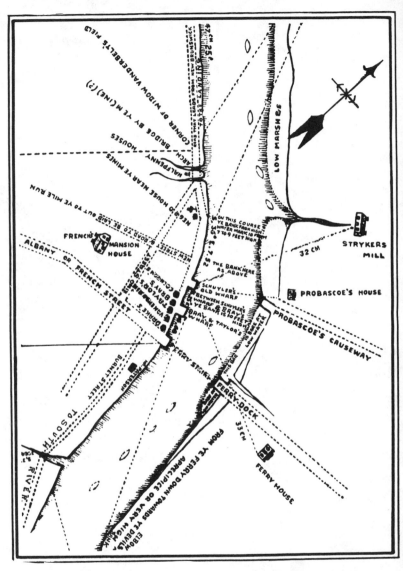

Earliest Existing Map of New Brunswick, New Jersey

in the section; for most of those we saw could easily have been lifted up by a single man. In another place we discovered a number of little round pebbles, but we did not find either mountains or rocks.

About noon we arrived at *New Brunswick*, (situated about thirty miles from Trenton and sixty from Philadelphia), a pretty little town in the province of New Jersey, in a valley on the west side of the river Raritan. On account of its low location, it cannot be seen (coming from Pennsylvania) before you get to the top of the hill, which is quite close to it. The town extends north and south along the river. The German inhabitants have two churches, one of stone and the other of wood. The English church is likewise of the latter material, but the Presbyterians are building one of stone. The Town Hall makes a good appearance. Some of the other houses are built of brick, but most of them are made either wholly of wood, or of brick and wood. The wooden buildings are not made of strong timber, but merely of boards or planks, which are within joined by laths. Houses built of both wood and brick have only the wall towards the street made of the latter, all the other sides being boards. This peculiar kind of ostentation would easily lead a traveller who passes through the town in haste to believe that most of the houses are built of brick. The houses are covered with shingles. Before each door is a veranda to which you ascend by steps from the street; it resembles a small balcony, and has benches on both sides on which the people sit in the evening to enjoy the fresh air and to watch the passers-by. The town has only one street lengthways, and at its northern extremity there is a cross street: both of these are of a considerable length.

The river Raritan passes close by the town, and is deep enough for large sailing vessels. Its breadth near the town is about the distance of a common gun shot. The tide comes up several miles beyond the town, which contributes not a little to the ease and convenience of securing vessels which dock along the bridge. The river has generally very high and steep banks on both sides, but near the town there are no such banks, because it is situated in a low valley. One of the streets is almost entirely inhabited by Dutchmen who came hither from Albany, and for that reason it is called Albany Street. These Dutch people keep company only with themselves, and seldom or never go amongst the other inhabitants, living as it were quite separate from them.

New Brunswick belongs to New Jersey; however, the greatest part or rather all its trade is with New York, which is about forty English miles away. To that place they send grain, flour in great quantities, bread, several other necessaries, a great quantity of linseed, boards, timber, wooden vessels and all sorts of carpenters' work. Several small boats pass every day back and forth between these two towns. The inhabitants likewise get a considerable profit from the travellers, who every hour pass through on the highroad.

The steep banks consist of red limestone, which I have before described. It is here plainly visible that the strata are not horizontal but considerably inclined especially towards the south. The weather and the air has in a great measure dissolved the stone here. I inquired whether it could not be made use of, but was assured that in building houses it was entirely useless, for, though it was hard and permanent under ground, yet on being dug out and exposed for a time to the air, it first crumbled into greater, then into lesser pieces, and at last crumbled into dust. An inhabitant of New Brunswick tried to build a house of this stone, but upon being exposed to the air, it soon began to change so much that the owner was obliged to put boards all over the wall to preserve it from falling to pieces. The people claim however that this stone makes a very good fertilizer, if it is scattered upon the soil in its crumbled state, for it is said to stifle the weeds: it is therefore utilized both on the fields and in gardens.[1]

Toward evening we continued our journey and were, together with our horses, ferried over the river Raritan. In a very dry summer, and when the tide has ebbed, it is by no means dangerous to ride through this river. On the opposite shore the red juniper tree was quite abundant. The country through which we now passed was well inhabited. In most places it was full of small pebbles.

We saw *Guinea hens* in many places where we passed by. They sometimes run about the fields, at a good distance from the farmhouses.

About eight English miles from New Brunswick, the road divided. We took the one on the left, for that on the right led to Amboy, the chief seatown in New Jersey. The country now had a charming appearance, some parts being high, others forming valleys, and all

---

[1] Probably it is a stone marl; a blue and reddish species of this kind is used with good success in the country of Bamff in Scotland.—F.

First Reformed Church, (Built about 1714)
New Brunswick, New Jersey

Original Christ Church

of them well cultivated. From the hills you had a prospect of houses, farms, gardens, tilled land, forests, lakes, islands, roads, and pastures.

In most of the places where we travelled to-day the color of the ground was reddish. I have no doubt there were strata of the before-mentioned red limestone underneath. Sometimes the ground looked very much like a cinnabar ore.

Woodbridge is a small village on a plain, consisting of a few houses: we stopped there to rest our horses a little. The houses were most of them built of boards; the walls had a covering of shingles on the outside and these were cut square and all in each row were of the same length. Some of the houses had an Italian roof, but the greatest number had roofs with pediments, and the majority of them were covered with shingles. In most places we saw wells and apparatus for drawing up water.

Elizabethtown is a small place, about twenty miles from New Brunswick: we arrived there immediately after sunset. Its houses are for the most part scattered, but well built, generally of boards, with a roof and walls of shingles. There are likewise a few stone buildings. A little rivulet passes through the town from west to east; it is almost reduced to nothing at low tide, but at high water it is navigable for small boats. Here are two fine churches, either one of which makes a better appearance than any in Philadelphia. That belonging to the Church of England is built of brick, has a steeple with bells, and a balustrade around it from which there is a view of the country. The meetinghouse of the Presbyterians is built of wood, but has both steeple and bells and is, like the other houses, covered with shingles. The Town Hall makes likewise a good appearance, and has a belfry. The banks of the river are red from the reddish limestone. In and about the town are many gardens and orchards, and it might truly be said that Elizabethtown is situated in a garden, the ground in this neighborhood being even and well cultivated.

The *geese* in some of the places by which we passed that day and the next wore three or four little sticks a foot in length about their necks. They were fastened crossways, to prevent them from creeping through half-broken fences. They looked extremely awkward, and it was very diverting to see them in this attire.

At night we took up our lodgings at Elizabethtown Point, at an inn about two English miles from the town, and the last house on this road belonging to New Jersey. The man who had taken the lease of it, together with that of the ferry near it, told us that he paid a hundred and ten pounds of Pennsylvania currency to the owner annually.

OCTOBER THE 30TH

We were ready to proceed on our journey at sunrise. Near the inn where we had passed the night, we were to cross a river, and we were brought over, together with our horses, in a wretched, half-rotten ferry. This river penetrated quite a distance inland, and small vessels could easily sail up it. This was a great advantage to the inhabitants of the neighboring country, giving them an opportunity of sending their goods to New York with great ease; and they even made use of it for trading with the West Indies. The country was low on both sides of the river, and consisted of meadows. But there was no other grass except that which commonly grows in swampy grounds; for as the tide came up the river the low plains were sometimes flooded. The people hereabouts were said to be troubled in summer with immense swarms of gnats or mosquitoes which stung them and their cattle. This was ascribed to the low swampy meadows, on which these insects deposited their eggs.

Staten Island. As soon as we had crossed the river, we were upon Staten Island, which is entirely surrounded with salt water. This is the beginning of the province of New York. Most of the people settled here were Dutch, or such as came hither while the Dutch were yet in possession of this place. But at present they were scattered among the English and other European inhabitants and spoke English for the greatest part. The appearance of the country here is extremely pleasing, as it is not so much intercepted by woods, but offers more cultivated fields to view. Hills and valleys alternated.

The farms were near each other. Most of the houses were wooden; however, some were built of stone. Near every farmhouse was an orchard with apple trees: the fruit had for the greatest part been gathered already. Here, and on the whole journey before, I observed a cider press at every farmhouse, made in different ways, by which the people had already pressed the juice out of the apples, or were just busy with that work. Some people made use of a wheel made of thick oak planks, which turned upon a wooden axis by means of a horse drawing it, much in the same manner as in crushing woad, except that here the wheel runs upon planks.—Cherry trees stood along the fences round the cultivated fields. The latter were excellently situated, and sown with either wheat or rye. They had no ditches along their sides, but (as is usual in England) only furrows, drawn at greater or less distances from each other.

In one place we observed a water mill, so situated that when the tide flowed the water ran into a pond: but when it ebbed the floodgate was drawn up and the mill driven by the water flowing out of the pond.

Province of New York. About eight o'clock in the morning we arrived at the place where we were to take the ferry for New York. We left our horses there and went on board the boat: we were to go nine English miles by sea, and landed about eleven o'clock in the morning at New York. We saw a kind of wild duck in immense numbers upon the water: the people called them "blue bills" and they seemed to be the same as our pintail ducks, or Linné's *Anas acuta,* but they were very shy. On the shore of the continent we saw some very fine sloping cultivated fields, which at present looked quite green, the grain having already come up. We saw many boats in which the fishermen were busy catching oysters. For this purpose they made use of a kind of rake with long iron teeth bent inwards; these they used either singly or two tied together in such a manner that the teeth were turned towards each other.

OCTOBER THE 31ST

Oysters. About New York they find innumerable quantities of excellent oysters, and there are few places which have oysters of such an exquisite taste and of so great a size. They are pickled and sent to the West Indies and other places, which is done in the following manner. As soon as the oysters are caught, their shells are opened. and the fish washed clean; some water is then poured into a pot, the oysters are put into it, and they are boiled for a while; the pot is then taken off the fire again and the oysters taken out and put upon a dish till they are almost dry. Then some nutmeg, allspice and black pepper are added, and as much vinegar as is thought sufficient to give a sourish taste. All this is mixed with half the liquor in which the oysters are boiled, and put over the fire again. While boiling great care should be taken to skim off the thick scum. At last the whole pickling liquid is poured into a glass or earthen vessel, the oysters are put into it, and the vessel is well stopped to keep out the air. In this manner, oysters will keep for years, and may be sent to the most distant parts of the world.

The merchants here buy up great quantities of oysters about this time, pickle them in the above-mentioned manner, and send them to the West Indies, by which they frequently make a considerable profit; for the oysters, which cost them five shillings of their currency, they commonly sell for a pistole, or about six times as much as they give for them, and sometimes they get even more. The oysters which are thus pickled have a very fine flavor, but cannot be fried. The following is another way of preserving oysters: they are taken out of the shells, fried in butter, put into a glass or earthen vessel with the melted butter over them, so that they are fully covered with it and no air can get to them. Oysters prepared in this manner have likewise an agreeable taste, and are exported to the West Indies and other parts.

Oysters are here reckoned very wholesome, and some people assured us that they had not felt the least inconvenience after eating a considerable quantity of them. It is also a common rule here that oysters are best in those months which have an "r" in their names such as September, October, etc. but that they are not so good in other months. However, there are poor people who live all year long upon nothing but oysters and a little bread.

The sea near New York yields annually a great quantity of oysters. They are found chiefly in a muddy ground, where they lie in the slime and are not so frequent in a sandy bottom: a rocky and stony bottom is seldom found here. The oyster shells are gathered in great heaps and burnt into a lime which by some people is used in building houses, but it is not reckoned so good as that made of limestone. On our journey to New York we saw high heaps of oyster shells near the farmhouses upon the seashore, and about New York we observed the people had scattered them upon the fields which were sown with wheat. However, they were whole and not crushed.

The Indians who inhabited the coast before the arrival of the Europeans made oysters and other shell fish their chief food, and at present whenever they come to salt water where oysters are to be gotten, they are very active in catching them, and in selling them in great quantities to other Indians who live further inland. For this reason one sees immense quantities of oyster and mussel shells piled up near the places where it is certain that the Indians formerly built their huts. This circumstance ought to make us cautious in maintaining that in all places on the seashore, or further back in the country where such heaps of shells are to be met with, that the latter have lain there ever since the time when those places were overflowed by the sea.

Lobsters are also caught in great numbers hereabouts, pickled in much the same way as oysters, and sent to several places. I was told of a very remarkable circumstance about these lobsters, and I have afterwards frequently heard it mentioned. The coast of New York had European inhabitants for a considerable time, yet no lobsters were to be found on that coast, and though the people fished ever so often, they could never find any signs of lobsters being in this part of the sea. They were therefore continually brought in great well boats from New England, where they were plentiful. But it happened that one of these well boats broke in pieces near Hellgate, about ten English miles from New York, and all the lobsters in it were lost in the sea. Since that time they have so multiplied off this part of the coast that they are now caught in great abundance.

November the 1st

Fever and Remedies. A kind of cold fever, which the English in this country call "fever" and "ague," is very common in several parts of the English colonies. There are, however, other parts, where the people have never felt it. I shall later describe the symptoms of this disease at large. Several of the most prominent inhabitants of this town assured me that the disease was not nearly so common in New York as in Pennsylvania, where ten were seized by it to one in the former province. Therefore they were of the opinion that it was occasioned by the vapors arising from stagnant fresh water, from marshes, and from rivers,[1] for which reason those provinces situated on the seashore could not be so much affected by

---

[1] The spread of malaria by mosquitoes was of course unknown in 1748. Later Kalm devotes a great deal of space to the disease known as ague, fever, etc., and from its description we must conclude that it was or resembled the modern malaria.

it. However, the carelessness with which people eat quantities of melons, watermelons, peaches and other juicy fruit in summer, was reckoned to contribute much towards the progress of this fever, and repeated examples confirmed the truth of this opinion. The Jesuits' bark was reckoned a good remedy for it. It has, however, often been found to have operated contrary to expectation, though I am ignorant whether it was adulterated or some mistake had been committed in the manner of taking it. Mr. Davis van Horne, a merchant, told me that he cured himself and several other people of this fever by the leaves of the common garden sage, or *Salvia officinalis* of Linné. The leaves are crushed or pounded in a mortar and the juice is pressed out of them. This is continued till they get a spoonful of the liquid, which is mixed with lemon juice. This draught is taken about the time the cold chill comes on, and after taking it three or four times the fever generally disappears.

The bark of the white oak was reckoned the best remedy which had as yet been found against dysentery. It is reduced to a powder, and then taken. Some people assured me that in cases where nothing would help, this remedy had given a certain and speedy relief. The people in this place also make use of the bark (as is usually done in the English colonies) to dye wool brown, which looks like that of bohea tea, and does not fade by being exposed to the sun.

Clams. Among the numerous shell-fish which are found on the seashore, there are some which by the English are called clams and which bear some resemblance to the human ear. They have considerable thickness, and are chiefly white, excepting the pointed end, which both without and within is of a bluish color, between purple and violet. They are found in vast numbers on the seashore of New York, Long Island and other places. The shells contain a large amount of meat which is eaten both by the Indians and Europeans settled here.

A considerable commerce is carried on in this fish product with such Indians as live further up the country. When these people inhabited the coast, they were able to dig their own clams, which at that time constituted the great part of their food; but at present this is the business of the Dutch and English, who live on Long Island and other maritime provinces. As soon as the clams are dug, the soft part is taken out of the shells, drawn upon a wire, and hung up in the open air in order to dry by the heat of the sun.

When this is done, the fish is put into proper vessels and carried to Albany upon the river Hudson; there the Indians buy them, and reckon them one of their best dishes. Besides the Europeans, many of the native Indians come annually down to the seashore in order to get clams, proceeding with them afterwards in the manner I have just described. They are ordinarily prepared like oysters. Sometimes they are baked in their shells, sometimes stewed in butter; other times they are boiled and placed in meat-soups. They are often served on a platter with steaks or other meat. No matter how prepared they make a palatable food. I have often eaten them during my travels, but they seemed a little hard for the stomach to digest.

Wampum. The shells of these clams are used by the Indians as money and make what they call their wampum; they likewise serve their women as ornaments, when they intend to appear in full dress. This wampum is properly made of the purple part of the shells, which the Indians value more than the white part. A traveller who goes to trade with the Indians and is well stocked with it, may become a considerable gainer, but if he takes gold coin, or bullion, he will undoubtedly be a loser, for the Indians who live farther back in the country put little or no value upon these metals which we reckon so precious, as I have frequently observed in the course of my travels. The Indians formerly made their own wampum, though not without great difficulties, but at present it is made mostly by the Europeans, especially by the inhabitants of Albany, who make a considerable profit by it. Later I intend to relate the manner of making the wampum.

November the 2nd

The Jews. Besides the different sects of Christians, many Jews have settled in New York, who possess great privileges. They have a synagogue, own their dwelling-houses, possess large country-seats and are allowed to keep shops in town. They have likewise several ships, which they load and send out with their own goods. In fine, they enjoy all the privileges common to the other inhabitants of this town and province.

A daughter of one of the richest Jews had married a Christian after she had renounced the Jewish religion. Her sister did not wish either to marry a Jew, so went to London to get a Christian husband.

During my residence in New York, both at this time and for the next two years, I was frequently in company with Jews. I was informed among other things that these people never boiled any meat for themselves on Saturday, but that they always did it the day before, and that in winter they kept a fire during the whole Saturday. They commonly eat no pork; yet I have been told by several trustworthy men that many of them (especially the young Jews) when travelling, did not hesitate the least about eating this or any other meat that was put before them, even though they were in company with Christians. I was in their synagogue last evening for the first time, and to-day at noon I visited it again, and each time I was put in a special seat which was set apart for strangers or Christians. A young rabbi read the divine service, which was partly in Hebrew and partly in the Rabbinical dialect. Both men and women were dressed entirely in the English fashion; the former had their hats on, and did not once take them off during the service. The galleries, I observed, were reserved for the ladies, while the men sat below. During prayers the men spread a white cloth over their heads, which perhaps is to represent sackcloth. But I observed that the wealthier sort of people had a much richer cloth than the poorer ones. Many of the men had Hebrew books, in which they sang and read alternately. The rabbi stood in the middle of the synagogue and read with his face turned towards the east; he spoke however so fast as to make it almost impossible for any one to understand what he said.[1]

New York. New York, the capital of a province of the same name, is situated forty degrees and forty minutes north latitude and forty-seven degrees and four minutes western longitude from London, and is about ninety-seven English miles from Philadelphia. The location of the city is extremely advantageous for trade, for it stands a short distance from the sea upon a point which is formed by two bays, into one of which the river Hudson empties not far from the town. New York is therefore on three sides surrounded by water. The ground it is built on is level in some parts and hilly in others. The place is generally reckoned very healthy.

[1] As there were very few Jews in Sweden at the time, Kalm was a stranger to their manners and religious customs and therefore relates them as a kind of novelty.

The town was first settled by the Dutch. This, it is said, was done in the year 1623, when they were yet masters of the country: They called it New Amsterdam, and the country itself New Holland. The English, towards the end of the year 1664, taking possession of it under the command of Sir Cartes [1] and keeping it by virtue of the next treaty of peace, gave the name of New York to both the town and the province belonging to it. In size it comes next to Boston and Philadelphia, but with regard to fine buildings, opulence, and extensive commerce, it vies with them for supremacy. At present it is about half as large again as Gothenburg in Sweden.

The streets do not run so straight as those of Philadelphia, and sometimes are quite crooked; however, they are very spacious and well built, and most of them are paved, except in high places, where it has been found useless. In the chief streets there are trees planted, which in summer give them a fine appearance, and during the excessive heat at that time afford a cooling shade. I found it extremely pleasant to walk in the town, for it seemed like a garden. The trees which are planted for this purpose are chiefly of two kinds: the water beech, or Linné's *Platanus occidentalis*, which is very plentiful and gives an agreeable shade in summer by its great and numerous leaves; and the locust tree, or Linné's *Robinia Pseud-Acacia,* which is also frequent. The latter's fine leaves and fragrant scent which exhales from its flowers, make it very suitable for planting in the streets near the houses and in gardens. There are likewise lime trees and elms along these walks, but they are not by far so frequent as the others. One seldom met with trees of the same sort next to each other, they being in general planted alternately.

Tree Toads. Besides birds of all kinds which make these trees their abode, there is a kind of frog which frequents them in great numbers in summer. It is Dr. Linné's *Rana arborea*, and especially the American variety of this animal. They are very clamorous in the evening and in the night (especially when the days have been

---

[1] Kalm here must refer to *Carteret*. Sir George Carteret (d. 1680) was together with Lord Berkeley in 1664 granted, by King James, all the territory between the Hudson and the Delaware, and Sir George sent his kinsman Philip Carteret (1639-1682) to act as governor of the newly formed province of New Jersey. But Carteret did not conquer New Amsterdam. This had been done by Col. Richard Nicolls on August 29, 1664, before the new appointee arrived.

hot, and a rain was expected) and in a manner drown the singing of the birds. They frequently make such a noise that it is difficult for a person to make himself heard.

Most of the houses are built of brick, and are generally strong and neat, and several stories high. Some had, in the old style, turned the gable end toward the street; but the new houses were altered in this respect. Many of the houses had a balcony on the roof, on which the people used to sit evenings in the summer season; and from thence they had a pleasant view of a great part of the town, and likewise of a part of the adjacent water and the opposite shore. The roofs are commonly covered with tiles or shingles, the latter of which are made of the white fir tree, or *Pinus strobus* L., which grows further north in the country. The inhabitants are of the opinion that a roof made of these shingles is as durable as one made in Pennsylvania of the white cedar, or *Cupressus thyoides* L. The walls were whitewashed within, and I did not anywhere see wall paper, with which the people of this country seem in general to be but little acquainted. The walls were covered with all sorts of drawings and pictures in small frames. On each side of the chimneys they had usually a sort of alcove, and the walls under the windows were wainscoted and had benches placed near them. The cupboards and all the wood work were painted with a bluish gray color.

There are several churches in the town which deserve mention. 1. The *English Church*, built in the year 1695, at the west end of the town, is built of stone, and has a steeple with a bell. 2. The *New Dutch Church*, which is likewise built of stone, is pretty large, and is provided with a steeple; it also has a clock, the only one in the town. This church stands almost due north and south. In several instances here no particular point of the compass has been considered in erecting sacred buildings. Some churches stand, as is usual, east and west; others south and north, and others in still different positions. In the Dutch church, there is neither altar, vestry, choir, sconces, nor paintings. Some trees are planted round it, which make it look as if it were built in a wood. 3. The *Old Dutch Church* is also built of stone. It is not so large as the new one. It was painted on the inside, though without images, and adorned with a small organ, of which governor Burnet made them a present. The men for the most part sit in the gallery, and the women below.

4. The *Presbyterian Church*, which is pretty large and was built but lately. It is of stone, and has a steeple and bell in it.

5. The *German Lutheran Church*.

6. The *German Reformed Church*.

7. The *French Church*, for Protestant refugees.

8. The *Quakers' Meeting House*.

To these may be added the *Jewish Synagogue* which I mentioned before.

Toward the sea, on the extremity of the promontory, is a tolerably good fortress, called Fort George, which entirely commands the port, and can defend the town, at least from a sudden attack on the sea side. Besides that, it is also secured on the north, or the land side, by a pallisade, which however (since for a considerable time the people have had nothing to fear from an enemy) is in many places in a very poor condition.

There is no good water in the town itself, but at a little distance away there is a large spring which the inhabitants use for their tea and for other kitchen purposes. Those people however who are less particular in this matter use the water from the wells in town, though it be very bad. This want of good water is hard on strangers' horses that come to the place, for they do not like to drink the well water.

The *Port* is a good one: ships of the greatest tonnage can lie in it, close to the bridge; but its water is very salt as the sea continually washes into it, and therefore is never frozen, except in extraordinarily cold weather. This is of great advantage to the city and its commerce; for many ships enter or leave the port at all times of the year, unless the winds be contrary, a convenience, which as I have before observed, is wanting at Philadelphia. It is secured from all violent hurricanes from the southeast by Long Island, which is situated just in front of the town; therefore only the storms from the southwest are dangerous to the ships which ride at anchor here, because the port is open only on that side. The entrance however has its faults: one of them is that no men-of-war can pass through it; for though the water is pretty deep, it is not sufficiently so for great ships. Sometimes even merchant ships of a large size have by the rolling of the waves and by sinking down between them slightly touched the bottom, though without any bad consequences. Besides this, the channel is narrow, and therefore many ships have been lost here, because they may easily be cast upon a sandbar if the ship is not well piloted. Some old people, who had constantly sailed upon this channel, assured me that it was neither deeper nor shallower at present than in their youth.

The common difference between high and low water at New York amounts to about six feet, English measure. But at a certain time in every month, when the tide is higher than usual, the difference is seven feet.

Trade in New York. New York probably carries on a more extensive commerce than any town in the English North American provinces, at least it may be said to equal them. Boston and Philadelphia however come very close to it. The trade of New York extends to many places, and it is said they send more ships from there to London than they do from Philadelphia. They export to that capital all the various sorts of skins which they buy of the Indians, sugar, logwood, and other dyeing woods, rum, mahogany and many other goods which are the produce of the West Indies, together with all the specie which they get in the course of trade. Every year several ships are built here, which are sent to London and there sold; and of late years a quantity of iron has been shipped to England. In return for all these, cloth is imported from London and so is every article of English growth or manufacture, together with all sorts of foreign goods. England, and especially London, profits immensely by its trade with the American colonies; for not only New York but likewise all the other English towns on the continent import so many articles from England that all their specie, together with the goods which they get in other countries must all go to Old England to pay their accounts there, for which they are, however, insufficient. Hence it appears how much a well regulated colony contributes to the increase and welfare of its mother country.

New York sends many ships to the West Indies with flour, grain, biscuit, timber, tuns, boards, meat and pork, butter, and other provisions, together with some of the few fruits that grow here. Many ships go to Boston in New England with grain and flour, and take in exchange, meat, butter, timber, different sorts of fish, and other articles, which they carry further to the West Indies.

They now and then carry rum from Boston, which is distilled there in great quantities, and sell it here at a considerable advantage. Sometimes they send vessels with goods from New York to Philadelphia; and at other times they are sent from Philadelphia to New York, which is only done, as appears from the gazettes, because certain articles are cheaper at one place than at the other. They send ships to Ireland every year, laden with all kinds of West India goods, but especially with linseed, which is collected in this country. I have been assured that in some years no less than ten ships have been sent to Ireland, laden with nothing but linseed, because it is said the flax in Ireland does not give good seed. But probably the true reason is that the people of Ireland, in order to have the better flax, make use of the plant before the seed is ripe, and therefore are obliged to send for foreign seed. It becomes thus one of the chief articles of trade. At this time a bushel of linseed is sold for eight shillings of New York currency.

For the goods which are sold in the West Indies either ready money is accepted or West India goods, which are either first brought to New York or immediately sent to England or Holland. If a ship does not choose to take West India goods on its return to New York, or if nobody will freight it, it often goes to Newcastle in England to take on coal for ballast, which when brought home sells for a pretty good price. In many parts of the town coal is used both for kitchen fires and in other rooms, because it is considered cheaper than wood, which at present costs thirty shillings of New York currency per fathom, of which measure I have before made mention. New York has likewise some trade with South Carolina, to which it sends grain, flour, sugar, rum, and other goods, and takes rice in return, which is almost the only commodity exported from South Carolina.

The goods in which the province of New York trades are not numerous. It exports chiefly the skins of animals, which are bought of the Indians about Oswego; great quantities of boards, coming for the most part from Albany; timber and casks from that part of the country which lies about the Hudson River; and lastly wheat, flour, barley, oats and other kinds of grain, which are brought from New Jersey and the cultivated parts of this province. I have seen vessels from New Brunswick, laden with wheat which lay loose on board, with flour packed up in barrels, and also with great quantities of linseed. New York also exports pork and other meat from its own province, but not in any great amount; nor is the quantity of peas which the people about Albany bring very large. Iron, however, may be had more plentifully, as it is found in several parts of this province and is of a considerable value; but all other products of this country are of little account.

Most of the wine which is drunk here and in the other colonies is brought from the Isle of Madeira and is very strong and fiery.

No manufactures of note have as yet been established here; at present they get all manufactured goods, such as woolen and linen cloth, etc., from England, and especially from London.

The Hudson River is very convenient for the commerce of this city, as it is navigable for nearly a hundred and fifty English miles into the country, and flows into the bay, a little west of the town. During eight months of the year this river is full of greater and lesser vessels, either going to New York or returning from there, laden either with native or foreign goods.

I cannot make a true estimate of the ships that annually come to this town or sail from it. But I have learned in the Pennsylvania gazettes that from the first of December in 1729 to the fifth of December of the following year 211 ships entered the port of New York and 222 cleared it; and since that time there has been a great increase of trade here.

The country people come to market in New York twice a week, much in the same manner as they do at Philadelphia, with this difference, that the markets are here kept in several places, and one has to go from one to another sometimes to get what one needs.

Government in New York. The *governor* of the province of New York resides here, and has a palace in the fort. Among those who have been entrusted with this post, William Burnet deserves a perpetual remembrance. He was one of the sons of Dr. Thomas Burnet (so celebrated on account of his learning) [1] and seemed to have inherited the knowledge of his father. But his great assiduity in promoting the welfare of this province is what constitutes the principal merit of his character. The people of New York therefore still reckon him the best governor they ever had, and feel they cannot praise his services too much. The many astronomical ob-

---

[1] See p. 72.

servations which he made in these parts are inserted in several English works. In the year 1727, at the accession of king George the II. to the throne of Great Britain, he was appointed governor of New England. In consequence of this he left New York and went to Boston, where he died universally lamented on the seventh of September, 1729.

An *assembly of deputies* from all the different districts of the province is held at New York once or twice every year. It may be looked upon as a parliament or diet in miniature. Everything relating to the good of the province is here debated. The governor calls the assembly, and dissolves it at pleasure. This is a power which he ought only to make use of, either when no farther debates are necessary or when the members are not so unanimous in the service of their king and country as is their duty. It frequently happens, however, that led aside by caprice or by self-interested views, he exerts it to the prejudice of the province. The colony has sometimes had a governor, whose quarrels with the inhabitants have induced their representatives, or the members of the assembly, through malice to oppose indifferently everything he proposed, whether it was beneficial to the country or not. In such cases the governor has made use of his power, dissolving the assembly, and calling another soon after, which however he again dissolved upon the least mark of their ill humor. By this means he so tired them, by the many expenses which they were forced to bear in so short a time, that they were at last glad to unite with him in his endeavors for the good of the province. But there have likewise been governors who have called assemblies and dissolved them soon after, merely because the representatives did not act according to their whims or would not give their assent to proposals which were perhaps dangerous or hurtful to the common welfare.

The king appoints the governor according to his royal pleasure; but the inhabitants of the province make up his excellency's salary. Therefore a man entrusted with this position has greater or lesser revenues, according to his ability of gaining the confidence of the inhabitants. There are examples of governors in this and other provinces of North America, who by their dissensions with the inhabitants of their respective provinces have lost their whole salary, his Majesty having no power to make them pay it. If a governor had no other resources in such circumstances he would be obliged either to resign his office, to be content with an income too small for his dignity, or else to conform in everything to the inclinations of the inhabitants. But there are several stated profits which in some measure make up for this.

1. No one is allowed to keep a public house without the governor's leave, which is only to be obtained by the payment of a certain fee, according to the circumstances of the person. Some governors therefore, when the inhabitants refuse to pay them a salary, have hit upon the expedient of doubling the number of inns in their province.

2. Few people who intend to be married, unless they be very poor, will have their banns published from the pulpit; so instead of this they get licenses from the governor, which empower any minister to marry them. Now for such a license the governor receives about half a guinea, and this collected throughout the whole province amounts to a considerable sum.

3. The governor signs all passports, and especially of such travellers as go to sea, and this gives him another means of supplying his expenses. There are several other advantages allowed to him, but as they are very trifling I shall omit them.

At the above assembly the old laws are reviewed and amended, and new ones are made, and the regulation and circulation of the coinage together with all other affairs of that kind are there determined. For it is to be observed that each English colony in North America is independent of the other, and that each has its own laws and coinage, and may be looked upon in several lights as a state by itself. Hence it happens that in time of war things go on very slowly and irregularly here; for not only the opinion of one province is sometimes directly opposite to that of another, but frequently the views of the governor and those of the assembly of the same province are quite different; so that it is easy to see that, while the people are quarrelling about the best and cheapest manner of carrying on the war, an enemy has it in his power to take one place after another. It has usually happened that while some provinces have been suffering from their enemies, the neighboring ones have been quiet and inactive, as if it did not in the least concern them. They have frequently taken up two or three years in considering whether or not they should give assistance to an oppressed sister colony, and sometimes they have expressly declared themselves against it. There are instances of provinces which were not only neutral in such circumstances, but which even carry on a great trade with the power which at that very time is attacking and laying waste some other provinces.

The French in Canada, who are but an unimportant body in comparison with the English in America, have by this position of affairs been able to obtain great advantages in times of war; for if we judge from the number and power of the English, it would seem very easy for them to get the better of the French in America.

It is however of great advantage to the crown of England that the North American colonies are near a country, under the government of the French, like Canada. There is reason to believe that the king never was earnest in his attempts to expel the French from their possessions there; though it might have been done with little difficulty. For the English colonies in this part of the world have increased so much in their number of inhabitants, and in their riches, that they almost vie with Old England. Now in order to keep up the authority and trade of their mother country and to answer several other purposes they are forbidden to establish new manufactures, which would turn to the disadvantage of the British commerce. They are not allowed to dig for any gold or silver, unless they send it to England immediately; they have not the liberty of trading with any parts that do not belong to the British dominion, excepting a few places; nor are foreigners allowed to trade with the English colonies of North America. These and some other restrictions occasion the inhabitants of the English colonies to grow less tender for their mother country. This coldness is kept up by the many foreigners such as Germans, Dutch and French, who live among the English and have no particular attachment to Old England. Add to this also that many people can never be contented with their possessions, though they be ever so large. They will always be desirous of getting more, and of enjoying the pleasure which arises from a change. Their extraordinary liberty and their luxury often lead them to unrestrained acts of selfish and arbitrary nature.

I have been told by Englishmen, and not only by such as were born in America but also by those who came from Europe, that the English colonies in North America, in the space of thirty or fifty years, would be able to form a state by themselves entirely independent of Old England.[1] But as the whole country which lies along the seashore is unguarded, and on the land side is harassed by the French, these dangerous neighbors in times of war are sufficient to prevent the connection of the colonies with their mother country from being quite broken off. The English government has therefore sufficient reason to consider means of keeping the colonies in due submission. But I have almost gone too far from my purpose; I shall therefore finish my observations on New York.

The declination of the magnetic needle, in this town was observed by Philip Wells, the chief engineer of the province of New York in the year 1686, to be eight degrees and forty-five minutes to the westward. But in 1723 it was only seven degrees and twenty minutes according to the observations of governor Burnet. From this we may conclude that in thirty-eight years the magnet has approached about one degree and twenty-five minutes nearer to the true north; or, which is the same thing, about two minutes annually. Mr. Alexander,[2] a man of great knowledge in astronomy and in mathematics, assured me from several observations, that in the year 1750, on the eighteenth of September, the deviation was to be reckoned six degrees and twenty-two minutes west.

There are two printers in the town and every week some English papers are published, which contain news from all parts of the world.

The winter is much more severe here than in Pennsylvania, it being nearly as cold as in some of the provinces of Sweden: its season, however, is much shorter than with us. The spring is very early and the autumn very late, and the heat in summer is excessive. For this reason the melons planted in the fields are ripe at the beginning of August, whereas we can hardly bring them so soon to maturity under glass and in hotbeds. The cold of the winter I cannot justly determine, as the thermometric observations which were sent to me were all calculated after incorrect thermometers that were placed inside the houses and not in the open air. The snow lies for several months upon the ground, and sleighs

---

[1] As it happened, the American Colonies became independent just about thirty years after this date.

[2] James Alexander (1691-1756), Scotch-born lawyer, politician and statesman, who with Franklin was one of the founders of the American Philosophical Society. He had been trained early as an engineer officer, was well versed in mathematics, and carried on an extensive correspondence with the scientists of his day.

are made use of here as in Sweden, but they are rather bulky. The Hudson River is about an English mile and a half broad at its mouth; the difference between the highest flood and the lowest ebb is between six and seven feet; and the water is very brackish. Yet the ice may remain in it for several months and has sometimes a thickness of more than a foot.

Mosquitoes. The inhabitants are sometimes greatly troubled with mosquitoes. They either follow the hay which is made near the town, in the low meadows saturated with salt water, or they accompany the cattle at night when they are brought home. I have myself experienced, and have observed in others, how much these little animals can disfigure a person's face during a single night; for the skin is sometimes so covered over with little swellings from their stings that people are ashamed to appear in public.

Watermelons. The watermelons which are cultivated near the town grow very large: they are extremely delicious, and are better than in other parts of North America, though they are planted in the open fields and never in a hotbed. I saw a watermelon at Governor Clinton's in September, 1750, which weighed forty-seven English pounds, and at a merchant's store in town another of forty-two pounds weight; but they were reckoned the biggest obtainable in this locality.

In the year 1710 five chiefs or sachems of the Iroquois went from here to England in order to engage Queen Anne to make an alliance with them against the French. Their names, dress, reception at court, speeches to the Queen, opinion of England and of the European manners, and several other particulars about them are sufficiently known from other writings; it would therefore be unnecessary to enlarge upon them here. The sachems of the Indians have commonly no greater authority over their subjects than constables in a meeting of the inhabitants of a parish, and hardly that much. On my travels through the country of these Indians I had never any occasion to go and call upon the chiefs, for they always came into my habitation without being asked. These visits they usually paid in order to get a glass or two of brandy, which they value above anything they know. One of the five sachems mentioned above died in England; the others returned safe.

The *first colonists* in New York were Dutchmen. When the town and its territories were taken by the English and left to them by the next peace in exchange for Surinam,[1] the old inhabitants were allowed either to remain at New York, and enjoy all the privileges and immunities which they were possessed of before, or to leave the place with all their goods. Most of them chose the former; and therefore the inhabitants both of the town and of the province belonging to it are still for the greatest part Dutch, who still, and especially the old people, speak their mother tongue.

They were beginning however by degrees to change their manners and opinions, chiefly indeed in the town and in its neighborhood; for most of the young people now speak principally English, go only to the English church, and would even take it amiss if they were called Dutchmen and not Englishmen.

Treatment of Germans. Though the province of New York has been inhabited by Europeans much longer than Pennsylvania, yet it is not by far so populous as that colony. This cannot be ascribed to any particular discouragement arising from the nature of the soil, for that is pretty good, but I was told of a very different reason which I shall mention here. In the reign of Queen Anne, about the year 1709, many Germans came hither, who got a tract of land from the government on which they might settle. After they had lived here for some time, and had built houses and churches and cultivated fields and meadows, their liberties and privileges were infringed upon, and under several pretences they were repeatedly deprived of parts of their land. This at last roused the Germans; they returned violence for violence, and beat those who thus robbed them of their possessions. But these proceedings were looked upon in a very bad light by the government: the leading Germans being imprisoned, they were very roughly treated and punished with the utmost rigor of the law. This however so exasperated the rest, that the greater part of them left their houses and fields and went to settle in Pennsylvania. There they were exceedingly well received, got a considerable tract of land, and were granted great privileges in perpetuity. The Germans not satisfied with being themselves removed from New York, wrote to their relations and friends and advised them if ever they intended to come to America not to go to New York, where the government had shown itself so inequitable. This advice had such influence that the Germans, who afterwards emigrated in great numbers to North America,

---

[1] Surinam is Dutch Guiana.

constantly avoided New York and kept going to Pennsylvania. It sometimes happened that they were forced to go on board such ships as were bound for New York; but they had scarcely got on shore, when they hastened on to Pennsylvania, right before the eyes of all the inhabitants of New York.

The Dutch Settlers. But the lack of people in this province may likewise be accounted for in a different manner. As the Dutch, who first cultivated this section, obtained the liberty of staying here by the treaty with England, and of enjoying all their privileges and advantages without the least limitation, each of them took a very large piece of ground for himself, and many of the more powerful heads of families made themselves the possessors and masters of a country of as great territory as would be sufficient to form one of our moderately-sized, and even one of our large, parishes. Most of them being very rich, their envy of the English led them not to sell them any land, but at an excessive rate, a practice which is still punctually observed among their descendants. The English therefore, as well as people of other nations, have but little encouragement to settle here. On the other hand, they have sufficient opportunity in the other provinces to purchase land at a more moderate price, and with more security to themselves. It is not to be wondered then, that so many parts of New York are still uncultivated, and that it has entirely the appearance of a frontier-land. This instance may teach us how much a small mistake in a government can hamper the settling of a country.

NOVEMBER THE 3RD

About noon we set out from New York on our return, and continuing our journey we arrived at Philadelphia on the fifth of November.

Cider. In the neighborhood of this capital (of Pennsylvania) the people had a month ago made their cider, which they were obliged to do, because their apples were so ripe that they dropped from the trees. But on our journey through New York we observed the people still employed in pressing out the cider. This is a plain proof that in Pennsylvania the apples are ripe sooner than in New York; but whether this be owing to the nature of the soil or to a greater heat of the summer in Philadelphia, or to some other cause, I know not. However, there is not the slightest advantage in making cider so early, for long experience has taught the husbandmen that it is worse for being made early in the year, the great heat in the beginning of autumn being said to effect unfavorably the fermentation of the juice.

Skunks. There is a certain quadruped which is pretty common not only in Pennsylvania but likewise in other provinces both of South and North America, and goes by the name of polecat among the English. In New York they generally call it a skunk. The Swedes here by way of nickname called it *fiskatta*, on account of the horrid stench it sometimes causes, as I shall presently show. The French in Canada, for the same reason call it *bête puante* or stinking animal, and *enfant du diable* or child of the devil. Some of them likewise call it *pekan.* Catesby in his *Natural History of Carolina*,[1] has described it by the name of *Putorius Americanus striatus* and drawn it. Dr. Linné calls it *Viverra putorius.* This animal, which is very similar to the marten, is of about the same size and commonly black. On the back it has a longitudinal white stripe and two others, one on each side, parallel to the former. Sometimes, but very seldom, some are seen which are entirely white. On our return to Philadelphia we saw one of these animals not far from town near a farmer's house, killed by dogs. And afterwards I had during my stay in these parts several opportunities of seeing it and of hearing of its qualities. It keeps its young ones in holes in the ground and in hollow trees; for it does not confine itself to the ground, but climbs up trees with the greatest agility. It is a great enemy to birds; since it breaks their eggs and devours their young ones; and if it can get into a hen roost it soon destroys all its inhabitants.

This animal has a particular quality by which it is principally known; when it is pursued by men or dogs it runs at first as fast as it can, or climbs a tree; but if it is closely beset by its pursuers, it squirts its urine upon them.[2] This, according to some, it does by wetting its tail with the urine whence by a sudden motion it

---

[1] Vol. 2, p. 62, plate 62.

[2] The skunk or polecat does not eject its urine but "an offensively odorous secretion produced in two muscular-walled perineal glands." (Webster's Dictionary).

Kalm seriously considered bringing a living skunk back to Sweden with him, but was dissuaded from his purpose lest "both he and the skunk be thrown overboard on the way." (Letter to Linné of Dec. 5, 1750).

scatters it abroad; but others believe that it could send its urine equally far without the help of its tail; I find the former of these accounts to be the most likely. For some creditable people assured me that they had had their faces wetted with it all over, though they stood more than eighteen feet from the animal. The urine has so horrid a stench that nothing can equal it: it is something like that of the cranesbill or Linné's *Geranium robertianum,* but infinitely stronger. If you come near a polecat when it spreads its stench, you cannot breathe for a while, and it seems as if you were stifled; and in case the urine comes into the eyes of a person he is likely to be blinded. Many dogs that in a chase pursue the polecat very eagerly, run away as fast as they can when they are wetted. However, if they be of the true breed, they will not give over the pursuit till they have caught and killed the polecat; but they are obliged now and then to run their noses in the ground in order to relieve themselves.

Clothes which have been wetted by this animal retain the smell for more than a month; unless they be covered with fresh dirt, and suffered to remain under it for twenty-four hours; then it will in a great measure be removed. Those likewise who have gotten any of this liquid upon their face and hands, rub them with loose earth; and some even hold their hands in the ground for an hour as washing will not help them so soon. A certain man of rank, who had by accident been wetted by the polecat, stunk so ill that on going into a house the people either ran away as from a pest or on his opening the door told him to go to the devil. Dogs that have hunted a polecat are so offensive for several days afterwards that they cannot be borne in the house. At Philadelphia I once saw a great number of people on a market day throwing things at a dog that was so unfortunate as to have been engaged with a polecat just before, and to carry about him the tokens of its displeasure. Persons when travelling through a forest are often troubled with the stink which this creature makes; and sometimes the air is so much infected that it is necessary to hold one's nose. If the wind blows from the place where the polecat has been, or if it be quite calm, as at night, the smell is stronger and more disagreeable.

In the winter of 1749 a polecat, tempted by a dead lamb, came one night near the farmhouse where I then slept. Being immediately pursued by some dogs, it had recourse to its usual expedient in order to get rid of them. The attempt succeeded, the dogs not choosing to continue the pursuit. The stink was so extremely penetrating that though I was at some distance it affected me in the same manner as if I had been stifled, and it was so disagreeable to the cattle that it made them bellow loudly. However, by degrees it vanished. Towards the end of the same year one of these animals got into our cellar, but no stench was observed, for it only vented that when it was pursued. The cook however found for several days in succession that some of the meat which was kept there had been eaten, and suspecting that it had been done by the cat she shut up all avenues in order to prevent him getting at it. But the next night, being awakened by a noise in the cellar, she went down, and though it was quite dark, saw an animal with two shining eyes, which seemed to be all on fire; she however resolutely killed it, but not before the polecat had thoroughly perfumed the cellar. The maid was sick of it for several days; and all the bread, meat and other provision kept in the cellar were so saturated with it, that we could not make the least use of them and were forced to throw them all away.

From an accident that happened at New York to one of my acquaintances, I conclude that the polecat either is not always very shy or that it sleeps very hard at night. This man coming home out of a wood in a summer evening, thought that he saw a plant standing before him; stooping to pluck it, he was to his cost convinced of his mistake by being all of a sudden covered with the secretion from a polecat, whose tail, as it stood upright, the good man had taken for a plant. The creature had taken its revenge so effectually that the man was much at a loss how to get rid of the stench.

However, though these animals play such disagreeable tricks, yet the English, the Swedes, the French, and the Indians in these parts tame them. They follow their masters like domestic animals, and never make use of their urine, except they be very much beaten or angered. When the Indians kill such a polecat, they always eat its flesh, but when they pull off its skin, they take care to cut away the bladder, that the flesh may not get a taste from it. I have spoken with both Englishmen and Frenchmen, who assured me that they had eaten of it, and found it very good meat, and not much unlike the flesh of a pig. The skin which is pretty coarse, and has long hair, is not used by the Europeans, but the Indians prepare it with the hair on, and make tobacco pouches of it, which they carry before them.

NOVEMBER THE 6TH

In the evening I went out of town to Mr. Bartram's. I found a man with him who lived in Carolina, and I obtained several particulars about that province from him, a few of which I shall here mention.

Carolina. Tar, pitch and *rice* are the chief products of Carolina. The soil is very sandy, and therefore many pines and firs grow in it, from which they make tar. The firs which are taken for this purpose are usually such as are dried up of themselves. The people here in general do not know how to prepare the firs by taking the bark off on one, or on several sides, as they do in Österbotten (Finland). In some parts of Carolina they likewise make use of the branches. The manner of burning or boiling, as the man described it to me, is entirely the same as in Finland. Pitch is made thus: they dig a hole into the ground and smear the inside well with clay, into which they pour the tar, and make a fire round it, which is kept up till the tar has got the consistency of pitch. They make two kinds of tar in the North American colonies: one is the common tar, which I have above described, and which is made of the stems, branches, and roots of such firs as are already considerably dried out before. This is the most common method in this country. The other way is by peeling the bark from the firs on one side and afterwards letting them stand another year during which the resin comes out between the cracks of the stem. The tree is then felled and burnt for tar; and the tar thus made is called "green tar"; not that there is that difference of color in it, for in this respect they are both pretty much alike; but the latter is called so from being made of green and fresh trees, whereas "common tar" is made of dead trees. The burning is done in the same manner as in Finland. They use only black firs, for the white firs will not serve their purpose, though they are excellent for boards, masts, etc. Green tar is dearer than common tar. It is already a pretty general complaint that the fir woods are almost wholly destroyed by this practice.

Rice is planted in great quantity in Carolina. It succeeds best in marshy and swampy grounds, which may be flooded, and likewise it ripens there the soonest. Where these cannot be had, they must choose a dry soil; but the rice produced here will be much inferior to the other. The land on which it is cultivated must never be manured. In Carolina they sow it in the middle of April, and it is ripe in September. It is planted in rows like peas and usually fifteen inches space is left between the rows. As soon as the plants come up, the field is laid under water. This not only greatly forwards the growth of the rice, but likewise kills all weeds, so as to render weeding unnecessary. The straw of rice is said to be excellent food for cattle, which eat it very greedily. Rice requires a hot climate, and therefore it will not succeed well in Virginia, the summer there being too short and the winter too cold; and much less will it grow in Pennsylvania. They are as yet ignorant in Carolina of the art of making arrack from rice: it is chiefly South Carolina that produces the greatest quantity of rice; and on the other hand they make the most tar in North Carolina.

November the 7th

The stranger from Carolina, whom I have mentioned before, had found many oyster shells at the bottom of a well, seventy English miles distant from the sea, and four from a river. They lay at a depth of fourteen English feet from the surface of the earth; the water in the well was brackish, but that in the river was fresh. The same man had at the building of a saw-mill, a mile and a half from a river, found, first sand, and then clay filled with oyster shells. Under these he found several bills of sea birds as he called them, which were already petrified: they were probably *Glossopetræ*.

Foxes. There are two species of foxes in the English colonies, one gray, and the other red; but later I shall show that there are others which sometimes appear in Canada. The gray foxes are here constantly, and are very common in Pennsylvania and in the southern provinces. In the northern ones they are pretty scarce, and the French in Canada call them Virginian foxes on that account. In size they do not quite come up to our foxes. They do no harm to lambs, but they prey upon all sorts of poultry, whenever they can get at them. They do not however seem to be looked upon as animals that cause a great deal of damage, for there is no reward given for killing them. Their skin is greatly sought by hatters, who employ the fur in their work. People have their clothes lined with it sometimes, and the grease is used for all sorts of rheumatic pains. These foxes are said to be less nimble than the red ones. They are sometimes tamed, though they are not allowed to run about, but are tied up. Mr. Catesby has drawn and described this animal in his *Natural History of Carolina*, under the name of the gray American fox.[1] A skin of it was sold in Philadelphia for two shillings and sixpence in Pennsylvania currency.

The red foxes are very scarce here: they are exactly the same as the European sort. Mr. Bartram and several others assured me, that according to the unanimous testimony of the Indians, this kind of fox never was in the country before the Europeans settled in it. But of the manner of their coming over I have two different accounts: Mr. Bartram and several other people were told by the Indians that these foxes came into America soon after the arrival of the Europeans, after an extraordinary cold winter, when the whole sea to the northward was frozen, from which they inferred that they had perhaps gotten over to America upon the ice from Greenland or the northern parts of Europe and Asia. But Mr. L. Evans and some others assured me that the following account was still known by the people. A gentleman of fortune in New England, who had a great inclination for hunting, brought over a great number of foxes from Europe, and let them loose in his territories, that he might be able to indulge his passion for hunting. This is said to have happened almost at the very beginning of New England's colonization by Europeans. These foxes were believed to have so multiplied, that all the red foxes in the country were their offspring. At present they are reckoned among the noxious creatures in these parts; for they are not contented, as the gray foxes, with killing fowl, but also attack lambs. In Pennsylvania therefore there is a reward of two shillings for killing an old red fox, and of one shilling for killing a young one. And in all the other prov-

---

[1] Vol. 2, p. 78.

inces there are likewise rewards offered for killing them. Their skin is in great demand, and is sold as dear as that of the gray foxes, that is, two shillings and sixpence, in Pennsylvania currency.

Wolves. There are two varieties of wolves here, which however seem to be of the same species. For some of them are yellowish, or almost pale gray, and others are black or dark brown. All the old Swedes related, that during their childhood, and still more at the arrival of their fathers, there were excessive numbers of wolves in the country, and that their howling and yelping might be heard all night. They also frequently tore in pieces sheep, hogs, and other young and small cattle. About that time or soon after, when the Swedes and the English were quite settled here, the Indians were attacked by the smallpox. This disease they got from the Europeans, for they knew nothing of it before. It killed many hundreds of them, and most of the Indians of the section, then called New Sweden, died of it. The wolves then came, attracted by the stench of so many corpses, in such great numbers that they devoured them all, and even attacked the poor sick Indians in their huts, so that the few healthy ones had enough to do to drive them away. But since that time they have disappeared, so that they are now seldom seen, and it is very rarely that they commit any disorders. This is attributed to the greater cultivation of the country, and to their being killed in great numbers. But further up the country, where it is less inhabited, they are still very abundant. On the coasts of Pennsylvania and New Jersey, the sheep stay all night in the fields, without the people fearing the wolves. However, to prevent their multiplying too much there is a reward of twenty shillings in Pennsylvania, and of thirty in New Jersey, for bringing in a dead wolf, and the person that brings it may keep the skin. But for a young wolf the reward is only ten shillings of Pennsylvania currency. There are instances of these wolves being made as tame as dogs.

Wild Oxen. The wild oxen have their habitat principally in the woods of Carolina, which are far up in the country. The inhabitants frequently hunt them, and salt their flesh like common beef, which is eaten by servants and the lower class of people. But the hide is of little use, having too large pores to be valuable for shoes. However, the poorer people in Carolina spread these hides on the ground instead of beds.

Plants in Carolina. The *Viscum filamentosum,* or fibrous mistle-toe, is found in abundance in Carolina; the inhabitants use it for straw in their beds, and to adorn their houses. The cattle are very fond of it, and it is likewise employed in packing goods.

The *Spartium scoparium* grew in Mr. Bartram's garden from English seeds. He said that he had several bushes of it, but that the frost in the cold winters here had killed most of them. They grow wild in several places in Sweden.

Truffles. Mr. Bartram had some truffles, or Linné's *Lycoperdon tuber,* which he had got out of a sandy soil in New Jersey, where they are abundant. These he showed to his friend from Carolina, and asked him whether they were the tuckahoo of the Indians. But the stranger said they were not, and added that though these truffles were very common in Carolina, yet he had never seen them used any other way but in milk, to cure dysentery; and he gave us the following description of the tuckahoo. It grows in several swamps and marshes and is often plentiful. The hogs greedily dig up its roots with their noses in such places; and the Indians in Carolina likewise gather them in their rambles in the woods, dry them in the sunshine, grind them and bake bread of them. While the root is fresh it is harsh and acrid, but being dried it loses the greatest part of its acrimony. To judge by these qualities the tuckahoo may very likely be the *Arum Virginicum.* Compare with this account, what will be related later of the *Tahim* and *Tuckah.*

After dinner I again returned to town.

NOVEMBER THE 8TH

Bees. Several English and Swedish farmers kept beehives, which afforded their possessors profit; for bees succeed very well here. The wax was for the most part sold to tradesmen; but the honey they made use of in their own families, in different ways. The people unanimously asserted that the common bees were not in North America before the arrival of the Europeans, but that they were first brought over by the English who settled here. The Indians also generally declare that their fathers had never seen any bees, either in the woods or anywhere else, before the Europeans had been here for a number of years. This is further confirmed by the name which the Indians give them; for having no particular name for them in their language they call them English flies, because the English first brought them over; but at present they fly plentifully about the woods of North America. However, it has been observed that always when the bees swarm, they spread to the southward, and never northward. It seems as though they did not find the latter direction so good for their constitution. Therefore they cannot stay in Canada, and all that have been carried thither died in winter. It seemed to me as though the bees in America were somewhat smaller than ours in Sweden. They have not yet been found in the woods on the other side of the Blue Mountains, which confirms the opinion of their being brought to America recently. A man told Mr. Bartram that on his travels in the woods of North America he had found another sort of bees, which, instead of separating their wax and honey, mixed both together in a great bag. But this account probably needs more particulars and factual certainty.

NOVEMBER THE 9TH

Birds. All the old Swedes and Englishmen born in America whom I ever questioned asserted that there were not nearly so many edible birds at present as there used to be when they were children, and that their decrease was visible. They even said that they had heard the same complaint from their fathers who were born in this locality. In their youth the bays, rivers and brooks were quite covered with all sorts of water fowl, such as wild geese, ducks, and the like. But at present there was sometimes not a single bird upon them. About sixty or seventy years ago, a single person could kill eighty ducks in a morning; but at present you frequently waited in vain for a single one. A Swede above ninety years old [1] assured me that he had in his youth killed twenty-three ducks at a shot (hunting party?). This good luck nobody is likely to have at present, as you are forced to ramble about for a whole day, without getting a sight of more than three or four. Cranes [2] at that time

---

[1] Undoubtedly the same one, whom we have met twice before.

[2] When Captain Amadas, the first Englishman that ever landed in North America, set foot on shore (to use his own words) "such a flocke of cranes (the most part white) arose under us with such a cry, redoubled by many echoes, as if an armie of men had shouted together."—F.

came hither by hundreds in the spring: at present there are very few. The wild turkeys, and the birds which the Swedes in this country call partridges and hazelhens, were seen in large flocks in the woods. But at this time a person gets tired with walking before he can start a single bird.

The cause of this diminution is not difficult to find. Before the arrival of the Europeans, the country was uncultivated and full of great forests. The few Indians that lived here seldom disturbed the birds. They carried on no trade among themselves, iron and gun powder were unknown to them. One hundredth part of the fowl which at that time were so plentiful here, would have sufficed to feed the few inhabitants. And considering that they cultivated their small maize fields, caught fish, hunted stags, beavers, bears, wild cattle, and other animals whose flesh was delicious to them, it will soon appear how little they disturbed the birds. But since the arrival of great crowds of Europeans, things are greatly changed; the country is well peopled, and the woods are cut down. The people, increasing in this country, have by hunting and shooting in part extirpated the birds, in part frightened them away. In spring the people still steal eggs, mothers and young indifferently, because no regulations are made to the contrary. And if any had been made, the spirit of freedom which prevails in the country would not suffer them to be obeyed. But though the eatable birds have been diminished greatly, yet there are others which have rather increased than decreased in number since the arrival of the Europeans. This can most properly be said of a species of daws which the English call blackbirds and the Swedes "corn thieves"; Dr. Linné calls them *Gracula quiscuta*. And with them the several varieties of squirrels, among the quadrupeds, have increased also: for these and the former live chiefly upon corn, or prefer it to anything else. As the population increases, the cultivation of corn increases, and of course the food of the above-mentioned animals is more plentiful. To this must be added, that these latter are rarely eaten, and therefore they are more at liberty to multiply their kind. There are likewise other birds which are not eaten, of which at present there are nearly as many as there were before the arrival of the Europeans. On the other hand I heard great complaints of the great decrease of eatable fowl, not only in this province but in all parts of North America where I have been.

Fish. Aged people had experienced with the fish the same conditions which I have just mentioned in regard to birds. In their youth, the bays, rivers and brooks, had such quantities of fish that at one draught in the morning they caught as many as a horse was able to carry home. But at present things are greatly altered, and they often work in vain all night long with their fishing tackle. The causes of this decrease of fish are partly the same as those of the diminution in the number of birds. They are of late caught by a greater variety of contrivances, and in different manners than before. The numerous mills on the rivers and brooks likewise contribute to it in part; for it has been observed here that the fish go up the river in order to spawn in shallow water in the spring; but when they meet with waterfalls that prevent their proceeding, they turn back, and never return.[1] Of this I was assured by a man of fortune at Boston; his father used to catch a number of herrings throughout the winter and almost always in summer, in a river, upon his countryseat; but when he built a mill with a dam in this water, they were lost. Consequently they complained here and everywhere of the decrease of fish. Old people asserted the same in regard to oysters at New York, for though they are still taken in considerable quantity and are as big and as delicious as can be wished, yet all the oyster fishermen own that the number diminishes greatly every year. The most natural cause of it is probably the immoderate catching of them at all times of the year.

Many old people said that the difference in the quantity of fish in their youth in comparison with that of to-day was as great as between day and night.

Herring. Mr. Franklin told me that in that part of New England where his father lived, two rivers flowed into the sea, in one of which they caught great numbers of herring, and in the other not one. Yet the places where these rivers discharged themselves into the sea were not far apart. They had observed that when the herrings came in spring to deposit their spawn, they always swam up one river, where they used to catch them, but never came into the other. This circumstance led Mr. Franklin's father, who had settled between the two rivers, to try whether it was not possible to make

---

[1] To-day it is well known of course, that unless the waterfalls be very high such fish as salmon and trout can ascend a waterfall by jumping.

the herrings also live in the other river. For that purpose he put out his nets, as they were coming up for spawning, and he caught some. He took the spawn out of them, and carefully carried it across the land to the other river. It was hatched, and the consequence was that every year afterwards they caught more herring in that river, and this is still the case. This leads one to believe that the fish always like to spawn in the same place where they are hatched, and from which they first put out to sea, being as it were accustomed to it. The following is another peculiar observation: it has never formerly been known that codfish were to be caught off Cape Henlopen; they were always caught at the mouth of the Delaware. But at present they are numerous in the former place. From this it may be concluded that fish likewise change their places of abode, of their own accord.

Greenland. A sea captain, who had been in Greenland, asserted from his own experience that on passing the seventieth degree of north latitude the summer heat was there much greater than it is below that degree. Hence he concluded that the summer heat at the pole itself must be still more excessive, since the sun shines there for such a long space of time without ever setting. The same account with similar consequences Mr. Franklin had heard from sea captains in Boston, who had sailed to the most northern parts of this hemisphere. But still more astonishing is the account he got from Captain Henry Atkins, who still lives at Boston. He had for some time been fishing along the coasts of New England, but not catching as much as he wished, he sailed north as far as Greenland. At last he went so far that he discovered people who had never seen Europeans before (and what is more astonishing) who had no idea of the use of fire, which they had never employed, and if they had known it, they could have made no use of their knowledge, as there were no trees in the country. But they ate the birds and fish raw. Captain Atkins got some very rare skins in exchange for some trifles.

It is already known from several accounts of voyages, that to the northward neither trees nor bushes, nor any ligneous plants are to be found fit for burning. But is it not probable that the inhabitants of so desolate a country, like other northern nations we know, burn the train oil of fishes, and the fat of animals in lamps, in order to boil their meat, to warm their subterraneous caves in winter, and to light them in the darkest season of the year? Otherwise they would be living in close proximity to eternal darkness.

NOVEMBER THE 11TH

The Moose. In several writings we read of a large animal, which is to be found in New England and other parts of North America. They sometimes dig very long and branched horns out of the ground in Ireland, and nobody in that country or anywhere else in the world knows an animal that has such horns. This has induced many people to believe that it is the Moose-deer so famous in North America, and that the horns found were of animals of this kind which had formerly lived in that island but had gradually become extinct. It has even been concluded that Ireland in distant ages either was connected to North America or that a number of little islands which are lost at present made a chain between them. This led me to inquire whether an animal with such excessive great horns as are ascribed to the Moose-deer had ever been seen in any part of this country. Mr. Bartram told me that notwithstanding he had carefully inquired to that purpose, yet there was no person who could give him any information which could be relied upon, and therefore he was entirely of the opinion that there was no such animal in North America. Mr. Franklin related that he had, when a boy, seen two of the animals which they call Moose-deer, but he well remembered that they were not of such a size as they must have been, if the horns found in Ireland were to fit them. The two animals which he saw, were brought to Boston in order to be sent to England to Queen Anne. Anyone who wanted to see them had to pay twopence. A merchant paid for a number of school-boys who wanted to see them, among whom was Franklin. The height of the animal up to the back was that of a pretty tall horse, but the head and its horns were still higher. Mr. Dudley [1] has given a description of the Moose-deer which is found in North America. On my travels in Canada I often inquired of the Frenchmen whether there had ever been seen so large an animal in this country,

---

[1] Paul Dudley (1675-1751), American jurist and naturalist. He wrote several pamphlets dealing with natural history and was elected to membership in the Royal Society.

as some people say there is in North America, and with such great horns as are sometimes dug up in Ireland. But I was always told that they had never heard of it, much less seen it. Some added that if there was such an animal, they certainly must have met with it in some of their excursions in the woods. There are elks here, which are either of the same sort as the Swedish or a variety of them. Of these they often catch some which are larger than usual, whence perhaps the report of the very large animal with excessive horns in North America first had its origin. These elks are called "Originals" by the French in Canada, which name they have borrowed from the Indians: perhaps Dudley, in describing the Moose-deer, meant no other animals than these large elks.[1]

Rocks and Minerals. Mr. Franklin gave me a piece of stone, which on account of its indestructibility in fire is used in New England for making smelting furnaces and forges. It consists of a mixture of *Lapis ollaris* or serpentine stone and of asbestos. The greatest part of it is a gray serpentine which is fat and smooth to the touch and is easily cut and worked. Here and there are some glittering speckles of that sort of asbestos whose fibres come from a center-like ray, or "star asbestos." This stone is not found in strata or solid rocks, but is here and there scattered on the fields.

Another stone is called *soapstone* by many of the Swedes, being as smooth as soap on the outside. They make use of it for rubbing spots out of their clothes. It might be called *Saxum talcosum particulis spataceis granatisque immixtis,* or a talc with mixed particles of spar and garnets. A more exact description I reserve for another work. At present I only add that the ground color is pale green, with some dark spots, and sometimes a few of a greenish hue. It is very smooth to the touch, and is formed in waves. I have seen large stones of it which were a fathom and more long, proportionably broad, and commonly six inches or a foot deep. But I cannot determine anything of their original size, as I have

[1] What gives still more weight to Mr. Kalm's opinion of the elk being the moose-deer, is the name Musu which the Algonquins give to the elk, as Mr. Kalm himself observes in the sequel of his work; and this circumstance is the more remarkable, as the Algonquins before the Iroquois or five nations got so great a power in America, were the most powerful nation in the northern part of this continent; in so much, that though they be now reduced to an inconsiderable number, their language is however a kind of universal language in North America; so that there is no doubt, that the elk is the famous Moose-deer.—F.

not been at the place where they are dug, and have only seen the stones at Philadelphia, which are brought there ready cut. The particles of talc in this stone are about thirty times as many as those of spar and garnet. It is found in many parts of the country, for example in the neighborhood of Chester in Pennsylvania. The English likewise call it soapstone and it is likely that the Swedes have borrowed the name from them.

This stone was chiefly employed in the following manner. First, the people took spots out of their clothes with it. But for this purpose the whole stone is not equally useful, for it includes in its clear particles some dark ones which consist wholly of serpentine stone, and may easily be cut with a knife. Some of the loose stone is scraped off like a powder, and strewed upon a greasy spot, on silk or any other material. This absorbs the grease, and after rubbing off the powder the spot disappears. As this stone is also very durable in fire, the country people make their hearths out of it, especially the place where the heat is the greatest, for the stone stands the strongest fire. If the people can get a sufficient quantity of this stone, they use it in laying steps before the houses instead of bricks, which are generally used for that purpose. The walls round the courtyards, gardens, burying places, and those for the cellar doors sloping towards the street, which are usually built of brick, are covered with a coping of this stone, for it holds excellently against all the effects of the sun, air, rain and storm, and does not decay but protects the bricks. On account of this quality, people commonly get the door posts in which their hinges are fastened made of this stone; and in several public buildings, such as the house of assembly for the province, the whole lower wall and the cornerstones are built of it.

The *salt* which is used in the English North American colonies is brought from the West Indies, but it is more corrosive than the European. The Indians have in some places salt springs from which they get salt by boiling. I shall later have occasion to describe some of them. Mr. Franklin was of the opinion that the people in Pennsylvania could easier make good salt of sea water than in New England, where sometimes salt is made of the sea water on their coast, though their location is more northerly.

Lead ore has been discovered in Pennsylvania, but as it is not found in any quantity, nobody has ever attempted to use it. Loadstones of good quality have likewise been found, and I myself possess several pretty pieces of them.

Iron is dug in such great quantities in Pennsylvania and in the other American provinces of the English, that they could provide with that commodity not only England but almost all Europe, and perhaps the greater part of the globe. Commonly the ore is here mined infinitely easier than our Swedish ore. For in many places with a pick ax, a crow-foot and a wooden club, it is obtained with the same ease with which a hole can be made in a hard soil. In many places the people know nothing of boring, blasting and firing; and the ore is likewise very fusible. Of this iron they get such quantities that not only the numerous inhabitants of the colonies themselves have enough of it but great amounts are sent to the West Indies, and they have lately begun even to trade with Europe in it. This iron is reckoned better for ship building than our Swedish iron, or any other, because salt water does not corrode it so much. Some people believe that without reckoning the freight they can sell their iron in England at a lower rate than any other nation, especially when the country becomes better peopled and labor cheaper.

Asbestos. The mountain flax, or that kind of stone which Bishop Brovallius in his lectures on mineralogy published in 1739 calls *Amiantus fibris separabilibus molliusculis*, or the amiant with easily separable soft fibres, is found abundantly in Pennsylvania. Some pieces are very soft, others pretty tough. Mr. Franklin told me that twenty and some odd years ago, when he made a voyage to England, he had a little purse with him, made of the mountain flax of this country, which he presented to Sir Hans Sloane.[1] I have likewise seen paper made of this stone, and I have received some small pieces of it which I keep in my cabinet. Mr. Franklin had been told by others that on exposing this mountain flax to the open air in winter, and leaving it in the cold and wet, it would grow together and become tougher and more suitable for spinning. But he did not venture to determine how well this opinion was founded. On this occasion he related a very amusing incident which happened

---

[1] Sir Hans Sloane (1660-1753), physician and naturalist, scientific writer and President of the Royal Society. His botanical work in the West Indies is still valuable. In 1696 he published in London his *Catalogus Plantarum quae in Insula Jamaica sponte proverriunt aut vulgo voluntur.*

to him with this mountain flax. He had several years ago gotten a piece of it, which he gave to one of his journeymen printers in order to get it made into a sheet at the paper mill. As soon as the fellow brought the paper, Mr. Franklin rolled it up and threw it into the fire, telling the journeyman he would see a miracle, a sheet of paper which did not burn. The ignorant fellow insisted upon the contrary, but was greatly terrified upon seeing himself convinced. Mr. Franklin then explained to him, though not very clearly, the peculiar qualities of the paper. As soon as he was gone, some of his acquaintances came in, who immediately recognized the paper. The journeyman thought he would show them a great curiosity and astonish them. He accordingly told them that he had curiously made a sheet of paper which would not burn, though it were thrown into the fire. They pretended to think it impossible, and he as strenuously maintained his assertion. At last they laid a wager about it, but while he was busy with stirring up the fire the others slyly besmeared the paper with fat. The journeyman, who was not aware of it, threw it into the fire and that moment it was all in flames. This astonished him so much that he was almost speechless, upon which they could not help laughing, and so disclosed the whole artifice.

Ants. In several houses of the town, a number of little ants run about, living under ground and in holes in the wall. The length of their bodies is one geometrical line [1/12th inch]. Their color is either black or dark red, and they have the custom of carrying off sweet things, if they can find them, in common with the ants of other countries. Mr. Franklin was much inclined to believe that these little insects could by some means communicate their thoughts or desires to each other, and he confirmed his opinion by some examples. When an ant finds some sugar, it runs immediately under the floor to its hole, where having stayed a little while a whole army comes out, unites and marches to the place where the sugar is, and carries it off by pieces. If an ant meets with a dead fly, which it cannot carry alone, it immediately hastens home, and soon after some more come out, creep to the fly and carry it away. Some time ago Mr. Franklin put a little earthen pot with treacle in it into a closet. A number of ants got into the pot and devoured the treacle very quietly. But as he observed it he shook them out, and tied the pot with a thin string to a nail which he had fastened

in the ceiling, so that the pot hung down by the string. A single
ant by chance remained in the pot: this ant ate till it was satisfied;
but when it wanted to get away it was under great concern to
find its way out. It ran about the bottom of the pot, but in vain.
At last it found after many attempts the way to get to the ceiling
by the string. After it had come there, it ran first to the wall and
then to the floor. It had hardly been away for half an hour, when
a great swarm of ants came out, climbed up to the ceiling, crept
along the string into the pot, and began to eat again. This they
continued till the treacle was all eaten. In the meantime one swarm
kept running down the string and the other up all day long.

NOVEMBER THE 12TH

Weather Prophecy. An honorable man who has long been in
this province asserted that from twenty years experience he had
found a confirmation of what other people had observed with regard
to the weather, *viz.* that the weather in winter was commonly foretold by that on the first of November, old style, or twelfth new
style. If that whole day were fair, the next winter would bring
but little rain and snow along with it; but if the first half of the
day were clear, and the other cloudy, the beginning of winter
would accordingly be fair, while its end and spring would turn
out rigorous and disagreeable.—I have heard similar signs of the
weather in other places, but meteorological observations and common sense have shown sufficiently how infinitely often these proph-
ecies have failed.

Springs. Pennsylvania abounds in springs, and you often find
a spring of clear water on one or the other side, and sometimes on
several sides, of a mountain. The people near such springs, besides
making the usual use of spring water, also conduct it into a little
stone building near the house, where they can confine it. In
summer they place their milk, bottles of wine and other liquors in
the water, where they keep cool and fresh. In many country houses
the kitchen or buttery was so situated that a rivulet ran under it,
and thus provided a ready supply of water.

Not only people of rank but even others that had some possessions,
frequently had fish ponds in the country near their homes. They
always took care that fresh water might run into their ponds,

which is very salutary for the fish; for that purpose the ponds were placed below a spring on a hill.

NOVEMBER THE 13TH

Irrigation of Meadows. I saw in several parts of this province a ready method of getting plenty of grass to grow in the meadows. Here it must be remembered what I have before mentioned about the springs, which are sometimes found on the sides of hills and sometimes in the valleys. The meadows lie usually in the dales between the hills; if they are too swampy and wet, the water is carried off by several ditches. But the summer in Pennsylvania is very hot, and the sun often burns the grass so much that it dries up entirely. The husbandmen therefore are very attentive to prevent this in their meadows. To that purpose they look for all the springs in the neighborhood of a meadow; and instead of letting the rivulets flow by the shortest way into the valleys, they conduct the water as much as possible and necessary to the higher part of the meadow, and then make several narrow channels from the brook down into the plain, so that it is entirely watered by it. When there are some very low places, the farmers frequently lay wooden gutters across them, through which the water flows to the other side, and thence it is again by very narrow ditches carried to all the places where it seems necessary. To raise the water and spread it more, there are high dikes built near the springs, between which the water rises till it is high enough to run down where the people want it. Industry and ingenuity go further; when a brook runs in a wood, in a direction away from a meadow, and it has been found by levelling, and taking an exact survey of the land between the meadow and the rivulet, that the latter can be conducted towards the former, a dike is made, which hems the course of the brook, and the water is led round the sides of many hills, sometimes for the space of an English mile and further, partly across valleys in wooden pipes, till at last it is brought to where it is wanted, and where it can be circulated as above-mentioned. One that has not seen it himself, cannot believe how great a quantity of grass there is in such meadows, especially near the little channels; while others, which have not been thus managed, look wretched. The meadows commonly lie in the valleys, and one or more of their sides have a

Sauerkraut Machine
(From Swedish original)

declivity, so that the water can easily be brought to run down them. These meadows which are so carefully watered, are usually mowed three times every summer. But it is likewise to be observed, that summer continues for seven months here. The inhabitants seldom fail to use a brook or spring in this manner, if it is not too far from the meadows.

The leaves have at present fallen from all the trees, both from the oaks and from all those which have deciduous leaves, and they cover the ground in the woods six inches deep. The great quantity of them which drop annually would necessarily seem to increase the upper black mould [or humus] greatly. However, it is not above three or four inches thick in the woods, and under it lies a brick-colored clay mixed with a sand of the same color. It is remarkable that a soil which in all probability has not been disturbed since the Flood should be covered with so little black mould; but I shall speak of this later.

Sauerkraut—Machine for Making it. At the house of a German I saw a peculiar machine used for preparing sauerkraut from cabbage, and since the method of slicing the cabbage is much more efficient than with the ordinary knives used for the purpose, and which are shaped like the letter "S," I shall here give a short description of it:—a tray was made of boards with a flat even bottom about three feet long and seven inches wide and with two-inch sides. In the middle of the bottom of the tray was a large, square opening, extending across the tray and being about four inches wide. Across this were placed three knives parallel to one another. The width of each knife was one and a half inches. Their edges were set aslant, like the blade in a plane, and in such a way that the edge of one slightly overlapped the back of the other. The cabbage was grated by these knives. On the tray was placed a moveable, bottomless box about the same width as the tray (10 in. long and 6 in. deep). The box was shoved back and forth across the knives when the cabbage was to be cut. The tray was placed over a barrel or other suitable receptacle so that the knives were over the opening of the barrel. When the cabbage was placed in the box and pushed back and forth over the knives the shredded cabbage fell into the barrel below. It is a very quick way to shred cabbage. On the long sides of the box there is, of course, as much of the lower edge cut off as is necessary to clear the knives. This machine can be called a cabbage plane except that in this case the plane is stationary.

NOVEMBER THE 14TH

Squirrels. The squirrels which run about in large numbers in the woods are of different species. I here intend to describe the most common sorts, more accurately.

Gray squirrels are very plentiful in Pennsylvania and in the other provinces of North America. Their shape corresponds to that of our Swedish squirrel; but they differ from them in keeping their gray color all year and by being somewhat bigger in size. The woods in all these provinces, and chiefly in New Sweden, consist of deciduous trees and in such these squirrels like to live. Ray in his *Synopsis Quadrupedum*, and Catesby in his *Natural History of Carolina* (vol. 2) call it the Virginian greater gray squirrel; and the latter has added a sketch of it from life. The Swedes call it *grå ekorre,* which is the same as the English "gray squirrel." Their nests are usually in hollow trees, and are made of moss, straw, and other soft things. Their food consists chiefly of nuts, as hazel nuts, chinquapins, chestnuts, walnuts, hickory nuts, and the acorns of the different sorts of oak which grow here; but they like corn best. The ground in the woods is in autumn covered with acorns and all kinds of nuts which drop from the numerous trees; of these the squirrels gather great stores for winter, which they lay up in holes dug by them for that purpose. They likewise carry a great quantity of them into their nests.

As soon as winter comes the snow and cold confines them to their holes for several days, especially when the weather is very severe. During this time they consume the little store which they have brought to their nests. As soon as the weather grows milder, they creep out and dig out part of the store which they have laid up in the ground. Of this they eat some on the spot and carry the rest into their nests in the trees. We frequently observed that in winter, at the eve of a great frost, when there had been some temperate weather, the squirrels, a day or two before an intensely cold spell, ran about the woods in greater numbers than usual, partly in order to eat their fill, and partly to store their nests with a new provision for the ensuing great cold, during which they did not venture to come out, but lay snug in their homes. Therefore seeing them run in the woods in larger numbers than ordinary was a safe forecast of cold weather.

The hogs which are here driven into the woods, while there is yet no snow in them, often do considerable damage to the poor squirrels by rooting up their store-holes and stealing their winter provisions. Both the Indians and the European Americans take great pains in ferreting out these store-holes, whether in trees or in the ground, as all the nuts they contain are choice and not only quite ripe but untouched by worms. The nuts and acorns which the dormice, or *Mus cricetus* L., store up in autumn are all in the same condition. The Swedes relate that during the long severe winter of 1741 there fell such a quantity of snow that the squirrels could not get to their store, and many of them starved to death.

The damage which these animals do in the corn fields, I have already described. They do the more harm, as they do not eat all the corn, but only the inner and sweet part, and, as it were, remove the husks. In spring towards the end of April, when the oaks were in full bloom I once observed a number of squirrels on them, sometimes five, six, or more in a tree, who bit off the stalks a little below the flowers, and dropped them on the ground. Whether they eat anything of them or make use of them for some other purpose I know not; but the ground was quite covered with oak flowers, to which a part of the stalk adhered. For this reason the oaks do not bear as much fruit by far—for feeding hogs and other animals— as they would otherwise do.

Of all the wild animals in this country the squirrels are some of the easiest to tame, especially when they are taken young for that purpose. I have seen them so tamed that they would follow the boys into the woods and run about everywhere, and when tired would sit on their shoulders. Sometimes they only ran a little way into the wood, and then returned home again to the little hole that had been fitted up for them. When they ate they sat almost upright, held their food between their forefeet and their tail bent upwards. When the tame ones got more than they could eat at a time, they carried the remainder to their nests and hid it amongst the wool which they lay upon. Such tame squirrels showed no fear of strangers, and would suffer themselves to be touched by everybody without offering to bite. They sometimes would leap upon a stranger, crawl into his clothes, and lie there in order to sleep. In the farmhouses where they were kept they played with cats and dogs. They also ate bread.

The wild gray squirrels hold up their tails when sitting. As soon as they perceive a man, they continually wave their tails and begin to gnash their teeth and make a great noise, which they keep up for a long time. Those who hunt birds and other animals are therefore very angry at them, as this noise betrays them and alarms the game. Though a gray squirrel does not seem to be very shy, it is very difficult to kill; for when it sees a man, it climbs up a tree, and commonly chooses the highest in the neighborhood. It then tries to hide itself behind the trunk, so that the hunter may not see it, and though he goes ever so fast round the tree, the squirrel changes its place just as quickly, if not quicker. If the tree has leafy branches, the squirrel presses himself down in the middle of one of them, and so close that it is hardly visible. You may then shake the tree, throw sticks and stones at the place where it lies, or shoot at it, yet it will never stir. If three branches join, it takes refuge between them, and lies as close to them as possible, and then it is sufficiently safe. Sometimes it escapes on a tree where there are old nests of squirrels or of large birds. It slips into them, and cannot be gotten out, either by shooting, throwing, or anything else; for the gray squirrels seldom leap from one tree to another, unless extreme danger compels them. They commonly run directly up the trees and down the same way, head first. Several of them which I shot in the woods had great numbers of fleas.

I have already mentioned that these squirrels are among the animals which at present are more plentiful than they formerly were, and that the infinitely greater cultivation of corn, which is their favorite food, is the cause of their multiplication. However, it is peculiar that in some years a greater number of squirrels come down from the northern countries into Pennsylvania and other English colonies than at other times. They commonly come in autumn, and are then very busy in the woods gathering nuts and acorns, which they carry into hollow trees or their store-holes, in order to be sufficiently provided with food for winter. They are so diligent in their storing up of provisions, that though the nuts have been extremely plentiful in a certain year, it is difficult to get a large quantity of them, for the squirrels have carried them away to their hidingplaces. The people here pretend from their own experience to know, that when the squirrels come down in such numbers from the more northern parts of the country the winter ensuing will be uncommonly rigorous and cold, and for that reason they always look upon their coming as a sure sign of a severe winter. Yet this does not always prove true, as I experienced in the autumn of the year 1749. At that time a great number of squirrels came south into the colonies, yet the winter was very mild and no colder than usual. But it appeared that their migration was occasioned by the scarcity of nuts and acorns, which occurred that year in the northern parts of the land, and obliged them to come hither for their food. Therefore they generally return the next year to the place from which they come.

Some people consider squirrel flesh a great dainty, but most do not care for it. The skin is of little value, yet small straps are sometimes made of it, as it is very tough. Others use it for fur lining, for want of a better. Ladies' shoes are occasionally made of it.

Snake Witchery. The rattlesnake often devours the squirrels, notwithstanding all their agility. This unwieldy creature is said to catch so agile a one merely by fascination. I have never had an opportunity of seeing how it is done; but so many reliable people assured me of the truth of the fact, and asserted that they were present and paid peculiar attention to it, that I am almost forced to believe their unanimous accounts. The spell is effected in the following manner: the snake lies at the bottom of the tree upon which the squirrel sits; its eyes are fixed upon the little animal, and from that moment it cannot escape; it begins a doleful outcry, which is so well known, that a person passing by, on hearing it, immediately knows that it is charmed by a snake. The squirrel runs up the tree a little way, comes downwards again, goes up, and then comes lower again. On that occasion it has been observed that the squirrel always goes down farther than it goes up. The snake still continues at the root of the tree, with its eyes fixed on the squirrel, with which its attention is so entirely taken up that a person accidentally approaching may make a considerable noise without the snake so much as turning about. The squirrel as before-mentioned comes constantly lower, and at last leaps down to the snake, whose mouth is already wide open for its reception. The poor little animal then with a piteous cry runs into the snake's jaws, and is swallowed at once, if it be not too big. But if its size will not allow it to be swallowed at once the snake licks it several times with its tongue, and smoothes it, and by that means makes it fit for swallowing. Everything else remarkable in this enchantment I have described in a treatise inserted in the *Memoirs of the Royal Swedish Academy of Sciences*, in the volume for the year 1753. I therefore shall not repeat the facts here. The same power of enchanting is ascribed to the kind of snake which is commonly called the black snake in America, and it is said to catch and devour squirrels in the same manner as the former.[1]

But these little animals do considerable damage to the corn, not only upon the stalk, as I have before observed, but even when it is brought home into the barns; for if they can get to it without any obstacle, they can in a few nights bring a whole bushel away into their secret holes. The government in most of the North American colonies has therefore been obliged to offer a certain premium, to be paid out of the common treasury, for the head of a squirrel. It seems inconceivable what a sum of money has been paid for gray and black squirrels' heads, in the province of Pennsylvania only, from the first of January, 1749, to the first of January, 1750; for when the deputies from the several districts of the province met, in order to deliberate upon the affairs of the province, each of them complained that their treasuries were exhausted by paying so much for squirrels; for at that time the law had appointed a reward of threepence for each squirrel's head. So great was the vengeance taken upon these little creatures, i. e. upon the gray and black squirrels, that it was found by inspecting accounts that in one year eight thousand pounds of Pennsylvania currency had been expended in paying rewards: this I was assured of by a man who had looked over the accounts himself. Many people, especially young men, left all other employment and went into the woods to shoot

---

[1] It has been observed, that only such squirrels and birds as have their nests near the place where such snakes come to, make this pitiful noise, and are so busy in running up and down the tree and the neighboring branches, in order to draw off the attention of the snake from their brood, and often they come so very near in order to fly away again, that being within reach of the snakes, they are at last bitten, poisoned and devoured; and this will I believe, perfectly account for the powers of fascinating birds and small creatures in the snakes.—F.

Forster apparently does not believe in snake enchantments.

squirrels. But the government having experienced how much three-pence per head took out of the treasury, settled half that sum upon each squirrel's head.

Flying squirrels are a peculiar kind, which seem to be the same as those which inhabit Finland, and which Dr. Linné in his *Fauna Svecica* calls *Sciurus volans*. The American flying squirrel is assuredly only a variety of that which we have in Finland. Catesby in his *Natural History of Carolina* has described it, and drawn it from life.[1] He likewise calls it *Sciurus volans*. They are found in the woods, but not very frequently. They are scarcely ever seen in the daytime, unless they are forced out by men who have discovered their nests. They sleep in the daytime, but as soon as it grows dark they come out and run about almost all night. They live in hollow trees and by cutting one down, seven or more flying squirrels are frequently found in it. By the additional skin with which Providence has provided them on both sides, they can sail from one tree to another. They expand their skins like wings, and contract them again as soon as they can grasp the opposite tree. Some people say that they fly in a horizontal line; but others asserted that they first flew a little downwards, and then rose again when they approached the tree toward which they were flying. They cannot fly further than four or five fathoms. Among all the squirrels in this country, these are the most easily tamed. The boys carry them to school, or wherever they go, without their ever attempting to escape. Even if they put their squirrel aside, it leaps upon them again immediately, creeps either into their bosom or their sleeve, or any fold of the clothes, and lies down to sleep. Its food is the same as that of the gray squirrel.

Chipmunks. There is small species of squirrel abounding in the woods, which the English call ground squirrels. Catesby has described and drawn them from life, in his *Natural History of Carolina*,[2] and Edwards in his *Natural History of Birds*. He and Dr. Linné call it *Sciurus striatus* [now *Tamias S.*], or the streaked squirrel. These do not properly live in trees, as others of this genus, but dig holes in the ground (much in the same manner as rabbits) in which they live, and whither they take refuge when they perceive any danger. Their holes go deep, and commonly further inside

---

[1] Vol. II, pp. 76, 77.
[2] Vol. II, p. 75.

divide into many branches. They are also cunning enough sometimes to make an opening or hole to the surface of the ground from one of these branches. The advantage they have from this is that when they stroll about for food, and the hole through which they went out is stopped up, they may not expose themselves to capture, but presently find the other hole into which they may retreat. But in autumn, when the leaves fall from the trees, or sometimes after, it is a diversion to see the consternation they are sometimes in when pursued; for their holes being easily covered with the great fall of leaves, or by the wind, they have great trouble to find them quickly. They then run back and forth, as if they had lost their way. They seem to know the places where they have made their subterraneous passages, but cannot conceive where the entrances are. If they be then pursued and you clap your hands, they know no other refuge than that of climbing up a tree; for it is to be observed that these squirrels always live under ground, and never climb trees unless pursued and unable in the hurry to find their holes. This kind of squirrel is much more numerous in Pennsylvania than in any other province of North America through which I have travelled. Its length is commonly six inches, without the little tail; and it is very narrow. The skin is ferruginous, or of a reddish brown, and marked with five black streaks, one of which runs along the back and two on each side. Their food consists of all sorts of grain, as rye, barley, wheat, corn, acorns, nuts, etc. They gather their winter provisions in autumn, like the common gray squirrels, and keep them in their holes under ground. If they get into a granary, they do as much mischief as mice and rats. It has often been observed that if, after eating rye, they find some wheat, they throw up the former, which they do not like so well as the wheat, in order to fill their belly with the latter. When the corn is harvested in the fields they are very busy biting off the ears and filling their cheeks with corn, so that they seem quite blown up. With this booty they hasten into their holes in the ground.

When a Swede once pretty late in autumn was digging for a mill dam in a neighboring hill, he came upon a subterraneous passage belonging to these squirrels. He followed it for some time, and discovered a walk on one side like a branch from the chief stem. It was nearly two feet long, and at its end had a quantity of choice acorns of the white oak, which the little foresighted animal had stored up for winter. Soon after he found another passage on the side like the former, but containing a fine store of corn; the next had hickory nuts; and the last and most hidden one contained some excellent chestnuts, to an amount which might have filled two hats.

In winter these squirrels are seldom seen, for during that season they live in their subterraneous holes upon the provisions which they have stored up there. However, on a very fine and clear day they sometimes come out. They frequently dig through the ground into cellars in which the country people lay up their apples, which they partly eat and partly spoil, so that the master has little or nothing left. They handle the corn stores fully as roughly as the apples. But the cats are their great enemies: they devour them or bring them home to their young. Their flesh is not eaten by men, and their skin is not used.

Of all the squirrels in the country these are the most difficult to be tamed; for though they be caught very young it is dangerous to touch them with naked hands, as they may bite very severely. Many boys, who had lost a good deal of time in trying to tame these squirrels, owned that they knew of no way to domesticate them, at least they were never so well tamed as the other species. In order to do anything towards taming them they must be caught when they are very small. Some people kept them in that state in a cage, because they looked very pretty.

I shall take another opportunity of speaking of the black and ferruginous squirrels, which likewise inhabit this country.

NOVEMBER THE 15TH

Hurricanes. In the morning I returned to Philadelphia. Mr. Cock told me to-day, and on some other occasions afterwards, of an accident which happened to him, and which seemed greatly to confirm a peculiar sign of an imminent hurricane. He sailed to the West Indies in a small yacht, and had an old man on board who had for a considerable time sailed on this sea. The old man sounding the depth, called to the mate to tell Mr. Cock to launch the boats immediately, and to put a sufficient number of men into them in order to tow the yacht during the calm that they might reach the island before them as soon as possible, as within twenty-four hours there would be a strong hurricane. Mr. Cock asked him what reasons he had to think so. The old man replied, that on sounding he saw the lead in the water at a distance of many fathoms more than he had seen it before; that therefore the water had become clear all of a sudden, which he looked upon as a certain sign of an impending hurricane in the sea. Mr. Cock likewise saw the excessive clearness of the water. He therefore gave immediate orders for launching the boats and towing the yacht, so that they arrived before night in a safe harbor. But before they had quite reached it the water acted as though it were boiling, though no wind was perceptible. In the ensuing night the hurricane came on, and raged with such violence that not only many ships were lost, and the roofs were torn off from the houses, but even Mr. Cock's yacht and other ships, though they were in safe harbors, were by the wind and violence of the sea washed so far up on shore that several weeks elapsed before they could be floated.

An old Dutch skipper said that he had once caught a dogfish in the Bay of New York, which when cut open was found to have a quantity of eels in his stomach.

NOVEMBER THE 18TH

Indian Pottery. Mr. Bartram showed me an earthen pot, which had been found in a place where the Indians formerly lived. He who first dug it out kept grease and fat in it with which to smear his shoes, boots and all sorts of leather. Mr. Bartram bought the pot of that man; it was yet entire and not damaged. I could perceive no glaze or color upon it, but on the outside it was very much ornamented and upon the whole well made. Mr. Bartram showed me several pieces of broken earthen vessels which the Indians formerly made use of. It plainly appeared in all these that they were not made of mere clay, but that different materials had been mixed with it, according to the nature of the places where they were made. Those Indians, for example, who lived near the seashore, pounded the shells of snails and mussels, and mixed them with the clay. Others who lived further back in the country, where mountain crystals could be found, pounded them and mixed them with their clay; but how they proceeded in making the vessels, is entirely unknown. It was plain that they did not bake them much, for they were so soft that they might be cut into pieces with a knife. The workmanship, however, seems to have been very good, for at present they find whole vessels or pieces in the ground which are not damaged at all, though they have lain in the earth many centuries. Before the Europeans settled in North America the Indians had no other vessels to boil their meat in than these earthen pots of their own making; but since their arrival the savages have always bought pots, kettles and other necessary vessels of the Europeans, and no longer take the pains of making any, so that this art is entirely lost among them. Such vessels of their own construction are therefore a great rarity, even among the Indians. I have seen such old pots and pieces of them consisting of a kind of serpentine stone, or Linné's *Talcum*.

Slate. Mr. Bartram likewise showed me little pieces of a black slate, which is plentifully found in some parts of the river Schuylkill. There are fragments to be found which are four feet or more square. The color and shape is the same as in the table slate (*Schistus tabularis* L.),[1] except that this is a little thicker. The inhabitants of the country thereabouts (in the neighborhood of the Schuylkill) cover their roofs with it. Mr. Bartram assured me that he had seen a whole roof composed of four such pieces of slate. The rays of the sun, heat, cold and rain do not act upon this stone.

Mr. Bartram further related that in several parts of the country caves or holes were to be found going deep into the mountains. He had been in several of them and had often found a number of stalactites, Linné's *Stalactites stillatitius*,[2] of different dimensions at the top. They differed in color, but the greatest curiosity was that in some of the caves Mr. Bartram had found stalactites whose outward side was spiraled from top to bottom. He had sent some pieces of it to London and had none at present.

NOVEMBER THE 19TH

Apple-dumplings. One apple dish which the English prepare is as follows: take an apple and pare it, make a dough of water, flour and butter. Roll it thin and enclose apple in it. This is then bound in a clean linen cloth, put in a pot and boiled. When done it is

---

[1] *Syst. Nat.* 3, p. 37.
[2] *Ibid.*, p. 183.

taken out, placed on table and served. While it is warm, the crust is cut on one side. Thereupon they mix butter and sugar, which is added to the apples; then the dish is ready. They call this apple dumpling, sometimes apple pudding. It tastes quite good. You get as many dumplings as you have apples.

November the 20th

The Environs of Raccoon, New Jersey. This morning I set out in company with a friend, on a journey to Raccoon in New Jersey,[1] where many Swedes live and who own their own church. We had three miles to go before we came to the ferry which was to bring us over the Delaware. The country here was very low in some places. The plains on the banks of the river were flooded at every high water or flowing of the tide, and at the ebbing they were left dry again. However the inhabitants of the country hereabouts met this situation, for they had in several places thrown up walls or dykes of earth near the river to prevent its overflowing the land which they made use of as meadows. On them the water-beeches (*Platanus occidentalis* L.) were planted in great numbers on both sides of the road, quite close together. These in summer afforded a pleasant shade, on account of the abundance and size of their leaves, and made the road extremely delightful, as it resembled a fine shady avenue. The Delaware has nearly the same breadth here as it has near Philadelphia. Near the landing several pretty houses had been built on both sides, where travellers might get all kinds of refreshment. On our journey from Pennsylvania to New Jersey we were brought over the Delaware in a ferry belonging to, and kept in repair by, the Pennsylvania men; but on our return we were obliged to take the ferry belonging to the New Jersey side. As soon as we had crossed the river we were in a different province, for the Delaware constitutes the division between Pennsylvania and New Jersey, so that everything to the west of it belongs to the former, and all to the east, to the latter province. Both these provinces have in most everything entirely different laws and currency.

We now pursued our journey further, and soon observed that the country on this side appeared very different from that on the other; for in Pennsylvania the ground consists of more clay and black

---

[1] Now Swedesboro.

(From J. W. Barber and Henry Howe, *Historical Collections of the State of New Jersey*, 1844)
Swedesboro (Raccoon), New Jersey

mould, and is very fertile, while in New Jersey it is more sandy and very poor, so that the horses sank deep into the sand in several parts of the road. Near the ferry landing and a little way along the shore was a thick fir wood; the trees were not very high, but vigorous looking. Between them appeared here and there some small oaks. But after travelling about three English miles, the fir wood ended, and we saw no more trees of this kind till we came to the church in Raccoon. In all the parts of Pennsylvania where I have been, I have found but few fir trees; on the other hand, they are abundant in New Jersey, and especially in the lower part of that province. Afterwards, all day, we found only deciduous trees; most of these were oaks of different sorts, and of considerable height, but they stood everywhere far enough apart to admit a chaise to pass through the wood without any inconvenience, there being seldom any shrubs or undergrowth between the trees to obstruct the way. The leaves had all fallen and covered the ground to a thickness of more than a hand's breadth. This had an appearance of increasing the upper black soil greatly. In several places flowed a small rivulet. The country was commonly level, but sometimes formed a few hills with an easy incline, though no high mountains appeared, and in a few places we found some small stones not bigger than a fist. Single farmhouses were scattered in the country, and in one place only was a small village. The country was yet more covered with forest than cultivated, and most of the time we were in a wood.

This day and the next we passed several kills, or small rivulets, which flowed out of the country into the Delaware with a gentle descent and rapidity. When the tide came up in the Delaware, it also rose in some of these rivulets a good way. Formerly they must have spread to a considerable breadth by the flowing of the tide, but at present there were meadows on their banks, formed by throwing up strong dikes as close as possible to the water, to keep it from overflowing. Such dikes were made along all rivers here to confine their water, and therefore when the tide was highest the water in the rivers was much higher than the meadows. In the dikes were gates through which the water could be drawn off or led into the meadows. They were sometimes placed on the outward side of the wall, in such a way that the water in the meadows would force them open while the river water would shut them.

In the evening we came to the house of the Swede named Peter Rambo, and we stayed there over night.

Pines. The pines which we have seen today and which I have mentioned before are of the kind which has double leaves and oblong cones covered with aculeated scales. The English, to distinguish it, call it the "Jersey pine." Commonly there are only two needles or leaves in one fascicle, as in our common Swedish pines, but sometimes three. The cones have long spines, so that they are difficult to handle. These pines at a distance look wholly like the Swedish ones, so that if the cones are not examined they may easily be taken for the same species. Of these pines they make a great quantity of tar, of which I shall speak later, but as most of them are only small they are good for nothing else; for if they be employed as posts or poles in the ground, they are in a short time rendered useless by rotting. As soon as they are cut down the worms are very greedy for them. They soon eat through the wood, only a few weeks after it is cut down. However, it is used as fuel where no other wood is to be gotten. In several places they make charcoal of it, as I intend to mention later. There is another thing which deserves mentioning in regard to these trees, and which several people besides myself have noticed. In the great heat of the summer the cattle prefer to stand under the evergreen trees rather than under the oak, hickory, walnut, water-beech and other trees of this kind, whose foliage is thicker and affords a better shade; and if there be but a single pine in a wood, as many cattle from the herd as can stand under it throng there. Some people infer from this that the resinous exhalations of these trees are beneficial to the cattle, and make them more inclined to be near firs and pines than any other trees.

Laurel Trees. The spoon tree, which never grows to a great height, was seen to-day in several places. The Swedes here have named it thus, because the Indians used to make their spoons and trowels of its wood. In my cabinet of natural curiosities I have a spoon made of this wood by an Indian, who had killed many stags and other animals on the very spot where Philadelphia afterwards was built, for in his time that spot was yet covered with trees and shrubs. The English call this tree a "laurel" because its leaves resemble those of the *Laurocerasus.* Dr. Linné, because of the peculiar friendship and kindness with which he has always honored me has been pleased to call this tree, *Kalmia foliis ovatis, corymbis terminalibus,* or *Kalmia latifolia.* It succeeds best on the side of hills, especially on the north side, where a brook passes by; therefore on meeting with some steep places (on hills) towards a brook, or on steep hillsides facing marshes, you are sure to find the *Kalmia.* But it frequently stands among beech trees. The further up the north side of a mountain the *Kalmias* stand, the shorter they are. I have seen them not only in Pennsylvania and New Jersey but also in New York, but there they are scarcer.[1] I never found them beyond the forty-second degree of north latitude, though I took ever so great care to look for them. They have the quality of preserving their fine green leaves throughout the winter, so that when all other trees have lost their ornaments, and stand quite naked, these adorn the woods with their green foliage. About the month of May they begin to flower in these parts, and then their beauty rivals that of most of the known trees in nature. The flowers are innumerable, and grow in great clusters. Before they open they have a fine red color, but as they expand, the sun bleaches them so that some are almost white; many keep the color of roses. Their shape is singular, for they resemble ancient cups. Their scent, however, is none of the most agreeable. In some places it is customary to decorate the churches on Christmas Day or New Year's Day with the fine branches of this tree.

But these trees are known for another remarkable quality; their leaves are poison to some animals, and food for others. Experience has taught the people that when sheep eat of these leaves, they either die immediately or fall very sick, and recover with great difficulty. The young and more tender sheep are killed by a small portion, but the older ones can bear a stronger dose. Yet these leaves will prove fatal to them too, if they eat much of them. The same noxious effect is observed on calves, if they eat too much of the foliage: they either die or do not recover easily. I can remember that in the summer of the year 1748 some calves ate of the leaves, but fell very sick, swelled, foamed at the mouth, and could hardly stand. However they were cured by giving them gunpowder and

---

[1] The mountain laurel is very common in Connecticut also, where it is the state flower. It is also the state flower of Pennsylvania.

other medicines. The sheep are most exposed to the temptation of these leaves in winter; for after having been kept in stables for some months they are greedy for all green food, especially if the snow still lies upon the fields, and therefore the green but poisonous leaves of the *Kalmia* are to them very tempting. Horses, oxen and cows which have eaten them, have likewise been very ill after the meal, and though none of them ever died from eating the laurel, most people believed that if they consumed too great a quantity death would certainly be the result. It has been observed that even if these animals eat only small amounts they suffer bad effects. On the other hand, the leaves of the *Kalmia* are the food of stags, when the snow covers the ground and hides all other provisions from them. Therefore, if they be shot in winter, their bowels are found filled with these leaves; and it is very extraordinary that if those bowels are then given to dogs, they become stupid, act as if drunk, and often fall so sick that they seem to be at the point of death. But the people who have eaten the venison have not felt the slightest indisposition. The leaves of the *Kalmia* are likewise the winter food of those birds which the Swedes in North America call "hazel-hens," and which stay here all winter, for when they are killed their crops are found filled with them.

The wood of the *Kalmia* is very hard, and some people on that account make the axes of their pulleys of it. Weaver's shuttles are made chiefly of it, and the weavers are of the opinion that no wood in this country is better for the purpose, for it is compact, may be made very smooth, and does not easily crack or burst. The joiners and turners here use it in all kinds of work which requires the best wood; they use chiefly the root because it is yellowish. The wood has a very suitable hardness and fineness, and from the center, spread, as it were, small rays, which are at some distance from each other. When the leaves of the *Kalmia* are thrown into the fire they crackle like salt. The chimney sweepers make brooms in winter of the branches with the leaves on them, since they cannot get others in that season. In the summer of the year 1750, a certain kind of worm devoured the leaves of almost all the trees in Pennsylvania; yet they did not venture to attack those of the *Kalmia*. Some people asserted that when a fire broke out in the woods, it never went beyond the *Kalmias* or spoon trees.

November the 21st

Corn. The Swedes and all the other inhabitants of the country plant great quantities of corn, both for themselves and for their cattle. It was asserted that it is the best food for hogs, because it makes them very fat, and gives their flesh an agreeable flavor, preferable to all other meat. I have sent two detailed dissertations on this cereal to the Swedish Royal Academy of Sciences, which are found in their *Memoirs* for the years 1751 and 1752, and thither I refer my readers.

Cart Wheels. The wheels of the carts which are here used, are made of two different kinds of wood. The fellies are usually manufactured of what is known as Spanish oak, and the spokes of white oak.

Other Trees. The *sassafras* tree grows everywhere in this locality. I have already indicated several particulars in regard to it, and intend to add a few more here. On throwing some of the wood into the fire it causes a crackling as if salt had been cast on it. The wood is used for fence posts, for it is said to last a long time in the ground. But it is also maintained that there is hardly any kind of wood which is more often attacked by worms than this, when it is exposed to the air without cover, and that in a short time it is worm-eaten through and through. The Swedes related that the Indians who formerly inhabited these parts made bowls of it. On cutting the sassafras tree or its shoots, and holding it to the nose, it has a strong but pleasant smell. Some people peel the root, and boil the peel with the beer which they may be brewing, because they believe it wholesome. The peel is put into brandy, either while it is distilling or after it is made.

An old Swede remembered that his mother cured many people of the dropsy by a decoction of the root of sassafras in water drunk every morning. At the same time she used to cup the patient on the feet. The old man assured me he had often seen people cured by this means, who had been brought to his mother wrapped up in sheets.

When a part of a forest is destined for cultivation, the sassafras trees are commonly left upon it, because they have a very thick foliage and afford a cool shade for the cattle during the very hot weather. Several of the Swedes wash and scour the vessels in which they intend to keep cider, beer or brandy with water in which the sassafras root or its peel has been boiled, which they think renders all those liquors more wholesome. Some people have their bedposts made of sassafras wood to expel the bed bugs, for its strong scent, it is said, prevents vermin from settling in them.[1] For two or three years this has the desired effect, or about as long as the wood keeps its strong aromatic smell, but after that time it has been observed to lose its effect. A joiner showed me a bed which he had made for himself, the posts of which were of sassafras wood, but as it was ten or twelve years old there were so many bugs in it that it seemed unlikely they would let him sleep peaceably. Some Englishmen related, that some years ago it had been customary in London, to drink a kind of tea of the flowers of sassafras, because it was looked upon as very healthful, but upon recollecting that the same potion was much used against the venereal disease, it was soon left off, lest those who used it should be looked upon as infected with that disease. In Pennsylvania some people put chips of sassafras into their chests, where they keep woolen stuffs, in order to expel the moths which commonly settle in them in summer. The root keeps its smell for a long while. I have seen one which had lain five or six years in the drawer of a table, and still preserved the strength of its scent.

A Swede named Rambo related that the Indians formerly dyed all sorts of leather with the bark of the chestnut oak.

Some old people remembered that in the year 1697 there had been so rigorous a winter that the ice on the Delaware River was two feet thick.

We left the place a little before noon and went to the house of a deacon of the Swedish Church, Eric Ragnilsson, where we remained for a few days.

NOVEMBER THE 22ND

Grass. Åke Helm was one of the most important Swedes in this place and his father came over to this country along with the Swedish Governor Printz;[2] he was upwards of seventy years of age. This old man told us, that in his youth there was grass in the

---

[1] Some farmers still make hen-roosts out of sassafras wood to keep vermin away.
[2] Johan Printz (1592-1663), governor of New Sweden from 1643-1653. He established on Tinicum Island the first permanent seat of government in Pennsylvania.

woods which grew very thick, and was everywhere two feet high, but that it was so much thinner at present that the cattle could hardly find food enough, and that therefore four cows now gave no more milk than one at that time. The causes for this change are easy to find. In the younger days of old Helm the country was little inhabited, and hardly a tenth part of the cattle kept which is there at present. A cow had therefore as much food at that time as ten now have. Further, most grasses here are annuals, and do not for several years in succession shoot up from the same root as our Swedish grasses. They must sow themselves every year, because the last year's plant dies away every autumn. The great numbers of cattle hinder this sowing, as the grass is eaten before it can produce flowers and seed. We need not therefore wonder that the grass is so thin on fields, hills and pastures in these provinces. This is likewise the reason why travellers in New Jersey, Pennsylvania and Maryland find many difficulties, especially in winter, to travel with their horses, for the grass in these provinces is not very abundant, the cattle having eaten it before it goes to seed. But farther to the north, as in Canada, there is a sufficient quantity of perennial grasses; so wisely has the Creator regulated everything. The cold parts of the earth naturally bring forth a more durable grass, because the inhabitants need more hay to feed their cattle with, on account of the length of the winter. The southern provinces again have less perennial grass, as the cattle may feed in the fields all winter. However, foresighted farmers have procured seeds of perennial grasses from England and other European states, and sowed them in their meadows, where they seem to thrive exceedingly well.

The *Persimmon* (*Diospyros Virginiana*) is quite common here. I have mentioned that before, but I intend now to add some particulars. Some of its fruit begins to ripen and becomes fit for eating about this time, for after it is frost bitten in autumn the people eat it like other fruit. It is very sweet and glutinous, yet has a little astringency. I frequently used to eat a great quantity of it without feeling the slightest discomfort. From the persimmon several Englishmen and Swedes brew a very palatable liquor in the following manner. Late in autumn after the fruit has been touched by the frost, a sufficient quantity is gathered, which is very easy, as each tree is well stocked with it. These persimmon apples are put into a dough of wheat or other flour, formed into cakes, and put into an oven, in which they remain till they are baked and sufficiently dry, when they are taken out again. Then, in order to brew the liquor, a pot full of water is put on the fire and some of the cakes are put in. These become soft by degrees as the water grows warm, and crumble to pieces. The pot is then taken from the fire, and the water in it well stirred so that the cakes may mix with it. This is then poured into another vessel, and they continue to steep and break up as many cakes as are necessary for the brewing. Then malt is added and one proceeds as usual with the brewing. Beer thus prepared is reckoned much preferable to other beer. Also brandy is prepared from this fruit as follows. Having collected a sufficient quantity of persimmons in autumn, they are put into a vessel, where they lie for a week till they are quite soft. Then one pours water on them, and in that state they are left to ferment of themselves, without promoting the fermentation by any addition. The brandy is then made in the ordinary way, and is said to be very good, especially if grapes (in particular those of the sweet sort), which are wild in the woods, be mixed with the persimmon fruit.[1] Some persimmons are ripe at the end of September, but most of them later, and some not before November and December, when the frost ripens them. The wood of this tree is very good for joiner's instruments, such as planes, handles to chisels, etc., but if after being cut down it lies exposed to sunshine and rain, it is the first wood which rots, and in a year's time there is nothing left that is useful. When the persimmon trees once get into a field, they are not easily removed, as they spread so much. I was told that if you cut off a branch and put it into the ground it strikes root; but in very severe winters these trees often die from frost, and they, together with the peach trees, bear cold the least of any.

November the 23rd

Pumpkins. Several kinds of pumpkins and melons are cultivated here. Originally they were in part cultivated by the Indians and in part brought over by Europeans. Of the pumpkins there is a kind which is crooked at the end and oblong otherwise, and is therefore called a crookneck. They keep almost all winter. There is yet another species of pumpkins which has the same quality. Others

---

[1] Cf. *Dialect Notes,* VI, Part VII, 360-361, about *Simmon beer.*

again are cut in slices, drawn upon thread, and dried. They keep all the year long, and are then boiled or stewed. All sorts of pumpkins are prepared for eating in different manners, as is customary in Sweden. Many farmers have a whole field of them.

Squashes are a kind of pumpkin which the Europeans got from the Indians, and I have already mentioned them. They are eaten boiled, either with meat or by themselves. In the first case, they are put on the edge of the dish round the meat; they require little care, for in whatever ground they are planted they grow and succeed well. If the seed is put into the fields in autumn it brings squashes the following spring, though during winter it has suffered from frost, snow and wet.

The *calabashes* are a kind of gourd, which are planted in quantities by the Swedes and other inhabitants, but they are not fit for eating, and are used for making all sorts of vessels. They are more difficult to raise than the squashes, for they do not always ripen here except when the weather is very warm. In order to make vessels of them, they are first dried well. The seeds, together with the pulpy and spongy matter in which they lie, are afterwards taken out and thrown away. The shells are scraped very clean within, and then large spoons or ladles, funnels, bowls, dishes and the like may be made of them. They are particularly fit for holding seeds which are to be sent over sea; for seeds keep their power of vegetating much longer if they be put in calabashes than by any other means. Some people scrape the outside of them before they are opened, dry them and then clean them within. This makes them as hard as bone. They are sometimes washed to keep their white color.

Buckwheat. Most of the farmers in this country sow buckwheat in the middle of July. It must not be sown later, for in that case the frost ruins it; and if it is sown before July, it flowers all summer long, though the flowers drop and no seed is generated. Some people plow the ground twice where they intend to sow buckwheat; others, only once, about two weeks before they sow it. As soon as it is sown the field is harrowed. It has been found by experience that in a wet year buckwheat is most likely to succeed. It stands on the fields till the frost comes. When the crop is favorable farmers get twenty, thirty, and even forty bushels from a bushel of seed. The Swedish churchwarden Ragnilsson, in whose house we stayed at this time, had obtained such a crop. From the flour they make buckwheat cakes and pudding. The cakes are usually made in the morning, and are baked in a frying pan or on a stone, are buttered and then eaten, when still warm, with tea or coffee, instead of toasted bread with butter, or toast, which the English usually eat at breakfast. The buckwheat cakes are very good, and are common at Philadelphia and in other English colonies, especially in winter. In Philadelphia there were some people who baked them and in the morning carried them around while still warm, to be sold. Buckwheat is an excellent food for fowls: they eat it eagerly, and lay more eggs than they do from other food. Hogs are likewise fattened with it. Buckwheat straw is of no use; it is therefore left upon the field in the places where it has been thrashed, or it is scattered in the orchards in order to serve as a manure by decaying. Neither cattle nor any other animal will eat it, except in the greatest necessity, when the snow covers the ground and nothing else is to be had. But though buckwheat is so common in the English colonies, yet the French had no knowledge of it in Canada, and it was never cultivated among them.

Glowworms. Towards night we found some glowworms in the wood. Their body was linear, consisting of eleven segments, a little pointed before and behind; the length from head to tail was five and a half geometrical lines [½ in.]. The color was brown and the segments joined in the same manner as in the onisci or woodlice. The antennæ or feelers were short and filiform, or thread-shaped; and the feet were fastened to the foremost segments of the body. When the insect crept, its hindmost segments were dragged on the ground, and helped its motion by pushing. The extremity of the tail contains a matter which shines in the dark [1] with a greenish light: the insect could draw it in so that it was invisible. It had rained considerably all day, yet they crept in great numbers along the bushes, so that the ground seemed as if it were sown with stars. I shall later have occasion to mention another kind of insect or fly which shines in the dark when flying in the air.

November the 24th

Holly. Holly, or *Ilex aquifolium*, grows in wet places, scattered

---

[1] The light is emitted from some of the abdominal segments, including the tail.

in the forest, and belongs to the rare trees. Its leaves are green both in summer and winter. The Swedes dry its leaves, crush them in a mortar, boil them in small beer, and take them to cure pleurisy or pains in the side.

Dyes. Red is dyed with brazilwood, and also with a kind of moss which grows on the trees here. Blue is dyed with indigo, but to get black, the leaves of the common field sorrel (*Rumex acetosella*) are boiled with the material to be dyed, which is then dried and boiled again with logwood and copperas. The black thus produced is said to be very durable. The people spin and weave a great part of their every-day apparel and dye it in their houses. Flax is cultivated by many people and succeeds very well, but hemp is not used here.

Rye, wheat and *buckwheat* are cut with the sickle, but oats are mown with a scythe. The sickles which are here used are long and narrow, and their sharp edges have close teeth on the inner side. The field lies fallow for a year, and during that time the cattle may graze on it.

All the inhabitants of this place, from the highest to the lowest, have their own orchards, which are larger or smaller according to their wealth. The trees in them are chiefly peach, apple and cherry. Compare with this what I have already said upon this subject.

A little before noon we left this place and continued our journey past the Swedish Church in Raccoon to Pilesgrove. The country along the sides of this road is very sandy in many places and pretty nearly level. Here and there appear single farms, yet they are very scarce, and large extensive pieces of ground are still covered with forests, which consist chiefly of various species of oak and hickory. However, we could pass with ease through these woods, as there were few bushes (or undergrowth) and stones to be found. It was not only easy to ride in every part of the wood on horseback, but in most places there was sufficient room even for a small coach or a cart. Sometimes a few windfall trees which had been hurled on the ground by a hurricane or had fallen down through great age caused some hindrance.

November the 25th

Receding Waters. During my stay at Raccoon, at this time and all the ensuing winter, I endeavored to get all possible information from the old Swedes about the increase of land and the decrease of water in these parts. I shall therefore here insert the answers which I have received to my questions. I reproduce them in the form I got them, and I shall only add a few remarks which may serve to explain things. The reader therefore is left at liberty to draw his own inferences and conclusions.

One of the Swedes, named King, who was over fifty years of age, was convinced that at this season the little lakes, brooks, springs and rivers had much less water than they had had when he was a boy. He could mention several lakes on which the people went rowing in large boats in his youth, and had sufficient water even in the hottest summers; but now, they were either partially or entirely dried up. He himself had seen the fish dying in them, and he was apt to believe that at this time it did not rain so much in summer as it did when he was young. One of his relations, who lived on a hill near a brook about eight miles from the river Delaware, had got a well dug in his court yard. At the depth of forty feet, he found a quantity of shells of oysters and mussels, and likewise a great quantity of reed and pieces of broken branches. I asked to what causes they ascribed what they had discovered, and I was answered that some people believed these things had lain there ever since the Deluge, and others, that the ground itself grew.

Peter Rambo, a man who was near sixty years of age, assured me that in several places at Raccoon where wells had been dug, or any other work carried deep into the ground, he had seen great quantities of mussel shells and other marine animals. On digging wells the people had sometimes met with logs or wood at the depth of twenty feet, some of which were decayed and others appeared burnt. They once found a large spoon in the ground at this depth. Now is it not probable that the burnt wood which had thus been dug up, had only been blackened by a subterraneous mineral vapor? People however have concluded from this, that America had had inhabitants before the Flood. This man (Peter Rambo) further told me, that bricks had been found deep in the ground; but may not the brick-colored clay (of which the ground here chiefly consists, and which is a mixture of clay and sand) in a hard state have had the appearance of bricks? I have seen such hardened clay, which at first sight is easily mistaken for brick. He likewise asserted that the water in rivers was still as high as it used to be, as far back as memory could reach, but that in little lakes, ponds, and marshes the amount of water had visibly decreased, and that many of them had dried up completely.

Måns Keen,[1] a Swede over seventy years old, asserted that on digging a well he had seen at the depth of forty feet a great piece of chestnut wood, together with roots and stalks of reed, and a clayey earth like that which commonly covers the shores of salt water bays and coves. This clay had a similar smell and a saline taste. Måns Keen and several other people inferred therefrom that the whole country where Raccoon and Penn's Neck are situated was once entirely covered by the sea. They likewise knew that at a great depth in the ground such a trowel as the Indians use had been found.

Sven Lock and William Cobb, both over fifty years of age, agreed that in many places hereabouts, where wells had been dug, they had seen a great quantity of reed, mostly rotten, at the depth of twenty or thirty feet or more.

As Cobb made a well for himself, the workmen after digging down twenty feet came upon so thick a branch that they could not go deeper until it had been cut in two places. The wood was still very hard. It was very common to find near the surface of the earth quantities of all sorts of partially decayed leaves. On making a dike some years ago along the river on which the church at Raccoon stands, and upon cutting through a bank, it was found full of oyster shells, though this place is above a hundred and twenty English miles from the nearest seashore. These men, and all the inhabitants of Raccoon, concluded from this circumstance (of their own accord and without being led to the thought) that this tract of land had been a part of the sea many centuries ago. They likewise asserted that many little lakes, which in their youth had been full of water, even in the hottest season, now hardly formed a narrow stream in summer, except after heavy rains; but it did not appear to them that the rivers had lost any water.

Åke Helm, found (on digging a well) first sand and small stones to the depth of eight feet; next a pale colored clay and then a black one. At the depth of fifteen feet he found a piece of hard wood,

---

[1] A well-known descendant of the Delaware Swedes. The original name was *Kyn*. The noted American surgeon William Williams Keen, who recently died, was of the same family.

and several pieces of pyrites. He told me that he knew some places in the Delaware, where the people went in boats when he was young, but which at present were changed into little islands, some of which were near an English mile in length. These islands derive their origin from a sandbar or bank in the river; on this the water washes some clay, in which rushes come up and thus the rest is generated by degrees.

At a meeting of the oldest Swedes in the parish of Raccoon, I obtained the following answers to the questions which I asked them on the subject. Whenever they dig a well in this neighborhood they always find at the depth of twenty or thirty feet great numbers of oyster and clam shells. The clams are, as above-mentioned, a kind of large mussel which is found in bays, and of which the Indians make their money. In many places, on digging wells, a quantity of rushes and reeds have been found almost wholly undamaged; and once on such an occasion a whole bundle of flax was brought up, which had been discovered between twenty and thirty feet underground. All looked at it with astonishment, as it was beyond conception how it could have gotten there. But did not the good people mistake some American plants, such as the wild Virginian flax, or *Linum Virginianum*, and the *Antirrhinum Canadense*, for the common flax? Yet it is remarkable that the bundle was said to have been tied together. The Europeans on their arrival in America found our common flax neither growing wild nor cultivated by the Indians, how then could this bundle have gotten into the ground? Can it be supposed that past ages had seen a nation here so early acquainted with the use of flax? I would rather abide by the assumption that the above American plants, or other similar ones, had been taken for flax. Charcoal and firebrands had often been found under ground. The Swedish churchwarden, Eric Ragnilsson, told me that he had seen a quantity of them which had been brought up at the digging of a well. On such occasions, people had often (at a depth of between twenty and fifty feet) found great branches and blocks. There were some spots where twenty feet under the surface of the earth people had found such trowels as the Indians use. From these observations they all concluded, that this tract of land had formerly been the bottom of the sea. It is to be observed that most of the wells which had hitherto been made, had been dug in new settlements, where the forest was yet standing, and had probably stood for centuries. From the observations which have hitherto been mentioned, and to which I shall add similar ones later, we may, with a considerable degree of certainty, conclude that a great portion of the province of New Jersey, in ages unknown to posterity, was part of the bottom of the sea, and was afterwards formed by the slime and mud, and the many other things which the river Delaware carries down along with it from the higher parts of the country. However, Cape May seems to give some occasion for doubts, of which I shall speak further on.

NOVEMBER THE 27TH

The American *Evergreens* are:
1. *Ilex aquifolium,* holly.
2. *Kalmia latifolia,* the spoon tree.
3. *Kalmia angustifolia,* another species of it.
4. *Magnolia glauca,* the beaver tree. The young trees of this kind only keep their leaves, the others drop them.
5. *Viscum album,* or mistletoe; this commonly grows upon the *Nyssa aquatica,* or tupelo tree, upon the *Liquidambar styraciflua,* or sweet gum tree, the oak and lime tree, so that their whole crowns were frequently quite green in winter.
6. *Myrica cerifera,* or the candleberry tree; of this, however, only some of the youngest shrubs preserve any leaves; most of them had already lost them.
7. *Pinus abies,* the pine.
8. *Pinus sylvestris,* the fir.
9. *Cupressus thyoides,* the white cedar.
10. *Juniperus Virginiana,* the red cedar.

Several oaks and other trees shed their leaves here in winter, which however keep ever green a little more to the south and in Carolina.

NOVEMBER THE 30TH

Teeth. It has been observed that the Europeans in North America, whether they were born in Sweden, England, Germany, Holland, or in North America, of European parents, always lost their teeth much sooner than usual. The women especially were subject to this disagreeable fact; the men did not suffer so much from it. Girls not above twenty years old frequently had lost half of their teeth, without any hopes of getting new ones. I have attempted to determine the causes of this early loss of the teeth, but I know not whether I have hit upon the true one. Many people are of the opinion that the air of this country hurts the teeth. So much is certain, that the weather can nowhere be subject to more frequent and sudden changes; for the end of a hot day often turns out piercing cold and *vice versa*. Yet this change of weather, cannot be looked upon as having any effect upon the shedding of the teeth, for the Indians prove the contrary. They live in the same air, and always keep their teeth in a fine, white condition as long as they live. This I have seen myself and have been assured of by everybody. Others ascribe it to the great quantities of fruit and sweetmeats which are here eaten. But I have known many people who never eat any fruit, and still have hardly a tooth left.

I then began to suspect the tea, which is drunk here in the morning and afternoon, especially by women, and is so common at present that there is hardly a farmer's wife or a poor woman who does not drink tea in the morning. I was confirmed in this opinion when I took a journey through some parts of the country which were still inhabited by Indians. For Major General Johnson [1] told me at that time, that several of the Indians who lived close to the European settlements had learned to drink tea. And it has been observed that such of the Indian women as accustomed themselves too much to this beverage had in the same manner as the European women lost their teeth prematurely, though they had formerly been quite sound. Those again who had not used tea had preserved their teeth strong and sound to a great age.

I found afterwards that the use of tea could not entirely cause this condition. Several young women who lived in this country but were born in Europe, complained that they had lost most of their teeth after they had come to America. I asked whether they did not think that it arose from the frequent use of tea, as it was known that strong tea, as it were, entered into and corroded the teeth. But

---

[1] Undoubtedly Sir William Johnson (1715-1774), colonial superintendent of Indian affairs. He was in 1748 commissioned colonel of a regiment of militia. It was largely through his influence that the Six Nations did not go over to the French in 1745 when hostilities broke out upon the New York frontier.

they answered that they had lost their teeth before they had begun to drink tea. Continuing my inquiries I found at last a sufficient cause for it. Each of these women owned that they were accustomed to eat everything hot, and nothing was good in their opinion unless they could eat it as soon as it had come from the fire. This was likewise the case with the women in the country who lost their teeth much sooner and more generally than the men. They drank tea in greater quantity and much oftener, in the morning, and even at noon, when the employment of the men would not allow them to sit at the tea-table. Besides that, the Englishmen [at that time] cared very little for tea, and a bowl of punch was much more agreeable to them. When the English women drank tea, they never poured it out of the cup into the saucer to cool it, but drank it as hot as it came from the teapot. The Indian women in imitation of them, swallowed the tea in the same manner. On the contrary, those Indians whose teeth were sound never ate anything hot, but took their meat either cold or only just luke-warm.

I asked the Swedish churchwarden in Philadelphia, Mr. Bengtson, and a number of old Swedes, whether their parents and countrymen had likewise lost their teeth as soon as the American colonists, but they told me that they had preserved them to a very great age. Bengtson assured me that his father at the age of sixty had cracked peach stones and black walnuts with his teeth, notwithstanding their great hardness, which at this time nobody dares to venture at that age. This confirms what I have before said, for at that time the use of tea was not yet known in North America.

Malaria. No disease is more common here than that which the English call "fever and ague," which is intermittent or recurrent. But it often happens that a person who has had chills and fever, after being free from them for a week or two, has had them again every other second or third day. The fever often attacks the people at the end of August, or beginning of September, and commonly continues during autumn and winter till towards spring, when it ceases entirely. It generally begins with a headache followed by chills and fever. Often the chill is so great that both the patient, the bed upon which he lies, and everything else, shakes violently. During the fever, and also between the intervals of it, the afflicted one has a severe headache and also occasionally, during the fever, a pain under his heart.

Strangers who arrive here are commonly attacked by this sickness the first or second year after their arrival, and it acts more violently upon them than upon the natives, so that they sometimes die of it. But if they escape the first time, they have the advantage of not being visited again the next year, or perhaps ever. It is commonly said here that strangers get the fever to accustom them to the climate. The natives of European offspring have annual fits of this ague in some parts of the country. Some however are soon delivered from it, while in others, on the contrary, it continues for six months; and others are afflicted with it till they die. The Indians also suffer from it, but not so violently as the Europeans. No age is safe against it. In those places where it rages annually you see old men and women attacked by it, and even children in the cradle, sometimes not above three weeks old. It is a pity to see these poor children tormented when a chill comes upon them and to hear how they cry and suffer. It is the same with them as with the older people. This autumn the augue was more violent here than it was said to be ordinarily. People who are afflicted with it look as pale as death and are greatly weakened, but in general are not prevented from doing their work in the intervals. It is remarkable that every year there are great parts of the country where this fever rages, and others where scarcely a single person has been taken ill. It is likewise worthy of notice that there are places where the people cannot remember having heard of its ravages, though at present it begins to grow more common. Yet there is no visible difference between the various locations.

All the old Swedes, Englishmen, Germans, etc., unanimously asserted that the fever had never been so violent, or of such continuance when they were boys, as it is at present. They were likewise of the opinion that about the year 1680 there were not so many people afflicted with it as at this time. However, others, equally old, were of the opinion that the fever was proportionably as common formerly as it is at present, but that it could not at that time be so easily perceived on account of the scarcity of inhabitants and the great distance of their settlements from each other. It is therefore probable that the effects of the fever have at all times been the same.

Causes of the Ague. It would be difficult to determine the true causes of this disease; they seem to be numerous, and not always alike: sometimes, and I believe often, several of them unite. I have taken all possible care to find the opinions of the physicians here on that subject and I here offer them to the reader.

Some of them think that the peculiar qualities of the air of this country cause this fever; but most of them assert that it is generated by the standing and putrid water, which seems confirmed by experience. For it has been observed in this country, that such people as live in the neighborhood of morasses or swamps, or in places where a stagnant, stinking water is to be found, are commonly infested with the fever and ague every year, and get it more readily than others. This chiefly happens at a time of the year when those stagnant waters are most evaporated by the excessive heat of the sun, and the air is filled with the most noxious vapors. The fever likewise is very violent in all places which have a low location, and where salt water comes up with the tide twice in twenty-four hours, and unites with the stagnant, fresh water in the country.[1] On travelling in summer over such low places, where fresh and salt water unite, the nauseous stench arising from them often forces the traveller to hold his nose. On that account most of the inhabitants of Penn's Neck and Salem in New Jersey, where the ground has the above-mentioned quality, are annually infested with the fever to a much greater degree than an inhabitant of the higher part of the country, where the people are free from the fever. If a settler moves into the lower regions, he may be well assured that the fever will attack him at the usual time, and that he will get it again every year, as long as he continues in that low country. People of the liveliest complexion, on coming into the low parts of the country and continuing there for some time, have lost their color entirely and become pale. However, this cannot be the sole cause of the fever, as I have been in several parts of the country which had a low elevation and stagnant waters near them, where the people declared they seldom suffered from this sickness. But these places were about two or three degrees further north.

Others were of the opinion that diet had much to do with it, and chiefly laid the blame upon the inconsiderate and intemperate consumption of fruit. This is particularly the case with the Europeans who come into America, and are not used to its climate and its fruit,

---

[1] It is interesting at least that one hundred and eighty-five years ago observers connected what is now known as malaria with low lands and stagnant water.

for those who are born here can bear more, yet are not entirely free
from the bad effects of eating too much. I have heard many Englishmen, Germans, and others speak from their own experience on
this point. They owned that they had often tried, and were certain,
that after eating a watermelon once or twice before they had breakfasted they would have the fever and ague a few days after. Yet it is
remarkable that the French in Canada told me the fevers were less
common in that country, though they consumed as many watermelons as the English colonies, and it had never been observed that
they occasioned a fever; but that on coming in the hot season to the
Illinois, an Indian nation which lives in nearly the same latitude as
New Sweden, they could not eat a watermelon without feeling the
shaking fits of an ague, and that the Indians therefore warned them
not to eat of so dangerous a fruit. Does not this lead us to believe
that the greater heat in Pennsylvania and the territory of the Illinois,
which are both five or six degrees more southerly than Canada,
makes fruit in some measure more dangerous? In the English North
American colonies every countryman plants a number of watermelons which are eaten while the people make hay, or during the
harvest when they have nothing upon their stomachs, in order to
cool them during the great heat, as that juicy fruit seems very suitable for refreshment. In the same manner melons, cucumbers,
pumpkins, squashes, mulberries, apples, peaches, cherries, and such
fruit are eaten here in summer, and altogether contribute to the
attacks of the ague.[1]

But that the manner of living contributes greatly towards it may
be concluded from the unanimous accounts of old people concerning
the times of their childhood, according to which the inhabitants of
these parts were at that time not subject to so many diseases as they
are at present, and people were seldom sick. All the old Swedes
likewise agreed that their countrymen who first came to North America attained to a great age, and their children nearly to the same; but
that their grandchildren and great grandchildren did not reach the
age of their ancestors, and were not nearly so vigorous or healthy.
But the Swedes who first settled in America lived very frugally; they
were poor, and could not buy rum, brandy, or other strong liquors,

---

[1] We shall refrain from commenting on these speculative but honestly sought
causes of the "ague" or malaria. They speak for themselves and represent a groping
but frantic search for the truth.

which they seldom distilled themselves, as few of them had a distilling vessel. However, they sometimes had a good strong beer. They did not understand the art of making cider, which is now so common in the country. Tea, coffee, chocolate, which at present constitute even the country people's daily breakfast, were wholly unknown to them: most of them had never tasted sugar or punch. The tea which is now drunk is either very old, or mixed with all sorts of herbs, so that it no longer deserves the name of tea. Therefore it cannot have any good effect upon those who use it plentifully. Besides, it cannot fail in relaxing the bowels, as it is drunk both in the morning and in the afternoon quite boiling hot. The Indians, the offspring of the first inhabitants of this country, are a proof of what I have said. It is well known that their ancestors, at the time of the first arrival of the Europeans, lived to a very great age. According to common accounts, it was not then unusual to find people among the Indians who were over a hundred years old. They lived frugally and drank pure water. Brandy, rum, wine, and all the other strong liquors, were utterly unknown to them; but since the Christians taught them to drink these liquors, and the Indians found them so palatable, those who could not resist their appetites, hardly reached half the age of their parents.

Lastly, some people pretended that the loss of many odoriferous plants, with which the woods were filled at the arrival of the Europeans, but which the cattle have now destroyed, might be looked upon as a cause of the greater progress of the fever at present. The number of those strong plants occasioned a pleasant scent to rise in the woods every morning and evening. It is therefore not unreasonable to imagine that the noxiousness of the effluvia from decaying substances was then prevented, so that they were not so dangerous to the inhabitants.

Remedies for the Ague. Several remedies are employed against this disease. Jesuits' bark was formerly a certain one, but at present it is not always effective, though it is genuine and selected. Many people accused it of leaving something noxious in the body. Yet it was commonly observed that when the bark was good, and was taken as soon as the fever made its appearance, and before the body was weakened, it was almost sure to conquer, so that the chills never returned, and no pain or stiffness remained in the limbs. But when the disease is well established and has considerably weakened the patients, or they are naturally very weak, the fever leaves them after using the Jesuits' bark, but returns again in a week or fortnight's time, and obliges them to take the bark again; but the consequence frequently is a pain and stiffness in their limbs, and sometimes in their bowels, which almost hinders them from walking. This pain often continues for several years, and even accompanies some to the grave. This bad effect is partly attributed to the bark, which can seldom be got unadulterated here, and partly to the little care which the patients take in using it. A man of my acquaintance was particularly dexterous in expelling the ague by the use of the Jesuits' bark. His manner of proceeding was as follows: when it was possible, the patient must use the remedy as soon as the fever began and before it got a good hold. Before he took the medicine, he was to take a diaphoretic remedy, as that had been found very salutary, and as the fever is frequently of such a nature here as not to make the patient sweat, even when the attack of fever is upon him, a perspiration was to be brought about by some other means. To that purpose the patient took a dose on the day when he had his chill, and was not allowed to eat anything at night. The next morning he remained in a warm bed, drank a quantity of tea, and was well covered that he might perspire plentifully. He continued so till the perspiration ceased, and then left the bed in a hot room and washed his body with lukewarm water in order to cleanse it from the impurities that settled on it from the perspiration, and to prevent their stopping up of the pores. The patient was then dried again, and at last he took the bark several times in one day. This was repeated twice or thrice on the days after he had the chill and it commonly left him without returning, and most people recovered so well, that they did not look pale after their sickness.

The bark of the root of the Tulip tree, or *Liriodendron tulipifera,* taken in the same manner as Jesuits' bark, sometimes had a similar effect.

Several people peeled the roots of the *Cornus florida,* or dogwood, and gave this peel to the patients; and some people, who could not be cured by the Jesuits' bark, recovered by the help of this. I also saw people cured of the fever by taking powdered sulphur mixed with sugar every night before they went to bed and every morning before they got up. They took it three or four times in between, and at each time drank some warm liquor to wash the powder down.

However, others that tried the same remedy did not find much relief from it.

Some people collected the yellow bark of the peach tree, especially that which is on the root, and boiled it in water till half of it had evaporated. Of this decoction the patient took every morning about a wineglass full, before he had eaten anything. This liquor had a disagreeable taste, and contracted the mouth and tongue like alum; yet several persons at Raccoon who had tried many remedies in vain were cured by this.

Others boiled the leaves of the *Potentilla reptans* or of the *Potentilla Canadensis* in water and made the patients drink it before the chill came on, and it is well known that several persons recovered by this means.

Some used no medicines at all, and let the disease be until it passed away of itself.

A woman who had suffered a long time from chills and had used several remedies for it had at the advice of an old woman placed some spider's web inside a baked apple which she ate. She repeated this process twice with no effect, but the third time she tried it she became so inconceivably ill that everyone thought she would die. She fell into such a coma that she looked dead. Finally after two or three hours she regained consciousness, her sickness left her and she recovered and became well, but it was both a severe and adventurous cure.

The people who have settled on the Mohawk River in New York, both Indians and Europeans, collect the root of the *Geum rivale*, and pound it. This powder some of them boil in water till it is a pretty strong decoction; others add only cold water to it and leave it so for a day; others mix it with brandy. Of this medicine the patient takes a wineglass full on the morning of the day when the fever does not come, before he has eaten anything. I was assured that this was one of the surest remedies, and more certain than the Jesuits' bark.

People who lived near the iron mines declared that they were seldom if ever visited by the fever and ague; but when they had the fever they drank the water of such springs as came from the iron districts and had a strong metallic taste; and they assured me that this remedy was infallible. Other people therefore, who did not live very far from such springs, went to them for a few days when they had the fever in order to drink the water, which commonly cured them.

I have already shown above, under the date of November first, that sage mixed with lemon juice has been found very salutary for the ague. It is however universally held that that which cures one person of it may have no effect upon another.

Pleurisy is likewise a disease which the people of this country are much subject to. The Swedes of this province call it "pricks and burning" (*Stick och bränna*) and they always mean pleurisy whenever they mention those words. Many of the old Swedes told me that they had heard very little of it when they were young, and that their parents had known still less of it in their childhood; but that it was so common now that many people died every year of it. Yet it has been observed that in some years this disease has been very moderate and taken few people, while in other years it has exacted a great toll. It is more violent in some places than in others.

In the autumn of the year 1728 it swept away many at Penn's Neck, a place below Raccoon and nearer to the Delaware, where a number of Swedes have settled. Almost all of the old Swedes there died of it, though they were very numerous. Hence it happened that their children who were left at a very tender age, and grew up among the English children, forgot their mother tongue, so that few of them understand it at present. Since that time, though the pleurisy every year killed a few people at Penn's Neck, yet it did not carry off any considerable number. It rested as it were till the autumn of the year 1748; but then it began to cause a dreadful havoc, and every week six or ten of the old people died. The disease was so violent that when it attacked a person he seldom lived above two or three days, and of those who were taken ill with it very few recovered. When the pleurisy got into a house it killed most of the old people in it. It was a true pleurisy, but it had the peculiarity that it commonly began with a great swelling under the throat and in the neck, and with a difficulty of swallowing. Some people looked upon it as contagious, and others declared seriously that when it came into a family not only those who lived in the same house suffered from it but even such relations as lived far off. There were several people at Penn's Neck, who, without visiting their sick friends, got the pleurisy and died of it.—I do not dispute the truth of this, though I do not agree with the conclusion.—The pleurisy was the most violent in November; yet some old people died of it during the next winter, though children were pretty free from it. The physicians did not know what to make of it, nor how to cure it.

It is difficult to determine the causes of such violent diseases. An old English surgeon who lived here gave the following reason. The inhabitants of this country drink great quantities of punch and other strong liquors in summer, when it is very hot; by that means the veins in the diaphragm contract, and the blood grows thick. Towards the end of October and the beginning of November, the weather is apt to alter very suddenly, so that heat and cold change several times a day. When the people during this changeable weather are in the open air, they commonly get this disease. It is likewise certain that the air is more unwholesome one year than another, which depends upon the heat and other circumstances. This peculiar quality of the air must of course produce pleurisy. It is remarkable that both in the year 1728 and at present when so many people died at Penn's Neck, which lies remarkably low, very few took sick or died at Raccoon which is situated pretty high. Though other conditions are much the same, the people in the former place have settled between marshes and swamps, in which the water stagnates and putrifies, and often these places are in addition covered with trees, which preserve the moisture still more, and near such marshes are the houses. Lastly the water at Penn's Neck is not reckoned so good as that in Raccoon, but has a bad taste. It likewise becomes brackish in several little rivers when the Delaware, at high tide, runs up into them. On the banks of these rivulets live many of the Swedes and use the water from them.

DECEMBER THE 3RD

This morning I set out for Philadelphia, where I arrived in the evening.

Wild grapes are very abundant in the woods, and are of various kinds. A species of them which is remarkable for size grows in the marshes, and is greedily eaten by the raccoon. They are therefore called marsh grapes, but the English call them "fox grapes." They are seldom eaten because of the disagreeable taste. The people make use of a small kind of wild grape which grows on a dry soil. Late in autumn when they are ripe they are eaten raw, and have a very good flavor, being a mixture of sweet and acid. Some people dry these grapes when gathered and bake them in tarts, etc. They likewise use them for dried sweetmeats. The Swedes formerly made a pretty good wine from them, but have now given up the practice. However, some of the English still press an agreeable liquor from these grapes, which they assured me was as good as the best claret, and that it would keep for several years.

The manner of preparing this sort of wine has been described at large in an almanack of this country for the year 1743, and is as follows: the grapes are collected from the twenty-first of September to about the eleventh of November, that is, as they grow ripe. They must be gathered in dry weather, and after the dew is gone off. The grapes are cleared of the cobwebs, dry leaves and other things adhering to them. Next a great hogshead is prepared which has either had treacle or brandy in it. It is washed very clean, one of the bottoms knocked out and the other placed on a stand for the purpose, or on pieces of wood in the cellar, or else in a warm room, about two feet above the ground. The grapes are put into this hogshead, and as they sink lower in three or four days' time more are added. A man with naked feet gets into the hogshead and treads the grapes, and in about half an hour's time the juice is forced out. The man then turns the lowest grapes uppermost, and treads them for about a quarter of an hour. This is sufficient to squeeze the good juice out of them, for an additional pressure would even crush the unripe grapes, and give the whole a disagreeable flavor. The hogshead is then covered with a thick blanket; but if there is no cellar, or it is very cold, two are spread over it. Under this covering the juice is left to ferment for the first time, and in the next four or five days it ferments and works very strongly. As soon as the fermentation ceases, a hole is made about six inches from the bottom and some of the juice is tapped off about twice a day. As soon as this is clear and settled, it is poured into a keg of sufficient size. From twenty bushels of grapes they get about as many gallons of juice. The keg remains untouched and the juice in it ferments a second time. Now it is necessary that the keg be quite full. The scum which settles at the bunghole must be taken off, and the keg kept filled up with more juice, which is held ready for the purpose. This is continued till Christmas, when the keg may be sealed up. At last, in February, the wine is ready and bottled. It is also usual here to put some of the ripe grapes into a vessel in order to make vinegar, and that which is gotten by this means is very good. Several people make brandy from these grapes, which has a very pleasant taste, but is still more pleasant if the fruit of the persimmon is mixed with it.

The wood of the grapevines is of no use, it is so brittle that it cannot be used for thongs. On cutting into the stem, a white, insipid resin comes out a few hours after the wound is made. In many gardens vines are planted for the purpose of making arbors for which they are indeed excellent, as their large and numerous leaves form a very close cover against the scorching heat of the sun. When the vines flower here in May and June, the flowers exhale a strong, but exceedingly pleasant and refreshing smell, which is perceptible even at a great distance. Therefore on coming into the woods about that time, you may judge from the sweet perfume in the air, arising from the flowers of the vines, that you are near them, though you do not see them. Though the winters be ever so severe, they do not affect the vines. Each grape is about the size of a pea, but further southward they are said to be of the size of common raisins, and of a finer flavor. Further up in the country, during a part of autumn, they are the chief food of bears, who climb up the trees after them. People are of the opinion that if the wild vines were cultivated with more care, the grapes would grow larger and more palatable.

A Swedish lady told me that one of the best remedies for cough, whooping-cough, and pain in the chest, was to take a piece of steel, heat it to a red heat, put it into a bowl of fresh milk and drink this as soon as it became cool enough. This was repeated several times. Her children had had such a severe case of whooping-cough that they had almost died. She was then advised by an acquaintance to use this remedy with the assurance that her children would improve within two or three days. She claimed that her experience proved that it was true and that she had later helped many others with the same remedy.

December the 5th

Weather Forecasting. I shall here mention two means of forecasting the weather, which were greatly valued here. Some people pretended to foretell that the ensuing winter would not be a severe one.

This they conjectured from having seen wild geese and other migratory birds go to the south in October, but return only a few days ago in great numbers, and even pass on further to the north. Indeed, the ensuing winter was one of the most temperate ones.

Several persons likewise assured us that we should have rain before tomorrow night. The reason they gave for this conjecture was that this morning at sunrise they had from their windows seen everything very plainly on the other side of the river, so that it appeared much nearer than usual, and that this commonly foreboded rain.—In the main this forecast was fulfilled.

Minerals. The Indians before the arrival of the Europeans, had no notion of the use of iron, though that metal was abundant in their country. However, they knew in some measure how to make use of copper. Some Dutchmen who lived here still preserved the old account among them, that their ancestors on their first settling in New York had come upon many of the Indians who had tobacco pipes of copper, and who made them understand by signs that they got them in the neighborhood. Afterwards a fine copper mine was discovered, upon the second river between Elizabethtown and New York. On digging in this mine the people found holes made in the mountain, out of which some copper had been taken, and they found even some tools, which the Indians probably made use of when they endeavored to get the metal for their pipes. Such holes in the mountains were also found in some parts of Pennsylvania, *viz.* below Newcastle, towards the coast, and there were always some traces of a copper ore along with them. Some people conjectured that the Spaniards, after discovering Mexico, sailed along the coasts of North America and landed now and then in order to inquire whether any gold or silver was to be found, and that they perhaps made these holes in the mountains. But supposing them to have made such a voyage along the coasts, they could not immediately have found the copper mines, and they probably did not stop to blast this ore, as they were bent only upon gold and silver. It is therefore almost certain that the Indians dug these holes. Or may we be allowed to suspect that our old Northmen, long before the discoveries of Columbus, came into these parts and met with such veins of copper, when they sailed to what they called *Vinland,*[1] of which

---

[1] There is no doubt that the Norsemen visited the coasts of North America in the beginning of the eleventh century, and probably later, but it is not at all certain that

our ancient traditional records called Sagas speak, and which undoubtedly was North America? But in regard to this I shall have occasion later to explain my sentiments better. It was remarkable that in all the places where such holes had lately been found in the mountains, and which manifestly seemed to have been dug by men, the pits were always covered with a great quantity of earth, as though they were intended to remain hidden from strangers.

DECEMBER THE 6TH

Poisonous Fish. On long voyages the sailors sometimes catch such fish as are unknown to any of the ship's company; but as they are very eager for fresh provisions they seldom abstain from eating them. However, it often proves too risky, experience having shown that their want of caution has sometimes cost them their lives, when poisonous fish have been caught. But there is a method of detecting them, as I have heard from several ship captains. It is usual when such unknown fish is boiled to put a silver button or some other piece of silver into the kettle. If the fish is poisonous the silver will turn black, otherwise it will not change. Some of the seamen referred to their own repeated experiences in attesting the infallibility of this method.

Mrs. Rebeson, one of the Swedish women in the city, always had a butter which in quality and palatability seemed to surpass most of its kind in this vicinity. The butter churned in winter was just as good as the summer variety, and the old as good as the new. It was said to keep longer than other butter. She had learned the art of making it from a Quaker's wife. Besides the fact that the cows are fed with good hay and the butter is made from good cream, the chief reason for its excellent quality is that after churning the milk is never washed out of the butter with water but forced out through kneading. This method requires more time and trouble than any other, but the excellence of the product richly pays for the work. Anyone may verify this by trying the method.

---

they came as far south as Pennsylvania. Some students believe that the Northmen entered Chesapeake Bay, but most scholars localize their operations somewhere between New York harbor and Nova Scotia.

It was Kalm who first acquainted Benjamin Franklin with the facts of the early Norse discovery of America.

Mr. Franklin and several other gentlemen frequently told me that an Indian, who owned Rhode Island, had sold it to the English for a pair of spectacles. It is large enough for a principality, and has a separate government at present.—This Indian knew how to set a true value upon a pair of glasses, for if they were scarce, they would undoubtedly on account of their great usefulness have the same value as diamonds.

The *education* of children among the English in this country was well established in many ways. They had separate schools for small boys and girls. When a child was a little over three it was sent to school both morning and afternoon. They probably realized that such little children would not be able to read much, but they would be rid of them at home and thought it would protect them from any misbehavior. Also they would acquire a liking for being with other children. Englishmen used only one kind of letters, i. e. Latin. So the youth could learn them easier than is the case in Sweden where children have to learn both the Latin and the Swedish alphabet. We ought to do the same in Sweden, because in general the letters are more even and look better. They are also more readable and heavier in type, a great advantage for weak or old eyes. Ordinarily the English write a very clear and readable script. In fact a great number of their women also write very neatly, at least not in such a scrawl as some of our Swedish men and women. It is probably just as easy to learn to write well as carelessly. One reason for the superiority of English penmanship is the character of the children's copy books, which are made of clean paper, on every leaf of which are letters neatly and legibly engraved in copper. These copy books are, also, reasonable in price. We (Swedes) ought to imitate this method more.

Servants. The servants which are employed in the English-American colonies are either free persons or slaves, and the former, again, are of two different classes.

1. Those who are entirely free serve by the year. They are not only allowed to leave their service at the expiration of their year, but may leave it at any time when they do not agree with their masters. However, in that case they are in danger of losing their wages, which are very considerable. A man servant who has some ability gets between sixteen and twenty pounds in Pennsylvania currency, but those in the country do not get so much. A maidservant gets eight

or ten pounds a year. These servants have their food besides their wages, but they must buy their own clothes, and whatever they get of these as gifts they must thank their master's generosity for.

Indenture. 2. The second kind of free servants consists of such persons as annually come from Germany, England and other countries, in order to settle here. These newcomers are very numerous every year: there are old and young of both sexes. Some of them have fled from oppression, under which they have labored. Others have been driven from their country by religious persecution, but most of them are poor and have not money enough to pay their passage, which is between six and eight pounds sterling for each person. Therefore, they agree with the captain that they will suffer themselves to be sold for a few years on their arrival. In that case the person who buys them pays the freight for them; but frequently very old people come over who cannot pay their passage, they therefore sell their children for several years, so that they serve both for themselves and for their parents. There are likewise some who pay part of their passage, and they are sold only for a short time. From these circumstances it appears that the price on the poor foreigners who come over to North America varies considerably, and that some of them have to serve longer than others. When their time has expired, they get a new suit of clothes from their master and some other things. He is likewise obliged to feed and clothe them during the years of their servitude. Many of the Germans who come hither bring money enough with them to pay their passage, but prefer to be sold, hoping that during their servitude they may get a knowledge of the language and character of the country and the life, that they may the better be able to consider what they shall do when they have gotten their liberty. Such servants are preferable to all others, because they are not so expensive. To buy a negro or black slave requires too much money at one time; and men or maids who get yearly wages are likewise too costly. But this kind of servant may be gotten for half the money, and even for less; for they commonly pay fourteen pounds, Pennsylvania currency, for a person who is to serve four years, and so on in proportion. Their wages therefore are not above three pounds Pennsylvania currency per annum. These servants are, after the English, called *servingar* by the Swedes. When a person has bought such a servant for a certain number of years, and has an intention to sell him again, he is at liberty to do so, but is obliged, at the expiration of the term of servitude, to provide the usual suit of clothes for the servant, unless he has made that part of the bargain with the purchaser. The English and Irish commonly sell themselves for four years, but the Germans frequently agree with the captain before they set out, to pay him a certain sum of money, for a certain number of persons. As soon as they arrive in America they go about and try to get a man who will pay the passage for them. In return they give according to their circumstances, one or several of their children to serve a certain number of years. At last they make their bargain with the highest bidder.

3. The *negroes* or blacks constitute the third kind. They are in a manner slaves; for when a negro is once bought, he is the purchaser's servant as long as he lives, unless he gives him to another, or sets him free. However, it is not in the power of the master to kill his negro for a fault, but he must leave it to the magistrates to proceed according to the laws. Formerly the negroes were brought over from Africa, and bought by almost everyone who could afford it, the Quakers alone being an exception. But these are no longer so particular and now they have as many negroes as other people. However, many people cannot conquer the idea of its being contrary to the laws of Christianity to keep slaves. There are likewise several free negroes in town, who have been lucky enough to get a very zealous Quaker for their master, and who gave them their liberty after they had faithfully served him for a time.

At present they seldom bring over any negroes to the English colonies, for those which were formerly brought thither have multiplied rapidly. In regard to their marriage they proceed as follows: in case you have not only male but likewise female negroes, they may intermarry, and then the children are all your slaves. But if you possess a male negro only and he has an inclination to marry a female belonging to a different master, you do not hinder your negro in so delicate a point, but it is of no advantage to you, for the children belong to the master of the female. It is therefore practically advantageous to have negro women. A man who kills his negro is, legally, punishable by death, but there is no instance here of a white man ever having been executed for this crime. A few years ago it happened that a master killed his slave. His friends and even the magistrates secretly advised him to make his escape, as otherwise they could not avoid taking him prisoner, and then he would be condemned to die according to the laws of the country, without any hopes of being saved. This leniency was granted toward him, that the negroes might not have the satisfaction of seeing a master executed for killing his slave. This would lead them to all sorts of dangerous designs against their masters, and to value themselves too much.

The negroes were formerly brought from Africa, as I mentioned before, but now this seldom happens, for they are bought in the West Indies, or American Islands, whither they were originally brought from their own country. It has been found that in transporting the negroes from Africa directly to these northern countries, they have not such good health as when they come gradually, by shorter stages, and are first carried from Africa to the West Indies, and from thence to North America. It has frequently been found, that the negroes cannot stand the cold here so well as the Europeans or whites; for while the latter are not in the least affected by the cold, the toes and fingers of the former are frequently frozen. There is likewise a material difference among them in this point; for those who come immediately from Africa, cannot bear the cold so well as those who are either born in this country, or have been here for a considerable time. The frost easily hurts the hands or feet of the negroes who come from Africa, or occasions violent pains in their whole body, or in some parts of it, though it does not at all affect those who have been here for some time. There are frequent examples that the negroes on their passage from Africa, if it happens in winter, have some of their limbs frozen on board the ship, when the cold is but very moderate and the sailors are scarcely obliged to cover their hands. I was even assured that some negroes have been seen here who had excessive pain in their legs, which afterwards broke in the middle, and dropped entirely from the body, together with the flesh on them. Thus it is the same case with men here as with plants which are brought from the southern countries, before they accustom themselves to a colder climate.

The price of negroes differs according to their age, health and ability. A full grown negro costs from forty pounds to a hundred of Pennsylvania currency. There are even examples that a gentleman has paid a hundred pounds for a black slave at Philadelphia and refused to sell him again for the same money. A negro boy or girl of two or three years old, can hardly be gotten for less than eight or fourteen pounds in Pennsylvania money. Not only the Quakers but also several Christians of other denominations sometimes set their negroes at liberty. This is done in the following manner: when a gentleman has a faithful negro who has done him great services, he sometimes declares him independent at his own death. This is however very expensive; for they are obliged to make a provision for the negro thus set at liberty, to afford him subsistence when he is grown old, that he may not be driven by necessity to wicked actions, or that he may fall a charge to anybody, for these free negroes become very lazy and indolent afterwards. But the children which the free negro has begot during his servitude are all slaves, though their father be free. On the other hand, those negro children which are born after the parent was freed are free. The negroes in the North American colonies are treated more mildly and fed better than those in the West Indies. They have as good food as the rest of the servants, and they possess equal advantages in all things, except their being obliged to serve their whole lifetime and get no other wages than what their master's goodness allows them. They are likewise clad at their master's expense. On the contrary, in the West Indies, and especially in the Spanish Islands, they are treated very cruelly; therefore no threats make more impression upon a negro here than that of sending him over to the West Indies, in case he will not reform. It has likewise been frequently found by experience that when you show too much kindness to these negroes, they grow so obstinate that they will no longer do anything but of their own accord. Therefore a strict discipline is very necessary, if their master expects to be satisfied with their services.

In the year 1620 some negroes were brought to North America in a Dutch ship, and in Virginia they bought twenty of them. These are said to have been the first that came hither. When the Indians, who were then more numerous in the country than at present, saw these black people for the first time, they thought they were a real breed of devils, and therefore they called them *manito* for a long while. This word in their language signifies not only god but also devil. Some time before that, when they saw the first European ship on their coasts, they were quite convinced that God himself was in the ship. This account I got from some Indians, who preserved it among them as a tradition which they had received from their ancestors. Therefore the arrival of the negroes seemed to them to have confused everything; but since that time, they have entertained less disagreeable notions of the negroes, for at present many live among them, and they even sometimes intermarry, as I myself have seen.

The negroes have therefore been upwards of a hundred and thirty years in this country. As the winters here, especially in New England and New York, are as severe as our Swedish winter, I very carefully inquired whether the cold had not been observed to affect the color of the negroes, and to change it, so that the third or fourth generation from the first that came hither became less black than their ancestors. But I was generally answered that there was not the slightest difference of color to be perceived; and that a negro born here of parents who were likewise born in this country, and whose ancestors, both men and women had all been blacks born in this country, up to the third or fourth generation, was not at all different in color from those negroes who were brought directly from Africa. Hence many people concluded that a negro or his posterity did not change color, though they continued ever so long in a cold climate; but the union of a white man with a negro woman, or of a negro man with a white woman had an entirely different result. Therefore to prevent any disagreeable mixtures of the white people and negroes, and to hinder the latter from forming too great opinions of themselves, to the disadvantage of their masters, I am told there was a law passed prohibiting the whites of both sexes to marry negroes, under pain of almost capital punishment, with deprivation and other severer penalties for the clergyman who married them. But that the whites and blacks sometimes copulated, appears from children of a mixed complexion, which are sometimes born.

It is likewise greatly to be pitied that the masters of these negroes in most of the English colonies take little care of their spiritual welfare, and let them live on in their pagan darkness. There are even some who would be very ill pleased [with negro enlightenment], and would in every way hinder their negroes from being instructed in the doctrines of Christianity. To this they are led partly by the conceit of its being shameful to have a spiritual brother or sister among so despicable a people; partly by thinking that they would not be able to keep their negroes so subjected afterwards; and partly through fear of the negroes growing too proud on seeing themselves upon a level with their masters in religious matters.

Several writings are well known which mention that the negroes in South America have a kind of poison with which they kill each other, though the effect is not sudden, and takes effect a long time after the person has taken it. The same dangerous art of poisoning is known by the negroes in North America, as has frequently been experienced. However, only a few of them know the secret, and they likewise know the remedy for it; therefore when a negro feels himself poisoned and can recollect the enemy who might possibly have given him the poison, he goes to him, and endeavors by money and entreaties to move him to deliver him from its effects. But if the negro is malicious, he not only denies that he ever poisoned him, but likewise that he knows an antidote for it. This poison does not kill immediately, as I have noted, for sometimes the sick person dies several years afterward. But from the moment he has the poison he falls into a sort of consumption state and enjoys but few days of good health. Such a poor wretch often knows that he is poisoned the moment he gets it. The negroes commonly employ it on such of their brethren as behave well [toward the whites], are beloved by their masters, and separate, as it were, from their countrymen, or do not like to converse with them. They have likewise often other reasons for their enmity; but there are few examples of their having poisoned their masters. Perhaps the mild treatment they receive, keeps them from doing it, or perhaps they fear that they may be discovered, and that in such a case, the severest punishments would be inflicted on them.

They never disclose the nature of the poison, and keep it inconceivably secret. It is probable that it is a very common article, which may be had anywhere in the world; for wherever the blacks are they can always easily procure it. Therefore it cannot be a plant, as several learned men have thought, for that is not to be found everywhere. I have heard many accounts here of negroes who have been killed by this poison. I shall only mention one incident which happened during my stay in this country. A man here had a negro who was exceedingly faithful to him, and behaved so well that he would not have exchanged him for twenty other negroes. His master likewise showed him a peculiar kindness, and the slave's conduct equalled that of the best servant. He likewise conversed as little as possible with the other negroes. On that account they hated him to excess, but as he was scarcely ever in company with them they had no opportunity of conveying the poison to him, which they had often tried. However, on coming to town during the fair (for he lived in the country) some other negroes invited him to drink with them. At first he would not, but they pressed him till he was obliged to comply. As soon as he came into the room, the others took a pot from the wall and pledged him, desiring him to drink likewise. He drank, but when he took the pot from his mouth, he said: "what beer is this? It is full of . . ." I purposely omit what he mentioned, for it seems undoubtedly to have been the name of the poison with which the malicious negroes do so much harm, and which is to be met with almost everywhere. It might be too much employed to wicked purposes, and it is therefore better that it remains unknown. The other negroes and negro-women began laughing at the complaints of their hated country-man, and danced and sang as if they had done an excellent thing, and had at last won the point so much wished for. The innocent negro went away immediately, and when he got home asserted that the other negroes had certainly poisoned him: he then fell into a decline, and no remedy could prevent his death.

DECEMBER THE 7TH

In the morning I undertook a little journey again to Raccoon, New Jersey.

Large Families. It does not seem difficult to find out the reasons why the people multiply faster here than in Europe. As soon as a person is old enough he may marry in these provinces without any fear of poverty. There is such an amount of good land yet un-cultivated that a newly married man can, without difficulty, get a spot of ground where he may comfortably subsist with his wife and children. The taxes are very low, and he need not be under any concern on their account. The liberties he enjoys are so great that he considers himself as a prince in his possessions. I shall here demonstrate by some plain examples what these conditions accomplish.

Måns Keen, one of the [above-mentioned] Swedes in Raccoon, was now near seventy years old. He had many children, grand-children, and great-grandchildren; so that of those who were yet alive he could count forty-five persons. Besides them, several of his children and grandchildren died young, and some at a mature age. He had, therefore, been uncommonly well blessed. Yet his happiness is not comparable to that which is to be seen in the following examples, which I have taken from the Philadelphia newspapers.

In the year 1732, January the 24th, there died at Ipswich, in New England, Mrs. Sarah Tuthil, a widow, aged eighty-six years. She had brought thirteen children into the world, and from seven of them only she had seen one hundred and seventy-seven grandchildren and great-grandchildren.

In the year 1739, May the 30th, the children, grandchildren and great-grandchildren, of Mr. Richard Buttington, in the parish of Chester, in Pennsylvania, were assembled in his house. They made together one hundred and fifteen persons. Mr. Buttington was born in England, and was then entering his eighty-fifth year. He was at that time quite healthy and active, and had a good memory. His eldest son, then sixty years old, was the first Englishman born in Pennsylvania.

In the year 1742, on the 8th of January, there died at Trenton, in New Jersey, Mrs. Sarah Furman, a widow, aged ninety-seven years. She was born in New England; and left five children, sixty-one grandchildren, one hundred and eighty-two great-grandchildren, and twelve great-great-grandchildren, who were all alive when she died.

In the year 1739, on the 28th of January, there died at South Kingston, in New England, Mrs. Maria Hazard, a widow, in the hundredth year of her age. She was born in Rhode Island, and was a grandmother of the then vice-governor of that colony, Mr. George Hazard. She could count altogether five hundred children, grandchildren, great-grandchildren, and great-great-grandchildren. When she died, two hundred and five of them were alive; a granddaughter of hers had already been grandmother near fifteen years.

In this manner the usual wish or blessing in our liturgy, that newly married couples may see their grandchildren till the third and fourth generation, has been literally fulfilled in regard to some of these persons.[1]

----

[1] Mr. Kalm speaks here of the Swedish Liturgy.—F.

DECEMBER THE 9TH

Injurious Insects. In every country we commonly meet with a number of insects. Many of them, though they be ever so small and contemptible, can do considerable damage to the inhabitants. Of these dangerous insects there are some in North America also. Some are peculiar to that country, others are common to Europe as well.

I have already mentioned the mosquitoe, as a kind of disagreeable gnat; and another noxious insect, the *Bruchus pisi,* which destroys whole fields of peas.[1] I shall here add some more [pests].

Seventeen-year Locusts. There is a kind of locust which about every seventeenth year comes in incredible numbers. The insects come out of the ground in the middle of May, and make, for six weeks, such a noise in the trees and woods that two persons who meet in such places cannot hear each other unless they speak louder than the locusts can chirp. During that time they make holes in the soft bark of the small branches with their tail. This ruins these branches. They do no other harm to the trees or plants. In the interval between the years when they are so numerous, only a straggler is seen or heard occasionally in the woods.

Caterpillars. There is also a kind of caterpillar in these provinces which eats the leaves from the trees. They are innumerable in some years. In the intervals there are but few of them; but when they do come, they strip the trees so entirely of their leaves, that in the middle of summer they are as naked as in winter. They eat all kinds of leaves, and very few trees are left untouched by them. About that time of the year the heat is very excessive. Stripping the trees of their leaves has the fatal consequence that they cannot withstand the heat any longer, and dry up entirely. In this manner great forests are often ruined entirely. The Swedes who live here showed me, here and there, great tracts in the woods, where young trees were now growing instead of the old ones which some years ago had been destroyed by the caterpillars. These caterpillars afterwards change into moths, or *phalænæ,* which will be described below in their proper place.

Grass-worms. In some years the grass-worms[2] do a great deal

---

[1] See pp. 76 and 91 ff.
[2] Probably the larvæ of a noctuid moth, *Laphygma frugiperda,* which is very destructive to grass, especially in the southern United States.

of damage in several places, both in the meadows and cornfields. For these are at certain times overrun with great armies of these worms, as well as with the other insects. It is very fortunate that these many plagues do not come all together. For in those years when the locusts are numerous, the caterpillars and grass-worms are few, and it so happens that only one of the three kinds comes at a time. Then there are several years when they are very scarce. The grass-worms have been observed to settle chiefly in a rich soil; but as soon as careful husbandmen discover them, they dig narrow ditches with almost perpendicular sides round the field which the worms have infested. As they creep along they all fall into the ditch, and cannot get out again. I was assured by many persons that these three sorts of insects followed each other pretty closely; and that the locusts came in the first year, the caterpillars in the second, and the grass-worms in the last: I have likewise found by may own experience that this is partly true.

Moths. Moths or *Tineæ*, which eat clothes, are likewise abundant here. I have seen cloth, worsted gloves, and other woolen stuffs, which had hung all the summer locked up in a clothespress and had not been taken care of, so damaged by these worms that whole pieces fell out. Sometimes they were so spoiled that they could not be mended again. Furs which had been kept in the garret were frequently so ruined by moths that the hair came off by the handfuls. I am not certain however whether these insects were originally in the country, or whether they were brought over from Europe.

Fleas. Fleas are likewise to be found in this part of the world. Many thousands were undoubtedly brought over from other countries; yet immense numbers of them have certainly been here since time immemorial. I have seen them on the gray squirrels, and on the hares which have been killed in wild parts of this country, where no human creature ever lived. As I afterwards came further north into the country and was obliged to lie at night in the huts and beds of the Indians, I was so plagued by the immense quantities of fleas, that I imagined I had been put in a burning fire. They drove me from the bed, and I was very glad to sleep on the benches below the roof of the huts. But it is easy to conceive that the many dogs which the Indians keep, breed fleas without end. Dogs and men lie promiscuously in the huts; and a stranger can hardly lie down and shut his eyes before he is in danger either of being squeezed to death or stifled by a dozen or more dogs, which lie round him, and upon him, in order to have a good resting place. For I imagine they do not expect that strangers will venture to beat them or throw them off, as their masters and mistresses commonly do.

Crickets. The noisy crickets (*Gryllus domesticus*) which are sometimes met with in the houses in Sweden I have not perceived in any part of Pennsylvania or New Jersey, and other people whom I have asked could not assert that they had ever seen any. In summer there is a kind of black cricket in the fields, which makes exactly the same chirping noise as our house crickets. But it sticks to the fields, and is silent as soon as winter or the cold weather sets in. They say it sometimes happens that these field crickets take refuge in houses, and chirp continually there while it is warm weather, or where the rooms are warm; but as soon as it grows cold they are silent. In some parts of the province of New York, and in Canada, every farmhouse and most of the houses in the towns swarm with so many that no Swedish farmer's house could be in a worse condition from them. They continue their music there throughout the whole winter and summer.

Bedbugs (*Cimex lectularius*) are very plentiful in this part of the world too. I have against my will been sorely tormented by them in many places, both among Englishmen and Frenchmen, but I do not remember having seen any among the Indians, during my stay at Fort Frédéric (Crown Point). The commander there, Mr. de Lusignan, told me that none of the Illinois or other Indians of the western parts of North America knew anything of these vermin. And he added, that he could with certainty say this from his own experience, having been among them a great deal. Yet I cannot determine whether bedbugs were first brought over by the Europeans, or whether they were in the country originally. Many people looked upon them as natives of this country, and as a proof of it said, that under the wings of bats people had often found some which had eaten very deep into the flesh. It was therefore believed that the bats had got them in some hollow tree, and had afterwards brought them into the houses, as they commonly planted themselves close to the walls, and crept into cracks. But as I have never seen any bedbugs upon bats, I cannot say anything upon that subject. Perhaps a louse or a tick (*Acarus*) had been taken for a bedbug. Or, if a real one were found upon a bat's wing, it is very easy to conceive that it had fixed itself on the bat, while the latter was in the chinks of a house stocked with them.

As the people here could not bear these vermin, any more than we can in Sweden, they endeavored to exterminate them by various means. I have already remarked above [1] that the beds for that reason were made of sassafras wood, but that they were only temporary remedies. Some persons assured me that they had found from their own experience and by repeated trials, that no remedy was more effectual towards the expulsion of bedbugs than injecting boiling water into all the cracks where they were, and washing all the wood of the beds with it. This being twice or thrice repeated, the bugs were wholly destroyed. If there are bugs in neighboring houses, they will cling to one's clothes, and thus be brought over into other houses.

I cannot say whether these remedies are good or not, as I have not tried them; but by repeated experiments I have been convinced that sulphur, if it be properly employed, entirely destroys bugs and their eggs in beds and walls, though they are ten times more numerous than ants in an ant-hill.[2]

Cockroaches. The millbeetles or cockroaches do not feel that they either should be excluded from the New World and so are likewise a plague of North America, and are found in many of its provinces. The learned Dr. Colden [3] was of the opinion that these insects were really natives of the West Indies, and that those that were found in North America were brought over from those islands. To confirm his opinion he said that it was still daily seen how ships coming with goods from the West Indies to North America brought millbeetles with them in great numbers. But from the observations which I have made in this country, I have reason to believe that these insects have been in North America since time immemorial. Yet notwithstanding this I do not deny their being brought over

---

[1] See page 180.

[2] A still more infallible remedy, is to wash all the furniture, infested with that vermin, with a solution of arsenic.—F.

[3] Cadwallader Colden (1688-1776), Scotch-American physician, botanist, mathematician, politician and lieutenant-governor of New York, 1761-1766. He introduced the Linnean system into America and furnished Linné with descriptions of several hundred American plants. He wrote a *History of the Five Indian Nations of Canada* (1727) and several medical works.

from the West Indies. They are in almost every house in the city of New York; and those have undoubtedly come over with ships. But how can that be said of those millbeetles, which are found in the midst of the woods·and wildernesses?

The English also call the millbeetles cockroaches, and the Dutch give them the name of *kackerlack*. The Swedes in this country call them *brödätare* or "breadeaters," on account of the damage they do to the bread, which I am going to describe. Dr. Linné calls them *Blatta Orientalis*. Many of the Swedes call them *kackerlackor*. They are not only seen in houses, but in summer they appear often in the woods, and run around tree stumps. On bringing in all sorts of old rotten blocks of wood for fuel in February, I discovered several cockroaches in them. They appeared at first quite torpid, but after lying in the room for a while they became very lively, and began to run about. I afterwards found very often, that when old rotten wood was brought home in winter and cut in pieces for fuel, the cockroaches hibernated in it in large numbers. In the same winter a fellow cut down a large dry tree, and was about to split it. I then observed in a crack, some fathoms above the ground, several live cockroaches together with the common ants. They had, it seems, crept up a great way, in order to find winter-quarters. On travelling in the middle of October, 1749, through the uninhabited country between the English and French colonies, and making a fire at night near a thick, half-rotten tree, on the shore of lake Champlain, hordes of cockroaches came out of the decaying wood, being wakened by the smoke and the fire, and driven out of their holes. The Frenchmen, who were then in my company, did not know them, and could not give them any name. In Canada the French did not remember seeing any in the houses. In Pennsylvania, I am told, they are found in immense numbers about the sheaves of corn, during the harvest. At other times they live commonly in the houses in the English settlements, and hide in the crevices, especially in the cracks of those beams which support the ceiling and are nearest to the chimney. They do a great deal of damage by eating the soft parts of bread. If they have once made a hole into a loaf, they will in a little time eat all the soft part in it, so that on cutting the loaf nothing but the crust is left. I am told that they eat other food also. Sometimes they bite people's noses or feet, while they are asleep. An old Swede, called Sven Lock, a grandson of the Rev. Mr. Lockenius,[1] one of the first Swedish clergymen that came to New Sweden, told me that he had in his younger years been once very much frightened on account of a cockroach which crept into his ear while he was asleep. He awoke suddenly, jumped out of bed, and felt that the insect, probably out of fear, was endeavoring with all its strength to get deeper. These attempts of the cockroach were so painful to him that he thought his head would burst, and he became almost mad. However, he hastened to the well, and bringing up a bucket full of water, threw some into his ear. As soon as the cockroach found itself in danger of being drowned, it endeavored to save itself, and pushed backwards out of the ear, with its hind feet, and thus happily delivered the poor man from his fears.

Woodlice are disagreeable insects, which in a manner are worse than the preceding, but as I have already described them in a separate article, which is printed among the memoirs of the Royal Swedish Academy of Sciences for the year 1754, I refer my readers to that account.

DECEMBER THE 11TH

This morning I made a little excursion to Penn's Neck, and across the Delaware to Wilmington. The country round Penn's Neck had the same qualities as that about other places in this part of New Jersey. The ground consists chiefly of sand, covered with a thin stratum of black soil. It is not very hilly and in most places is covered with open woods of hardwood trees, especially oak. Now and then you see a single farm, and a little cultivated field around it. Here and there are little marshes or swamps, and sometimes a sluggish brook.

Trees. The woods of these parts consist of all sorts of trees, but chiefly of oak and hickory. They have certainly never been cut down, and have always grown without hindrance. It might therefore be expected that there are trees of an uncommonly great age to be

---

[1] Lars Lock, Latinized and Anglicized into Laurence Lockenius or Laurence Lock, came to America in the time of Governor Printz. He preached at Tinicum and Christina. For many years he was the only clergyman the Swedes had. He died in 1688.

found in them. But it happens otherwise, and there are very few
trees three hundred years old. Most of them are only two hundred
years and this convinced me that trees have the same quality as
animals, and die after they have arrived at a certain age. We find
great forests here, but when the trees in them have stood a hundred
and fifty or a hundred and eighty years, they are either rotting
within, or losing their crown. Sometimes their wood becomes quite
soft, or their roots are not longer able to draw in sufficient nourishment. Therefore when storms blow, the trees are broken off either
just at the root or further up. Several trees are likewise torn out by
their roots in the powerful winds. The storms thus cause great
devastations in these forests. Everywhere you see trees felled by
the winds, after they are too much weakened by one or the other
of the above-mentioned causes. Fire, too, breaks out often in the
woods and burns the trees half way through or more at the root,
so that they are easily broken off by the wind.

Windfalls. On travelling through these woods, I purposely tried
to find out, by the position of the trees which had fallen down, which
winds are the strongest hereabouts. But I could not conclude anything with certainty, for the trees fell on all sides and lay towards
all the points of the compass. I therefore judged, that any wind
which blows from that side where the roots of the tree are weakest
and shortest, and where it can make the least resistance, must root
it up and hurl it down. In this manner the old trees die continually,
and are succeeded by a younger generation. Those which are thrown
down lie on the ground and putrify, sooner or later, and by that
means increase the black soil [humus], into which the leaves are
likewise finally changed. The leaves drop abundantly in autumn,
are blown about by the winds for some time, but are finally heaped
up, and lie on both sides of the trees, which have fallen down. It
requires several years before a tree is entirely reduced to dust. When
the winds tear up a tree by the roots, a quantity of loose soil commonly comes up with and sticks to them for a time, but at last it
drops off and forms a little hillock, which is afterwards augmented
by the leaves. In this way many holes and mounds are formed.

Some trees are more inclined to decay than others. The tupelo tree
(*Nyssa*), the tulip tree (*Liriodendron*), and the sweet gum tree
(*Liquidambar*), I learned became rotten in a short time. The hickory
did not take much time, and the black oak fell to pieces sooner than

the white oak; but this was owing to circumstances. If the bark remained on the wood, it was for the greatest part rotten and entirely eaten by worms within the space of six, eight, or ten years, so that nothing was to be found but a reddish brown dust. But if the bark was taken off, the oaks would often lie twenty years before they were completely rotten. The suddenness of a tree's growth, the large pores, and the frequent changes of heat and humidity in summer, cause it to rot sooner. To this must be added, that all sorts of insects make holes into the trunks of the fallen trees, and by that means the moisture and the air get into the tree, which must of course forward putrefaction. Most of the trees here are deciduous. Many of them begin to rot while they are yet standing and blooming. This forms the hollow trees, in which many animals make their nests and have their places of refuge.

The breadth of the Delaware directly opposite Wilmington is reckoned two and a half English miles, yet to look at it it did not seem to be so great. Here the depth of the river, in the middle, is said to be from four to six fathoms.

December the 12th

The joiners say that among the trees of this country they use chiefly the black walnut, the wild cherry, and the curled maple. Of the black walnut (*Juglans nigra*) there is yet a sufficient quantity, but careless people are trying to destroy it, and some peasants even use it as fuel. The wood of the wild cherry tree (*Prunus Virginiana*) is very good, and looks exceedingly well; it has a yellow color, and the older the furniture is, which is made of it, the better it looks. But it is already scarce, for people cut it everywhere without replanting. The curled maple (*Acer rubrum*) is a species of the common red maple, and likewise very difficult to obtain. You may cut down many trees without finding the wood which you want. The wood of the sweet gum tree (*Liquidambar*) is also used in joiner's work, such as tables and other furniture. But it must not be brought near the fire, because it warps. The pines and the white cedars (*Cupressus thyoides*) are likewise made use of by the joiners for different sorts of work.

The millers said that the axle-trees of the mill wheels were made of white oak, and that they lasted for three or four years, but that pine did not keep so well. The cogs of the mill wheel and the pullies were made of the white walnut because it was said to be the hardest which could be found here. The wood of mulberry trees was of all others reckoned the most excellent for pegs and plugs in ships and boats.

At night I crossed the Delaware, from Wilmington to the New Jersey side.

DECEMBER THE 13TH

In the morning I returned to Raccoon.

Growths on Trees. On many trees either on the trunk or on the branches are greater or lesser growths or excrescences. Sometimes there is only a single one on a tree. In size there is considerable difference, for some of these knobs are as big, or bigger, than a man's head, others are only small. They project above the surface of the tree, like a tumor. Sometimes a tree is quite covered with them. They do not lie on one side only, but often form a circle round a branch, and even round the trunk itself. The trees which have these protuberances are not always large ones, but may be not above a fathom high. The knobs commonly consist of the same material as the tree itself, and look within like curled wood. Some of them are hollow. When a knob or hump on a little tree is cut open, we commonly find a number of little worms in it, which are sometimes also common in the larger knobs. This shows the origin of the excrescenses in general. When the tree is stung by insects, which lay their eggs under the bark, the sap runs out and gradually condenses into a protuberance. Only the trees with deciduous leaves have these knobs, and among them chiefly the oak. The black and Spanish oak especially have the greatest abundance of swellings. The ash trees (*Fraxinus excelsior*), and the red maple (*Acer rubrum*) also have them. Formerly the Swedes, and more especially the Finlanders, who have settled here, made dishes, bowls, etc. of the knobs which were on the ash trees. These vessels, I am told, were very pretty, and looked as if they were made of curled wood. The oak knobs cannot be employed in this manner, as they are usually worm-eaten and rotten within. At present the Swedes no longer make use of such bowls and dishes but of earthen ware or vessels made of other wood. Some swellings are of an uncommon size and give a tree a grotesque appearance. Such trees are very common in this country.[1]

Roads. The roads are good or bad according to the condition of the ground. In a sandy soil the roads are dry and good; but in a clayey one they are bad. The people here are very careless in repairing them. If a rivulet is not very big, they do not make a bridge over it, and travellers may get over as best they can. Therefore many people are in danger of being drowned in places where the water has risen by a heavy rain. When a tree falls across the road it is seldom cut off to keep the road clear but the people go round it. This they can easily do, since the ground is very even, and without stones. It has no underwood or shrubs, and the trees on it stand far apart. Hence the roads here have many turnings.

When one is travelling on a public highway one never finds a gate or a cultivated field to pass through, because the fields lie often on both sides of the road, with fences on both sides. So that one travels as if through a lane. If one wishes to visit a farm it may happen that one has to cross a cultivated field right near the farmhouse; but then no gates are used but only an opening in the fence.

The farms are most of them single, and you seldom meet even with two together, except in towns, or places which are intended for towns; therefore there are but few villages. Each farm has its cornfields, its woods, its pastures and meadows. This may perhaps have contributed something towards the extermination of wolves, since they encountered houses everywhere, and people fired at them. Two or three farmhouses sometimes have a pasture or a wood in common, but there are seldom more together, and most of them have their land entirely separate from that of others.

---

[1] In Siberia, and in the province of Wiatka, in the government of Cazan, in Russia, the inhabitants make use of the knobs, which are pretty frequently found in birches, to make bowls and other domestic utensils thereof. They are turned, made pretty thin, and covered with a kind of varnish, which gives them a pretty appearance; for the utensil looks yellow, and is marbled quite in a picturesque manner, with brown veins. The best kind of these vessels are made so thin that they are semidiaphanous, and when put into hot water they grow quite pliant, and may be formed by main force, quite flat, but when again left to themselves, and grown cold, they return to their original shape. This kind of wood is called, in Russia, *Kap,* and the vessels made of it, *kappowie Tchashki,* and are pretty high in price, when they are of the best kind, and well varnished.—F.

DECEMBER THE 18TH

Marriages. All persons who intend to be married, must either have their banns published three times from the pulpit in the same way as in Sweden, or get a license from the governor, which they give to the clergyman who keeps it permanently. If the clergyman does not observe this well but marries people on his own responsibility, he is liable to a large fine if a complaint is made against him. The banns of the poorer people only are published, and all those who are a little above them get a license. In that license he declares that he has examined the affair and found no obstacles to hinder the marriage, and therefore allows it. The license is signed by the governor; but, before he delivers it, the bridegroom must come to him in company with two creditable and well known men, who answer for him that there really is no lawful obstacle to his marriage. These men must sign a certificate, in which they make themselves answerable for, and engage to bear all the damages of, any complaints made by the relations of the persons who intend to be married, by their guardians, their masters, or by those to whom they may have been promised before. For the governor cannot possibly know all these circumstances. They further certify that nothing hinders the intended union, and that nothing is to be feared on that account. For a license they pay (at Philadelphia) five and twenty shillings in Pennsylvania money. The governor keeps twenty shillings, or one pound, and the remaining five shillings belong to his secretary. The license is directed only to protestant clergymen, because the Catholic priests have no right to marry anyone here. The Quakers have a special permission for their marriages. According to English law a man must be twenty-one and a girl eighteen before they can marry; but then they can do so whenever they wish without asking their parents' permission. Before this time, however, they have to get consent from their parents or guardians. But as it would be very troublesome, especially for those who live far from the governor's residence, to come up to town for every license, and to bring the men with them who are to answer for them, the clergymen in the country commonly obtain from the governors as many printed licenses and certificates as they think they will need in the immediate future, with blanks left for the names. When a candidate for marriage comes to the minister with his sponsors, he obtains a license as soon as the sponsors have signed the above-mentioned guarantee. The minister gets the twenty-five shillings intended for the governor and his secretary, plus his own fee. The money that is collected is delivered to the governor as soon as the officiating clergyman comes to town, together with the certificates, which are signed by two men, as above-mentioned. The minister then brings back again as many licenses as he thinks sufficient for the interim between this and the next visit to the governor. Hence we may conceive that the governors in the English North American colonies, besides their salaries, have very considerable revenues.

There is a great mixture of people of all sorts in these colonies, partly of such as have lately come over from Europe, and partly of such as have not yet any settled place of abode. Hence, it frequently happens that when a clergyman has married such a couple, the bridegroom says he has no money at present, but will pay the fee at the first opportunity. Then he goes off with his wife and the clergyman never gets his due. This proceeding has given occasion to a custom which is now common in Maryland. When the clergyman marries a very poor couple, he breaks off in the middle of the ceremony, and cries out, "Where is my fee?" The man must then give the money, and the clergyman proceeds; but if the bridegroom has no money, the clergyman defers the marriage till another time, when the man is better provided. People of fortune, from whom the clergyman is sure to get his due, need not fear this disagreeable question when they are married.

However, though the parson has a license to marry a couple, yet if he be not very careful, he may get into difficulties, for in many parts of the country there is a law made, which, notwithstanding the governor's license, greatly limits a clergyman in some cases. He is not allowed to marry a couple who are not yet of age, unless he be certain of the consent of their parents. He cannot marry such strangers as have bound themselves to serve a certain number of years, in order to pay off their passage from Europe, without the consent of their masters. If he acts without their consent, or in opposition to it, he must pay a penalty of fifty pounds, Pennsylvania currency, though he has the license and the certificate of the two men who are to answer for any objection. But parents or masters give themselves no concern about these men, and hold the clergyman responsible. He is at liberty, however, to prosecute those sponsors who signed the certificate, and to get his damages repaid. With the consent of the parents and masters he may marry people without danger to himself. No clergyman is allowed to marry a negro with one of European extraction, lest he must pay a penalty of one hundred pounds, according to the laws of Pennsylvania.

A Curious Marriage Custom. There is a very amusing custom here in regard to marrying. When a man dies, and leaves his widow in great poverty, or so that she cannot pay all the debts with what little she has left, and notwithstanding all that, there is a person who is willing to marry her, she must be married in no other habit than her nightgown. By that means she leaves to the creditors of her deceased husband her clothes and everything which they find in the house. But she is not obliged to pay them anything more, because she has left them all she had, even her clothes, keeping only a nightgown to cover her, which the laws of the country cannot refuse her. As soon as she is married, and no longer belongs to the deceased husband, she puts on the clothes which the second husband has given her. The Swedish clergymen here have often been obliged to marry a woman in a dress which is very poor and scanty. This appears from the registers kept in the churches, and from the accounts given by the clergymen themselves. I have likewise often seen accounts of such marriages in the English gazettes, which are printed in these colonies; and I particularly remember the following account: a woman went, with no other dress than her nightgown, out of the house of her deceased husband to that of her bridegroom, who met her half way with fine new clothes, and said, before all who were present, that he lent them to his bride, and put them on her with his own hands. It seems he said that he *lent* the clothes, for if he had said he *gave* them, the creditors of the first husband might have come and claimed them, pretending that she was looked upon as the relict of her first husband until she was married to the second.

DECEMBER THE 21ST

Old Wells. It seems very probable, from the following observations, that long before the arrival of the Swedes, there had been other Europeans in this province; and later we shall give more confirmations of this opinion. The same old Måns Keen, whom I have already mentioned, told me repeatedly, that on the arrival of the Swedes in the last century, and on their making a settlement, called Helsing-burg,[1] on the banks of the Delaware, somewhat below the place where Salem is now situated, they found, at the depth of twenty feet, some wells, inclosed by walls. This could not be a work of the native Americans or Indians, as bricks were entirely unknown to them when the Europeans first came here at the end of the fifteenth century; and they knew still less how to make use of them. The wells were, at that time, on the land; but in such a place, on the banks of the Delaware, that was sometimes under water and some-times dry. But since that time the ground has so washed away that the wells are entirely covered by the river, the water being seldom low enough to show the wells. When the Swedes afterwards dug new wells for themselves, at some distance from the former, they discovered some broken earthen vessels in the ground and some good whole bricks. They have often run into them in the ground when plowing.

From these relics, it seems, we may conclude that in earlier times either Europeans or other people of the then civilized parts of the world, were carried hither by storms, or other accidents, settled here, on the banks of the river, burnt bricks, and made a colony here, but that they afterwards were absorbed by the Indians, or were killed by them. They may gradually, by conversing with the Indians, have learnt their manners and way of thinking. The Swedes themselves were accused of being already half Indians when the English arrived in the year 1682. And we still see that the French, English, Germans, Dutch, and other Europeans, who have lived for several years in distant provinces, near and among the Indians, grow so like them in their behavior and thoughts that they can only be distinguished by the difference of their color. But history, together with the tradi-tion among the Indians, assures us, that the above-mentioned wells and bricks cannot have been made at the time of Columbus's expedi-tion, nor soon after, as the traditions of the Indians say that those wells were made long before that epoch. This account of the wells, which had been inclosed with bricks, and of such bricks as have been found in several places in the earth, I have afterwards heard repeated by many other old Swedes.

-----

[1] One of the Swedish forts on the Delaware, Elfsborg, was erected at "Elsingburg Ford Point," which recalls the Swedish "Helsingborg," hence the name and spelling.

DECEMBER THE 22ND

An old farmer foretold a decided change of the weather because the air was very warm this day at noon, though the morning had been very cold. This he likewise concluded from having observed the clouds gathering about the sun. The meteorological observations annexed to the end of this work will prove that his observation was correct.

DECEMBER THE 31ST

Toothache. The remedies for toothache are almost as numerous as days in the year. There is hardly an old woman but can tell you three or four scores of them, and she is perfectly certain that they are as infallible and speedy in giving relief as a month's fasting, with bread and water, is to a heavy stomach. Yet it happens often, nay too frequently, that this painful disease eludes all this formidable army of cures. However, I cannot forbear giving the following remedies, which have sometimes in this country been found effectual against the toothache.

When the pains come from the hollowness of the teeth, the following remedy is said to have had a good effect; a little cotton is put at the bottom of a tobacco pipe; the tobacco is put on top of it, and lighted; and you smoke till it is almost burnt up. By smoking, the oil of the tobacco gets into the cotton, which is then taken out and applied to the tooth as hot as it can be borne.

The chief remedy of the Iroquois for the toothache caused by hollow teeth I heard from Captain Lindsey's wife at Oswego. She assured me that she knew from her own experience that the remedy was effectual.—They take the seed capsules of the Virginian anemone, as soon as the seed is ripe, and rub them in pieces. It will then be rough, and look like cotton. This cotton-like substance is dipped into strong brandy, and then put into the hollow tooth, which commonly ceases to ache soon after. The brandy is biting or sharp, and the seeds of the anemone, as most seeds of the *Polyandria polygynia* class of plants, are bitter. They therefore, both together, help to assuage the pain; and this remedy is much like the former. Besides that we have many seeds which have the same qualities as the American anemone.

The following treatment was much in vogue against the tooth-ache which is attended with a swelling: they boil gruel, or corn-meal and milk; to this they add, while it is yet over the fire some lard or other suet, and stir it well, that everything may mix equally. The gruel is put on a piece of cloth and applied as hot as possible to the swelled cheek, where it is kept till it is cool again. I have found that this remedy has been very efficacious against a swelling; as it lessens the pain, abates the swelling, opens a gathering, if there be any, and procures a good discharge of pus.

Sometimes toothache is caused by hollow and decayed teeth; in such a case it was said to be best to have them pulled. But one should then practise caution because the pains like to move to the next tooth. Therefore the patient, before the tooth is pulled, should take something to cause a sweat. Against toothache caused by cold the sweating was one of the best cures.

I have seen the Iroquois boil the inner bark of the *Sambucus Canadensis*, or Canada elder, and put it on that part of the cheek in which the pain was most violent. This, I am told, often diminishes the pain.

Among the Iroquois, or Five Nations, on the Mohawk River, I saw a young Indian woman, who by frequent drinking of tea had gotten a violent toothache. To cure it she boiled the *Myrica asplenii folia*, and tied it, as hot as she could bear it, on the whole cheek. She said that remedy had often cured the toothache before.

JANUARY THE 2ND, 1749

Before the Europeans under the direction of Columbus came to the West Indies, the savages or Indians (who had lived there since antiquity) were entirely unacquainted with iron, which appears very strange to us, as North America, almost every part of it, contains a number of iron mines. They were therefore obliged to supply this want with sharp stones, shells, claws of birds and wild beasts, pieces of bones, and other things of that kind, whenever they intended to make hatchets, knives, and similar instruments. Hence it appears that they must have led a very wretched life. The old Swedes who lived here, and had had an intercourse with the Indians when they were young, and at a time when they were yet very numerous in these parts, could tell a great many things concerning their manner of living. At this time the people find accidentally, by plowing and digging in the ground, several of the instruments which the Indians employed, before the Swedes and other Europeans had provided them with iron tools. For it is observable that the Indians at present make use of no other tools than such as are made of iron and other metals, and which they always get from the Europeans. Of this I shall give more particulars at a later time. But having already had an opportunity of seeing and partly collecting a great many of the ancient Indian tools, I shall here describe some of them.

Indian tools and weapons. Their hatchets are commonly made of stone. Their shape is similar to that of the wedges with which we cleave our wood, about half a foot long, and narrow in proportion; they are made like a wedge, sharp at one end, but rather blunter than our wedges. As this hatchet must be fixed on a handle, there is a notch made all round the thick end. To fasten it, they split a stick at one end, and put the stone between it, so that the two halves of the stick come into the notches of the stone; then they tie the two split ends together with a rope or something like it, almost in the same way as smiths fasten the wedges with which they cut off iron to a split stick. Some of these stone-hatchets are not notched or furrowed at the upper end, and it seems they only hold those in their hands in order to hew or strike with them, and do not make handles for them. Most of the hatchets which I have seen consist of a hard rock-stone; but some are made of a fine, hard, black, apyrous stone. When the Indians intend to fell a thick strong tree, they cannot use their hatchets, but for want of proper instruments employ fire. They set fire to a great quantity of wood at the roots of the tree, and make it fall by that means. In order that the fire does not reach higher than they would have it, they fasten some rags to a pole, dip them into water, and keep continually washing the tree, a little above the fire. Whenever they intend to hollow out a thick tree for a canoe, they lay dry branches all along the felled trunk of the tree as far as it must be hollowed out. They then put fire to those dry branches, and as soon as they are burnt, they are replaced by others. While these branches are burning, the Indians are busy with wet rags, pouring water upon the tree to prevent the fire from spreading too far on the sides and at the ends. The tree being burnt hollow as far as they find it sufficient, or as far as it can be without damaging the canoe, they take the above described stone hatchets, or sharp pieces of flint and quartz, or sharp shells, and scrape off the burnt part of the wood, and smoothen the boats within. By this means they likewise give it what shape they please. Instead of cutting with a hatchet such a piece of wood as is necessary for making a canoe, they employ fire. A canoe is commonly between thirty and forty feet long. The chief use of their hatchets is, according to the unanimous accounts of all the Swedes, to make good fields for corn plantations; for if the ground where they intend to do this planting is covered with trees, they cut off the bark all round them with their hatchets, especially at the time when the sap is running. By that means the tree becomes dry, and cannot take any more nourishment, and the leaves can no longer obstruct the rays of the sun from passing. The smaller trees are then pulled out by main force, and the ground is turned up a little with crooked or sharp branches.

Instead of knives they were satisfied with little sharp pieces of flint or quartz, or else some other hard kind of a stone, or with a sharp shell, or with a piece of a bone which they had sharpened.

At the end of their arrows they fastened narrow angulated pieces of stone; they made use of these, since they had no iron or wood of sufficient hardness for the purpose. The points were commonly flint or quartz, but were sometimes made from another kind of stone. Some employed the bones of animals or the claws of birds and beasts. Some of these ancient harpoons were very blunt, and it looked as though the Indians could kill birds and small quadrupeds with them; but whether they could enter deep into the body of a great beast or of a man by the velocity which they got from the bow, I cannot ascertain. Yet some have been found very sharp and well made.

The Indians had stone pestles, about a foot long, and as thick as a man's arm. They consisted chiefly of a black sort of stone, and were formerly employed by the Indians for pounding corn, which has since olden times been their chief and almost only grain. They had neither windmills, water-mills, nor hand-mills, to grind it, and did not so much as know a mill before the Europeans came into the country. I have spoken with old Frenchmen, in Canada, who told me, that the Indians had been astonished beyond expression, when the French set up the first windmill. They came in numbers, even from the most distant parts, to view this marvel and could sit for days staring at it. They were long of the opinion that it was not driven by the wind, but by the spirits who lived within it. They were almost equally astonished when the first water-mill was built. They formerly pounded all their corn or grain in hollow trees, with the above-mentioned stone pestles. Many Indians had only wooden ones. The blackish stone, of which the hatchets and pestle are sometimes made, is very good for grindstones, and therefore both the English and the Swedes employed the hatchets and pestles chiefly for those implements whenever they could get them.

The old boilers or kettles of the Indians, are either made of clay or of different kinds of potstone (*Lapis ollaris*). The former consist of a dark clay, mixed with grains of white sand or quartz, and burnt in the fire. Many of these kettles have two holes in the upper margin, one on each side, through which the Indians put a stick, and hold the kettle over the fire, as long as is needed. Most of the kettles have no feet. It is remarkable that no pots of this kind have been found glazed, either on the outside or the inside. A few of the oldest Swedes could yet remember seeing the Indians boil their meat in these pots. They are very thin, and of different sizes; they are made sometimes of a greenish, and sometimes of a grey potstone, and some are made of another species of fire-proof stone; the bottom and the margin are frequently above an inch thick. The Indians, notwithstanding their being unacquainted with iron, steel, and other metals, have learned to hollow out these pots or kettles of potstone very ingeniously.

The old tobacco pipes of the Indians are likewise made of clay, potstone, or serpentine. The first sort is shaped like our tobacco pipes, though much coarser and not so well made. The tube is thick and short, hardly an inch long, but sometimes as long as a finger; its color comes nearest to that of our tobacco pipes which have been long used. Their pipes of potstone are made of the same material as their kettles. Some of them are pretty well made, though they have neither iron nor steel. But besides these kinds of tobacco pipes, we find another kind which is made with great ingenuity of a very fine, red potstone or a variety of serpentine marble. Such pipes are very scarce and are seldom used by any other than the Indian sachems or elders. The fine red stone of which these pipes are made is, also, very rare, and is found only in the country of those Indians who are called Aiouez,[1] and who, according to Father Charlevoix, live on

---

[1] Probably the Ajoues tribe, south of the Missouri.

the other side of the Mississippi.[1] The Indians themselves commonly value a pipe of this kind as much as a piece of silver of the same size, and sometimes they hold it still more valuable. Of the same kind of stone, commonly, is their pipe of peace, which the French call *calumet de paix*, and which they use in their treaties of peace and alliances. Most authors who have written of these nations mention this article, and I intend to speak of it further when an opportunity offers.

The Indians employ hooks made of bone, or bird's claws, instead of fishing hooks. Some of the oldest Swedes here told me, that when they were young a great number of Indians had been in this part of the country and had caught fish in the river Delaware with these hooks.

They made fire by rubbing one end of a hard piece of wood continually against another soft and dry one, till the wood began to smoke, and afterwards to burn.

The early stone implements of the Indians are still found now and then in the ground when digging or plowing, but no longer in as great numbers as in the days of the old settlers, partly because they have been forced down too deep in the cultivation of the fields, partly because time has hidden them deeper, and in part because some of them have disintegrated with age.

Such were the tools of the ancient Indians, and the use which they made of them before the Europeans invaded this country and before they (the Indians) were acquainted with the advantages of iron. North America abounds in iron mines, and the Indians lived all about the country before the arrival of the Europeans, so that several places can be shown in this country where at present there are iron mines, and where, not a hundred years ago, stood great towns or villages of the Indians. It is therefore very remarkable that the Indians did not know how to make use of a metal or ore which was always under their eyes, and on which they could not avoid treading every day. They even lived upon the very spots where iron ores were afterwards found, and yet they often went many miles in order to get a wretched hatchet, knife, or the like, as above described. They were forced to employ several days in order to sharpen their tools, by rubbing them against a rock, or other stones, though the advantage was far from being equal to the labor. They could never cut

---

[1] See his *Journal historique d'un voyage de l'Amérique*, Tome V.

down a thick tree with their hatchets, and only with difficulty could they fell a small one. They could not hollow out a tree with their hatchets, or do a hundredth part of the work which we can perform with ease by the help of our iron tools. Thus we see how disadvantageous the ignorance and inconsiderate contempt of useful arts is. Happy is the country which knows their full value!

JANUARY THE 5TH

Christmas Day was celebrated today by the Swedes and English, for they still use the old calendar. The members of the English church hardly made any bigger preparations for Christmas than for any Sunday, and when it came on a week-day it was not celebrated any more than a Swedish Apostle's Day. The Quakers paid still less attention to it, for since they observe no other holiday except Sunday they work on Christmas Day just as on any other day, unless it comes on Sunday. Formerly the Swedes here had candles in the church on Christmas Day and celebrated it as we do in Old Sweden, but now no candles are used and its observance has been much curtailed.

JANUARY THE 6TH

Hares. There are a great number of hares in this country, but they differ from our Swedish ones in their size, which is very small, being but little bigger than that of a rabbit. They keep almost the same gray color both in summer and winter, which our Northern hares have in summer only. The tip of their ears is always gray, and not black; the tail is likewise gray on the upper side, at all seasons. They breed several times a year: in spring they lodge their young ones in hollow trees, and in summer, in the months of June and July, they breed in the grass. When they are frightened they commonly take refuge in hollow trees, out of which they are taken by means of a crooked stick, or by cutting a hole into the tree, opposite to the place where they lie; or by smoke, which is occasioned by making a fire on the outside of the tree. On all these occasions the dogs must be at hand. These hares never bite, and can be touched without any danger. In daytime they usually lie in hollow trees, and hardly ever stir unless they be disturbed by men or dogs; but in the night they come out and seek their food. In bad weather, or when it snows, they lie close for a day or two and do not venture to leave their retreats. They do a great deal of mischief in the cabbage fields, but apple trees suffer infinitely more from them, for they peel off all the bark next to the ground. The people here agreed that the hares are fatter in a cold and severe winter than in a mild and wet one, for which they could give me several reasons from their own conjectures. Whenever they caught one alive they generally killed it by seizing his back legs with the left hand, holding it a short distance away and giving it a blow on the neck with the right hand, whereupon it died immediately. The skin is useless, because it is so delicate that when you attempt to separate it from the flesh, you need only pull at the fur, and the skin follows. These hares cannot be tamed. They are at all times, even in the midst of winter, plagued with a number of common fleas.[1]

JANUARY THE 16TH

Mice. The common mice were in great abundance in the towns and in the country; they do as much mischief as in the old countries. Oldmixon in his book, the *British Empire in America*,[2] writes that North America had neither rats nor mice before European ships brought them over. How far this is true I know not. It is true that in several wild places, where no man ever lived, I have seen and killed common mice, in crevices of stones or mountains; and is it probable that all such mice as are spread in this manner throughout the inland parts of the country derive their origin from those which were brought over from Europe?

Rats. Rats likewise may be ranked among those animals which do great damage in this country. They live both in the cities and in the country, and destroy provisions. Their size is the same as that of our common rats, but their color differs, for they are gray or blue-gray. I inquired of the Swedes whether these rats had been here prior to the arrival of the Europeans, or whether they came over in the ships? But I could not get a reliable answer. All agreed that a

---

[1] This account sufficiently proves that these hares are a species distinct from our European reddish gray kind, and also different from that species or variety only, which in the northern parts of Europe and Asia is white in winter, with black tipped ears, and has a gray coat in summer. Upon a closer examination naturalists will perhaps find more characteristics to distinguish them more accurately.—F.

[2] Vol. 1, p. 444.

number of these dangerous and mischievous animals were every year brought to America by ships from Europe and other countries. But Mr. Bartram maintained that before the Europeans settled here, rats had been in the country; for he saw a great number of them on the high mountains which are commonly called the Blue Mountains, where they live among stones and in the subterraneous grottoes which are found there. They always hide in the daytime, and you hardly ever see one out; but at night they come forth and make a terrible noise. When the cold is very violent they seem quite torpid, for during the continuance of the cold weather, one cannot hear the least noise, or shrieking, occasioned by them. It is to be recalled that neither the Swedes nor the English have any dampers in their houses, scarcely any ceilings, but only loose boards. The walls in the wooden houses are frequently not closed, even with moss; so that the rooms, though they have fires in them, are no warmer than barns. The rooms where the servants sleep never have any fire in them, though the winter is pretty severe sometimes. The rats have, therefore, little or no warmth to attract them in winter; but as soon as a milder season makes its appearance, they come out again. We observed several times this winter that the rats were very active, and made an unusual noise all night just before a severe cold. It seems they had some sensation of cold weather being at hand; and that they therefore ate sufficiently, or stored up provisions. In mild weather they used to carry away apples and other food. Therefore, we could always conclude with certainty, when the rats made an uncommon noise at night, or were extremely greedy, that a severe cold would ensue. I have already noted [1] that the gray squirrels in this country have the same behavior. When these and the common mice eat corn, they do not consume the whole grains, but only the loose, sweet and soft kernel, and leave the rest.

January the 21st

Winter. The cold now equalled that of Sweden, though this country is so much more southerly. The Celsian (Centigrade) or Swedish thermometer was this morning twenty-two degrees ($7\frac{3}{4}°$ below zero F.) below the freezing point. As the rooms here are without any shutters, the cracks in the walls without moss, and since some-

---

[1] Pages 164 ff.

times there is no fireplace or chimney in the room, the winters here must be very disagreeable to one who is used to our warm Swedish, winter rooms. But the greatest comfort here is that the cold is of a very short duration. Some days of this month the room in which I lodged was such that I could not write two lines before the ink froze on my pen. When I did not write I could not leave the ink-stand on the table, but was forced to put it upon the hearth or into my pocket. Yet, notwithstanding it was so cold, as appears from the meteorological observations at the end of this work,[1] and though it snowed sometimes for several days and nights, and the snow lay near six inches high upon the ground, yet all the cattle were obliged to stay day and night in the fields, during the whole winter. For neither the English nor the Swedes had any stables; but the Germans and Dutch had preserved the custom of their country, and generally kept their cattle in barns during the winter. Almost all the old Swedes say, that on their first arrival in this country they made barns for their cattle, as is usual in Sweden; but as the English came and settled among them, and left their cattle in the fields all winter, as is customary in England, they left off their former custom, and adopted the English one. They owned, however, that the cattle suffered greatly in winter, when it was very cold, especially when it froze after a rain; and that some cattle were killed by it in several places in the long winter of the year 1741. About noon the cattle went out into the woods where there were yet some leaves on the young oaks; but they did not eat the leaves, and only bit off the extremities of the branches and the tops of the youngest oaks. The horses went into the cornfields, and ate the dry leaves on the few stalks which remained. The sheep ran about the woods and in fields. The chickens perched on the trees of the gardens at night, for they had no henhouses. The hogs were likewise exposed to the roughness of the weather within a small enclosure.

The Snowbird. A small kind of bird, which the Swedes call "snow-bird" and the English "chick bird," came into the houses about this time. At other times it sought its food along the roads. It is seldom seen except when it snows. Catesby in his *Natural History of Carolina,* calls it *Passer nivalis;* and Dr. Linné, in his *Systema Naturæ,* calls it *Emberiza hyemalis.* [It is now classified as the *Junco hyemal.*]

---

[1] See p. 748. It was —22° C.

The river Delaware was now covered with ice opposite Philadelphia and even somewhat lower, and the people could walk over it; but nobody ventured to ride over on horseback.

<h3 style="text-align:center">JANUARY THE 22ND</h3>

Partridges. There are partridges in this country, but they are not the same kind as ours. The Swedes sometimes call them *rapphöns* (partridges), and sometimes *åkerhöns* (quails). Some of the English also call them partridges, others quails. Their shape is almost the same as that of the European partridges, and their characteristics are the same. I mean they run and hide themselves when pursued. But they are smaller, and entirely different in color. In this work I cannot insert, at large, the descriptions which I have made of birds, insects, quadrupeds, and plants; because it would swell my volumes too much. I only observe that the feet are naked and not hairy; the back is spotted with brown, black, and white; the breast is dark yellow; and the belly whitish, with black edges on the tips of the feathers. The size is nearly that of a hazel-hen, or *Tetrao bonasia.* Above each eye is a narrow streak of whitish yellow. These birds are numerous in New Sweden, i. e. in this part of the country. On going but a little way you meet with great coveys of them, except near the towns, where they are either extirpated or frightened away by frequent shooting. They are always in lesser or greater coveys, do not fly very much, but run in the fields, and keep under the bushes and near the fences where they seek their food. They are considered a very delicious food, and the people here prepare them in different ways. For that purpose they are caught and shot in great numbers. They are caught by putting up a sieve or a square open box, made of boards, in the places they frequent. The people strew some oats under the sieve, and lift it up on one side by a stick; and as soon as the partridges come under the sieve, in order to pick up the oats, it falls, and they are caught alive. Sometimes they get several partridges at once. When they run in the bushes you can come very near them, without starting them. When they sleep at night they lie together in a heap. They scratch in the bushes and upon the field like common chickens. In spring they make their nests either under a bush or in the cornfields, or on the hills under the open sky. They scratch some hay together, into which they lay about thirteen white eggs. They eat several sorts of grain and grass seeds. They have likewise been seen eating the berries of the sumach or *Rhus glabra*. Some people have taken them young and kept them in a cage till they were tame; then they let them go, and they followed the chickens and never left the farm.

Fences. The fences built in Pennsylvania and New Jersey, but especially in New York, are those which on account of their serpentine form resembling worms are called "worm-fences" [1] in English. The rails which compose this fence are taken from different trees, but they are not all of equal duration. The red cedar is reckoned the most durable of any, for it lasts over thirty years; but it is very scarce, and grows only in a single place hereabouts, so that no fences can be made of it. It is true, the fences about Philadelphia (which however are wholly different from the worm fences) are all made of red cedar; but it has been brought by water from Egg Harbor, where it grows in abundance. In the Philadelphia fences the posts stuck into the ground are made of the white cedar, or *Cupressus thyoides*, and the rails or poles which are laid between them of the red cedar or *Juniperus Virginiana*. Next to the cedar wood, oak and chestnut are reckoned best. Chestnut is commonly preferred, but it is not so plentiful as to be made into fences; in its stead they make use of several sorts of oak. In order to make rails the people do not cut down the young trees, as is common with us, but they fell here and there large trees, cut them in several places, leaving the pieces as long as it is necessary, and split them into rails of the desired thickness; a single tree affords a multitude of rails. Several old men in this country told me that the Swedes on their arrival here made such fences as are usual in Sweden, but they were forced to leave off in a few years time, because they could not get posts enough. They had found by experience that a post put into the ground would not last above four or six years before the part underground was entirely rotten. But the chief thing was that they could not get any switches to tie them together; [2] they made some of hickory, which

---

[1] The well-known zigzag fence of rails crossing at the ends. It is also called "snake fence" or "Virginia rail fence." Kalm's detailed description of it is here omitted as unnecessary.

[2] In this type of Swedish fence, the rails rest one end on the ground and the other on horizontally-attached switches, which are suspended between and, like thongs, hold together two upright, slender posts driven into the ground, the rails being sandwiched in between them. The rails, though parallel to each other and as close together

is one of the toughest trees in this country, and of white oak; but in the space of a year or two the switches were rotten, the fence fell to pieces of itself, and they were forced to give up making such fences. Several of the newcomers again attempted, but with the same bad success, to make fences with posts and switches. The Swedish way of fencing therefore will not succeed here. Thus the worm fence is one of the most useful sorts of inclosures, especially as they cannot get any posts made of the wood of this country to last above six or eight years in the ground without rotting. The rails in this country are very heavy, and the posts cannot sustain them very well, especially when it blows a storm; but the worm fences are easily put up again, when they are forced down. Experience has shown that a fence made of chestnut or white oak seldom hold out over ten or twelve years, before the rails and posts are so thoroughly rotten that they are good only for fuel. When the poles or rails are made of other wood the fences hardly last six or eight years. Considering how much more wood the worm-fences require (since they zigzag) than other fences which go in straight lines, and that they are so soon useless, one may imagine how the forests will be consumed, and what sort of an appearance the country will have forty or fifty years hence, in case no change is made. Also, an incredible amount of wood is really squandered in this country for fuel; day and night all winter, or for nearly half of the year, in all rooms, a fire is kept going.

FEBRUARY THE 8TH

The *muskrats*, so called by the English in this country on account of their scent, are pretty common in North America; they always live near the water, especially on the banks of lakes, rivers and brooks. On travelling to places where they are, you see the holes which they have dug in the ground just at the water's edge, or a little above its surface. In these holes they have their nests, and there they stay whenever they are not in the water in pursuit of food. The Swedes call them *desmansråttor*,[1] and the French *rats musqués*.

as the thickness of the switches will allow, are laid obliquely with reference to the surface of the ground. The binding switches or thongs are usually made of spruce branches.

[1] *Desman* signifies musk in Swedish and in some provincial dialects of the German language; consequently *Desman-rat* is nothing but muskrat, and from hence Mr. de Buffon has formed his desman or Russian Muskrat.—F.

Linné calls this animal the *Castor zibethicus*. Their food is chiefly the mussels which lie at the bottom of lakes and rivers. You see a number of such shells near the entrance of their holes. I am told they likewise eat several kinds of roots and plants. They differ from the European muskrat, or Linné's *Castor moschatus*. The teeth are the same in both; the tail of the American variety is compressed on the sides so that one sharp edge goes upwards and the other downwards. The hind feet are not palmated, or joined by a moveable skin, but are peculiar for having on both sides of the feet, long, white, close, pectinated, offstanding hair, besides the short hair with which the feet are covered. Such hairs are on both sides of the toes, and do the same service in swimming as a web. Their size is that of a small cat, or to be more accurate, the length of the body is about ten inches, and the tail of the same length. The color of the head, neck, back, sides, and of the outside of the thighs, is blackish brown; the hairs are soft and shining. Under the neck, on the breasts, and on the inside of the thighs, they are gray and the belly reddish-brown. They make their nests in the dikes that are erected along the banks of rivers to keep the water from the adjoining meadows; but they often do a great deal of damage by spoiling the dikes with digging and opening passages for the water to come into the meadows; whereas beavers stop up all the holes in a dike or bank. They make their nests of twigs and such things externally, and carry soft stuff into them for their young ones to lie upon. The Swedes asserted that they could never observe a diminution in their number, but believed that they were as numerous at present as formerly. As they damage the banks so considerably, the people are endeavoring to destroy them when they can find their nests. The skin is sold and this is an inducement to catch the animal. The skin of a muskrat formerly cost but threepence, but at present they bring from sixpence to ninepence in the market. The skins are chiefly used by hatters, who make hats of the hair, which are said to be nearly as good as beaver hats. The muskrats are commonly caught in traps, with apples as bait. In the country of the Iroquois I saw the Indians following the holes of the muskrats by digging till they came to their nests, where they killed them all. Nobody here eats their flesh; I do not know whether the Indians eat it, for they are commonly not fastidious in the choice of meat. The muskbag is put between the clothes in order to preserve them against moths. It is very difficult to extirpate these rats when they are once settled in a bank. A Swede, however, told me that he had freed his dam or piece of dike along the river from them in the following manner: he sought and found their holes, stopped them all up with earth, excepting one, on that side from whence the wind came. He put a quantity of sulphur into the open entrance, set fire to it, and then closed the hole, leaving but a small one for the wind to pass through. The smoke of the sulphur then entered their most remote nests, and stifled all the animals. As soon as the sulphur was burnt, he was obliged to dig up part of the ground in the bank where they had their nests; and he found them lying dead in heaps. He sold the skins, and they paid him for his trouble, not to mention the advantage he got by clearing his dam of the muskrats.

Beavers were formerly abundant in New Sweden, as all the old Swedes here told me. At that time they saw one dam after another raised in the rivers and brooks by beavers. But after the Europeans had come over in such great numbers and cultivated the country more, many beavers had been killed. Many, too, had just died out and some had moved further into the country, where the people were not so numerous. Therefore there is but a single place in Pennsylvania where beavers are to be seen; their chief food is the bark of the beaver tree, of *Magnolia glauca*, which they prefer to any other. So the Swedes used to put branches of this tree into traps near the beaver dams which they laid for the beavers, while they were plentiful; and they could be almost certain of good success. Some persons in Philadelphia have tamed beavers, so that they go fishing with them, and they always come back to their masters. Major Roderfort, in New York, related that he had a tame beaver over half a year in his house, where he went about loose, like a dog. The Major gave him bread, and sometimes fish, which he liked very much. He got as much water as he wanted in a bowl. All the rags and soft things he found he dragged into a corner where he used to sleep, and made a bed of them. The cat in the house, having kittens, took possession of his bed, and he did not hinder her. When the cat went out, the beaver often took a kitten between his forepaws and held it to his breast to warm it, and doted upon it; as soon as the cat returned he gave her the kitten again. Sometimes he grumbled but never did any harm or attempted to bite.

Mink. The English and the Swedes gave the name of mink

(*Putorius vison*) to an animal of this country, which also lives either in the water or very near it. I have never had an opportunity to see any more than the skin of this animal. But the shape of this skin, and the unanimous accounts I have heard of it, make me conclude with certainty that it belonged to the genus of weasels or *mustelæ*. The greatest skin I ever saw, was two feet, a smaller one was about one foot, and about three and one-third inches broad, before it was cut. The color was dark brown and sometimes almost black. The tail was bushy, like that of a marten. The hair was very close; and the ears short, with short hair. The length of the feet belonging to the smaller skin was about two inches long. I am told this animal is so similar to the American polecat, or *Viverra putorius,* that the two are hardly distinguishable.[1] I have had the following accounts given me of its way of living; it seldom appears in daytime, but at night it comes out of the hollow trees, on the banks of the rivers. Sometimes it lives in the rocks and bridges, at Philadelphia, where it is a cruel enemy to the rats. Sometimes it gets into the courtyards at night and creeps into the chicken-house, through a small hole, where it kills all the poultry and sucks their blood, but seldom eats one. If it comes upon geese, fowls, ducks, or other birds on the road, it kills and devours them. It lives upon fish and birds. When a brook is near the houses it is not easy to keep ducks and geese, for the mink, which lives near rivers, kills the young ones. It first kills as many as it can seize, and then it carries them off and feasts upon them. In banks and dams near the water it likewise does mischief by digging. To catch it the people put up traps, into which they put heads of birds, fishes or other meat. The skin is sold in the towns, and at Philadelphia they give twenty pence and even two shillings apiece for them, according to their size. Some of the ladies get muffs made of these skins; but for the greatest part they are sent over to England, from which they are distributed to other countries. The old Swedes told me that the Indians formerly used to eat all kinds of flesh, except that of the mink.

Raccoon. I have already mentioned something of the raccoon;[2] I shall here add more of the nature of this animal and its mode of living in its habitat, in a place which is properly its native country.[3]

---

[1] This seems incredible when we remember that the "American polecat" is the skunk.

[2] See pages 53 and 111.

[3] The village of Raccoon. The name is from the Algonquin word *arakun.*

The English call it everywhere by the name of raccoon, which name they have undoubtedly taken from one of the Indian nations; the Dutch call it *hespan*, the Swedes, *espan*, and the Iroquois, *attigbro*. It commonly lodges in hollow trees, lies close in the daytime, never going out except on a dark, cloudy day; but at night it rambles and seeks its food. I have been told by several people that in bad weather, especially when it snows and blows a storm, the raccoon lies in its hole for a week without coming out once; during that time it lives by sucking and licking its paws. Its food consists of the several sorts of fruit, and corn, while the ears are soft. In gardens it often does a great deal of damage to the apples, chestnuts, plums, and wild grapes, which are its favorite food; to the poultry it is very cruel. When it finds the hens on their eggs, it first kills them, and then eats the eggs. It is caught by dogs, which trace it back to its nest in hollow trees, or by snares and traps, in which a chicken, some other bird, or a fish is put for bait. It generally brings forth its two or three young in May when it prepares its nest. Some people eat its flesh. It leaps with all its feet at once; on account of this and of several other qualities many people here reckoned that it belonged to the genus of bears. The skin is sold for eighteen pence at Philadelphia. I was told that the raccoons were not nearly so numerous as they were formerly; yet in the more inland parts they were abundant. I have mentioned before the use which the hatters make of their furs, that they are easily tamed, and that they like sweetmeats, etc. Of all the North American wild quadrupeds none can be tamed as easily as this one.

FEBRUARY THE 10TH

In the morning I started for Philadelphia, where I arrived in the evening. On my arrival at the ferry on the Delaware River, I found the river quite covered with drifting ice, which at first prevented our crossing. After waiting about an hour, and making an opening near the ferry, I, together with many other passengers, got over before any more ice came down. As it began to freeze very hard soon after the twelfth of January (or New Year, according to the old style) the river was covered with ice, which by the intenseness of the cold grew so strong that the people crossed the river with horses at Philadelphia. The ice continued till the eighth of February, when it began to get loose, and the violent hurricane, which came that night, broke it, and it was driven down so fast that on the twelfth of February not a single piece was left, excepting a cake or two near the shore.

Crows flew in great numbers together to-day and settled on the tops of trees. During the whole winter we hardly observed one, though they are said to winter here. All spring they used to sit on the tops of trees in the morning; yet not crowded together but scattered in several trees. They belong to the injurious birds in this part of the world, for they live chiefly upon all kinds of grain. After the corn is planted or sown, they scratch the kernels out of the ground and eat them. When the corn begins to ripen, they peck a hole into the involucrum which surrounds the ear, by which means the corn is spoiled, as the rain passes through the hole which they have made and occasions the putrefaction of the corn. Besides eating grain they also steal chickens. They are very fond of dead carcasses. Some years ago the government of Pennsylvania gave threepence, and that of New Jersey fourpence, premium for every head of a crow; but this law has now been repealed, as the expenses became too great. I saw boys in several places having crows that were very tame. They had clipped their wings. The latter hopped about the fields near the farmhouses where they belonged, but always returned again, without endeavoring to escape on any occasion. These American crows are only a variety of the Royston crow, or Linné's *Corvus cornix*.

FEBRUARY THE 12TH

In the afternoon I returned to Raccoon from Philadelphia.

Trees. On my journey to Raccoon I attentively observed the trees which still had dry leaves left. They were pale, but had not all dropped from the trees enumerated:

The beech tree (*Fagus sylvatica*), whether great or small, always kept a considerable part of its leaves during the whole winter, even till spring. The large trees kept the lower leaves.

The white oak (*Quercus alba*). Most of the young trees which were less than a quarter of a yard in diameter had the greatest part of their leaves left, but the old trees had lost most of theirs, except in some places where they had new shoots. The color of the dry leaves was much paler in the white oak than in the black one.

Quercus rubra. Most of the young trees still held their dried leaves. Their color was reddish brown, and darker than that of the white oak.

The Spanish oak is a mere variety of the black oak. The young trees of this kind also still retained their leaves.

A scarce species of oak which is known by its leaves having a triangular apex or top, whose angles terminate in a short bristle, and whose leaves are smooth below but woolly above.[1] The young oaks of this species still had their leaves.—In short, whenever I came into any wood where the above kinds of oaks prevailed, of only twenty years or less in age, I always found the leaves on them.

It seems that an omniscient Providence has, besides other provisions, also aimed to protect several sorts of birds—it being very cold and stormy about this time, by preserving the dry leaves on these trees. I have this winter at several times seen birds hiding in the old leaves during a severe cold or storm.

February the 13th

Insects. As I began to dig a hole to-day, I found several insects which had crept deep into the ground in order to pass the winter. As soon as they came to the air they moved their limbs a little, but had not strength sufficient for creeping, except the black ants, which moved a little, though slowly. The insects found thus were the following:

Formica nigra, or the black ants, were pretty numerous, and somewhat lively. These lay about ten inches below the surface.

Carabus latus. Some of these [ground beetles] lay at the same depth with the ants. This is a very common insect in all North America.

Scarabæus castaneus; chestnut-colored, with a hairy thorax; the elytræ shorter than the abdomen, with several longitudinal lines, beset with hair. It is something similar to the cockchafer, but differs in many respects. I found it very abundant in the ground.

Gryllus campestris, or the field cricket: they lay ten inches deep; they were quite torpid, but as soon as they came into a warm place they revived and were quite lively. In summer I have found these crickets in great numbers in all parts of North America where I have

---

[1] This seems to be nothing but a variety of the *Quercus rubra* L.—F.

been. They leaped about on the fields, and made a noise like that of our common house crickets, so that it would be difficult to distinguish them by their chirping. They sometimes made so great a noise that it caused pain in the ears and two people talking could hardly hear each other. In places where the rattlesnakes lived, the field crickets were very annoying, and in a manner dangerous, for their violent chirping drowned out the warning which that horrid snake gives with its rattle, and thus deprived one of the means of avoiding it. I have already mentioned that they wintered sometimes in chimneys.[1] Here they lay all winter in the ground, but at the beginning of March, as the air grew warm, they came out of their holes and began their music, though at first it was but very faint and rarely heard. When we were forced on our travels to sleep in uninhabited places the crickets got into the folds of our clothes, so that we were obliged to stop awhile every morning and thoroughly shake our clothes before we could get rid of them.

The red ants (*Formica rufa*), which in Sweden make the big ant hills, I likewise found on this and on the following days. They were not in the ground, for when my servant Jungström cut down old dry trees, he met with a number of them in the cracks of the tree. These cracks were at the height of many yards and the ants had crept so high in order to find their winter habitation. As soon as they came into a warm place, they began to stir about very briskly.

FEBRUARY THE 14TH

Birds. The Swedes and the English gave the name of "blue bird" to a very pretty little bird, which was of a fine blue color. Linné calls it *Motacilla scalis.* Catesby has drawn it in his *Natural History of Carolina,*[2] and described it by the name of *Rubecula Americana cœrulea;* and Edwards [3] has represented it in his *Natural History of Birds.*[4] In my own journal I called it *Motacilla cœrulea nitida, pectore rufo, ventre albo.* In Catesby's plate, I must point out, the color of the breast should be a cloudy red or ferruginous; the tibiæ and feet

---

[1] See p. 215.

[2] Vol. I, pl. 47.—F.

[3] George Edwards (1693-1773), English naturalist. His *History of Birds* appeared in 1745-1751.

[4] Plate and page 26.—F.

black as jet; the bill too should be real black; the blue color in general ought to be much deeper, more lively and shining. No bird in Sweden has so shining and deep a blue color as this one. The jay has perhaps a plumage like it. The food of the blue bird is not merely insects, he likewise feeds upon plants; therefore in winter, when no insects are to be found he comes to the farmhouses in order to subsist on the seeds of hay, and other small grains.

The *red bird* is another species of a small bird. Catesby has likewise drawn it.[1] Dr. Linné calls it the *Loxia cardinalis*. It belongs to that class of birds which are enemies of bees, lying in wait for them and eating them. I fed a male for five months in a cage. It ate both corn and buckwheat, for I gave it nothing else. By its song it attracted others of its species to the courtyard, and after we had put some corn on the ground under the window where I had it, the others came there every day to get their food; it was then easy to catch them by means of traps. Some of them, especially old ones, both males and females, would die with grief on being put into cages. Those on the other hand which had grown tame began to sing exceedingly sweet. Their note very nearly resembles that of our European nightingale, and on account of their agreeable song, they are sent abundantly to London, in cages. They have such strength in their bill that when you hold your hand to them they pinch it so hard as to cause the blood to issue forth. On spring mornings they sit warbling on the tops of the highest trees in the woods. But in cages they can sit still for an hour; the next moment they hop up and down, singing; and so they go on alternately all day.

FEBRUARY THE 17TH

Cranes (*Ardea Canadensis*) were sometimes seen flying in the daytime, to the northward. They usually stop here early in spring for a time, but they do not make their nest here since they proceed further north. Certain old Swedes told me that in their younger years, when the country was but little cultivated, an incredible number of cranes were here every spring, but that at present they were not so numerous. Several settlers here eat them when they can shoot them. They are said to do no harm to grain or anything else.

---

[1] See Catesby's *Natural History*, vol. I, pl. 38. *Coccothraustes rubra*.

This morning I went down to Penn's Neck, and returned in the evening.

Snow still lay in several parts of the woods, especially where the trees stood very thick, and the sun could not penetrate. However, it was not more than four inches deep. All along the roads was ice, especially in the woods, so that it was very difficult to ride horses which were not well shod. The people who have settled here know little of sleighs, but ride to church on horseback all winter, though the snow is sometimes near a foot deep. It seldom stays more than a week before it melts, when some fresh snow may fall.

Blackbirds. A species of bird, called by the Swedes, "corn thieves," [1] do the greatest mischief in this country. They have given them that name because they eat corn, both publicly and privately, just after it is planted and when it is ripe. The English call them blackbirds. There are two species of them, both described and drawn by Catesby; one is called *Monedula purpurea* and the other *Sturnus niger*. Though they are very different, yet there is so great a friendship between them that they frequently accompany each other in mixed flocks. However, in Pennsylvania one species is more numerous than the other and members of it often fly by themselves without any of the second, a red-winged variety. The first sort, or the purple daws, bear in many points so great a likeness to the daw, the starling and the thrush that it is difficult to determine to which genus they belong; but they seem to come nearest to the starling, for the bill is exactly the same as that of the thrush, though the tongue, the flight, their manner of perching, their song and shape, make them starlings entirely. At a distance they look almost black, but close by they have a very blue or purple cast, but not so much as in Catesby's print.

As these blackbirds are so noted for their misbehavior, I will here give a short description of them. Their size is that of a starling. The bill is conic, almost subulated, straight, convex, naked at the base, black, with almost equal mandibles, the upper being only very little angulated, so as to form almost squares; they are placed obliquely at the base of the bill and have no hair. There is a little horny knob, or a small prominence on the upper side of them. The tongue is sharp

---

[1] See p. 153.

and bifid at the point. The iris of the eyes is pale; the forehead, the crown, the nucha, the upper part and the sides of the head under the eyes are dark blue; all the back and coverts of the wings are purple; the upper coverts of the tail are not of so conspicuous a purple color, but are, as it were, blackened with soot; the nine primary quill feathers are black; the other secondary ones are likewise black, but their outward margin is purple; the twelve tail feathers have a blackish purple color and their tips are round; those on the outside are the shortest, and the middle extremely long. When the tail is spread, it looks round towards the extremity. The throat is bluish green, and shining; the breast is likewise black or shining green, according as you turn it to the light; the belly is blackish, and the vent feathers are obscurely purple-colored; the parts of the breast and belly which are covered by the wings are purple-colored; the wings are black below, or rather sooty; and the thighs have blackish feathers. The legs (*tibiæ*) and the toes are of a shining black. It has four toes, as most birds have. The claws are black, and that on the back toe is longer than the rest. Dr. Linné calls this bird the *Gracula quiscuta*.

A few of these birds are said to winter in swamps, which are overgrown with thick woods; and they appear only in mild weather. But the greatest number go south at the approach of winter. To-day I saw them for the first time this year. They were already flying in great flocks. Their chief and most agreeable food is corn. They come in great swarms in spring, soon after the corn is planted. They scratch up the kernels of corn and eat them. As soon as the leaf comes up, they take hold of it with their bills, and pull it up, together with the corn or grain; and thus they give a great deal of trouble to the country people, even so early in spring. To lessen their greediness for corn, some people dip the grains of that plant in a decoction of the root of the *veratrum album*, or white hellebore, (of which I shall speak later) and plant them afterwards. When the corn thief eats a grain or two, which are so prepared, his head is made dizzy and he falls down: this frightens his companions, and they dare not venture to the place again. But they repay themselves amply near autumn when the corn is ripe; for at that time, they are continually feasting. They assemble by thousands in the corn fields and exact a heavy tax. They are very bold, for when they are disturbed they only go and settle in another part of the field. In that manner they always pass from one end of the field to the other, and do not leave it till they are satisfied. They fly in incredibly large flocks in autumn; and it can hardly be conceived whence such immense numbers of them can come. When they rise in the air they darken the sky, and make it look almost black. They are then in such great numbers and so close together, that it is surprising how they find room to move their wings. I have known a person to shoot a great number of them on one side of a corn field, which did not frighten the rest; for they only just took flight and dropped at about the distance of a musket-shot in another part of the field, and kept changing their place when their enemy approached. They tired the sportsman before he could drive them from the corn, though he killed a great many of them at every shot. They likewise eat the seeds of the aquatic grass (*Zizania aquatica*) late in autumn, after the corn has been harvested. I am told they also eat buckwheat and oats. Some people say that they even eat wheat, barley, and rye, when pressed by hunger; yet, from the best information I could obtain, they have not been found to do any damage to these species of grain. In spring they sit in flocks on the trees, near the farms; and their note is quite pleasing. As they are so destructive to corn, the odium of the inhabitants against them is carried so far that the laws of Pennsylvania and New Jersey have settled a premium of threepence a dozen for dead corn thieves. In New England the people are still greater enemies to them; for Dr. Franklin told me in the spring of the year 1750, that by means of the premiums which had been paid for killing them in New England they had been so thoroughly extirpated, that they were very rarely seen, and in a few places only. But in the summer of the year 1749 an immense quantity of worms appeared on the meadows, which devoured the grass, and did great damage, so the people repented of their enmity against the corn thieves for they thought they had observed that those birds lived chiefly on such worms before the corn was ripe, and consequently exterminated them, or at least prevented their increasing too much. They seem therefore to be entitled, as it were, to some reward for their trouble. But after these enemies and destroyers of the worms (the corn thieves) were killed off, the worms were of course more at liberty to multiply, and therefore they grew so numerous that they did more mischief now than the birds did before. In the summer of 1749 the worms left so little grass in New England that the inhabitants were forced to get hay from Pennsylvania, and even from Old England. The corn thieves, too, have enemies besides the human species. A species of a small hawk lives on them, and on other little birds. I saw some of these hawks pursuing the corn thieves and catching them in the air. Nobody eats the flesh of the purple corn thieves or daws (*Gracula quiscuta*), but that of the red-winged blackbirds or starlings (*Oriolus phœniceus*) are sometimes eaten. Some old people have told me that this part of America still contained as many starlings as it did formerly. The cause of this they derive from the corn which is now sown in much greater quantity than formerly; therefore the birds can get their food with more ease at present.

The American cranberry, Vaccinium hispidulum, is extremely abundant in North America and grows in such places as we commonly find our cranberries in Sweden. Those in America are probably bigger, but in qualities so like the Swedish that many people would take them to be just another variety. The English call them cranberries, the Swedes *tranbär*, and the French in Canada *atopa*, (atocas) which is a name they have borrowed from the Indians. They are brought to market every Wednesday and Saturday at Philadelphia, late in autumn. They are boiled and prepared in the same manner as we do our red *lingon*, or *Vaccinium vitis idœa*, and they are used during winter and part of summer in tarts and other kinds of pastry. But as they are very sour, they require a great deal of sugar. That is not very dear, however, in a country where the sugar plantations are near by. Quantities of these berries are sent over, preserved, to Europe and to the West Indies.

MARCH THE 2ND

Mytilus anatinus, a kind of mussel, was found abundantly in little furrows, which crossed the meadows here. The shells were frequently covered on the outside with a thin crust of particles of iron, when the water in the furrows came from an iron mine. Englishmen and Swedes who lived here seldom made any use of them; but the Indians broiled and ate them. A few Europeans ate them occasionally.

The snow still remained in some parts of the wood, where it was very shady, but the fields were quite free from it. Cows, horses, sheep and hogs went into the woods to seek their food, which was as yet very scanty.

March the 3rd

The Swedes call a species of little birds *snöfågel,* and the English call it snowbird. This is Dr. Linné's *Emberiza hyemalis.*[1] The reason why it is called snowbird is because it never appears in summer, but only in winter when the fields are covered with snow. In some winters they come in as great numbers as the corn thieves, fly about the houses and barns and into the gardens eating the corn and the grass seeds which they find scattered on the hills.

At eight o'clock at night we observed a meteor, commonly called a snow-fire.[2] I have described this meteor in the memoirs of the Royal Swedish Academy of Sciences.[3]

Wild pigeons (*Columba macroura* or *migratoria*) flew in the woods in numbers beyond conception, and I was assured that they were more plentiful than they had been for several years past. They came this week and continued here for about a fortnight, after which they all disappeared, whence they came. I shall speak of them more particularly in another place.

March the 7th

Several people told me that it was a certain sign of bad weather here when a thunderstorm rose in the south or southwest and spread to the east and afterwards to the north: but that on the contrary, when it did not spread at all, or when it moved both east and west, though it should rise in the south or southwest, it would then prognosticate fair weather. To-day it was heard in the southwest, but it did not spread at all. (See the meteorological observations, at the end of this work.)

Till now the frost had continued in the ground, so that if any one had a mind to dig a hole he was forced to cut through with a pickax. However, it had not penetrated above four inches. And to-day it had almost gone out. This made the soil so soft that on riding, even in the woods, the horse sank in very deep.

Damage by Frost. I often inquired among the old Englishmen and Swedes whether they had found that any trees were killed in very

---

[1] See above p. 236.
[2] Probably nothing but an Aurora borealis.
[3] See vol. for the year 1752.

AMERICAN MIGRATORY PIDGEON.

From English edition

PURPLE JACKDAW.

RED-WINGED STARE.

From English edition

severe winters, or had received much damage. I was answered that young hickory trees were often killed in very cold weather, and that the young oaks likewise suffered in the same manner. Nay sometimes black oaks, five inches in diameter, were killed by the frost in a severe winter, and sometimes, though very seldom, a mulberry tree was killed. Peach trees very frequently died in a cold winter, and often all the peach trees in a whole district were killed by a severe frost. It had been found repeatedly, with regard to these trees, that they could stand the frost much better on hills, than in valleys; so much, that when the trees in a valley were killed by frost those on a hill were not hurt at all. They assured me that they had never observed that the black walnut tree, the sassafras or other trees, had been hurt in winter. In regard to a frost in spring they had observed at different times, that a cold night or two happened often after the trees had pretty large leaves and that by this frost most of them were killed. But the leaves thus ruined had always been supplied by fresh ones. It is remarkable that when such cold nights occur in May and June the frost acts chiefly upon the more delicate trees, and in such a manner that all the leaves to the height of seven and even of ten feet from the ground are killed by the frost, and all the top remains unhurt. Several old Swedes and Englishmen assured me that they had made this observation, and the attentive engineer, Mr. Lewis Evans, demonstrated it to me from his notes. Such a cold night happened here in the year 1746, in the night between the fourteenth and fifteenth of June, new style, attended with the same effect as described in Mr. Evans's observations. The trees which were then in blossom had lost both their leaves and their flowers in these parts which were nearest the ground; sometime afterwards they got fresh leaves, but no new flowers. Further it is observable that the cold nights which happen in spring and summer never do any hurt to high grounds, damaging only the low and moist ones. They are likewise very perceptible in such places where limestone is found, and though all the other parts of the country be not visited by such cold nights in a summer, yet those where limestone lies have commonly one or two every summer. Frequently these places are situated on a high ground; but they suffer notwithstanding their location, while a little way off in a lower ground, where no limestone is to be found, the effects of the cold nights are not felt. Mr. Evans was the first who made this observation, and I have had occasion at different times to see the truth of it, on my travels, as I shall mention further on. The young hickory trees have their leaves killed sooner than other trees on such a cold night, and the young oaks next; this has been observed by other people, and I found it to be true in the years 1749 and 1750.

MARCH THE 11TH

Woodpeckers. Of the genus of woodpeckers we find here all those which Catesby in his first volume of the *Natural History of Carolina*, has drawn and described. I shall only enumerate them, and add one or two of their qualities; but their description at large I defer for another occasion.

Picus principalis, the king of the woodpeckers, is found here, though very seldom, and only at a certain season.

Picus pileatus, the crested woodpecker; this I have already mentioned.

Cuculus auratus, the gold-winged woodpecker. This species is plentiful here, and the Swedes call it *hittock*, and *piut*, both names of which have a relation to its note. It is almost continually on the ground, and is not observed to pick in the trees; it lives chiefly on insects, but sometimes becomes the prey of hawks; it is commonly very fat, and its flesh is very palatable. As it stays all year, and cannot easily get insects in winter, it must doubtless eat some kinds of seed in the fields. Its form, and some of its qualities, make it resemble a cuckoo.

Picus Carolinus, the Carolina woodpecker. It lives here also, and the color of its head is a deeper and more shining red than Catesby has represented it.[1]

Picus villosus, the spotted, hairy, middle-sized woodpecker is abundant here; it destroys the apple trees by pecking holes in them.

Picus erythrocephalus, the red-headed woodpecker. This bird was seen often in the country, and the Swedes called it merely *hackspett* or woodpecker. They give the same name to all the birds which I now enumerate, the gold-winged woodpecker excepted. This species is destructive to corn fields and orchards, for it pecks through the ears of corn and eats apples. In some years they are very numerous, especially where sweet apples grow, which they eat until nothing but the mere peels remain. Some years ago there was a premium of two-

---

[1] Vol. I of *op. cit.*, p. 19.

pence per head, paid from the public funds, in order to extirpate this pernicious bird, but this law has been repealed. They are likewise very fond of acorns. At the approach of winter they travel to the southward. But when they stay in flocks in the woods, at the beginning of winter, the people look upon it as a sign of a mild winter.

Picus varius, the lesser, spotted, yellow-bellied woodpecker. These birds are much more numerous than many people like; for this, as well as the preceding and succeeding species, are very hurtful to apple trees.

Picus pubescens, or the least spotted woodpecker. This species abounds here. Of all the woodpeckers it is the most dangerous to orchards, because it is the most daring. As soon as it has pecked a hole in a tree, it makes another close to the first, in a horizontal direction, proceeding till it has pecked a circle of holes round the tree. Therefore the apple trees in the orchards here have several rings round their trunks, which lie very close to each other, frequently only an inch apart. Sometimes these birds peck the holes so close that the tree dries up. This bird, as Catesby remarks, is so like the lesser spotted woodpecker in regard to its color and other qualities, that they would be taken for the same bird, were not the former (the *Picus pubescens*) a great deal smaller. They are alike in the bad habit of pecking holes into apple trees.

Frogs. Rana ocellata is a type of frogs here, which the Swedes call *sill-hoppetåssor*, i. e. herring-hoppers, and which now begin to peep in the evening and at night in swamps, pools, and ponds. The name which the Swedes give them is derived from their beginning to make their noise in spring at the same time when the people here go catching what are called herrings, which however differ greatly from the true European herrings. These frogs have a peculiar note, which is not like that of our European frogs, but rather corresponds with the chirping of some large birds, and can nearly be expressed by "piiit, piiit." With this noise they continue throughout a great part of spring, beginning their noise soon after sunset and finishing it just before sunrise. The sound is sharp, but yet so loud that it can be heard at a great distance. When they expect rain they cry much worse than usual, and begin in the middle of the day, or when it grows cloudy, and the rain comes usually six hours after.

When it snowed on the sixteenth of this month, and blew very violently all day, there was not the least sign of the frogs at night, and during the whole time that it was cold and while the snow lay on the fields, the frost had so silenced them that we could not hear one; but as soon as the mild weather returned, they began their noise again. They were very timorous, and it was difficult to catch them; for as soon as a person approached the place where they lived they became silent without any one of them appearing. It seems that they hide themselves entirely under water, except the tip of the snout, when they peep. For when I stepped to the pond where they were, I could not observe a single one hopping into the water. I could not see any of them before I had emptied a whole pool, where they lodged. Their color is a dirty green, variegated with spots of brown. When they are touched they make a noise and moan, and then sometimes assume a form, as if they had blown up the hind part of the back, so that it makes a high elevation; and then they do not stir, though touched. When they are put alive into spirits of wine, they die within half a minute.

MARCH THE 12TH

The Robin. The bird which the English and Swedes in this country call robin redbreast, is found here all the year round. It is a very different bird from that which in England bears the same name. It is Linné's *Turdus migratorius.* It sings very melodiously, is not very shy, but hops on the ground quite close to the houses. In Philadelphia it is kept in a cage for its singing.

Plants. The hazels (*Corylus avellana*) were now blossoming. They succeeded best in a rich mould, and the Swedes reckoned it a sign of a good soil where they found them growing.

MARCH THE 13TH

The alder (*Betula alnus*) was just blossoming.

Skunk Cabbage. The *Dracontium foetidum* [1] grew plentifully in the marshes and was beginning to flower. Among the stinking plants this is the most foetid; its nauseous scent was so strong that I could hardly examine the flower; and when I smelled it a little too long my

---

[1] This must be the skunk cabbage or *Spathyrma foetida.*

head ached. The Swedes call it *björnblad* (bear's-leaf) or *björn-rötter* (bear's root). The English call it polecat-root [polecat weed], because its effluvia are as nauseous and foetid as those of the polecat, which I have mentioned before. The flowers are purple-colored. When they are in full flower the leaves begin to come out of the ground; in summer the cattle do not touch it. Dr. Colden [1] told me that he had employed the root in all cases where the root of the arum is used, especially against the scurvy, etc. The Swedish name it got from the fact that the bears, when they leave their winter quarters in spring, are fond of it. It is a common plant in all North America.

The *Draba verna* was abundant here, and was now in bloom.

The *Veratrum album* (hellebore) was very common in the marshes, and in low places over all North America. The Swedes here call it *dock*, *dockor* (dolls), or *dockrötter*, that is, purple root [for dolls] because the children make dolls of its stalks and leaves. The English call it itch-weed or hellebore. It is a poisonous plant, and therefore the cattle never touch it. However, it sometimes happens that the cattle are deceived in the beginning of spring, when the pastures are bare, and eat of the fine broad green leaves of this plant, which come up very early; but such a meal frequently proves fatal to them. Sheep and geese have likewise often been killed by it. By means of its roots the corn is preserved from the greediness of voracious birds in the following manner: the roots are boiled in water, into which the corn is put as soon as the water is quite cool; the corn must lie all night in it, and is then planted as usual. Then when the starlings, crows, or other birds, pick up or pluck out the grains of corn, their heads grow delirious, and they fall, which so frightens the rest that they never venture on the field again. When those which have tasted the grains recover, they leave the field, and are no more tempted to visit it again. By thus preparing corn, one must be very careful that no other creatures touch it; for when ducks or fowls eat a grain or two of the corn which is thus steeped, they become very sick, and if they swallow a considerable quantity they die. When the root is thrown away raw, no animal eats it; but when it is put out boiled, its sweet taste tempts the beasts. Dogs have been seen to eat a little of it, and have become very sick; however, they have recovered after a vomit, for

---

[1] See page 216, note.

when animals cannot free themselves of it by this means, they often die. Some people boil the root for medicinal purposes, washing scorbutic parts with the water or decoction. This is said to cause some pain, and even a plentiful discharge of urine, but the patient is said to be cured thereby. When the children here are plagued with vermin, the women boil this root, put the comb into the decoction, and comb the head with it, and this kills the lice most effectually.

March the 17th

Indians. At the first arrival of the Swedes in this country, and long after that time, it was filled with Indians. But as the Europeans proceeded to cultivate the land, the Indians sold their land, and went further into the country. But in reality few of the Indians really left the country in this manner; most of them ended their days before, either by wars among themselves, or by the small-pox, a disease which the Indians were unacquainted with before their commerce with the Europeans, and which since that time has killed incredible numbers of them. For though they can heal wounds and other external hurts, yet they do not understand fever or other internal diseases. One can imagine how ill they would succeed with the cure of the small-pox when as soon as the pustules appeared they leaped naked into the cold of the rivers, lakes, or fountains, and either dived over head into it, or poured it over their body in great abundance in order to cool the heat of the fever. In the same manner they carried their children, when they had the small-pox, into the water and ducked them.[1] But brandy is said to have killed most

---

[1] Professor Kalm wrote this when the truly laudable method of treating the small-pox with a cold regimen, was not yet adopted; and he thought therefore, the way in which the Americans treated this disease, was the cause of its being so deleterious. But when the Khalmucks, in the Russian dominions, get the small-pox, it has been observed that very few escape. Of this I believe no other reason can be alleged than that the small-pox is always dangerous, either when the open pores of the human skin are too numerous, which is caused by opening them in a warm water bath; or when they are too much closed, which is the case with all the nations, that are dirty and greasy. All the American Indians rub their body with oils, the Khalmucks never wash themselves and rub their bodies and their fur coats with grease; the Hottentots are, I believe, known to be patterns of filthiness, their bodies being richly anointed with their ornamental greasy sheep guts; this shuts up all the pores; hinders perspiration entirely, and makes the small-pox always lethal among these nations; to which we may yet add the too frequent use of spirituous, inflammatory liquors, since their acquaintance with the Europeans.—F.

of the Indians. This liquor was, also, entirely unknown to them before the Europeans came hither, but after they had tasted it, they could never get enough of it. A man can hardly have a greater desire of a thing than the Indians have for brandy. I have heard them say that to die by drinking brandy was a desirable and an honorable death; and indeed it was a very common thing to kill themselves by drinking this liquor to excess.

The food of these Indians was very different from that of the inhabitants of the other parts of the world. Wheat, rye, barley, oats, and rice, were then quite unknown in this part of the world. In the same manner it was with regard to the fruits and herbs which were eaten in the old countries. The corn, some varieties of beans, and melons, comprised almost the whole of the Indian agriculture and gardening; and dogs were the only domestic animals in North America. But as their agriculture and their gardening were very trifling, and they could hardly live two months in a year upon their produce, they were forced to resort to hunting and fishing, which at that time, and even at present, furnish their chief subsistence; and they also in part had to rely on the products of plants and trees. Some of the old Swedes were yet alive who in their younger years had had intercourse with the Indians and had thoroughly observed their manners. I was therefore desirous of knowing which of the native herbs they made use of for food at that time; and all the old men agreed that the following plants were what they used chiefly in New Sweden.

Hopniss or *Håpniss* was the Indian name of a wild plant, which they ate at that time. The Swedes still call it by that name and it grows in the meadows in good soil. The roots resemble potatoes, and are boiled by the Indians, who eat them instead of bread. Some of the Swedes at that time likewise ate this root for want of bread. Some of the English still eat it instead of potatoes. Mr. Bartram told me that the Indians who live farther in the country do not only eat these roots, which have as good taste as potatoes, but likewise take the peas which lie in the pods of this plant, and prepare them like common peas.[1] Dr. Linné calls the plant *Glycine apios*.

Sagittaria. Katniss is another Indian name of a plant, the root of which they were also accustomed to eat, when they lived here. The Swedes still preserve this name. It grows in low, muddy and very

---

[1] A relative of this root is the soy bean.

wet ground. The root is oblong, commonly an inch and a half long, and one inch and a quarter broad in the middle; but some of the roots are as big as a man's fist. The Indians either boiled this root or roasted it in hot ashes. Some of the Swedes ate it with much relish at the time when the Indians were so near the coast; but at present none of them make any use of the roots. Nils Gustafson told me that he had often eaten these roots when he was a boy, and that he liked them very well at that time. He added that the Indians, especially the women, travelled to some islands, at about Whitsuntide, dug out the roots, and brought them home; and while they had them, they desired no other food. They were said to have been almost destroyed by hogs, which were exceedingly greedy for them. The cattle are very fond of their leaves. I afterwards got some of these roots roasted, and in my opinion they tasted good, though they were rather dry. The taste was nearly the same as that of potatoes. When the Indians come down to the coast and see the turnips of the Europeans, they likewise give them the name of *katniss*. The *katniss* is an arrow-head or *Sagittaria,* and is only a variety of the Swedish arrow-head or *Sagittaria sagittifolia,* for the plant above the ground is entirely the same, while the root underground is much greater in the American than in the European variety. Mr. Osbeck in his *Voyage to China* [1] mentions that the Chinese plant a *Sagittaria* and eat its roots. This seems undoubtedly to be a variety of this *katniss*. Further in the north of this part of America I met with the other species of *Sagittaria* which we have in Sweden.

Arum Virginicum. Taw-ho and *Taw-him* was the Indian name of another plant, the root of which Indians eat. Some of them likewise call it *Tuckáh*; but most of the Swedes still knew it by the name of *Taw-ho*. It grows in moist ground and swamps. Hogs are very fond of the roots, and grow very fat by feeding on them. Therefore, they often visit the [wet] places where these roots grow, and they are frequently seen rooting up the mud, and falling with their whole body into the water, so that only a little of the back part is out of the water. It is therefore very plain, that these roots must have been extirpated in places which are frequented by hogs. The roots often grow to the thickness of a man's thigh. When they are fresh, they have a pungent taste, and are reckoned a poison in that fresh state. Nor did the Indians ever venture to eat them raw, but prepared

---

[1] Vol. I p. 334, of the Swedish edition.

them in the following manner: they gathered a great heap of these roots, dug a great long hole, sometimes two or three fathoms and upwards in length, into which they put the roots, and covered them with the earth that had been taken out of the hole; they made a great fire above it, which burnt till they thought proper to remove it; and then they dug up the roots and consumed them with great avidity. These roots, when prepared in this manner, I am told, taste like potatoes. The Indians never dry or preserve them, but always dig them fresh out of the marshes, when they want them. This *Taw-ho* is the *Arum Virginicum*, or Virginian wake-robin. It is remarkable that the arums, with the plants next akin to them, are eaten by men in different parts of the world, though their roots, when raw, have a fiery pungent taste, and are almost poisonous in that state. How can men have learned that plants so extremely opposite to our nature were eatable, and that their poison, which burns on the tongue, can be conquered by fire? Thus the root of the *Calla palustris*, which grows in the north of Europe, is sometimes used instead of bread in an emergency. The North American Indians eat this species of arum. Those of South America, and of the West Indies, eat other species of arums. The Hottentots at the Cape of Good Hope in Africa, prepare bread from a species of arum or wake-robin, which is as burning and poisonous as the other species of this plant. In the same manner, they employ the roots of some kinds of arum as a food, in Egypt and Asia. Probably, that severe but sometimes useful teacher, necessity, has first taught men to find out a food, which the first taste would have rejected as useless. This *Taw-ho* seems to be the same as the *Tuckahoo,* of the Indians in Carolina.[1]

Golden Club. Taw-kee is another plant, so named by the Indians who eat it. Some of them call it *Taw-kim*, and others *Tackvim*. The Swedes always call it by the name of *Taw-kee*. The plant grows in marshes, near moist and low grounds, and is very plentiful in North America. The cattle, hogs and stags, are very fond of the leaves in spring; for they are some of the earliest. The leaves are broad, like those of the *Convallaria,* or the Lily of the Valley, green on the upper side, and covered with very minute hair, so that they look like fine velvet. The Indians pluck the seeds, and keep them for eating. They cannot be eaten fresh or raw, but must be dried. The Indians

---

[1] Cf. page 151.

were forced to boil them repeatedly in water, before they were fit for use; and then they ate them like peas. When the Swedes gave them butter or milk, they boiled or broiled the seeds in it. Sometimes they employ these seeds instead of bread; and they taste like peas. Some of the Swedes likewise ate them; and the old men among them told me, they liked this food better than any of the other plants which the Indians formerly made use of. The *Taw-kee* is the *Orontium aquaticum*.

Huckleberries. Bilberries were likewise a very common dish among the Indians. They are called huckleberries by the English here, and belong to the various species of *Vaccinium*, which are all of them different from our Swedish bilberry bush, though their berries, in regard to color, shape, and taste, are so similar to the Swedish bilberry that they are with difficulty distinguished from each other. The American ones grow on shrubs, which are from two to four feet high; and there are some species which are above six feet in height. The Indians formerly plucked them in abundance every year, dried them either in the sunshine or by the fireside, and afterwards prepared them for eating in different manners. These huckleberries are still a dainty dish among the Indians. On my travels through the country of the Iroquois, they offered me, whenever they designed to treat me well, fresh corn bread, baked in an oblong shape, mixed with dried huckleberries, which lay as close in it as the raisins in a plum pudding. I shall write more about it later. The Europeans also used to collect a quantity of these berries, to dry them in ovens, to bake them in tarts, and to employ them in several other ways. Some preserve them with treacle. They are likewise eaten raw, either with or without fresh milk.

I shall on the twenty-seventh of March find occasion to mention another dish, which the Indians ate formerly, and still eat, at formal ceremonies.

MARCH THE 18TH

Weather. During almost the whole of this spring, the weather and the winds were calm in the morning at sunrise. At eight o'clock the wind began to blow pretty hard, and continued so all day, till sunset, when it ceased, and all the night was calm. This was the regular course of the weather; but sometimes the winds raged, without intermission, for two or three days. At noon it was usually most violent. But ordinarily the wind decreased and increased as follows: at six in the morning we had a calm; at seven, a very gentle western breeze, which grew stronger at eight; at eleven the wind was much stronger; but at four in the afternoon it was no stronger than it was at eight o'clock in the morning; and thus it went on decreasing till it was quite calm just before sunset. The winds this spring blew generally from the west, as appears from the observations at the end of this work.

I was told, that it was a very certain forecast of bad weather, when you saw clouds on the horizon in the southwest, about sunset, and that if those clouds sank below the horizon a while later it would rain the next day, though all the forenoon were fair and clear. But if some clouds were seen in the southwest on the horizon at sunset, and they rose some time later, one could expect fair weather the next day.

March the 20th

An old Swede foretold a change in the weather, because it was calm today; for when there had been wind for a few days and a calm followed, they said there would be rain or snow, or some other change in the weather. I was likewise told that some people here were of the opinion that the weather commonly altered on Friday. In case it had rained or blown hard all week and a change was to happen, it would commonly fall on Friday.—How far the forecast has been true appears from my own observations of the weather, to which I refer.

March the 21st

The red maple (*Acer rubrum*) and the American elm (*Ulmus Americana*) now began to flower, and some of the latter were already in full blossom.

March the 24th

I walked quite far today in order to see whether I could find any plants in blossom. But the cloudy weather and the great rains which had lately fallen had allowed little or nothing to grow up. The leaves now began to grow green. The plants which I have just mentioned were now in full bloom.

The noble liverwort, or *Anemone hepatica,*[1] was now everywhere in flower. It was abundant, and the Swedes called it *blå-blomster* or blue-flower. They did not know any use for it.

Near all the fields through which I walked to-day, I did not see a single ditch, though many needed them. But the people generally followed the English way of making no ditches along the fields, whether they needed them or not. The consequence was that the late rain had in many places washed away great pieces of the grounds sown with wheat and rye. There were no ridges left between the fields, except very narrow ones near the fence, which were entirely overgrown with the sumach, or *Rhus glabra,* and with blackberry bushes, so that there the cattle could find very little or no food. The grain fields were laid out in rectilinear sections, or divided into lots which were nearly twenty feet wide, and separated from each other by furrows. These pieces were flat and not elevated in the middle.

Meloe majalis, a species of oil-beetle, crept about on the hills.

Papilio antiopa, or willow butterfly, flew in the woods to-day, and was the first butterfly which I saw this year.

Papilio Euphrosyne, or the April butterfly, was one of the scarce species. The other American insects, which I saw and described this day and the days immediately following, I shall mention on some other occasion. In the meantime I shall only mention those which seem remarkable for some peculiar qualities.

Haystacks. The haystacks were commonly made here after the Swedish manner, that is, in the shape of a thick and short cone, without any cover over it. When the people wanted any hay, they cut some of it loose, by a specially made cutter. However, many people, especially in the environs of Philadelphia, had haystacks with roofs which could be moved up and down. Near the surface of the ground were some poles laid, on which the hay was put, that the air might pass freely through it. I have mentioned before that the cattle had no stables in winter or summer and were obliged to graze in the open air during the whole year. However, in Philadelphia, and in a few other places, I saw that those people who made use of the latter kind of haystacks, *viz.* that with movable roofs, com-

---

[1] Probably the flower now named liver leaf, *Hepatica triloba.*

monly had built them so that the hay was put a fathom or two above the ground, on a floor of boards, under which the cattle could stand in winter when the weather was very bad. Under this floor were partitions of boards on all the sides, which however stood far enough from each other to afford the air a free passage.

MARCH THE 27TH

New Sweden. In the morning I went to speak with the old Swede, Nils Gustafson, who, [as I have noted] was ninety-one years of age. I intended to get some account of the former state of New Sweden. The country which I now passed through was the same as that which I had found in those parts of North America I had already seen. It was diversified with a variety of little hills and valleys: the former consisted of a very pale brick-colored earth, composed, for the greatest part, of a fine sand, mixed with some mould. I saw no mountains and no stones, except some little ones not above the size of a pigeon's or hen's egg lying on the hills and commonly consisting of white quartz, which was generally smooth and polished on the outside. At the bottom, along the valleys, ran sometimes creeks of crystalline water, the bottom of which were covered with white pebbles. Now and then I came upon swamps in the valleys. Sometimes there appeared, though at considerable distances from each other, some farms, frequently surrounded on all sides by grain fields. Almost on every field there yet remained the stumps of trees, which had been cut down, a proof that this country had not been long cultivated, having been overgrown with trees forty or fifty years ago. The farms did not lie together in villages, or so that several of them were near each other in one place, but were all unconnected. Each countryman lived by himself, and had his own ground about the house separated from the property of his neighbor. The greatest part of the land, between these farms so far apart, was overgrown with woods, consisting of tall trees. However, there was a fine space between the trees, so that one could ride on horseback without inconvenience in the woods, and even with a cart in most places; and the ground was very flat and uniform at the same time. Here and there appeared some fallen trees, thrown down by the wind; some were torn up by the roots; others broken straight across the trunk. In some parts of the country the trees were thick and tall, but in others I found large tracts covered with young trees, only twenty, thirty, or forty years old. On these tracts, I am told, the Indians formerly had their little plantations. I did not yet see any marks of the leaves coming out, and I did not meet with a flower in the woods, for the cold winds, which had blown for several days successively, had hindered this. The woods consisted chiefly of several species of oak, and of hickory. The swamps were filled with red maple which was now all in flower and made these places look real red at a distance.

A Nonagenarian. The old Swede, whom I came to visit, seemed to be still pretty healthy and could walk by the help of a cane, but he complained of having felt in these later years some pains in his back and limbs, and confessed that he now could keep his feet warm in winter only by sitting near the fire. He said he could very well remember the state of this country at the time when the Dutch possessed it, and in what circumstances it was in before the arrival of the English. He added, that he had brought a great deal of timber to Philadelphia at the time that it was built. He still remembered to have seen a great forest on the spot where Philadelphia now stands. The father of this old man had been one of the Swedes who were sent over from Sweden in order to cultivate and inhabit this country. He gave me the following answers to the questions I asked him.

The Swedish Settlers. Query, Whence did the Swedes, who first came hither, get their cattle? The old man answered, that when he was a boy, his father and other people had told him that the Swedes brought their horses, cows and oxen, sheep, hogs, geese, and ducks, over with them. There were but few of a kind at first, but they multiplied greatly here afterwards. He said that Maryland, New York, New England, and Virginia, had been earlier inhabited by Europeans than this part of the country; but he did not know whether the Swedes ever got cattle of any kind from any of these provinces, except from New York. While he was yet very young, the Swedes, as far as he could remember, had already a sufficient stock of all these animals. The hogs had propagated so much at that time, there being so great a plenty of food for them, that they ran about wild in the woods, and that the people were obliged to shoot them when they wanted them. The old man likewise recollected, that horses ran wild in the woods, in some places; but he could not tell whether any other kind of cattle turned wild. He thought that the cattle grew as big at present as they did when he was a boy, provided they received as much food as they needed. For in his younger years food for all kinds of cattle was so plentiful and abundant that the cattle were extremely fat. A cow at that time gave more milk than three or four do at present; but she got more and better food at that time than three or four get now; and, as the old man said, the scanty allowance of grass which the cattle get in summer is really very pitiful. The causes of this scarcity of grass have already been mentioned.

Query, Whence did the English in Pennsylvania and New Jersey get their cattle? They bought them chiefly from the Swedes and Dutch who lived here, and a small number were brought over from Old England. The physical form of the cattle and the unanimous accounts of the English here confirmed what the old man had said.

Query, Whence did the Swedes here settled get their several sorts of grain and likewise their fruit trees and kitchen herbs? The old man told me that he had frequently heard when he was young, that the Swedes had brought all kinds of grain and fruits and herbs or seeds of them with them. For as far as he could recollect, the Swedes here were plentifully provided with wheat, rye, barley and oats. The Swedes, at that time, brewed all their beer of malt made of barley, and likewise made good strong beer. They had already got distilling apparatus, and when they intended to distil they lent their apparatus to one another. At first they were forced to buy corn of the Indians, both for sowing and eating. But after continuing for some years in this country, they extended their corn plantations so much that the Indians were obliged some time after to buy corn of the Swedes. The old man likewise assured me that the Indians formerly, and about the time of the first settling of the Swedes, were more industrious, but that now they had become very lazy in comparison. While he was young the Swedes had a great quantity of very good white cabbage. Winter cabbage, or kale, which was left on the ground during winter, was also abundant. They were likewise well provided with turnips. In winter they kept them in holes under ground. But the old man did not like that method, for when they had lain too long in these holes in winter they became spongy.

He preferred that method of keeping them which is now commonly adopted, and which consists in the following particulars. After the turnips have been taken out of the ground in autumn, and exposed to the air for a while, they are put in a heap upon the field, covered with straw at the top, and on the sides, and with earth over the straw. By this means they stand the winter very well here, and do not become spongy. The Indians were very fond of turnips, and called them sometimes *hopniss*, sometimes *katniss*. Nobody around here had ever heard of *rutabagas* or Swedish turnips. The Swedes likewise cultivated carrots, in the old man's younger years. Among the fruit trees were apple trees. They were not numerous, and only some of the Swedes had little orchards of them, while others had not a single tree. None of the Swedes made cider, for it had come into use but lately. The Swedes brewed beer and that was their common drink. But at present there are very few who brew beer, for they commonly make cider. Cherry trees were abundant when Nils Gustafson was a boy. Peach trees were at that time more numerous than at present and the Swedes brewed beer of the fruit. The old man could not tell from whence the Swedes first of all got the peach trees.

Indians. During the younger years of this old man, the Indians were everywhere in the country. They lived among the Swedes. The old man mentioned Swedes who had been killed by the Indians, and he mentioned two of his countrymen who had been scalped by them. They stole children from the Swedes, and carried them off, and they were never heard of again. Once they came and killed some of them and took their scalps; on that occasion they scalped a little girl, and would have killed her, if they had not perceived a boat full of Swedes, making towards them, which obliged them to flee. The girl's scalp afterwards healed, but no hair grew on it; she was married, had many children, and lived to a great age. At another time the Indians attempted to kill the mother of this old man, but he vigorously resisted them until a number of Swedes came up, who frightened the Indians and made them run away. Nobody could ever find out to what nation these savages belonged; for in general they lived very peaceably with the Swedes.

The *Indians* had their little plantations of corn in many places. Before the Swedes came into this country, the Red Men had no other hatchets than those made of stone; in order to make corn plantations they cut out the trees and prepared the ground in the manner I have mentioned before.[1] They planted but little corn, for they lived chiefly by hunting, and throughout the greatest part of the summer *hopniss, katniss, taw-ho*, and whortleberries were their chief food. They had no horses or other cattle which could be employed in their agriculture, and therefore did all the work with their own hands. After they had reaped the corn, they kept it in holes under ground during winter; they seldom dug these holes deeper than a fathom, and often not so deep; at the bottom and on the sides they put broad pieces of bark. If bark could not be had, the *Andropogon bicorne*, a grass which grows in great plenty here, and which the English call Indian grass and the Swedes wildgrass, supplied the want of the former. The ears of corn were then thrown into the hole and covered to a considerable thickness with this grass, and the whole again covered by a sufficient quantity of earth. Corn was kept extremely well in those holes, and each Indian had several such subterraneous stores, where his corn lay safe, though he travelled far from it. After the Swedes had settled here and planted apple trees and peach trees, the Indians, and especially their women, sometimes stole the fruit in great quantity; but when the Swedes caught them, they gave them a severe drubbing, took the fruit from them, and often their clothes too. In the same manner it happened sometimes that as the Swedes had a great increase of hogs, and they ran about in the woods, the Indians killed some of them privately and ate them: but there were likewise some Indians who bought hogs of the Swedes and raised them. They taught them to follow them like dogs, and whenever they moved from one place to another their pigs always went with them. Some of those savages got such numbers of these animals, that they afterwards gave them to the Swedes for a trifle. When the Swedes arrived in America the Indians had no domestic animals, except a species of little dog. The Indians were extremely fond of milk and drank it with pleasure when the Swedes gave it to them. They likewise prepared a kind of liquor like milk by gathering a great number of hickory and black walnuts, dried and crushed them. Then they took out the kernels, pounded them as fine as flour, and mixed this with water so that it looked like milk and was almost as sweet. They had tobacco pipes of clay, manufactured by themselves, at the time that the Swedes arrived here;

---

[1] See page 230.

they did not always smoke true tobacco, but made use of another plant instead of it, which was unknown to the old Swedes. It was not the common mullein, or *Verbascum thapsus,* which is generally called Indian tobacco here.

Religion among the Indians. As to their religion the old man thought it very trifling, and even believed that they had none at all. When they heard loud claps of thunder they said that the evil spirit was angry. Some of them said that they believed in a god, who lives in heaven. The old Swede once walked with an Indian, and they encountered a red-spotted snake on the road: the old man therefore went to seek a stick in order to kill it, but the Indian begged him not to touch it, because it was sacred to him. Perhaps the Swede would not have killed it, but on hearing that it was the Indian's deity, he took a stick and killed it, in the presence of the Indian, saying: "Because thou believest in it, I think myself obliged to kill it." Sometimes the Indians came into the Swedish churches, looked around, listened and went away again. One day as this old Swede was at church and did not sing, because he had no psalmbook, one of the Indians, who was well acquainted with him, tapped him on the shoulder, and said: "Why dost thou not sing with the others, Tantánta! Tantánta! Tantánta?" On another occasion, as a sermon was preached in the Swedish church at Raccoon, an Indian came in, looked about him, and after listening awhile to the preacher, he said: "Ugh! A lot of prattle and nonsense, but neither brandy nor cider!" and went out again. For it is to be observed that when an Indian makes a speech to his companions, in order to encourage them to war, or to anything else, they all drink immoderately on those occasions.

At the time when the Swedes arrived, they bought land at a very small price. For a piece of baize, or a pot full of brandy or the like, they could get a piece of ground, which at present would be worth more than four hundred pounds, Pennsylvania currency. When they sold a piece of land, they commonly signed an agreement; and though they could neither read nor write, they scribbled their marks, or signatures, at the bottom of it. The father of old Nils Gustafson bought a piece of ground from the Indians in New Jersey. As soon as the agreement was drawn up, and the Indians were about to sign it, one of them, whose name signified a beaver, drew a beaver; another of them drew a bow and arrow; and a third a mountain, instead of his name. Their canoes were made of thick trees, which they hollowed out by fire, and made them smooth again with their hatchets, as has been mentioned before.

Weather. The following account the old man gave me in answer to my questions with regard to the weather and its changes. It was his opinion that the weather had always been pretty uniform ever since his childhood; that there happened as great storms at present as formerly; that the summers now were sometimes hotter and sometimes colder than they were at that time; that the winters were often as cold and as long as formerly; and that there still often falls as great a quantity of snow as in former times. However, he thought that no cold winter came up to that which reigned after the summer when the Swedish clergymen Rudman and Björk came here, in the year 1697, which is often mentioned in the almanacks of this country; and which I have mentioned before.[1] For in that winter the river Delaware was so thickly covered with ice that the old man brought many wagons full of hay over it near Christina; and that it was passable on sledges even lower down. No cattle, as far as he could recollect, were frozen to death during the cold winters, except, in later years, such cattle as were lean, and had no stable in which to seek protection. It commonly rains in summer as it did formerly, excepting that, during the last years the summers have been drier. Nor could the old Swede find a diminution of water in the brooks, rivers and swamps. He allowed, as a very common and certain fact, that wherever you dug wells you would strike oyster shells in the ground.

Old Gustafson was of the opinion, that intermitting fevers were as frequent and violent formerly as they are now; but believed that they seemed more uncommon, because there were fewer people at that time here. When he got this fever, he was not yet full grown. He got it in summer, and had it till the ensuing spring, which is almost a year; but it did not hinder him from doing his work, either within or out of doors. Pleurisy likewise attacked one or two of the Swedes formerly; but it was not nearly so common as it is now. The people in general were very healthy at that time.

Some years ago the old Swede's eyes were so much weakened that he was forced to make use of a pair of spectacles. He then got a fever, which was so violent that it was feared he would not re-

---

[1] See page 180. Eric Björk settled at Christina.

cover. However, he became quite well again, and at the same time got new strength in his eyes, so that he has been able to read without spectacles ever since.

Houses in New Sweden. The houses which the Swedes built when they first settled here were very poor. The whole house consisted of one little room, the door of which was so low, that one was obliged to stoop in order to get in. As they had brought no glass with them, they were obliged to be content with little holes, before which a moveable board was fastened. They found no moss, or at least none which could have been serviceable in stopping up holes or cracks in the walls. They were therefore forced to close them, using clay, both inside and out. The chimneys were masoned in a corner, either of gray stone, or (in places where there were no stone) of mere clay, which they laid very thick in one corner of the house. The ovens for baking were likewise inside. So far as we know the Swedes never used any dampers [as I have already mentioned], perhaps because they had none of iron and did not feel that the winters here were either cold or long enough [to need them], and also because in the beginning they had an abundance of fuel.

Dress in New Sweden. Before the English came to settle here, the Swedes could not get as many clothes as they needed, and were therefore obliged to get along as well as they could. The men wore waistcoats and breeches of skins. Hats were not in fashion, and they made little caps, provided with flaps; some made fur caps. They had worsted stockings. Their shoes were of their own making. Some of them had learned to prepare leather, and to make common shoes, with heels; but those who were not shoemakers by profession took the length of their feet and sewed the leather together accordingly, taking a piece for the sole, one for the hind-quarters, and another one for the uppers. These shoes were called *kippaka.* At that time, they likewise sowed flax here, and wove linen cloth. Hemp was not to be had; and they made use of linen and wild hemp for fishing tackle. The women were dressed in jackets and petticoats of skins. Their beds, excepting the sheets, were skins of various animals; such as bears, wolves, etc.

Tea, coffee, and chocolate, which are at present universally in use here, were then [1] wholly unknown. Bread and butter, and other

---

[1] Before the English settled here.—F.

substantial food, was what they breakfasted upon; and the above-mentioned superfluities have only been lately introduced, according to the account of the old Swede. Sugar and molasses they had in abundance, so far back as he could remember. Rum could formerly be had for a more moderate price than at present.

English Customs Replace the Swedish. From the accounts of this old man I concluded, that before the English settled here they followed wholly the customs of Old Sweden; but after the English had been in the country for some time, the Swedes began gradually to follow theirs. When this Swede was but a boy, there were two Swedish smiths here, who made hatchets, knives, and scythes, exactly like the Swedish ones, and made them sharper than they can be gotten now. The hatchets now in use are often the English style, with a broad edge, and their handles are very narrow. They had no jack-knives. Almost all the Swedes had bath-houses and they commonly bathed every Saturday, but now these bath-houses are done away with. They celebrated Christmas with several sorts of games, and with various special dishes, as is usual in Sweden; all of which is now, for the greatest part, given up. In the younger years of this Swede, they made a strange kind of cart here. They sawed off round cross sections of thick liquidambar logs, and used two of them for the front wheels and two more for the back wheels. With those carts they brought home their wood. Their sledges were at that time made almost as they are now, being about twice as broad as the true Swedish ones. Timber and great beams of wood were carried upon a dray. They baked great loaves, such as they do now. They never had any hard, crackerhole-bread or *knäckebröd* [now called "health bread" in the United States], though the clergymen who came from Sweden commonly had some baked.

The English on their arrival here bought large tracts of land of the Swedes for almost nothing. The father of the old Swede sold an estate to the English, which at this time would be reckoned worth three hundred pounds, for which he got a cow, a sow, and a hundred pumpkins.

With regard to the decrease in the number of birds and fish, he was wholly of that opinion which I have already mentioned.[1] This was the account which the old man gave me of the former state of

---

[1] See pages 152 ff.

the Swedes in this country. I shall speak more particularly of it later.

Storms are sometimes very violent here, and often tear up great trees. They proceed as it were along peculiar tracts or lines. In some places, especially in a hurricane's tract, all the trees are struck down, and it looks afterwards as if the woods had been hurled down intentionally; but close to the tract the trees are not damaged. Such is the place which was shown to me to-day. It is dangerous to go into the woods where the hurricanes blow, for the trees fall before one has time to guard oneself or make the least provision for escape.

The Pennsylvanian aspen was now in full blossom. But neither this tree nor any of those kin to it showed their leaves.

An old countryman asserted that he commonly sowed a bushel of rye on an acre of ground and got twenty bushels in return; but from a bushel of barley he got thirty bushels. However, in that case the ground must be well prepared. Wheat returns about as much as rye. The soil was a clay mixed with sand and mould. In the evening I returned.[1]

MARCH THE 28TH

Insects. I found a black beetle [2] (*Scarabæus*) with a pentagonal oval *clypeus* or shield, on the head a short blunt horn, and a bibbous, or hump-backed *thorax* or corselet. This beetle is one of the bigger sort here. I found here and there holes on the hills, which were so wide that I could put my finger into them. On digging them up I always found these beetles lying at the bottom, about five inches under ground. Sometimes there were short whitish worms, about as thick as one's finger, which lay with the beetles; and perhaps they were related to them. There were likewise other insects who made their home in such holes, as the black cricket (*Gryllus niger*), spiders, earth-beetles (*Carabi*) and others. This beetle had a scent exactly like the *Trifolium melilotus cærulea*, or the blue melilot. It was entirely covered with oblong pale ticks (*Acari*). Its feet were as strong as those of the common dung-beetle (*Scarabæus stercorarius*).

---

[1] From Nils Gustafson, the old Swede.—F.

[2] The beetle here described seems to be the *Scarabæ Carolinus*, L. *Syst. Nat.* p. 545, and of Drury Illustrations of *Nat. Hist.* tab. 35., f. a. It is common in New York, New Jersey, Pennsylvania, Maryland, and Carolina.—F.

APRIL THE 4TH

A *Cicindela,* or shining beetle, with a gold-green head, thorax, and feet, and a blue-green abdomen or belly, flew everywhere about the fields, and was hunting other insects. It is very common in North America, and seems to be a mere variety of the *Cicindela campestris.*

Cimex lacustris, a kind of water-bug, hopped in great numbers on the surface of waters which had a slow course.

Dytiscus piceus, or the great water-beetle, swam sometimes in the water.

About sixty years ago, the greatest part of this country was covered with tall and large trees, and the swamps were full of water. But it has undergone so great a change, as few other places have undergone in so short a time. At present the forests are cut down in most places, the swamps drained by ditches, the country cultivated, and changed into grain fields, meadows, and pastures. Therefore, it seems very reasonable to suppose, that so sudden a change has likewise had some effect upon the climate. I was therefore desirous of hearing from the old Swedes, who have lived the longest in this country, and have been inhabitants of this place during the whole time of the change mentioned, whether the present state of the weather was in some particulars remarkably different from that which they noticed in their younger years. The following is an account which they unanimously gave me in answer to this question.

Change of Climate. The winter came sooner formerly than it does now. Mr. Isaac Norris,[1] a wealthy merchant, who has a considerable share in the government of Pennsylvania, confirmed this by a particular account. His father [Isaac Norris, Mayor of Philadelphia, d. 1735], one of the first English merchants in this country, observed that in his younger years, the river Delaware was commonly covered with ice about the middle of November, old style, so that the merchants were obliged to bring down their ships in great haste before that time, for fear of their being obliged to lie idle all

---

[1] Isaac Norris (1707-1766), party leader, was a member of the Pennsylvania Assembly, with few interruptions from 1734 to 1764. He was the leader of the Peace Party. In 1745 he had been appointed by the governor to treat with the Indians in a matter of land ownership and in 1755 was sent to Albany on a similar mission. Norris was a man of liberal education, and his library of 1500 volumes was left to Dickinson College.

winter. On the contrary, this river seldom freezes over at present, before the middle of December, old style.

It snowed much more in winter formerly than it does now; but the weather in general was likewise more constant and uniform; and when the cold set in, it continued to the end of February, or till March, old style, when it commonly began to grow warm. At present, it is warm, even the very next day after a severe cold; and sometimes the weather changes several times a day.[1]

Most of the old people here were of the opinion that spring came much later at present than formerly, and that it was now much colder in the latter end of February and during the whole month of March than when they were young. Formerly the fields were as green and the air as warm towards the end of February as it is now in March, or in the beginning of April, old style. The Swedes at that time made use of the phrase: *Påsk bittida, Påsk sent, alltid gräs*, that is, we have always grass at Easter, whether it be soon or late in the year. But perhaps we can account as follows for the opinion which the people here have, that vegetation used to appear earlier than it does now. Formerly the cattle were not so numerous as now; however, the woods were full of grass and herbs, which, according to the testimony of all the old people here, grew to half the height of a man. At present a great part of the annual grasses and plants have been entirely extirpated by the continual grazing of cattle. These annual grasses were probably green very early in spring, and (being extirpated) might lead the people to believe that everything came on sooner formerly than it does at present.

It used to rain more abundantly than it does now; during the harvest especially the rains fell so heavily that it was very difficult to bring home the hay and grain. Some of the last years have been extremely dry. However, a few people were of the opinion that it rained as much now as formerly.

All the people agreed that the weather was not by far so inconstant, when they were young as it is now. For at present it happens at all times of the year that when a day has been warm, the next is very cold, and *vice versa*. It frequently happens that the weather alters several times in one day; so that when it has been a pretty

---

[1] There seems to have been some difference of opinion as to the change of climate on Colonial territory. Cf. p. 271.

warm morning, the wind blows from N. W. about ten o'clock and brings cold air with it; yet a little after noon it may be warm again. My meteorological observations sufficiently confirm the reality of these sudden changes of weather, which are said to cause in a great measure the people to be more unhealthy at present than they were formerly.

I likewise found everybody agreed in asserting, that the winter betwixt the autumn of the year 1697 and the spring of the year 1698 was the coldest and the severest which they had ever felt.

APRIL THE 6TH

Plants. Sanguinaria Canadensis, which is here called bloodroot, because the root is large and red, and when cut, looks like the root of a red beet, and the *Epigæa repens*, which some call the creeping ground laurel, were both beginning to flower. The former grew in a rich mould, the other in a poorer soil. The *Laurus æstivalis*, which some people call spicewood, likewise began to blossom about this time; its leaves had not yet appeared; it liked a moist soil in the woods.

APRIL THE 9TH

Apocynum cannabinum was by the Swedes called Indian hemp (wild hemp),[1] and grew plentifully in old grain grounds, in woods, on hills, and in high glades. The Swedes have given it the name Indian hemp, because the Indians formerly and even now apply it to the same purposes as the Europeans do hemp; for the stalk may be divided into filaments, and is easily prepared. When the Indians were still living among the Swedes, in Pennsylvania and New Jersey, they made ropes of this *Apocynum*, which the Swedes bought, and used them as bridles, and for nets. These ropes were stronger and kept longer in water than such as were made of common hemp. The Swedes usually got thirty feet of these ropes for one piece of bread. Many of the Europeans still buy such ropes because they last so well. The Indians also make several other articles of their hemp, such as various sizes of bags, pouches, quilts and linings. On

---

[1] *Wilskt Hampa.*—F.

my journey through the country of the Iroquois I saw the women employed in the manufacture of this hemp. They made use neither of spinning wheels nor distaffs, but rolled the filaments upon their bare thighs, and made thread and strings of them, which they dyed red, yellow, black, etc. and afterwards worked them into goods with a great deal of ingenuity. The plant is perennial, which renders the annual planting of it altogether unnecessary. Out of the root and stalk of this plant, when it is fresh, comes a white milky juice, which is somewhat poisonous. Sometimes the fishing equipment of the Indians consists entirely of this hemp. The Europeans make no use of it that I know of.

Cat-tail. Flax or *cat-tail* are names given to a grass which grows in bays, rivers, and in deep whirlpools, and which is known to botanists by the name of *Typha latifolia*. Its leaves are here twisted together, and formed into great oblong rings, which are put upon the horse's neck, between the mane and the collar, in order to prevent the horse's neck from being hurt by the collar. The bottoms of chairs (now called rush-bottom chairs) were frequently made of these leaves, twisted together. Formerly the Swedes employed the down which surrounds its seeds and put it into their beds instead of feathers; but as it coalesces into lumps after the beds have been used for some time, they have left off making use of them. I omit the use of this plant in medicine, that being the peculiar province of the physicians.

Garlic. A species of leek very much like that which appears only in woods on hills in Sweden, grows at present on almost all sandy grain fields. The English here call it garlic. On some fields it grows in great abundance. When the cattle graze on such fields and eat the garlic, their milk, and the butter which is made of it, taste so strongly of it that they are scarcely consumable. Sometimes they buy butter in the Philadelphia markets which tastes so strongly of garlic that it is useless. On this account they do not allow milk cows to graze on fields where garlic abounds: this they reserve for other species of cattle. When the cattle eat much of this garlic in summer, their flesh likewise gets such a strong flavor that it is unfit for eating. This kind of garlic appears early in spring, and the horses always pass by it without ever touching it.

It would take too much room in my Journal and render it too prolix, were I to mark down the time when every wild plant in this country was in blossom, when it got ripe seeds, what soil was peculiar to it, besides other circumstances. Some of my readers would be but little amused with such a botanical digression. I intend therefore to reserve all this for another work, which will give a particular account of all the plants of North America; [1] and I shall only mention such trees and plants here, which deserve to be made known for some peculiar quality.

APRIL THE 12TH

This morning I went to Philadelphia and the places adjacent, in order to learn whether more plants had lately sprung up there than at Raccoon and in New Jersey in general. The wet weather which had set in during the preceding days had made the roads very bad in low and clayey places.

Reckless Burning. The leaves which dropped last autumn had covered the ground three or four inches in depth. As this seemed to hinder the growth of the grass, it was customary to burn it in March, or at the end of that month (according to the old style), in order to give the grass the opportunity of growing up. I found several spots burnt in this manner to-day; but if it be useful one way, it does a great deal of damage in another. All the young shoots of several trees were burnt with the dead leaves, which diminishes the wood and timber considerably; and in places where the dead leaves had been burnt for several years in succession the old trees only were left, which being cut down, there remained nothing but a large field, and without any wood. At the same time all sorts of trees and plants were consumed by the fire, or at least deprived of their power of budding. Now, a great number of the plants and most of the grasses here are annuals; their seeds fall between the leaves, and by that means are burnt. This is another cause of universal complaint that grass is much scarcer at present in the woods than it was formerly. A great number of dry and hollow trees are burnt at the same time, though they could serve as fuel in the houses, and by that

---

[1] Kalm later prepared such a work, but the publication of it had to be given up, presumably because of the lack of funds. His master and teacher Linné, however, incorporated many of Kalm's observations and descriptions in his scientific writings, notably the *Species Plantarum* (1st ed. 1753; 2nd ed. 1761-1762) so that the contribution was not lost to the world.

means spare part of the forests. The upper mould likewise burns away in part by that means, not to mention several other inconveniences with which this burning of the dead leaves is attended. To this purpose the government of Pennsylvania has lately published an edict which prohibits this burning; but every one does as he pleases and this prohibition meets with a general censure.

Ticks. There are vast numbers of woodlice in the woods about this time; they are a very disagreeable insect, for as soon as a person sits down on an old stump or log, or on the ground itself, a whole army of woodlice creep upon his clothes and imperceptibly come upon the naked body. I have given a full account of their bad qualities and of other circumstances relating to them in the *Memoirs* of the Swedish Royal Academy of Sciences.[1]

I had a piece of *petrified wood* given me to-day, which was found deep in the ground at Raccoon. In this wood the fibres and annual rings appeared very plainly; it seemed to be a piece of hickory, for it was similar to it in every respect as if it had just been cut from a hickory log.

More Old Clam Shells. I was likewise presented with some shells to-day which the English commonly call clams, and whereof the Indians make their ornaments and money,[2] which I shall take an opportunity of speaking about further on. These clams were not fresh, but such as are everywhere found in New Jersey, on digging deep into the ground; the live shells of this kind are found only in salt water and on the seacoasts. But these gift clams had been discovered at Raccoon, about eight or nine English miles from the Delaware River, and almost a hundred from the nearest seashore.

At night I went to Mr. Bartram's estate.

APRIL THE 13TH

I employed this day in several observations relative to botany.

Wasps. Two nests of wasps hung in a high maple tree, over a brook. Their form was wholly the same as that of our wasp nests, but they exceeded them in size. Each nest was ten inches in diameter; in each nest were three combs, one above another, of which the nethermost was the biggest, and the two uppermost smaller in

---

[1] See the volume for the year 1754, p. 19.
[2] See also above, p. 129.

proportion. There were some eggs in them. The diameter of the lowest comb was about six and one quarter inches, and that of the uppermost three inches and three quarters. The cells in which the eggs or the young ones were deposited were hexagonal, and the color of the nest gray. I was told that the wasps make this kind of nest out of the gray splints, which stick to old fences and walls. A dark brown wasp with black antennæ, two black rings on the belly, and purple wings, flew about the trees, and was perhaps an inhabitant of these parts.

Another kind of wasp which is larger than this one, makes its nest open. It consists merely of one comb which has no covering, and is made of bits of twigs. The cells are horizontal, and when the eggs or the young larvæ lie in them, they have lids or coverings, that the rain may not get into them. But whither the old wasps retreat during storms is a mystery to me, unless they creep into the crevices of rocks. That side of the comb which is uppermost is covered with some oily particles, so that the rain cannot penetrate. The cells are hexagonal, from five to seven lines deep and two lines in diameter. Mr. Bartram observed that these nests are built of two sorts of materials, *viz.* (1) the splints which are found upon old pales, or fences, which are flaked off by the air. The wasps have often been observed to sit on such old wood and to gnaw away these splints. (2) The sides and the lids or covers of the cells are made of an animal substance, or glutinous matter, thrown up by the wasps, or prepared in their mouths. When this substance is thrown into the fire, it does not burn, but is only singed and smells like burnt hair or horn. But the bottom of the nest when put into the fire burns like linen or half-rotten wood, and leaves a smell of burnt wood. The wasps, whose nests I have now described, have three elevated black shining points on the forehead,[1] and a pentagonal black spot on the thorax. Toward the end of autumn these wasps creep into the crevices of mountains, where they lie torpid during winter. In spring, when the sun begins to operate, they come out during daytime, but return towards night when it grows cold. I saw them early in spring during sunshine in and about some crevices in the moun-

---

[1] These three points are common to most insects, and ought therefore not to be made characteristics of any particular species. They are called *Stemmata,* and are a kind of eye which serve the insects for looking at distant objects, as the compound eyes do for objects near at hand.—F.

tains. I was told of another species of wasps, which make their nests underground.

Gyrinus natator (*Americanus*), or the whirl-beetles.[1] These were found dancing in great numbers on the surface of the waters.

April the 14th

Sawmills. This morning I went down to Chester. In several places on the road are sawmills, but those which I observed to-day had no more than one saw. I likewise perceived that the woods and forests of these parts had been very roughly treated. It is customary here, when they erect sawmills, grist mills, or iron works, to direct the water by a different course almost horizontally until they come to a place suitable for building. This was not done with boards like a flume but by ditches. The dam itself was provided with sluice gates.

April the 16th

Swallows. This morning I returned to Raccoon. This country has several kinds of swallows, *viz.* such as live in barns, in chimneys, and under ground; there are likewise martins.

The *barn swallows*, or, as some Swedes called them, house swallows, are those with a furcated tail. They are Linné's *Hirundo rustica*. I found them in all parts of North America which I traversed. They correspond very nearly to the European house swallow in regard to their color, but there seems to be a small difference in the note. I took no notice this year when they arrived; but the following year, 1750, I observed them for the first time on the 10th of April (new style). The next morning I saw great numbers of them sitting on posts and planks, and they were as wet as if they had just come out of the sea. They build their nests in houses and under the roofs on the outside; I likewise found their nests built on the lower side of overhanging rocks. They build, too, under the edges of perpendicular rocks; and this shows where the swallow made their nests before the Europeans settled and built houses here; for it is

---

[1] Now called the whirligig beetle, *Gyrinus borealis*.

well known that the huts of the Indians could not serve the purpose of the swallows. A very reliable lady and her children told me the following story, assuring me that they were eye-witnesses to it: a couple of swallows built their nest in the stable belonging to the lady; the female swallow laid eggs in it, set on them and was about to brood them. Some days afterward the people saw the female still sitting on the eggs but the male flying about the nest and sometimes settling on a nail and uttering a very plaintive note, which betrayed his uneasiness. On a nearer examination the female was found dead in the nest and the people flung her away. The male then went to sit upon the eggs, but after being about two hours on them, and thinking the business too troublesome for him, he went out and returned in the afternoon with another female, which sat upon the eggs and afterwards fed the young ones till they were able to provide for themselves. The people differed here in their opinions about the abode of swallows in winter. A number of the Swedes thought that they lay at the bottom of the sea; others, not only Swedes and Englishmen but also the French in Canada, thought that they migrated southward in autumn and returned in spring. I have likewise been credibly informed in Albany, that they have been found sleeping in deep holes and clefts or rocks during winter.

The *chimney swallows* are the second species, and they derive their name from building their nests in chimneys which are not used in summer. Sometimes when the fire is not very large they do not mind the smoke, and remain in the chimney. I did not see them this year till late in May, but in the ensuing year, 1750, they arrived on the third of May, for they appear much later than other swallows. It is remarkable that each feather in their tail ends in a stiff sharp point, like the end of an awl; they apply the tail to the side of the wall in the chimneys, hold themselves with their feet, and the stiff tail serves to keep them up. They cause a great thundering noise all day long by flying up and down in the chimneys; and as they build their nests in chimneys only, and it is well known that the Indians have not so much as a hearth made of masonry, much less a chimney, but make their fires on the ground in their huts, it is a natural question to ask: where did these swallows build their nests before the Europeans came and made houses with chimneys? It is probable that they formerly made them in great hollow trees.

This opinion was accepted by Mr. Bartram and many others here. Catesby has described the chimney swallow and drawn it,[1] and Dr. Linné calls it *Hirundo palasgia*.

The *ground swallows* or sand martins (Linné's *Hirundo riparia*) are seen everywhere in America; they make their nests in the ground on the steep shores of rivers and lakes.

Purple Martins. The purple martins have likewise been described and drawn in their natural colors by Catesby.[2] Dr. Linné calls them *Hirundo purpurea*. They are less common here than the former species. I have seen in several places little bird houses made of boards and attached to the outside of the walls, for the purpose of inducing these martins to make their nests in them, for the people are very desirous of having them near their houses. They drive away both hawks and crows as soon as they see them, and by their anxious note alarm the poultry of the approach of their enemies. The chickens run to shelter as soon as they are warned by the martins.

April the 17th

Leatherwood. The *Dirca palustris,* or moose-wood, is a little shrub which grows on hills, near swamps and marshes, and is now in full blossom. The English in Albany call it leatherwood, because its bark is as tough as leather. The French in Canada call it *bois de plomb*, or lead wood, because the wood itself is as soft and as tough as lead. The bark of this shrub was used for ropes, baskets, etc. by the Indians, while they lived among the Swedes. And it is really very fit for that purpose on account of its remarkable strength and toughness, which is equal to that of the lime tree bark. The English and the Dutch in many parts of North America, and the French in Canada, employ this bark in all cases where we in Europe use the lime tree bark, especially for binding purposes. The tree itself is very tough, and one cannot easily separate a branch from the trunk without the help of a knife. Some people use switches of this tree for whipping their children.

---

[1] *Hirundo, caudâ aculeatâ, Americana.* Catesby, *Carol.* vol. III, t. 8.—F.
[2] *Hirundo purpurea. Nat. Hist. of Carol.* vol. I, t. 51.—F.

APRIL THE 18TH

Sorrel. Both the Swedish and English settlers are in the spring accustomed to prepare greens from various plants of which the following are the most important; *Rumex crispus* L. is a kind of sorrel which grows at the edge of cultivated fields and elsewhere in rather low land. Farmers choose a variety which has green leaves instead of pale colored ones. All sorrel is not suitable for greens, for the leaves of some are very bitter. These green leaves are gathered at this time everywhere and used by some people in the same way that Swedes prepare spinach. But they generally boil the leaves in the water in which they had cooked meat. Then they eat it alone or with the meat. It is served on a platter and eaten with a knife, which is different from the Swedish custom. Here also vinegar is placed in a special container on the table to be used on the kale. I must confess that this dish tastes very good.

In some places hereabouts there was a plant, the *Chenopodium album* L., which grew in great quantities in rich soil and was the second plant used as kale. Only the young plants a few inches in height were used and prepared like the preceding one.

Poke. The third plant used here in springtime for the same purpose is the *Phytolacca*, or the poke. It is prepared in the same way as the preceeding two. It must be used when very young for it becomes poisonous when old. It may become two or more feet tall. I have eaten of it several times without any bad effects.

APRIL THE 20TH

To-day I found the strawberries in bloom for the first time this year. The fruit is commonly larger than that in Sweden; but it seems to be less sweet and palatable.

Abundance of Food. The annual harvest, I am told, always affords plenty of bread for the inhabitants, though one year may be better than the rest. A venerable septuagenarian Swede, Åke Helm, [whom we have met before] assured me that in his lifetime there had been no failure of crops but that the people had always had plenty. It is likewise to be observed that the people eat their bread or corn, rye, or wheat, quite pure and free from chaff and other impurities. Many aged Swedes and Englishmen confirmed this account, and said that they could not remember any crop so bad as to make the people suffer in the least, much less that anybody had starved to death, while they were in America. Sometimes the price of grain rose higher in one year than in another, on account of a great drought or bad weather, but still there was always sufficient for the consumption of the inhabitants. Nor is it likely that any great famine can happen in this country, unless it please God to afflict it with extraordinary punishments. The weather is well known from more than sixty years experience. Here are no nights cold enough to hurt the seeds. The rainy periods are of short duration and the drought is seldom or never severe. But the chief thing is the great variety of grain. The people sow the different kinds, at different times and seasons, and though one crop turns out bad, yet another succeeds. The summer is so long that of some species of grain they may get two crops. There is hardly a month from May to October or November, inclusive, in which the people do not reap some kind of cereal, or gather some sort of fruit. It would indeed be a very great misfortune if a bad crop should happen; for here, as in many other places, they lay up no stores, and are contented with living from hand to mouth, as the saying goes.

The peach trees were now everywhere in blossom; their leaves had not yet come out of the buds and therefore the flowers showed to greater advantage; their beautiful pale red color had a very fine effect; and they grew so close that the branches were completely covered with them. The other fruit trees were not yet in bloom, but the apple blossoms were beginning to appear.

Currants. The English and the Swedes of America give the name of currants [1] to a shrub which grows in wet ground and near swamps, and which is now in blossom; its fiowers are white, have a very agreeable fragrance and grow in oblong bunches; the fruit is very good eating, when it is ripe; the style is thread-shaped (*filiformis*), and shorter than the stamen; it is divided in the middle, into five parts, or stigmata. Dr. Linné calls it *Cratægus*,[2] and Dr. Gronovius calls it a *Mespilus*.[3]

---

[1] It must be carefully distinguished from what is called currants, in England, which is the *Ribes rubrum.*—F.

[2] *Cratægus tomentosa,* L. *Spec. Pl.* p. 682.—F.

[3] *Mespilus inermis, foliis ovato-oblongis, serratis, subtus tomentosis.* Gronov. *Fl. Virgin.* 55.—F.

The Whippoorwill. The Swedes give the name of whipperiwill, and the English that of whippoorwill, to a kind of nocturnal bird, whose voice is heard in North America, almost throughout the whole night. Catesby and Edwards both have described and drawn it.[1] Dr. Linné calls it a variety of the *Caprimulgas Europæus,* or goat-sucker: its shape, color, size, and other qualities, make it difficult to distinguish it from other species. But the peculiar note of the American one distinguishes it from the European, and from all other birds: it is not found here during winter, but returns with the beginning of summer. I heard it to-day for the first time, and many other people said that they had not heard it before this summer; its English and Swedish name is taken from its note; but accurately speaking, it does not call "whipperiwill," nor "whíp-poor-wíll," but rather "whipperiwhip," so that the first and last syllables are accented, and the intermediate ones but slightly pronounced. The English change the call of this bird into "whip-poor-will," that it may have some kind of signification: it is neither heard nor seen in daytime; but soon after sunset it begins to call, and continues for a good while, as the cuckoo does in Europe. After it has continued calling in one place for a time it removes to another and begins again. It commonly comes several times in a night, and settles close to the houses; I have seen it coming late in the evening, settling on the steps of the house in order to sing its song. It is not very shy, and when a person stands still it will settle close by him and begin to call. It comes to the houses in order to get its food, which consists of insects; and those always abound near houses at night. When it sits calling its "whipperiwhip" and sees an insect passing, it flies up and catches it, and settles again. Sometimes you hear four or five, or more near each other, calling as it were for a wager, and raising a great noise in the woods. They are seldom heard in towns, being either extirpated there or frightened away by frequent shooting. They do not like to sit in trees, but are commonly on the ground or very low bushes or on the lower rails of fences. They always fly near the ground: they continue their calling

---

[1] *Caprimulgus minor Americanus.* Catesby. *Nat. Hist. of Carolina,* vol. III, t. 16. Edward's *Nat. Hist. of Birds.* t. 63.—F.

To-day it is classified as the *Antrostomus vociferus.*—Ed.

at night till it grows dark. They are silent till the dawn of day comes on, and then they call till the sun rises. The sun seems to stop their mouths, or dazzle their eyes so as to make them keep still. I have never heard them call in the middle of the night, though I have hearkened very attentively on purpose to hear it; and many others have done the same. I am told they make no nests but lay two eggs in the open fields. My servant shot at one which sat on a bush near the house, and though he did not hit it it fell down through fear, and lay for some time as if dead, though it recovered afterwards. It never attempted to bite when it was held in the hands, only endeavoring to get loose by stirring itself about. Above and close under the eyes were several black long and stiff bristles, as in other nocturnal birds. The Europeans eat it. Mr. Catesby says the Indians affirm that they never saw these birds, or heard them, before a certain great battle in which the Europeans killed a great number of Indians. Therefore they suppose that these birds, which are restless and utter their plaintive note at night, are the souls of their ancestors who died in battle.

April the 24th

Trees. To-day the cherry trees began to show their blossoms; they had already pretty large leaves. The apple trees also began to bloom, but the cherry trees were earlier and got a greenish hue from their leaves. The mulberry trees were still quite bare, and I was sorry to find that this tree is one of the latest to get leaves, though one of the first to develop fruit.

April the 26th

This morning I travelled to Penn's Neck. The tulip trees, especially the tall ones, looked very green, being covered with leaves; this tree is therefore one of the earliest to acquire foliage.

To-day I saw the flowers of the sassafras tree (*Laurus sassafras*). The leaves had not come out. The flowers have a fine smell.

The Lupine. The *Lupinus perennis* is abundant in the woods, and grows equally in good soil and in poor. I often found it thriving on very poor sandy fields, and on heaths, where no other plants will grow. Its flowers, which commonly appear in the middle of May, make a fine show by their purple hue. I was told, that the cattle eat these flowers very greedily; but I was sorry to find very often that they were not so fond of it, as it is represented, especially when they had anything else to eat; and they seldom touched it notwithstanding its fine green color and its softness. The horses eat the flowers, but leave the stalks and leaves. If ever the cattle eat this plant in spring it is because of necessity and hunger, which give it a relish. This country does not afford any green pastures like the Swedish ones; the woods are the places where the cattle must collect their food. The ground in the woods is quite even with gently rising knolls. The trees stand far apart, but the ground between them is not covered with greensod, for there are but few kinds of grass in the woods, and the blades of it stand single and scattered. The soil is very loose, partly owing to the dead leaves which cover the ground during a great part of the year. Thus the cattle find very little grass in the forests and are forced to be satisfied with all kinds of plants which come in their way, whether they be good or bad food. I saw all spring long how the cattle bit off the tops and shoots of young trees and ate them; for no plants had come up and they stood in general but very thin, scattered here and there, as I have just mentioned. Hence you may easily imagine that hunger compels the cattle to eat plants which they would not touch, were they better provided for. However, I am of the opinion that it would be worth while to make use of this lupine to improve dry sandy heaths, and, I believe, it would not be absolutely impossible to find out the means of making it agreeable to the cattle.

Various Observations. The oaks here have qualities similar to the European ones. They keep their dead leaves almost the whole winter, and are very backward in getting fresh ones. They had no leaves as yet, and were only just beginning to show a few.

The humming bird which the Swedes call "king's bird,"[1] and which I have mentioned before, appeared hereabouts to-day for the first time this spring. Nor had anyone else in this vicinity seen it this year.

A number of oil beetles, (*Meloë proscarabæus*) sat on the leaves of wild hellebore, (*Veratrum album*) and feasted on them. I watched them a great while, and they devoured a leaf in a very few

---

[1] *Kungsfågel.* See page 112. To-day the *kungsfågel* is our golden-crested wren.

minutes. Some of them had already eaten so much that they could hardly creep. Thus this plant, which is almost certain death to other animals, is their dainty food.

The fireflies appeared at night, for the first time this year, and flew about between the trees in the woods, It seemed in the dark as if sparks of fire flew up and down. I will give a more particular account of them in another place.

Towards night I went to Raccoon.

MAY THE 1ST

Last night was so cold that the ground at sunrise was as white as snow from the hoary frost. The Swedish thermometer was a degree and a half below the freezing point (29°3/10 F.). We observed no ice in the rivers or waters of any depth; but upon such only as were about three inches deep the ice lay to the thickness of one third part of a line. The evening before, the wind was south, but the night was calm. The apple trees and cherry trees were in full blossom. The peach trees had almost stopped blooming. Most of the forest trees had already gotten new and tender leaves, and most of them were in flower, as almost all kinds of oaks, the dogwood, (*Cornus florida*), hickory, wild plums, sassafras, horn-beam, beeches, etc.

Frost Damage. The plants which were found damaged by the frost, were the following.

1. The hickory. Most of the young trees of this kind had their leaves killed by the frost, so that they looked all black in the afternoon; the leaves were touched by frost everywhere in the fields, near the marshes, and in the woods.

2. The black oak. Several of these trees had their leaves damaged by the frost.

3. The white oak. Some very young trees of this species had lost their leaves for the same cause.

4. The blossoms of the cherry trees were harmed in several locations.

5. The flowers of the English walnut tree were spoiled entirely by the frost.

6. The *Rhus glabra*. Some of these [sumach] trees had already gotten leaves, and they were killed by the cold.

7. The *Rhus radicans*; the tender young sprouts of this climber suffered from the frost, and had their leaves partly killed.

8. The *Thalictra*, or meadow rues, had both their flowers and leaves hurt by the frost.

9. The *Podophyllum peltatum*. Of this [May apple] plant there was not above one in five hundred hurt by the frost.

10. The ferns. A number of them, which had just come up, were damaged.—I should add several other plants which were spoiled, but I could not distinguish them entirely on account of their tender growth.

I visited several places to-day.

The *Bartsia coccinea* grew in great abundance on several low meadows. Its flower buds were already tinged with a beautiful scarlet which adorned the meadows. It is not yet applied to any use but that of delighting the sight.

Various Observations on Trees. One of the·Swedes here had planted an English walnut tree (*Juglans regia*) in his garden, and it was now about two fathoms high. It was in full blossom, and had already large leaves, whereas the black walnut trees which grow wild in every part of this country had not yet any leaves or flowers. Last night's frost had killed all the leaves of the European variety. Dr. Franklin told me afterwards that there had been some English walnut trees in Philadelphia which thrived very well for a while, but that they had finally been killed by the frost.

I looked about me for the trees which had not yet gotten fresh leaves, and I found the following ones:

Juglans nigra, or the black walnut tree.
Fraxinus excelsior, or the ash.
Acer negundo, called the white ash here.
Nyssa aquatica, the tupelo tree.
Diospyros Virginiana, or the persimmon.
Vitis labrusca, or the fox grapes; and
Rhus glabra, or the sumach.

The trees whose leaves were coming out, were the following:

Morus rubra, the mulberry tree.
Fagus castanea, the chestnut tree.
Platanus occidentalis, or the water beech.
Laurus sassafras, the sassafras tree.
Juglans alba, the hickory.

Some trees of this kind had already large leaves, but others had none at all; the same difference I believe exists among the other species of hickory.

The Virginian cherry tree grows here and there in the woods and glades. Its leaves are already pretty large, but the flowers are not yet entirely open.

The sassafras tree was now everywhere in flower; but as I have just noted, its leaves had not yet quite opened.

Sweet Gum Tree. The *Liquidambar styraciflua,* or sweet gum tree, grows in the woods, especially in wet soil, in and near pools. Its leaves had now sprouted on the top. This tree grows to a great thickness, and its height rivals that of the tallest firs and oaks. As it grows higher, the lower branches die and drop, and leave the trunk smooth and straight, with a great crown at the very summit. The seeds are contained in round, dentated cones, which drop in autumn. It is therefore not particularly pleasant to dance barefooted under these trees. As the tree is very tall, the high winds carry the seeds away to a great distance. I have already given an account of the use of this tree to which I must add the following.

The wood can be made very smooth, because its veins are extremely fine; but it is not hard. You can carve letters on it with a knife, which will seem to be engraved. Mr. Lewis Evans told me that from his own experience no wood in this country was more fit for making brass casting moulds than this. Otherwise carpenters agreed unanimously that this wood had the same properties as tulip wood, namely, that it expands from moisture and shrinks from drought. I inquired of Mr. Bartram whether he on this tree had found the resin, which is so much praised in medicine. He told me that a very pleasant-smelling resin always flows out of any cut or wound which is made in the tree, but that the quantity here was too small to pay for the labor of collecting it. This resin or gum first gave rise to the English name. The further south you go, the greater quantity of gum the tree yields, so that it is easy to collect it. Mr. Bartram was of the opinion, that this tree was properly created for the climate of Carolina and that it was brought by various means as far north as New York. In the southern countries the heat of the sun fills the tree with gum, but in the northern ones it does not.

May the 2nd

This morning I travelled down to Salem, in order to see the country.

Description of Country. The sassafras trees stood widely scattered in the woods, and along the fences round the fields. It was now distinguishable at a distance for its fine yellow flowers. The leaves had not yet fully come out.

In some meadows the grass had already grown up pretty high; but these meadows were marshy, and no cattle had been on them this year. They are mown twice a year, *viz.* in May, and the end of August, or the beginning of September, old style. I saw some meadows of this kind to-day in which I saw grass which was now almost ready to be mown; and many meadows in Sweden do not have such grass at harvest time as these had now. They lay in marshes and valleys, where the sun had very great power. The grass consisted merely of a cyperus grass or sedge.

The wild plum trees were now everywhere in flower; they grow here and there in the woods, but commonly near marshes and in wet ground. They can be seen at a distance by their white flowers. The fruit when ripe is edible.

The *Cornus florida,* or dogwood, grows in the forests, on hills, on plains, in valleys in marshes, and near brooks. I cannot therefore say which is its native soil. However, it seems that in a low but not a wet soil it succeeds best. It was now adorned with its great snowy *involucra*, which rendered it conspicuous even at a distance. At this time it is a pleasure to travel through the woods, so much are they beautified by the blossoms of this tree. The flowers which are within the *involucra* began to open today. The tree does not grow to any considerable height or thickness, but is about the size of our mountain ash (*Sorbus aucuparia*). There are three species of this tree in the woods; one with great white *involucra*, another with small white ones, and a third with reddish ones.

The woods were now full of birds. I saw the lesser species everywhere hopping on the ground or in bushes, without any great fear. It is therefore very easy for all kinds of snakes to approach and bite them. I believe that the rattlesnake has nothing to do but to lie still, and before long some little bird or other will pass by or run directly upon it, giving the snake an opportunity of catching it, without any enchantment.

Salem. Salem is a little trading town, situated a short distance from the Delaware River. The houses stand far apart; they are partly stone and partly wood. A rivulet passes by the town and flows into the Delaware. The inhabitants live by their several trades, as well as they can. In the neighborhood of Salem are some very low and swampy meadows; and therefore it is reckoned a very unwholesome place. Experience has shown, that those who came hither from other places to settle got a very pale and sickly look, though they arrived in perfect health, and with a very fresh color. The town is very easily distinguished about this time by the disagreeable stench which arises from the swamps. The vapors of the putrid water are carried to those inhabitants which live next to the marshes, and enter the body along with the air, and through the pores, and are thus hurtful to health. At the end of every summer, the intermittent fevers are very frequent. I know two young men who came with me from England to America: soon after their arrival at Philadelphia they went to Salem, in perfect health; but a few weeks after they fell sick, and before the winter was half over they had both died.

Saffron. Many of the inhabitants plant saffron; but it is not so good or so strong as the English and French variety. Perhaps it is better after being stored for a number of years, like tobacco.

The *Gossypium herbaceum,* or cotton plant, is an annual, and several of the inhabitants of Salem had begun to sow it. Some had procured the seeds from Carolina, where they have large plantations of cotton; but others got it out of cotton which they had bought. They said it was difficult at first to get ripe seeds from the plants which were planted here, for the summer in Carolina, whence their first seed had come, is both longer and hotter than it is here. But after the plants had become more used to the climate they seemingly hastened more than formerly, so that the seeds were ripe on time.

MAY THE 4TH

The Crab Apple. Crab trees[1] are a species of wild apple trees, which grow in the woods and glades, but especially on little hillocks, near rivers. In New Jersey the tree is rather scarce; but in Pennsyl-

---

[1] *Malus coronaria* perhaps, or *Malus sylvestris,* Linné called it *Pyrus coronaria.* There are several species.

vania it is plentiful. Some people had planted a single tree of this kind near their farms, on account of the fine smell which its flowers afford. Some of its flowers opened about a day or two ago; however, most of them were not yet open. They are exactly like the blossoms of the common apple trees, except that the color is a little more reddish in the crab trees, though some kinds of the cultivated trees have flowers which are almost as red. But the smell distinguishes them plainly, for the wild trees have a very pleasant smell, somewhat like the raspberry. The apples or crabs, are small, sour, and unfit for anything but vinegar. They lie under the trees all winter, and acquire a yellow color. They seldom begin to rot before spring.

Native Trees versus Foreign. I cannot omit an observation here. The crab trees opened their flowers only yesterday and to-day, whereas the cultivated apple trees did not blossom before the twelfth of May; on the other hand, the cultivated European ones, had already opened their blossoms on the twenty-fourth of April. The black walnut trees of this country had neither leaves nor flowers, when the European variety had large leaves and blossoms. Hence it appears that trees brought over from Europe, of the same kind as the wild trees of America, flower much sooner than the latter. I cannot say what the reason is for this forwardness of the European trees in this country, unless they bring forth their blossom as soon as they get a certain degree of warmth, which they had had in their native country. It seems that European trees do not expect, after a certain degree of warmth, any such cold nights as will kill their flowers; for in the cold countries there seldom appear any hot days succeeded by such cold nights as will harm the flowers. On the contrary the wild trees in this country are directed by experience, (if I may so speak) not to trust to the first heat, but wait for a greater one, when they are tolerably safe from cold nights. Therefore it happens often that the flowers of the European trees are killed by the frosts here; but the native trees are seldom damaged, though they be of the same kind as the European. This is a manifest proof of the wisdom of the Creator.

MAY THE 5TH

Rapaapo. Early this morning I went to Rapaapo, which is a very large village, whose farms, however, lie quite scattered. It was inhabited only by Swedes, and not a single Englishman or people of any other nation had settled it. Therefore they have preserved their native Swedish tongue there better than elsewhere, and mixed but few English words with it. The intention of my journey was partly to see the place, and to collect plants and other natural curiosities there, and partly to find the places where the white cedar, or *Cupressus thyoides*, grows.

Wild Honeysuckle. The mayflowers, as the Swedes call them, were plentiful in the woods wherever I went to-day; especially on a dry soil, or one that is somewhat moist. The Swedes have given them this name, because they are in full blossom in May. Some of the Swedes and the Dutch call them *Pinxterbloem* (Whitsunday flowers), as they really are in blossom about Whitsuntide. The English call them wild honeysuckle, and at a distance they have some similarity to the honeysuckle or *Lonicera*. Dr. Linné and other botanists call it an *Azalea*.[1] Its flowers were now open, and added a new ornament to the woods, being little inferior to the flowers of the honeysuckle and *Hedysarum*. They are arranged in a circle round the stem's extremity, and have either a dark red or a light red color; but, by standing for some time the sun bleaches them, and at last they become whitish. I know not why Colden calls them yellow.[2] The height of the bushes is not always the same. Some are as tall as a full grown man, and taller, others but low, and some are not above a palm from the ground, though all full of flowers. The people have not yet found that this plant may be applied to any practical use; they only gather the flowers and put them in pots, because they are so beautiful. They have some smell, but I cannot say it is very pleasant.[3] However, the beauty of the color entitles them to a place in every flower garden.

To-day I saw the first ear of this year's rye. In Sweden, rye begins to show its ears about Ericmas, that is, about the eighteenth of May, old style.[4] But in New Sweden, the people said they always saw the ears of rye in April, old style, whether the spring began late

---

[1] *Azalea nudiflora.* Linné *Spec. Plant.* p. 214. *Azalea ramis infra flores nudis.* Gron. *Virg.* 21.

[2] *Azalea erecta, foliis ovatis, integris, alternis, flore luteo, pilos, præcoci.* Cold. *Ebor.,* 25.

[3] The editor disagrees, if it is the flower known as the wild honeysuckle in Connecticut, *Azalea nudiflora* or Pinxterflower.

[4] Accordingly about the twenty-ninth of May, new style.

or early. However, in some years the ears come early, and in others late in April. This spring was reckoned one of the late ones.

Frogs. Bullfrogs are a large species of frog, which I had an opportunity of hearing and seeing to-day. As I was riding out I heard a roaring before me; and I thought it was a bull in the bushes on the other side of the dike, though the sound was rather more hoarse than that of a bull. I was however afraid that a bad goring bull might be near me, though I did not see him; and I continued to think so till some hours after, when I talked with some Swedes about the bullfrogs, and by their account I immediately found that I had heard them, for the Swedes told me that there were numbers of them in the dike. I afterwards hunted them. Of all the frogs in this country this is doubtless the biggest. I am told that towards autumn, as soon as the air begins to grow a little cool, they hide themselves under the mud which lies at the bottom of ponds and stagnant waters, and lie there torpid during the winter. As soon as the weather grows mild, towards summer, they begin to get out of their holes, and croak. If the spring, that is, if the mild weather, begins early, they appear about the end of March, old style; but if it comes later, they stay under water till April. They live in ponds and bogs of stagnant water; they are never in any flowing water. When many of them croak together, they make an enormous noise. Their croak resembles exactly the roaring of an ox or bull, which is somewhat hoarse. They croak so loud that two people talking by the side of a pond cannot understand each other. It seems as if they had a captain among them: for when he begins to croak, all the others follow; and when he stops, the others are all silent. When this captain gives the signal for stopping, you hear a note like "poop" coming from him. In daytime they seldom make any great noise, unless the sky is cloudy. But the night is their croaking time; and, when all is calm you may hear them, though you are near a mile and a half off. When they croak, they are commonly near the surface of the water, under the bushes, and have their heads out of the water. Therefore, by going slowly, one may get close up to them before they disappear. As soon as they are completely under water, they believe themselves safe, though the water be shallow.

Sometimes they sit at a good distance from the pond; but as soon as they suspect any danger, they hasten with great leaps into the water. They are very expert at hopping. A full grown bullfrog

takes nearly three yards at a hop. I have often been told the following story by the old Swedes, an occurrence which happened here at
the time when the Indians lived with the Swedes. It is well known,
that the Indians are excellent runners; I have seen them at Colonel
Johnson's [1] equal the best horse in its swiftest course, and almost
pass by it. Therefore, in order to try how well the bullfrogs could
leap, some of the Swedes laid a wager with a young Indian that he
could not overtake the frog, provided it had two leaps handicap.
They carried a bullfrog, which they had caught in a large pond,
upon a field, and burnt his posterior with a torch, and then let him
go.[2] The fire, and the Indian who endeavored to keep close to the
frog, had such an effect upon it, that it made its long hops across
the field as fast as it could. The Indian began to pursue the frog
with all his might at the proper time: the noise he made in running
frightened the poor frog; undoubtedly it was afraid of being tor-
tured with fire again, and therefore it redoubled its leaps, and by
that means it reached the pond before the Indian could overtake it.

Frog's Legs. In some years bullfrogs are more numerous than
others. Nobody could tell, whether the snakes had ever ventured to
eat them, though they eat all the lesser kinds of frogs. The women
are no friends to these beasts, because they kill and eat young duck-
lings and goslings. Sometimes they carry off chickens that come too
near the ponds. I have not observed that they bite when they are
held in the hands, though they have some small teeth. When they
are beaten, they cry out like children. I was told that some people
eat the thighs and hind legs, and that they are very palatable.

"White Cedar." A tree which grows in the swamps here, and in
other parts of America, goes by the name of the white juniper tree.
Its trunk indeed looks like one of our old tall, straight juniper trees
in Sweden: but the leaves are different, and the wood is white. The
English call it white cedar, because the boards which are made of
the wood, are like those made of cedar. But neither of these names
is correct, for the tree is of the cypress variety.[3] It always grows in
wet ground or swamps: it is therefore difficult to get to it, because
the ground between the little hillocks is full of water. The trees
stand both on the hillocks and in the water: they grow very close

---

[1] Later Sir William Johnson. See page 190, note.

[2] Did Mark Twain know this story when he wrote his "Jumping Frog"?

[3] *Cupressus thyoides.* L. *Spec. Pl.* p. 1422. *Cypressus Americana, fructo minimo.*
Miller's *Gard. Dictionary.*—F.

together, and have straight, thick and tall trunks; but their numbers
have been greatly reduced. In places where they are left to grow up,
they grow as tall and as thick as the tallest fir trees. They preserve
their green leaves both in winter and summer; the tall ones have no
branches on the lower part of the trunk.

Its Habitat and Uses. The marshes where these trees grow are
called cedar swamps, which are numerous in New Jersey, and likewise in some parts of Pennsylvania and New York. The most
northerly place where it has been found hitherto is near Goshen in
New York, under forty-one degrees and twenty-five minutes north
latitude, as I am informed by Dr. Colden. To the north of Goshen it
has not been found in the woods. The white cedar is one of the
trees which resist decay the most; and when it is put above ground,
it will last longer than underground. Therefore it is employed for
many purposes; it makes good fence rails, and also posts which are
to be put into the ground; but in this point, the red cedar is still preferable to the white. It likewise makes good canoes. The young
trees are used for hoops round barrels, turns, etc., because they are
thin and pliable; the thick, tall trees afford timber and wood for
cooper's work. Houses which are built of it surpass in duration
those which are built of American oak. Many of the houses in
Rapaapo were made of this white cedar; but the chief article which
the white cedar produces is the shingle, which is the best for several reasons; first, it is more durable than any other made of
American wood, the red cedar shingles excepted; secondly it is
very light, so that no strong beams are required to support the roof.
For the same reason it is unnecessary to build thick walls, because
they do not have to support heavy roofs. When fires break out it is
less dangerous to go under or along the roofs, because the shingles
being very light can do little hurt by falling. They suck the water,
being somewhat spongy, so that the roofs can easily be wetted in
case of fire. On the other hand, their oiliness prevents the water
from damaging them, for it evaporates easily. When they burn and
are carried about by the wind, they constitute what is commonly
called a dead coal, which does not easily set fire where it alights. The
roofs made of these shingles can easily be cut through, if required
because they are thin, and not very hard. For these qualities the
people in the country and in the towns are very desirous of having
their houses covered with white cedar shingles, if the wood can be

gotten. Therefore all churches, and the houses of the more substantial inhabitants of the towns, have shingle roofs. In many parts of New York province where the white cedar does not grow, the people however, have their houses roofed with cedar shingles, which they get from other parts. To that purpose great quantities of shingles are annually transported from Eggharbor and other parts of New Jersey, to the town of New York, whence they are distributed throughout the province. A quantity of white cedar wood is likewise exported every year to the West Indies, for shingles, pipe staves, etc. Thus the inhabitants here, are not only lessening the number of these trees, but are even extirpating them entirely. People are here (and in many other places) in regard to wood, bent only upon their own present advantage, utterly regardless of posterity. By these means many swamps are already quite destitute of cedars, having only young shoots left; and I plainly observed by counting the circles of the trunk, that they do not grow up very quickly, but require a great deal of time before they can be cut for timber. It is well known that a tree gets only one circle every year; a trunk eighteen inches in diameter, had one hundred and eight circles round the thicker end; another seventeen inches in diameter, had a hundred and sixteen; and another, two feet in diameter had one hundred and forty-two circles upon it. Thus nearly eighty years growth is required before a white cedar raised from seed can be used for timber. Among the advantages which the white cedar shingles have over others, [as I have just noted] the people reckon their lightness. But this good and useful quality may in the future turn out very disadvantageous to Philadelphia and other places where the houses are roofed with cedar shingles; for as the roofs made of these shingles are very light, and bear but a trifling weight on the walls, the people have made the latter very thin. I measured the thickness of the wall of several houses here, three stories high (cellar and garret not included), and found most of them nine inches and a half, and some ten inches thick. Therefore it is by no means surprising that violent hurricanes sometimes make the brick gable ends vibrate perceptibly, especially in such houses as have a very open location. And since the cedar trees will soon be gone in this country, and the present roofs when rotten must be replaced with heavier ones, of brick or of other wood, it is more than probable that the thin walls will not be able to bear such an additional weight, and will either break or require support by props: or else the whole house must be pulled down and rebuilt with thicker walls. This observation has already been made by myself and others. Some of the people here make use of the chips [or shavings] of white cedar instead of tea, assuring me that they preferred it in regard to its wholesomeness to all foreign tea. All the inhabitants here were of the opinion that the water in the cedar swamps is wholesomer than any other drink: it creates a great appetite, which they endeavored to prove by several examples. They ascribed this quality to the water itself, which is filled with the resin of the trees, and to the exhalations which come from the trees, and can easily be smelled. The people also thought that the yellowish color of the water, which stands between the cedar trees, was due to the resin which comes out of the roots of these trees. They likewise all agreed that this water was always very cold in the hottest season, which might be partly owing to the continual shade it is in. I knew several people who were resolved to go to these cedar swamps, and use the waters for the recovery of their appetite. Mr. Bartram had planted a white cedar in a dry soil, but it could not succeed there. He then put it into a swampy ground, where it got, as it were, new life, and thrived very well; and though it was not taller than a man, yet it was full of cones. Another thing is very remarkable with regard to the propagation of this tree: Mr. Bartram had cut off its branches in spring two years in succession and put them into the swampy soil, where they took root and succeeded very well. I have seen them myself.

Red Cedar. The red juniper is another tree which I have mentioned very frequently in the course of my account. The Swedes have given it the name of red juniper because the wood is very red and fine within. The English call it red cedar, and the French *Cèdre rouge.* However, the Swedish name is the most correct, as the tree belongs to the junipers.[1] At its first growth it has a good deal of similarity to the Swedish juniper,[2] but after it is grown up it gets quite different leaves. The berry resembles exactly that of the Swedish juniper, in regard to its color and shape; however, it is not so big, though the red cedar grows very tall. At Raccoon these trees stood single, and were not so tall; but at other places I have seen them standing together in groves. They like the same ground

---

[1] *Juniperus Virginiana.* L. *Spec. Pl.* p. 114.—F.
[2] *Juniperus communis.* L. *Spec. Pl.* p. 1470.—F.

as the common Swedish juniper, especially the rising banks of rivers and other high land, in a dry and frequently poor soil. I have seen them growing in abundance, as thick and tall as the tallest fir trees, on poor dry and sandy heaths. Near Canada, or in the most northerly places where I have seen them, they usually choose the steep sides of the mountains, and there they grow promiscuously with the common juniper. The most northerly places where I have found them wild in the woods is in Canada, eighteen French miles south of the Fort Saint Jean, or St. John, in about 44°35' north latitude. I have likewise seen it growing very well in a garden, on the island of Magdalene, in the river St. Lawrence, close by the town of Montreal, in Canada, belonging to the then governor of Montreal, Monsieur le Baron de Longueuil. But it had come from the south and was transplanted here. Of all the woods in this country, this is without exception the most durable, and withstands weathering longer than any other. It is therefore employed in all cases where wood is most liable to rot, especially for all kinds of posts which are to be put into the ground. Some people say that if an iron be put into the ground along with a pole of cedar, the iron will be half corroded by rust in the same time that the wood will be rotten. In many places both the rails and the posts belonging to them are made of red cedar. The best canoes, consisting of a single piece of wood, are made of red cedar; for they last longer than any others, and are very light. In New York I have seen quite large yachts built of this wood. Several yachts which go from New York to Albany, up the river Hudson, are built in a different manner, as I have mentioned in the first volume.[1] In Philadelphia they cannot make any yachts or other boats of red cedar, because the quantity and the size of the trees will not allow it. For the same reason they do not roof their houses with red cedar shingles; but in such places where it is plentiful it makes excellent roofs. The heart of this cedar is of a fine red color, and whatever is made of it looks very fine, and has a very agreeable and wholesome smell. But the color fades by degrees, or the wood would be very suitable for cabinet work. I saw a parlor in the countryseat of Mr. Isaac Norris, one of the members of the Pennsylvania House of Assembly, wainscoted many years ago

---

[1] See vol. I, page 62 of this edition. "The lower part of the yachts, which is continually under water, is made of black oak; the upper part is built of red cedar, because it is sometimes above and sometimes in the water."—F

with boards of red cedar. Mr. Norris assured me that the cedar looked exceedingly well in the beginning, but it was quite faded when I saw it, and the boards looked very shabby. The boards near the window especially had entirely lost their color, so that Mr. Norris had been obliged to put mahogany in their stead. However, I am told that the wood will keep its color if a thin varnish is put upon it while it is fresh, and just after it has been planed, and if care is taken that the wood is not afterwards rubbed or scratched. At least it makes the wood keep its color much longer than it would otherwise. Since it has a very pleasant smell, when fresh, some people put the shavings and chips of it among their linen to secure it against being worm eaten. Some likewise get bureaus, [chests], etc. made of red cedar, for the same reason. But it is only useful for this purpose as long as it is fresh, for it loses its smell after a time, and is then no longer good for keeping out insects. It is sometimes sent to England as timber and brings a good price. In many places round Philadelphia on the country estates there are avenues of red cedars leading from the high road to the house. The lower branches are cut and only a fine crown left. In winter, when most other trees have lost their leaves, this looks very attractive. This tree has likewise a very slow growth, for a trunk thirteen inches and a quarter in diameter had one hundred and eighty-eight rings, or annual circles, and another, eighteen inches in diameter, had at least two hundred and fifty, for a great number of the rings were so fine that they could not be counted. This tree is propagated in the manner as the common juniper tree is in Sweden, *viz.* chiefly by birds, which eat the berries and emit the seeds whole. To encourage the planting of this useful tree, a description of the method of doing it, written by Mr. Bartram, was inserted in a Pennsylvania almanack, called *Poor Richard Improved*, for the year 1749. In it was explained the manner of planting and augmenting the number of these trees, and mention is made of some of the purposes to which they may be employed.

In the evening I returned to Raccoon.

MAY THE 6TH

The *Mulberry tree* (*Morus rubra*) began to blossom about this time, but its leaves were yet very small. The people divided the

trees or flowers into male and female, and said that those which
never bore any fruit were males, and those which did, females.

Smilax laurifolia was superabundant in all the swamps near this
place. Its leaves were now beginning to come out, for it sheds them
all every winter. It climbs up along trees and shrubs, and runs
across from one tree or bush to another. By this means it shuts up
the passage between the trees, fastening itself everywhere with its
cirrhi or tendrils, so that it is with the utmost difficulty one can
force a passage in the swamps and woods where it is plentiful. The
stalk towards the bottom is full of long spines, which are as strong
as those of a rosebush. They catch hold of clothes and tear them.
This troublesome plant may sometimes bring you into imminent
danger when botanizing or going into the woods, for, not to mention that one's clothes must be absolutely ruined by its numberless
spines, it occasions a deep shade in the woods, by crossing from
tree to tree so often. This forces you to stoop, and even to creep on
all fours through the little passages which are left close to the
ground, and then you cannot be careful enough to prevent a snake
(of which there are large numbers here) from darting into your
face. The stalk of the plant is the same color as that of young rose-
bushes. It is green and smooth between the spines, so that a stranger
might take it to be a kind of thorn in winter, when the leaves are
gone. It is therefore called "green thorn" by the Swedes.

MAY THE 8TH

Caterpillars. The trees here were now overrun with innumerable
caterpillars; one kind especially was worse than all the others. They
immediately formed great white webs between the branches of the
trees, so that they could be seen, even at a distance. In each of these
webs were thousands of caterpillars, which crept out of them afterwards and spread chiefly upon the apple trees. They consumed the
leaves, and often left not one on a whole branch. I was told that
some years ago they did so much damage that the apple trees and
peach trees hardly bore any fruit at all, because they had consumed
all the leaves and exposed the naked trees to the intense heat of the
sun which caused several of the trees to die. The people took the
following method of killing these caterpillars. They fixed some
straw or flax on a pole, set it on fire, and held it under the webs or

nests, so that a part was burnt, and a part fell to the ground. However, large numbers of the caterpillars crept up the trees again,
which could have been prevented, if they had been trod upon or
killed in some other way. I called chickens to the places where they
crept on the ground in such numbers, but they would not eat them.
Nor did the wild birds like them; for the trees were full of these
webs, though whole flocks of little birds had their nests in the
gardens and orchards.

MAY THE 18TH

Sweden versus America. Though it was already quite late in May
the nights were very dark here. About an hour after sunset it was so
dark that it was impossible to read a book, though the type was ever
so large. About ten o'clock on a clear night the darkness had increased so much that it looked like one of the darkest though starlight
nights in an autumn in Sweden. It also seemed to me that though
the nights were clear, the stars did not give so great a light as they
did in Sweden. And as at this season the nights were usually dark,
and the sky covered with clouds, I could compare them only to dark
and cloudy Swedish winter nights. It was therefore, at this time of
the year, very difficult to travel during such cloudy nights; for neither
man nor horse could find his way. The nights here in general seem
very disagreeable to me, in comparison with the light and glorious
summer nights of Sweden. Ignorance sometimes makes us think
slightly of our native land, Sweden. If other countries have their
advantages, Sweden also has hers; and upon comparing the advantages and disadvantages of different places, Sweden will be found to
be not inferior to any of them.

Old and New Sweden Compared. I shall briefly mention in what
points I think Sweden is preferable to this part of America; and why
I, as Ulysses did with Ithaca, prefer Old Sweden to New Sweden.

Dark Nights and Changeable Weather. The nights are very dark
here all summer; and in winter they are quite as dark, if not darker,
than the winter nights in Sweden; for here is no Aurora Borealis, and
the stars give a very faint light. It is very remarkable if an Aurora
Borealis appears once or twice a year. The winters here bring no
snow, to make the nights clear and travelling more safe and easy.
The cold is, however, frequently as intense as in Old Sweden. The

snow which falls lies only a few days, and always goes off with a
great deal of wet. The rattlesnakes, horned snakes, red-bellied, green,
and other poisonous snakes, against whose bite there is frequently
no remedy, are numerous here. To these I must add the wood lice
with which the forests are so pestered that it is impossible to pass
through a bush or to sit down, though the place be ever so pleasant,
without having a whole swarm of them on your clothes. The inconvenience and trouble they cause, both to man and beast, I have
described in the *Memoirs* of the Royal Swedish Academy of Sciences.
The weather is so inconstant here that when a day is most excessively
hot, the next is often cold. This sudden change often happens in one
day; and few people can endure these changes, without impairing
their health. The heat in summer is intense, and the cold in winter
often more piercing. However, one can always secure one's self
against the cold; but when the great heat is of any duration, there is
hardly any protection against it. It tires one so that one does not
know which way to turn. It has frequently happened that people
who walked into the fields dropped down dead on account of the
violence of the heat. Several illnesses prevail here; and they increase
every year. Nobody is left unattacked by the intermitting fever; and
many people are forced to suffer it every year, together with other
diseases. Peas cannot be sown, on account of the insects which consume them. There are worms in the rye seed, and myriads of them
in the cherry trees. The caterpillars often eat all the leaves from the
trees, so that they cannot bear fruit that year; and large numbers die
every year, both of fruit trees and forest trees. The grass in the
meadows is likewise consumed by a kind of worm; another species
causes the plums to drop before they are half ripe. The oak here
affords not nearly so good timber as the European oak. The fences
cannot last above eighteen years. The houses are of no long duration.
The meadows are poor, and what grass they have is bad. The pasture for cattle in the forests consists of such plants as they do not like,
and which they are compelled to eat by necessity; for it is difficult to
find any grass in the great forests where the trees stand far apart, and
where the soil is excellent. For this reason, the cattle are forced, during almost the whole winter and part of the summer, to live upon
the young shoots and branches of trees, which sometimes have no
leaves: therefore, the cows give very little milk, and decrease in size
every generation. The houses are extremely unfit for winter habita-

tion. Hurricanes are frequent, which overthrow trees, carry away roofs, and sometimes houses, and do a great deal of damage. Some of these inconveniences might be remedied by diligence; but others cannot, or only with difficulty. Thus every country has its advantages, and its defects: happy is he who can content himself with his own lot! [1]

Careless Agriculture. The rye grows very poorly in most of the fields, which is chiefly owing to the carelessness in agriculture, and to the poorness of the fields, which are seldom or never manured. After the inhabitants have converted a tract of land into a tillable field, which has been a forest for many centuries, and which consequently has a very fine soil, the colonists use it as such as long as it will bear any crops; and when it ceases to bear any, they turn it into pastures for the cattle, and take new grain fields in another place, where a rich black soil can be found and where it has never been made use of. This kind of agriculture will do for a time; but it will afterwards have bad consequences, as every one may clearly see. A few of the inhabitants, however, treated their fields a little better: the English in general have carried agriculture to a higher degree of perfection than any other nation. But the depth and richness of the soil found here by the English settler (as they were preparing land for plowing, which had been covered with woods from times immemorial) misled them, and made them careless husbandmen. It is well known that the Indians lived in this country for several centuries before the Europeans came into it; but it is likewise known, that they lived chiefly by hunting and fishing, and had hardly any agriculture. They planted corn and some species of beans and pumpkins; and at the same time it is certain that a plantation of such vegetables as serve an Indian family during one year take up no more ground than a farmer in our country takes to plant cabbage for his family. At least, a farmer's cabbage and turnip ground, taken together, is always as extensive, if not more so, than all the corn fields and kitchen gardens of an Indian family. Therefore, the Indians could hardly subsist for one month upon the produce of their gardens and fields. Commonly, the little villages of Indians are about twelve or eighteen

---

[1] In comparing his native land with America Kalm rises to commendable heights, for here profound feeling, intelligence and conviction dictate the style. The same is true in the following section where the practical scientist lashes the Colonial farmer for his carelessness and shortsightedness.

miles distant from each other. Hence one may judge how little ground was formerly employed for planting; and the rest was overgrown with large, tall trees. And though they cleared (as is yet usual) new ground, as soon as the old one had lost its fertility, such little pieces as they made use of were very inconsiderable, when compared to the vast forest which remained. Thus the upper fertile soil increased considerably, for centuries; and the Europeans coming to America found a rich, fine soil before them, lying as loose between the trees as the best bed in a garden. They had nothing to do but to cut down the wood, put it up in heaps, and to clear the dead leaves away. They could then immediately proceed to plowing, which in such loose ground is very easy; and having sown their grain, they got a most plentiful harvest. This easy method of getting a rich crop has spoiled the English and other European settlers, and induced them to adopt the same method of agriculture as the Indians; that is, to sow uncultivated grounds, as long as they will produce a crop without manuring, but to turn them into pastures as soon as they can bear no more, and to take on new spots of ground, covered since ancient times with woods, which have been spared by the fire or the hatchet ever since the Creation. This is likewise the reason why agriculture and its science is so imperfect here that one can travel several days and learn almost nothing about land, neither from the English, nor from the Swedes, Germans, Dutch and French; except that from their gross mistakes and carelessness of the future, one finds opportunities every day of making all sorts of observations, and of growing wise by their mistakes. In a word, the grain fields, the meadows, the forests, the cattle, etc. are treated with equal carelessness; and the characteristics of the English nation, so well skilled in these branches of husbandry, is scarcely recognizable here. We can hardly be more hostile toward our woods in Sweden and Finland than they are here: their eyes are fixed upon the present gain, and they are blind to the future. Their cattle grow poorer daily in quality and size because of hunger, as I have before mentioned. On my travels in this country I observed several plants, which the horses and cows preferred to all others. They were wild in this country and likewise grew well on the driest and poorest ground, where no other plants would succeed. But the inhabitants did not know how to turn this to their advantage, owing to the little account made of Natural History, that science being here (as in other parts of the world) looked upon as a mere trifle, and the pastime of fools. I am certain, and my certainty is founded upon experience, that by means of these plants, in the space of a few years, I should be able to turn the poorest ground, which would hardly afford food for a cow, into the richest and most fertile meadow, where great flocks of cattle would find superfluous food, and grow fat. I own that these useful plants are not to be found on the grounds of every planter: but with a small share of natural knowledge, a man could easily collect them in the places where they are to be had. I was astonished when I heard the country people complaining of the badness of the pastures; but I likewise perceived their negligence, and often saw excellent plants growing on their own grounds, which only required a little more attention and assistance from their unexperienced owner. I found everywhere the wisdom and goodness of the Creator; but too seldom saw any inclination to make use of them or adequate estimation of them, among men.

> *O fortunatos nimium, sua si bona norint,*
> *Agricolas!* Virg.

I have been led to these reflections, which may perhaps seem foreign to my purpose, by the bad and neglected state of agriculture in every part of this continent, and because I wanted to show the reason why this journal is so thinly stocked with economical advantages in several branches of husbandry. [There were so few of them]. I do not however deny, that I have here and there found skilful farmers, but they were very scarce.

Poultry Enemies. Birds of prey which pursue the poultry are found in abundance here and are even more plentiful than in Sweden. They enjoy great security here, as there are still great forests in many places, whence they can come unawares upon chickens and ducks. To the birds of prey it is quite indifferent whether the woods consist of good or bad trees, provided they have protection there. At night the owls, which are very numerous, endanger the safety of the tame fowls. They live chiefly in the marshes, give a disagreeable shriek at night, and attack the chickens, which commonly roost at night in the apple, peach, and cherry trees in the garden. But since people are clearing this country of woods, as we are in Sweden and Finland, it may result in exposing the birds of prey more than at present, and in depriving them of the opportunities of doing mischief with so much ease.

Stags. The thick forests of America contain numerous stags; they do not seem to be a different species from the European stags. An Englishman was possessed of a tame hind. It is observed that though these creatures are very shy when wild in the woods and the cedar swamps, which are very much frequented by them, they can be tamed to such a degree, if taken young, that they will come of their own accord to people, and even to strangers: this hind was caught when it was very small. The color of its whole body was a dirty reddish brown, the belly and the under side of the tail excepted, which were white; the ears were gray; the head, towards the nose, was very narrow, but upon the whole the creature looked slim and trim. The hair lay close together, and was quite short; the tail reached almost to the bend of the knee, near which, on the inside of each hind-foot, was a knob or callus. The owner of the hind said that he had tamed several stags, by catching them while they were very young, and had owned this one three years. She was now with calf. It had a little bell hung about its neck, so that by walking in the woods the people might know it to be tame and take care not to shoot it. It was at liberty to go where it pleased, and to keep it confined would have been a pretty hard task, as it could leap over the highest fences. Sometimes it went far into the woods, and frequently staid away a night or two, but afterwards returned home like other cattle. When it went into the woods, it was often accompanied by wild stags, and decoyed them back even into the very houses, especially in rutting time, giving its master numerous opportunities of shooting the wild stags almost at his door. Its scent was excellent, and when it had turned toward the wind, I often saw it rising and looking in that direction; though I did not see any people on the road, they commonly appeared about an hour after. As soon as the wild deer have the scent of a man they make off. In winter the man fed the hind with grain and hay; but in summer it went out into the woods and meadows, seeking its own food, eating both grass and other plants. It was now kept in a meadow; it ate chiefly clover and the leaves of hickory, the *Andromeda paniculata,* and the *Geranium maculatrum.* It was also fond of the leaves of the common plantain, or *Plantago,* grasses and several other plants. The owner of this hind sold stags to people in Philadelphia who sent them as curiosities to other places. He got twenty-five, thirty, or forty shillings apiece for them. The food of the wild stags in summer is grass and various herbs; but in winter when they cannot get them, they eat the shoots and young buds of branches. I have already mentioned [1] that they eat without any danger the spoon tree, or *Kalmia latifolia*, which is poisonous to other animals. In the long, severe winter, which commenced here upon the nineteenth of December, 1740, and continued to the thirteenth of March, old style, during the course of which there fell a great quantity of snow, the deer were found dead, but chiefly further north, where the snow was deeper. Nobody could determine whether their death was the consequence of the great quantity and depth of the snow, which hindered their getting out, whether the cold had been too severe and of too long duration, or whether they were short of food. The old people likewise relate, that vast numbers of stags came down in the year 1705, when there was a heavy fall of snow, near a yard deep, and that they were afterwards found dead in the woods, in great numbers, because the snow was deeper than they could pass through. Innumerable birds were likewise found dead at that time. That same winter, a stag came to Matsong into the stables and ate hay together with the cattle. It was so pinched by hunger, that it grew tame immediately and did not run away from the people. It afterwards remained on the farm as any other domesticated animal. All aged persons asserted that formerly this country had had more stags than it has at present. It was formerly not uncommon to see thirty or forty of them in a herd together. The reason of their decrease is chiefly owing to the increase of population, the destruction of the woods, and the number of people who kill and frighten the stags at present. However, farther north where there are great forests and frontier conditions there are yet great numbers of them. Among their enemies are the lynxes of this country, which are the same as the Swedish ones. [2] They climb up the trees, and when the stag passes beneath, they dart down upon him, get a firm grip, bite and suck the blood, and never give up till they have killed him.

Dung Beetles. I saw several holes in the ground, both on hills and on fields, and fallow grounds; they were round, and commonly about

---

[1] See p. 178.

[2] *Varglo; Felis Lynx,* L. The Swedes mention two kinds of lynx, the one is called the *Varglo,* or wolf-lynx, and the other the *Kattlo,* or cat-lynx. The Germans make the same distinction, and call the former *Wolf-luchs,* and the latter *Katz-luchs:* the former is the biggest, of a brownish red, mixed with gray and white, on its back, and white towards the belly, with brownish spots; the latter is smaller, and has a coat which is more white, and with more spots.—F.

an inch wide; they went almost perpendicularly into the earth, and were made by dung beetles, or by angle worms. The dung beetles had dug very deep into the ground, through horse dung, though it lay on the hardest road, so that a great heap of earth lay near it. These holes were afterwards occupied by other insects, especially grasshoppers, (*Grylli*) and *Cicadæ*; for by digging these holes up I usually found one or more young ones of these insects which had not yet attained their full size.

MAY THE 19TH

Kalm Leaves Raccoon. This morning I left Raccoon, the parish in the country called New Sweden, and which is yet chiefly inhabited by Swedes, in order to proceed in my travels toward the north. I first intended to set out at the beginning of April, but for several reasons this was not advisable. No leaves had come out at that time, and hardly any flowers appeared. I did not know what plants grew here in spring; for the autumnal plants are different from the vernal ones. The Swedes had this winter told me the agricultural and medical uses of many plants, to which they gave names unknown to me: they could not then show me those plants on account of the season and by their deficient and erroneous descriptions I was not able to guess what plants they meant. By going away so early as the beginning of April, I would have remained in uncertainty in regard to these things. It was therefore fit, that I should spend a part of the spring at Raccoon, especially as I had still time enough left for my tour to the north.

The Black Snake. On the road we saw a black snake, which we killed, and found to be just five feet long. Catesby has described it and its qualities, and also drawn it.[1] The fullgrown black snakes are commonly between five and six feet long, but very slender. The thickest I ever saw was in the broadest part hardly three inches thick. The back is black, shining, and smooth; the chin white and smooth; the belly whitish turning into blue, shining, and very smooth. There is probably more than one variety of this snake. One which was forty-five inches long, had a hundred and eighty-six abdominal scutes (*Scuta abdominalia*) and ninety-two caudal plates (*Squamæ*

---

[1] *Anguis niger.* See Catesby's *Nat. Hist. of Carol.* II, t. 48.

subcaudales)[1], which I found to be true, by a repeated counting of them. Another, which was about forty-two inches in length, had a hundred and eighty-four abdominal scutes, and only sixty-four caudal scales. This I likewise assured myself of by counting the scutes over again. It is possible that the end of this last snake's tail was cut off, and the wound healed up again.[2]

The country abounds with black snakes. They are among the first that come out in spring, and often appear very early if warm weather comes, while if it grows cold again after that, they become numb and lie stiff and torpid on the ground or on the ice. When taken in this state and put before a fire, they revive in a short time. It has sometimes happened, when the beginning of January is very warm, that they come out of their winter habitations. They commonly appear about the end of March, old style.

This is the swiftest of all the snakes which are to be found here, for it moves so quick that a dog can hardly catch it. It is therefore almost impossible for a man to escape it, if pursued, but happily its bite is neither poisonous nor in any other way dangerous. Many people have been bitten by it in the woods, and have scarcely felt any more inconvenience than if they had been cut by a knife; the wounded place remains painful for only a short time. The black snakes seldom do any harm, except in spring when they copulate; but if anybody comes in their way at that time they are so vexed that they pursue the intruder as fast as they can, and if they then meet with a person who is afraid of them, he will be in great trouble. I am acquainted with several people, who have on such an occasion run hard enough to be quite out of breath in endeavoring to escape the snake, which moved with the swiftness of an arrow after them. If a person thus pursued can muster up courage enough to face the

---

[1] Kalm uses the terms *segmenta abdominis* and *semisegmenta caudæ*, respectively, for these scutes or scales.

[2] It has been found by repeated experience that the specific character employed by Dr. Linné for the distinction of the species of snakes, taken from their *Scuta abdominalia & caudalia*, or their *Squamæ subcaudales*, varies greatly in snakes of the same species, so that often the difference amounts to ten or more: the whole number of the *scuta* sometimes helps to find out the species; care ought however to be taken, that the snake may not by any accident have lost its tail, and that it be growing again; in which case, it is impossible to make use of this character. The character is not quite so good and decisive, as may be wished, but neither are the marks taken from colors, spots, stripes, etc., quite constant; and so it is better to make use of an imperfect character, than none at all. Time, and greater acquaintance with this class of animals, may perhaps clear up their natural characters.—F.

snake with a stick or anything else, when it either passes by him or when he steps aside to avoid it, it will turn back and seek refuge in flight. It is, however, sometimes bold enough to rush up to a man, and not to depart before it has received a good blow. I have been assured by several, that when it overtakes a person, who has tried to escape it, and who has not had courage enough to oppose it, it winds itself around his feet, so as to make him fall down; it then bites him several times in the leg, or whatever part it can get hold of, and goes off again. I shall mention two circumstances, which confirm what I have said. During my stay in New York, Dr. Colden told me, that in the spring, 1748, he had several workmen at his countryseat, and among them one lately arrived from Europe, who of course knew very little of the qualities of the black snake. The other workmen seeing a great black snake copulating with its female, engaged the newcomer to go and kill it, which he intended to do with a little stick. But on approaching the place where the snakes lay, they perceived him, and the male, maddened by the intrusion, left his pleasure to pursue the fellow with amazing swiftness. The latter little expected such courage in the snake, and flinging away his stick began to run as fast as he was able. The snake pursued him, overtook him, and twisting itself several times round his feet, threw him down, and frightened him almost out of his senses; he could not get rid of the snake, till he took a knife and cut it through in two or three places. The other workmen enjoyed this fight, and laughed at it, without offering to help their companion. Many people at Albany told me of an accident which happened to a young lady, who went out of town in summer, together with many other girls, attended by her negro. She sat down in the wood, in a place where the others were running about, and before she was aware, a black snake being disturbed in its amours, ran under her petticoats and twisted itself round her waist so that she fell backwards in a swoon occasioned by her fright or by the compression which the snake caused. The negro came up to her and suspecting that a black snake might have hurt her, on making use of a remedy to bring his lady to herself again, he lifted up her clothes and really found the snake wound tight about her body. The negro was not able to tear it away and therefore cut it, and the girl regained consciousness. But she was so embarrassed about it that she finally pined away and died. At other times of the year this snake is more apt to run away than to attack people. However, I have heard it asserted frequently that even in summer when its time of copulation is passed, it pursues people, especially children, if it finds that they are afraid and run away from it._ Several people also assured me from their own experience, that it may be provoked to pursue people if they throw something at it, and then run away. I cannot well doubt this, as I have heard it said by large numbers of creditable people. But, [personally] I was never successful in irritating the black snakes. I always ran away on seeing one, or flung something at it, and then took to my heels, but I could never bring the snakes to pursue me: I know not for what reason they shunned me, unless they took me for an artful seducer. They have always tried to run away with the swiftness of an arrow.

Most people in this country ascribed to this snake a power of fascinating birds and squirrels, as I have described elsewhere in my journal.[1] When the snake lies under a tree, and has fixed his eyes on a bird or squirrel above, it obliges them to come down and to go directly into its mouth. I cannot account for this, for I never saw it done. However, I have a list of more than twenty persons, among which are some of the most reliable people, who have all unanimously, though living far distant from each other, asserted the same thing. They assured me upon their honor, that they have seen (several times) these black snakes bewitching squirrels and birds which sat on the tops of trees, the snake lying at the foot of the tree, with its eyes fixed upon the bird or squirrel, which sits above it and utters a doleful note, from which it is easy to conclude with certainty that it is about to be charmed, though you cannot see it. The bird or squirrel runs up and down along the tree continuing its plaintive song, and always comes nearer the snake, whose eyes are unalterably fixed upon it. It seemed as if these poor creatures endeavored to escape the snake, by hopping or running up the tree; but there appears to be a power which holds them. They are forced downwards, and each time that they turn back, they approach nearer their enemy, till they are at last forced to leap into its mouth, which stands wide open for that purpose. Numbers of squirrels and birds are continually running and hopping fearless in the woods on the ground where the snakes lie in wait for them, and can easily give these poor creatures a mortal bite. Therefore it seems that this fas-

---

[1] See pages 34-35 and 167 ff.

cination might be thus interpreted, that the creature has first got a mortal wound from the snake, which is sure of its bite, and lies quiet, being assured that the wounded creature has been poisoned, or at least feels pain from the violence of the bite, and that it will at last be obliged to come down into its mouth. The plaintive note is perhaps occasioned by the acuteness of the pain which the wound gives the creature. But to this it may be objected, that the bite of the black snake is not poisonous. It may further be objected, that if the snake could come near enough to a bird or squirrel to give it a mortal bite, it might as easily keep hold of it, or as it sometimes does with poultry, twist round and strangle or stifle it. But the chief objection against this interpretation is the following account which I received from the most creditable people, who have assured me of the truth of it. The squirrel being upon the point of running into the snake's mouth, the spectators have not been able to let it come to that pitch, but killed the snake, and as soon as it had been given a mortal blow, the squirrel or bird destined for destruction got away, and left off its mournful note, as if it had broken loose from a net. Some say that if they only touched the snake, so as to draw off its attention from the squirrel, it went off quickly, not stopping till it had gotten to a great distance. Why do the squirrels or birds go away so suddenly, and why no sooner? If they had been poisoned or bitten by the snake before, so as not to be able to get from the tree, and to be forced to approach the snake more and more, they could however not get new strength by the snake being killed or diverted. Therefore, it seems that they are enchanted only while the snake has its eyes fixed on them. However, this looks odd and unaccountable, though many of the worthiest and most reputable people have related it, and it is so universally believed here that to doubt it would be to expose one's self to general laughter.

The black snakes kill the smaller species of frogs and eat them. If they get at eggs of poultry or of other birds, they make holes in them and suck the contents. When the hens are sitting on the eggs, they creep into the nest, wind round the birds, stifle them and suck the eggs. Mr. Bartram asserted that he had often seen this snake creep up into the tallest trees, after bird's eggs or young birds, always with the head foremost when descending. A Swede told me that a black snake had once got the head of one of his hens in its mouth, and had wound himself several times round the body, when he came and killed the snake. The hen afterwards became as well as ever.

The snake is very fond of milk, and it is difficult to keep it out, when it has once gone into a cellar where milk it kept. It has been seen drinking milk out of the same dish with children, without biting them, though they often gave it blows with the spoon upon the head when it was overgreedy. I never heard it hiss. It can raise more than one half of its body from the ground, in order to look about. It changes its skin every year, and its skin is said to be a remedy for the cramp, if continually worn about the body.

The rye was now beginning to bloom.

Soil and River Banks. I have often observed with astonishment on my travels the great difference between the plants and the soil on the two opposite banks of brooks. Sometimes a brook, which one can stride over, has plants on one bank widely different from those on the opposite bank. Therefore, whenever I come to a large brook or a river, I expect to find plants which I have not seen before. Their seeds are carried down with the stream from distant parts. The soil is likewise very often different on the different sides of a rivulet, being rich and fertile on the one, and dry, barren and sandy on the other. But a large river can make still greater differences. Thus we see a great disparity between the provinces of Pennsylvania and New Jersey, which are only divided by the river Delaware. In Pennsylvania the soil consists of a mould mixed with sand and clay, and is very rich and fertile; and in the woods, the ground is mountainous and stony. On the other hand, in the province of New Jersey the soil is poor and dry, and not very fertile, some parts excepted. You can hardly find a stone in New Jersey, much less mountains. In Pennsylvania you scarcely ever see a fir tree, and in New Jersey there are whole woods of it.

This evening I arrived at Philadelphia.

MAY THE 22ND

The *locusts* began to creep out of their holes in the ground last night, and continued to do so to-day. As soon as their wings were dry, they began their song, which is almost sufficient to make one deaf, when travelling through the woods. This year there was an immense number of them. I have given a minute account of them, of their food, qualities, etc. in the *Memoirs* of the Swedish Royal Academy of Sciences; [1] it is therefore needless to repeat it here, and I refer the reader to the quoted place.

MAY THE 25TH

The *tulip tree* (*Liriodendron tulipifera*) was now in full blossom. The flowers have a resemblance to tulips, look pretty, and though they have no smell to delight the nose, yet the eye is pleased to see trees as tall as full grown oaks covered with tulip-like flowers.

On the flowers of the tulip tree was an olive colored chafer (*Scarabæus*) without horns (*muticus*), the suture and borders of his wing shells (*elytræ*) were black, and his thighs brown. I cannot with certainty say whether he was collecting the pollen of the flower or whether he was mating. Later in summer I saw the same kind of beetle make deep holes into the ripe mulberries, either to eat them or to lay eggs in them. I likewise found it abundant in the leaves of the *Magnolia glauca,* or beaver tree.

The strawberries were now ripe on the hills.

The country people were already bringing ripe cherries to town; but there were only a few to satisfy curiosity, yet we may form a judgment of the climate from this.

MAY THE 26TH

A Tornado. A peculiar kind of storm called a "travat" or "travade" [2] (tornado), raged today. In the evening about ten o'clock, when the sky was clear, a thick, black cloud came rushing suddenly from the southwest. The air was quite calm, and we could not feel any breeze. But the approach of this cloud was perceived from the strong rushing noise in the woods to the southwest, and which increased in proportion as the cloud came nearer. As soon as it had come up to us, it was attended by a violent gust of wind, which in its course threw down the weaker fences, carried them a good way along with it, and broke down several trees. It was then followed by a hard shower of rain, which put an end to the storm, and everything was calm as before. These travadoes are frequent in

[1] See Bibliography at end of this work, item No. 22.
[2] From the Portuguese *travados*, pl., a tornado.

summer, and have the quality of cooling the air. However, they often do considerable damage. They are commonly attended by thunder and lightning; as soon as they have passed over, the sky is as clear as it was before.

MAY THE 28TH

The *Magnolia glauca* was now in full bloom. Its flowers have a very pleasant fragrance, which refreshes the travellers in the woods, especially towards the evening. The flowers of the wild grapevine afterwards took the place of the magnolia. Several other flowers contributed likewise towards perfuming the ambient air.

Dwarf Laurel. The *Kalmia angustifolia* was now everywhere in flower. It grows chiefly on sandy heaths, or on dry poor grounds, where few other plants thrive; it is common in Pennsylvania, but particularly in New Jersey and the province of New York; it is scarce in Canada; its leaves stay all winter; the flowers are a real ornament to the woods; they grow in bunches like crowns and are of a fine bright purple; at the bottom is a circle of deep purple, and within it a grayish or whitish color. The flowers grow in bunches round the end of the stalk, and make it look like a decorated pyramid. The English at New York call this plant the "dwarf laurel." Its qualities are the same as those of the *Kalmia latifolia, viz.* that it kills sheep and other smaller animals, when they eat plentifully of it. I do not know whether it is poisonous to the larger cattle. It is not of any known use, and only serves to attract the eye while in bloom.

Mountain Laurel. The *Kalmia latifolia* was also in blossom. It rivals the preceding one in the beauty of its color; yet though it is conspicuous in regard to the color and shape of its flowers, it is in no way remarkable for smell, such as the magnolia is, for it has little or no smell at all. So equally and justly does nature distribute her gifts; no part of the creation has them all, each has its own, and none is absolutely without a share of them.

MAY THE 30TH

The Origin of Indians and Greenlanders. The Moravian Brethren, who in May arrived in great numbers from Europe, at New York, brought two converted Greenlanders with them. The Moravians who had already settled in America, immediately sent some of their brethren from Philadelphia to the newcomers, in order to welcome them. Among these deputies were two North American Indians, who had been converted to their doctrine, and likewise two South American Indians, from Surinam. These three kinds of converted heathens accordingly met at New York. I had no opportunity of meeting them; but all those who had seen them, and with whom I conversed, thought that they had plainly perceived a similarity in their features and shape, the Greenlanders being only somewhat smaller. They concluded therefore that all these three kinds of pagans were the posterity of one and the same descendant of Noah, or that they were perhaps yet more nearly related. How far their guesses are to be relied upon, I cannot determine.

To-day I ate ripe cherries for the first time this year; they now began to be more common and therefore not so expensive.

Yams are a species of roots, which are cultivated in the hottest parts of America, for eating, as we do potatoes. It has not yet been attempted to plant them here, and they are brought from the West Indies in ships; therefore they are reckoned a rarity here, and as such I ate them at Dr. Franklin's to-day. They are white and taste like common potatoes, but are not quite so palatable; and I think it would not be worth while to plant them in Sweden, though they might bear the climate. The species these roots belong to is the *Dioscorea alata.*

Cheese. The inhabitants here make plenty of cheese, but it is not reckoned as good as the English kind. However, some take it to be just as good when old, and so it seemed to me. A man from Boston in New England told me that they made very good cheese there; but that they took care to keep the cattle from salt water, especially those who live near the seacoasts; for it has been found that the cheese will not become so good when the cows graze near salt water as it will when they have fresh water. This, however, wants closer examination in my opinion.

MAY THE 31ST

About noon I left Philadelphia and went on board a small yacht which sails continually up and down upon the river Delaware, be-

tween Trenton and Philadelphia. We sailed up the river with fair
wind and weather. Sturgeons leaped often a fathom into the air.
We saw them continuing this exercise all day, till we came to Trenton. The banks on the Pennsylvania side were low, and those on the
New Jersey side steep and sandy, but not very high. On both sides
we perceived forests of tall deciduous trees.

During the course of this month the forenoon was always calm;
but immediately after noon it began to blow gently, and sometimes
pretty strongly. This morning was likewise fair; and in the afternoon it was cloudy, but did not rain.

The banks of the river were sometimes high, and sometimes low.
We saw some small houses near the shore, in the woods, and now
and then a good stone house. The river now grew narrower. About
three o'clock this afternoon we passed Burlington.

Burlington, New Jersey. Burlington, the chief town in the province
of New Jersey and the residence of the governor, is but a small town,
about twenty miles from Philadelphia, on the eastern side of the
Delaware. The houses are built chiefly of stone, though they stand
far apart. The town has a good location, since ships of considerable
tonnage can sail close up to it: but Philadelphia prevents its carrying on an extensive trade; for the proprietors of that place [1] have
granted owners large privileges by which it is increased so as to
swallow all the trade of the adjacent towns. The house of the governor at Burlington is only a small one: it is built of stone, close by
the riverside, and is the first building in the town as you come from
Philadelphia. It has been observed that at the full moons, when the
tides are highest, and the high water at Cape Henlopen comes at
nine o'clock in the morning, it will be at Chester, on the river
Delaware, about ten minutes after one o'clock; at Philadelphia,
about ten minutes after two o'clock; and at Burlington about ten
minutes after three o'clock; for the tide in the river Delaware comes
up to Trenton. These observations were communicated to me by
Mr. Lewis Evans.

The banks of the river were now for the most part high and steep
on the New Jersey side, consisting of a pale brick-colored soil. On
the Pennsylvania side, they were gently sloping and consisted of a
blackish rich mould, mixed with particles of glimmer (mica). On
the New Jersey side appeared some firs, but seldom on the other,

---

[1] William Penn, Esq., and his heirs after him.—F.

except in a few places where chance had accidentally brought them over from New Jersey.

At Trenton. Towards night, after the tide had begun to ebb and the wind had subsided, we could not proceed, but dropped our anchor about seven miles from Trenton, and passed the night there. The woods were full of fireflies, (*Lampyris*) which flew like sparks of fire between the trees, and sometimes across the river. In the marshes the bullfrogs now and then began their hideous croaking; and more than a hundred of them croaked together. The whip-poorwill was also heard everywhere.

JUNE THE 1ST

We continued our voyage this morning, after the rain was over. The river Delaware was very narrow here, and the banks the same as we found them yesterday, after we had passed Burlington. About eight o'clock in the morning we arrived at Trenton.[1]

JUNE THE 2ND

This morning we left Trenton, and proceeded towards New York. We rode in an ordinary open wagon which in stony places came near shaking liver and lungs out of you, otherwise the better class people travel with their own horses whether they ride in a wagon or chaise, or on horseback; the latter is the more common method of travelling. The country I have described before.[2] The fields were sown with wheat, rye, corn, oats, hemp, and flax. In several places, we saw very large pieces of ground planted with hemp. We saw an abundance of chestnut trees in the woods. They often stood in excessively poor ground, which was neither too dry nor too wet. Tulip trees did not appear on the road; but the people said there were some in the woods. The beaver tree (*Magnolia glauca*) grew in the swamps. It was now in flower, and the fragrance of its blossoms had so perfumed the air, that one could enjoy it before one approached the swamps; and this fine smell likewise showed that a beaver tree was near us, though we did not always see it.

The *Phlox glaberrima* grows abundantly in the woods, and makes

---

[1] See page 117.
[2] See pages 118 ff.

a grand show with its red flowers. It grows in such soil here as in Europe, is occupied by the *Lychnis viscaria* and *Lychnis dioica,* or red catchfly and campion. The *Phlox maculata* grows abundantly in wet ground, and has fine red and odoriferous flowers. It grows on low meadows, where in Europe the meadow-pinks, or *Lychnis flos cuculi,* are found. By adding to these flowers the *Bartsia coccinea,* the *Lobelia cardinalis,* and the *Monarda didyma,* which grow wild in this country, we must own that the land is undoubtedly adorned with the finest red imaginable.

The sassafras tree was abundant in the woods and near the fences.

The houses which we passed by were most of them wooden. In one place I saw the people building a house with walls of mere clay, which is also employed in making ovens for baking.

Buckwheat was already coming up in several places. We saw single plants of it all day in the woods and in the fields, but always by the side of the road, whence it may be concluded that they spring up from lost and scattered seeds.

Late in the evening we arrived at New Brunswick.[1]

June the 3rd

At noon we went on board a boat bound for New York, and sailed down the [Raritan] river, which had at first very high, steep banks of red limestone on each side which I have mentioned before.[2] Now and then, there was a farmhouse on the high shore. As we came lower down, we saw on both sides great fields and meadows close to the water. We could not sail at random for the river was often shallow in some places, and sometimes in the very middle. For that reason the course which we were to take was marked out by branches of leaves. At last we reached the sea, [at Perth Amboy], which bounded our view on the south; but on the other side we were continually in sight of land at some distance. On coming to the mouth of the river, we had a choice of two roads to New York: *viz.* either inside or outside of Staten Island. The inhabitants are determined in their choice by the weather; for when it is stormy and cloudy, or dark, they do not venture to sail out into the open sea. We took that course now, it being very pleasant weather; and

---

[1] See an account of this place on pp. 121 ff.
[2] See p. 122.

though we ran aground once or twice, we got loose again and arrived at New York about nine o'clock. Of this town I have already given an account.[1]

June the 4th

I found grapevines in several gardens, which had first been brought from the old countries. They bear almost annually a quantity of excellent grapes. When the winters are very severe they are killed by the frost, but the next spring new shoots spring up from the root.

Wild strawberries were now sold in abundance about the town every day. They are eaten in the same way as in Sweden, either with or without fresh milk, or with a small amount of wine and sugar. They are also used in place of candy in the usual way. An Englishman from Jamaica asserted that in that island there were no wild strawberries. In North America the strawberries have the character of clinging fast to the calyx so that they do not loosen so quickly as our Swedish ones. Snakes are said to be very fond of these berries. Although there is a fairly large quantity in some places here, their amount is not the same as in Finland and Sweden.

Red clover was sown in a few places on the hills outside. The country people were now employed in mowing the meadows. Some had already been mown, and the dry clover put in haycocks to be carted away at the first opportunity.

Cherry trees were planted in great quantities before the farmhouses and along the highroads, from Philadelphia to New Brunswick; but beyond that place they became more scarce. On coming to Staten Island, in the province of New York, I found them very common again, near the gardens. Here are not so many varieties of cherries as there are in Pennsylvania. I seldom saw any of the black sweet cherries [2] in New York, but commonly the sour red ones. All travellers are allowed to pluck ripe fruit in any garden which they pass by, provided they do not break any branches; and not even the most covetous farmer hindered them from so doing. It was a common custom, and any countryman knew that if the farmer tried to prevent it, he would be abused in return. Between New Brunswick

---

[1] See pages 130 ff.
[2] Commonly called black-heart cherries.—F.

and Staten Island, are a few cherry gardens, but proportionately more apple orchards.

JUNE THE 6TH

Several gentlemen and merchants, between fifty and sixty years of age, asserted that during their life they had plainly found several kinds of fish decrease in number every year, and that they could not get near so many fish now as they could formerly.

Rum, a brandy prepared from sugar cane, and in great use with all the English North American colonists, is reckoned much wholesomer than brandy made from wine or grain. In confirmation of this opinion they say that if you put a piece of fresh meat into rum and another into brandy, and leave them there for a few months, that in the rum will keep as it was, but that in the brandy will be eaten and full of holes. But this experiment is said to be unreliable. Major Roderfort told me that upon the Canada expedition he had observed that most of his soldiers who drank brandy for a time died; but those who drank rum were not hurt, though they got drunk with it every day for a considerable time.

Long Island is the name of an island opposite the town of New York. The northern part of it is much more fertile than the southern. Formerly a number of Indians lived there; and there are still a few left who, however, decrease every year, because they leave the place. The soil of the southern part of the island is very poor; but this deficiency is made up by a vast quantity of oysters, lobsters, crabs, several kinds of fish, and numbers of water fowl, all of which are far more abundant there than on the northern shores of the island. Therefore the Indians formerly chose the southern part to live in, because they subsisted on oysters and other products of the sea. When the tide is out, it is very easy to fill a whole cart with oysters that have been driven on shore by one flood tide. The island is strewed with oyster shells and other shells, which the Indians have left there; these shells now serve as good manure for the fields. The southern part of Long Island is turned into meadows, and the northern part into arable land. The winter is more constant on the northern part, and the snow in spring lies longer there than on the southern one. The people are very prolific here, and commonly tall and strong.

JUNE THE 10TH

The Hudson River. At noon we left New York, and sailed with a gentle wind up the Hudson River, in a boat bound for Albany. All afternoon we kept seeing a whole fleet of little boats returning from New York, whither they had brought provisions and other goods for sale, which on account of the extensive commerce of this town and the great number of its inhabitants find a good market. The Hudson River runs from north to south here, except for some high pieces of land which sometimes project far into it, and alter its direction. Its breadth at the mouth is reckoned about a mile and a quarter. Some porpoises played and tumbled in the river. The eastern shore, or the New York side, is at first very steep and high, but the western very sloping and covered with woods. There appeared farmhouses on both sides surrounded with plowed land. The soil of the steep shores is of a pale brick color and some little rocks of gray sandstone are seen here and there. About ten or twelve miles from New York, the western shore appears quite different from what it is further south; it consists of steep mountains with perpendicular sides towards the river, and they are exactly like the steep sides of the mountains of Hall and Hunnebärg in Västergötland, Sweden. Sometimes a rock projects out like the pointed angle of a bastion. The tops of these mountains are covered with oaks and other trees. A number of stones of all sizes lie along the shore, having rolled down from the mountains.

These high and steep mountains continued for a few English miles on the western shore; but on the eastern side the land was diversified with hills and valleys, which were commonly covered with hardwood trees, amongst which there appeared a farm now and then in a glade. The hills were covered with stones in some places. About twelve miles from New York we saw sturgeons [1] (*Acipenser sturio*) leaping up out of the water, and on the whole passage we met porpoises in the river. As we proceeded we found the eastern banks of the river very much cultivated, and a number of pretty farms surrounded with orchards, and fine plowed fields presented themselves to our view. About twenty-two miles from New York, the high

---

[1] The New York sturgeons, which I saw this year brought over, had short blunt noses, in which particular they are different from the English ones, which have long noses.—F.

mountains which I have before mentioned left us, and made as it were a high ridge here from east to west across the country. This altered the face of the land on the western shore of the river: from mountainous it became interspersed with little valleys and round hillocks, which were scarcely inhabited at all; but the eastern shore continued to afford us a delightful prospect. After sailing a little while in the night, we cast our anchor and lay here till the morning, especially as the tide was ebbing with great force.

June the 11th

This morning we continued our voyage up the river, with the tide and a faint breeeze. We now passed the Highland mountains, which were to the east of us. They consisted of gray sandstone, were very high, quite steep, and covered with deciduous trees, firs and red cedars. The western shore was rocky, which however did not come up to the height of the mountains on the opposite shore. The tops of these eastern mountains were cut off from our sight by a thick fog which surrounded them. The country was unfit for cultivation, being so full of rocks, and accordingly we saw no farms. The distance from these mountains to New York is computed at thirty-six English miles.

Ascending the Hudson. A thick fog now rose from the high mountains like the thick smoke of a charcoal kiln. For the space of a few English miles we had hills and rocks on the western banks of the river, and a variation of lesser and greater mountains and valleys covered with young firs, red cedars and oaks, on the eastern side. The hills close to the riverside were usually low, but their height increased as they were further from the river. Afterwards we saw for some miles nothing but high rolling mountains and valleys, both covered with woods; the valleys were in reality nothing but low rocks, and stood perpendicular towards the river in many places. The breadth of the river was sometimes two or three musket shots, but commonly not over one. Every now and then we saw several kinds of fish leaping out of the water. The wind vanished after about ten o'clock in the morning, and forced us to go forwards with our oars, the tide being almost spent. In one place on the western shore we saw a wooden house painted red, and we were told that there was a sawmill further up. But besides this we did not see a farm or any cultivated grounds all forenoon.

The Course of Rivers. The water in the river has here no longer a brackish taste; yet I was told that the tide, especially when the wind is south, sometimes carries the salt water further north with it. The color of the water is likewise altered, for it appears darker here than before.—To account for the original course of rivers is very difficult, if not wholly impossible. Some rivers may have come from a great reservoir of water, which being considerably increased by heavy rains or other circumstances overflows old bounds down to the lower countries through the places where it meets with the least resistance. This is perhaps the reason why some rivers have such a winding course through fields of soft earth, and where mountains, rocks and stones divert their passage. However, it seems that some rivers derive their origin from the Creation itself, and that Providence then pointed out their course; for their existence can, in all probability, not be owing to the accidental eruption of water alone. Among these rivers we may rank the Hudson. I was surprised on seeing its course and the varied character of its shores. It rises a long way north of Albany, and descends to New York, in a direct line from north to south, which is a distance of about a hundred and sixty English miles, and perhaps more, for the little windings which it makes are of no signification.[1] In many places between New York and Albany, are ridges of high mountains running west and east. But it is remarkable that they go on undisturbed till they come to the Hudson, which cuts directly across them, and frequently their sides stand perpendicular towards the river. There is an opening left in the chain of mountains, as broad as the river, for it to pass through, and the mountains go on as before, on the other side, in the same direction. It is likewise remarkable, that the river in such places where it passes through the mountains is as deep, and often deeper, than in the other places. The perpendicular rocks on the sides of the river are surprising, and it appears that if no passages had been opened by Providence for the river to pass through, the mountains in the upper part of the country would have been inundated, since these mountains, like so many dikes, would have hindered the water from going on. Query, why does this river go on in a direct line for

---

[1] The Hudson River is about 300 miles long.

so considerable a distance? Why do the many passages, through which the river flows across the mountains, lie along the same meridian? Why are there no rapids at some of these openings, or at least shallow water with a rocky bed?

We now perceived excessively high and steep mountains on both sides of the river, which echoed back each sound we uttered. Yet notwithstanding they were so high and steep, they were covered with small trees. The Blue Mountains,[1] which reared their towering summits above all the other mountains, were now seen before us towards the north, but at a great distance. The skipper told us that often on one of the mountains on the west side of the river one can see a light at night which people claim is a carbuncle. The country began here to look more cultivated and less mountainous. The last of the high western mountains was called Butterhill; beyond it the country between the mountains grew more level. The farms became very numerous, and we had a view of many grain fields between the hills. Before we passed the latter we had the wind in our face, and we could only get forward by tacking, which took much time, as the river was hardly a musket shot in breadth. Afterwards we cast anchor, because we had both wind and tide against us. While we waited for the return of the tide and the change of the wind we went on shore.

Sassafras trees and chestnut trees grow here in great abundance. I found a tulip tree in some parts of the wood, and also the *Kalmia latifolia*, which was now in full blossom, though the flowers were already fading.

Some time after noon a fair wind arose from the southwest, so we weighed anchor and continued our voyage. The place where we lay at anchor was just at the end of those steep and amazingly high mountains. They consisted of gray rock, and close to them, on the shore, lay a vast number of little stones. As soon as we had passed these mountains, the country became more level and higher. The river likewise increased in breadth, so as to be nearly an English mile broad. After sailing for some time we found no more mountains along the river; but to the eastward was a high chain of mountains [of the Berkshire system] whose sides were covered with woods up

---

[1] Here presumably southern parts of the Catskill mountain system. The name in Kalm's day was probably loosely descriptive primarily, and applied to the whole eastern chain of mountains.

to more than half of their height. The summits however were quite
barren; for I suppose that nothing would grow there on account of
the great degree of heat,[1] dryness, and the violence of the wind, to
which that part was exposed. The eastern side of the river was much
more cultivated than the western, where we seldom saw a house.
The land was covered with woods though it was in general very level.
About fifty-six English miles from New York the country is not
very high; yet it is everywhere covered with woods, except for some
pioneer farms which were scattered here and there. The high mountains which we left in the afternoon, now appeared above the woods
and the countryside. These mountains, which were called the Highlands, did not extend further north than the other on the opposite
side, in the place where we anchored. Their sides (not those towards
the river) were seldom perpendicular, but sloping, so that one could
climb up to the top, though not without difficulty.

On some of the higher grounds near the river, the people burnt
lime. The master of the boat told me that they quarry a fine bluish-
gray limestone in the highlands along both sides of the river, for the
space of several English miles, and burn lime out of it. But for
several miles distance there is no more limestone and they find also
none on the banks till they come to Albany.

We passed by a little neck of land which projected on the western
side in the river and was called Dance. The name of this place is
said to derive its origin from a festival which the Dutch celebrated
here in former times, and at which they danced and diverted themselves; but once there came a number of Indians who killed them all.

We cast anchor late at night, because the wind ceased and the tide
was ebbing. The depth of the river is twelve fathoms here. The
fireflies flew over the river in large numbers at night and often settled
upon the rigging.

June the 12th

This morning we proceeded with the tide, but against the wind.
The river was here a musketshot broad. The country in general was

---

[1] Mr. Kalm was certainly mistaken, by thinking the summits of these mountains
without wood, "on account of the great degree of heat": for it is a general notion,
founded on experience, that the sun operates not so much on the tops of mountains,
as in plains or valleys, and the cold often hinders the increase of wood on the summits
of high mountains.—F.

low on both sides, consisting of low rocks and stony fields, which were however covered with woods. It is so rocky, stony, and poor that nobody can settle on it, or inhabit it, there being no spot of ground fit for cultivation. The country continued to have the same appearance for the space of a few miles, and we never perceived a single settlement. At eleven o'clock this morning we came to a little island, which lies in the middle of the river, and is said to be half-way between New York and Albany. The shores were still low, stony and rocky, as before. But at a greater distance we saw high mountains, covered with woods, chiefly on the western shore, raising their summits above the rest of the country; and still further off, the Blue Mountains rose up above them. Towards noon it was quite calm, and we went on very slow. Here the land was well cultivated, especially on the eastern shore, and full of great plowed fields; yet the soil seemed sandy. Several villages lay on the eastern side, and one of them, called Strasburg,[1] was inhabited by a number of Germans. To the west we saw several cultivated places. The Blue Mountains could be seen very plainly from here. They appeared through the clouds and towered above all other mountains. The river was fully an English mile broad opposite Strasburg.

They use a yellow *Agaricus*, or fungus, which grows on maple trees, for tinder. That which is found on the red-flowering maple (*Acer rubrum*) is reckoned the best, and next in goodness is that of the sugar maple (*Acer saccarinum*), which is sometimes reckoned as good as the former.

Rhinebeck is a place located a short distance from Strasburg, further from the river. It is inhabited by many Germans, who have a church there. Their clergyman at present was the Rev. Mr. Hartwig, who knew some Swedish, having been at Gothenburg for a time. This little town is not visible from the riverside.

At two in the afternoon it began again to blow from the south, which enabled us to proceed. The country on the eastern side was high, and consisted of well cultivated soil. We had fine plowed fields, well-built farms and good orchards in view. The western shore was likewise somewhat high, but still covered with woods, and we now and then, though seldom, saw one or two little settlements. The river was more than an English mile broad in most places, and

---

[1] Modern Staatsburg (Statesburg)?

came in such a straight line from the north that the water vanished from view.

JUNE THE 13TH

The wind favored our voyage during the whole night, so that I had no opportunity of observing the nature of the country. This morning at five o'clock we were but nine English miles from Albany. The country on both sides of the river was low and covered with woods, only here and there were a few little scattered settlements. On the banks of the river were wet meadows, covered with sword grass (*Carex*), and they formed several little islands. We saw no mountains and hastened towards Albany. The land on both sides of the river was chiefly low, and more carefully cultivated as we came nearer to Albany. Here we could see everywhere the type of haystacks with movable roofs which I have described before.[1] As to the houses which we saw, some were of wood, others of stone. The river was seldom above a musketshot broad, and in several parts of it were sandbars which required great skill in navigating the boats. At eight o'clock in the morning we arrived at Albany.

Arriving at Albany. All the boats which ply between Albany and New York belong to Albany. They go up and down the Hudson River as long as it is open and free from ice. They bring from Albany boards or planks, and all sorts of timber, flour, peas, and furs, which they get from the Indians, or which are smuggled from the French. They come home almost empty, and only bring a few kinds of merchandise with them, the chief of which is rum. This is absolutely necessary to the inhabitants of Albany. They cheat the Indians in the fur trade with it; for when the Indians are drunk they are practically blind and will leave it to the Albany whites to fix the price of the furs. The boats are quite large, and have a good cabin, in which the passengers can be very commodiously lodged. They are usually built of red cedar or of white oak. Frequently the bottom consists of white oak, and the sides of red cedar, because the latter withstands decay much longer than the former. The red cedar is likewise apt to split when it hits against anything, and the Hudson is in many places full of sand and rocks, against which the keel of the boat sometimes strikes. Therefore people choose white oak

---

[1] See p. 264.

for the bottom, as being the softer wood, and not splitting so easily. The bottom, being continually under water, is not so much exposed to weathering and holds out longer.

Canoes. The canoes which the boats always have along with them are made of a single piece of wood, hollowed out: they are sharp on both ends, frequently three or four fathoms long, and as broad as the thickness of the wood will allow. The people in it do not row sitting, but usually a fellow stands at each end, with a short oar in his hand, with which he controls and propels the canoe. Those which are made here at Albany are commonly of white pine. They can do service for eight or twelve years, especially if they be tarred and painted. At Albany they are made of pine since there is no other wood fit for them; at New York they are made of the tulip tree, and, in other parts of the country of red or white cedars: but both these trees are so small in the neighborhood of Albany that they are unfit for canoes. There are no seats in them, for if they had any, they would be more liable to be upset, as one could not keep one's equilibrium so well. One has to sit in the bottom of these canoes.

Battoes [1] are another kind of boats which are much in use in Albany: they are made of boards of white pine; the bottom is flat, that they may row the better in shallow water. They are sharp at both ends, and somewhat higher towards the end than in the middle. They have seats in them, and are rowed as common boats. They are long, yet not all alike. Usually they are three and sometimes four fathoms long. The height from the bottom to the top of the board (for the sides stand almost perpendicular) is from twenty inches to two feet, and the breadth in the middle about a yard and six inches. They are chiefly made use of for carrying goods along the river to the Indians, that is, when those rivers are open enough for the battoes to pass through, and when they need not be carried by land a great way. The boats made of the bark of trees break easily by knocking against a stone, and the canoes cannot carry a great cargo, and are easily upset; the battoes are therefore preferable to them both. I saw no boats here like those in Sweden or other parts of Europe.

Temperature at Albany. Frequently the cold does a great deal of damage at Albany. There is hardly a month in summer during which

---

[1] From the French *bateaux* (boats).—F.

a frost does not occur. Spring comes very late, and in April and May are numerous cold nights which frequently kill the flowers of trees and kitchen herbs. It was feared last May that the blossoms of the apple trees had been so severely damaged by the frost that next autumn there would be but very few apples. Even the oak blossoms in the woods are very often killed by the cold. The autumn here is of long continuance, with warm days and nights. However, the cold nights commonly commence towards the end of September, and are frequent in October. The people are forced to keep their cattle in stables from the middle of November till March or April, and must find them hay during that time.[1]

During summer, the wind blows mostly from the south and brings a great drought along with it. Sometimes it rains a little, and as soon as it has rained the wind veers to northwest, blowing for several days from that point and then returning to the south. I have had frequent opportunities of seeing this condition of wind happen precisely, both this year and the following.

JUNE THE 15TH

The fences were made of pine boards, of which there is an abundance in the extensive woods; and there are many saw mills to cut it into boards.

Fruit Trees. The several sorts of apple trees were said to grow very well here, and bear as fine fruit as in any other part of North America. Each farm has a large orchard. They have some apples here which are very large, and very palatable; they are sent to New York and other places as a rarity. People make excellent cider in the autumn in the country round Albany. From the seed which I gathered of these large apples and planted in Åbo, Finland, a number of trees have come up which seem to thrive well and have not been injured in the least by our winters; but since they have not yet blossomed I cannot say whether the fruit which they bear will resemble that grown in Albany. Pear trees do not succeed here. This was complained of in many other parts of North America. But I fear that they do not take sufficient care in the management and planting of them, for I have seen fine pears in several parts of Pennsylvania. Peach trees have often been planted here, and never suc-

---

[1] The reader must reckon all this according to the old style.—F.

ceed well. This was attributed to a worm which lives in the ground and eats through the root, so that the tree dies. Perhaps the severity of the winter contributes much to it. They plant no other fruit trees at Albany besides these I have mentioned.

Grains. They sow as much hemp and flax here as they want for home consumption. They sow corn in great abundance; a loose soil is reckoned the best for this purpose, for it will not thrive in clay. From half a bushel they reap a hundred bushels. They reckon corn a very suitable kind of crop, because the young plant recovers after being hurt by the frost. They have had instances here of the plants freezing off twice in the spring, close to the ground, and yet surviving and yielding an excellent crop. Corn has likewise the advantage of standing much longer against a drought than wheat. The larger sort of corn which is commonly sown here ripens in September.

Wheat is sown in the neighborhood of Albany to great advantage. From one bushel they get twelve sometimes; if the soil is good, they get twenty bushels. If their crop amounts only to a ten-fold yield, they think it a very mediocre one. The inhabitants of the country round Albany are Dutch and Germans. The Germans live in several great villages, and sow great quantities of wheat which is brought to Albany, whence they send many boats laden with flour to New York. The wheat flour from Albany is reckoned the best in all North America, except that from Sopus (Esopus) or King's Town (Kingston), a place between Albany and New York. All the bread in Albany is made of wheat. At New York they pay for the Albany flour with a few shillings more per hundred weight than for that from other places.

Rye is likewise sown here, but not so generally as wheat. They do not sow much barley, because they do not reckon the profits very great. Wheat is so plentiful that they make malt of that. In the neighborhood of New York, I saw great fields sown with barley. They do not sow more oats than are necessary for their horses.

Peas. The Dutch and Germans who live hereabouts sow peas in great abundance; they grow very well, and are annually carried to New York in great quantities. They were free from insects for a considerable time. But of late years the same pest which destroys the peas in Pennsylvania, New Jersey, and the lower parts of the province of New York,[1] has likewise appeared destructive among the

---

[1] I have mentioned them before. See pp. 91-93.

peas here. It is a real loss to this town, and to the other parts of North America, which used to get so many peas from here for their own consumption and that of their sailors. It had been found that if they procured good peas from Albany and sowed them near King's Town, or the lower part of the province of New York, they succeeded very well the first year, but were so full of worms the second and following years that nobody could or would eat them.—Some people put ashes into the pot, among the peas, when they will not boil or soften well; but whether this is wholesome and agreeable to the palate, I do not know.

Potatoes are planted by almost everyone. Some people preferred ashes to sand for keeping them in during winter. Some people in Ireland are said to have the custom in autumn of placing the potatoes in an oven and drying them a bit, when they are said to keep better over winter; but these potatoes cannot later be planted, only eaten. The Bermuda potatoes have likewise been planted here, and succeed pretty well. The greatest difficulty is to keep them during winter, for they generally rot in that season.

The humming bird comes to this place sometimes, but is rather a scarce bird.

The *shingles* with which the houses are covered are made of the white pine, which is reckoned as good and as durable and sometimes better than the white cedar. It is claimed that such a roof will last forty years. The white pine is found abundant here, in such places where common pines grow in Sweden. I have never seen them in the lower parts of the province of New York, nor in New Jersey or Pennsylvania. A vast quantity of lumber from the white pine is prepared annually on this side of Albany, which is brought down to New York and exported.

Grapevines. The woods abound with grapevines, which likewise grow on the steep banks of the river in surprising quantities. They climb to the tops of trees on the bank and bend them by their weight. But where they find no trees they hang down along the steep shores and cover them entirely. The grapes are eaten after the frost has touched them, for they are too sour before. They are not much used in any other way.

Gnats. The vast woods and uninhabited grounds between Albany and Canada contain immense swarms of gnats which annoy the travellers. To be in some measure secured against these insects some besmear their face with butter or grease, for the gnats do not like to settle on greasy places. The great heat makes boots very uncomfortable; but to prevent the gnats from stinging the legs they wrap some paper round them, under the stockings. Some travellers wear caps which cover the whole face, and some have gauze over the eyes. At night they lie in tents, if they can carry any with them, and make a great fire at the entrance so that the smoke will drive the pests away.

JUNE THE 16TH

The porpoises seldom go higher up the Hudson River than the salt water goes; after that, the sturgeons fill their place. It has however sometimes happened that porpoises have gone clear up to Albany.

There is a report that a whale once came up the river to this town.

The fireflies (*Lampyris*), which are the same as those that we find in Pennsylvania during summer, are seen here in abundance every night. They fly up and down in the streets of the town. They come into the houses if the doors and windows are open.

JUNE THE 19TH

Several of the Pennsylvania trees are not to be seen in these woods: *viz.*

Magnolia glauca, the beaver tree;
Nyssa aquatica, the tupelo tree;
Liquidambar styraciflua, the sweet-gum tree;
Diospyros Virginiana, the persimmon;
Liriodendron tulipifera, the tulip tree;
Juglans nigra, the black walnut tree;
Quercus——, the swamp oak; [1]
Cercis Canadensis, the sallad tree;
Robinia pseudacacia, the locust tree;
Gleditsia triacanthos, the honey-locust tree;
Annona muricata, the papaw tree;
Celtis occidentalis, the nettle tree;

and a number of shrubs, which are never found here. The more northerly location of the place, the height of the Blue Mountains, and the course of the rivers, which flow here southward into the sea, and accordingly carry the seeds of plants from north to south

---

[1] *Quercus aquatica?*

and not the contrary way, are chiefly the causes that several plants which grow in Pennsylvania cannot be found here.

An Island near Albany. This afternoon I went to see an island which lies in the middle of the river about a mile below the town. This island is an English mile long, and not above a quarter of a mile broad. It is almost entirely turned into plowed fields, and is inhabited by a single planter, who besides possessing this island is the owner of three more. Here we saw no woods, except a few trees which were left round the island on the shore and formed as it were a tall, large hedge. The red maple (*Acer rubrum*) grows in abundance in several places. Its leaves are white or silvery on the under sides, and, when agitated by the wind, they make the tree appear as if it were full of white flowers. The water beech (*Plantanus occidentalis*) grows to a great height and is one of the best shade trees here. The water poplar is the most common tree hereabouts, grows exceedingly well on the shores of the river, and is as tall as the tallest of our aspens. In summer it affords the best shade for men and cattle against the scorching heat. On the banks of rivers and lakes it is one of the most useful trees, because it holds the soil by its extensively branched roots, and prevents the water from washing it away. The water beech and the elm tree (*Ulmus*) serve the same purpose. The wild plum trees are plentiful here and full of unripe fruit. Its wood is not made use of, but its fruit is eaten. Sumach (*Rhus glabra*) is plentiful here, as also the wild grapevines which climb up the trees and creep along the high shores of the river. I was told that the grapes ripen very late, though they are already pretty large. The American elm tree (*Ulmus Americana*) forms several high hedges. The soil of this island is a rich mould, mixed with sand, which is chiefly employed in corn plantations. There are likewise large fields of potatoes. The whole island was leased for one hundred pounds of New York currency. The person who had taken the lease again let some greater and smaller lots of ground to the inhabitants of Albany for kitchen gardens, and by that means reimbursed himself. Portulaca (*Portulaca oleracea*) grows spontaneously here in great abundance and looks very well.

JUNE THE 20TH

The Hudson River at Albany. The tide in the Hudson goes about eight or ten English miles above Albany, and consequently runs one hundred and fifty-six English miles from the sea. In spring when the snow melts there is hardly any high tide near this town, for the great quantity of water which comes from the mountains during that season occasions a continual ebbing. This likewise happens after heavy rains.

The cold is generally reckoned very severe here. The ice in the Hudson is commonly three or four feet thick. On the third of April some of the inhabitants crossed the river with six pairs of horses. The ice commonly dissolves about the end of March, or beginning of April. Great pieces of ice come down about that time, which sometimes carry with them the houses that stand close to the shore. The water is very high at that time because the ice stops sometimes and piles up in places where the river is narrow. The water often has been observed to rise three fathoms higher than it commonly is in summer. The ground is frozen here in winter to the depth of three, four or five feet. On the sixteenth of November the boats are put up, and about the beginning or middle of April they are in motion again. People here are unacquainted with stoves, and their chimneys are so wide that one could almost drive through them with a horse and sleigh.

Drinking Water. The water of several wells in this town was very cool about this time, but had a kind of acid taste, which was not very agreeable. On a nearer examination, I found an abundance of little insects in it, which were probably *monoculi*.[1] Their length varied; some were a geometrical line and a half, others two, and others four lines long. They were very narrow and of a light color. The head was blacker and thicker than the other parts of the body, and about the size of a pin's head. The tail was divided into two branches, and each branch terminated in a little black globule. When these insects swam they proceeded in crooked or undulated lines, almost like tadpoles. I poured some of this water into a bowl, and added nearly a fourth part of rum to it. The *monoculi*, instead of being affected by it, swam about as briskly as they had done before. This shows that if one makes punch with this water it must be very strong to kill the *monoculi*. I think this water is not very wholesome for people who are not used to it, though the inhabitants of Albany, who drink it every day, say they do not feel the least discomfort from it. I have

---

[1] Now called Cyclops, a genus of minute crustaceans. They have one median eye (though a double one), hence the name.

been several times obliged to drink water here, in which I have plainly seen *monoculi* swimming; but the next day I generally felt something like a pea in my throat, or as if I had a swelling there, and this continued for about a week. I felt such swellings this year, both at Albany and in other places. My servant Jungström likewise got a great pain in his chest, and a sensation as from a swelling, after drinking water with *monoculi* in it; but whether these insects occasioned it or whether it came from some other cause, I cannot ascertain. However, I have always endeavored, as much as possible to do without such water as had *monoculi* in it. I have found them in very cold water, taken from the deepest wells, in different parts of this country. Perhaps many of our diseases arise from waters of this kind, which we do not sufficiently examine. I have frequently observed an abundance of minute insects in water which has been remarkable for its clearness. Almost each house in Albany has its well, the water of which is applied to common use; but for tea, brewing, and washing, they commonly take the water of the Hudson, which flows close by the town. This water is generally quite muddy and very warm in summer; and, on that account, it is kept in cellars, in order that the sediment may settle to the bottom, and that the water may cool a little.

We lodged with a gunsmith, who told us that the best charcoal for the forge was made of black pine. The next best in his opinion was that made of beech. The best and most expensive stocks for his muskets were made of wild cherry, and next to these he valued most those of the red maple. They scarcely make use of any other wood for this purpose. The black walnut tree affords excellent wood for stocks, but it does not grow in the neighborhood of Albany.

June the 21st

Description of Albany. Next to New York Albany is the principal town, or at least the most wealthy, in the province of New York. It is situated on the slope of a hill, close to the western shore of the Hudson River, about one hundred and forty-six English miles from New York. The town extends along the river, which flows here from N.N.E. to S.S.W. The high mountains in the west, above the town, bound the view on that side. There are two churches in Albany, one

(From J. W. Barber, *Historical Collections of the State of New York*, 1851)

Old Dutch Church, Albany

English and the other Dutch. The Dutch church stands a short distance from the river on the east side of the market. It is built of stone and in the middle it has a small steeple with a bell. It has but one minister who preaches twice every Sunday. The English church is situated on the hill at the west end of the market, directly under the fort. It is likewise built of stone but has no steeple. There is no service at the church at this time because they have no minister, but all the people understand Dutch, the garrison excepted. The minister of this church has a settled income of one hundred pounds sterling, which he gets from England. The town hall lies to the south of the Dutch church, close by the riverside. It is a fine building of stone, three stories high. It has a small tower or steeple, with a bell, and a gilt ball and vane at the top of it.

The houses in this town are very neat, and partly built of stones covered with shingles of white pine. Some are slated with tile from Holland, because the clay of this neighborhood is not considered fit for tiles. Most of the houses are built in the old Frankish way, with the gable-end towards the street, except a few, which were recently built in the modern style. A great number of houses are built like those of New Brunswick, which I have described,[1] the gable-end towards the street being of bricks and all the other walls of boards. The outside of the houses is never covered with lime or mortar, nor have I seen it practised in any North American towns which I have visited; and the walls do not seem to be damaged by the weather. The eaves on the roofs reach almost to the middle of the street. This preserves the walls from being damaged by the rain, but it is extremely disagreeable in rainy weather for the people in the streets, there being hardly any means of avoiding the water from the eaves. The front doors are generally in the middle of the houses, and on both sides are porches with seats, on which during fair weather the people spend almost the whole day, especially on those porches which are in the shade. The people seem to move with the sun and the shade, always keeping in the latter. When the sun is too hot the people disappear. In the evening the verandas are full of people of both sexes; but this is rather troublesome because a gentleman has to keep his hat in constant motion, for the people here are not Quakers whose hats are as though nailed to the head. It is considered very impolite

---

[1] See page 121.

not to lift your hat and greet everyone. The streets are broad, and some of them are paved. In some parts they are lined with trees. The long streets are almost parallel to the river, and the others intersect them at right angles. The street which goes between the two churches is five times broader than the others and serves as a marketplace. The streets upon the whole are very dirty because the people leave their cattle in them during the summer nights. There are two marketplaces in town, to which the country people come twice a week. There are no city gates here but for the most part just open holes through which people pass in and out of the town.

The fort lies higher than any other building on a high steep hill on the west side of the town. It is a great building of stone surrounded with high and thick walls. Its location is very bad, as it can serve only to keep off plundering parties without being able to sustain a siege. There are numerous high hills to the west of the fort, which command it, and from which one may see all that is done within it. There is commonly an officer and a number of soldiers quartered in it. They say the fort contains a spring of water.

Trade. The location of Albany is very advantageous in regard to trade. The Hudson River which flows close by it is from twelve to twenty feet deep. There is not yet any quay made for the better landing of the boats, because the people fear it will suffer greatly or be entirely carried away in spring by the ice which then comes down the river. The vessels which are in use here may come pretty near the shore in order to be loaded, and heavy goods are brought to them upon canoes tied together. Albany carries on a considerable commerce with New York, chiefly in furs, boards, wheat, flour, peas, several kinds of timber, etc. There is not a place in all the British colonies, the Hudson's Bay settlements excepted, where such quantities of furs and skins are bought of the Indians as at Albany. Most of the merchants in this town send a clerk or agent to Oswego, an English trading town on Lake Ontario, to which the Indians come with their furs. I intend to give a more minute account of this place in my Journal for the year 1750. The merchants from Albany spend the whole summer at Oswego, and trade with many tribes of Indians who come with their goods. Many people have assured me that the Indians are frequently cheated in disposing of their goods, especially when they are drunk, and that sometimes they do not get one half or even one tenth of the value of their goods. I have been a witness to several transactions of this kind. The merchants of Albany glory in these tricks, and are highly pleased when they have given a poor Indian, a greater portion of brandy than he can stand, and when they can, after that, get all his goods for mere trifles. The Indians often find when they are sober again, that they have for once drunk as much as they are able of a liquor which they value beyond anything else in the whole world, and they are quite insensible to their loss if they again get a draught of this nectar. Besides this trade at Oswego, a number of Indians come to Albany from several places especially from Canada; but from this latter place, they hardly bring anything but beaver skins. There is a great penalty in Canada for carrying furs to the English, that trade belonging to the French West India Company. Notwithstanding that the French merchants in Canada carry on a considerable smuggling trade. They send their furs by means of the Indians to their agent at Albany, who purchases them at the price which they have fixed upon with the French merchants. The Indians take in return several kinds of cloth, and other goods, which may be bought here at a lower rate than those which are sent to Canada from France.

The greater part of the merchants at Albany have extensive estates in the country and a large property in forests. If their estates have a little brook, they do not fail to erect a sawmill upon it for sawing boards and planks, which many boats take during the summer to New York, having scarcely any other cargo.

Many people at Albany make wampum for the Indians, which is their ornament and money, by grinding and finishing certain kinds of shells and mussels. This is of considerable profit to the inhabitants. I shall speak of this kind of money later. The extensive trade which the inhabitants of Albany carry on, and their sparing manner of living, in the Dutch way, contribute to the considerable wealth which many of them have acquired.

The Dutch in Albany. The inhabitants of Albany and its environs are almost all Dutchmen. They speak Dutch, have Dutch preachers, and the divine service is performed in that language. Their manners are likewise quite Dutch; their dress is however like that of the English. It is well known that the first Europeans who settled in the province of New York were Dutchmen. During the time that they were the masters of this province, they seized New Sweden of which they were jealous. However, the pleasure of possessing this conquered land and their own was but of short duration, for towards the end of 1664 Sir Robert Carr,[1] by order of King Charles the second, went to New York, then New Amsterdam, and took it. Soon after Colonel Nicolls [2] went to Albany, which then bore the name of Fort Orange, and upon taking it, named it Albany, from the Duke of York's Scotch title. The Dutch inhabitants were allowed either to continue where they were, and under the protection of the English to enjoy all their former privileges, or to leave the country. The greater part of them chose to stay and from them the Dutchmen are descended who now live in the province of New York, and who possess the greatest and best estates in that province.

The avarice, selfishness and immeasurable love of money of the inhabitants of Albany are very well known throughout all North America, by the French and even by the Dutch, in the lower part of New York province. If anyone ever intends to go to Albany it is said in jest that he is about to go to the land of Canaan, since Canaan and the land of the Jews mean one and the same thing, and that Albany is a fatherland and proper home for arch-Jews, since the inhabitants of Albany are even worse. If a real Jew, who understands the art of getting forward perfectly well, should settle amongst them, they would not fail to ruin him. For this reason nobody comes to this place without the most pressing necessity; and therefore I was asked in several places, both this and the following year, what induced me to make the pilgrimage to this New Canaan. I likewise found that the judgment which people formed of them was not without foundation. For though they seldom see any strangers, (except those who go from the British colonies to Canada and back again) and one might therefore expert to find victuals and accommodation for travellers cheaper than in places where they always resort, yet I experienced the contrary. I was here obliged to pay for everything twice, thrice and four times as much as in any part of North America which I have passed through. If I wanted their assistance, I was

---

[1] British commissioner to New England in 1664. With Nicolls he took New Amsterdam from the Dutch, 1664, as Kalm relates, and named it New York. He died in 1667.

[2] Sir Richard Nicolls (1624-1672), chief military director in the capture of New Amsterdam, and the first English colonial governor of New York.

obliged to pay them very well for it, and when I wanted to purchase anything or be helped in some case or other, I could at once see what kind of blood ran in their veins, for they either fixed exorbitant prices for their services or were very reluctant to assist me. Such was this people in general. However, there were some among them who equalled any in North America or anywhere else, in politeness, equity, goodness, and readiness to serve and to oblige; but their number fell far short of that of the former. If I may be allowed to declare my conjectures, the origin of the inhabitants of Albany and its neighborhood seems to me to be as follows. While the Dutch possessed this country, and intended to people it, the government sent a pack of vagabonds of which they intended to clear their native country, and sent them along with a number of other settlers to this province. The vagabonds were sent far from the other colonists, upon the borders towards the Indians and other enemies, and a few honest families were persuaded to go with them, in order to keep them in bounds. I cannot in any other way account for the difference between the inhabitants of Albany and the other descendants of so respectable a nation as the Dutch, who are settled in the lower part of New York province. The latter are civil, obliging, just in prices, and sincere; and though they are not ceremonious, yet they are well meaning and honest and their promises may be relied on.

The behavior of the inhabitants of Albany during the war between England and France, which ended with the peace of Aix la Chapelle, has, among several other causes, contributed to make them the object of hatred in all the British colonies, but more especially in New England. For at the beginning of that war when the Indians of both parties had received orders to commence hostilities, the French engaged theirs to attack the inhabitants of New England, which they faithfully executed, killing everybody they met with, and carrying off whatever they found. During this time the people of Albany remained neutral, and carried on a great trade with the very Indians who murdered the inhabitants of New England. Articles such as silver spoons, bowls, cups, etc. of which the Indians robbed the houses in New England, were carried to Albany, for sale. The people of that town bought up these silver vessels, though the names of the owners were engraved on many of them, and encouraged the Indians to get more of them, promising to pay them well, and whatever they would demand. This was afterwards interpreted by the inhabitants of New England to mean that the colonists of Albany encouraged the Indians to kill more of the New England people, who were in a manner their brothers, and who were subjects of the same crown. Upon the first news of this behavior, which the Indians themselves spread in New England, the inhabitants of the latter province were greatly incensed, and threatened that the first step they would take in another war would be to burn Albany and the adjacent parts. In the present war it will sufficiently appear how backward the other British provinces in America are in assisting Albany, and the neighboring places, in case of an attack from the French or Indians.[1] The hatred which the English bear against the people at Albany is very great, but that of the Albanians against the English is carried to a ten times higher degree. This hatred has subsisted ever since the time when the English conquered this section, and is not yet extinguished, though they could never have gotten larger advantages under the Dutch government than they have obtained under that of the English. For, in a manner, their privileges are greater than those of Englishmen themselves.

In their homes the inhabitants of Albany are much more sparing than the English and are stingier with their food. Generally what they serve is just enough for the meal and sometimes hardly that. The punch bowl is much more rarely seen than among the English. The women are perfectly well acquainted with economy; they rise early, go to sleep very late, and are almost superstitiously clean in regard to the floor, which is frequently scoured several times in the week. Inside the homes the women are neatly but not lavishly dressed. The children are taught both English and Dutch. The servants in the town are chiefly negroes. Some of the inhabitants wear their own hair very short, without a bag or queue, because these are looked upon as the characteristics of Frenchmen. As I wore my hair in a bag the first day I came here from Canada, I was surrounded with children, who called me a Frenchman, and some of the boldest offered to pull at my French head dress, so I was glad to get rid of it.

Their food and its preparation is very different from that of the English. Their breakfast is tea, commonly without milk. About thirty or forty years ago, tea was unknown to them, and they break-

---

[1] Mr. Kalm published his third volume just during the time of the last war.—F.

fasted either upon bread and butter, or bread and milk. They never put sugar into the cup, but take a small bit of it into their mouths while they drink. Along with the tea they eat bread and butter, with slices of dried beef. The host himself generally says grace aloud. Coffee is not usual here. They breakfast generally about seven. Their dinner is buttermilk and bread, to which they add sugar on special occasions, when it is a delicious dish for them, or fresh milk and bread, with boiled or roasted meat. They sometimes make use of buttermilk instead of fresh milk, in which to boil a thin kind of porridge that tastes very sour but not disagreeable in hot weather. With each dinner they have a large salad, prepared with an abundance of vinegar, and very little or no oil. They frequently drink buttermilk and eat bread and salad, one mouthful after another. Their supper consists generally of bread and butter, and milk with small pieces of bread in it. The butter is very salt. Sometimes too they have chocolate. They occasionally eat cheese at breakfast and at dinner; it is not in slices, but scraped or rasped, so as to resemble coarse flour, which they pretend adds to the good taste of cheese. They commonly drink very weak beer, or pure water.

A Conference with the Indians. The governor of New York often confers at Albany with the Indians of the Five Nations, or the Iroquois, (Mohawks, Senekas, Cayugaws, Onondagoes, and Oneidas), especially when they intend either to make war upon, or to continue a war against the French. Sometimes, also, their deliberations turn upon their conversion to the Christian religion, and it appears by the answer of one of the Indian chiefs or sachems to Governor Hunter,[1] at a conference in this town, that the English do not pay so much attention to a work of so much consequence as the French do, and that they do not send such able men to instruct the Indians, as they ought to do.[2] For after Governor Hunter had presented these In-

---

[1] Robert Hunter (d. 1734). He returned to England in 1719, and was later governor of Jamaica. See Richard L. Beyer's article in the *Dictionary of American Biography.*

[2] Mr. Kalm is, I believe, not rightly informed. The French ecclesiastics have allured some few wretched Indians to their religion and interest, and settled them in small villages; but by the accounts of their behavior, in the several wars of the French and English, they were always guilty of the greatest cruelties and brutalities; and more so than their heathen countrymen; and therefore it seems that they have been rather perverted than converted. On the other hand, the English have translated the Bible

dians, by order of Queen Anne, with many clothes and other presents, of which they were fond, he intended to convince them still more of her Majesty's good-will and care for them, by adding *that their good mother, the Queen, had not only generously provided them with fine clothes for their bodies, but likewise intended to adorn their souls by the preaching of the gospel; and that to this purpose some ministers should be sent to them to instruct them.* The governor had scarce ended, when one of the oldest sachems got up and answered *that in the name of all the Indians, he thanked their gracious good queen and mother for the fine clothes she had sent them; but that in regard to the ministers, they had already had some among them,* (who he likewise named) *who instead of preaching the holy gospel to them had taught them to drink to excess, to cheat, and to quarrel among themselves because in order to get furs they had brought brandy along with which they filled the Indians and deceived them.* He then entreated the governor to take from them these preachers, and a number of other Europeans who resided amongst them, for, before they came among them the Indians had been an honest, sober, and innocent people, but now most of them had become rogues. He pointed out that they formerly had the fear of God, but that they hardly believed his existence at present; that if he (the governor) would do them any favor, he should send two or three blacksmiths amongst them, to teach them to forge iron, in which they were inexperienced. The governor could not forbear laughing at this extraordinary speech. I think the words of St. Paul not wholly unapplicable on this occasion: For your sake the name of God is blasphemed amongst the heathens (Gentiles).[1]

---

into the language of the Virginian Indians, and converted many of them to the true knowledge of God; and at this present time, the Indian charity schools, and missions, conducted by the Rev. Mr. Eleazar Wheelock, have brought numbers of the Indians to the knowledge of the true God. The society for propagating the gospel in foreign parts, sends every year many missionaries, at their own expense, among the Indians. And the Moravian Brethren are also very active in the conversion of Gentiles; so that if Mr. Kalm had considered all these circumstances, he would have judged otherwise of the zeal of the British nation, in propagating the gospel among the Indians.—F.

The editor has purposely left this note by Forster for whatever it may be worth. After reading both sides the reader may judge for himself. Of Kalm's sincerity and veracity there is no doubt, but there was undoubtedly much to be censured on both sides. Neither the French, Dutch, nor the English were angels in their treatment of the Indians, and to-day we are not so certain that the English settled this land in order to save the souls of the savages.

[1] Romans II, 24.—F.

June the 21st (P.M.)

Leaving Albany. About five o'clock in the afternoon we left Albany, and proceeded towards Canada. We had two men with us, who were to accompany us to the first French place, which is Fort St. Frédéric, or as the English call it, Crown Point. For this service each of them was to receive five pounds of New York currency, in addition to food and drink. This is the common price here, and he that does not choose to conform to it, is obliged to travel alone. We were forced to use a canoe, as we could get neither battoes, nor boats of bark; and as there was a good road along the west side of the Hudson, we left the men to follow in the canoe, and we went along on the shore, that we might be able to examine it and its natural characteristics with greater accuracy. It is very inconvenient to row in these canoes; for one stands at each end and pushes the boat forwards. They commonly keep close to the shore, that they may be able to reach the ground easily. The rowers [or paddlers] are forced to stand upright while they row in a canoe.[1] We kept along the shore all evening. It consisted of large hills, and next to the water grew the trees, which I have mentioned above,[2] and which likewise are to be seen on the shores of the isle, in the river, situated below Albany. The easterly shore of the river is uncultivated, woody, and hilly; but the western is flat, cultivated, and chiefly turned into plowed fields, which has no drains, though they need them in some places. It appears very plainly here that the river has formerly been broader, for there is a sloping bank on the grain fields, at about thirty yards distance from the river, which always runs parallel. From this it sufficiently appears that the rising ground formerly was the shore or the river, and the fields its bed. As a further proof, it may be added that the same shells which abound on the present shore of the river, and are not applied to any use by the inhabitants, lie plentifully scattered on these fields. I cannot say whether this change was occasioned by the diminishing of the water in the river, or by its washing some earth down the river and carrying it to its sides, or by the river's cutting deeper in on the sides.

Agriculture. All land was plowed very even, as is usual in the

---

[1] This must have been a hazardous method. To-day, of course, the navigators of a canoe generally sit while propelling the boat.

[2] See p. 338.

Swedish province of Uppland. Some fields were sown with yellow and others with white wheat. Now and then we saw great fields of flax, which was now beginning to flower. In some parts it grew very well, and in others it was poor. The excessive drought which had continued throughout this spring had parched all the grass and plants on hills and higher grounds, leaving no other green plant than the common mullein (*Verbascum thapsus* L.) which I saw in several places, on the driest and highest hills, growing in spite of the parching heat of the sun, and though the pastures and meadows were excessively poor and afforded scarcely any food at all; yet the cattle never touched the mullein. Now and then I found fields with peas, but the charlock (wild mustard) (*Sinapis arvensis* L.) kept them quite under. The soil in most of these fields is a fine black mould, which goes down pretty deep.

The wild vines cover all the hills along the rivers, on which no other plants grow, and on those which are covered with trees they climb to the tops and wholly cover them, making them bend down with their weight. They had already large grapes; we saw them in abundance all day, and during all the time that we kept to the Hudson, on the hills, along the shores, and on some little islands in the river.

The white-beaked corn thieves (*blackbirds*) appeared now and then, flying amongst the bushes; their note is pleasant and they are not so large as the black corn thieves, (*Oriolus Phœniceus*). We saw them near New York, for the first time.

We found a water beech tree (*Platanus occidentalis*) cut down near the road, measuring about five feet in diameter.

This day and for some days afterwards we met islands in the river. The larger ones were cultivated and turned into grain fields and meadows. We walked about five English miles along the river to-day, and found the ground, during that time, very uniform and consisting of pure earth. I did not see a single stone on the fields. The red maple, the water beech, the water aspen, the wild plum tree, the sumach, the elm, the wild grape-vines, and some species of willows were the trees which we found on the rising shores of the river, where some asparagus (*Asparagus officinalis*) grew wild.

North of Albany. We passed the night about six miles from Albany in a countryman's cottage. On the west side of the river we saw several houses, one after another, inhabited by the descendants

Cohocs Falls

From original and English editions

of the first Dutch settlers, who lived by cultivating their grounds. About an English mile beyond our lodgings was a place where the tide stops in the Hudson, there being only small and shallow streams above it. At that place they catch a good many kinds of fish in the river.

The barns were generally built in the Dutch way, as I have before described them; [1] for in the middle was the threshing floor, above it a place for the hay and straw, and on each side stables for horses, cows, and other animals. The barn itself was very large. Sometimes the buildings of a farm consisted only of a small cottage with a garret above it, together with a barn upon the above plan.

JUNE THE 22ND

This morning I followed one of our guides to the waterfall near Cohoes, in the Mohawk River, before it empties into the Hudson River. These falls are about three English miles from the place where I passed the night. The country as far as the falls is a plain, and is hilly only about the fall itself. The wood is cleared in most places, and the ground cultivated and interspersed with pretty farmhouses.

The *Cohoes Falls* are among the greatest in this locality. They are in the Mohawk, before it unites with the Hudson. Above and below the falls both sides and the bottom are of rock, and there is a cliff at the fall itself, running everywhere equally high, and crossing the river in a straight line with the side which forms the fall. It represents, as it were, a wall towards the lower side, which is not quite perpendicular, wanting about four yards. The height of this wall, over which the water rolls, appeared to me about twenty or twenty-four yards. I had noted this height in my diary, and afterwards found it agreed pretty well with the account which that ingenious engineer, Mr. Lewis Evans, gave me at Philadelphia. He said that he had geometrically measured the breadth and height of the falls, and found it nine hundred English feet broad and seventy-five feet high. There was very little water in the river and it only ran over the falls in a few places. In some spots where the water had rolled down before, it had cut deep holes below into the rock, sometimes to the depth of two or three fathoms. The bed of the river,

---
[1] See page 118.

below the falls, was quite dry, there being only a channel in the middle fourteen feet broad, and a fathom or somewhat more deep, through which the water passed which came over the falls. We saw a number of holes in the rock below the falls, which bore a perfect resemblance to those in Sweden which we call giants' pots, or mountain kettles. They differed in size, there being large deep ones, and small shallow ones. We had clear uninterrupted sunshine, not a cloud above the horizon, and no wind at all. However, close to this fall, where the water was in such a small quantity, there was a continual drizzling rain, occasioned by the vapors which rose from the water during its fall, and were carried about by the wind. Therefore, in coming within a musketshot of the falls, against the wind, our clothes became wet at once, as from rain. The whirlpools, which were in the water below the falls, contained several kinds of fish; and they were caught by some people, who amused themselves with angling. The rocks hereabouts consist of the same black stone which forms the hills about Albany. When exposed to the air, it is apt to split into horizontal flakes, as slate does.

I here saw a kind of fence which we had not seen before, but which was used all along the Hudson where there was a quantity of woods. It can be called a timber fence, for it consisted of long, thick logs, and was about four feet high. It was made by placing the long logs at right angles to and upon short ones and fitting them together by having suitable crescent-shaped hollows in the short logs [in the manner of building log cabins]. Such a fence is possible only where there is plenty of trees.

En Route for Canada. At noon we continued our journey to Canada in the canoe, which was pretty long and made of a white pine. Somewhat beyond the farm where we lay at night, the river became so shallow that the men could reach bottom everywhere with their oars, it being in some parts not above two feet, and sometimes but one foot deep. The shore and bed of the river consisted of sand and pebbles. The river was very rapid, and against us, so that our men found it hard work to propel themselves forward against the stream. The hills along the shore consisted merely of earth, and were very high and steep in places. The breadth of the river was generally near two musketshots.

Sturgeons abound in the Hudson River. We saw them all day long leaping high up into the air, especially in the evening. Our guides, and the people who lived hereabouts, asserted that they never see any sturgeons in winter time, because these fish go into the sea late in autumn, but come up again in spring and stay in the river all summer. They are said to prefer the shallowest places in the river, which agreed pretty well with our observations, for we never saw them leap out of the water except in shallow spots. Their food is said to be several kinds of *confervæ*,[1] which grow plentifully in some places on the bottom of the river, for these weeds are found in their bellies when they are opened. The Dutch who settled here, and the Indians, fish for sturgeon, and every night of our voyage upon this river we observed several boats with people who speared them with harpoons. The torches which they employed were made of that kind of pine which they call the black pine here. The nights were exceedingly dark, though they were now the shortest, and though we were in a country so much to the south of Sweden. The shores of the river lay covered with dead sturgeons, which had been wounded with the harpoon, but escaped, and died afterwards; they occasioned an insupportable stench during the excessive heat of the day.

Indians. As we went further up the river we saw an Indian woman and her boy sitting in a boat of bark, and an Indian wading through the river, with a great cap of bark on his head. Near them was an island which was temporarily inhabited by a number of Indians who had gone there for sturgeon fishing. We went to their huts to learn if we could get one of them to accompany us to St. Anne and help us make a bark canoe to get to Fort St. Frédéric. On our arrival we found that all the men had gone into the woods hunting this morning, and we persuaded their boys to go and look for them. They demanded bread for payment, and we gave them twenty little round loaves; for as they found that it was of great importance to us to speak with the Indians, they raised difficulties and would not go till we gave them what they wanted. The island belonged to the Dutch, who had cultivated it. But they had now leased it to the Indians who planted their corn and several kinds of melons on it. The latter built their huts or wigwams on this island, on a very simple plan. Four posts were put into the ground perpendicularly, over which they had placed poles, and made a roof of bark upon them. The huts had either no walls at all, or they consisted of branches with leaves, which were fixed to the poles. The beds con-

---

[1] Water plants, probably of the filamentous green alga type.

sisted of deerskins which were spread on the ground. The utensils were a couple of small kettles, two ladles, and a bucket or two, of bark, made tight enough to hold water. The sturgeons were cut into long slices, and hung up in the sunshine to dry, to be ready to serve as food in winter. The Indian women were sitting at their work on the hill, upon deer-skins.—They never use chairs, but sit on the ground. However, they do not sit crosslegged, as the Turks do, but between their feet, which though they be turned backwards, are not crossed, but bent outwards. The women wear no headdress, and have black hair. They have a short blue petticoat which reaches to their knees, the edge of which is bordered with red or other-colored ribbons. They wear their blouses over their petticoats. They have large ear-rings, and their hair is tied behind and wrapped in ribbons. Their wampum, or pearls, and their money, which is made of shells, are tied round the neck and hang down on the breast. This is their whole dress. They were now making several articles of skins, to which they sewed the quills of the American porcupine, having dyed them black or red or left them in their original color.

Towards evening we went from there to a farm close to the river where we found only one man looking after the corn and the fields, the inhabitants not having yet returned from [King George's] war.

The little brooks here contain crawfish, which are exactly the same as ours,[1] with this difference only that they are somewhat smaller; however, the Dutch settlers will not eat them.

June the 23rd

Bargaining with Indians. We waited a good while for the Indians, who had promised to come home in order to show us the way to Fort St. Anne, and to assist us in making a boat of bark to continue our voyage. About eight o'clock three of the men arrived. Their hair was black, and cut short; they wore light gray pieces of woolen cloth over their shoulders, shirts which covered their thighs, and pieces of cloth or skins which they wrapped round the legs and parts of the thighs in place of stockings. They had neither hats, caps nor breeches. Two of them had painted the upper part of their foreheads and their cheeks with vermilion. Round their necks were ribbons from which bags hung down to the breast, containing a knive.

---

[1] *Cancer Astacus* L.—F.

They promised to accompany us for thirty shillings, but soon after changed their minds and went with an Englishman who gave them more. Thus we were obliged to make this journey quite alone. The Indians, however, were honest enough to return us fifteen shillings, which we had paid them beforehand.

Our last night's lodging was about ten English miles from Albany. During the war which had just ended, the inhabitants had all retreated from thence to Albany, because the French Indians had taken or killed all the people they met with, set the houses on fire, and cut down the trees. Therefore, when the inhabitants returned, things looked wretched; they found no houses, and were forced to lie under a few boards which they propped up against each other.

The river was almost a musketshot broad, and the ground on both sides cultivated. The hills near the river were steep, and the earth a pale color. The American elder (*Sambucus occidentalis* [1]) grows in incredible quantities on those hills, which appear white from the abundance of flowers on the tree.

All day long we had one group of rapids after another, full of rocks, which were great obstacles to our getting ahead. The water in the river was very clear, and generally shallow, being only from two to four feet deep, running very violently against us in most places. The shore was covered with pebbles and a grey sand. The hills consisted of earth, and were high and abrupt. The river was near two musketshots broad. On both sides the land was sometimes cultivated, and sometimes it was covered with woods.

The hills near the river abound with red and white clover. We found both these kinds plentiful in the woods. It is therefore difficult to determine whether they were brought over by the Europeans, as some people think, or whether they were originally in America, which the Indians deny.

We found purslane (*Portulaca oleracea*) growing plentifully in a dry sandy soil. In gardens it was one of the worst weeds.

We found people returning everywhere to their habitations, which they had been forced to leave during the war.

The farms were commonly built close to the river, on the hills. Each house had a little kitchen garden and a still lesser orchard. Some farms, however, had large gardens. The kitchen gardens yielded several kinds of pumpkins, watermelon and kidney beans.

---

[1] *Sambucus Canadensis* L.—F.

This year the trees had few or no apples on account of the frosty nights which had come in May and the drought which had continued throughout this summer.

Houses north of Albany. The houses hereabouts are generally built of beams and of unburnt bricks dried by the sun and the air. The beams are first erected, and upon them a gable with two walls, and the spars. The wall on the gable is made of nothing but boards. The roof is covered with shingles of fir. They make the walls of unburnt bricks, between the beams, to keep the rooms warmer; and that they might not easily be destroyed by rain and air they are covered with boards on the outside. There is generally a cellar beneath the house. The fireplaces among the Dutch were always built in, so that nothing projected out, and it looked as though they made a fire against the wall itself.

The farms are either built close to the riverside or on the high grounds, and around them are large fields of corn.

Muskrats. We saw great numbers of muskrats (*Castor zibethicus* L.) on the shores of the river, where they had many holes, some on a level with the surface of the water. These holes were large enough to admit a kitten. Before and in the entrance to the holes, lay a quantity of empty shells, the animals of which had been eaten by the muskrats.[1] They are caught in traps placed along the waterside and baited with corn or apples.

Sassafras trees abound here but never grow to any considerable height. Chestnut trees appear now and then. The cockspur hawthorn (*Cratægus crus galli* L.) grows in the poorest soil, and has very long spines, which shows that it may be very advantageously planted in hedges, especially in a poor soil.

This night we lodged with a farmer, who had returned to his farm after the war was over. All his buildings except the big barn had been burnt.

JUNE THE 24TH

The farm where we passed the night was the last in the province of New York, towards Canada, which had been left standing and which was now inhabited. Further on we met other inhabitants,

---

[1] This appears to be a new observation, as Linné, De Buffon, and Sarrasin pretend, they only feed on the *Acorus*, or reeds, and other roots.—F.

but they had no houses and lived in huts of boards, the houses having been destroyed during the war.

As we continued our journey we observed the country on both sides of the river was generally flat, but sometimes hilly, and large tracts of it covered with fir trees. Now and then we found some parts turned into plowed fields and meadows; however, the greater part was covered with woods. From Albany almost halfway to Saratoga the river ran very rapidly, and it was difficult to make headway northward. But afterwards it became very deep for the space of several miles and the water moved very slowly. The shores were very steep, though they were not very high. The river was two musketshots broad. In the afternoon it changed its direction, for hitherto it had been from north to south, but now it came from N.N.E. to S.S.W. and sometimes from N.E. to S.W.

Anthills are very scarce in America and I do not remember seeing a single one before I came to the Cohoes Falls. We observed a few in the woods today. The ants were the same species as our common red ones (*Formica rufa* L.). The anthills consisted chiefly of the slate-like crumbled stone which abounds here, there being no other material for them.

The Flora of Northern New York. Chestnut trees grew scattered in the woods. We were told that mulberry trees (*Morus rubra* L.) likewise grew wild here, but were rather scarce. This is the most northerly place where they grow in America; at least, they have not been observed further north. We met with wild parsnips every day but usually where the land was or had been cultivated. Hemp grew wild and in great abundance near old plantations. The woods abounded with woodlice, which were extremely troublesome to us. The *Thuya occidentalis* L. appeared along the shores of the river. I had not seen it there before. The trees which grew along the shores and on the adjacent hills, within our sight to-day were elms, birches, white firs, alders, dogwood trees, lime trees, red willows and chestnut trees. The American elder (*Sambucus Canadensis* L.) and the wild grapevines appeared only in places where the ground had been somewhat cultivated, as if they were desirous of being the companions of men. The lime trees and white walnut trees were the most numerous. The horn beams, with inflated cones, (*Carpinus ostrya* L.) appeared now and then; but the water beech and water poplar never came within sight any more.

We frequently saw squirrels and black squirrels in the woods.

A little distance from Saratoga, we met two Indians in their boats of bark, which could scarce contain more than one person.

Near Saratoga the river became shallow and rapid again. The ground had here been turned into grain fields and meadows, but on account of the war it lay waste.

Saratoga was a fort built of wood by the English to stop the attacks of the French Indians upon the English inhabitants in these parts, and to serve as a rampart to Albany. It was situated on a hill, on the west side of the Hudson River, and was built of thick posts driven into the ground, close to each other, in the manner of palisades, forming a square, the length of whose sides was within the reach of a musketshot. At each corner were the houses of the officers, and within the palisades the barracks, all of timber. This fort had been kept in order and was garrisoned till the last war, when the English themselves in 1747 set fire to it, not being able to defend themselves in it against the attacks of the French and their Indians; for as soon as a party of them went out of the fort, some of these enemies lay concealed and either took them all prisoners or shot them.

I shall only mention one out of many artful tricks which were played here, and which both the English and French who were present here at that time told me repeatedly: a party of French with their Indians, concealed themselves one night in a thicket near the fort. In the morning some of their Indians, as they had previously resolved, went to have a nearer view of the fort. The English fired upon them as soon as they saw them at a distance. The Indians pretended to be wounded, fell down, got up again, ran a little way, and dropped again. Above half the garrison rushed out to take them prisoners, but as soon as they had come up with them, the French and the remaining Indians came out of the bushes between the fortress and the English, surrounded them and took them prisoners. Those who remained in the fort had hardly time to shut the gates, nor could they fire upon the enemy, because they equally exposed their countrymen to danger, and they were vexed to see their enemies take and carry them off in their sight and under their cannon. Such French artifices as these made the English weary of their ill-planned fort. We saw some of the palisades still in the ground. There was an island in the river, near Saratoga, much better situated for a fortification. The country is flat on both sides of the river near Saratoga and its soil fertile. The wood round about was generally cut down. The shores of the river were high, steep and consisted of mould. We saw some hills in the north beyond the distant forests. The inhabitants are of Dutch extraction, and bear an inveterate hatred to all Englishmen.

We lay over night in a little hut of boards erected by the people who had come to live here.

June the 25th

Several sawmills had been built here before the war, which were very profitable to the inhabitants on account of the abundance of wood which grows here. The boards were easily brought to Albany and thence to New York in rafts every spring with the high water; but all the mills were burnt at present.

This morning we proceeded up the river, but after we had advanced about an English mile, we encountered a waterfall which cost us a deal of pains before we could get our canoe over it. The water was very deep just below the rapids, owing to its hollowing the rock out by the fall. In every place where we met with rocks in the river we found the water very deep, from two to four fathoms and upwards; because by finding a resistance it had worked a deeper channel into the ground. Above the falls the river was very deep again, the water sliding along silently, and increasing its speed suddenly near the shores. On both sides up to Fort Nicholson the shore was covered with tall trees. After rowing several miles we passed another waterfall, which was longer and more dangerous than the preceding one.

Giants' pots,[1] which I have described in the *Memoirs* of the Royal Swedish Academy of Sciences, are abundant near the fall of the rock which extends across the river. The rock was almost dry at present because the river contains very little water at this season of the year. Some of the giants' pots were round, but in general they were oblong. At the bottom of most of them lay either stones or grit in abundance. Some were fifteen inches in diameter, but some were less. Their depth was likewise different and some that I observed were above

---

[1] This is the literal meaning of the Swedish word *jätte-grytor*. See the *Memoirs of the Swedish Academy of Sciences* for the year 1743, p. 122, and Kalm's *Resa*, vol. I, pp. 135-136.

two feet deep. It is plain that they owed their origin to the whirling of the water round a pebble, which by that means was put in motion, together with the sand.

Through the Wilderness on Foot. We had intended to proceed close up to Fort Nicholson in the canoe, which would have been a great convenience to us; but we found it impossible to get over the upper falls, the canoe being heavy, and there being scarcely any water in the river except in one place where it flowed over the rock, and where it was impossible to get up on account of the steepness and the violence of the fall. We were accordingly obliged to leave our canoe here, and to carry our baggage through unfrequented woods to Fort Anne, on the river Woodcreek, which is a space from forty-three to fifty English miles, during which we were quite exhausted through the excess of heat. Sometimes we had no other way of crossing deep rivers than by cutting down tall trees, which stood on their banks, and throwing them across the water. All the land we passed over this afternoon was rather level, without hills or stones, and entirely covered with a tall and thick forest in which we continually met trees which had fallen down, because no one made the least use of the woods. We passed the next night in the midst of the forest, plagued with mosquitoes, gnats and woodlice, and in fear of all kinds of snakes.

June the 26th

Early this morning we continued our journey through the woods, along the Hudson River. There was an old path leading to Fort Nicholson, but it was so overgrown with grass that we discovered it only with great difficulty. In some places we found plenty of raspberries, some of which were already ripe.

Fort Nicholson is the place on the eastern shore of the Hudson where a wooden fortification formerly stood. We arrived there some time before noon and rested a while. Colonel Lydius [1] resided here till the beginning of the last war, chiefly with a view of carrying on a greater trade with the French Indians; but during the war, they burnt his house and took his son prisoner. The fort was situated on a plain, but at present the place is all overgrown with a thicket. It

---

[1] Probably John Henry Lydius, an Englishman.

was built in the year 1709, during the war which Queen Anne carried on against the French, and it was named after the brave English general Nicholson.[1] It was not so much a fort as a storehouse to Fort Anne. In the year 1711, when the English naval attempt upon Canada miscarried, the English themselves set fire to this place. The soil hereabouts seems to be pretty fertile. The river Hudson passed close by here.

Some time in the afternoon we continued our journey. We had hitherto followed the eastern shore of the Hudson and gone almost due north; but now we left it, and went E.N.E. or N.E. across the woods, in order to come to the upper end of the river Woodcreek, which flows to Fort St. Frédéric, where we might go in a boat from the former place. The ground we passed over this afternoon was generally flat and somewhat low. Now and then we passed rivulets, which were generally dried up during this season. Sometimes we saw a little hill, but neither mountains nor stones, and the country was everywhere covered with tall and thick forests. The trees stood close and afforded a fine shade, but the pleasure which we enjoyed from it was lessened by the incredible quantity of gnats which fill the woods. We found several plants here, but they were far from each other (as in our woods where the cattle have destroyed them), though no cattle ever came here. The ground was everywhere thickly covered with leaves of the last autumn. In some places we found the ground overgrown with great quantities of moss. The soil was generally very good, consisting of a deep mould in which the plants thrive very well. Therefore it seems that it would respond very well if it were cultivated. However, flowing waters were very scarce hereabouts; and if the woods were cleared, how great would be the effects of the parching heat of the sun, which might then act with its full force!

We lodged this night near a brook, in order to be sufficiently supplied with water, which was not everywhere at hand during this season. The mosquitoes, punchins or gnats and the woodlice were very troublesome. Our fear of snakes and especially of the Indians made the night's rest very uncertain and insecure.

Punchins, as the Dutch call them, are the little gnats (*Culex puli-*

---

[1] Presumably Sir Francis Nicholson (1660-1728), British colonial official. He served against Canada in 1705, and became governor of Acadia in 1713. He was given the title of lieutenant-general, 1725.

caris L.) which abound here. They are very minute, and their wings grey, with black spots. They are ten times worse than the larger ones (*Culex pipiens* L.) or mosquitoes, for their size renders them next to imperceptible; they are everywhere careless of their lives, suck their fill of blood and cause a burning pain.

We heard several great trees fall of themselves in the night, though it was so calm that not a leaf stirred. They made a dreadful cracking noise.

JUNE THE 27TH

We continued our journey in the morning. We found the country like that which we passed over yesterday, except for a few hills. Early this morning we plainly heard a waterfall or some rushing rapids in the Hudson River. In every part of the forest we found trees thrown down either through storms or age; but none were cut down, there being no inhabitants. And though the wood was very fine, nobody made use of it. We found it very difficult to get over such trees, because they had blocked all the paths, and close to them was the chief retreat of rattlesnakes during the intense heat of the day.

Arrival at Fort Anne. About two o'clock this afternoon we arrived at Fort Anne. It lies on the Woodcreek River, which is here at its source no bigger than a little brook. We stayed here all day, and the next, in order to make a new boat of bark, because there was no possibility to go down the river to Fort St. Frédéric without one. We arrived in time, for one of our guides fell ill that morning and could not have gone any further with his load. If he had been worse, we should have been obliged to stop on his account, which would have put us under great difficulties, as our provisions would soon have been exhausted, and from the wilderness where we were we could not have arrived at any inhabited place in less than three or four days. Happily we reached the wished-for place, and the sick man had time to rest and recover.

Around Fort Anne we found a number of common mice. They were probably the offspring of those which were brought to the fort in the soldier's provisions, at the time when it was kept in a state of defense.—We saw some apple and plum trees, which were certainly planted when the fort was in a good condition.

JUNE THE 28TH

The American elm grew in abundance in the forest hereabouts. There were two kinds of it. One was called the white elm, on account of the inside of the tree being white. It was more plentiful than the other species, which was called the red elm, because the color of the wood was reddish. The boats here were commonly made of the white bark. With the bark of hickory, which was employed as bast, they sewed the elmbark together, and with the bark of the red elm they joined the ends of the boat so close as to keep the water out. They beat the bark between two stones or, for want of them, between two pieces of wood.

The Making of a Bark Boat. The making of the boat took up half our time yesterday and all to-day. To make such a boat they pick out a thick tall elm, with a smooth bark, and with as few branches as possible. This tree is cut down, and great care is taken to prevent the bark from being hurt by falling against other trees or against the ground. With this view some people do not fell the trees, but climb to the top of them, split the bark and strip it off, which was the method our carpenter took. The bark is split on one side, in a straight line along the tree, as long as the boat it intends to be. At the same time the bark is carefully cut from the trunk a little way on both sides of the slit, that it may more easily separate. It is then peeled off very carefully, and particular care is taken not to make any holes in it. This is easy when the sap is in the trees, and at other seasons they are heated by fire for that purpose. The bark thus stripped off is spread on the ground in a level place, [with the smooth side down, later] turning the inside upwards. To stretch better, some logs of wood or stones are carefully put on it, which press it down. Then the sides of the bark are gently bent upwards in order to form the sides of the boat. Some sticks are then fixed into the ground, at the distance of three or four feet from each other, in a curved line, which the sides of the boat are intended to follow, supporting the bark intended for them. The sides are then bent in the form which the boat is to have, and according to that the sticks are either put nearer or further off. The ribs of the boat are made of thick branches of hickory, these being tough and pliable. They are cut into several flat pieces, about an inch thick, and bent into the

form which the ribs require, according to their places in the broader
or narrower part of the boat. Being thus bent, they are put across
the boat, upon the bark, or its bottom, pretty close together, about a
span or ten inches from each other. The upper edge on each side of
the boat is made of two thin strips of the length of the boat, which
are put close and flat against the side of the boat, where they are to
be joined. The edge of the bark is put between these two strips and
sewed up with threads of bast, of the mouse wood [1] or other tough
bark, or with roots. But before it is thus sewed up, the ends of the
ribs are likewise put between the strips on each side, taking care to
keep them at some distance from each other. After this is done, the
strips are sewed together, and being bent properly, both their ends
join at each end of the boat, where they are tied together with ropes.
To prevent the widening of the boat at the top, three or four trans-
verse bands are put across it, from one edge to the other, at a distance
of thirty or forty inches from each other. These bands are commonly
made of hickory, on account of its toughness and flexibility, and have
a good length. Their extremities are put through the bark on both
sides, just below the strips, which form the edges. They are bent up
above those strips and twisted round the middle part of the bands,
where they are carefully tied by ropes. As the bark at the two ends of
the boat cannot be put so close together as to keep the water out, the
crevices are stopped up with the crushed or pounded bark of the red
elm which in that state looks like oakum. Some pieces of bark are
put upon the ribs in the boat, otherwise the foot would easily pierce
the thin and weak bark below, which forms the bottom of the boat.
For the better security some thin boards are commonly laid at the
bottom, which may be trod upon with more safety. The side of the
bark which has been upon the wood, thus becomes the outside of the
boat, because it is smooth and slippery and cuts the water with less
difficulty than the other. The building of these boats is not always
quick; for sometimes it happens that after peeling the bark off an
elm, and carefully examining it, it is found pierced with holes and
splits, or it is too thin to venture one's life in. In such a case another
elm must be found; and it sometimes happens that several elms must
be stripped of their bark, before one is found fit for a boat. The boat

---

[1] Probably "moosewood" which Kalm took for "mousewood"; in that case it is
either "leatherwood" (*Dirca palustris*) or the "striped maple" (*Acer Pennsylvanicum*
L.).

which we made was big enough to carry four persons, with our baggage, which weighed somewhat more than a man.

All possible precautions must be taken in rowing on the rivers and lakes of these parts with a bark boat. For as the rivers, and even the lakes, contain a number of fallen trees, which are commonly hidden under the water, the boat may easily run against a sharp branch which will tear half the boat away, if one is rowing energetically, exposing the people in it to great danger, where the water is very deep, and especially if such a branch also holds the boat fast.

The boarding of such a frail vessel must be done with great care, and for the greater safety, without shoes. For with the shoes on, and still more with a sudden leap into the boat, the heels may easily pierce the bottom of it, which might sometimes result in dire consequences, especially when the boat is near a rock or close to deep water; and such places are common in the lakes and rivers here.

I never saw the *mosquitoes* more plentiful in any part of America than they are here. They were so eager for our blood that we could not rest all night, though we had surrounded ourselves with fire.

Woodlice (*Acarus Americanus* L.) abound here, and were more plentiful than on any part of the journey. Scarcely had a person sat down before a whole swarm of them crept upon his clothes. They caused us as much inconvenience as the gnats had the previous night, and continued to do so during the whole, short time we stayed here. Their bite is very disagreeable and they would prove very dangerous if any one of them should creep into a man's ear, from whence it is difficult to extract them. There are examples of people whose ears were swelled to the size of a fist on account of one of these insects creeping into them and biting. More is said about them in the description which I have given to the Royal Swedish Academy of Sciences.[1]

The whipperiwill, or whippoorwill, cried all night on every side. Fireflies flew in large numbers through the woods at night.

A Colonial "Pork-Barrel"! Fort Anne, where we now encamped, derives its name from Queen Anne; for in her time it served as a fortification against the French. It lies on the western side of the river Woodcreek, which is here as small as a brook, of a fathom's breadth, and may be waded through in any part during this season.

---

[1] See the *Memoirs of the Royal Academy* for the year 1754, pp. 19-31; also Bibliography in this work, item No. 21.

The fort is built in the same manner as the forts Saratoga and Nicholson, that is to say, of palisades, within which the soldiers were quartered, and at the corner of which were blockhouses providing lodgings for the officers. The whole consisted of wood, because it was erected only with a view for protection against wandering marauders. It is built on a little rising ground which runs obliquely to the river. The country round about it is partly flat, partly hilly, and partly marshy, but it consists merely of earth, and not a stone could be found there even if you would pay for it. General Nicholson built this fort in the year 1709; but at the conclusion of the war against the French it shared the fate of Saratoga and Fort Nicholson, being in 1711 burnt by the English themselves. The facts were these: in 1711 the English resolved to attack Canada, by land and by sea, at the same time. A powerful English fleet sailed up the St. Lawrence to besiege Quebec, and General Nicholson, who was the greatest promoter of this expedition, lead a large army to this place by land, to attack Montreal simultaneously; but a great part of the English fleet was shipwrecked in the St. Lawrence, and obliged to return to New England. The news of this misfortune was immediately communicated to General Nicholson, who was advised to retreat. Captain [Walter] Butler, who commanded Fort Mohawk during my stay in America, told me that he had been at Fort Anne in 1711 and that General Nicholson was about to leave it and go down the river Woodcreek in boats ready for that purpose, when he received the accounts of the disaster which had befallen the fleet. He was so enraged that he endeavored to tear his wig, but it being too strong for him he flung it to the ground and trampled on it, crying out, "Roguery, treachery!" He then set fire to the fort and returned. We saw the remains of the burnt palisades in the ground, and I asked my guides why the English had gone to such great expenses in erecting the fort, and why they had afterwards burnt it without any previous consideration? They replied that it was done to have another opportunity to extract money from the government; for the latter would appropriate a large sum for the rebuilding of the fort, the biggest proportion of which would perhaps reach the pockets of a few of the promoting authorities, who would then again erect only a wretched fort. They further told me that some of the richest people in Albany had promoted their poor relations to the places for supplying the army with bread etc., with a view to patch up their broken fortunes, and that they had acquired such wealth as rendered them equal to the richest inhabitants of Albany.

Excessive Heat. The heat was excessive to-day, especially in the afternoon when it was quite calm. We were on the very spot where Fort Anne formerly stood; it was a little place free from trees, but surrounded with them on every side, where the sun had full liberty to heat the air. After noon it grew as warm as in a hot bath, and I never felt a greater heat. I found it difficult to breathe and it seemed to me as if my lungs could not draw in a sufficient quantity of air. I was more eased when I went down into the valleys, and especially along the Woodcreek. I tried to fan the air to me with my hat, but it only increased the difficulty of breathing and I received the greatest relief when I went to the water, and in a shady place frequently sprinkled some water in the air before me. My companions also suffered a great deal, but they did not find such difficulty in breathing as I had experienced. However, towards evening the air became somewhat cooler.

JUNE THE 29TH

Having finally completed our boat, after a great deal of labor and trouble, we continued our journey this Sunday morning. Our provisions, which were much diminished, urged us to make great haste, for by being obliged to carry everything on our backs through this wilderness to Fort Anne, we had not been able to take a great quantity of provisions with us, having been compelled to include in our baggage several other very necessary things; nevertheless we always ate very heartily. As there was very little water in the river, and several trees had fallen across it, which frequently stopped the boat, I left the men in it and walked along the shore with Jungström. The ground on both sides of the river was so low that it must be under water in spring and autumn. The shores were covered with several sorts of trees which stood at moderate distances from each other, and a great deal of grass grew between them. The trees afforded the fine shade which is so necessary and agreeable in this hot season: but the pleasure it gave was considerably lessened by the number of gnats which we encountered. The soil was extremely rich.

Beaver Dams. As we came lower down the river, the dams, which the beavers had made in it, produced new difficulties. These laborious animals had carried together all sorts of boughs and branches and placed them across the river, putting mud and clay in betwixt them, to stop the water. They had bit off the ends of the branches as neatly as if they had been chopped off with a hatchet. The grass about these places was trod down by them, and in the neighborhood of the dams we sometimes came upon paths in the grass, where the beavers probably had dragged trees along. We found a row of dams before us, which stopped us a considerable while, as we could not get forwards with the boat till we had cut through them.

A Discovery. As soon as the river was more open we got into the boat again, and continued our journey in it. The breadth of the river, however, did not exceed eight or nine yards, and frequently it was not above three or four yards broad, and generally so shallow that our boat got on with difficulty. Sometimes it acquired such a sudden depth that we could not reach the ground with sticks of seven feet in length. The stream was very rapid in some places and very slow in others. The shores were low at first, but afterwards remarkably high and steep, and now and then a rock projected into the water which always caused a great depth in such places. The rocks consisted here of a gray quartz, mixed with gray limestone, lying in strata. The water in the river was very clear and transparent, and we saw several little paths leading to it from the woods, said to be made by beavers and other animals, which came here to drink. After going a little more than three English miles we came to a place where a fire was yet burning and we could see from the trodden grass that people had been lying there the night before, and then we little thought that we had narrowly escaped death the night before, as we heard this evening. Now and then we ran into trees lying across the river, and some beaver dams, which obstructed our way.

How Kalm Escaped Death. Towards night we met a French sergeant, and five French soldiers, who had been sent by the commander of Fort St. Frédéric, to accompany three Englishmen to Saratoga, and to defend them in case of necessity against six French Indians who had gone to be revenged on the English for killing the brother of one of them in the last war. The peace had already been concluded at that time, but as it had not yet been proclaimed in Canada the Indians thought they could take this step; therefore they silently got away, contrary to the order of the Governor of Montreal, and proceeded towards the English plantations. We here had occasion to perceive the care of Providence for us, in escaping these savage barbarians. We had found the grass trod down all the day long, but had had no thoughts of danger, as we believed that everything was quiet and peaceable. We were afterwards informed, that these Indians had trod the grass down, and passed the last night in the place where we found the burning brands in the morning. The usual route which they were to have taken was by Fort Anne, but to shorten their journey they had gone an unfrequented path. If they had gone on towards Fort Anne, they would have met us without doubt, and looking upon us all as Englishmen, for whose blood they were thirsting, they could easily have surprised and shot us all, and by that means have been rid of the trouble of going any further to satisfy their cruelty. We were not a little agitated when the Frenchmen told us how near death we had been to-day. We passed the night here, and though the French repeatedly advised and desired me not to venture any further with my English company, but to follow them to the first English settlement, and then back to Fort St. Frédéric, yet I resolved, with the protection of the Almighty, to continue my journey the next day.

Wild Pigeons. We saw immense numbers of the wild pigeons which I have previously described [1] flying in the woods, and which sometimes come in incredible flocks to the southern English colonies, without the inhabitants there knowing where they come from. They have their nests in the trees here, and almost all night make a rustling, whirring noise and cooing in the trees where they roost. The French shot a great number of them, and gave us some, in which we found a great quantity of the seeds of the elm, which evidently demonstrated the care of Providence in supplying them with food; for in May the seeds of the red maple, which abounds here, are ripe, and drop from the trees and are eaten by the pigeons during that time: afterwards, the seeds of the elm ripen, which then become their food, till other seeds mature for them. Their flesh is the most palatable of any bird's flesh I have ever tasted.

Falling Trees. Almost every night we heard some trees crack and fall while we lay here in the wood, though the air was so calm that not a leaf stirred. The reason for this breaking I am totally unac-

---

[1] See page 252.

quainted with. Perhaps the dew or something else loosens the roots
of trees more at night; or perhaps something falls too heavily on the
branches on one side of the tree. It may be that the above-mentioned
wild pigeons settle in such quantities on one tree as to weigh it down;
or perhaps the tree begins to bend more and more to one side from
its center of gravity, making the weight continuously greater for the
roots to support till it comes to the point when it can no longer keep
upright, which may as well happen in the midst of a calm night
as at any other time. When the wind blows hard it is reckoned very
dangerous to sleep or walk in the woods on account of the many
trees which fall in them; and even when it is very calm there is some
danger in passing under very large and old trees. I was told, in several parts of America that the storms or hurricanes sometimes pass
over only a small part of the woods and tear down the trees in it;
and I have had opportunities of confirming the truth of this observation by finding places in the forests where almost all the trees had
crashed down, and lay in one direction.

Tea, which is brought in great quantities from China, is differ-
ently esteemed by different people, and I think we would be as well,
and our purses much better, if we were without both tea and coffee.
However, I must be impartial, and mention in praise of tea that if
it be useful it must certainly be so in summer on such journeys as
mine through a vast wilderness, where one cannot carry wine or
other liquors and where the water is generally unfit for use, being
full of insects. In such cases it is very refreshing when boiled and
made into tea, and I cannot sufficiently describe the fine taste it has
under such circumstances. It relieves a weary traveller more than
can be imagined, as I myself have experienced, and as have also a
great many others who have travelled through the primeval forests
of America. On such journeys tea is found to be almost as necessary
as food.[1]

JUNE THE 30TH

This morning we left our boat to the Frenchmen, who used it to
carry their provisions; for we could not make any further use of it

---

[1] On my travels through the desert plains, beyond the river Volga, I have had
several opportunities of making the same observations on tea; and every traveller, in
the same circumstances, will readily allow them to be very just.—F.

on account of the number of trees which the French had thrown across the river during the last war to prevent the attacks of the English upon Canada. The Frenchmen gave us leave to make use of one of their birch canoes, which they had left behind them, about six miles from the place where we passed the last night. Thus we continued our journey on foot, along the river, and found the country flat, with some little vales here and there. It was everywhere covered with tall deciduous trees, among which the beech, the elm, the American lime tree, and the sugar maple were the most numerous. The trees stood some distance from each other, and the soil in which they grew was extremely rich.

Description of the Country. After we had walked about a Swedish mile, or six English miles, we came to the place where the six Frenchmen had left their bark boats, of which we took one, and rowed down the river, which was now between nineteen and twenty yards broad. The banks on both sides were very smooth and not very high. Sometimes we found a hill consisting of gray quartz, mixed with small fine grains of gray spar. We likewise observed black stripes of it; but they were so small that I could not determine whether they were mica or another kind of stone. The hills were frequently divided into strata, lying one above another and of the thickness of five inches. The strata went from north to south, and were not quite horizontal but dipping to the north. As we went further on we saw high steep hills on the riverside, partly covered with trees; but in other parts the banks consisted of a swampy turf ground, which gave way when it was walked upon and had some similarity to the sides of our marshes which my countrymen are now about to drain. In those parts where the ground was low and flat we did not see any stones either on the ground or on the softer shore, and both sides of the river, when they were not hilly, were covered with tall elms, American lime trees, sugar maples, beeches, hickory trees, some water beeches and white walnut trees.

On our left we saw an old fortification of stones laid above one another; but nobody could tell me whether the Indians or the Europeans had built it.

Kalm's Party Goes Astray. We had rowed very fast all afternoon in order to make a good distance, and we thought that we were upon the right road, but found ourselves greatly mistaken; for towards night we observed that the reeds in the river bent towards us, which was a mark that it flowed towards us; whereas, if we had been on the right body of water it should have run in the direction we were going. We likewise observed from the trees which lay across the river, that nobody had lately passed that way, though we should have seen the tracks of the Frenchmen in the grass along the shore, when they brought their boat over these trees. At last we plainly saw that the river flowed against us, and we were convinced that we had gone twelve English miles and upwards upon a wrong river, which obliged us to return, and to row till very late at night. We sometimes thought, through fear, that the Indians, who had gone to murder some English, would unavoidably meet us. Though we rowed very fast, we were not able to-day to get halfway back to the place where we first left the right river.

The most odoriferous effluvia sometimes came from the banks of the river at nightfall, but we could not determine what flowers diffused them. However, we supposed they chiefly arose from the *Asclepias syriaca,* and the *Apocynum· androsæmifolium.*

The muskrats could likewise be smelled at night. They had many holes in the shores even with the surface of the water.

We passed the night on an island, where we could not sleep on account of the gnats. We did not venture to make a fire, for fear the Indians should find us out and kill us. We heard several of their dogs barking in the woods, at a great distance from us, which added to our uneasiness in this wilderness.

July the 1st, 1749

At daybreak we got up, and rowed a good while before we got to the place where we left the true course. The country which we passed was the poorest and most disagreeable imaginable. We saw nothing but a row of amazingly high mountains covered with woods, steep and rough on their sides, so that we found it difficult to get to an open place in order to land and boil our dinner. In many places the ground, which was very smooth, was under water, and looked like the sections of our Swedish morasses which are being drained; for this reason the Dutch in Albany call these parts the "Drowned Lands." [3] Some of the mountains run from S.S.W. to

---

[3] *De verdronkene landen.*—F.

N.N.E., and along the river they form perpendicular shores, and are full of rocks of different sizes. The river flows for several miles from south to north.

The wind blew from the north all day, and made it very hard work for us to travel on, though we all rowed as hard as we could, for what little we had left of our provisions was eaten to-day at breakfast. The river was frequently an English mile and more broad, then it became narrow again, and so on alternately; but upon the whole it kept a good breadth, and was surrounded on both sides by high mountains.

Approaching Fort St. Frédéric. About six o'clock in the evening we arrived at a point of land about twelve English miles from Fort St. Frédéric. Behind this point the river is converted into a spacious bay, and as the wind still kept blowing pretty strong from the north, it was impossible for us to proceed, since we were extremely tired. We were therefore obliged to pass the night here, in spite of the remonstrances of our hungry stomachs.

It is to be attributed to the peculiar grace of God towards us that we met the above-mentioned Frenchmen on our journey, and that they gave us leave to take one of their bark boats. It hardly happens once in three years that the French take this route to Albany; for they commonly pass over Lake St. Sacrement, or, as the English call it, Lake George, which is the nearer and better way, and everybody wondered why they took this troublesome one. If we had not gotten their large, strong boat, and had been obliged to keep the one we had made, we would in all probability have been very ill off; for to venture upon the great bay during the least wind with so wretched a vessel, would have been a great piece of temerity, and we should have been in danger of starving if we had waited for a calm. For being without fire-arms, and these wildernesses having but few quadrupeds, we would have been obliged to subsist upon frogs and snakes, which (especially the latter) abound in these parts. I can never think of this journey, without reverently acknowledging the peculiar care and providence of the merciful Creator.

JULY THE 2ND

At Crown Point. Early this morning we set out on our journey again, it being moonlight and calm, and we feared lest the wind should change and become unfavorable to us if we stopped any longer. We all rowed as hard as possible, and happily arrived about eight in the morning at Fort St. Frédéric, which the English call Crown Point. Monsieur Lusignan,[1] the governor, received us very politely. He was about fifty years old, well acquainted with polite literature, and had made several journeys into this country, by which he had acquired an exact knowledge of several things relative to it.

Drought. I was informed that during the whole of this summer, a continual drought had been here, and that they had not had any rain since last spring. The excessive heat had retarded the growth of plants, and on all dry hills the grass and a vast number of plants had dried up. The small trees which grew near rocks, heated by the sun, had withered leaves and the grain in the fields bore a very wretched appearance. The wheat had not yet eared, nor were the peas in blossom. The ground was full of wide, deep cracks, into which the little snakes retired and hid when pursued, as into an impregnable asylum.

Forest Fires. The country hereabout, it is said, contains vast forests of firs of the white, black and red varieties, which formerly had been still more extensive. One of the chief reasons for their decrease is the numerous fires which happen every year in the woods, through the carelessness of the Indians, who frequently make great fires when they are hunting, which spread over the fir woods when everything is dry.

French Advancement of the Natural Sciences. Great efforts are made here for the advancement of Natural History, and there are few places in the world where such good regulations are made for this purpose, all of which is chiefly owing to the care and zeal of a single person. From this it appears how well a useful science is received and advanced when the leading men of a country are its patrons. The governor of the fort was pleased to show me a long article which the then governor-general of Canada, the Marquis de la Galissonnière[2] had sent him. It was the same Marquis, who some years after, as a French admiral, engaged the English fleet under

---

[1] See diary of July 5th.

[2] Rolland-Michel Barrin, Marquis de la Gallissonnière (1693-1756), French naval officer, who was governor-general of Canada from June, 1747, to the fall of 1749, during the English captivity of the Marquis de la Jonquière who had first been appointed. Gallissonnière returned to France in 1749. See L. W. Marchand's French translation of Kalm's *Travels*, vol. 2, p. 240, note.

Admiral Byng[1] the consequence of which was the conquest of Minorca. In this above-mentioned work a number of trees and plants are mentioned, which grow in North America, and deserve to be collected and cultivated on account of their useful qualities. Some of them are described, among which is the *Polygala senega,* or rattle-snake root, and with several of them the places where they grow are mentioned. It is further requested that all kinds of seeds and roots be gathered here; and, to assist such an undertaking, a method of preserving the gathered seeds and roots, is prescribed, so that they may grow and be sent to Paris. Specimens of all kinds of minerals are required, and all the places in the French settlements are mentioned, where any useful or remarkable stone, earth, or ore has been found. There is likewise a manner of making observations and collections of curiosities in the animal kingdom. To these requests it is added, to inquire and get information in every possible manner as to what purpose and in what manner the Indians employ certain plants and other products of nature, in medicines or in any other field of usefulness. This useful paper was drawn up by order of the Marquis de la Galissonnière, by Mr. Gauthier,[2] the royal physician at Quebec, and afterwards corrected and improved by the marquis's own hand. He had several copies made of it which he sent to all the officers in the forts and also to other learned men who travelled in the country. At the end of the work is an injunction to the superior officers, to let the governor-general know which of the common soldiers had used the greatest diligence in the discovery and collection of plants and other natural curiosities, that he might be able to promote them when an opportunity occurred to places adapted to their respective capacities, or to reward them in some other manner. I found that the people of distinction had here in general a much greater taste for natural history and other learning than in the English colonies, where it was everybody's sole care

---

[1] John Byng (1704-1757), the unfortunate English admiral, was sent in 1756 from the Mediterranean, where he was commanding, to the West Indies to prevent the French from taking Minorca. He handled his ships unskillfully, it was claimed, and was defeated on May 20. He was recalled, tried by court martial, sentenced to death for neglect of duty, and shot, January 27, 1757.

[2] This physician, whose name was Jean François *Gauthier* (1711-1756), (but whose name in the records is spelled in various ways), was Kalm's cicerone and travelling companion in Canada. Kalm refers to him several times in his *Travels.* (See article on Kalm in the *Dictionnaire Générale du Canada*). Gauthier, a "modest savant," succeeded Sarrazin as royal physician at Quebec.

and employment to scrape a fortune together, and where the sciences were held in universal contempt.[1] It was still complained of here, that those who studied natural history, did not sufficiently inquire into the medicinal use of the plants of Canada.

The Health of the French Canadians. The French who are born in France are said to enjoy better health in Canada than in their native country, and to attain to a greater age than the French born in Canada. I was likewise assured that the European Frenchmen can do more work, and perform more journeys in winter, without prejudice to their health than those born in this country. The intermittent fever which attacks the Europeans on their arrival in Pennsylvania, and which as it were, hardens them to the climate,[2] is not known here, and the people are as well after their arrival as before. The English have frequently observed that those who

---

[1] It seems Mr. Kalm has forgotten his own assertions in the first volume. Dr. Colden, Dr. Franklin, and Mr. Bartram, have been the great promoters and investigators of nature in this country; and how would the inhabitants of Old England have gotten the fine collections of North American trees, shrubs, and plants, which grow at present almost in every garden, and are as if they were naturalized in Old England, had they not been assisted by their friends, and by the curious in North America. One need only cast an eye on Dr. Linné's new edition of his *Systema*, and the repeated mention of Dr. Garden, in order to be convinced that the English in America have contributed a greater share towards promoting natural history, than any nation under heaven, and certainly more than the French, though their learned men are often handsomely pensioned by their great Monarque: on the other hand the English study that branch of knowledge, from the sole motive of its utility, and the pleasure it affords to a thinking being, without any of those mercenary views, held forth to the learned of other countries. And as to the other parts of literature, the English in America are undoubtedly superior to the French in Canada, witness the many useful institutions, colleges, and schools founded in the English colonies in North America, and so many very considerable libraries now erecting in this country, which contain such a choice of useful and curious books, as were very little known in Canada, before it fell into the hands of the English; not to mention the productions of original genius written by American born.—F.

We leave this interesting note as it is, for whatever value it may have. The reader will easily understand Kalm's meaning. Forster's defense of the English is very touching and commendable, though there will be a difference of opinion as to what nation is the most mercenary and which one shows the greatest interest in science for culture's sake. Kalm means, in part, that in the English colonies there was no concerted scientific study directed from high governmental sources. The interest in natural sciences was, after all, limited to a few individuals from the rank and file. Men like Franklin were the great *private* exceptions, who, as one might contend, proved the rule of the *general* lack of scientific investigations in the English colonies. Cf. letter by Kalm to Linné of December 5, 1750 in *Bref och Skrifvelser af och till Carl von Linné*, I, viii, 60 (Uppsala, 1922, edited by J. M. Hulth).

[2] See pages 191 ff.

are born in America of European parents can hardly ever bear sea voyages or move to the different southern parts of America with as healthy results as those born in Europe. The French born in Canada have the same constitutions, and when any of them go to the West India Islands, such as Martinique, Domingo, etc., and make a long stay there, they commonly fall sick and die soon after: those who fall ill there seldom recover, unless they are brought back to Canada. On the contrary, those who go from France directly to those islands can more easily bear the climate, and attain a great age there; this I heard confirmed in many parts of Canada.

JULY THE 5TH

Indian Revenge. While we were at dinner we heard several times a repeated, disagreeable, bloodcurdling outcry, some distance from the fort, in the river Woodcreek: Mr. Lusignan,[1] the commander, told us this cry was ominous, because he could conclude from it that the Indians, whom we escaped near Fort Anne, had completed their design of avenging the death of one of their brethren upon the English, and that their shouts showed that they had killed an Englishman. As soon as I came to the window, I saw their boat, with a long pole at the front, at the extremity of which they had put a bloody human scalp. As soon as they had landed, we heard that they, being six in number, had continued their journey (from the place where we saw marks of their passing the night) till they had gotten within the English boundaries, where they found a man and his son employed in harvesting. They crept on towards this man and shot him dead. This happened near the very village where the English, two years before, killed the brother of one of these Indians, who had then gone out to attack them. According to their custom they cut off the scalp of the dead man and took it with them, together with his clothes and his son, who was about nine years old. As soon as they came within a mile of Fort St. Frédéric, they put the scalp on a pole in the fore part of the boat, and shouted as a sign of their success. They were dressed in shirts, as usual, but some of them had put on the dead man's

---

[1] Paul-Louis Lusignan (1691; d. sometime after 1752), French army officer, had been promoted to a captaincy in 1744. He was, as Kalm asserts, at this time commander of Fort Saint-Frédéric.

clothes; one his coat, the other his breeches, another his hat, etc. Their faces were painted with vermilion, with which their shirts were marked across the shoulders. Most of them had great rings in their ears, which seemed to be a great inconvenience to them, as they were obliged to hold them when they leaped or did anything which required a violent motion. Some of them had girdles of the skins of rattlesnakes, with the rattles on them; the son of the murdered man had nothing but his shirt, breeches and cap, and the Indians had marked his shoulders with red. When they got on shore they took hold of the pole on which the scalp was put, and danced and sung at the same time. Their object of taking the boy was to carry him to their tent, to bring him up instead of their dead brother, and afterwards to marry him to one of their relations so that he might become one of them. Notwithstanding they had perpetrated this act of violence in time of peace, contrary to the command of the governor in Montreal, and to the advice of the governor of St. Frédéric, the latter could not at present deny them provisions and whatever they wanted for their journey, because he did not think it advisable to exasperate them; but when they came to Montreal, the governor called them to account for this action, and took the boy from them, whom he afterwards sent to his relations. Mr. Lusignan asked them what they would have done to me and my companions, if they had met us in the wilderness. They replied that as it was their chief intention to take their revenge on the Englishmen in the village where their brother had been killed, they would have let us alone. But it would have depended on the humor they were in when we first came in sight. However, the commander and all the Frenchmen said that what had happened to me was infinitely safer and better.

Huge Skeleton Found. Some years ago a skeleton of an amazingly large animal had been found in that part of Canada where the Illinois live. One of the lieutenants in the fort assured me that he had seen it. The Indians who were there, had found it in a swamp. They were surprised at the sight of it, and when they were asked what they thought it was, they answered that it must be the skeleton of the chief or father of all the beavers. It was of a prodigious bulk, and had thick white teeth, about ten inches long. It was looked upon as the skeleton of an elephant. The lieutenant assured me that the figure of the whole snout was yet to be seen though it was half mouldered. He added that he had not observed whether any of the bones had been taken away, but thought the skeleton lay intact there. I have heard people talk of this monstrous skeleton in several parts of Canada.[1]

Bears are plentiful hereabouts, and they kept a young one about three months old at the fort. He had exactly the same shape and qualities as our common bears in Europe, except the ears, which seemed to be longer in proportion, and the hairs which were stiffer; his color was deep brown, almost black. He played and wrestled every day with one of the dogs. A vast number of bear skins are annually exported to France from Canada. The Indians prepare an oil from bear's fat, with which in summer they daub their faces, hands, and all naked parts of their body, to secure them from the bite of the gnats. With this oil they likewise frequently smear the body, when they are excessively cold, tired with labor, hurt, and in other cases. They believe it softens the skin, and makes the body pliant, and promotes longevity.

Dandelion Greens. The common dandelion (*Leontodon taraxacum* L.) grew in abundance in the pastures and on the borders of the grain fields, and was now in flower. In spring when the young leaves begin to come up, the French dig up the plants, take their roots, wash them, cut them, and prepare them in vinegar as a common salad; but they have a bitter taste. It is not usual here to make use of the leaves for eating.

JULY THE 6TH

Veterans' Cottages. The soldiers who had been paid off after the war had built houses round the fort on the grounds allotted to them; but most of these habitations were no more than wretched cottages, no better than those in the most wretched places of Sweden. There was that difference, however, that their inhabitants here were rarely oppressed by hunger, and could eat good and pure wheat bread.[2] The huts which they had erected consisted of boards, standing perpendicularly close to each other. The roofs were of wood too. The crevices were stopped up with clay to keep

---

[1] The country of the Illinois is on the river Ohio, near the place where the English have found some bones which are supposed to have belonged to elephants.—F.

[2] Very little wheat bread was consumed in Sweden or Finland in Kalm's day.

the room warm. The floor was usually clay, or a black limestone, which is common here. The hearth was built of the same stone, except the place where the fire was to burn, which was made of gray stones, which for the greatest part consisted of particles of quartz. In some hearths the stones quite close to the fireplace were of limestone. However, I was assured that there was no danger of fire, especially if the stones which were most exposed to the heat were of a large size. Dampers had never been used here and the people had no glass in their windows.

The *Fences* were just like the most common ones in Sweden only that the distance between the slender upright posts was sometimes as much as eighteen feet. For binding the pairs of posts hickory was used; a circle was made of it and then tied.

JULY THE 8TH

The *Galium tinctorium* is called *Tisavojaune rouge* by the French throughout all Canada, and abounds in the woods round this place, growing in a moist but fine soil. The roots of this plant are employed by the Indians in dying the quills of the American porcupines red, which they put into several pieces of their work; and air, sun or water seldom change this color. The French women in Canada sometimes dye their clothes red with these roots, which are small, like those of the *Galium luteum,* or yellow bedstraw.

The horses are left out of doors during the winter, and find their food in the woods, living upon nothing but dry plants, which are very abundant; however, they do not lose weight by this food, but look very fine and plump in the spring.

JULY THE 9TH

The skeleton of a whale was found a few French miles from Quebec, and one French mile [1] from the St. Lawrence River, in a place where there is no flowing water at present. This skeleton was of a very great size, and the governor of the fort said he had spoken with several people who had seen it.

---

[1] A French mile (lieue) seems to be about three English miles. The value of it varied with the time and locality. To-day, in France, it is the equivalent of 4 km. or 2.49 miles.

JULY THE 10TH

Boats. The boats which are here used, are of three kinds. 1. Bark boats, made of the bark of trees, with ribs of wood; 2. Canoes, consisting of a single piece of wood, hollowed out, which I have already described. These are here made of the white fir, and of different sizes. They are not propelled by rowing but by paddling, by which method not half the strength can be applied which is used in rowing. A single man might, I think, row as fast as two of them could paddle.[1] 3. The third kind of boats are bateaux. They are always made very large here, and used for large cargoes. They are flat-bottomed, and the bottom is made of red, but more commonly of white, oak which shows better resistance when it runs against a stone than other wood. The sides are made of white fir, because oak would make the bateau too heavy. They make plenty of tar and pitch here.

A Soldier's Rations. The soldiery enjoy such advantages here as they are not allowed in any part of the world. Those who formed the garrison of this place had a very plentiful allowance from their government. They get every day a pound and a half of wheat bread, which is almost more than they can eat. They likewise get plenty of peas, bacon, and salt or dried meat. Sometimes they kill oxen and other cattle, the flesh of which is distributed among the soldiers. All the officers kept cows, at the expense of the king, and the milk they gave was more than sufficient to supply them. The soldiers had each a small garden outside the fort, which they were allowed to attend and to plant in it whatever they liked. Some of them had built summerhouses in them and planted all kinds of vegetables. The governor told me that it was a general custom to allow the soldiers a plot of ground for kitchen gardens, at such of the French forts hereabouts as were not situated near great towns, from whence they could be supplied. In time of peace the soldiers have very little guard duty when at the fort; and as the lake close by was full of fish, and the woods abounded with birds and animals, those amongst them who chose to be diligent could live extremely well and like a lord in regard to food. Each soldier got a new coat every two years; but annually, a waistcoat, cap, hat, breeches, cravat, two pair of stockings, two pair of shoes, and as much wood as he had occasion for in

---

[1] Kalm forgets that in this case the shape of the boat makes it more suitable for paddling.

winter. They likewise got five *sols* [1] apiece every day, which is augmented to thirty sols when they have any particular labor for the
king. When this is considered it is not surprising to find the men
are very healthy, well fed, strong and lively here. When a soldier
falls sick he is brought to the hospital, where the king provides him
with a bed, food, medicine, and people to take care of and serve him.
When some of them asked leave to be absent for a day or two to go
away it was generally granted them if circumstances would permit,
and they enjoyed as usual their share of provisions and money, but
were obliged to get some of their comrades to mount guard for them
as often as it came to their turns, for which they gave them an
equivalent. The governor and officers were duly honored by the
soldiers; however, the soldiers and officers often spoke together as
comrades, without any ceremonies, and with a very becoming freedom. The soldiers who are sent hither from France commonly serve
till they are forty or fifty years old, after which they are honorably
discharged and allowed to settle upon and cultivate a piece of
ground. But if they have agreed on their arrival to serve no longer
than a certain number of years, they are dismissed at the expiration
of their term. Those who are born here commonly agree to serve
the crown during six, eight, or ten years, after which they are [hon-
orably] discharged and settle down as farmers in the country. The
king presents each discharged soldier with a piece of land, being
commonly 40 arpents [2] long and but three broad, if the soil be of
equal goodness throughout; but they get somewhat more, if it be
poorer. As soon as a soldier settles to cultivate such a piece of land,
he is at first assisted by the king, who supplies him, his wife and
children with provisions during the first three or four years. The
king likewise gives him a cow and the most necessary instruments
for agriculture. Some soldiers are sent to assist him in building a
house, for which the king pays them. These are of great help to a
poor man who begins to keep house, and it seems that in a country
where the troops are so highly distinguished by royal favor, the king
cannot be at a loss for soldiers. For the better cultivation and popu-

---

[1] See *infra,* p. 399, note.

[2] The value of the *arpent* varied with the locality. According to Kalm, as given in
the original, there were 84 arpents in a French *lieue* or mile. The latter is 4444 1/2
meters, which makes an arpent about 175 feet. To-day the Canadian *arpent* is supposed to be 12 rods or 198 feet. The *arpent* of Kalm's day was probably somewhere
between the two.

lation of Canada, a plan was proposed some years ago for sending three hundred men over from France every year, by which means the old soldiers might always be retired, marry and settle in the country. The land which was allotted to the soldiers about this place, was very good, consisting throughout of a deep mould, mixed with clay.

The *food* which the better classes of Frenchmen ate was as follows: for dinner, clear soup, with slices of wheat bread and various kinds of relishes; then a dish of cooked meat, sometimes fried after being cooked; occasionally beef or mutton, squabs or fowl. It was almost always fresh. Often the third course consisted of green peas and occasionally fried fish. The wheat bread used was quite good, but ordinarily, according to my taste, too salt. The salt was a gray, finely powdered variety. No cheese was served and very little butter, which had little salt in it. Milk was seldom used and generally it was boiled milk with slices of wheat bread in it, or fresh milk with berries similar to our blackberries. Occasionally pancakes were to be had. For a beverage the Frenchmen either used pure wine, usually red wine, mixed with water, or else just water or spruce beer. In the evening there were served two dishes of meat, both fried, sometimes a fricassee or fried pigeons, also fried fish, and now and then milk with berries. The third course in the evening was almost always a salad prepared in the usual manner.

July the 11th

The harrows which they make use of here are made entirely of wood and of a triangular form. The plows seem to be less serviceable. The wheels upon which the plow beam is placed, are as thick as the wheels of a cart, and all the wood work is so clumsily made that it requires a horse to draw the plow over the surface of the field.

Minerals. Stones of different sorts lay scattered on the fields. Some were from three to five feet high, and about three feet broad. They were very much alike in regard to the kind of stone; however I observed three different species of them.

1. Some consisted of quartz, whose brown color resembled sugar candy, and which was mixed with a black small-grain mica, a black hornstone, and a few minute grains of a brown spar. The quartz was most abundant in the mixture; the mica was likewise in great

quantity, but the amount of spar was insignificant. The several kinds of stones were well mixed, and though the eye could distinguish them yet no instrument could separate them. The stone was very hard and compact, and the grains of quartz looked very fine.

2. Some pieces consisted of gray particles of quartz, black glimmer, and hornstone, together with a few particles of spar, which made a very close, hard, and compact mixture, differing from the former only in color.

3. A few of the stones were of a mixture of a light quartz and black mica, to which some red grains of quartz were added. The spar was most predominant in this mixture, and the mica appeared in large flakes. This stone was not so well mixed as the former, and was by far not so hard and so compact, being easily broken. [All are varieties of granite].

On the mountains where Fort St. Frédéric is built, and on those where the above-mentioned kinds of stone are found there was also a black limestone, lying in lamellæ as slate does, and it might be called a kind of slate, which can be turned into quicklime by burning.[1] This limestone is black inside, and when broken, appears to be of an exceedingly fine texture. There are some grains of a dark spar scattered in it, which together with some other irregularities, form veins. The strata which lie uppermost in the mountains consist of a gray limestone, which is seemingly no more than a variety of the preceding. The black limestone is constantly found filled with petrifications of all kinds, and chiefly the following:

Pectinites, or petrified *Ostreæ Pectines.* These pertrified shells are more abundant than any others that have been found here, and sometimes whole strata are found consisting only of a quantity of shells of this sort, fused together. They are generally small, never exceeding an inch and a half in length. They are found in two different states of petrification: one always shows the impressions of the elevated and hollow surfaces of the shells, without any vestige of the shells themselves. In the other appears the real shell sticking in the stone, and by its light color is easily distinguishable from the rock, which is black. Both these kinds are plentiful in the stone; however, the impressions are more in number than the real shells.

---

[1] *Marmor schistosum,* L. *Syst.* III, p. 40. *Marmor unicolor nigrum,* Wall. *Min.* p. 61, n. 2. Lime-slates, *schistus calcareu,* Forst. *Introd. to Min.* p. 9.—F.

Some of the shells are very convex, especially in the middle, where they form as it were a hump; others again are concave in the middle; but in most of them the outward surface is remarkably elevated. The furrows always run longitudinally, or from the top, diverging to the margin.

Petrified Cornua Ammonis. These are also frequently found, but are not equal to the former in number; like the *pectinitæ* they are found both petrified and in impressions. Amongst them were some petrified snails. Several of these *Cornua ammonis* were remarkably big, and I do not remember having seen their equal, for they measured above two feet in diameter.

Different kinds of *corals* could be plainly seen in, and separated from, the stone in which they lay. Some were white and ramose, or Lithophytes; others were starry corals, or Madrepores; the latter were rather scarce.

I must give the name of "stone balls" to a kind of stone foreign to me, which is found in great numbers in some of the rocks.[1] They were globular, one half of them projecting generally above the rock, and the other remaining in it. They consisted of nearly parallel fibres, which radiated from the bottom as from a center, and spread over the surface of the ball and were gray in color. The outside of the balls was smooth, but had a number of small pores, which externally appeared to be covered with a pale gray crust. They were from one inch to an inch and a half in diameter.

Sand. Among some other kinds of sand which are found on the shores of Lake Champlain, two are very peculiar, and usually lie in the same place; the one is black, and the other reddish brown, or garnet-colored. The black sand, always lying uppermost, consists of very fine grains, which, when examined by a microscope, appear to have a dark blue color, like that of a smooth iron not attacked by rust. Some grains are roundish, but most of them angular, with shining surfaces; and they sparkle when the sun shines. All the grains of this sand without exception are attracted by the magnet. Amongst these black or deep blue grains, they meet with a few of a red or garnet-colored variety, which is the same as the red sand that lies immediately under it, and which I shall now describe. This red or garnet-colored sand is very fine, but not so fine as the black sand.

---

[1] Amygdaloid rocks containing small cavities which were later filled by material different from that of the mass, such as agate, or quartz.

Its grains not only are the color of garnets, but they are really nothing but pounded garnets. Some grains are round, others angulated; all shine and are semipellucid; but the magnet has no effect on them, and they do not sparkle so much in sunshine. This red sand is seldom found very pure, being commonly mixed with a white variety, consisting of particles of quartz. The black and red sand is not found in every part of the shore, but only in a few places, in the order before mentioned. The uppermost or black sand lay about a quarter of an inch deep; when it was carefully taken off the sand under it became of a deeper red the deeper it lay, and its depth was commonly greater than that of the former. When this was carefully removed, the white sand of quartz appeared mixed very much at top with the red sand, but growing purer the deeper it lay. This white sand was above four inches deep, had round grains, which made it entirely like pearl sand. Below this was a pale gray angulated quartz sand. In some places the garnet-colored kind lay uppermost, and this gray angulated one immediately under it, without a grain of either the black or the white sand.

I cannot determine the origin of the black or steel-colored sand, for it was not known here whether there were iron mines in the neighborhood or not. But I am rather inclined to believe they may be found in these parts, as they are common in different parts of Canada, and as this sand is found on the shores of almost all the lakes and rivers in Canada, though not in equal quantities. The red or garnet-colored sand has its origin hereabouts, for although the rocks near Fort St. Frédéric contained no garnets, yet there are stones of different sizes on the shores, quite different from those which form the rocks; these stones are very full of grains of garnet, and when pounded there is no perceptible difference between them and the red sand. In the more northerly parts of Canada, or below Quebec, the mountains themselves contain a great number of garnets. The garnet-colored sand is very common on the shores of the St. Lawrence River. I shall leave out several observations which I made upon the minerals hereabouts, as uninteresting to most of my readers.

Plants. The *Apocynum androsæmifolium* grows in abundance on hills covered with trees, and is in full flower about this time. The French call it *Herbe à la puce.* When the stalk is cut or torn, a white milky juice comes out. The French attribute the same qual-

ities to this plant as the poison tree, or *Rhus vernix,* has in the English colonies; that its poison is noxious to some persons and harmless
to others. The milky juice, when spread upon the hands and body,
has no bad effect on some persons, whereas others cannot come near
it without being blistered. I saw a soldier whose hands were blistered
all over, merely by plucking the plant in order to show it me; and
it is said its exhalations affect some people, when they come within
reach of them. It is generally allowed here that the lactescent juice
of this plant, when spread on any part of the human body not only
swells the part but frequently corrodes the skin, at least there are
few examples of persons on whom it had no effect. As for my part,
it has never hurt me, though in the presence of several people I
touched the plant and rubbed my hands with the juice till they were
white all over, and I have often rubbed the plant in my hands till it
was quite crushed without feeling the least inconvenience or change
on my hand. Cattle never touch this plant.

JULY THE 12TH

The burdock, or *Arctium lappa,* grows in several places about the
fort and the governor told me that its tender shoots are eaten in
spring as radishes, after the exterior peel is taken off.

The *Sison Canadense* abounds in the woods of all North America.
The French call it *cerfeuil sauvage,* and make use of it in spring, in
green soups, like chervil. It is universally praised here as a wholesome, antiscorbutic plant, and as one of the best which can be had
here in spring.

The *Asclepia syriaca,* or, as the French call it, *le cotonnier,* grows
abundant in the country, on the sides of hills which lie near rivers,
as well as in a dry and open place in the woods and in a rich loose
soil. When the stalk is cut or broken it emits a lactescent juice, and
for this reason the plant is reckoned in some degree poisonous. The
French in Canada nevertheless use its tender shoots in spring, preparing them like asparagus, and the use of them is not attended with
any bad consequences, as the slender shoots have not yet had time to
suck up anything poisonous. Its flowers are very fragrant, and,
when in season, they fill the woods with their sweet exhalations
and make it agreeable to travel in them, especially in the evening.
The French in Canada make a sugar of the flowers, which for that

purpose are gathered in the morning, when they are covered with dew. This dew is pressed out, and by boiling yields a very good brown, palatable sugar. The pods of this plant when ripe contain a kind of wool, which encloses the seed and resembles cotton, whence the plant has gotten its French name. The poor collect it and with it fill their beds, especially their children's, instead of feathers. This plant flowers in Canada at the end of June and the beginning of July, and the seeds are ripe in the middle of September. The horses never eat this plant.

JULY THE 16TH

West of Lake Champlain. This morning I crossed Lake Champlain to the high mountain on its western side, in order to examine the plants and other curiosities there. From the top of the rocks, a little distance from Fort St. Frédéric, a row of very high mountains appear on the western shore of Lake Champlain, extending from south to north. On the eastern side of this lake is another chain of high mountains, running in the same direction. Those on the eastern side are not close to the lake, being about ten or twelve miles from it. The country between them is low and flat, and covered with woods, which likewise clothe the mountains, except in such places as the fires, which destroy the forest here, have reached and burnt them down. These mountains have generally steep sides, but sometimes they are found gradually sloping. We crossed the lake in a canoe, which could only contain three persons, and as soon as we landed we walked from the shore to the top of one of the mountains. Their sides are very steep, and covered with earth, and some great rocks lay on them. All the mountains are covered with trees, but in some places the forests have been destroyed by fire. After a great deal of trouble we reached the top of one of the mountains, which was covered with a dusty mould. It was not one of the highest, and some of those which were at a greater distance were much higher, but we had no time to go to them, for the wind increased and our boat was but a little one. We found no curious plants or anything remarkable there.

When we returned to the shore we found the wind had risen to such an intensity that we did not venture to cross the lake in our boat, and for that reason I left a man to bring it back, as soon as the

wind had subsided and we walked round the bay, which was a distance of about seven English miles. I was followed by my servant,
and for want of a road we kept close to the shore, where we passed
over mountains and sharp stones, through thick forests and deep
marshes, all of which were known to be inhabited by numberless
rattlesnakes, of which we happily saw none at all. The shore is full
of stones in some places, and covered with large angulated rocks,
which are sometimes roundish and their edges as if worn off. Now
and then we came upon a small sandy spot, covered with gray, but
chiefly with the fine red sand which I have before mentioned; and
the black iron sand likewise occurred sometimes. We found stones
of red mica of a fine texture on the mountains. Occasionally these
mountains with the trees on them stood perpendicular at the waterside, but in some places the shore was marshy.

I saw a number of petrified *Cornua ammonis* in one place, near
the shore, among a number of stones and rocks. The rocks consist of
gray limestone, which is a variety of the black, and lies in strata, as
that does. Some of them contain a number of petrifications, with
and without shells. In one place we found prodigiously large *Cornua
ammonis,* about twenty inches in breadth. In some places the water
had worn off the stone, but could not have the same effect on the
petrifications which lay elevated above and in a manner glued on
to the stones.

The mountains near the shore are amazingly high and large, consisting of compact gray rock, which does not lie in strata as the limestone. Its chief constituent parts are a gray quartz and a dark mica.
This rock reached down to the water in places where the mountains
stood close to the shore, but where they were a short distance from
it there were strata of gray and black limestone, which reached to
the water side, and which I never have seen covered with gray rocks.

The *Zizania aquatica* grew in the mud, and in the most rapid
parts of brooks, and was in full bloom about this time.

JULY THE 17TH

Diseases Common in Canada. The diseases which ravage the Indians are rheumatism and pleurisy, which arise from their being
obliged frequently to lie in the wet parts of the woods at night, from
the sudden changes of heat and cold, to which the air is exposed

here, and from their being frequently loaded with too great a quantity of strong liquor, in which case they commonly lie down naked in the open air, without any regard to the season or the weather. Of these diseases, pleurisy especially is likewise very common among the French here. The governor told me he had once had a very violent fit of the latter, and that Dr. Sarrazin had cured him in the following manner, which has been found to succeed best here. He gave him sudorifics, which were to operate in eight or ten hours. He was then bled, and the sudorifics repeated. He was bled again, and that effectually cured him.

Dr. Sarrazin was the royal physician at Quebec,[1] and a correspondent of the Royal Academy of Sciences at Paris. He was possessed of great knowledge in the practice of medicine, anatomy, and other sciences, and very agreeable in his behavior. He died at Quebec, of a malignant fever, which had been brought to that place by a ship, and with which he was infected at a hospital, where he visited the sick. He left a son, who likewise studied medicine, and went to France to make himself more perfect in the practical part of it, but he died there.

The intermitting fever sometimes appears amongst the people here, and venereal disease is common. The Indians are likewise infected with it; many of them have had it, and some still have it; but they are possessed of an infallible art of curing it. There are examples of Frenchmen and Indians, infected all through the body with this disease, who have been "radically" and perfectly cured by the Indians within five or six months. The French have not been able to find out this remedy, though they know that the Indians employ no mercury, but that their chief remedies are roots, which are unknown to the French. I afterwards heard what these plants were and gave an account of them to the Royal Swedish Academy of Sciences.[2]

We are very well acquainted in Sweden with the pain caused by the *tæniæ,* or a kind of tape worm. They are less abundant in the

---

[1] Michel Sarrazin (1659-1734), army surgeon and all-round scientific scholar, came to Quebec about 1685; went back to France later to get his doctor's degree, and returned to Canada in 1697. He was a botanist, chemist and biologist. See article on Dr. Sarrazin in the *Dictionnaire Générale du Canada.* His descendants are still living in the vicinity of Quebec.

[2] See the *Memoirs* of that Academy, for the year 1750, p. 284. The *Stillingia Sylvatica* is probably one of these roots.—F. See also Bibliography, item 15.

British North American colonies, but in Canada they are very common. Some of these worms, which have been evacuated by a person, have been several yards long. It is not known whether the Indians are afflicted with them or not. No particular remedies against them are known here, and no one can tell whence they come, though the eating of some fruits contributes, as is believed, in creating them.

JULY THE 19TH

At Crown Point. Fort St. Frédéric [1] is a fortification on the southern extremity of Lake Champlain, situated on a neck of land, between that lake and the river, which arises from the union of the river Woodcreek, and Lake St. Sacrement. The breadth of this river is here about a good musketshot. The English call this fortress Crown Point, but its French name is derived from the French secretary of state, Frédéric Maurepas,[2] in whose hands the direction and management of the French court of admiralty was at the time of the erection of this fort. It is to be observed that the government of Canada is subjected to the court of admiralty in France, and the governor-general is always chosen from that court. As most places in Canada bear the names of saints, custom has made it necessary to prefix the word Saint to the name of the fortress. The fort is built on a rock, consisting of black lime or slate, as mentioned before. It is nearly square, has high, thick walls made of the same limestone, of which there is a quarry about half a mile from the fort. On the eastern part of the fort, is a high tower, which is proof against bombshells, provided with very thick and substantial walls, and well stored with cannon from the bottom almost to the very top; and the governor lives in the tower. On one side of the fort is a pretty little church, and on the other side, houses of stone for the officers and soldiers. There are sharp rocks on all sides towards the land, beyond a cannon shot from the fort, and among them are some which are as high as the walls of the fort and very near them.

The Englishmen insist that this fort is built on their territory and that the boundary between the French and English colonies in this locality lies between Fort St. Jean and the Prairie de la Madeleine;

---

[1] See diary for July the 2nd.
[2] Jean-Frédéric Phélypeaux, Comte de Maurepas (1701-1781), minister of France. See article in *Nouvelle Biographie Générale.*

on the other hand, the French maintain that the boundary runs through the woods, between Lake St. Sacrement and Fort Nicholson.

The soil about Fort St. Frédéric is said to be very fertile, on both sides of the river, and before the last war a great many French families, especially old soldiers, settled there, but the king obliged them to go into Canada, or to settle close to the fort, and sleep in it at night. A great number of them returned at this time, and it was thought that about forty or fifty families would go to settle here this autumn. Within one or two musketshots to the east of the fort, is a windmill, built of stone with very thick walls, and most of the flour which is needed to supply the fort is ground here. This windmill is so constructed as to serve the purpose of a redoubt, and at the top of it are four or five small pieces of cannon. During the last war, a number of soldiers was quartered in this mill, because they could from there look a great way up the river, and observe whether the English boats approached, which could not be done from the fort itself. This was a matter of great consequence, as the English might (if this guard had not been placed here) have gone in their little boats close under the western shore of the river, and then the hills would have prevented their being seen from the fort. Therefore the fort ought to have been built on the spot where the mill stands, and all those who come to see it, are immediately struck with the absurdity of its location. If it had been erected in the place of the mill, it would have commanded the river, and prevented the approach of the enemy; and a small ditch cut through the loose limestone, from the river (which comes out of the Lake St. Sacrement) to Lake Champlain, would have surrounded the fort with flowing water, because it would have been situated at the extremity of the neck of land. In that case the fort would always have been sufficiently supplied with fresh water, and at a distance from the high rocks which surround it in its present situation. We prepared to-day to leave this place, having waited several days for the arrival of the boat, which plies constantly all summer between the forts Saint Jean and Fort St. Frédéric. During our stay here, we had received many favors. The governor of the fort, Mr. Lusignan, a man of learning and of great politeness, heaped kindness upon us, and treated us with as much civility as

if we had been his relations. I had the honor of eating at his table
during my stay here, and my servant was allowed to eat with his.
We had our rooms, etc. to ourselves, and at our departure the
governor supplied us with ample provisions for our journey to Fort
Saint Jean. In short he did us more favors than we could have expected from our own countrymen, and the officers were likewise
particularly obliging to us.[1]

On Lake Champlain. About eleven o'clock in the morning we
set out with a fair wind. On both sides of the lake are high chains
of mountains, with the difference which I have before observed,
that on the eastern shore is a low piece of ground covered with a
forest, extending between nine to twelve English miles, after which
the mountains begin, and the country beyond them belongs to New
England. This chain consists of high mountains, which are to be
considered as the boundaries between the French and English possessions in these parts of North America.[2] On the western shore of
the lake, the mountains reach to the waterside. The lake at first is
but a French mile broad, but keeps increasing in size afterwards.
The country is inhabited within a French mile of the fort, but after
that it is covered with a thick forest. At a distance of about ten
French miles from Fort St. Frédéric, the lake is four such miles
broad, and we perceived some islands in it. The captain of the
boat said there were some of considerable size. He assured me that
the lake was in most parts so deep that a line of two hundred yards
could not fathom it, and close to the shore, where a chain of mountains generally runs across the country, it frequently has a depth
of eighty fathoms. Fourteen French miles from Fort St. Frédéric
we saw four large islands in the lake, which is here about six French
miles broad. This day the sky was cloudy, and the clouds, which
were very low, seemed to surround several high mountains near
the lake with a fog. From many mountains the fog rose, as the
smoke of a charcoal kiln. Now and then we saw a little river which

---

[1] From the article on Kalm in the *Dictionnaire Générale du Canada* we learn that
French scientists had always been guests of the Swedish government when travelling
in Sweden; hence as a matter of reciprocal hospitality and courtesy, Kalm became
the guest of the French government while in Canada, and was treated accordingly.
This made him very kindly disposed toward the French there. Forster felt that
Kalm was prejudiced in favor of the Canadians. Cf. below, p. 771, note.
[2] Obviously the Green Mountains of Vermont.

emptied into the lake. The country behind the high mountains, on the western side of the lake, is, as I am told, covered for many miles with tall forests, intersected by many rivers and brooks with marshes and small lakes, and is very suitable for habitations. The shores are sometimes rocky and sometimes sandy here. Towards night the mountains decreased gradually. The lake was very clear, and we observed neither rocks nor shallows in it. Late last night the wind abated, and we anchored close to the shore, and spent one night here.

JULY THE 20TH

This morning we proceeded with a fair wind. The place where we passed the night was more than half way to Fort Saint Jean, for the distance of that place from Fort St. Frédéric, across Lake Champlain, is computed to be forty-one French miles. That lake is here about six English miles in breadth. The mountains were now out of sight, and the country low, plain and covered with trees. The shores were sandy, and the lake appeared now from four to six miles broad. It was really broader but the islands made it appear narrower.

Indians. We often saw Indians in bark boats, close to the shore, which was, however, not inhabited, for the Indians came here only to catch sturgeons, wherewith this lake abounds, and which we often saw leaping up into the air. These Indians lead a very singular life. At one time of the year they live on the small store of corn, beans, and melons, which they have planted; during another period, or about this time, their food is fish, without bread or any other meat; and another season they eat nothing but game, such as stags, roes, beavers, etc., which they shoot in the woods and rivers. They, however, enjoy long life, perfect health, and are more able to undergo hardships than other people. They sing and dance, are joyful, and always content, and would not for a great deal exchange their manner of life for that which is preferred in Europe.

When we were yet ten French miles from Fort Saint Jean, we saw some houses on the western side of the lake, in which the French had lived before the last war, and which they then abandoned, as it was by no means safe. They now returned to them again. These were the first houses and settlements which we saw after we had left those about Fort St. Frédéric.

An Old Fort. There formerly was a wooden fort or redoubt on the eastern side of the lake, near the waterside, and the place where it stood was shown to me; at present it is quite overgrown with trees. The French built it to prevent the incursions of the Indians over this lake, and I was assured that many Frenchmen had been slain in these places. At the same time the Canadians told me that they numbered four women to one man in Canada, because annually several Frenchmen were killed on their expeditions which they undertook for the sake of trading with the Indians.

A *windmill*, built of stone, stood on the east side of the lake on a projecting piece of ground. Some Frenchmen lived near it; but they left it when the war broke out, and have not yet come back to it. From this mill to Fort Saint Jean they considered it eight French miles. The English, with their Indians, had burnt the houses here several times, but the mill remained unhurt.

The boat in which we went to Saint Jean was the first that was built here, and employed on Lake Champlain, for formerly they made use of bateaux to send provisions over the lake. The Captain was a Frenchman born in this country. He had built it in order to find out the true course, between Fort Saint Jean and Fort Saint Frédéric. Opposite the windmill the lake was about three fathoms deep, but it grew more and more shallow, the nearer it came to Fort Saint Jean.

We now perceived houses on the shore again. The captain had otter skins in the cabin, which in color and species were just like the European ones. Otters are said to be very abundant in Canada.

Seals. Sealskins were here made use of to cover boxes and trunks, and they often made provision bags and hand bags of them in Canada. The common people had their tobacco pouches made of the same skins and the shape was like those of the same material used in western Sweden (Gothenburg and Bohuslän) and in Norway. They folded them together when they carried them about or laid them aside. The fur was on the outside. The common people were accustomed to smoke tobacco a good deal on their journeys and at their work; but I never saw anyone chew it here as the English and Dutch sailors are accustomed to do. The seals here are the same as the Swedish, which are gray with black spots. They are said to be plentiful at the mouth of the St. Lawrence River, below Quebec, and to go up that river as far as the water is salt. They have not been found in any of the great lakes of Canada. The French call them *Loups marins*, or sea-wolves.

Prayers. The French, in their colonies, spend much more time in prayer and external worship than the English and Dutch settlers in the British colonies. The latter have neither morning nor evening prayer in their ships and boats, and no difference is made between Sundays and other days. They never, or very seldom, say grace at dinner. On the contrary, the French here have prayers every morning and night on board their ships, and on Sundays they pray more than commonly. They regularly say grace both before and after their meals and cross themselves. The captain kneels in prayer at his bed in the morning, and every one says prayers in private as soon as he gets up. At Fort St. Frédéric all the soldiers assemble together for morning and evening prayers. The only fault is that most of the prayers are read in Latin, which a great part of the people do not understand.

Below the afore-mentioned windmill, the breadth of the lake is about a musketshot, and it looks more like a river than a lake. The country on both sides is low and flat, and covered with deciduous trees. We saw at first a few scattered cottages along the shore, but a little further on, the country was uninhabited without interruption. The lake was here from six to ten feet deep, and had several islands. During the whole course of this voyage, the direction of the lake was always directly from S. S. W. to N. N. E.

In some parts of Canada are great tracts of land belonging to individual persons. From these lands, pieces of forty arpens long and four wide are allotted to each discharged soldier who intends to settle here; but after his household is established, he is obliged to pay the owner of the land six French francs annually.

The lake was now so shallow in several places that we were obliged to trace the way for the boat by sounding the depth with branches of trees. In other places opposite it was sometimes two fathoms deep.

In the evening, about sunset, we arrived at Fort Saint Jean, or St.

John, having had a continual change of rain, sunshine, wind and calm all afternoon.

Fort St. John. St. John is a wooden fort, which the French built in 1748 on the western shore of the mouth of Lake Champlain, close to the waterside. It was intended to protect the country round about it, which they were then going to people, and to serve as a magazine for provisions and ammunition, which were usually sent from Montreal to Fort St. Frédéric; because they might go in boats from here to the last mentioned place, which is impossible lower down, as about two gunshots further there is a shallow place full of stones and very rapid water in the river, over which they can pass only in bateaux or flat vessels. Formerly Fort Chambly, which lies four French miles lower, was the magazine of provisions; but since they were forced first to send them hither in bateaux, and then from here in boats, and the route to Fort Chambly from Montreal being by land, and round about, this fort was erected. It has a low location and lies on a sandy soil, and the country about it is likewise low, flat and covered with woods. The fort is square and includes the space of one arpent square. In each of the two corners which look towards the lake is a wooden building, four stories high, the foundation of which is of stone to the height of about a fathom and a half. In these buildings which are polyangular, are holes for cannon and lesser fire-arms. In each of the two other corners towards the country is only a little wooden house, two stories high. These buildings are intended for the habitation of the soldiers and for the better defense of the place. Between these houses there are poles, two fathoms and a half high, sharpened at the top and driven into the ground close to one another. They are made of the *thuya* tree, which is here considered the best wood for withstanding rot, and is much preferable to fir in that point. Lower down the palisades are double, one row within the other. For the convenience of the soldiers, a broad elevated platform of more than two yards in height is made in the inside of the fort all along the palisades, with a balustrade. On this platform the soldiers stand and fire through the holes upon the enemy, without being exposed to their fire. In the last year, 1748, two hundred men were in garrison here, but at this time there were only a governor, a commissary, a baker, and six soldiers to take care of the fort and buildings, and to superintend the provisions which are carried to this place. The person who now commanded at the fort, was the Chevalier de Gannes, a very agreeable gentleman, and brother-in-law of Mr. Lusignan, the governor of Fort St. Frédéric. The ground about the fort, on both sides of the water, is rich and has a very good soil, but it is still uninhabited, though it is said people expected to settle here soon.

The French in all Canada call the gnats *marangoins*, which name, it is said, they have borrowed from the Indians. These insects are in such prodigious numbers in the woods round Fort St. John, that it would have been more properly called Fort de Marangoins. The marshes and the low nature of the country, together with the extent of the woods, contributed to bring about this condition. Sometimes when the forest is cut down, the water drained, and the country cultivated, they probably will decrease in number and vanish at last, as they have done in other places.

The Rattlesnake, according to the unanimous accounts of the French, is never seen in this neighborhood, nor further north near Montreal and Quebec. The mountains which surround Fort St. Frédéric are the most northerly part on this side where they have been seen. Of all the snakes which are found in Canada to the north of these mountains none is poisonous enough to do any great harm, and all without exception run away when they see a man. My remarks on the nature and properties of the rattlesnake, I have communicated to the Royal Swedish Academy of Sciences,[1] and thither I refer my readers.

JULY THE 22ND

This evening some people arrived with horses from Prairie in order to fetch us. The commander had sent for them at my desire, because there were not yet any horses near Fort St. John, the place being only a year old, and the people had not had time to settle near

---

[1] See the *Memoirs* for the years 1752 and 1753. For details consult Bibliography, item 20.

it. Those who led the horses, brought letters to the governor from the governor-general of Canada, the Marquis de la Galissonnière, dated at Quebec the fifteenth of this month, and from the vice-governor of Montreal, the Baron de Longueuil, dated the twenty-first of the same month. They mentioned that I had been particularly recommended by the French court, and that the commander should supply me with everything I wanted and forward my journey.[1] At the same time he received two little casks of wine for me, which they thought would relieve me on my journey. At night we drank to the health of their Majesties, the Kings of France and Sweden, under a salute from the cannon of the fort, and to the health of the governor-general and others.

JULY THE 23RD

En Route to Laprairie. This morning we set out on our journey to Prairie [Laprairie], whence we intended to proceed to Montreal. The distance of Prairie from Fort. St. John, by land, is six French miles, and from there to Montreal two and a half, by the St. Lawrence River. At first we kept along the shore, so that we had on our right the Rivière de St. Jean (St. John's River). This is the name of the mouth of Lake Champlain, which flows into the St. Lawrence River, and is sometimes called Rivière de Champlain (Champlain River). After we had travelled about a French mile, we turned to the left from the shore. The country was always low, woody, and pretty wet, though it was in the midst of summer; so that we found it difficult to proceed. But it is to be observed that Fort St. John was built only last summer, when this road was first made, and consequently it could not yet have acquired a proper degree of solidity. Two hundred and sixty men were three months at work in making this road, for which they were fed at the expense of the government, and each received thirty sols a day. I was told that they would again resume the work next autumn. The country hereabouts is low and woody, and of course the breeding place of millions of gnats and flies, which were very troublesome to us. After we had gone about three French miles we came out of the woods and the ground seemed to have been formerly a marsh which

---

[1] Cf. page 510, note.

was now dried up. From here we had a pretty good view on all sides. On our right at a great distance we saw two high mountains, rising remarkably above the rest and they were not far from Fort Champlain. We could likewise from here see the high mountain which lies near Montreal, and our road went on nearly in a straight line. Soon after we got again upon wet and low grounds, and after that into a wood which consisted chiefly of the fir with leaves which have a silvery underside (*Abies foliis subtus argenteis*). We found the soil which we passed over to-day very fine and rich, and when the woods are cleared and the ground cultivated, it will probably prove very fertile. There are no rocks, and hardly any stones near the road.

About four French miles from Fort St. John, the country had quite another appearance. It was all cultivated, and a continual variety of fields with excellent wheat, peas and oats, presented itself to our view; but we saw no other kind of grain. The farms stood scattered, and each of them was surrounded by its corn fields and meadows. The houses were built of wood and were very small. Instead of moss, which cannot be found here, they used clay for stopping up the crevices in the walls. The roofs were made at a very steep angle, and covered with straw. The soil was good, flat, and divided by several rivulets, and only in a few places were there any little hills. The view is very fine from this part of the road, and as far as I could see the country it was cultivated. All the fields were covered with grain, and they generally used summer wheat here.— The ground is still very fertile, so that there is no occasion for leaving it fallow. The forests are pretty much cleared, and it is to be feared that there will be a time when wood will become very scarce. Such is the appearance of the country nearly up to Prairie, and to the St. Lawrence River, which at last we now had always in sight. In a word this country was, in my opinion, one of the finest of North America that I had hitherto seen.

About dinner time we arrived at Prairie which is situated on a little rising ground near the St. Lawrence River. We stayed here to-day, because I intended to visit the places in this neighborhood before I went on.

Laprairie de la Madeleine is a small village on the eastern side of the St. Lawrence River about two French miles and a half from

Montreal, which lies N. W. from here on the other side of the river. All the country round Prairie is quite flat, and has hardly any hills. On all sides are large grain fields, meadows and pastures. On the western side the St. Lawrence River passes by and has here a breadth of a French mile and a half, if not more. Most of the houses in Prairie are built of timber with sloping wooden roofs, and the crevices in the walls are stopped up with clay. There are some little buildings of stone, chiefly of black limestone or of pieces of rock; in using the latter material the encasement of the doors and windows was made of the black limestone. In the midst of the village is a pretty church of stone, with a steeple at the west end of it, furnished with bells. Before the door is a cross, together with ladders, tongs, hammers, nails, etc. which are to represent all the instruments made use of at the crucifixion of our Savior, and perhaps many others besides. The village is surrounded with palisades, from four to five yards high, put up formerly as a barrier against the incursions of the Indians. Without these palisades are several little kitchen and pleasure gardens, but there are very few fruit trees. There are few rising grounds along the river here. In this place there lives a priest, and a captain, who assumes the name of governor. The grain fields round the place are extensive and sown with summer wheat, but rye, barley and corn are never seen. To the southwest of this place is a great fall in the St. Lawrence River, and the noise which it causes may be plainly heard here. When the water in spring increases in the river on account of the ice which then begins to dissolve, it sometimes happens to rise so high as to overflow a great part of the fields, and, instead of fertilizing them as the river Nile fertilizes the Egyptian fields by its inundations, it does them much damage by depositing a number of grasses and plants on them, the seeds of which spread the worst kind of weeds and ruin the fields. These inundations oblige the people to take their cattle a great way off, because the water covers a great tract of land, but happily it never stays on it above two or three days. The cause of these inundations is generally the blocking of the water by ice in some part of the river.

The *Zizania aquatica*, or *Folle avoine* [wild rice], grows plentiful in the rivulet or brook, which flows a short distance below Prairie.

July the 24th

At Montreal. This morning I went from Prairie in a bateau to
Montreal on the St. Lawrence River. The river is very rapid, but
not very deep near Prairie, so that the boats cannot go higher than
Montreal, except in spring with the high water, when they can come
up to Prairie, but not further. The town of Montreal may be seen
at Prairie and all the way down to it. On our arrival there we
found a crowd of people at the gate of the town where we were to
pass through. They were very desirous of seeing us, because they
were informed that some Swedes were to come to town, people of
whom they had heard something, but whom they had never seen;
and we were assured by everybody, that we were the first Swedes
that ever had come to Montreal. As soon as we had landed, the
governor of the town sent a captain to me, who desired that I would
follow him to the governor's house, where he introduced me to him
in a room where the governor was with some friends. Baron Longueuil was as yet vice-governor, but he daily expected his promotion
from France. He received me more civilly and generously than I
can well describe, and showed me letters from the governor-general
at Quebec, the Marquis de la Galissonnière, who mentioned that he
had received orders from the French court to supply me with whatever I should want, as I was to travel in this country at the expense
of his most Christian majesty. In short Governor Longueuil loaded
me with greater favors than I could expect or even imagine, both
during that stay and on my return from Quebec.

July the 25th

French Manners and Customs. The difference between the manners and customs of the French in Montreal and Canada, and those
of the English in the American colonies, is as great as that between
the manners of those two nations in Europe. The women in general are handsome here; they are well bred and virtuous, with an
innocent and becoming freedom. They dress up very fine on Sundays; about the same as our Swedish women, and though on the
other days they do not take much pains with other parts of their

dress, yet they are very fond of adorning their heads. Their hair is always curled, powdered and ornamented with glittering bodkins and aigrettes. Every day but Sunday they wear a little neat jacket, and a short skirt which hardly reaches halfway down the leg, and sometimes not that far. And in this particular they seem to imitate the Indian women. The heels of their shoes are high and very narrow, and it is surprising how they can walk on them. In their domestic duties they greatly surpass the English women in the plantations, who indeed have taken the liberty of throwing all the burden of housekeeping upon their husbands, and sit in their chairs all day with folded arms.[1] The women in Canada on the contrary do not spare themselves, especially among the common people, where they are always in the fields, meadows, stables, etc. and do not dislike any work whatsoever. However, they seem rather remiss in regard to the cleaning of the utensils and apartments, for sometimes the floors, both in the town and country, are hardly cleaned once in six months, which is a disagreeable sight to one who comes from amongst the Dutch and English, where the constant scouring and scrubbing of the floors is reckoned as important as the exercise of religion itself. To prevent the thick dust, which is thus left on the floor from being noxious to the health, the women wet it several times a day, which lays the dust, and they repeat this as often as the dust is dry and begins to rise again. Upon the whole, however, they are not averse to the taking part in all the business of housekeeping, and I have with pleasure seen the daughters of the better sort of people and of the governor himself, not too finely dressed, going into kitchens and cellars to see that everything was done as it ought to be. And they also carry their sewing with them, even the governor's daughters.

The men are extremely civil and take their hats off to every person whom they meet in the streets. This is difficult for anyone whose duties demand that he be out doors often, especially in the

---

[1] It seems, that for the future, the fair sex in the English colonies in North America, will no longer deserve the reproaches Mr. Kalm stigmatizes them with repeatedly, since it is generally reported, that the ladies of late have vied one with another, in providing their families with linen, stockings, and homespun cloth of their own making, and that a general spirit of industry prevails among them at this present time.—F.

evening when every family sits outside their door, near the street. It is customary to return a visit the day after you have received one, even though one should have several scores to pay in one day.

Animals in Canada. I have been told by some Frenchmen, who had gone beaver hunting with the Indians to the northern parts of Canada, about fifty French miles from Hudson Bay, that the animals whose skins they endeavor to get, and which are there in great abundance, are beavers, wild cats, or lynx, and martens. These animals are the more valued, the further they are caught to the north, for their skins have better hair, and look better or worse the further they come from the north or south.

White partridges (*Perdrix blanches*) is the name which the French in Canada give to a kind of bird abounding during winter near Hudson Bay, and which undoubtedly are our *ptarmigans*, snow-hens (*Tetrao lagopus*). They are very plentiful at the time of a great frost, and when a considerable quantity of snow happens to fall. The greater the cold or snow, the greater the number of birds. They were described to me as having rough white feet, and being white all over, except for three or four black feathers in the tail; and they are considered very fine eating. From Edward's *Natural History of Birds*[1] it appears that the ptarmigans are common about Hudson Bay.

Hares are likewise said to be plentiful near Hudson Bay, and they are abundant even in Canada, where I have often seen and found them, corresponding perfectly with our Swedish hares. In summer they have a brownish gray and in winter a snowy white color, as with us.

The Trades. Mechanical trades, such as architecture, cabinet work, turning, and brick making, are not yet so advanced here as they ought to be, and the English in that particular outdo the French. The chief cause of this is that scarcely any other people than dismissed soldiers come to settle here, who have not had any opportunity of learning a mechanical trade, but have sometimes accidentally, and through necessity been obliged to do it. There are, however, some who have a good skill in mechanics, and I saw

---

[1] Page 72.—F.

a person here who made very good clocks and watches, though he had had but very little instruction.

July the 27th

The *common houseflies* (*Musca domestica*) were observed in this country about one hundred and fifty years ago, as I have been assured by several persons in this town, and in Quebec. All the Indians assert the same thing, and are of the opinion that the common flies first came over here with the Europeans and their ships, which were stranded on this coast. I shall not dispute this; however, I know, that while I was on the frontier between Saratoga and Crown Point, or Fort St. Frédéric, and sat down to rest or to eat, a number of our common flies always came and settled on me. It is therefore dubious whether they have not been longer in America than the term above-mentioned, or whether they have been imported from Europe. On the other hand, it may be urged that the flies were left in that wilderness at the time when Fort Anne was yet in a good condition, and when the English often travelled there and back again; not to mention that several Europeans, both before and after that time, had travelled through those places and carried the flies with them, which had been attracted by their provisions.

Wild cattle were abundant in the southern parts of Canada, and have been there since times immemorial. They were particularly plentiful in those parts where the Illinois Indians lived, which were nearly in the same latitude with Philadelphia; but further to the north they are seldom observed. I saw the skin of a wild ox to-day; it was as big as one of the largest ox hides in Europe, but had better hair. This was dark brown like that on a brown bearskin. That which was close to the skin is as soft as wool. This hide was not very thick and in general was not considered so valuable (in France) as a bearskin. In winter it is spread on the floor to keep the feet warm. Some of these wild cattle, as I am told, have a long and fine wool, as good if not better than sheep wool. They make stockings, cloth, gloves, and other pieces of worsted work of it, which look as well as if they were made of the best sheep wool.

The Indians employ it for several uses. The flesh is as good and fat as the best beef. Sometimes the hides are thick, and may be used as cowhides are in Europe. The wild cattle in general are said to be stronger and bigger than European cattle, and of a brownish red color. Their horns are short, though very thick close to the head. These and several other qualities, which they have in common with and in greater perfection than the domestic cattle, have induced some to endeavor to tame them, by which means they would obtain the advantages arising from their good hair, and, on account of their great strength, could employ them successfully in agriculture. With this view some have repeatedly gotten young wild calves and brought them up in Quebec and other places among the tame cattle, but they have usually died in three or four years time; and though they have seen people every day, they have always retained a natural ferocity. They have constantly been very shy, pricked up their ears at the sight of a man, and have trembled or run about, so that the art of taming them has not hitherto been successful. Some have been of the opinion that these cattle cannot bear the cold well, as they never go north of the place I mentioned, though the summers be very hot, even in those northern parts. They think that when the country about the Illinois is better peopled it will be more easy to tame these cattle, and that afterwards they may more easily be accustomed to the northerly climates.[1] The Indians and French in Canada, make use of the horns of these creatures to put gunpowder in. I have briefly mentioned the wild cattle in the former parts of this journey.[2]

The peace, which was concluded between France and England, was proclaimed to-day.[3] The soldiers were under arms, the artillery on the walls was fired off, and some salutes were given by the small firearms. All night fireworks were exhibited, and the whole town was illuminated. All the streets were crowded with people till late at night. The governor invited me to supper and to partake of the joy of the inhabitants. There were present a number of officers

---

[1] But by this means they would lose that superiority, which in their wild state they have over the tame cattle; as all the progenies of tamed animals degenerate from the excellence of their wild and free ancestors.—F.

[2] See pp. 110 and 150.

[3] The peace of Aix-la-Chapelle of October 18, 1748, ending King George's War.

and other persons of distinction, and they drank merrily far into the night.

JULY THE 28TH

On the Isle of Madeleine. This morning I accompanied the governor, Baron de Longueuil,[1] and his family to a little island called *Madeleine,* which is his own property. It lies in the St. Lawrence River directly opposite the town on the eastern side. The governor had here a very neat house, though it was not very large, a fine, extensive garden, and a yard. The river passes between the town and this island, and is very rapid. Near the town it is deep enough for large boats, but towards the island it grows more shallow, so that they are obliged to push the boats forward with poles. There was a mill on the island, turned by the mere force of the stream, without an additional milldam. In the mill I noticed that the stones did not consist of one single piece but were made of several pieces. The upper millstone was quite large, made of eight separate parts which were joined very close together and bound with a thick iron band. The lower stone was the same. The upper one had been imported from France but the other was native. The trough of the funnel through which the grain ran down was shaken in this manner: the upper part of the pinion-axle was joined above the millstone to a square piece of hard wood about four inches square. When this turned its four corners turned against the end of the hopper which projected out on one side, so the funnel was shaken and the grain ran down. The wheels and axle were made of white oak, but the cogs of the wheel and other parts of the machinery were made of the sugar maple or wild cherry. Still, the former was said to be the most in use, because it was considered hard wood, especially if it had grown in dry places.

Trees and Plants. The smooth sumach, or *Rhus glabra,* grows abundantly here. I have nowhere seen it so tall as in this place, where it had sometimes the height of eight yards and a proportional thickness.

---

[1] Charles le Moyne, second Baron de Longueuil, became governor of Montreal in 1749, and "commandant-general" of the Colony in 1752. See note by Marchand in his French edition, vol. 2, p. 48.

Sassafras is planted here, for it is never found wild in these parts, Fort Anne being the most northerly place where I have found it wild. Those shrubs which were on the island had been planted many years ago; however, they were only small shrubs, from two to three feet high, and scarcely that much. The reason is that the main stem is killed every winter almost down to the very root, and must produce new shoots every spring, as I have found from my own observations here; and so it appeared to be near the forts Anne, Nicholson, and Oswego. It will therefore be useless to attempt to plant sassafras in a very cold climate.

The *Mulberry trees* (*Morus rubra* L.) are likewise planted here. I saw four or five of them about five yards high, which the governor told me had been twenty years in this place and brought from more southerly parts, since they do not grow wild near Montreal. The most northerly place where I have found it growing wild is about twenty English miles north of Albany. I had this confirmed by the country people who live in that place, and who at the same time informed me that it was very scarce in the woods. When I came to Saratoga I inquired whether any of these mulberry trees had been found in that neighborhood, but everybody told me that they were never seen in those parts, and that the before-mentioned place twenty miles above Albany is the most northern point where they grow. The mulberry trees that were planted on this island succeed very well, though they are placed in poor soil. Their foliage was large and thick, but they did not bear any fruit this year. I was informed that they can bear a considerable degree of cold.

The *waterbeech* had been planted here in a shady place, and had grown to a great height. All the French hereabouts call it *cotonnier*.[1] It is never found wild near the St. Lawrence River, nor north of Fort St. Frédéric, where it is now very scarce.

The *red cedar* is called *cèdre rouge* by the French, and that was also planted in the governor's garden, whither it had been brought from more southern parts, for it is not to be found in the forests hereabouts. However, it grew very well here.

About half an hour after seven in the evening we left this pleas-

---

[1] Cotton tree. Mr. Kalm mentions before that this name is given to the *Asclepias Syriaca*. See page 387.

ant island, and an hour after our return the Baron de Longueuil received two agreeable pieces of news at once. The first was that his son who had been two years in France, had returned; and the second, that he had brought with him the royal patents for his father, by which he was appointed governor of Montreal and the country belonging to it.

People make use of *fans* here which are made of the tails of the wild turkeys. As soon as the birds are shot, their tails are spread like fans, and dried, by which means they keep their shape. The ladies and the men of distinction in town carry these fans, when they walk in the streets during the intense heat.

All the *grass* on the meadows round Montreal consists of a species of meadow grass, or the *Poa capillaris* L. This is a very slender grass which grows very close and succeeds even on the driest hills. It is however not rich in foliage; but the slender stalk is used for hay. We have numerous kinds of grasses in Sweden which are much more useful than this.

JULY THE 30TH

The *wild plum trees* grow in great abundance on the hills along the rivulets about the town. They were so loaded with fruit that the boughs were bent downwards by the weight. The fruit was not yet ripe, but when it comes to that perfection it has a red color and a fine taste, and preserves are sometimes made of it.

Black currants (*Ribes nigrum* L.) are plentiful in the same places, and its berries were ripe at this time. They are very small, and not by far so fine as those in Sweden.

Parsnips grow in great abundance on the rising banks of rivers, along the grain fields, and in other places. This led me to think that they were original natives of America, and not first brought over by the Europeans. But on my journey into the country of the Iroquois, where no European ever had a settlement, I never once saw it, though the soil was excellent; and from this it appears plain enough that it was transported hither from Europe, and is not originally an American plant. Therefore it is in vain sought for in any part of this continent except among the European settlements.

The governor-general of Canada commonly resides at Quebec, but occasionally he goes to Montreal, and generally spends the winter there. In summer he resides chiefly at Quebec on account of the king's ships which arrive there during that season and bring him letters which he must answer; besides, there is much other business at that time. During his residence in Montreal he lives in the castle, as it is called, which is a large house of stone, built by the former governor-general Vaudreuil, that still belongs to his family, who rents it to the king. General de la Galissonière is said to like Montreal better than Quebec, and indeed the location of the former is by far the more agreeable one.

Canada Money. Canada had scarcely any other money but paper currency. I hardly ever saw any coin, except French sols, consisting of copper, with a very small mixture of silver. They were quite thin by constant circulation, and were valued at a sol and a half. The bills were not printed but written. Their origin is as follows. The French king having found it very dangerous to send money for the pay of troops and other purposes over to Canada, on account of privateers, shipwrecks, and other accidents, ordered that instead of it the intendant, or king's steward at Quebec, or the commissary at Montreal, should write bills for the value of the sums which are due to the troops, and which he distributes to each soldier. On these bills is incribed that they bear the value of such or such a sum, till next October, and they are signed by the intendant or the commissary, and in the interval they bear the value of money. In the month of October, at a certain stated time, every one brings the bills in his possession to the intendant at Quebec, or the commissary at Montreal, who exchanges them for bills of exchange upon France, which are paid in lawful money, at the king's exchequer, as soon as they are presented. If the money is not yet wanted the bill may be kept till October of the following year, when it may be exchanged by one of those gentlemen for a bill upon France. The paper money can only be delivered in October and exchanged for bills upon France. They are of different values, and some do not exceed a livre, and perhaps some are still less. Towards autumn when the merchant ships come in from France, the merchants endeavor to get as many bills as they can and change them for bills upon the French treasury. These bills are partly printed, spaces being left for the name, sum, etc. But the first bill, or paper currency, is all written, and is therefore subject to be counterfeited, which has sometimes been done; but the great punishments which have been inflicted upon the authors of these forged bills, and which generally are capital, have deterred people from attempting it again; so that examples of this kind are very scarce at present. As there is a great want of small coin here, the buyers or sellers are frequently obliged to suffer a small loss and can pay no intermediate prices between one livre and two.[1] For example if I wanted to buy something for which the price was ten livres, and I did not have the bills for the exact amount, I was obliged to pay one or two livres extra. In this transaction the one who was anxious to buy or sell was generally the sufferer.

Wages. On the farms the wages for servants and day laborers was ordinarily a little less than in the cities. They commonly give one hundred and fifty livres a year to a faithful and diligent man servant, and to a maid servant of the same character one hundred livres. A journeyman to an artist gets three or four livres a day, and a common laboring man gets thirty or forty sols a day. The scarcity of laboring people occasions the wages to be so high; for almost everybody finds it easy to be a farmer in this uncultivated country where he can live well, and at so small an expense that he does not care to work for others.

Montreal is the second town in Canada, in regard to size and wealth; but it is the first on account of its fine location and mild climate. A short distance above the town the St. Lawrence River divides into several branches, and by that means forms several islands, among which the isle of Montreal is the greatest. It is ten French miles long, and near four broad, in its broadest part. The town of Montreal is built on the eastern side of the island, and close to one of the largest branches of the St. Lawrence River; and thus it has a very pleasant and advantageous location. The town has a square form, or rather it is a rectangular parallelogram, the long and

---

[1] The sol is the lowest coin in Canada, and is about the value of a penny in the English colonies. A livre, or franc, (for they are both the same) contains twenty sols; and three livres or francs make an écu or crown.—F.

eastern side of which extends along the great branch of the river.
On the other side it is surrounded with excellent grain fields, charming meadows and delightful woods. It has the name of Montreal
from a great mountain about half a mile westwards of the town,
which lifts its head far above the woods. M. Cartier,[1] one of
the first Frenchmen who surveyed Canada more accurately, named
this mountain on his arrival on this island, in the year 1535, when
he visited the mountain and the Indian town Hochelaga near it.
The priests who, according to the Roman Catholic way, would call
every place in this country after some saint or other, called Montreal, Ville Marie, but they have not been able to make this name
general, for it has always kept its first name. It is pretty well forti-
fied, and surrounded with a high and thick wall. On the east side
it has the St. Lawrence River, and on all the other sides a deep
ditch filled with water, which secures the inhabitants against all
danger from sudden incursions of the enemy's troops. However,
it cannot long stand a regular siege, because it requires a great
garrison on account of its extent and because it consists chiefly of
wooden houses. Here are several churches, of which I shall only
mention that belonging to the friars of the order of St. Sulpicius,
that of the Jesuits, that of the Franciscan Friars, that belonging to
the nunnery, and that of the hospital. Of these the first is by far
the finest, both in regard to its outward and inward ornaments,
not only in this place but in all Canada. The priests of the seminary
of St. Sulpicius have a fine large house, where they live together.
The college of the Franciscan Friars is likewise spacious, and has
good walls, but it is not so magnificent as the former. The college
of the Jesuits is small, but well built. To each of these three buildings are annexed fine large gardens, for the amusement, health and
use of the communities to which they belong. Some of the houses
in the town are built of stone, but most of them are of timber,
though very neatly built. Each of the better sort of houses has a
door towards the street, with a seat on each side of it, for amusement and recreation in the morning and evening. The long streets
are broad and straight and divided at right angles by the short ones:
some are paved but most of them are very uneven. The gates of
the town are numerous; on the east side of the town towards the

---

[1] Jacques Cartier (1494-about 1552 or later), famous French navigator.

river are five, two great and three lesser ones, and on the other side are likewise several. The governor-general of Canada, when he is at Montreal, resides in the so-called castle, which the government hires for that purpose of the family of Vaudreuil; but the governor of Montreal is obliged to buy or hire a house in town; though I was told that the government contributed towards paying the rent.

In the town is a *Nunnery,* and outside its walls half of another, for though the last was ready, it had not yet been confirmed by His Holiness the Pope. In the first they do not receive every girl that offers herself, for their parents must pay about five hundred écus or crowns for them. Some indeed are admitted for three hundred écus, but they are obliged to serve those who pay more than they. No poor girls are taken in.

The Hospital. The King has erected a hospital for sick soldiers here. The sick person is there provided with everything he wants, and the king pays twelve sols every day for his keep, attendants, etc. The surgeons are paid by the king. When an officer is brought to this hospital, who has fallen sick in the service of the crown, he receives victuals and attendance gratis; but if he has gotten a sickness in the execution of his private concerns, and comes to be cured here, he must pay it out of his own purse. When there is room enough in the hospital, they likewise take in some of the sick inhabitants of the town and country. They have the medicines and the attendance of the surgeons gratis, but must pay twelve sols per day for meat, etc.

Every Friday is a *market day* when the country people come to the town with provisions, and those who want them must supply themselves on that day because it is the only market day in the whole week. On that day, too, a number of Indians come to town to sell their goods and buy others.

The *declination of the magnetic needle* was here ten degrees and thirty-eight minutes, west. M. Gillion, one of the priests here, who had a particular taste for mathematics and astronomy, had drawn a meridian in the garden of the seminary which he said he had examined repeatedly by the sun and stars, and found it to be very exact. I compared my compass with it, taking care that no iron was near it, and found its declination just the same as that which I have before mentioned. According to M. Gillion's observations, the

latitude of Montreal is forty-five degrees and twenty-seven minutes.
Temperature at Montreal. M. Pontarion, another priest, had
made thermometrical observations in Montreal, from the beginning
of this year, 1749. He made use of the Reaumur thermometer,
which he placed sometimes in a window half open, and sometimes
in one entirely open, and accordingly it will seldom mark the
greatest degree of cold in the air. However, I shall give a short
abstract of his observations for the winter months. In January the
greatest cold was on the 18th day of the month, when the Reau-
murian thermometer was 23° below the freezing point (—19¾°F.).
The least degree of cold was on the 31st of the same month when
it was just at the freezing point, but most of the days of this month
it was from 12° to 15° below freezing. In February the greatest
cold was on the 19th and 25th when the thermometer was 14° below, and the least was on the 3rd day of that month when it rose 8°
above the freezing point (50° F.); but it was generally 11° below
it. In March the greatest cold was on the 3rd when it was 10°
below the freezing point (9.5° F.) and on the 22nd, 23rd, and 24th
it was mildest, being 15° above it (65¾° F.): in general it was
4° below it (23° F.). In April the greatest degree of cold happened
on the 7th, the thermometer being 5° below the freezing point
(20¾° F.); the 25th was the mildest day, it being 20° above the
freezing point (77° F.); but in general it was 12° above it (59° F.).
These are the contents chiefly of M. Pontarion's observations
during those months; but I found, by the manner he made his
observations, that the cold had every day been from 4° to 6° (9 to
13½° F.) greater than he had marked it. He had likewise marked
in his journal that the ice in the St. Lawrence River broke on the
3rd of April at Montreal, and only on the 20th day of that month
at Quebec. On the 3rd of May some trees began to flower at Montreal, and on the 12th the hoarfrost was so great that the trees were
covered with it as with snow. The ice in the river close to this town
is over a French foot thick every winter, and sometimes it is two
feet, as I was informed by all whom I consulted on that head.

Several of the friars here told me that the summers had been
remarkably longer in Canada since its cultivation than they used
to be before; it begins earlier and ends later. The winters, on the
other hand, are much shorter, but the friars were of the opinion

that they were as severe as formerly, though they were not of the same duration, and the summer at present was no hotter than it used to be. The coldest winds at Montreal are those from the north and northwest.

AUGUST THE 2ND

En Route for Quebec. Early this morning we left Montreal and went in a bateau on our journey to Quebec in company with the second major of Montreal, M. de Sermonville. We went down the St. Lawrence River, which was here pretty broad on our left; on the northwest side was the isle of Montreal, and on the right a number of other isles, and the shore. The isle of Montreal was closely inhabited along the river; it was very flat and the rising land near the shore consisted of pure earth and was between three or four yards high. The woods were cut down along the riverside for the distance of an English mile. The dwelling houses were built of wood or stone indiscriminately, and whitewashed on the outside. The other buildings, such as barns, stables, etc. were all of wood. The ground next to the river was turned either into grain fields or meadows. Now and then we perceived churches on both sides of the river, the steeples of which were generally on that side of the church which looked towards the river, because they are not obliged here to put the steeples on the west end of the churches as in Sweden. Within six French miles of Montreal we saw several islands of different sizes in the river, and most of them were inhabited. Those without houses were sometimes turned into grain fields, but generally into grazing land. We saw no mountains, hills, rocks or stones to-day, the country being flat throughout, and consisting of pure earth.

All the *farms in Canada* stand separate from one another, so that each farmer has his possessions entirely separate from those of his neighbor. Each church, it is true, has a little village near it; but that consists chiefly of the parsonage, a school for the boys and girls of the place, and of the houses of tradesmen, but rarely of farmhouses; and if that was the case, their fields were still separated. The farmhouses hereabouts are generally all built along the rising banks of the river, either close to the water or at some distance from

it, and about three or four arpens from each other. To some farms
are annexed small orchards but they are in general without them;
however, almost every farmer has a kitchen-garden.

Peach Trees. I have been told by all those who have made journeys to the southern parts of Canada and to the Mississippi River
that the woods there abound with peach trees which bear excellent
fruit, and that the Indians of those parts say that the trees have
been there since time immemorial.

The *farmhouses* are generally built of stone, but sometimes of
timber, and have three or four rooms. The windows, seldom of
glass, are most frequently of paper. They have iron stoves in one
of the rooms and fireplaces in the rest, always without dampers.
The roofs are covered with boards, and the crevices and chinks are
filled up with clay. Other farm buildings are covered with straw.
The fences are like our common ones.

Road Shrines. There are several crosses put up by the roadside,
which is parallel to the shores of the river. These crosses are very
common in Canada, and are put up to excite devotion in the travellers. They are made of wood, five or six yards high, and proportion-
ally broad. In that side which faces the road is a square hole, in
which they place an image of our Savior, the Cross, or of the Holy
Virgin with the Child in her arms, and before that they put a piece
of glass, to prevent its being spoiled by the weather. Everyone who
passes by crosses himself, raises his hat or does some other bit of
reverence. Those crosses which are not far from churches, are very
much adorned, and they put up about them all the instruments
which they think the Jews employed in crucifying our Savior,
such as a hammer, tongs, nails, a flask of vinegar, and perhaps many
more than were really used. A figure of the cock, which crowed
when St. Peter denied our Lord, is commonly put at the top of the
cross.

The country on both sides was very delightful to-day, and the fine
state of its cultivation added greatly to the beauty of the scene. It
could really be called a village, beginning at Montreal and ending
at Quebec, which is a distance of more than one hundred and eighty
miles, for the farmhouses are never above five arpens and sometimes but three apart, a few places excepted. The prospect is exceedingly beautiful when the river flows on for several miles in a

straight line, because it then shortens the distances between the houses, and makes them form one continued village.

Women's Dress. All the women in the country without exception, wear caps of some kind or other. Their jackets are short and so are their skirts, which scarcely reach down to the middle of their legs. Their shoes are often like those of the Finnish women, but are sometimes provided with heels. They have a silver cross hanging down on the breast. In general they are very industrious. However I saw some, who, like the English women in the colonies, did nothing but prattle all day. When they have anything to do within doors, they (especially the girls) commonly sing songs in which the words *amour* and *coeur* are very frequent. In the country it is usual that when the husband receives a visit from persons of rank and dines with them, his wife stands behind and serves him, but in the town the ladies are more distinguished, and would willingly assume an equal if not a superior position to their husbands. When they go out of doors they wear long cloaks, which cover all their other clothes and are either grey, brown or blue. Men sometimes make use of them when they are obliged to walk in the rain. The women have the advantage of being in a *déshabillé* under these cloaks, without anybody's perceiving it.

We sometimes saw *windmills* near the farms. They were generally built of stone, with a roof of boards, which together with its wings could be turned to the wind.

The breadth of the river was not always the same to-day; in the narrowest place it was about a quarter of an English mile broad; in other parts it was nearly two English miles. The shore was sometimes high and steep, and sometimes low or sloping.

At three o'clock this afternoon we passed by a river [the Chambly] which empties into the St. Lawrence and comes from Lake Champlain, in the middle of which is a large island. The large boats which go between Montreal and Quebec pass on the southeast side of this island because it is deeper there; but some boats prefer the northwest side because it is nearer and yet deep enough for them. Besides this island there are several more hereabouts which are all inhabited. Somewhat further on, the country on both sides of the river is uninhabited near the shore till we come to the Lac St. Pierre, because it is so low as to be overflowed at certain times of the year. But still further inland, where the country is more elevated, it was said to be just as well populated as those places that we had passed previously to-day.

Lac St. Pierre is a part of the St. Lawrence River which is so broad that we could hardly see anything but sky and water before us, and I was everywhere told that it is seven French miles long and three broad. From the middle of this lake, as it is called, you see a large high country in the west which appears above the woods. In the lake are many places covered with a kind of rush, or *Scirpus palustris* L. There are no houses in sight on either side of the lake, because the land is rather too low there, and in spring the water rises so high that they can go with boats between the trees. However, at some distance from the shores where the ground is higher, the farms are close together. We saw no islands in the lake this afternoon, but the next day we came upon some.

Late in the evening we left Lac St. Pierre and rowed up a little river called Rivière de Loup, in order to come to a house where we might pass the night. Having rowed about an English mile we found the country inhabited on both sides of the river. Its shores are high but the country in general is flat. We passed the night in a farmhouse. The territory of Montreal extends to this place, but here begins the jurisdiction of the governor of Trois Rivières, which they reckon eight miles from here.

AUGUST THE 3RD

At five o'clock in the morning we set out again, and first rowed down the little river till we came into the Lac St. Pierre, down which we proceeded. After we had gone a good way, we perceived a high chain of mountains in the northwest, which were very much elevated above the low, flat country. The northwest shore of the lake was now in general very closely inhabited; but on the southeast side we saw no houses, and only a country covered with woods, which is sometimes said to be under water, but behind which there are, as I am told, a great number of farms. Towards the end of the lake, the river went into its proper bounds again, being not even a mile and a half broad, and afterwards it grew still narrower.

From the end of Lac St. Pierre to Trois Rivières, they reckon three French miles, and at about eleven o'clock in the morning we arrived at the latter place, where we attended divine service.

Trois Rivières is a little market town which had the appearance of a large village. It is, however, numbered among the three great towns of Canada, which are Quebec, Montreal, and Trois Rivières. It is said to lie in the middle between the two first, and is thirty French miles distant from each. The town is built on the north side of the St. Lawrence River on a flat, elevated sandbar and its location is very pleasant. On one side the river passes by, and it is here an English mile and a half broad. On the other side are fine grain fields, though the soil is very sandy. In the town are two churches of stone, a nunnery, and a house for the friars of the order of St. Francis. This town is likewise the seat of the third governor in Canada, whose house is also of stone. Most of the other houses are of timber, a single story high, tolerably well built, and stand very much apart. The streets are crooked. The shore here consists of sand, and the rising grounds along it are pretty high. When the wind is very violent here, it raises the sand, and blows it about the streets, making it very troublesome to walk in them. The nuns, who are about twenty in number, are very ingenious in all kinds of needlework. This town formerly flourished more than any other in Canada, for the Indians brought their goods to it from all sides; but since that time they have gone to Montreal and Quebec, and to the English, on account of their wars with the Iroquois, or Five Nations, and for several other reasons, so that this town is at present very much reduced by it. Its present inhabitants live chiefly by agriculture, though the neighboring ironworks may serve in some measure to support them. About an English mile below the town a great river flows into the St. Lawrence River, but first divides into three branches, so that it appears as if three rivers emptied themselves there. This has given occasion to call the river and this town, Trois Rivières (the Three Rivers).

The tide goes about a French mile above Trois Rivières, though it is so trifling as to be hardly observable. But about the equinoxes, and at the new moons and full moons in spring and autumn, the difference between the highest and lowest water is two feet. Accordingly the tide in this river goes very far up, for from the above-mentioned place to the sea they reckon about a hundred and fifty French miles.[1]

While my companions were resting I went on horseback to view the ironworks. The country which I passed through was pretty high, sandy, and generally flat. I saw neither stones nor mountains here.

The *ironworks* which is the only one in this country, lies three miles to the west of Trois Rivières. Here are two great forges, besides two lesser ones under the same roof. The bellows were made of wood, and everything else is as in the Swedish forges. The melting furnaces stand close to the forges, and are the same as ours. The ore is gotten two French miles and a half from the ironworks, and is carried thither on sledges in the winter. It is a kind of moor ore, which lies in veins within six inches or a foot from the surface of the ground. Each vein is from six to eighteen inches deep, and below it is a white sand. The veins are surrounded with this sand on both sides and covered at the top with a thin earth. The ore is pretty rich and lies in the veins in loose lumps the size of two fists, though there are a few which are nearly eighteen inches thick. These lumps are full of holes, which are filled with ochre. The ore is so soft that it may be crushed betwixt the fingers. They make use of a gray limestone which is quarried in the neighborhood for promoting the smelting of the ore. To that purpose they likewise employ a clay marle, which is found near this place. Charcoal is to be had in great abundance here, because all the country round this place is covered with woods, which have never been disturbed except by storms and old age. The charcoal from evergreen trees, that is, from the fir, is best for the forge, but that of deciduous trees is best for the smelting oven. The iron which is here made was to me described as soft, pliable, and tough, and is said to have the quality of not being attacked by rust as easily as other iron. In this point there appears a great difference between the Spanish iron and this, in shipbuilding. This smeltery was first founded in 1737, by private persons, who afterwards ceded it to the king. They cast cannon and mortars here of different sizes, iron stoves which are in use all over Canada, kettles, etc. not to mention the bars which are made here. They have likewise tried to make steel here,

---

[1] By the "sea" here is meant the Gulf of St. Lawrence, not the Atlantic Ocean.

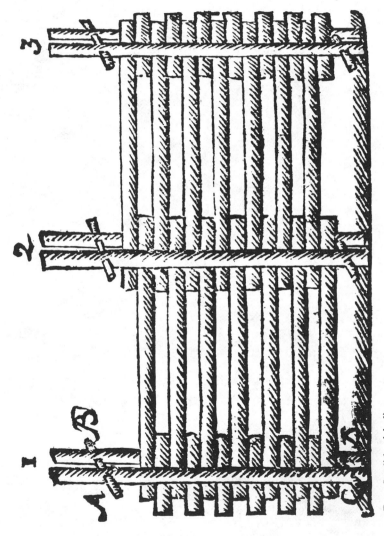

(From Swedish original)

Type of Fence

but cannot bring it to any great perfection because they are unacquainted with the best manner of preparing it. Here are many officers and overseers who have very good houses, built on purpose for them. It is agreed on all sides that the revenues of the ironworks do not pay the expenses which the king must every year have for maintaining them. They lay the fault on the bad state of the population and say that the inhabitants in the country are few, and that these have enough to do in attending to their agriculture, and that it therefore costs large sums to get a sufficient number of workmen. But however plausible this may appear, yet it is surprising that the king should be a loser in carrying on this work, for the ore is easily broken, very near the furnaces, and it is very fusible. The iron is good, and can be very conveniently transported over the country. These are, moreover, the only ironworks in the country from which everybody must supply himself with iron tools and what other iron he wants. But the officers and workmen belonging to the smeltery appear to be in very affluent circumstances. A river runs down from the ironworks into the St. Lawrence River, by which all the iron can be sent in boats throughout the country at a low rate.—In the evening I returned again to Trois Rivières.

[At this point Kalm introduces a long description of a prevalent type of Canadian fence. It does not seem necessary to include the verbal details here, and Forster and all his imitating translators omit it entirely, but as a middle course, and to illustrate Kalm's curious interest and versatility (as well as his sense of completeness) we are reproducing the drawing of the fence from the original.]

August the 4th

At the dawn of day we left this place and went on towards Quebec. We found the land on the north side of the river somewhat elevated, sandy, and closely inhabited along the shore. The southeast shore, we were told, is equally well inhabited, but the woods along that shore prevented our seeing the houses which are built further back in the country, the land close to the river being so low as to be subject to annual inundations. Near Trois Rivières, the river grows somewhat narrow but it enlarges again as soon as you come a little below that place, and has the breadth of more than two English miles.

Prayers. The French in their travels generally read a *Kyrie eleison* every morning before they started off. Every time when I followed them on water they chose the *Kyrie eleison*, which is found in their prayer books for Saturday and is almost wholly directed to the Holy Virgin. Almost every word of it is in Latin and although women, common people, and in fact most of the higher classes in Canada hardly understood a word of it, the whole morning prayer and the benediction were in this language. If there were women in the company the foremost of them was elected to read this litany in a very loud voice and to enumerate all the titles of honor which in it are attributed to the Holy Virgin. But in the absence of women it was done by one of the most distinguished of the men. At the end of every sentence the others present would answer *Ora pro nobis.* When priests were present they conducted the service. The women knew this Latin litany perfectly, so that they didn't miss a word. Some of the titles of the Holy Virgin were *Mater divinae gratiae, Virgo potens, Virgo clemens, Rosa mystica, Domus aurea, Regina angelorum,* etc. At everyone of these expressions everyone exclaimed *Ora pro nobis.* It was both strange and amusing to see and hear how eagerly the women and soldiers said their prayers in Latin and did not themselves understand a word of what they said. When all the prayers were ended the soldiers cried *Vive le Roi!* and that is about all they understood of the prayer proceedings. I have noticed in the papal service that it is directed almost entirely toward the external; the heart representing the internal is seldom touched. It all seems to be a ceremony. In the meantime the people are very faithful in these observances, because everyone tries by these means to put God under some obligation and intends by it to make himself deserving of some reward.

As we went on we saw several churches of stone, and often very well built ones. The shores of the river are closely inhabited for about three quarters of an English mile back in the country, but beyond that the woods and the wilderness increase. All the rivulets joining the St. Lawrence River are likewise well inhabited on both sides. I observed throughout Canada that the cultivated lands lie only along the St. Lawrence River and near the other rivers in the country, the environs of towns excepted, round which the country is all cultivated and inhabited within the distance of twelve or eighteen English miles. The great islands in the river are likewise inhabited.

The shores of the river now became higher, more oblique and steep; however, they consisted chiefly of earth. Here and there some rivers or large brooks flow into the St. Lawrence River, among which one of the largest is the Rivière Puante, which unites on the southeast side with the St. Lawrence, about two French miles below Trois Rivières, and has on its banks, a little way from its mouth, a town called Becancourt which is wholly inhabited by Abenakee Indians who have been converted to the Roman Catholic religion, and have Jesuits among them. At a great distance, on the northwest side of the river we saw a chain of very high mountains, running from north to south, elevated above the rest of the country, which is quite flat here without any remarkable hills.

Here were several limekilns along the river; and the limestone employed in them is quarried in the neighboring hills. It is compact and gray, and the lime it yields is quite white.

The fields here are generally sown with wheat [in this country they ate only wheat bread] oats, corn and peas. Pumpkins and watermelons are planted in abundance near the farms.

A humming bird (*Trochilus colubris*) flew among the bushes in a place where we landed to-day. The French call it *Oiseau mouche,* and say it is pretty common in Canada; and I have seen it since several times at Quebec.

About five o'clock in the afternoon we were obliged to take our night's lodgings on shore, the wind blowing very strong against us and being attended with rain. I found that the nearer we came to Quebec, the more open and free from woods was the country. The place where we passed the night is twelve French miles from Quebec.

Fish Traps. They have a very peculiar method of catching fish near the shore here. They place hedges along the shore, made of twisted oziers, so close that no fish can get through them, and from one foot to a yard high, according to the different depth of the water. For this purpose they choose places where the water runs off during the ebb, and leaves the hedges quite dry. Within this inclosure they place several weels, or wickerwork fish traps, in the form of cylinders, but broader at the base. They are placed upright, and are about a yard high and two feet and a half wide: on one side near the bottom is an entrance for the fishes, made of twigs, and sometimes of yarn made into a net. Opposite to this entrance, on the other side of the weel, facing the lower part of the river, is another entrance, like the first, and leading to a box of boards about four feet long, two deep and two broad. Near each of the weels is a hedge, leading obliquely to the long hedge, and making an acute angle with it. This hedge is made in order to lead the fish directly into the trap, and it is placed on that end of the long hedge which points towards the upper part of the river. When the tide comes up the river, the fish, and chiefly the eels, go up with it along the river side; when the water begins to ebb, the fish likewise go down the river, and meeting with the hedges, they swim along them, till they come through the weels into the boxes of boards [or eelpots], at the top of which there is a hole with a cover, through which the fish can be taken out. This apparatus is made chiefly for catching eels. In some places hereabouts they place nets instead of the hedges of twigs.

The shores of the river consisted no longer of pure loam; but of a species of slate. They are very steep and nearly perpendicular here, and the slate of which they consist is black, with a brown cast. The slate is divisible into thin shivers, no thicker than the blade of a knife. This slate moulders as soon as it is exposed to the open air, and the shore is covered with fine grains of sand which are nothing but particles of such mouldered slate. Some of the strata run horizontally, others obliquely, dipping to the south and rising to the north, and sometimes the contrary way. Sometimes they form bendings like large semicircles: sometimes a perpendicular line cuts off the strata to the depth of two feet, and the slates on both sides of the line form a perpendicular and smooth wall. In some places hereabouts, they find amongst the slate a stratum about four inches thick of a gray, compact but pretty soft limestone, of which the Indians for many centuries have made tobacco pipes and the French still make them.[1]

---

[1] This limestone seems to be a marle, or rather a kind of stone-marle for there is a whitish kind of it in the Krim Tartary, and near Stiva or Thebes, in Greece, which

AUGUST THE 5TH

Approaching Quebec. This morning we continued our journey rowing against the wind. The appearance of the shores was the same as yesterday; they were high, pretty steep or perpendicular, and consisted of the black slate before described. The country at the top was a plain without eminences, and was closely inhabited along the river for about the space of an English mile and a half inland. There are no islands in this part of the river, but several stone places, perceptible at low water only, which have several times proved fatal to travellers. The breadth of the river varies; in some parts it was a little more than three quarters of a mile, in others half a mile, and in some over two miles. The inhabitants made use of the same method of catching eels along the shores here as that which I described yesterday. In many places they make use of nets instead of the wicker traps.

Insects. Bedbugs (*Cimex lectularius*) abound in Canada; and I met with them in every place where I lodged, both in the towns and country, and the people know of no other remedy for them than patience.

The crickets (*Gryllus domesticus*) are also abundant in Canada, especially in the country where these disagreeable guests lodge in the chimneys; nor are they uncommon in the towns. They stay here both summer and winter, and frequently cut clothes in pieces for a pastime.

The cockroaches (*Blatta orientalis*) have never been found in the houses here.

Landing at Quebec. The shores of the river grow more sloping as you come nearer to Quebec. To the northward appears a high ridge of mountains. About two French miles and a half from Quebec the river becomes very narrow, the shores being within the reach of a musketshot from each other. The country on both sides was sloping, hilly, covered with trees, and had many small rocks. The shore was stony. About four o'clock in the afternoon we happily arrived at Quebec. The city does not appear till one is close to

---

is employed by the Turks and Tartars for making heads of pipes, and that from the first place is called Keffekil, and in the latter, Sea-Scum; it may be very easily cut, but grows harder in time.—F.

it, the view being intercepted by a high mountain on the south side. However, a part of the fortifications appears at a good distance, being situated on the same mountain. As soon as the soldiers, who were with us, saw Quebec, they called out that all those who had never been there before should be ducked, if they did not pay something to release themselves. This custom even the governor-general of Canada was obliged to submit to on his first journey to Quebec. We did not care when we came in sight of this town to be exempted from this old custom, which is very advantageous to the rowers, as it enables them to spend a merry evening on their arrival at Quebec after their troublesome labor.

Immediately after my arrival, the officer who had accompanied me from Montreal led me to the palace of the then vice-governor-general of Canada, the Marquis de la Galissonnière, a nobleman of uncommon qualities, who behaved towards me with extraordinary goodness during the time he stayed in this country.[1] He had already ordered some apartments to be got ready for me, and took care to provide me with everything I wanted, besides honoring me so far as to invite me to his table almost every day I was in town.

AUGUST THE 6TH

Quebec, the chief city in Canada, lies on the western shore of the St. Lawrence River, close to the water's edge on a neck of land bounded by that river on the east side, and by the St. Charles River on the north. The mountain, on which the town is built rises still higher on the south side and behind it begin great pastures. The same mountain also extends a good way westward. The city is divided into the lower and the upper section.[2] The lower lies on the river east of the upper. The neck of land I mentioned before, was formed by the dirt and filth, which had from time to time been accumulated there, and by a cliff which projects out at that point, not by any gradual diminution of the water. The upper city lies above the other on a high hill and takes up five or six times the space of the lower, though it is not quite so populous. The mountain on which the upper city is located reached above the houses of the lower city.

---

[1] The Marquis returned to France on September 24, 1749.
[2] *La haute Ville et la basse Ville.*

Notwithstanding, the latter are three or four stories high, and the view from the palace of the lower city (part of which is immediately under it) is amazing. There is only one easy way of getting to the upper city, and that is where a part of the mountain has been blown away. This road is very steep, although it is serpentine. However, people go up and down it in carriages and with wagons. All the other roads up the mountain are so steep that it is very difficult to climb to the top on them. Most of the merchants live in the lower city, where the houses are built very close together. The streets in it are narrow, very rough, and almost always wet. There is likewise a church and a small marketplace. The upper city is inhabited by people of quality, by several persons belonging to the different offices, by tradesmen and others. In this part are the chief buildings of the town, among which the following are worthy of particular notice.

I. The *Palace* is situated on the south or steepest side of the mountain, just above the lower city. It is not properly a palace but a large building of stone, two stories high, extending north and south. On the west side of it is a courtyard, surrounded partly by a wall and partly by houses. On the east side, or towards the river, is a gallery as long as the whole building, and about two fathoms broad, paved with smooth flags and protected on the outside by iron railings from which the city and the river exhibit a charming view. This gallery serves as a very agreeable walk after dinner, and those who come to speak with the governor-general wait here till he is at leisure. The palace is the lodging of the governor-general of Canada, and a number of soldiers stand guard before it, both at the gate and in the courtyard. When the governor or the bishop comes in or goes out they must present arms and beat the drum. The governor-general has his own chapel where he offers prayers. However, he often goes to mass at the church of the Recollets,[1] which is very near the palace.

II. The *churches* in this town are seven or eight in number and are all built of stone.

1. The *Cathedral* is on the right hand, coming from the lower to the upper city, somewhat beyond the bishop's house. The people were at present employed in ornamenting it. The organ had just

---

[1] A kind of Franciscan friars, called *Ordo St. Francisci strictioris observantiæ.*

been removed because of the improvements being made on it. On its west side was a round steeple of two stories, the lower one of which contained bells. The pulpit and some other parts within the church were of gilt. The seats were excellent.

2. The Jesuits' Church is built in the form of a cross and has a round steeple. This is the only temple that has a clock, and I shall mention it more particularly below.

3. The Recollets Church, or the Temple of the Barefooted Friars, is opposite the gate of the palace on the west side, looks well, and has a pretty high pointed steeple, with a compartment below for the bells.

4. The Church of the Ursulines has a round spire.

5. The Church of the Hospital.

6. The Bishop's Chapel.

7. The church in the lower city was built in 1690, after the town had been delivered from the English, and is called *Notre Dame de la Victoire*. It has a small steeple in the middle of the roof, square at the bottom and round at the top.

8. The little Chapel of the Governor-General may likewise be ranked amongst these churches.

III. The Bishop's House is the first on the right hand coming from the lower to the upper town. It is a fine large building, surrounded by an extensive courtyard and kitchen garden on one side, and by a wall on the other.

IV. The College of the Jesuits, which I shall describe more particularly, has a much more noble appearance in regard to its size and architecture than the Palace itself, and would be proper for a palace if it had a more advantageous location. It is about four times as large as the Palace, and is the finest building in town. It stands on the north side of a market, on the south side of which is the Cathedral.

V. The House of the Recollets lies to the west, near the Palace and directly over against it, and consists of a spacious building with a large orchard and kitchen garden. The house is two stories high. In each story is a narrow gallery with rooms and halls on one or both sides.

VI. The Hotel de Dieu, where the sick are taken care of, will be described more minutely later. The nuns that serve the sick are of the Augustine order.

VII. *Le Seminaire* or the house of the clergy is a large building on the northeast side of the cathedral. Here is on one side a spacious court, and on the other, towards the river, a large orchard with a kitchen garden. Of all the buildings in the town none has so fine a view as that in the garden belonging to this house, which lies on the high shore and commands a good distance down the river. The Jesuits on the other hand have the worst, and hardly any view at all from their college; nor have the Recollets any fine views from their home. In this building all the clergy of Quebec lodge with their superior. They have large pieces of land in several parts of Canada presented to them by the government, from which they derive a bountiful income so that they can live exceedingly well.

VIII. The Convent of the Ursuline Nuns will be described later.

These are the chief public buildings in town, but to the northwest, just before the town is:

IX. The house of the intendant, a public building whose size makes it fit for a palace. It is covered with tin, and stands in a second lower town, situated southward upon the St. Charles River. It has a large and fine garden on its north side. In this house all the deliberations concerning this province are held, and the gentlemen who have the management of the police and the civil power meet here and the intendant (mayor) generally presides. In affairs of great consequence the governor-general is also present. On one side of this building is the storehouse of the crown, and on the other the prison.

Most of the houses in Quebec are built of stone, and in the upper city they are generally but one story high, the public buildings excepted. I saw a few wooden houses in the town, but these may not be rebuilt when decayed. The houses and churches in the city are not built of bricks, but of the black "lime-slate" or [calcareous schist] of which the mountain consists and whereon Quebec stands. When these strata of slate are quarried deep in the mountain, they look very compact at first, and appear to have no fragments or *lamellæ* at all; but after being exposed a while to the air they separate into thin leaves. This slate is soft and easily cut, and the city walls as well as the garden ones consist chiefly of this material. The roofs of the public buildings are covered with common slate, which is brought from France because there is none in Canada.[1]

The *slated roofs* have for some years withstood the changes of air and weather without suffering any damage. The private houses have roofs of boards which are laid parallel to the spars, and sometimes to the eaves, or sometimes obliquely. The corners of houses are made of a gray small-grained limestone, which has a strong smell like stink-stone, and the windows are generally enchased with it. This limestone is more useful in those places than the lime-slate which always shivers in the air. The outsides of the houses are generally whitewashed. The windows are placed on the inner side of the walls; for they sometimes have double windows in winter. The middle roof has two, or at most three spars, covered with boards only. The rooms are warmed in winter by small iron stoves, which are removed in summer. There were no dampers anywhere. The floors are very dirty in every house and have all the appearance of being cleaned but once every year.

The Powder Magazine stands on the summit of the mountain on which the city is built, and south of the palace.

The streets in the upper city have a sufficient breadth, but are very rough on account of the rock on which it lies, and this renders them very disagreeable and troublesome, both to foot passengers and carriages. The black lime slabs basset out everywhere into sharp angles, which cut the shoes into pieces. The streets cross each other at all angles and are very crooked.

The many great orchards and kitchen gardens near the house of the Jesuits and other public and private buildings make the town appear very large, though the number of houses it contains is not very great. Its extent from south to north is said to be about six hundred toises,[2] and from the shore of the river along the lower town to the western wall between three hundred and fifty, and four hundred toises. It must be here observed that this space is not yet wholly inhabited, for on the west and south side, along the town

---

[1] In his French version of Kalm's *Travels,* Marchand points out that common slate is now found in several places in Canada, especially in the eastern sections. See vol. 2, p. 78, note.

[2] Fathoms.

walls, are large pieces of land without any buildings on them, and destined to be built upon in future time when the number of inhabitants will have increased in Quebec.

The bishop whose see is in the city is the only bishop in Canada. His diocese extends to Louisiana on the Mexican Gulf in the south and to the South Sea on the west.

No bishop, the pope excepted, ever had a more extensive diocese. But his spiritual flock is very small some distance from Quebec, and his sheep are often many hundred miles distant from each other.

Quebec as a Seaport. Quebec is the only seaport and trading town in all Canada, and from there all the produce of the country is exported. The port is below the town on the river, which is there about a quarter of a French mile broad, twenty-five fathoms deep, and its bottom is very good for anchoring. The ships are secured from all storms in this port; however, the northeast wind is the worst, because the town is more exposed in a storm from this direction. When I arrived here I counted thirteen large and small vessels, but in the evening before I left Quebec I counted twenty-three, and they expected more to come in. But it is to be remarked that no other ships than French ones can come into the port, though they may come from any place in France, or even from the French possessions in the West Indies. All foreign goods which are found in Montreal and other parts of Canada must first come from here. Similarly the French merchants from Montreal, after having spent six months among various Indian nations in order to purchase skins of beasts and furs, return about the end of August and go down to Quebec in September or October to sell their goods. The privilege of selling the imported goods should have vastly enriched the merchants of Quebec; but this is contradicted by others, who allow that there are a few in affluent circumstances, but that the majority possess no more than is absolutely necessary for their bare subsistence, and that several are very much in debt, which they say is owing to their luxury and vanity and to the fact that no one wanted to be poorer than the other. The merchants dress very finely, are extravagant in their repasts, and their ladies are every day in full dress and as much adorned as if they were to go to court.

The town is surrounded on almost all sides by a high wall, and especially towards the land. It was not quite completed when I was there and they were very busy finishing it. It is built of the above lamellated black limestone and of dark gray sandstone. For the corners of the gates they have employed gray limestone. They have not made any walls towards the water side, but nature seems to have worked for them by placing a rock there which it is impossible to ascend. All the rising land thereabouts is likewise so well planted with cannon that it seems impossible for an enemy's ships or boats to come to the town without running into imminent danger of being sunk. On the land side the town is likewise guarded by high mountains so that nature and art have combined to fortify it.

History of Quebec. Quebec was founded by its former governor, Samuel de Champlain, in the year 1608. We are informed by history that its rise was very slow. In 1629 towards the end of July it was taken by two Englishmen Lewis and Thomas Kirke [1] by capitulation and surrendered to them by the above-mentioned Champlain. At that time, Canada and Quebec were wholly destitute of provisions, so that they looked upon the English more as their deliverers than their enemies. The above-mentioned Kirkes were the brothers of the English admiral David Kirke, who lay with his fleet somewhat lower in the river. In the year 1632 the French got the town of Quebec and all Canada returned to them by the treaty of St. Germain-en-Laye. It is remarkable that the French were doubtful whether they should reclaim Canada from the English or leave it to them. The greater part were of the opinion that to keep it would be of no advantage to France, because the country was so cold, its expenses far exceeded the income, and France could not people so extensive a country without weakening herself, as Spain had done before; furthermore, that it was better to keep the people in France and employ them in all sorts of manufactures, which would oblige the other European powers who have colonies in America to bring their raw materials to French ports and take French manufactured goods in return. Those on the other hand who had more foresighted views knew that the climate was not so unfavorable as it had been represented. They likewise believed that that which caused the expenses was a fault of the Company, because it did not manage the country well. They would not have many people sent over at once,

---

[1] The capture of Quebec is generally credited to Sir David Kirke (1597-1656), the brother of Lewis and Thomas. See Kalm's account below.

but a few at a time, so that France might not feel it. They hoped that this colony would in future times make France powerful, for its inhabitants would become more and more acquainted with the herring, whale and cod fisheries, and likewise with the capturing of seals, and that by this means Canada would become a school for training seamen. They further mentioned the several sorts of furs, the conversion of the Indians, the shipbuilding, and the various uses of the extensive woods; and lastly, that it would be a considerable advantage to France even though it should reap no other benefit than to hinder by this means the progress of the English in America and of their increasing power, which would otherwise become insupportable to France, not to mention several other reasons. Time has shown that these reasons were the result of mature judgment and that they laid the foundation for the rise of France. It were to be wished that we had been of the same opinion in Sweden at a time when we were actually in possession of New Sweden, the finest and best province in all North America, or when we were yet in a condition to get possession of it. Wisdom and foresight do not only look upon the present times but also look into the future.

In the year 1663, at the beginning of February, the great earthquake was felt in Quebec and a great part of Canada, and there are still some vestiges of its effects at that time; however, no lives were lost.

On the 16th of October, 1690, Quebec was besieged by the English general, William Phipps,[1] who was obliged to retire in disgrace a few days afterward with great loss. The English have tried several times to repair their losses, but the St. Lawrence River has always been a very good defense for this country. An enemy, and one that is not acquainted with this river, cannot ascend without being ruined; for in the neighborhood of Quebec it abounds with hidden rocks, and has strong currents in some places so that the channel follows an extremely serpentine course. [In the last war, however, the Englishmen (1759) became masters of Quebec and a part of Canada.]

The name of Quebec is said to be derived from a Norman word, on account of its situation on a neck or point of land. For when one

---

[1] Sir William Phipps (1651-1695), later became governor of Massachusetts. He was in 1694 summoned to England to answer charges of inaction and failure in his campaigns against the French and Indians, but died before proceedings were undertaken.

comes up the river by l'Isle d'Orleans, that part of the St. Lawrence River which lies above the town does not come in sight and it appears as if the St. Charles River which lies just before were a continuation of the St. Lawrence. But on advancing further the true course of the river comes within sight, and has at first a great similarity to the mouth of a river or a great bay. This has given occasion for a sailor, who saw it unexpectedly, to cry out in his provincial dialect *Qué bec*,[1] that is, "what a point of land!" and from this it is thought the city obtained its name. Others derive it from the Algonquin word *Quebégo* or *Québec* signifying that which grows narrow because the river becomes narrower as it comes nearer to the town. The pronunciation of "Quebec" by the French was "Kebäk" almost without any accent; they gave each syllable a quantitative value. The word Canada was pronounced both by the French and English with the accent on the first syllable.

The St. Lawrence River is exactly a quarter of a French mile, or about three quarters of an English mile broad at Quebec. The salt water never comes up to the town and therefore the inhabitants can make use of the water in the river for their kitchens, drinking purposes, etc. All accounts agree that notwithstanding the breadth of this river and the violence of its course, especially during ebb, it is covered with ice during the whole winter, which is strong enough for walking, and a carriage may go over it. It is said to happen frequently that when the river has been open in May there are such cold nights in this month that it freezes again and will bear walking over. This is a clear proof of the intenseness of the cold here, especially when one considers what I shall mention immediately about the ebbing and flowing of the tide in this river. The greatest breadth of the river at its mouth is computed to be twenty-six French miles [or about seventy-five English miles], though the boundary between the sea and the river cannot well be ascertained, as the latter gradually looses itself in and unites with the former. The greatest part of the water contained in the numerous lakes of Canada, four or five of which are like large seas, is forced to empty into the ocean by means of this river alone. The navigation up this river from the sea is rendered very dangerous by the strength of the current, and by the number of sandbars which often arise in places where they never

---

[1] Meaning *Quel bec.*—F.

were before. The English encountered this formation of new sandbars once or twice when they intended to conquer Canada. Hence
the French have good reasons to look upon the river as a barrier to
Canada.[1]

The tide goes far beyond Quebec in the St. Lawrence River, as I
have mentioned above. The difference between high and low water
is generally between fifteen and sixteen feet, French measure; but
with the new and full moon, and when the wind is likewise favorable, the difference is seventeen or eighteen feet, which is indeed
considerable.

AUGUST THE 7TH

Ginseng is the current French name in Canada of a plant, the root
of which has a very great value in China.[2] It has been growing since
time immemorial in the Chinese Tartary and in Korea, where it
is annually collected and brought to China. Father Du Halde[3] says
it is the most precious and the most useful of all the plants in eastern
Tartary, and attracts every year a number of people into the deserts
of that country. The Mantchou-Tartars call it Orhota, that is, the
most noble or the queen of plants.[4] The Tartars and Chinese praise
it very much, and ascribe to it the power of curing several dangerous
diseases and that of restoring to the body new strength and supplying the loss caused by the exertion of the mental and physical fac-
ulties. An ounce of ginseng brings the surprising price of seven or
eight ounces of silver at Peking. When the French botanists in
Canada first saw a picture of it, they remembered to have seen a
similar plant in this country. They were confirmed in their conjecture by considering that several settlements in Canada lie in the same

---

[1] The river St. Lawrence, was no more a barrier to the victorious British fleets in
the last war, nor were the fortifications of Quebec capable to withstand the gallant
attacks of their land army, which disappointed the good Frenchmen in Canada of
their too sanguine expectations and at present they are rather happy at this change of
fortune, which has made them subjects of the British sceptre, whose mild influence they
at present enjoy.—F.

[2] Botanists know this plant by the descriptive name of *Panex quinque folium, foliis
ternatis quinatis* L.

[3] Jean-Baptiste Du Halde (1647-1743, French geographer. In 1735 he published
in four volumes his *Description Géographique, historique, chronologique, politique
et physique de l'Empire de la Chine et de la Tartarie chinoise.*

[4] Peter Osbeck's *Voyage to China*, Vol. I, p. 223.

latitude as those parts of the Chinese Tartary, and China, where the true ginseng grows wild. They succeeded in their attempt and found the same plant wild and abundant in several parts of North America, both in the French and English plantations, in level parts of the woods. It is fond of shade, and of a deep rich earth, and of land which is neither wet nor high. It is not common everywhere, for sometimes one may search the woods for the space of several miles without finding a single plant of it, but in those spots where it grows it is always found in great abundance. It flowers in May and June and its berries are ripe at the end of August. It bears transplanting very well, and will soon thrive in its new ground. Some people here who have gathered the berries and put them into their kitchen gardens told me that they lay one or two years in the ground without coming up. The Iroquois call the ginseng roots *Garangtoging* which it is said signifies a child, the roots bearing a faint resemblance to one: but others are of the opinion that they mean the thigh and leg by it, and the roots look very much like that. The French use this root for curing asthma, as a stomachic, and promoting fertility in women. The trade which is carried on with it here is very brisk, for they gather great quantities of it and send them to France, whence they are brought to China and sold there to great advantage.[1] It is said that the merchants in France met with amazing success in this trade at the first outset, but by continuing to send the ginseng over to China its price has fallen considerably there and consequently in France and Canada; however, they still find some profit in it. In the summer of 1748 a pound of ginseng was sold for six francs or livres at Quebec; but its common price here is one hundred sols or five livres. During my stay in Canada all the merchants at Quebec and Montreal received orders from their correspondents in France to send over a quantity of ginseng, there being an uncommon demand for it this summer. The roots were accordingly collected in Canada with all possible haste.[2] The Indians especially travelled about the

---

[1] Mr. Osbeck seems to doubt whether the Europeans reap any advantages from the ginseng trade or not, because the Chinese do not value the Canada roots so much as those of the Chinese Tartary and therefore the former bear scarce half the price of the latter. See Osbeck's *Voyage to China*, Vol. I, p. 223.—F.

[2] Marchand in the previously quoted *Voyage de Kalm en Amérique*, Vol. 2, p. 88, note 2, states that the ginseng trade in Canada was ruined by the merchants who, in order to hasten their profits, employed artificial heat to dry the plants instead

country in order to collect as much as they could and to sell it to the merchants at Montreal. The Indians in the neighborhood of this town were likewise so much taken up with this business that the French farmers were not able during that time to hire a single Indian, as they commonly do to help them in the harvest. Many people feared lest by continuing for several successive years to collect these plants without leaving one or two in each place to propagate their species, there would soon be very few of them left, which I think is very likely to happen, for by all accounts they formerly grew in abundance round Montreal, but at present there is not a single plant of it to be found, so effectually have they been rooted out. This obliged the Indians this summer to go far within the English boundaries to collect these roots. From the merchants in Montreal one received 40 francs a minot (39 liters) of these fresh roots. After the Indians have sold the fresh product to the merchants, the latter must take a great deal of pains with them. They are spread on the floor to dry, which commonly requires two months or more, according as the season is wet or dry. During that time they must be turned once or twice every day, lest they should spoil or moulder. Ginseng has never been found far north of Montreal. The father superior of the clergy here and several other people assured me that the Chinese value the Canada ginseng as much as the Tartarian [1] and that no one ever had been entirely acquainted with the Chinese method of preparing it. However, it is thought that amongst other preparations they dip the roots in a decoction of the leaves of ginseng. The roots prepared by the Chinese are almost transparent and look like horn inside, and the roots which are fit for use must be heavy, solid or compact inside.

"*Maiden Hair*". The plants which throughout Canada bear the name of *Herba capillaris* is likewise one of those with which a great trade is carried on in Canada. The English in their plantations call it "maiden hair"; it grows in all their North American colonies, which I travelled through and likewise in the southern parts of Canada; but I never found it near Quebec. It grows in the woods

---

of following the slow and natural method of dessication; that is, failure was due to greed and the method of preparation rather than to any deficiency in the quality of the Canadian plant itself.

[1] This is directly opposite to Mr. Osbeck's assertion. See the preceding page, note 1.—F.

in shady places and in a good soil.[1] Several people in Albany and Canada assured me that its leaves were very much used instead of tea, in consumption, cough, and all kinds of pectoral diseases. This they have learnt from the Indians who have made use of the plant for these purposes since ancient times. This American maiden hair is reckoned preferable in surgery to that which we have in Europe [2] and therefore they send a great quantity of it to France every year. The price varies and is regulated according to the grade of the plant, the care in preparing it, and the quantity which is to be gotten. For if it is brought to Quebec in great abundance, the price falls, and on the contrary it rises when the quantity gathered is but small. Usually the price at Quebec is between five and fifteen sols a pound. The Indians went into the woods about this time and travelled far above Montreal in quest of this plant.

Kitchen herbs succeed very well here. White cabbage is very fine but sometimes suffers greatly from worms. Onions (*Allium cepa*) are very much in use here together with other species of leeks. They likewise plant several species of pumpkins, melons, lettuce, wild chiccory or wild endive (*Cichorium intybus*), several kinds of peas, beans, Turkish beans, carrots, and cucumbers. They have plenty of red beets, horseradish, and common radishes, thyme, and marjoram. Turnips are sown in abundance and used chiefly in winter. Parsnips are sometimes eaten, though not very commonly. Few people took notice of potatoes, and neither the common (*Solanum tuberosum*) nor the Bermuda ones (*Convolvulus batatas*) were planted in Canada; only a few had any artichokes. When the French here are asked why they do not plant potatoes, they answer that they do not like them and they laugh at the English who are so fond of them. Throughout all North America the root cabbage [3] (*Brassica gongylodes* L.) is unknown to the Swedes, English, Dutch, Irish, Germans, and French. Those who have been employed in sowing and planting kitchen herbs in Canada and have had some experience in

---

[1] It is the *Adiantum pedatum* of L. *sp. pl.* p. 1557. Cornutus, in his *Canadens. plant. historia*, p. 7, calls it *Adiantum Americanum*, and gives together with the description, a figure of it, p. 6.—F.

[2] *Adiantum Capillus Veneris*. True Maiden Hair.

[3] This is a kind of cabbage with large round eatable roots, which grow out above the ground, wherein it differs from the turnip cabbage (*Brassica Napobrassica*) whose roots grow in the ground. Both are common in Germany, and the former likewise in Italy.

gardening told me that they were obliged to send for fresh seeds from France every year, because they commonly loose their strength here in the third generation and do not produce such plants as would equal the original ones in quality.

Antiquities. The Europeans have never been able to find any alphabetical characters, much less writings or books among the Indians, who have inhabited North America since time immemorial, and comprise several nations and dialects. These Indians have therefore lived in the greatest ignorance and darkness for several centuries; they are totally unacquainted with the state of their country before the arrival of the Europeans, and all their knowledge of it consists of vague traditions and mere fables. It is not certain whether any other nations possessed America before the present Indian inhabitants came into it, or whether any other nations visited this part of the globe, before Columbus discovered it.[1] It is equally unknown whether the Christian religion was ever preached here in former times. I conversed with several Jesuits who undertook long journeys in this extensive country and asked them whether they had met with any marks that there had formerly been any Christians among the Indians who lived here, but they all answered they had not found any. The Indians have always been as ignorant of architecture and manual labor as of science and writing. In vain does one seek for well-built towns and houses, artificially built fortifications, high towers and pillars, and such like among them, which the Old World can show from the most ancient times. Their dwelling places are wretched huts of bark, exposed on all sides to wind and rain. All their masonry work consists in placing a few gray stones on the ground round their fireplace to prevent the firebrands from spreading too far in their hut, or rather to mark out the space intended for the fireplace in it. Travelers do not enjoy a tenth part of the pleasure in traversing these countries which they must receive on their journeys through our old countries where they, almost every day, meet with some vestige or other of antiquity. Now an ancient celebrated town presents itself to view; here the remains of an old castle; there a field where many centuries ago the most powerful and the most

[1] Kalm here seems to have forgotten, temporarily, the well-known fact, which he mentions elsewhere, that the Norsemen visited North America "long before Columbus's time" (see pp. 202 and 443). Since Kalm wrote his *Travels,* a vast amount of literature has appeared on this subject, and it seems unnecessary to elaborate on the topic here.

skilful generals and the greatest kings fought a bloody battle; now the native spot and residence of some great or learned man. In such places the mind is delighted in various ways, and represents all past occurrences in living color to itself. We can enjoy none of these pleasures in America. The history of the country can be traced no further than from the arrival of the Europeans, for everything that happened before that period, is more like fiction or a dream than anything that really happened. In later times there have, however, been found a few marks of antiquity, from which it may be conjectured that North America was formerly inhabited by a nation more versed in science and more civilized than that which the Europeans found on their arrival here; or that a great military expedition was undertaken to this continent from the known parts of the world.

This is confirmed by an account which I received from Mr. de Verandrier (or de Vérendrye)[1] who commanded the expedition to the "South Sea" (Pacific Ocean) in person, of which I shall presently give an account. I have heard it repeated by others who were eyewitnesses of everything that happened on that occasion. Some years before I came to Canada, the then governor-general, Chevalier de Beauharnois,[2] gave Mr. de Vérendrye an order to go from Canada, with a number of people, on an expedition across North America to the South Sea in order to examine how far those two places were distant from each other and to find out what advantages might accrue to Canada or Louisiana, from a communication with that ocean. They set out on horseback from Montreal and went as far west as they could on account of the lakes, rivers and mountains which fell in their way. As they came far into the country beyond many nations they sometimes met with large tracts of land, free from wood, but covered with a kind of very tall grass for the space of some days' journey. Many of these fields were everywhere covered

---

[1] Probably Pierre Gaultier de Varenne, Sieur de La Vérendrye (1685-1749), noted Canadian pioneer and explorer. A condensed account of his expedition to the West is given in the *Dictionnaire Générale du Canada* under his name. He died soon after this conversation took place. He had several sons, some of whom accompanied him on his expeditions. One of these, Louis Joseph de La V. (1717-1761), is undoubtedly the one to whom Marchand refers in his *op. cit.*, 2, 94, note 1, as Lieut. de La Verandrière.

[2] Charles de la Boische, Marquis de Beauharnois (1670-1749), was appointed the fifteenth governor of New France in 1726 and was recalled in 1747. See *Dictionnaire Générale du Canada.*

with furrows, as if they had once been plowed and sown. It is
to be noticed that the nations which now inhabit North America
could not cultivate the land in this manner, because they never made
use of horses, oxen, plows, or any instruments of husbandry, nor had
they ever seen a plow before the Europeans came to them. In two
or three places, at a considerable distance from each other, our
travellers met with impressions of the feet of grown people and children in a rock but this seems to have been no more than a *Lusus
Naturæ*. When they came to the west where, to the best of their
knowledge, no Frenchmen, or European, had ever been, they found
in one place in the woods, and again on a large plain, great pillars
of stone, leaning upon each other. The pillars consisted of one
single stone each, and the Frenchmen could not but suppose that
they had been erected by human hands. Sometimes they have found
such stones laid upon one another, and, as it were, formed into a
wall. In some of those places where they found such stones, they
could not find any other sort. They have not been able to discover
any characters or writing upon any of these, though they have made
a very careful search for them. At last they met with a large stone,
like a pillar, and in it a smaller stone was fixed, which was covered
on both sides with unknown characters. This stone, which was
about a foot of French measure in length, and between four or five
inches broad, they broke loose, and carried to Canada with them,
whence it was sent to France to the secretary of state, the Count of
Maurepas. What became of it afterwards is unknown to them, but
they think it is yet preserved in his collection. Several of the Jesuits,
who had seen and handled this stone in Canada, unanimously affirm
that the letters on it are the same as those in books containing accounts of Tataria, are called Tatarian characters,[1] and that on com-

---

[1] This account seems to be highly probable, for we find in Marco Polo that Kublai-
Khan, one of the successors of Gengbizkhan, after the conquest of the southern part
of China, sent ships out to conquer the kingdom of Japan, or, as they call it, Nipan-
gri, but in a terrible storm the whole fleet was cast away, and nothing was ever
heard of the men of the fleet. It seems that some of these ships were cast to the
shores, opposite the great American lakes, between forty and fifty degrees north
latitude, and there probably erected these monuments, and were the ancestors of
some nations, who are called Mozemlecks, and have some degree of civilization.
Another part of this fleet it seems reached the country opposite Mexico, and there
founded the Mexican empire, which, according to their own records, as preserved by
the Spaniards, and in their painted annals, in Purchas's *Pilgrimage*, are very recent;
so that they can scarcely remember any more than seven princes before Motezuma

paring both together they found them just alike. Notwithstanding the questions which the French on the South Sea expedition asked the people there concerning the time when and by whom those pillars were erected, what their traditions and sentiments concerning them were, who had written the characters, what was meant by them, what kind of letters they were, in what language they were written, and other circumstances; yet they could never get the least explanation, the Indians being as ignorant of all those things as the French themselves. All they could say was that these stones had been in those places since ancient times. The places where the pillars stood were near nine hundred French miles westward of Montreal. The chief intention of this journey, *viz.* to come to the South Sea, and to examine its distance from Canada, was never attained on this occasion. For the people sent out for that purpose were induced to take part in a war between some of the most distant Indian nations, in which some of the French were taken prisoners, and the rest obliged to return. Among the last and most westerly Indians they were with, they heard that the South Sea was but a few days journey off; that they (the Indians) often traded with the Spaniards on that coast and sometimes also they went to Hudson

II. who was reigning when the Spaniards arrived there, 1519, under Fernanado Cortez; consequently the first of these princes, supposing each had a reign of thirty-three years and four months, and adding to it the sixteen years of Motezuma, began to reign in the year 1270, when Kublai Khan, the conqueror of all China and of Japan, was on the throne, and in whose time happened, I believe, the first abortive expedition to Japan, which I mentioned above, and probably furnished North America with civilized inhabitants. There is, if I am not mistaken, a great similarity between the figures of the Mexican idols, and those which are usual among the Tartars, who embrace the doctrines and religion of the Dalaï-Lama, whose religion Kublai-Khan first introduced among the Monguls, or Moguls. The savage Indians of North America, it seems, have another origin, and are probably descended from the Yukag·biri and Tchucktchi, inhabitants of the most easterly and northerly part of Asia, where, according to the accounts of the Russians, there is but a small traject to America. The ferocity of these nations, similar to that of the American, their way of painting, their fondness of inebriating liquors, (which the Yukaghiri prepare from poisonous and inebriating mushrooms, bought of the Russians) and many other things, show them plainly to be of the same origin. The Esquimaux seem to be the same nation with the inhabitants of Greenland, the Samoyedes and Lapponians. South America, and especially Peru, is probably peopled from the great unknown south continent, which is very near America, civilized, and full of inhabitants of various colors: who therefore might very easily be cast on the America continent in boats, or proas.—F.

I have let this note by Forster remain to show the character of the speculation on such a topic in the eighteenth century.—Ed.

Bay, to trade with the English. Some of these savages had houses which were made of earth. Many nations had never seen any Frenchmen; they were commonly clad in skins, but many were naked.

All those who had made long journeys in Canada to the south, but chiefly westward, agreed that there were many great plains destitute of trees, where the land was furrowed, as if it had been plowed. In what manner this happened, no one knew, for the grain fields of a great village or town of the Indians are scarce above four or six of our acres in extent; whereas those furrowed plains sometimes continued for several days' journey, except now and then a small smooth spot, and here and there some rising grounds.

I could not hear of any more relics of antiquity in Canada, notwithstanding my careful inquiries after them. In the continuation of my journey for the year 1750 [1] I shall find an opportunity of speaking of two other remarkable curiosities. Our Swedish scholar, Mr. George Westmann, A.M., has clearly and circumstantially shown, that our Scandinavians, chiefly the northern ones, long before Columbus's time, have undertaken voyages to North America; see his dissertation on that subject, which he read at Åbo in 1747 for obtaining his degree.[2]

AUGUST THE 8TH

A Convent. This morning I visited the largest nunnery in Quebec. Men are prohibited from visiting under very heavy punishments, except in some rooms, divided by iron rails, where the men and women that do not belong to the convent stand without and the nuns within the rails and converse with each other. But to increase the many favors which the French nation heaped upon me as a Swede, the governor-general got the bishop's leave for me to enter the convent and see its construction. The bishop alone has the power of granting this favor to men, but he does it very sparingly. The royal physician and a surgeon, are, however, at liberty to go in as often as they think proper. Mr. Gauthier,[3] a man of great knowl-

---

[1] This part was not published until 1929. See Introduction.

[2] Westmann's thesis was published in Åbo, 1757, and bore the title of *Itinera priscorum Scandianorum in Americam.* It was written under Kalm's direction. See Bibliography, item 25.

[3] See page 375, note 2.

edge in medicine and botany, was then the royal physician here, and accompanied me to the convent. We first saw the hospital, which I shall presently describe, and then entered the convent which forms a part of the hospital. It is a great building of stone, three stories high, divided on the inside into long corridors on both sides of which are cells, halls and rooms. The cells of the nuns are in the highest story, on both sides of the long corridor. They are only small, not painted on the inside, but hung with paper pictures of saints and of our Savior on the Cross. A bed with curtains and good bed clothes, a little narrow desk, and a chair or two, comprise the whole furniture of a cell. They have no fires in winter, and the nuns are forced to lie in the cold cells. In the long hall is a stove which is heated in winter, and as all the rooms are left open some warmth can by this means come into them. In the middle story are the rooms where they pass the day together. One of these is the work room. This is large, finely painted and decorated, and has an iron stove. Here nuns were at their needlework, embroidering, gilding and making flowers of silk, which bear a great similarity to the natural ones. In a word, they were all employed in such fine and delicate work as was suitable to ladies of their rank in life. In another hall they assemble to hold their deliberations. Another apartment contains those who are indisposed, but such as are more dangerously ill have rooms to themselves. The novices or newcomers are taught and instructed in another hall. Another is destined for their refectory, or dining room, in which are tables on all sides. On one side of it is a small desk, on which is laid a French book concerning the life of those saints who are mentioned in the New Testament. When they dine, all are silent. One of the eldest enters the pulpit and reads a part of the book before mentioned and when they have gone through it, they read some other religious book. During the meal they sit on that side of the table which is turned towards the wall. In almost every room is a gilt table on which are placed candles, the pictures of our Savior on the Cross and of some saint. Before these tables they say their prayers. On one side is the church and near it a large room divided from the church by rails, so that the nuns can only look into it. In this room they remain during divine service. The priest is in the church where the nuns hand him his sacerdotal garments through a hole, for they are not allowed to go into the vestry or to be in the same room with the priest. There are still several other rooms and halls here, the use of which I do not remember. The lowest story contains a kitchen, bake house, several butteries, etc. In the garrets they keep their grain and dry their linen. In the middle story is a balcony on the outside, almost round the whole building, where the nuns are allowed to take the air. The view from the convent is very fine on every side. The river, the fields and the meadows beyond the town appear there to great advantage. On one side of the convent is a large garden, in which the nuns are at liberty to walk about. It belongs to the convent, and is surrounded with a high wall. There is a quantity of all sorts of vegetables in it and a number of apple, cherry and white walnut trees and red currant bushes. This convent, they say, contains about fifty nuns, most of them advanced in years, scarcely any being under forty years of age. At this time there were two young ladies among them who were being instructed in those things which belong to the knowledge of nuns. They are not allowed to become nuns immediately after their entrance, but must pass through a noviciate of two or three years in order to learn whether they will be constant. For during that time it is in their power to leave the convent, if a monastic life does not suit their inclinations. But as soon as they are received among the nuns and have taken their vows, they are obliged to continue their whole life in it. If they appear willing to change their mode of life, they are locked up in a room from which they can never get out. The nuns of this convent never go further from it than to the hospital, which lies near it and even constitutes a part of it. They go there to attend the sick and to take care of them. Upon my leaving, the abbess asked me if I was satisfied with their institution, whereupon I told them that their convent was beautiful enough, though their mode of living was much circumscribed. Thereupon she told me that she and her sisters would heartily ask God to make me a good Roman Catholic. I answered her that I was far more anxious to be and remain a good Christian, and that as a recompense for their honors and prayers I would not fail earnestly to ask God that they too might remain good Christians, because that would be the highest degree of a true religion that a mortal could find. Thereupon she smilingly bade me farewell. I was told by several people here, some of which were ladies, that none of the nuns went into a convent till she had attained an age in which she had small hopes of ever getting a husband. The nuns of all the three convents in Quebec looked very old, by which it seems that there is some foundation for this assertion. All agree here that the men are much less numerous in Canada than the women, for the men die on their voyages. Many go to the West Indies and either settle or die there; many are killed in battles, etc. Hence there seems to be a necessity of some women going into convents.

The hospital, as I have before mentioned, forms a part of the convent. It consists of two large halls, and some rooms near the apothecary's shop. In the halls are two rows of beds on each side. The beds next to the wall are furnished with curtains, the outward ones are without them. In each bed are fine bed clothes with clean double sheets. As soon as a sick person has left his bed, it is made again to keep the hospital in cleanliness and order. The beds are two or three yards distant, and near each is a small table. There are good iron stoves, and fine windows in this hall. The nuns attend the sick people, and bring them food and other necessaries. Besides them there are some men who attend, and a surgeon. The royal physician is likewise obliged to come hither once or twice every day, look after everything and give prescriptions. They commonly receive sick soldiers into this hospital, who are very numerous in July and August, when the king's ships arrive, and in time of war. But at other times, when no great number of soldiers are sick, other poor people can take their places, as far as the number of empty beds will reach. The king provides everything here that is requisite for the sick persons, *viz.* provisions, medicines, fuel, etc. Those who are very ill are put into separate rooms, in order that the noise in the great hall may not be troublesome to them.

The civility of the inhabitants here is more refined than that of the Dutch and English in the settlements belonging to Great Britain. On the street they raised their hat only to acquaintances and to those of the upper classes. Young men often kept their hats on inside where there were women, but most of them, especially the older ones, took them off. The English, on the other hand, do not idle their time away in dressing as the French do here. The ladies, especially, dress and powder their hair every day, and put their locks in papers every night. This idle custom had not been introduced in the English settlements. The gentlemen generally wear their own hair, but some have wigs, and there were a few so distinguished that they had a queue. People of rank are accustomed to wear lace-trimmed clothes and all the crown officers carry swords. All the gentlemen, even those of rank, the governor-general excepted, when they go into town on a day that looks like rain, carry their cloaks on their left arm. Acquaintances of either sex, who have not seen each other for some time, on meeting again salute with mutual kisses.

Canadian Plants. The plants which I have collected in Canada and which I have partly described, I pass over as I have done before, that I may not tire the patience of my readers by a tedious enumeration. If I should crowd my journal with my daily botanical observations, and descriptions of animals, birds, insects, ores, and like curiosities, it would be swelled to six or ten times its present size. I therefore spare all these things, consisting chiefly of dry descriptions of natural curiosities, for a *Flora Canadensis* or a similar work.[1] The same I must say in regard to the observations I have made in medicine. I have carefully collected all I could on this journey, concerning the medicinal use of the American plants and the cures some of which they reckon infallible in more than one place. But medicine not being my principal study (though from my youth I always was fond of it) I may probably have omitted remarkable circumstances in my accounts of medicines and cures, though one cannot be too accurate in such remarks, or at least one would not find them as they ought to be. This will excuse me for avoiding as much as possible to mention such things as belong to that field and are above my knowledge. Concerning the Canadian plants, I can here add that the further you go northward, the more you find the plants are the same as the Swedish ones: thus, on the north side of Quebec, a fourth part of the plants, if not more, are the same as the wild plants in Sweden. A few plants and trees which have a particular quality or are applied to some particular use, shall, however, be mentioned in a few words later.

The reindeer moss (*Lichen rangiferinus*) grows plentifully in the woods round Quebec. M. Gauthier, and several other gentlemen, told me that the French, on their long journeys through the woods, on account of their fur trade with the Indians, sometimes boil this moss and drink the decoction for want of better food, when their provisions are at an end, and they say it is very nutritive.

---

[1] This work was never completed or published, though the essential results of Kalm's investigations were utilized by other botanists, chiefly Linné.

Several Frenchmen who have been in the Terra Labrador, where there are many reindeer (which the French and Indians here call cariboux), related that all the land there is in most places covered with this reindeer moss, so that the ground looks white as snow.

AUGUST THE 10TH

The Jesuit College. To-day I dined with the Jesuits. A few days before, I paid my visit to them, and the next day their president and another Father Jesuit called on me and invited me to dine with them to-day. I attended divine service in their church, which is a part of their house. It is very fine within, though it has no seats, for everyone is obliged to kneel down during the service. Above the church is a small steeple with a clock. The building the Jesuits live in is magnificently built, and looks exceedingly fine, both without and within, which makes it similar to a fine palace. It consists of stone, is three stories high, exclusive of the garret, is covered with slate and built in a square form like the new [royal] palace at Stockholm, including a large court. Its size is such that three hundred families would find room enough in it, though at present there were not above twenty Jesuits in it. Sometimes there is a much greater number of them, especially when those return who have been sent as missionaries into the country. There is a long corridor along all the sides of the square, in every story, on both sides of which are either cells, halls, or other apartments for the friars. There are also their library, apothecary shop, etc. Everything is very well regulated, and the Jesuits are well accommodated here. On the outside is their college which is on two sides surrounded with great orchards and kitchen gardens, in which they have fine walks. A part of the trees here are the remains of the forest which stood here when the French began to build this town. They have besides planted a number of fruit trees, and the garden is stocked with all sorts of plants for use in the kitchen. The Jesuits dine together in a great hall. There are tables placed all round it along the walls, and seats between the tables and the walls, but not on the other side. Near one wall is a pulpit at which one of the fathers stands during the meal in order to read some religious book, but this day it was omitted, all the time being employed in conversation. They dine very well, and their dishes are as numerous as at the greatest feasts.

In this spacious building you do not see a single woman; all are "fathers", or "brothers", the latter of which are young men, brought up to be Jesuits. They prepare the meal and bring it upon the table, for common servants are not admitted.

The Jesuits in this country were dressed as follows; they had their own hair, but it was cut short; at the top of the head they were shaved so that it was only a bare spot. The older ones had hoods of black cloth, the younger went bare-headed inside the house but sometimes had a hat on. All shaved their beard, like all other Frenchmen in Canada. The tie was black and often was but the collar of the coat: the latter was long, black and reached to the shoes. It buttoned tight to the body with the buttons in front and was besides bound about the waist with a black band. It was so covered with buttons down the entire front that only these and the shoes could be seen. Often slippers were used in place of shoes. Sometimes they wore a tight-fitting jacket which reached to the knees. They had no collars like clergymen nor did their shirt sleeves show. The oldest had black caps, shaped like a cone, at the top was a black tassel which they also wore on their calotte.

Clergymen in Canada. Besides the bishop there are three kinds of clergymen in Canada; *viz.* Jesuits, priests and recollets. The Jesuits are without doubt the most important; therefore they commonly say here, by way of proverb, that a hatchet is sufficient to cut out a recollet; a priest can be made with a pair of scissors, but a Jesuit requires a paint-brush [1] to show how much he surpasses the others. The Jesuits are usually very learned, studious and civil and agreeable in company. In their whole deportment there is something pleasing. It is no wonder therefore that they captivate the minds of people. They seldom speak of religious matters, and if it happens they generally avoid disputes. They are very ready to do anyone a service, and when they see that their assistance is wanted they hardly give one time to speak of it, falling to work immediately to bring about what is required of them. Their conversation is very entertaining and learned, so that one cannot be tired of their company. Among all the Jesuits I have conversed with in Canada, I have not found one who was not possessed of these qualities in a very eminent degree.

---

[1] *Pour faire un recolet il faut une hachette, pour un prêtre un ciseau, mais pour un Jesuite il faut un pinceau.*—F.

They have large possessions in this country which the French king gave them. At Montreal they have likewise a fine church and a neat little house with a small but pretty garden next to it. They do not care to become pastors of a congregation in the town or country; but leave their places, together with the emoluments arising from them, to the priests. All their business here is to convert the heathens; and with that view their missionaries are scattered over every part of this country. Near every town and village peopled by converted Indians are one or two Jesuits, who take great care that they may not return to paganism but live as Christians ought to do. Thus there are Jesuits with the converted Indians in Tadoussac, Lorette, Becancourt, St. François, Sault St. Louis, and all over Canada. There are likewise Jesuit missionaries with those who are not converted; so that there is commonly a Jesuit in every large village belonging to the Indians, whom he endeavors on all occasions to convert. In winter he goes on their great hunts where he is frequently obliged to suffer all imaginable inconveniences such as walking in the snow all day, lying in the open air all winter, being out both in good and bad weather, the Indians not fearing any kind of weather, and lying in the Indian huts, which often swarm with fleas and other vermin, etc. The Jesuits undergo all these hardships both for the sake of converting the Indians and also for political reasons. The Jesuits are of great use to their king, for they are frequently able to persuade the Indians to break their treaty with the English, to make war upon them, to bring their furs to the French and not to permit the English to come amongst them. But there is some danger attending these attempts, for when the Indians are drunk they sometimes kill the missionaries who live with them, calling them spies, or excusing themselves by saying that the brandy had killed them. These are accordingly the chief occupations of the Jesuits here. They do not go to visit the sick in the town, they do not hear confessions, and attend no funerals. I have never seen them go in processions in remembrance of the Virgin Mary and other saints. They seldom go into a house in order to get food, and if they are invited they do not like to stay except they be on a journey. Everybody sees that they are, as it were, selected from the people on account of their superior genius and qualities. They are here held a most cunning set of people, who generally succeed in their undertakings and surpass all others in acuteness of understanding. I have therefore several times observed that they have enemies in Canada. They never receive any others into their society but persons of very promising parts, so that there are no blockheads among them. On the other hand the priests take any kind of people into their order they can find; and in the choice of monks, they are even less careful. The Jesuits who live here have all come from France, and many of them return thither again after a stay of a few years here. Some (five or six of whom are yet alive) who were born in Canada, went over to France and were received among the Jesuits there, but none of them ever came back to Canada. I know not what political reason hindered them. During my stay in Quebec one of the priests with the bishop's leave, gave up his priesthood and became a Jesuit. The other priests were very ill pleased with this, because it seemed as if he looked upon their condition as too lowly for himself. Those congregations in the country that pay rents to the Jesuits, have, however, divine service performed by priests, who are appointed by the bishop; and the land rent belongs only to the Jesuits. Neither the priests nor the Jesuits carry on any trade with furs and skins, leaving that entirely to the merchants.

This afternoon I visited the building called the Seminary, where all the priests live in common. They have a great house built of stone with corridors in it, and rooms on each side. It is several stories high, and close to it is a fine garden full of all sorts of fruit trees and vegetables, and it is divided by walks. The view from here is the finest in Quebec. The priests of the Seminary are not much inferior to the Jesuits in civility, and therefore I spent my time very agreeably in their company.

The priests are the second and most numerous class of the clergy in this country, for most of the churches, both in towns and villages (the Indian converts excepted) are served by priests. A few of them are likewise missionaries. In Canada are two seminaries: one in Quebec, the other in Montreal. The priests of the seminary in Montreal are of the order of St. Sulpicius, and supply only the congregation on the isle of Montreal and the town of the same name. At all the other churches in Canada, the priests belonging to the Quebec seminary officiate. The former, or those of the order of St. Sulpicius, all come from France, and I was assured that they never allow a native of Canada to come among them. In the seminary at Quebec, the natives of Canada constitute the greater part. In order to fit the children of this country for orders, there are schools at Quebec and St. Joachim where the youths are taught Latin and instructed in the knowledge of those things and sciences, which have a more immediate connection with the business they are intended for. However, they are not very particular in their choice, and people of a middling capacity are often received among them. They do not seem to have made great progress in Latin; for notwithstanding the service is read in that language, and they read their Latin breviary and other books, every day, yet most of them find it very difficult to speak it. All the priests in the Quebec seminary are consecrated by the Bishop of Canada. The dress of the priests differed from that worn by the Jesuits in that the former wore collars, either yellow or light blue. When they travelled they always brought with them their breviary or prayer book in a little skin bag, which they hung about the neck or on their arm, and read so often in it, that it seemed to us they had been prescribed to read certain sections in it every day. Both the seminaries receive great revenues from the king; that in Quebec has above thirty thousand livres a year. All the country on the west side of the St. Lawrence River from the town of Quebec to St. Paul Bay belongs to this seminary, besides their other possessions in the country. They lease the land to the settlers for a certain rent, which, if it be annually paid according to their agreement, the children or heirs of the settlers may remain in an undisturbed possession of the lands. A piece of land three arpens [1] broad and thirty, forty, or fifty arpens long, pays annually an ecu [2] and a couple of chickens, or some other additional trifle. In such places as have convenient waterfalls they have built watermills, or sawmills, from which they annually get considerable sums. The seminary of Montreal possesses the whole ground on which that town stands, together with the whole Isle of Montreal. I have been assured that the ground rent of the town and isle is computed at seventy thousand livres, besides what they get for saying masses, baptizing, holding confessions, attending at marriages and funerals, etc. All the revenues of ground rent belong to the seminaries alone, and the priests in the country have

---

[1] A Canadian arpent, linear measure, as noted before, is said to be 12 rods. A square arpent is a little less than an acre. Its value varied greatly.

[2] A French coin, value about one English crown.—F.

no share in them. But as the seminary in Montreal, consisting only of sixteen priests, has greater revenues than it can expend, a large sum of money is annually sent over to France, to the chief seminary there. The land rents belonging to the Quebec seminary are employed for the use of the priests in it, and for the maintenance of a number of young people who are brought up to take the orders. The priests who live in the country parishes, get the tithes from their congregation, together with the extra pay on visiting the sick, etc. In small congregations the king gives the priests an additional sum. When a priest in the country grows old, and has done good services, he is sometimes allowed to come into the seminary in town and spend the rest of his days there. The seminaries are allowed to place the priests on their own estates; but the other places are appointed by the bishop.

The recollets or Barefooted Monks are the third class of clergymen in Canada. They have a fine large dwelling house here, and a fine church, where they officiate. Near it is a large and fine garden, which they cultivate with great application. In Montreal and Trois Rivières they are lodged almost in the same manner as here. They do not endeavor to choose the best fellows amongst them, but take all they can get. They do not torment their brains with much learning and I have been assured that after they have put on their monastic habit they do not study to increase their knowledge but forget even what little they knew before. Their dress consists of a long black frock of coarse cloth extending down to the heels. On the back of it near the collar hangs a hood like a bag fastened to it, which they pull over the head in bad weather. It is like the hoods which our women now use on their cloaks, a custom which probably first came from the monks. On their head they have small calottes; the hair is cut very short, reaching only to the ears. Around the waist they wear a narrow hemp rope which encircles the body several times. In the summer they go barelegged, with wooden shoes, but in winter they have stockings. They wear no linen shirts but a woolen coat next to the body. When they walk in processions they put a black mantle outside of their frock, extending down to the waist. At night they generally lie on mats or some other hard mattresses. However, I have sometimes seen good beds in the cells of some of them. They have no possessions here, having made vows of poverty, and live chiefly on the alms which people give them. To this purpose, the young monks, or brothers, go into the houses with a bag, and beg what they want. They have no congregations in the country, but sometimes go among the Indians as missionaries. In each fort, which contains forty men, the king keeps one of these monks who officiates there instead of a priest. The king gives him lodging, provisions, servants, and all he wants; besides two hundred livres a year. Half of it he sends to the community he belongs to; the other half he reserves for his own use. On board the king's ships are generally no other priests than these friars, who are therefore looked upon as people belonging to the king. When one of the chief priests in the country dies, and his place cannot immediately be filled, they send one of these friars there to officiate while the place is vacant. Some of these monks come from France and some are natives of Canada. There are no other monks in Canada besides these, except now and then one of the order of St. Augustine or some other, who come with one of the king's ships, but goes off with it again.

AUGUST THE 11TH

A Convent. This morning I took a walk out of town, with the royal physician M. Gauthier, in order to collect plants and to see a nunnery at some distance from Quebec. This cloister, which is built very magnificently of stone, lies in a pleasant spot surrounded with grain fields, meadows, and woods, and from which Quebec and the St. Lawrence may be seen. A hospital for poor old people, cripples, etc. makes up part of the cloister and is divided up into two halls, one for men, the other for women. The nuns attend both sexes, with this difference however, that they only prepare the meal for the men and bring it in to them, give them medicine, clear the table when they have eaten, leaving the rest for male servants. But in the hall where the women are, they do all the work that is to be done. The regulation in the hospital was the same as in that at Quebec. To show me a particular favor, the bishop, at the desire of the Marquis de la Galissonnière, governor-general of Canada, granted me leave to see this nunnery also, where no man is allowed to enter without his leave, which is an honor he seldom confers on anybody. The abbess led me and M. Gauthier through all the apartments, accompanied by a great number of nuns. Most of the nuns here are of noble families and one was the daughter of a governor. She had a grand air. Many of them are old, but there are likewise some very young ones among them, who looked very well. They all seemed to be more polite than those in the other nunnery. Their rooms are the same as in the last place except for some additional furniture in their cells. The beds are hung with blue curtains; there are a couple of small bureaux, a table between them, and some pictures on the walls. There are however no stoves in any cell. But those halls and rooms in which they are assembled together, and in which the sick ones lie, are supplied with an iron stove. I did not find out the number of nuns here, but I saw a great number of them. Here are likewise some probationers preparing for their reception as nuns. A number of little girls are sent hither by their parents to be instructed by the nuns in the principles of Christian religion and in all sorts of ladies' work, and when they are through their parents take them home again. The convent at a distance looks like a palace, and, as I am told, was founded by a bishop who they say is buried in a part of the church.

We botanized till dinner time in the neighboring meadows, and then returned to the convent to dine with a venerable old father recollet, who officiated here as a priest. The dishes were all prepared by nuns, and as numerous and various as on the tables of great men. There were likewise several sorts of wine and among the many dainties served at the end of a meal were these: white Canadian walnuts coated with sugar, pears and apples with syrup, apples preserved in spirits of wine, small sugared lemons from the West Indies, strawberry preserves and angelica roots. The revenues of this convent are said to be considerable. At the top of the building is a small belfry. Considering the large tracts of land which the king has given in Canada to convents, Jesuits, priests, and several families of rank, it seems he must have very little left for himself.

Our *common raspberries* are so plentiful here on the hills near grain fields, rivers and brooks, that the branches look quite red on account of the number of berries on them. They are ripe about this time and eaten as a dessert after dinner. They are served either with or without fresh milk and powdered sugar. Sometimes they are kept through the winter in glass jars with syrup.

The *mountain ash*, or sorb tree (*Sorbus aucuparia*) is pretty common in the woods hereabouts.

The *northeast wind* is considered the most piercing of all here. Many prominent people assured me that this wind when it is very violent in winter pierces through walls of a moderate thickness, so that the whole wall on the inside of the house is covered with snow or a thick hoarfrost, and that a candle placed near a thinner wall is almost blown out by the wind which continually comes through. This wind damages the houses which are built of stone, and forces the owners to repair them very frequently on the northeast side. The north and northeast winds are consequently considered very cold here. In summer the north wind is generally attended with rain.

The difference of climate between Quebec and Montreal is by all said to be very great. The wind and weather of Montreal are often entirely different from what they are at Quebec. The winter there is not nearly so cold as in the latter place. Several sorts of fine pears will grow near Montreal, but are far from succeeding at Quebec, where the frost frequently kills them. Quebec has generally more rainy weather, spring begins later and winter sooner than at Montreal, where all sorts of fruits ripen a week or two earlier than at Quebec.

August the 12th

Mixed Blood. This afternoon I and my servant went out of town, to stay in the country for a couple of days that I might have more leisure to examine the plants which grow in the woods here, and the nature of the country. In order to proceed the better, the governor-general had sent for an Indian from Lorette to show us the way and teach us what use they make of the wild plants hereabouts. This Indian was an Englishman by birth, taken by the Indians thirty years ago when he was a boy and adopted by them according to their custom in the place of a relation of theirs killed by the enemy. Since that time he had constantly stayed with them, become a Roman Catholic and married an Indian woman. He dressed like an Indian, spoke English and French and many of the Indian dialects. In the wars between the French and English in this country, the French Indians made many prisoners of both sexes in the English plantations, adopted them afterwards, and married them to people of the Indian nations. Hence the Indian blood in Canada is very much mixed with European blood, and a large number of the Indians now living owe their origin to Europe. It is also remarkable that a great number of the people they had taken during the war and incorporated with their nations, especially the young people, did not choose to return to their native country, though their parents and nearest relations came to them and endeavored to persuade them to, and though it was in their power to do so. The free life led by the Indians pleased them better than that of their European relations; they dressed like the Indians and regulated all their affairs in their way. It is therefore difficult to distinguish them, except by their color, which is somewhat whiter than that of the Indians. There are likewise examples of some Frenchmen going amongst the Indians and following their mode of life. There is on the contrary scarcely one instance of an Indian adopting the European customs; for those who were taken prisoners in the war always endeavored to return to their own people again, even after several years of captivity, though they enjoyed all the privileges that were ever possessed by the Europeans in America.

Geological Formations. The lands which we passed over were everywhere laid out into grain fields, meadows, or pastures. Almost all around us were presented to our view farms and farmhouses and excellent fields and grazing land. Near the town the land is pretty flat and intersected now and then by a clear rivulet. The roads are very good, broad, and lined with ditches on each side in low ground. Further from the town the land rises higher and higher, and consists as it were of terraces, one above another. This rising ground is, however, pretty smooth, chiefly without stones and covered with rich earth. Under that is the black lime stratum which is so common hereabouts, and is broken into small lamellæ and corroded by the air. Some of the strata were horizontal, others perpendicular. I have likewise found such perpendicular strata of lime lamellæ in other places in the neighborhood of Quebec. All the hills are cultivated and some are adorned with fine churches, houses and crop-bearing fields. The meadows are commonly in the valleys, though some are in higher places. Soon after we had a fine view from one of these hills. Quebec appeared very plainly to the eastward, and the St. Lawrence River could likewise be seen. Further away, on the southeast side of that river, appeared a long chain of high mountains running generally parallel to it, though many miles distant from it. To the west again, at some distance from the rising lands where we were, the hills changed into a long chain of very high mountains, lying very close to each other and running parallel likewise to the river, that is, nearly from south to north. These high mountains consisted of grey rock composed of several kinds of [conglomerate] stones which I shall mention later. These mountains seemed to prove that the lime slate strata were of as ancient a date as the gray rock and not formed in later times; for the amazingly large rocks lay on the top of the mountains, which consisted of black slate.

Grass and Meadows. The high meadows in Canada are excellent and by far preferable to the meadows round Philadelphia and in the other English colonies. The further I advanced northward here the finer were the meadows, and the turf upon them was better and closer. Almost all the grass here is of two kinds, *viz.* a species of the narrow leaved meadow grass (*Poa angustifolia* L.), its spikes containing either three or four flowers which are so exceedingly small that the plant might easily be taken for a bent grass (*Agrostis* L.), and its seeds have several small downy hairs at the bottom. The other plant, which grows in the meadows, is the white clover.[1] These two plants form the hay in the meadows. They stand close and thick together, and the meadow grass (*Poa*) is pretty tall, but has very thin stalks. At the root of the meadow grass the ground is covered with white clover, so that one cannot wish for finer meadows than are found here. Almost all have been formerly tilled fields, as appears from the furrows on the ground which still remain. They can be mown but once every summer, as spring commences very late.

Haymaking. Farmers were now busy making hay and getting it in and I was told they had begun about a week ago. The scythes are like our Swedish ones; the men mow and the women rake. The hay was prepared in much the same way as with us, but the tools are a little different. The head of the rake is smaller, has tines on both sides and is a little heavier. The hay is raked into rows; and they also use a kind of wooden fork for both pitching and raking. In so doing, however, a good deal of the hay is left on the field

---

[1] *Trifolium repens* L.

since this does not rake as clean as an ordinary rake. There were no hillocks on these meadows. The hay is taken away in four-wheeled carts drawn by either horses or oxen. The oxen are hitched in such a way as to pull with their horns instead of their shoulders. Some of the hay barns were out on the fields. They have haystacks near most of their meadows, and on the wet ones they make use of conic haystacks. Their grass lots are usually without fences, the cattle being in the pastures on the other side of the woods and cow-herds take care of them where they are necessary.

The *grain fields* are pretty large. I saw no ditches anywhere, though they seemed to be needed in some places. They are divided into ridges, of the breadth of two or three yards broad, between the shallow furrows. The perpendicular height of the middle of the ridge, from the level to the ground is near one foot. All the grain is summer sown, for as the cold in winter destroys the grain which lies in the ground, it is never sown in autumn. I found white wheat most common in the fields. There are likewise large fields with peas, oats, in some places summer rye, and now and then barley. Near almost every farm I found cabbages, pumpkins, and melons. The fields are not always sown, but lie fallow every two years. The fallow fields not being plowed in summer the weeds grow without restraint in them and the cattle are allowed to roam over them all season.[1]

There was a superabundance of fences around here, since every farm was isolated and the fields divided into small pastures. It will be difficult to obtain material for these fences when the woods are used up; in the future they will probably have to use hedges for their enclosures. It is a stroke of good fortune though, that there is a large amount of cockspur hawthorn growing in the neighborhood. Happy they who will think of it in time! (Here follows a description of a palisade type of fence which it seems unnecessary to reproduce).

The houses in the country are built of stone or wood. The stone

---

[1] Here follows, in the original, an account of the fences made use of near Quebec, which is intended only for the Swedes, but not for a nation that has made such progress in agriculture and husbandry as the English.—F.

This is a rather typical case of the patronizing air of the English translator toward Kalm and Sweden. While some of the matter is superfluous to us, a part has been reintroduced here.—Ed.

houses are not of bricks, as there is not yet any considerable quantity of brick made here. People therefore take what stones they can find in the neighborhood, especially the black slate. This is quite compact when quarried, but shatters when exposed to the air; however, this is of little consequence as the stones stick fast in the wall, and do not fall apart. For want of it they sometimes make their buildings of limestone or sandstone, and sometimes of gray stone. The walls of such houses are commonly two feet thick and seldom thinner. The people here can have lime everywhere in this neighborhood. The greater part of the houses in the country are built of wood, and sometimes plastered over on the outside. The chinks in the walls are filled with clay instead of moss. The houses are seldom above one story high. The windows are always set in the inner part of the wall, never in the outer, unless double windows are used. The panes are set with putty and not lead. In the city glass is used for the windows for the most part, but further inland they use paper. In opening the windows they use hooks as with us. The floors are sometimes of wood and sometimes of clay. The ceiling consists generally of loose boards without any filling, so that much of the internal heat is wasted. In every room is either a chimney or a stove, or both. The stoves have the form of an oblong square; some are entirely of iron, about two feet and a half long, one foot and a half or two feet high, and near a foot and a half broad. These iron stoves are all cast at the ironworks at Trois Rivières. Some are made of bricks or stones, not much larger than the iron stoves, but covered at the top with an iron plate. The smoke from the stoves is conveyed up the chimney by an iron pipe in which there are no dampers, so that a good deal of their heat is lost. In summer the stoves are removed. The roofs are always very steep, either of the Italian type or with gables. They are made of long boards, laid horizontally, the upper overlapping the lower. Wooden shingles are not used since they are too liable to catch fire, for which reason they are forbidden in Quebec. Barns have thatched roofs, very high and steep. The dwelling houses generally have three rooms. The baking oven is built separately outside the house, either of brick or stone, and covered with clay. Brick ovens, however, are rare.

This evening we arrived at Lorette, where we lodged with the Jesuits.

AUGUST THE 13TH

Botanizing. In the morning we continued our journey through the woods to the high mountains, in order to see what scarce plants and curiosities we could get there. The ground was flat at first and covered with a thick wood all round, except in marshy places. Nearly half the plants which are to be found here grow in the woods and morasses in Sweden.

We saw *wild cherry trees* here of two kinds which are probably mere varieties, though they differ in several respects. Both are pretty common in Canada and both have red berries. One kind which is called *cerisier* by the French tastes like our Alpine cherries and their acid contracts the mouth and cheeks. The berries of the other sort have an agreeable sourness and a pleasant taste.

The three-leaved hellebore (*Helleborus trifolius*) grows in great quantities in the woods, and in many places it covers the ground by itself. However, it commonly chooses mossy places that are not very wet, and the wood sorrel (*Oxalis acetosella* L.), with the mountain enchanter's nightshade (*Circæa alpina* L.) are its companions. Its seeds were not yet ripe and most of the stalks had no seeds at all. This plant is called *Tissavoyanne jaune* by the French in Canada. Its leaves and stalks are used by the Indians for giving a fine yellow color to several kinds of work which they make of prepared skins. The French who have learnt this from them, dye wool and other things yellow with this plant.

We climbed with a great deal of difficulty to the top of one of the highest mountains here, and I was vexed to find nothing at its summit but what I had seen in other parts of Canada before. We had not even the pleasure of a view, because the trees with which the mountain is covered obstructed it. The trees that grow here are a kind of hornbeam, or *Carpinus ostrya* L., the American elm, the red maple, the sugar maple, that kind of maple which cures scorched wounds (which I have not yet described), the beech, the common birch tree, the sugar birch (*Betula nigra* L.) the mountain ash, the Canada pine, called *perusse*, the mealy tree with dentated leaves (*Viburnum dentatum* L.), the ash, the cherry tree (*Cerisier*) just before described and the berry-bearing yew (*Taxus baccata*).

The gnats in this wood were more numerous than we could have wished. Their bite caused such swelling of the skin that it was difficult to shave, and the Jesuits at Lorette said the best preservative against their attacks was to rub the face and naked parts of the body with grease. Cold water they reckoned the best remedy against the bite, when the wounded places were washed with it immediately afterward.

At night we returned to Lorette, having accurately examined the plants of note which we found to-day.

AUGUST THE 14TH

Lorette is a village, three French miles to the west of Quebec, inhabited chiefly by Indians of the Huron nation converted to the Roman Catholic religion. The village lies near a little river which tumbles over a rock there with a great noise and turns a sawmill and a flourmill. When the Jesuits who are now with them, arrived among them, they lived in their usual huts, which are made like those of the Laplanders. They have since laid aside this custom and built all their houses after the French fashion. In each house are two rooms, *viz*. their bedroom and the kitchen. In one room is a small oven of stone, covered on top with an iron plate. Their beds are near the wall, and they put no other clothes on them than those which they are dressed in. Their other furniture and utensils look equally wretched. There is a fine little church here, with a steeple and a bell. The steeple is raised pretty high and covered with white tin plates. They pretend that there is some similarity between this church in its shape and plan and the Santa Casa at Loretto in Italy, whence this village has gotten its name. Close to the church is a house built of stone for the clergymen, two Jesuits, who constantly live here. The divine service is as regularly attended here as in any other Roman Catholic church, and it is a pleasure to hear the vocal skill and pleasant voices of the Indians, especially of the women, when singing all sorts of hymns in their own language. The Indians dress chiefly like the other adjacent Indian nations; the men, however, like to wear waistcoats, or jackets, like the French. The women keep exactly to the Indian dress. It is certain that these Indians and their ancestors long ago, on being converted to the Christian religion, made a vow to God never to drink strong liquors. This vow they have kept pretty inviolable hitherto, so that one seldom sees one of them drunk, though brandy and other strong liquors are goods which other Indians prefer to life itself.

The Indians of Lorette. These Indians have made the French their patterns in several things besides the houses. They all plant corn; and some have small fields of wheat and rye. Some of them keep cows. They plant our common sun-flower (*Helianthus annuus*), in their corn fields and mix the seeds of it into their sagamite or corn soup. The corn which they plant here is of the small sort, which ripens sooner than the other. Its kernels are smaller but give more and better flour in proportion. It commonly ripens here at the middle, sometimes however at the end, of August. The mills belong to the Jesuits who get paid for everything they grind.

The Swedish winter wheat and winter rye has been tried in Canada, to see how well it would succeed; for Canadians employ nothing but spring wheat or rye, since it has been found that French wheat and rye die here in winter, if it be sown in autumn. Dr. Sarrazin [1] therefore (as I was told by the eldest of the two Jesuits here) got a small quantity of winter wheat and rye from Sweden. It was sown in autumn, not hurt by the winter, and gave good results. The ears were not so large as those of the Canadian grain, but weighed nearly twice as much, and gave a greater quantity of finer flour, than the summer variety. Nobody could tell me why the experiments have not been continued. They cannot, I am told, bake such white bread here of the summer grain as they can in France of their winter wheat. Many people assured me that all the spring wheat now used here came from Sweden or Norway; for the French, on their arrival, found the winters in Canada too severe for the French winter seed, and their summer variety did not always ripen on account of the shortness of the summer. Therefore they began to look upon Canada as little better than a useless country where nobody could live, till they fell upon the idea of getting their spring grain from the most northern parts of Europe, which has succeeded very well.

To-day I returned to Quebec making botanical observations on the way.

---

[1] See pp. 390, note.

AUGUST THE 15TH

The New Governor Arrives. The new governor-general of all Canada, the Marquis de la Jonquière,[1] arrived last night in the river before Quebec; but it being late he reserved his public entrance for to-day. He had left France on the second of June, but could not reach Quebec before this time on account of the difficulty which great ships find in passing the sandbars in the St. Lawrence River. The ships cannot venture to go up without a fair wind, being forced to sail in many windings and frequently in a very narrow channel. To-day was another great feast on account of the Ascension of the Virgin Mary which is very highly celebrated in Roman Catholic countries. This day was accordingly doubly remarkable both on account of the holiday and of the arrival of the new governor-general, who is always received with great pomp, as he is really a viceroy here.

About eight o'clock the chief people in town assembled at the house of Mr. de Vaudreuil,[2] who had lately been nominated governor of Trois Rivières and lived in the lower town, and whose father had likewise been governor-general of Canada. Thither came likewise the Marquis de la Galissonnière, who had till now been governor-general, and was to sail for France at the first opportunity. He was accompanied by all the people belonging to the government. I was likewise invited to see this festivity. At half an hour after eight the new governor-general went from the ship into a barge covered with red cloth. A signal with cannons was given from the ramparts for all the bells in the town to be set ringing. All the people of distinction went down to the shore to salute the governor who, on alighting from the barge, was received by the Marquis de la Galissonnière. After they had saluted each other, the commandant of the town addressed the new governor in a very

[1] Pierre-Jacques de Taffanel, Marquis de la Jonquière (1685-1752). See long article on him in the *Dictionnaire Générale du Canada,* which quotes Kalm on Jonquière's personal appearance. See *infra,* p. 465.

[2] Pierre de Rigaud, Marquis de Vaudreuil-Cavagnol (1698-1778), last governor of New France (1755-1760), had been appointed governor of the Three Rivers in 1733, and was governor of Louisiana from 1742 to 1755. Obviously he did not spend much time in Louisiana if he still made his home in Trois Rivières. Or was the candidate "lately" nominated for the governorship of Three Rivers one of the other eleven Vaudreuil brothers, of whom Pierre was one? It was the latter who surrendered all Canada to the British in 1760.

elegant speech which he answered very concisely. Then all the cannon on the ramparts gave a general salute. The whole street up to the cathedral was lined with men in arms, chiefly drawn from among the burgesses. The governor then walked towards the cathedral, dressed in a suit of red, with an abundance of gold lace. His servants went before him in green, carrying firearms on their shoulders. On his arrival at the cathedral he was received by the Bishop of Canada and the whole clergy assembled. The bishop was arrayed in his pontifical robes, and had a long gilt tiara on his head, and a great crozier of massy silver in his hand. After the bishop had addressed a short speech to the governor-general, a priest brought a silver crucifix on a long stick (two priests with lighted tapers in their hands, going on each side of it) to be kissed by the governor. The bishop and the priests then went through the long aisle up to the choir. The servants of the governor followed with their hats on, and arms on their shoulders. At last came the governor and his suite and after them a crowd of people. At the beginning of the choir the governor and the General de la Galissonnière, stopped before a chair covered with red cloth and stood there during the whole time of the celebration of the mass, which was celebrated by the bishop himself. From the church he went to the palace where the gentlemen of note in the town afterwards went to pay their respects to him. The members of the different orders with their respective superiors likewise came to him to testify their joy on account of his happy arrival. Among the numbers that came to visit him none stayed to dine but those that were invited beforehand, among which I had the honor to be one. The entertainment lasted very long, and was as elegant as the occasion required.

The governor-general, Marquis de la Jonquière, was very tall, and at that time above sixty years old. He had fought a desperate naval battle with the English in the last war, but had been obliged to surrender, the English being, as it was told, vastly superior in the number of ships and men. On this occasion he was wounded by a ball which entered one side of his shoulder and came out at the other, so that he walked slightly bent over. He was very complaisant but knew how to preserve his dignity when he distributed favors.

Early Refrigeration. Many of the gentlemen present at this en-

tertainment asserted that the following method had been successfully employed to keep wine, beer, or water cool during the summer. The wine or other liquor is bottled; the bottles are well corked,
hung up in the air and wrapped in wet cloth. This cools the wine
in the bottles notwithstanding it was quite warm before. After a
little while the cloths are again made wet, with the coldest water
that is to be had and this continued. The wine, or other liquor, in
the bottles is then always colder than the water with which the
cloths are made wet. And though the bottles should be hung up in
the sunshine, the above way of proceeding will always have the
same effect.[1]

A Catholic Procession. The procession in memory of the Ascension of the Holy Virgin, which was held to-day in Quebec, was in
its way quite magnificent. Believing that She ascended on this
day, the Catholics marched from one church to the other through
the whole city. The people flocked to see the procession, as though
they had never seen it before. It was said that they were always
anxious to behold such sights. First came two boys who were constantly ringing bells, followed by a man carrying a banner with a
painting of Christ on the Cross on one side and one of Mary, Joseph
and the Christ-Child between them on the other. Then came another man bearing a painted wooden image of the Savior on the
Cross. Both the banner and the image were on long poles. Then
followed the Recollets, or Mendicant Friars, dressed in the costume
described above, walking far apart and two abreast; and since they
affected poverty they carried aloft a plain wooden cross. Next,
borne on a pole, part of which was of silver, came a silver image
of the crucified Savior. Then followed several pairs of boys, about
ten or twelve years old, clad in red tunics with white cottas over
them and wearing red, cone-shaped caps. Then came some more

---

[1] It has been observed by several experiments, that any [container of] liquid dipped
into another liquor, and then exposed into the air for evaporation, will get a remarkable degree of cold; the quicker the evaporation succeeds, after repeated dippings,
the greater the cold. Therefore spirit of wine evaporating quicker than water, cools
more than water; and spirit of sal ammoniac, made by quick-lime, being still more
volatile than spirit of wine, its cooling quality is still greater. The evaporation succeeds
better by moving the vessel containing the liquor, by exposing it to the air, and by
blowing upon it, or using a pair of bellows. See de Marian, *Dissertation sur la
Glace*, Prof. Richman in Nov. *Comment. Petrop. ad an.* 1747, & 1748. p. 284, and
Dr. Cullen in the *Edinburgh physical and literary Essays and Observations*, Vol. II.
p. 145.—F.

boys with black tunics and caps, followed by the priests. Some of the latter wore white chasubles, others long silk mantles of various colors, with black, cone-shaped hoods on their heads and bluish bands. A priest swinging a censer followed. Then came two priests bearing a silver image of the Virgin enclosed in a silver-plated shrine, surrounded on all sides by men carrying lanterns with wax-candles, and after came the most distinguished clergymen in their long robes. The bishop in full dress with his official crosier ended the religious group of the procession. After him marched the governor-general's private guard with guns on their shoulders, followed by the governor himself, de la Jonquière, and General de la Galissonnière, walking abreast. Last came a number of distinguished residents and a large group of common people. As the procession passed the castle the soldiers presented arms in salute, the drums were beaten, and the cannon roared from the forts, as was customary on such occasions. Those who stood nearest where the procession passed kneeled as the image of the Holy Virgin was carried by; but at the figure of Christ they remained standing. Onlookers who watched at some distance did not kneel. Thus the procession went on, the priests singing as they marched.[1]

AUGUST THE 16TH

The occidental *Arbor vitæ* (*Thuja occidentalis* L.), is a tree which grows very plentiful in Canada but not much further south. The most southerly place I have seen it in is a place a little to the south of Saratoga in the province of New York, and likewise near Casses [Cassius?] in the same province, which places are in forty-two degrees and ten minutes north latitude. Mr. Bartram, however, informed me that he had found a single tree of this kind in Virginia, near the falls in the James River. Doctor Colden likewise asserted that he had seen it in many places round his seat Coldingham, which lies between New York and Albany, about forty-one degrees thirty minutes north latitude. The French, all over Canada, call it *Cèdre blanc*. The English and Dutch in Albany also call it the white cedar. The English in Virginia have called a *Thuya* which grows with them a juniper. The places and the soil where it grows

---

[1] This passage, like several others, is omitted by Forster and subsequent translators, who did not use the original.

best are not always alike; however, it generally succeeds in ground where its roots have sufficient moisture. It seems to prefer swamps, marshes, and other wet places to all others, and there it grows pretty tall. Stony hills, and places where a number of stones lie together, covered with several kinds of mosses (*Lichen, Bryum, Hypnum*), seemed to be the next in order where it grows. When the seashores are hilly and covered with mossy stones, the *Thuya* seldom fails to grow on them. It is likewise seen now and then on the hills near rivers and other high grounds, which are covered with earth or mould; but it is to be observed that such places commonly have a sourish water in them, or receive moisture from the higher localities. I have however seen it growing in some pretty dry places; but there it never grows to any great size. It is found frequently in the clefts of mountains, but cannot grow to any remarkable height or thickness. The tallest trees I have found in the woods in Canada are from thirty to thirty-six feet high. A barked tree of exactly twelve inches diameter had ninety-two rings round the trunk; another of one foot and four inches in diameter had one hundred and forty-two rings.

Thuya Wood. The Canadians generally use this tree for the following purposes. Since it is thought the most durable wood in Canada, and that which best withstands rotting, so as to remain undamaged for over a man's age, fences of all kinds are scarcely made of any other than this wood. All the posts which are driven into the ground are made of the Thuya wood. The palisades round the forts in Canada are likewise made of the same material. The beams in the houses are made of it; and the thin narrow pieces of wood which form both the ribs and the bottom of the bark boats commonly used here are taken from this wood, because it is pliant enough for the purpose, especially while it is fresh. It is also very light. The Thuya wood is reckoned one of the best for the use in limekilns. Its branches are used all over Canada for brooms, and the twigs and leaves of it being naturally bent together seem to be very proper for the purpose. The Indians make such brooms and bring them to the towns to sell, nor do I remember having seen any made of any other wood. The fresh branches have a peculiar, agreeable scent which is pretty strongly smelled in houses where they make use of brooms of this kind.

Thuya in Medicine. This Thuya is used for several medicinal purposes. The commandant of Fort St. Frédéric, M. de Lusignan, could never sufficiently praise its excellence for rheumatic pains. He told me he had often seen it tried with remarkably good success upon several persons in the following manner. The fresh leaves are pounded in a mortar and mixed with hog's grease or any other grease. This is boiled together till it becomes a salve, which is spread on linen and applied to the part where the pain is. The salve gives certain relief in a short time. Against violent pains which move up and down in the thighs and sometimes spread all over the body, they recommend the following remedy. Take of the leaves of a kind of polypody [1] four-fifths, and of the cones of the Thuya one-fifth, both reduced to a coarse powder by themselves, and mixed together afterwards. Then pour milkwarm water on it so as to make a poultice, spread it on linen, and wrap it round the body: but as the poultice burns like fire, they commonly lay a cloth between it and the body, otherwise it would burn and scorch the skin. I have heard this remedy praised beyond measure, by people who said they had experienced its good effects. Among these was a woman who said that she had applied such a poultice for three days, after which her severe pain passed away entirely. An Iroquois Indian told me that a decoction of Thuya leaves was used as a remedy for a cough. In the neighborhood of Saratoga, they use this decoction in the intermittent fevers.

The Thuya tree keeps its leaves, and is green all winter. Its seeds are ripe towards the end of September, old style. The fourth of October of this year, 1749, some of the cones, especially those which stood much exposed to the heat of the sun, had already dropped their seeds, and all the other cones were opening in order to shed them. This tree has, in common with many other American trees, the quality of growing plentifully in marshes and thick woods, which may with certainty be called its native habitat. However, there is scarcely a single Thuya tree in those places which bears seeds; if, on the other hand, a tree accidentally stands on the outside of a wood, on the seashore, or in a field where the air can freely get to it, it is always full of seeds. I have found this to be the case with the Thuya on innumerable occasions. It is the same with the sugar maple, the maple which is good for healing scorched wounds, the white fir tree, the pine called *pérusse*, the mulberry tree, the sas-

---

[1] *Polypodium fronde pinnaia, pinnis alternis ad basin superne appendiculatis.*—F.

safras and several others. In England this tree is everywhere called *arbor vitæ* by the farmers.

The Ursuline Convent. To-day I went to see the nunnery of the Ursulines, which is located on nearly the same plan as the two other nunneries. It lies in the town and has a very fine church. The nuns are renowned for their piety and they go abroad less than any others. Men are not allowed to go into this convent either, except by the special license of the bishop, which is given as a great favor. The royal physician and the surgeon are alone entitled to go in as often as they please to visit the sick. At the desire of the Marquis de la Galissonière the bishop granted me leave to visit this nunnery together with the royal physician M. Gauthier. On our arrival we were received by the abbess, who was attended by a great number of nuns, for the most part old ones. We saw the church; and it being Sunday, we found some nuns on every side of it kneeling by themselves and saying prayers. As soon as we came into the church, the abbess and the nuns with her dropped on their knees and so did M. Gauthier and myself. We then went to an apartment or small chapel dedicated to the Virgin Mary, at the entrance of which they all fell on their knees again. In several places which we visited we found images and paintings and candles which burned in front of some of these. The nuns explained that these pictures and images were not kept here for worship, because God alone is to be worshipped, but only to arouse piety through them. We afterwards saw the kitchen, the dining hall and the apartment they work in, which is large and fine. They do all sorts of neat work there, gild pictures, make artificial flowers, etc. The dining hall is similar to those in the other two nunneries. Under the tables are small drawers for each nun to keep her napkin, knife and fork, and other things in. Their cells are small, and each nun has one to herself. The walls are not painted; a little bed, a table with a drawer, a crucifix and pictures of saints on it, and a chair, constitute the whole furniture of a cell. We were then led into a room full of young ladies about twelve years old and below that age, sent hither by their parents to be instructed in reading and in matters of religion. They are allowed to go to visit their relations once a day, but must not stay away long. When they have learned read-ing and have received instructions in religion they return to their parents again. Near the nunnery is a fine garden which is sur-rounded with a high wall. It belongs to this institution and is stocked with all sorts of kitchen herbs and fruit trees. When the nuns are at work, or during dinner, everything is silent in the rooms, unless some one of them reads to the others; but after din-ner they have leave to take a walk for an hour or two in the garden, or to divert themselves within doors. After we had seen everything remarkable here we took our leave.

A Mineral Spring. About a quarter of a Swedish mile to the west of Quebec is a well of mineral waters which contains a great deal of iron ochre and has a pretty strong taste. M. Gauthier said that he had prescribed it with success in costive cases and like dis-eases.

Snakes. I have been assured that there are no snakes in the woods and fields round Quebec whose bite is poisonous, so that one can safely walk in the grass. I have never found any that endeavored to bite, and all were frightened. In the southern parts of Canada it is not advisable to be off one's guard.

A very small species of black ant (*Formica nigra* L.) live in ant-hills, in high grounds and in woods; they look exactly like our Swedish ants, but are much smaller.

AUGUST THE 21ST

Indians. To-day there were representatives of three Indian nations in this country with the governor-general, *viz.* Huron, Mickmacks and Anies,[1] the last of which are a nation of Iroquois and allies of the English. They were taken prisoners in the last war.

The Hurons are some of the same Indians as those who live at Lorette, and have received the Christian religion. They are a tall, robust people, well shaped, and of a copper color. They have short black hair which is shaved on the forehead from one ear to the other. None of them wear hats or caps. Some have earrings, others not. Many of them have the face painted all over with cinnabar; others have only strokes of it on the forehead and near the ears; and

---

[1] Probably Oneidas.—F.

some paint their hair with the same material. Red is the color they chiefly use in painting themselves, but I have also seen some who had daubed their face with black. Many of them have figures on the face and on the whole body, which are stained into the skin, so as to be indelible. The manner of making them shall be described later. These figures are commonly black; some have a snake painted on each cheek, some have several crosses, some an arrow, others the sun, or anything else their imagination leads them to. They have such figures likewise on the breast, thighs and other parts of the body; but some have no figures at all. They wear a shirt which is either white or blue striped and a shaggy piece of cloth, which is either blue or white, with a blue or red stripe below. This they always carry over their shoulders, or let it hang down, in which case they wrap it round their middle. Round their neck they have a string of violet wampum, with some white wampum between them. These wampum are small, of the figure of oblong pearls, and made of the shells which the English call clams (*Venus mercenaria* L.) I shall make a more particular mention of them later. At the end of the wampum strings, many of the Indians wear a large French silver coin with the king's effigy on their breasts. Others have a large shell on the breast, of a fine white color, which they value very highly; others again have no ornament at all round the neck. They all have their breasts uncovered. In front hangs their tobacco pouch made of the skin of an animal with the hairy side turned outwards. Their shoes are made of skins, and bear a great resemblance to the heel-less shoes which the women in Finland use. Instead of stockings they wrap the legs in pieces of blue cloth, as I have seen the Russian peasants do.

The Mickmacks are dressed like the Hurons, but distinguish themselves by their long straight hair of jet-black. Almost all the Indians have black straight hair; however, I have met with a few, whose hair was quite curly. But it is to be observed that it is difficult to judge the true complexion of the Canada Indians, their blood being mixed with the European, either by the adopted prisoners of both sexes or by the Frenchmen who travel in the country and often contribute their share towards the increase of the Indian families, to which the women, it is said, have no serious objection. The Mickmacks are commonly not so tall as the Hurons. I have not seen any Indians whose hair is as long and straight as theirs.

Their language is different from that of the Hurons; therefore there is an interpreter here for them on purpose.

The Anies are the third kind of Indians which came hither. Fifty of them went to war, being allies of the English, in order to plunder in the neighborhood of Montreal. But the French, being informed of their scheme, laid an ambush, and killed with the first discharge of their guns forty-four of them so that only the four who were here to-day saved their lives, and two others who were ill at this time. They are as tall as the Hurons, whose language they speak. The Hurons seem to have a longer and the Anies a rounder face. The Anies have something cruel in their looks; but their dress is the same as that of the other Indians. They wear an oblong piece of white tin in the hair which lies on the neck. One of those I saw had taken a flower of the rose mallow, out of a garden where it was in full blossom at this time, and put it in the hair at the top of his head. Each of the Indians has a tobacco pipe of gray limestone which is blackened afterwards and has a stem of wood. There were no Indian women present at this interview. As soon as the governor came in and was seated in order to speak with them the Mickmacks sat down on the ground, like Laplanders, but the other Indians took chairs.

There is no printing press in Canada, nor has there formerly been any: but all books are brought from France and all the orders made in the country are written, which [as I have shown] extends even to the paper currency. They pretend that the press is not yet introduced here because it might be the means of propagating libels against the government and religion. But the true reason seems to lie in the poverty of the country, as no printer could make a sufficient number of books for his subsistence; and another reason may be that France now has the profit arising from the exportation of books hither.

Food. The meals here are in many respects different from those in the English provinces. This depends upon the difference of custom, taste, and religion, between the two nations. French Canadians eat three meals a day, *viz.* breakfast, dinner and supper. They breakfast commonly between seven and eight, for the French here rise very early, and the governor-general can be seen at seven o'clock, the time when he has his levee. Some of the men dip a piece of bread in brandy and eat it; others take a dram of brandy and eat a piece of bread after it. Chocolate is likewise very common for breakfast, and many of the ladies drink coffee. Some eat no breakfast at all. I have never seen tea used here, perhaps because they can get coffee and chocolate from the French provinces in America, in the southern part, but must get tea from China. They consider it is not worth their while to send the money out of their country for it. I never saw them have bread and butter for breakfast. Dinner is exactly at noon. People of quality have a great many dishes and the rest follow their example, when they invite strangers. The loaves are oval and baked of wheat flour. For each person they put a plate, napkin, spoon and fork. (In the English colonies a napkin is seldom or never used). Sometimes they also provide knives, but they are generally omitted, all the ladies and gentlemen being provided with their own knives. The spoons and forks are of silver, and the plates of Delft ware. The meal begins with a soup with a good deal of bread in it. Then follow fresh meats of various kinds, boiled and roasted, poultry, or game, fricassees, ragouts, etc. of several sorts, together with different kinds of salads. They commonly drink red claret at dinner, either mixed with water or clear; and spruce beer is likewise much in use. The ladies drink water and sometimes wine. Each one has his own glass and can drink as much as he wishes, for the bottles are put on the table. Butter is seldom served, and if it is, it is chiefly for the guest present who likes it. But it is so fresh that one has to salt it at the table. The salt is white and finely powdered, though now and then a gray salt is used. After the main course is finished the table is always cleared. Finally the fruit and sweetmeats are served, which are of many different kinds, *viz.* walnuts from France or Canada, either ripe or pickled; almonds; raisins; hazel-nuts; several kinds of berries which are ripe in the summer season, such as currants, red and black, and cranberries which are preserved in treacle; many preserves in sugar, as strawberries, raspberries, blackberries, and mossberries. Cheese is likewise a part of the dessert, and so is milk, which they drink last of all, with sugar. Friday and Saturday, the "lean" days, they eat no meat according to the Roman Catholic rites; but they well know how to guard against hunger. On those days they boil all sorts of vegetables like peas, beans and cabbage, and fruit, fish, eggs, and milk are prepared in various ways. They cut cucumbers into slices and eat them with cream, which is a very good dish. Sometimes they put whole cucumbers on the table and everybody that likes them takes one, peels and slices it, and dips the slices into salt, eating them like radishes. Melons abound here and are always eaten without sugar. In brief, they live here just as well on Fridays and Saturdays, and I who am not a particular lover of meats would willingly have had all the days so-called lean days. There is always salt and pepper on the table. They never put any sugar into wine or brandy, and upon the whole they and the English do not use half so much sugar as we do in Sweden, though both nations have large sugar plantations in their West Indian possessions. They say no grace before or after their meals, but only cross themselves, a custom which is likewise omitted by some. Immediately after dinner they drink coffee without cream. Supper is commonly at seven o'clock, or between seven and eight at night, and the dishes the same as at dinner. Pudding is not seen here and neither is punch, the favorite drink of the Englishmen, though the Canadians know what it is.

August the 23rd

The "Poor Man's Horse". In many places hereabouts they use their dogs to fetch water out of the river. I saw two great dogs to-day attached to a little cart, one before the other. They had neat harnesses, like horses, and bits in their mouths. In the cart was a barrel. The dogs were directed by a boy, who ran behind the cart, and as soon as they came to the river they jumped in of their own accord. When the barrel was filled, the dogs drew their burden up the hill again to their house. I frequently saw dogs employed in this manner, during my stay at Quebec. Sometimes they put but one dog before the watercart, which was made small on purpose. The dogs were not very large, hardly of the size of our common farmer's dogs. The boys that attended them had great whips with which they urged them on occasionally. I have seen them fetch not only water but also wood and other things. In winter it is customary in Canada, for travellers to put dogs before little sledges, made on purpose to hold their clothes, provisions, etc. Poor people commonly employ them on their winter journeys, and go on foot themselves.

Almost all the wood, which the poorer people in this country fetch out of the woods in winter, is drawn by dogs, which have therefore got the name of "the poor man's horse". They commonly place a pair of dogs before each load of wood. I have likewise seen some neat little sledges for ladies to ride in, in winter. They are drawn by a span of dogs, and go faster on a good road than one would think. A middle-sized dog is sufficient to draw a single person when the roads are good. I have been told by old people that horses were very scarce here in their youth and almost all the transportation was then effected by dogs. Several Frenchmen who have been among the Esquimaux on Terra Labrador have assured me that they not only make use of dogs for drawing drays, with their provisions and other necessaries, but are likewise employed for pulling the men themselves in little sledges.

AUGUST THE 25TH

The high hills to the west of the town abound with springs. These hills consist of the black slate, before mentioned, and are pretty steep, so that it is difficult to get to the top. Their perpendicular height is about twenty or four and twenty yards. Their summits are destitute of trees and covered with a thin crust of earth, lying on the lime slate, and are tilled or used for pastures. It seems inconceivable therefore that these naked hills can have so many running springs, which in some places gush out like torrents. Have these hills the quality of attracting the water out of the air in the day time or at night? Or are the lime slates more apt to do it than others?

Domestic Animals. All the horses in Canada are strong, well-built, swift, as tall as the horses of our cavalry, and of a breed imported from France. The inhabitants have the custom of docking the tails of their horses, which is a rather severe treatment, as they cannot defend themselves against the numerous swarms of gnats, gadflies, and horseflies. They put the horses in tandem before their carts, which has probably occasioned the docking of their tails, as the horses would hurt the eyes of those behind them, by moving their tails to and fro. The governor-general and a few of the chief people in town have coaches; the rest make use of open two-wheeled carts. It is a general complaint that the country people are beginning to keep too many horses, by which means the cows are kept short of food in winter.

The cows have likewise been imported from France, and are of the size of our common Swedish cows. Everybody agreed that the cattle, which are born of the original French breed, never grow to the same size as the parent stock. This they ascribe to the cold winters, during which they are obliged to put their cattle into stables and give them but little food. Almost all the cows have horns; a few, however, I have seen without them. A cow without horns would be reckoned an unheard-of curiosity in Pennsylvania. Is not this to be attributed to the cold? The cows give as much milk here as in France. The beef and veal at Quebec is reckoned fatter and more palatable than at Montreal. Some look upon the salty pastures below Quebec as the cause of this difference. But this does not seem sufficient, for most of the cattle which are sold at Quebec have no meadows with arrow-headed grass (*Triglochin*) on which they graze. In Canada the oxen draw with the horns, but in the English colonies they draw with their withers, as horses do. The cows vary in color; however, most of them are either red or black.

Every countryman commonly keeps a few sheep which supply him with as much wool as he needs to clothe himself with. The better sort of clothes are brought from France. The sheep degenerate here after they are imported from France, their wool becoming coarser, and their progeny is still poorer. The want of food in winter is said to cause this degeneration.

I have not seen any goats in Canada, and I have been assured that there are none. I have seen but very few in the English colonies, and only in their towns where they are kept on account of some sick people who drink the milk by the advice of their physicians.

The harrows are triangular; two of the sides are six feet and the third four feet long. The teeth and every other part of the harrows are of wood. The teeth are about five inches long and about as far apart.

The view of the country about a quarter of a Swedish mile north of Quebec on the west side of the St. Lawrence River is very fine. The country is very steep towards the river, and grows higher as you go further from the water. In many places it is naturally divided into terraces. From the heights one can see a great distance; Quebec appears very flat to the south and the St. Lawrence River is to the east, on which vessels sail up and down. To the west are high mountains where the slope to the river begins. All the country is cultivated and laid out in grain fields, meadows, and pastures; most of the fields are sown with wheat, many with white oats, and some with peas. Several fine houses and farms are interspersed all over the country and none are ever together. The dwelling-house is commonly built of black slate and generally whitewashed on the outside. Many rivulets and brooks flow down from the heights below the mountains, the latter consisting entirely of the black slate that shatters into pieces when exposed to the air. On the slate lies earth two or three feet in depth. The soil in the grain fields is always mixed with little pieces of the slate. All the rivulets cut their beds deep into the ground so that their shores are commonly of limestone. A dark gray variety is sometimes found among the strata, which, when broken, smells like stinkstone.

Shipbuilding. They were now building several ships below Quebec, at the king's command. However, before my departure an order arrived from France prohibiting the further building of ships of war, except those which were already on the stocks; because they have found that the ships built of American oak did not last so long as those of the European variety. Near Quebec is found very little oak, and what grows there is not fit for use, being very small; therefore they are obliged to fetch their oak timber from those parts of Canada which border upon New England. But all the North American oaks have the quality of lasting longer and withstanding rot better the further north they grow, and *vice versa.* The timber from the confines of New England is brought in floats or rafts on the rivers near those parts and near St. Pierre which empty into the great St. Lawrence River. Some oak is likewise brought from the country between Montreal and Fort St. Frédéric, or Fort Champlain; but it is not held so good as the first, and the place it comes from is further distant.

August the 26th

To-day they showed some green earth which had been brought to the general, Marquis de la Galissonnière, from the upper parts of

Canada. It was a clay which cohered very fast together, and was of a green color throughout like verdi-gris.[1]

All the brooks in Canada contain *crawfish,* the same kind as ours. The French are fond of eating them and say they have vastly decreased in number since they began to catch them.

The common people in the country seem to be very poor. They have the necessaries of life but little else. They are content with meals of dry bread and water, bringing all other provisions, such as butter, cheese, meat, poultry, eggs, etc. to town, in order to get money for them, for which they buy clothes and brandy for themselves and finery for their women. Notwithstanding their poverty, they are always cheerful and in high spirits.

AUGUST THE 29TH

Down the St. Lawrence. By the desire of the governor-general, Marquis de la Jonquière, and of Marquis de la Galissonnière, I set out with some French gentlemen, to visit the so-called silver mine, or the lead mine, near the Bay of St. Paul. I was glad to undertake this journey, as it gave me an opportunity of seeing a much greater part of the country than I should otherwise have done. This morning therefore we set out on our tour in a boat and went down the St. Lawrence River.

The harvest was now at hand, and I saw all the people at work in the grain fields. They had begun to reap wheat and oats a week ago.

The view near Quebec is very beautiful from the river. The town lies very high and all the churches and other buildings appear very conspicuous. The ships in the river below ornament the landscape on that side. The powder magazine, which stands at the summit of the mountain, on which the town is built, towers above all the other buildings.

The country we passed by afforded a no less charming sight. The St. Lawrence River flows nearly from south to north here; on both sides of it are cultivated fields, but more on the west than on the east side. The hills on both shores are steep and high. A number of fine elevations separated from each other, large fields which

---

[1] It was probably impregnated with particles of copper ore.—F.

looked quite white with the grain that covered them, and excellent woods of deciduous trees, made the country round us look very pleasant. Now and then we saw a stone church, and in several places brooks fell from the hills into the river. Where the brooks are large enough the people have erected sawmills and grist mills.

After rowing for the space of a French mile and a half we came to the Isle of Orleans, which is a large island, near seven French miles and a half long, and almost two of those miles broad, in the widest part. It lies in the middle of the St. Lawrence River, is very high, and has steep and very woody shores. There are some places without trees, which have farmhouses below, quite close to the shore. The Isle itself is well cultivated and nothing but fine houses of stone, large grain fields, meadows, pastures, woods of hardwood trees, and some churches built of stone, are to be seen on it.

We passed into that branch of the river which flows on the west side of the Isle of Orleans, it being the shortest. It is reckoned about a quarter of a French mile broad, but ships cannot take this route on account of the sandbars which lie here near the projecting points of land, the shallowness of the water and the rocks and stone at the bottom. The shores on both sides still kept the same appearance as before. On the west side, or on the mainland, the hills near the river consist throughout of black slate, and the houses of the peasants are made of this kind of stone, whitewashed on the outside. Some few houses are of a different kind of stone. The chain of high, large mountains which is on the west side of the river, and runs nearly from south to north, gradually comes nearer to the river; for at Quebec they are nearly two French miles distant from the shore, but nine French miles lower down the river they are close to the shore. These mountains are generally covered with woods, but in some places the woods have been destroyed by accidental fires. About seven and a half French miles from Quebec, on the west side of the river, is a church, called St. Anne, close to the shore. This church is remarkable because the ships from France and other parts, as soon as they have gotten far enough up the St. Lawrence River to get sight of it, give a general discharge of their artillery as a sign of joy that they have passed all danger in the river, and have escaped all the sandbars in it.

The water has a pale red color and was very dirty in those parts of the river which we saw to-day, though it was everywhere computed above six fathoms deep. Somewhat below St. Anne, on the west side of the St. Lawrence River, another river, called la Grande Rivière, or the Great River, empties into it. Its water flows with such violence as to make its way almost into the middle of the branch of the St. Lawrence River which runs between the shore and the Isle of Orleans.

About two o'clock in the afternoon the tide began to flow up the river, and the wind being likewise against us we could not proceed any farther till the tide began to ebb. We therefore took up our night lodgings at a great farm belonging to the priests in Quebec, near which is a fine church called St. Joachim, after a voyage of about eight French miles. We were exceedingly well received here. The king has given all the country round about this place to the Seminary, or the priests at Quebec, who have leased it to farmers. They have built houses on it. Here are two priests, and a number of young boys, whom they instruct in reading, writing, and Latin. Most of these boys are designed for the priesthood. Directly opposite this farm to the east is the northernmost point, or the extremity of the Isle of Orleans.

All the gardens in Canada abound with red currant bushes, which were at first brought over from Europe. Everywhere they were red with berries.

The wild grapevines (*Vitis labrusca & vulpina*) grow quite plentifully in the woods. In all other parts of Canada they plant them in the gardens, near arbors and summer houses. The latter are made entirely of laths, over which the vines climb with their tendrils, and cover them entirely with their foliage so as to shelter them entirely from the heat of the sun. They are very refreshing and cool in summer. The wheat was never cut here with a scythe but with a sickle which was larger than usual. These sickles were filed full of grooves on the lower side. The grooves were so deep that the edge too was full of them. There were two kinds of haystacks here on the meadows, one the ordinary cone-shaped, the other like our contrivances for drying peas. I saw no hay sheds out in the fields in this part of Canada.

The strong contrary winds obliged us to lie all night at St. Joachim.

AUGUST THE 30TH

This morning we continued our journey in spite of the wind, which was very violent against us. The water in the river begins to get a brackish taste when the tide is highest, somewhat below St. Joachim, and the further one goes down, the more the saline taste increases of course. At first the western shore of the river has fine but low grain fields, but soon after the high mountains run close to the riverside. Before they come to the river the hilly shores consist of black slate, but as soon as the high mountains appear on the river side the slate disappears. There the stone, of which the high mountains consist, is a chalky rock, mixed with mica and quartz (*Saxum micaceo quarzoso-calcarium*). The mica is black; the quartz partly violet, and partly gray. All the four constituent parts are so well mixed together that they can not be easily separated by an instrument, though plainly distinguishable with the eye. During our journey today the breadth of the river was generally three French miles. They showed me the channels where the ships are obliged to go, which seemed to be very difficult, as the vessels are obliged to bear away from either shore, as occasion requires, or as the rocks and sandbars in the river oblige them to do. We were often obliged on this journey to experience a kind of optical illusion with respect to distances. Often when a high mountain seemed but a short distance away we found that it was several miles distant. There was the same effect upon the land as upon the sea in judging the distance to an island. [This was probably due to the clear atmosphere.—Ed.]

For the distance of five French miles we had a very dangerous passage to go through, for the whole western shore, along which we rowed, consisted of very high and steep mountains, where we could not have found a single place to land with safety for the distance of five miles, in case a high wind had arisen. There are indeed two or three openings, or holes in the mountains, into which one could have drawn the boat in the greatest danger. But they are so narrow that in case the boat could not find them in a hurry, it would inevitably be dashed against the rocks. These high mountains are either quite bare, or covered with some small firs, standing far apart. In some places there are great clefts on the mountain slopes, in which trees grow very close together and are taller than on the other parts of the mountain; so that those places looked like hedges planted on the solid rock. A little while afterward we passed a small church and some farms round it. The place is called Petite Rivière, and they say its inhabitants are very poor, which seems very probable. They have no more land to cultivate than what lies between the mountains and the river, which in the widest part is not above three musketshots, and in most parts but one broad. About seventeen French miles from Quebec the water is so salty in the river that no one can drink it, our rowers therefore provided themselves with a kettle full of fresh spring water this morning. About five o'clock in the evening we arrived at St. Paul's Bay and took our lodgings with the priests, who have a fine large house here and who entertained us very hospitably.

Bay St. Paul is a small parish, about thirteen French miles below Quebec, lying some distance from the shore of a bay formed by the river, on a low plain. It is surrounded by high mountains on every side, one large gap excepted which is over near the river. All the farms are some distance from each other. The church is reckoned one of the most ancient in Canada, which seems to be confirmed by its bad architecture and want of ornaments, for the walls are formed of timber, erected perpendicularly about two feet from each other, supporting the roof. Between these pieces of timber, they have made the walls of the church of black slate. The roof is flat. The church has no steeple, but a bell fixed above the roof, in the open air. Almost all the country in this neighborhood belongs to the priests, who have leased it to the farmers. The inhabitants live chiefly upon agriculture and the making of tar, which is sold at Quebec.

Since this country is low and situated upon a bay of the river, it may be conjectured that this flat ground was formerly part of the bottom of the river, and was formed either by a decrease of water in the river or by an increase of earth, which was carried upon it from the land by the brooks or thrown on it by storms. A great part of the plants which are to be found here are likewise marine, such as glasswort, sea milkwort, and seaside peas (*Salicornia, Glaux, Pisum maritimum*). But when I asked the inhabitants whether they found shells in the ground by digging for wells, they always answered in the negative. I received the same answer from those who live in the low fields directly north of Quebec, and all agreed that they never found anything by digging except different kinds of earth and sand.

It is to be noted that there is generally a different wind in the bay from that in the river, which arises from the high mountains covered with tall woods that surround it on every side but one. For example, when the wind comes from the river, it strikes against one of the mountains at the entrance of the bay, it is reflected, and consequently takes a direction quite different from what it had before.

I found sand of three kinds upon the shore: one is a clear coarse sand, consisting of angulated grains of quartz, and is very common on the shore; another is a fine black sand, which I have likewise found in abundance on the shores of Lake Champlain [1] and which is common all over Canada. Almost every grain of it is attracted by the magnet. Besides this, there is a garnet-colored sand [2] which is likewise very fine. This may owe its origin to the garnet-colored grains of sand which are to be found in all the stones and mountains here near the shore. The sand may have arisen from the crumbled pieces of some stones, or the stones may have been composed of it. I have found both this and the black sand on the shores in several parts of this journey, but the black sand was always the most plentiful.

AUGUST THE 31ST

All the high hills in the neighborhood sent up a smoke this morning as from a charcoal kiln.

Gnats are innumerable here, and as soon as a person comes out of doors they immediately attack him; and they are still worse in the woods. They are exactly the same gnats as our common Swedish ones, being only somewhat smaller than the North American kind. Near Fort St. Jean, I have also seen gnats which were the same as ours, but they were somewhat bigger, almost of the size of our craneflies (*Tipula hortorum* L.). Those which abound here are immeasurably blood-thirsty. However, I comforted myself with the thought that the time of their disappearance was near at hand.

[1] See page 385.
[2] See page 386.

En Route to the Mines. This afternoon we went still lower down the St. Lawrence River, to a place where, we were told, there were silver or lead mines. Somewhat below Bay St. Paul we passed a neck of land which consists entirely of a gray, rather compact limestone, lying in tipping and almost perpendicular strata. It seems to be merely a variety of the slate. The strata incline to the southeast and basset out to the southwest. The thickness of each is from ten to fifteen inches. When the stone is broken it has a strong smell, like stinkstone. We kept, as before, to the western shore of the river, which consists of nothing but steep mountains and rocks. The river is not more than three French miles broad here. Now and then we could see stripes in the rock of a fine white, loose, semi-opaque spar. In some places of the river are boulders as big as houses, which have rolled down from the mountains in spring. The places they formerly occupied are plainly to be seen.

In several places they have eel traps in the river, like those I have before described.[1]

Algonquin Words. By way of amusement I wrote down a few Algonquin words which I learned from a Jesuit who has been a long time among the Algonquins. They call water, *nypi*; *mukuman*, knife; the head, *ustigon*; the heart, *uthä*; the body, *wihas*; the foot, *ushita*; a little boat, *ush*; a ship, *nabiḳoän*; fire, *sḳute*; hay, *masḳusu*; the hare, *whabus*; (they have a verb which expresses the action of hunting hare, derived from the noun); the marten, *whabistania*; the elk, *musu*[2] (but so that the final u is hardly pronounced); the reindeer, *atticḳu*; the mouse, *manitulsis*; beaver, *amisḳu*. The Jesuit who told me those particulars, likewise informed me that he had great reason to believe that if any Indians here owed their origin to Tartary, he thought the Algonquins certainly did; for their language is universally spoken in that part of North America which lies far to the west of Canada, towards Asia.

---

[1] See page 423.

[2] The famous "moose deer" is accordingly nothing but an elk; for no one can deny the derivation of moose deer from *moosu*. Considering especially that before the Iroquois or the Five Nations grew to that power, which they at present have all over North America, the Algonquins were then the leading nation among the Indians and their language was of course then a most universal language over the greater part of North America; and though they have been very nearly destroyed by the Iroquois, their language is still more universal in Canada than any of the rest.—F.

It is said to be a very rich language in vocabulary; as, for example, the verb "to go upon the ice", is entirely different in the Algonquin from "to go upon dry land", "to go upon the mountains", etc.

Late last night we arrived at *Terre d'Eboulement,* which is twenty-two French miles from Quebec, and the last cultivated place on the western shore of the St. Lawrence River. The country lower down is said to be so mountainous that nobody can live in it, there not being a single spot of ground which can be tilled. A little church belonging to this place stands on the shore near the water.

No walnut trees grow near this village, nor are there any varieties of them further north of this place. At Bay St. Paul, there are two or three walnut trees of that species which the English call butternut trees; but they are looked upon as great rarities, and there are no others in the neighborhood. Oaks of any kind will not grow near this place, either lower down or further north.

Wheat is the kind of grain which is sown in the greatest quantities here. The soil is pretty fertile, and they have sometimes gotten twenty-four or twenty-six bushels from one, though the harvest is generally ten or twelvefold. The bread here is whiter than anywhere else in Canada. The Canadians sow plenty of oats, and it succeeds better than wheat. They sow likewise a great quantity of peas, which yield a larger crop than any cereal and there are examples of them producing a hundredfold.

Here are but few birds, and those that pass the summer here migrate in autumn, so that there are no other birds than snowbirds, red partridges and ravens in winter. Even crows do not venture to expose themselves to the rigors of winter but take flight in autumn.

The bullfrogs live in the pools of this neighborhood. Fireflies are likewise to be found here.

Instead of candles they use lamps in country places, in which they burn the train-oil of porpoises, a common oil here. Where they have none of it, they use instead the train-oil of seals.

September the 1st

There was a woman with child in this village who was now in the fifty-ninth year of her age. She had not had her catamenia for eighteen years. In the year 1748 she got the small pox and now she was very large. She said she was very well and could feel the motions of the foetus. She looked healthy and her husband was living. Since her case was such an uncommon one she was brought to the royal physician, M. Gauthier, so that he might thoroughly observe her condition. He had accompanied us on this journey.

At half an hour after seven this morning we went down the river. The country near Terre d'Eboulement is high and consists of hills of a loose earth, which lies in three or four terraces above each other, and are all well cultivated and mostly turned into grain fields, though there are likewise meadows and pastures.

The great earthquake which happened in Canada, in February, 1663, and which is mentioned by Charlevoix,[1] had done considerable damage to this place. Many hills had tumbled down, and a great part of the grain fields on the lowest hills had been destroyed. They showed me several little islands which arose in the river on this occasion.

There are pieces of black limeslate scattered on those hills which consist of earth. For the space of eight French miles along the side of the river there is not a piece of slate to be seen; but instead of it there are high gray mountains, consisting of a rock which contains a purple and a crystalline quartz mixed with limestone and black mica. The face of these mountains go into the water. We now began to see the slate again.

The river is here computed to be about four French miles broad.

On the sides of the river, about two French miles inland, there are such terraces of earth as at Terre d'Eboulement; but soon after they are succeeded by high disagreeable-looking mountains.

Several brooks flow with a great noise into the river here over the steep shores. These are sometimes several yards high and consist either of earth or of rock.

A Mineral Spring. One of these brooks which flows over a hill of limestone contains a mineral water. It has a strong smell of sulphur, is very clear, and does not change its color when mixed with gall-apples. If it is poured into a silver cup, it looks as if the cup were gilt; and the water leaves a sediment of a crimson color at the bottom. The stones and pieces of wood which lie in the water are covered with a mud which is pale gray at the top and black at the bottom of the stone. This mud has not much pungency, but

---

[1] See his *Histoire de la Nouvelle France,* tom. II. p. m. 125.—F.

tastes like oil of tobacco. My hands had a sulphurous smell all day because I had handled some of the wet stones.

The slate now abounds again near the level of the water. It lies in strata, which are placed almost perpendicularly near each other, inclining a little towards W. S. W. Each stratum is between ten and fifteen inches thick. Most of them are split into thin slabs where they are exposed to the air, but in the inside, whither neither sun, air nor water can penetrate, they are close and compact. Some of these stones are not quite black but have a grayish cast.

About noon we arrived at *Cap aux Oyes*, or Geese Cape, which has probably gotten its name from the number of wild geese which the French found near it on their first arrival in Canada. At present we saw neither geese nor any kind of birds here, a single raven excepted. Here we were to examine the renowned metallic veins in the mountain; but found nothing more than small veins of a fine white spar, containing a few specks of lead ore. Cap aux Oyes is computed to be from twenty-two to twenty-five French miles distant from Quebec. I was much pleased in finding that most of the plants are the same as those which grow in Sweden; some of which are:

The sand reed (*Arundo arenaria* L.) which grows in abundance in the sand and prevents its being blown about by the wind.

The sea lyme grass (*Elymus arenarius* L.) likewise abounds on the shores. Both it and the preceding plant are called *Seigle de mer* (sea rye) by the French. I have been assured that these plants grow in great plenty in Newfoundland and on other North American shores, the places covered with them looking, at a distance, like grain fields, which might explain the passage in our northern accounts of the excellent Vinland,[1] which mentions that they had found whole fields of wheat growing wild.

The seaside plantain (*Plantago maritima* L.) is found very frequently on the shore. The French boil its leaves in a broth on their sea voyages or eat them as salad. It may likewise be pickled like samphire.

The bear berries (*Arbutus uva ursi* L.) grow in great abundance

---

[1] *Vinland det goda,* or the good wine-land, is the name which the old Scandinavian navigators gave to America, which they discovered long before Columbus. See Torfæi *Historia Vinlandiæ antiquae S. partis Americæ Septentrionalis,* Hafniæ 1715, 4to. and Mr. George Westmann's A. M. Dissertation on that Subject. Åbo, 1747.—F.

Westmann's thesis was not printed until 1757—Ed. see p. 776, item 35.

here. The Indians, French, English and Dutch, in those parts of North America which I have seen, call them *Sagáckhomi*, and mix the leaves with tobacco for their use. Even the children use only the Indian name for these berries.

Gale, or sweet willow (*Myrica gale* L.) is likewise abundant here. The French call it *laurier,* and some *poivrier*. They put the leaves into their broth to give it a pleasant taste.

The sea rocket (*Bunias cakile* L.) is, likewise, not uncommon. Its root is pounded, mixed with flour, and eaten here when there is a scarcity of bread.

The sorb tree, or mountain ash, the cranberry bush (*lingon*), the juniper tree, the seaside peas, the *Linnæa,* and many other Swedish plants, are likewise to be found here.

We returned to Bay St. Paul to-day. A gray seal swam behind the boat for some time, but was not near enough to be shot.

SEPTEMBER THE 2ND

Lead Ore. This morning we went to see the silver or lead veins. They lie a little on the south side of the mills, belonging to the priests. The mountain in which the veins lie has the same constituent parts as the other high gray rocks in this place, *viz.* a rock composed of a whitish or pale gray limestone, a purple or almost garnet-colored quartz, and black mica. The limestone is found in greater quantities here than the other parts, and it is so fine that the grain is hardly visible. It effervesces very strongly with *aqua fortis* (nitric acid). The purple or garnet-colored quartz is next in quantity; it lies scattered in exceedingly small grains, and strikes fire when struck with steel. The little black particles of mica follow next; and last of all, the transparent crystalline specks of quartz. There are some small grains of spar in the limestone. All the different kinds of stone are very well mixed together, except that the mica now and then forms little veins and lines. The stone is very hard, but when exposed to sunshine and the open air it changes so much as to look quite rotten, and becomes brittle; in that case its constituent particles grow quite indistinguishable. The mountain is full of perpendicular clefts in which the veins of lead ore run from E. S. E. to W. N. W. It seems the mountain had formerly gotten cracks here which afterwards filled up with a kind of stone in which the lead ore was generated. That stone which contains the lead ore is a soft, white and often semidiaphanous spar, which is very easily malleable. In it there are sometimes stripes of a snowy white limestone and almost always veins of a green kind of stone like quartz. This spar has many cracks, and divides into such pieces as quartz, but is much softer, never strikes fire with steel, does not effervesce with acids, and is not smooth to the touch. It appears to be a variety of Wallerius's [1] vitrescent spar.[2] There are sometimes small pieces of a grayish quartz in this spar, which emit strong sparks of fire when struck with steel. In these kinds of stone the lead ore is lodged. It commonly lies in little lumps of the size of peas, but sometimes in specks of an inch square, or bigger. The ore is very clear, and lies in little cubes.[3] It is generally very poor, a few places excepted. The veins of soft spar and other kinds of stone are very narrow, and commonly from ten to fifteen inches broad. In a few places they are twenty inches broad, and in one single place twenty-two and a half. The brook which intersects the mountain towards the mills, runs down so deep into the mountain that the distance from the edge to the bottom of the brook is nearly twelve yards. Here I examined the veins and found that they always kept the same breadth, not increasing near the bottom of the brook, and likewise that they were no richer below than at the top. Hence it may easily be concluded that it is not worth while sinking mines here. Of these veins there are three or four in this neighborhood at some distance from each other, but all of the same quality. The veins are almost perpendicular, sometimes deviating a little. When pieces of the green stone before mentioned lie in the running water, a great deal of the adherent white spar and limestone is consumed; but the green stone remains untouched. That part of the veins which is turned towards the air and rain has mouldered a great part of the spar and limestone, but the green stone has resisted their attacks. Sometimes deep holes are found in these veins which are filled with mountain crystals. The greatest quantity of lead or silver ore is to be found next to the rock. There

---

[1] Johan Gottschalk Wallerius (1709-1785), Swedish geologist. The Swedish edition of his *Mineralogia* had appeared in 1747.

[2] See Wallerius's *Mineralogy*, Germ. ed. p. 87. Forst. *Introd. to Mineralogy*, p. 13.—F.

[3] It is a cubic lead ore, or galena. Forster's *Introd. to Min.* p. 51.—F.

are now and then little grains of pyrites in the spar, which have a fine gold color. The green stone when pulverized and put on a red-hot shovel burns with a blue flame. Some say they can then observe a sulphurous smell, which I could never perceive, though my sense of smell is perfect. When this green stone has grown quite red-hot, it loses its green color and acquires a whitish one, but will not effervesce with *aqua fortis*.

The sulphurous springs (if I may so call them) are at the foot of the mountain which contains the silver or lead ore. Several springs join here and form a little brook. The water in those brooks is covered with a white scum and leaves a white, mealy matter on the trees and other bodies in its way; this matter has a strong sulphurous smell. Trees covered with this mealy matter, when dried and set on fire, burn with a blue flame and emit a smell of sulphur. The water does not change by being mixed with gall-apples, nor does it change blue paper into a different color. It makes no good lather with soap. Silver is tarnished and turns black if kept in this water for a little while. The blade of a knife was turned quite black after it had lain about three hours in it. It has a disagreeable smell which, they say, it spreads still more in rainy weather. A number of grasshoppers had fallen into it at present. The inhabitants used this water as a remedy against the itch.

Silver Ore. In the afternoon we went to see another vein which had been spoken of as silver ore. It lay about a quarter of a mile to the northeast of St. Paul's Bay, near a point of land called Cap au Corbeau, close to the shore of the St. Lawrence River. The mountain in which these veins lay consisted of a pale red vitrescent spar, a black mica, a pale limestone, purple or garnet-colored grains of quartz, and some transparent quartz. Sometimes the reddish vitrescent spar is the most abundant and lies in long stripes of small hard grains. Sometimes the fine black mica abounds more than the remaining constituent parts; and these two last kinds of stone generally run in alternate stripes. The white limestone, which consists of almost invisible particles, is mixed in among them. The garnet-colored quartz grains appear here and there, and sometimes form whole stripes. They are as big as pin's heads, round, shining, and strike fire with steel. All these stones are very hard, and the mountains near the sea are composed entirely of them. They sometimes lie in almost perpendicular strata of ten or fifteen inches thickness.

The strata, however, point with their upper ends to the northwest, and go upwards from the river, as if the water, which is close to the southeast side of the muontains, had forced the strata to lean toward that side. These mountains contain very narrow veins of a white and sometimes of a greenish, fine semidiaphanous, soft spar, which crumbles easily into grains. In this spar they very frequently find specks, which look like a calamine blend.[1] Now and then, but very seldom, there is a grain of lead ore. The mountains near the shore consist sometimes of a black fine-grained hornstone and a ferruginous limestone. The hornstone in that case is always present in three or four times as great a quantity as the limestone.

In this neighborhood there is likewise a *sulphurous spring*, having exactly the same qualities as that which I have before described. The broad-leaved reed mace (*Typha latifolia* L.) grows right in the spring, and succeeds extremely well. A mountain ash stood near it whose berries were of a pale yellow faded color, whereas on all other mountain ashes they have a deep red color.

Great quantities of tar are made at Bay St. Paul. We now passed near a place in which they burn tar during summer. It is exactly the same as ours in Österbotten, Finland, only somewhat smaller in size, though I have been told that there are sometimes very great manufactures of it here. The tar is made solely from *Pin rouge* or red pine. All other firs, of which there are several kinds here, are not fit for this purpose, because they do not give tar enough to repay the trouble of making it. People use only the roots which are full of resin, and which they dig out of the ground with about two yards of the trunk, just above the root, laying aside all the rest. They have not yet learned the art of drawing the resin from one side of the tree by peeling off the bark; at least they never use this method. The tar barrels are only about half the size of ours. Such a barrel holds forty-six pots,[2] and sells at present for twenty-five francs at Quebec. The tar is considered quite good.

The sand on the shore of the St. Lawrence River consists in some places of a kind of pearl sand. The grains are of quartz, small and translucent. In some places it consists of little particles of mica;

---

[1] Forster's *Introd. to Mineralogy*, p. 50. *Zincum sterilum* Linn. *Syst. Nat.* III, p. 126. Ed. XII.—F.

[2] A pot equals 1.81 quarts or 2 liters, French measure.

and there are likewise spots covered with the garnet-colored sand which I have before described and which abounds in Canada.

SEPTEMBER THE 4TH

The mountains hereabouts were covered with a very thick fog to-day, resembling the smoke of a charcoal kiln. Many of these mountains are very high. During my stay in Canada, I asked many people who have travelled much in North America, whether they ever met with mountains so high that the snow never melted on them, to which they always answered in the negative. They say that the snow sometimes stays on the highest, *viz.* on some of those between Canada and the English colonies during a great part of the summer, but that it melts as soon as the great heat begins.

Flax. Iron Ore. Every countryman sows as much flax as he wants for his own use. They had already harvested it some time ago, and spread it on the fields, meadows, and pastures, in order to bleach it. It was very short this year in Canada. They find iron ore in several places hereabouts. Almost a Swedish mile from Bay St. Paul up in the country there is a whole mountain full of iron ore. The country round it is covered with a thick forest, and has many rivulets of different sizes which seem to make the erection of iron works very easy here. But since the government has suffered very much by the ironworks at Trois Rivières, nobody ventures to propose anything further in that way.

SEPTEMBER THE 5TH

Early this morning we set out on our return to Quebec. We continued our journey at noon, notwithstanding the heavy rain and thunder we got afterwards. At that time we were just at Petite Rivière, with the tide beginning to ebb, and it was impossible for us to go against it; therefore we lay by here and went on shore.

Petite Rivière is a little village on the western side of the river St. Lawrence and lies on a little creek, from which it takes its name. The houses are built of stone and are dispersed over the country. Here is likewise a fine little church of stone. To the west of the village are some very high mountains, which causes the sun to set three or four hours sooner here than in other places. The St. Lawrence River annually cuts off a piece of land on the east side of the village so that the inhabitants fear they will in a short time lose all the land they possess here, which at most is but a musket-shot broad. All the houses here are full of children.

Schist. The slate schists on the hills are of two kinds. One is black, which I have often mentioned and on which the town of Quebec is built. The other is generally black but sometimes dark gray and seems to be a species of the former. It is called *Pierre à chaux* here. It is chiefly distinguished from the former by being cut very easily, giving a very white lime when burnt, and not easily mouldering into thin slabs in the air. The walls of the houses here are made entirely of this slate, and likewise the chimneys, those places excepted which are exposed to the greatest fire, where they place pieces of gray rock mixed with a deal of mica. The mountains near Petite Rivière consist merely of gray rock which is just the same as that which I described near the lead mines of Bay St. Paul. The foot of these mountains is composed of one of the lime-slates. A great number of the Canada mountains of gray rock rest on a calcareous schist, in the same manner as the gray rocks of Västergötland in Sweden.

September the 6th

Eels and porpoises are caught here at a certain season of the year, *viz.* at the end of September and during the whole month of October. The eels come up the river at that time and are caught in the manner I have before described. They are followed by the porpoises which feed upon them. The greater the quantity of eels, the greater the number of porpoises, which are caught in the following manner. When the tide ebbs in the river, the porpoises commonly go down along the sides of the river, catching the eels which they find there. The inhabitants of this place therefore stick little branches with leaves into the river, in a curved line or arch, the ends of which face the shore, but stand at some distance from it, leaving a passage. The branches stand about two feet distant from each other. When the porpoises come amongst them and perceive the rustling the water makes with the leaves, they dare not venture to proceed, fearing a snare or trap and endeavor to return.

Meanwhile the water has receded so much that in going back they light upon one of the ends of the arch, whose moving leaves frighten them again. In this confusion they swim backwards and forwards till all the water has ebbed off, and they lie on the bottom where the inhabitants kill them. They furnish a great quantity of train-oil.

Near the shore is a gray clay, full of ferruginous cracks, and pierced by worms. The holes are small, perpendicular, and big enough to admit a medium-sized pin. Their sides are likewise ferruginous and half petrified; and where the clay has been washed away by the water, the rest looks like ochre-colored stumps of tobacco pipe tubes.

At noon we left Petite Rivière, which lies in a little bay, and proceeded to St. Joachim. Between these two places the western shore of the St. Lawrence River consists of prominent mountains, between which there are several small bays. It has been found by long experience that there is always a wind on these mountains. And when the wind is pretty strong at the last mentioned place it is not advisable to go to Quebec in a boat, the wind and waves in that case being very strong near these mountains. We had at present an opportunity of experiencing it. In the creeks between the mountains, the water was calm, but on our coming near one of the points formed by the high mountains the waves increased and the wind was so strong that two people were forced to take care of the helm, and the mast broke. The waves are likewise greatly increased by the strong current near those points or capes.

SEPTEMBER THE 7TH

A little before noon, we continued our voyage from St. Joachim.

Tree fungi are used very frequently instead of tinder. Those which are taken from the sugar maple are reckoned the best; those of the red maple are next in quality; and next to them, those of the sugar birch. For want of these they make use of those which grow on the aspen tree or tremble.

There are no other evergreen trees in this part of Canada than the thuya, the yew, and some of the fir. The thuya is esteemed for resisting decay much longer than any other wood; and next in value to it is the pine, called *pérusse* here.

They make *cheese* in several places here. That of the Isle of Orleans is, however, reckoned the best. This kind is small, thin, and round, and four such cheeses weigh about a French pound. Twelve of them sell for thirty sols. A pound of salt butter costs ten sols at Quebec, and of fresh butter, fifteen sols. Formerly they could get a pound of butter for four sols here.

The grain fields slope towards the river; they are allowed to lie fallow and to be sown alternately. The sown ones looked yellow at a distance and the fallow ones green. The weeds are left on the latter all summer for the cattle to feed upon.

The ash tree furnishes the best hoops for barrels here; and for want of it, they take the thuya, little birch trees, wild cherry trees and others.

Black Lime Schist. The hills near the river, on the western side opposite the Isle of Orleans, are very high and pretty steep. They consist for the most part of black schist. There are likewise some spots which consist of a rock which at first sight looks like a sandstone, and is composed of gray quartz, a reddish limestone, a little gray limestone, and some pale gray grains of sand. These parts of the stone are small and pretty equally mixed with each other. The stone looks red, with a grayish cast, and is very hard. It lies in strata, one above another. The thickness of each stratum is about five inches. It is remarkable that there are both elevated and hollow impressions of pectinites on the surface where one likewise meets with the petrified shells themselves; but on breaking the stone it does not even contain the least vestige of an impression or petrified shell. All the fossils are small, about the length and breadth of an inch. The particles of quartz in the stone strike fire with steel, and the particles of limestone effervesce strongly with *aqua fortis*. The upper and lower surfaces of the strata consist of limestone and the inner parts of quartz. Large quantities of this stone are quarried in order to build houses, pave floors, and make stair-cases of it. Great quantities of it are sent to Quebec. It is to be noted that there are petrefactions in this stone, but never any in the black calcareous schist.

The women dye their woolen yarn yellow with seeds of gale (*Myrica gale* L.), which is called *poivrier* here, and grows abundantly in wet places.

This evening M. Gauthier and I went to see the waterfall at Montmorency. The country near the river is high and level and laid out in meadows. Above them the high and steep hills begin, which are covered with a crust of earth and turned into grain fields. In some very steep places and near the rivulets, the hills consist of mere black lime schist, which is often crumbled into small pieces, like earth. All the fields below the hills are full of such pieces of schist. When some of the larger pieces are broken, they smell like stinkstone. In some more elevated places, the earth consists of a pale red color; and the schists are likewise reddish.

The *waterfall near Montmorency* is one of the highest I ever saw. It is in a river whose breadth is not very great and falls over the steep side of a hill, consisting entirely of black lime schist. The fall is now at the head of the bay. Both sides of the bay consist merely of the same schist, which is very much cracked and eroded, and though sloping one could hardly walk up it. The hill of schist under the waterfall is quite perpendicular, and one cannot look without astonishment at the quantity of water going over it. The rain of the preceding days had increased the water in the river to such an extent that it was terrifying to see such an amount of water hurling itself over the precipice. The breadth of the fall is not above ten or twelve yards. Its perpendicular height M. Gauthier and I guessed to be between a hundred and ten and a hundred and twenty feet; and on our return to Quebec, we found our guess confirmed by several gentlemen who had actually measured the fall and found it to be nearly as we had estimated. The people who live in the neighborhood exaggerate in their accounts of it, absolutely declaring that it is three hundred feet high. Father Charlevoix[1] is too sparing in giving it only forty feet in height.[2] At the bottom of the fall, there is always a thick fog of vapors, spreading about the water, being resolved into them by its violent fall. This fog occasions almost perpetual rain here, which is more or less heavy, in proportion to its distance from the fall. M. Gauthier and myself, together with the man who showed us the way, were willing to come nearer to the falling water in order to examine more accurately how it came down from such a height and how the stone

---

[1] See his *Histoire de la Nouv. France*, tom. V. p. m. 100.—F.
[2] Actual height of falls is 250 feet.

behind the water looked. But, when we were about twelve yards away from the fall, a sudden gust of wind blew a thick spray upon us, which in less than a minute had wet us thoroughly as if we had walked for half an hour in a heavy shower. We therefore hurried as fast as we could and were glad to get away. The noise of the fall is sometimes heard at Quebec which is two French miles to the south; and this is a sign of a northeast wind. At other times it can be well heard in the villages a good way to the north, and it is then reckoned an undoubted sign of a southwest wind or of rain. The black lime schist on the sides of the fall lies in dipping and almost perpendicular strata. In these lime slate strata, are the following kinds of stone.

Fibrous gypsum. This lies in very thin leaves between the cracks of the schist. Its color is a snowy white. I have found it in several parts of Canada in the same black limestone.

Pierre à Calumet. This is the French name of a stone disposed in strata between the lime schist and of which they make almost all the tobacco pipe heads in the country. The thickness of the strata varies. I have seen pieces nearly fifteen inches thick, but they are commonly between four and five inches. When the stone is long exposed to the open air or heat of the sun it becomes a yellowish color, but on the inside it is gray. It is of such compactness that its particles are not distinguishable by the naked eye. It is pretty soft, and will bear cutting with a knife. From this quality the people likewise judge the suitability of the stone for tobacco heads; for the hard pieces of it are not so fit for use as the softer ones. I have seen some of these stones split into thin leaves on the outside, where they were exposed to the sun. All the tobacco pipe heads, which the common people in Canada use, are made of this stone, and ornamented in different ways. A great part of the gentry likewise use them, especially when they are on a journey.

The Indians have employed this stone for the same purposes for several ages past, and have taught it to the Europeans. The heads of the tobacco pipes are naturally of a pale gray color, but they are blackened while they are quite new to make them look better. People cover the head all over with grease and hold it over a burning candle or any other fire, by which means it gets a good black color, which is increased by frequent use. The tubes of the pipes are always made of wood[1] and a brass wire holds them to the head.

There is no coal near this fall, or in the steep hills close to it. However, the people in the neighboring village showed me a piece of coal, which, they said, had been found on one of the hills near the fall.

We arrived at Quebec very late at night.

SEPTEMBER THE 8TH

Intermittent fevers of any kind are very rare at Quebec, M. Gauthier affirms. On the contrary, they are very common near Fort St. Frédéric and near Fort Detroit, which is a French colony, and between Lake Erie and Lake Huron, in forty-three degrees north latitude. Most of the natives told me something which I have mentioned before concerning both Englishmen and Frenchmen, namely, that native Europeans live longer than native Canadians. They also assured me that the second generation born here, and still more the third, does not reach the age of the preceding one, though one reason was ascribed to the dangerous journeys of the men for furs among the savages. It was said to be very rare to see a centenarian in Canada, though here and there an octogenarian was to be found.

Some of the people of quality make use of ice cellars to keep beer cool during the summer and to preserve fresh meat in the great heat. These cellars are commonly built of stone, under the house. The walls of them are covered with boards, because the ice is more easily consumed by stones. In winter they fill them with snow, which is beat down with the feet and covered with water. They then open the cellar holes and the door to admit the cold.—It is customary in summer to put a piece of ice into the water or wine which is to be drunk.

All the salt which is used here is imported from France. They had also made good salt in Canada from sea water; but since

---

[1] All over Poland, Russia, Turkey, and Tartary, they smoke pipes made of a kind of stone-marle, to which they fix long wooden tubes; for which latter purpose, they commonly employ the young shoots of the various kinds of Spiræa, which have a kind of pith easily to be thrust out. The stone-marle is called generally sea-scum, being pretty soft; and by the Tartars, in Crimea, it is called *keffekil*. And as it cuts so easily, various figures are curiously carved in it, when it is worked into pipe-heads, which often are mounted with silver.—F.

This is apparently the modern meerschaum pipe.

France keeps the salt trade entirely to itself, they had not continued to make it.

The Esquimaux are a particular kind of American savage, who live only near the water and never far in the country, on Terra Labrador, between the most outward point of the mouth of the St. Lawrence River and Hudson Bay. I have never had an opportunity of seeing one of them. I have spoken with many Frenchmen who have seen them and had them on board their own vessels. I shall here give a brief history of them, according to their unanimous accounts.

The Esquimaux are entirely different from the Indians of North America in regard to their complexion and their language. They are almost as white as Europeans, and have little eyes: the men, also, have beards. The Indians, on the contrary, are copper-colored, and the men have no beards. The Esquimaux language is said to contain some European words.[1] Their houses are either caverns or clefts in the mountains or huts of turf above ground. They never sow or plant vegetables, living chiefly on various kinds of whales, on seals (*Phoca vitulina* L.) and walruses (*Trichechus rosmarus* L.). Sometimes they likewise catch land animals, on which they feed. They eat most of their meat raw. Their drink is water and people have seen them drinking the sea water which was like brine.

Their shoes, stockings, breeches, and jackets are made of seal skins well prepared, and sewed together with sinews of whales, which may be twisted like threads and are very tough. Their clothes, the hairy side of which is turned outwards, are sewed together so well that they can walk up to their shoulders in water without wetting their underclothes. Under their upper clothes they wear shirts and waistcoats made of sealskins, prepared so well as to be quite soft. I saw one of their women's dresses: a cap, a waistcoat and skirt, made all of one piece of sealskin well prepared, soft to the touch, and the hair on the outside. There was a long train,

---

[1] The Moravian brethren in Greenland, coming once over with some Greenlanders to Terra Labrador, the Esquimaux ran away at their appearance; but they ordered one of their Greenlanders to call them back in his language. The Esquimaux, hearing his voice, and understanding the language, immediately stopped, came back, and were glad to find a countryman, and wherever, they went, among the other Esquimaux, they gave out, that one of their brethren was returned. This proves the Esquimaux to be of a tribe different from any European nation, as the Greenland language has no similarity with any language in Europe.—F.

about six inches wide, on the back of the skirt. In front it scarcely reached to the middle of the thigh, but under it the women wore breeches and boots, all of one piece. The Esquimaux women are said to be handsomer than any women of the American Indians, and their husbands are accordingly more jealous.

A Kayak. I have likewise seen an Esquimaux boat. The outside of it consisted entirely of skins, the hair of which had been taken off, so that they felt as smooth as vellum. The boat was near six feet long, but very narrow, and very sharp-pointed at the extremities. On the inside of such a boat, they place two or three thin boards, which give a kind of form to the boat. It is covered with skins at the top, excepting near one end, where there is a hole big enough for a single person to sit and row in and keep his thighs and legs under the cover. The shape of the hole resembles a semicircle, the base or diameter of which is turned towards the larger end of the boat. The hole is surrounded by wood, on which a soft skin is fastened, with straps in its upper part. When the Esquimau makes use of his boat he puts his legs and thighs under the deck, sits down at the bottom of the boat, draws the skin before mentioned round his body, and fastens it well with the straps; the waves may then beat over his boat with considerable violence, and not a single drop comes into it; the clothes of the Esquimaux keep him dry. He has an oar in his hand, which has a blade or paddle at each end; it serves him for rowing and keeping the boat in equilibrium during a storm. The paddles of the oar are very narrow. The boat will hold but a single person. Esquimaux have often been found safe in their boats many miles from land, in violent storms, where ships found it difficult to save themselves. Their boats float on the waves like bladders, and they row them with an incredible velocity. I am told they have boats of different shapes. They have likewise larger boats of wood, covered with skins, in which several people may sit and in which their women commonly go to sea.

Arms. Bows and arrows, javelins and harpoons, constitute their arms. With the last they kill whales and other large marine animals. The points of their arrows and harpoons are sometimes made of iron, sometimes of bone, and sometimes of the teeth of the walrus. Their quivers are made of seals' skins. The needles with which they sew their clothes are likewise made of iron or of bone. All their iron they get by some means or other from the Europeans.

They sometimes go on board the European ships in order to exchange some of their goods for knives and other iron. But it is not advisable for Europeans to go on shore, unless they be numerous, for the Esquimaux are false and treacherous and cannot suffer strangers amongst them. If they find themselves too weak, they run away at the approach of strangers; but if they think they are an over-match for them, they kill all that come in their way, without leaving a single one alive. The Europeans, therefore, do not venture to let a greater number of Esquimaux come on board their ships than they can easily master. If they are ship-wrecked on the Esquimaux coasts, they may as well be drowned in the sea as come safe to the shore: this many Europeans have experienced. The European boats and ships which the Esquimaux get into their power are immediately cut to pieces and robbed of all their nails and other iron which they work into knives, needles, arrow-heads, etc. They make use of fire for no other purposes than the working of iron and the preparing of the skins of animals. Their meat is eaten raw. When they come on board a European ship and are offered some of the sailors' food, they will never taste of it till they have seen some Europeans eat it. Though nothing pleases other savage nations so much as brandy, yet many Frenchmen have assured me that they never could prevail on the Esquimaux to take a dram of it. Their mistrust of other nations is the cause of it; for they undoubtedly imagine that they are going to poison them, or do them some harm; and I am not certain whether they do not judge right. They have no earrings, and do not paint the face like the American Indians. For many centuries past they have had dogs whose ears are erect and never hang down. They use them for hunting and instead of horses in winter for drawing their goods on the ice. They themselves sometimes ride in sledges drawn by these dogs. They have no other domestic animal. There are, indeed, plenty of reindeer in their country; but it is not known that either the Esquimaux or any of the Indians in America have ever tamed them. The French in Canada, who are in a manner the neighbors of the Esquimaux, have taken a deal of pains to carry on some kind of trade with them and to endeavor to engage them in a more friendly intercourse with other nations. For that purpose they took some

Esquimaux children, taught them to read, and educated them in the best manner possible. The intention of the French was to send these children back to the Esquimaux, that they might inform them of the kind treatment the French had given them and thereby incline them to conceive a better opinion of the French. But unhappily all the children died of the small pox and the scheme was dropped. Many persons in Canada doubted whether the scheme would have succeeded, though the children had survived. For they say there was formerly an Esquimau taken by the French and brought to Canada, where he stayed a good while and was treated with great civility. He learnt French pretty well, and seemed to relish the French way of living. When he was sent back to his countrymen, he was not able to make the least impression on them in favor of the French, but was killed by his nearest relations as half a Frenchman and a foreigner. This inhuman proceeding of the Esquimaux against all strangers is the reason why none of the Indians of North America ever give quarter to the Esquimaux if they meet them, but kill them on the spot, though they frequently pardon their other enemies, and incorporate the prisoners with their nation.

For the use of those who are fond of comparing the languages of several nations, I have here inserted a few Esquimaux words, communicated to me by the Jesuit Saint Pié. One, *kombuc*; two, *tigal*; three, *ké*; four, *missilagat*; water, *willalokto*; rain, *killaluck*; heaven, *taktuck*, or *nabugakshe*; the sun, *shikonak*, or *sakaknuk*; the moon, *takock*; an egg, *mannejuk*; the boat, *kayack*; the oar, *pacotick*; an arrow, *katso*; the head, *niakock*; the ear, *tchiu*; the eye, *killik*, or *shik*; the hair, *nutshad*; a tooth, *ukak*; the foot, *itikat*. Some think that the Esquimaux are nearly of the same origin as the Greenlanders, or Skralingers,[1] and pretend that there is a great affinity in the language.[2]

---

[1] The early Scandinavians who visited America in the tenth and eleventh centuries called the native North Americans *skrælingar* or *skrælings;* Dan. *skrælling,* a wretch. Whether they were Indians or Esquimaux is not absolutely certain, but to-day the opinion seems to favor the Indian theory. The real meaning of the term *skræling* is also unknown, but may be connected with an Old Norse word for "screaming" or "making a noise" (mod. Sw. *skrälla* or *skråla*), as the savage Indians were wont to do when going into battle, and as the natives did who encountered the Northmen. To-day the vast majority of the Greenlanders are Esquimaux. Kalm evidently here thought of the natives met by his ancient ancestors as Esquimaux.

[2] The above account of the Esquimaux may be compared with Henry Ellis's *Ac-*

Plum trees of different sorts, brought over from France, succeed very well here. The present year they did not begin to flower till this month. Some of them looked very well, and I am told the winter does not hurt them.

SEPTEMBER THE 11TH

The Marquis de la Galissonnière is one of the three noblemen, who, above all others, have gained high esteem with the French admiralty in the last war. The three are the Marquis de la Galissonnière, de la Jonquière, and de l'Etenduere.[1] The first of these was now above fifty years of age, of a low stature, and somewhat humpbacked, but of a very agreeable appearance. He had been here for some time as governor-general, and was soon going back to France. I have already mentioned something concerning this nobleman; but when I think of his many great qualities, I can never give him a sufficient encomium. He has a surprising knowledge in all branches of science, and especially in natural history, in which he is so well versed that when he began to speak with me about it I imagined I saw our great Linné under a new form. When he spoke of the use of natural history, of the method of learning, and employing it to raise the state of a country, I was astonished to see him take his reasons from politics, as well as natural philosophy, mathematics and other sciences. I own that my conversation with this nobleman was very instructive to me; and I always drew a deal of useful knowledge from it. He told me several ways of employing natural history to the purposes of politics, the science of government, and to make a country powerful in order to weaken its envious neighbors. Never has natural history had a greater promoter in this country; and it is very doubtful whether it will ever have his equal here. As soon as he got the place of governor-general, he began to take those measures for getting information in natural

count of a Voyage to Hudson's Bay, by the Dobbs Galley and California, etc. and *The Account of a Voyage for the Discovery of a North West Passage by Hudson's Straights*, by the Clerk of the California. Two vols. 8vo. And lastly, with Crantz's *History of Greenland*. two vols. 8vo.—F.

[1] This should be l'Estanduere. See Marchand, *op. cit.* 2, 183, note. He distinguished himself in a naval battle with the English, under vice-admiral Hawke, the following year.

history which I have mentioned before. When he saw people who had for some time been in a settled place of the country, especially in the more remote parts, or had travelled in those parts, he always questioned them about the trees, plants, earths, stones, ores, animals, etc. of the place. He likewise inquired what use the inhabitants made of these things; in what state their husbandry was; what lakes, rivers, and passages there were; and a number of other particulars. Those who seemed to have clearer notions than the rest were obliged to give him circumstantial descriptions of what they had seen. He himself wrote down all the accounts he received; and by this great application, so uncommon among persons of his rank, he soon acquired a knowledge of the most distant parts of America. The priests, commandants of forts, and of several distant places were often surprised by his questions, and wondered at his knowledge, when they came to Quebec to pay their visits to him; for he often told them that near such a mountain or on such a shore, etc. where they often went hunting, there were some particular plants, trees, soils, ores, etc. for he had gotten a knowledge of those things before. Hence it happened, that some of the inhabitants believed he had a preternatural knowledge of things, as he was able to mention all the curiosities of places, sometimes nearly two hundred Swedish miles from Quebec, though he had never been there himself, and though the others, on the other hand, had lived there for years. A person who did not know this gentleman well enough would have considered him dry and only moderately pleasant, in social relations, especially for one who had not penetrated into the sciences. But the more one became acquainted with him the better his good qualities appeared and the greater became the cause for respecting a person who was characterized by everything big. Never was there a better statesman than he; and nobody could take better measures or choose more proper means for improving a country and increasing its welfare. Canada was hardly acquainted with the treasure it possessed in the person of this nobleman, when it lost him again. The king wanted his services at home and could not leave him so far off. He was going to France with a collection of natural curiosities, and a quantity of young trees and plants, in boxes full of earth. I cannot describe all the favors he showed me. It was greater than I could have expected in my own fatherland. I do not know whether the natives or the sciences will miss him most, because he was the tenderest of fathers for both, and for the latter the biggest patron and promoter that any place has been able to show. Happy the country that has such a chief! There it is not necessary to lament about egoistic and imaginary obstacles for promoting deeds of public welfare. Such a chief gives encouragement to all things that benefit a fatherland.

Black lime schist (*or slate*) has been repeatedly mentioned during the course of my journey. I will here give a more minute description of it. The mountain on which Quebec is built, and the hills along the St. Lawrence River consist of it for several miles on both sides of Quebec. About a yard from the surface this stone is quite compact and without any cracks, so that one cannot perceive that it is slate, its grain being imperceptible. It lies in strata, which vary from three or four inches to twenty inches thickness or more. In the mountains on which Quebec is built, the strata do not lie horizontal, but dip so as to be nearly perpendicular, the upper ends pointing northwest and the lower ones southeast. For this reason the corners of these strata always stand out at the surface into the streets and cut the shoes in pieces. I have likewise seen some strata inclining to the north but nearly as perpendicular as the former. Horizontal strata, or nearly such, have been found too. The strata are divided by narrow cracks, which are commonly filled with fibrous white gypsum that can sometimes be pried loose with a knife, if the layer of stratum of slate above it is broken in pieces. And in that case it has the appearance of a thin white leaf. The larger cracks are almost filled up with transparent quartz crystals of different sizes. One part of the mountain contains vast quantities of these crystals, from which that point of the mountain which lies to the S. S. E. of the palace has got the name of Pointe de Diamante, or Diamond Point. The small cracks which divide the stone run generally at right angles. The distances between them are not always equal. The outside of the stratum, or that which is turned towards the other stratum, is frequently covered with a fine, black, shining membrane which looks like a kind of pyrous hornstone. In it there is sometimes a yellow pyrites in the form of small grains. I never found fossils or any other kinds of stone in it besides those I have just mentioned. The whole mountain on which Quebec is situated, consists entirely of lime schist from top to bottom. When this stone is broken, or scraped with a knife, it gives a strong smell like stinkstone. That part of the mountain which is exposed to the open air, has crumbled into small pieces, which have lost their black color and turned a pale red instead. Almost all the public and private buildings at Quebec are built of this schist, and likewise the walls round the town, and round the monasteries and gardens. It is easily quarried, and cut to the size desired. But it has the property of splitting into thin slabs, parallel to the surface of the stratum whence they are taken, after lying during one or more years in the air, and exposed to the sun. However, this quality does no damage to the walls in which they are placed; for the stones are laid on purpose in such a position that the cracks always run horizontally, and upper stones press so much upon the lower ones that they can only get cracks and split on the outside without going further inwards. The slabs always grow thinner as the houses grow older.

The Quebec Climate. In order to give my readers some idea of the climate of Quebec and of the different changes of heat and cold at the several seasons of the year, I will here insert some particulars extracted from the meteorological observations of the royal physician, M. Gauthier. He gave me a copy of those which he had made from October, 1744, to the end of September, 1746. The thermometrical observations I shall omit, because I do not believe them accurate. M. Gauthier made use of de la Hire's thermometer, in which the degrees of cold cannot be exactly determined since all the quicksilver in it is compressed into the globe at the bottom before the intense cold sets in. The observations are made throughout the year, between seven and eight in the morning and two and three in the afternoon, but he has seldom made any after noon. His thermometer was likewise inaccurate, for it was placed in a bad location, sometimes in an open window, and sometimes in the sun. —The new style is used in dates.

THE YEAR 1745

January. The 29th of this month the St. Lawrence River was covered over with ice near Quebec. In the observations of other years it is found that the river is sometimes covered with ice in the beginning of January or the beginning of December.

February. Nothing remarkable happened during the course of this month.

March. They say this has been the mildest winter they ever felt; even the eldest persons could not remember one so mild. The snow was only two feet deep, and the ice in the river opposite Quebec had the same thickness. On the twenty-first there was a thunder-storm, a soldier was struck and hurt very much. On the 19th and 20th they began to make incisions into the sugar maple and to pre-pare sugar from its sap.

April. During this month they continued to extract the sap of the sugar maple for making sugar. On the 7th the gardeners began to make hotbeds. On the 20th the ice in the river broke loose near Quebec and went down. This rarely happens so soon, for the St. Lawrence River is sometimes covered with ice opposite Quebec on the 10th of May. On the 22nd and 23rd, there fell a quantity of snow. On the 25th they began to sow near St. Joachim. The same day they saw some swallows. The 29th they sowed grain all over the country. Ever since the 23rd the river had been clear at Quebec.

May. The third of this month the cold was so great in the morn-ing that the Celsius or Swedish [centigrade] thermometer was four degrees below the freezing point; however, it did not hurt the grain. On the 16th all the spring sowing was done. On the 5th the sanguinaria, narcissus, and violet began to bloom. The 17th the wild cherry trees, raspberry bushes, apple trees and lime trees, be-gan to put out their leaves. The strawberries were in flower about that time. On the 29th the wild cherry trees were in blossom. On the 26th, part of the French apple trees and plum trees bloomed.

June. By the 5th of this month all the trees had leaves. The ap-ple trees were in full flower. Ripe strawberries were to be had on the 22nd. Here it is noted that the weather was very fine for the growth of vegetables.

July. The grain began to form into ears on the 12th, and had ears everywhere on the 21st. (It is to be observed that they sow nothing but spring seed here.) Soon after it began to flower. Hay making began the 22nd. All this month the weather was excellent.

August. On the 12th there were ripe pears and melons at Mont-real. On the 20th the wheat was ripe round Montreal, and the harvest began there. On the 22nd the harvest began at Quebec. On

the 30th and 31st, there was a very light hoarfrost on the ground.
September. The harvest of all kinds of grain ended on the 24th
and 25th. Melons, watermelons, cucumbers, and fine plums were
very plentiful during the course of this month. Apples and pears
were likewise ripe, which is not always the case. On the last days
of this month they began to plow the land. The following is one
of the observations of this month: "The old people in this country
say that the grain was formerly seldom ripe till the 15th or 16th of
September, and sometimes on the 12th, but no sooner. They likewise assert that it was never perfectly ripe. But since the woods
have been sufficiently cleared the beams of the sun have had more
room to operate and the grain ripens sooner than before." It is
further remarked that the hot summers are always very fruitful
in Canada, but that in most years only one-tenth of the grain ever
arrives at perfect maturity.

October. During this month the fields were plowed, and the
weather was very fine all the time. There was a little frost for
several nights, and on the 28th it snowed. Towards the end of this
month the trees began to shed their leaves.

November. They continued to plow till the 10th of this month,
when the trees had shed all their leaves. Till the 18th the cattle
went out of doors, a few days excepted when bad weather had kept
them inside. On the 16th there was some thunder and lightning.
There was not yet any ice in the St. Lawrence on the 24th.

December. During this month it is observed that the autumn
has been much milder than usual. On the 1st a ship could still set
sail for France; but on the 15th the St. Lawrence was covered with
ice on the sides, though open in the middle. In the Charles River
the ice was thick enough for horses with heavy loads to pass over
it. On the 26th the ice in the St. Lawrence River was washed away
by a heavy rain; but on the 28th, part of that river was again covered with ice.

The next observations show that this winter, too, has been one
of the mildest. I now resume the account of my journey.

This evening I left Quebec with a fair wind. The governor-
general of Canada, the Marquis de la Jonquière, ordered one of the
king's boats and seven men to bring me to Montreal. The middle
of the boat was covered with blue cloth under which we were se-

cured from the rain. This whole journey I made at the expense of the French king.[1] We went three French miles to-day.

SEPTEMBER THE 12TH

We continued our journey all day.

The small kind of maize, which ripens in three months' time, was ripe about this date and harvested and hung up to dry.

The weather about this time was like the beginning of our August, old style. Therefore, it seems, autumn commences a whole month later in Canada than in the central part of Sweden.

Kitchen Gardens. Near each farm there is a kitchen garden in which onions are most abundant, because the French farmers eat their dinners of them with bread, on Fridays and Saturdays, or fasting days. However, I cannot say the French are strict observers of fasting, for several of my rowers ate meat to-day, though it was Friday. The common people in Canada may be smelled when one passes by them on account of their frequent use of onions. Pumpkins also are abundant in the farmers' gardens. They prepare them in several ways, but the most common is to cut them through the middle, and place each half on the hearth, open side towards the fire, till it is roasted. The pulp is then cut out of the peel and eaten. Better class people put sugar on it. Carrots, lettuce, Turkish beans, cucumbers, and currant shrubs, are planted in every farmer's little kitchen garden.

Tobacco. Every farmer plants a quantity of tobacco near his house, in proportion to the size of his family. It is necessary that one should plant tobacco, because it is so universally smoked by the common people. Boys of ten or twelve years of age, as well as the old people, run about with a pipe in their mouth. Persons of the better class do not refuse either to smoke a pipe now and then. In the northern parts of Canada they generally smoke pure tobacco; but further north and around Montreal, they take the inner bark of the red Cornelian cherry (*Cornus sanguinea* L.), crush it, and mix it with the tobacco, to make it weaker. People of both sexes,

---

[1] According to Jos. E. Roy in *Voyage de Kalm en Canada* (1900) Kalm's total expenses in Canada amounted to 2,182 livres, all paid by the French Government.

and of all ranks, use snuff very much. Almost all the tobacco which is consumed here is the product of the country, and some people prefer it even to Virginian tobacco: but those who pretend to be connoisseurs reckon the last kind better than the other.

Manners and Customs. Though many nations imitate the French customs, I observed, on the contrary, that the French in Canada in many respects follow the customs of the Indians, with whom they have constant relations. They use the tobacco pipes, shoes, garters, and girdles of the Indians. They follow the Indian way of waging war exactly; they mix the same things with tobacco; they make use of the Indian bark boats and row them in the Indian way; they wrap a square piece of cloth round their feet, instead of stockings, and have adopted many other Indian fashions. When one comes into the house of a Canadian peasant or farmer, he gets up, takes his hat off to the stranger, invites him to sit down, puts his hat on and sits down again. The gentlemen and ladies, as well as the poorest peasants and their wives, are called Monsieur and Madame. The peasants, and especially their wives, wear shoes which consist of a piece of wood hollowed out, and are made almost as slippers. Their boys and the old peasants themselves wear their hair behind in a queue, and most of them wear red woolen caps at home and sometimes on their journeys.

Food. The farmers prepare most of their food from milk. Butter is seldom seen, and what they have is made of sour cream, and therefore not so good as English butter. A good deal of this butter has a slight taste of tallow. Congealed sour milk is found everywhere, in stone vessels. Many of the French are very fond of milk, which they eat chiefly on fast days. However, they have not so many methods of preparing it as we have in Sweden. The common way is to boil it, and put bits of wheat bread and a good deal of sugar into it. The French here eat nearly as much meat as the English on those days when their religion allows it. For excepting the soup, the salads and the dessert, all their other dishes consist of meat variously prepared.

At night we slept at a farmhouse near the river called Petite Rivière, which entered here into the St. Lawrence River. This place was reckoned sixteen French miles from Quebec, and ten from Trois Rivières. The tide was still strong here. Here is the last place where the hills along the river consist of black lime schist; further on they are composed merely of earth.

Fireflies flew about the woods at night, though not in great numbers; the French call them *Mouches à feu.*

The houses in this neighborhood are all made of wood. The rooms are pretty large. The inner roof rests on two, three, or four large, thick spars, according to the size of the room. The chinks are filled with clay instead of moss. The windows are made entirely of paper. The chimney is erected in the middle of the room; that part of the room which is opposite the fire is the kitchen; that which is behind the chimney serves the people for sleeping and entertaining strangers. Sometimes there is an iron stove behind the chimney.

September the 13th

Near *Champlain*, which is a place about five French miles from Trois Rivières, the steep hills near the river consist of a yellow and sometimes ochre-colored sandy earth, in which a number of small springs arise. The water in them is generally filled with yellow ochre, which is a sign that these dry sandy fields contain a great quantity of the same iron ore which is dug at Trois Rivières. It is not conceivable whence that number of small rivulets arise, the ground above being flat, and exceedingly dry in summer. The lands near the river are cultivated for about an English mile into the country, but behind them there are thick forests and low grounds. The woods, which collect a quantity of moisture and prevent the evaporation of the water, force it to make its way under ground to the river. The shores of the latter are here covered with a great deal of black iron sand.

Towards evening we arrived at Trois Rivières, where we stayed no longer than was necessary to deliver the letters which we had brought with us from Quebec. After that we went a French mile further north before we took up our night's lodging.

Three Old People. This afternoon we saw three remarkable old people. One was an old Jesuit, called father Joseph Aubery, who had been a missionary to the converted Indians of St. François. This summer he ended the fiftieth year of his mission. He therefore returned to Quebec to renew his vows there, and he seemed to be healthy and in good spirits. The other two people were our landlord and his wife; he was above eighty years of age, and she was not much younger. They had now been fifty-one years married. The year before, at the end of their fiftieth year of their marriage, they had gone to church together and offered up thanks to God Almighty for the great grace he had given them. They were yet quite well, content, merry and talkative. The old man said that he was at Quebec when the English besieged it, in the yaar 1690, and that the bishop went up and down the streets, dressed in his pontifical robes, with a sword in his hand, in order to encourage the soldiers.

The old man said that he thought the winters were formerly much colder than they are now. There fell likewise a greater quantity of snow when he was young. He could remember the time when pumpkins, cucumbers, etc. were killed by the frost about mid-summer, and he assured me that the summers were warmer now than they used to be. About thirty and some odd years ago there had been such a severe winter in Canada that the frost killed many birds; but the old man could not remember the particular year. Everybody admitted that the summers in 1748 and 1749 had been warmer in Canada than they had been in many years.

The soil is considered quite fertile here, and wheat yields a nine or tenfold harvest. But when this old man was a boy and the country was new and rich everywhere, they could get a twenty or four-and-twentyfold crop. They sow but little rye here, nor do they sow much barley, except for the use of the cattle. They complain, however, that when they have a bad crop, they are obliged to bake bread of barley in place of wheat.

SEPTEMBER THE 14TH

This morning we got up early and pursued our journey. After we had gone about two French miles we reached Lake St. Pierre, which we crossed. Many plants which are common in our Swedish lakes swim on the top of this water. This lake is said to be covered every winter with such strong ice that a hundred loaded horses could go over it together with safety.

A crawfish, or river lobster, somewhat like a crab but quite minute, about two geometrical lines long and broad in proportion, was frequently drawn up by us with the aquatic weeds. Its color was a pale greenish white.

The cordate pontederia (*Pontederia cordata* L.) grows plentifully on the sides of a long and narrow canal or waterway, in the places frequented by our water lilies (*Nymphæa*). A great number of hogs wade far into this kind of strait and sometimes duck the greatest part of their bodies under water in order to get at the roots which they are very fond of.

As soon as we had passed over Lake St. Pierre, the face of the country was entirely changed, and became as agreeable as could be wished. The isles, and the land on both sides of us, looked like the prettiest pleasure gardens; and this continued till we neared Montreal.

Near every farm on the riverside there were some boats, hollowed out of the trunks of single trees but commonly neat and well made, having the proper shape of boats. In only one single place did I see a boat made of the bark of trees.

SEPTEMBER THE 15TH

We continued our journey early this morning. On account of the force of the stream which came down against us, we were sometimes obliged to let the rowers go on shore and draw the boat up. At four o'clock in the evening we arrived at Montreal; and our voyage was reckoned a happy one, because the violence of the river flowing against us all the way, and the changeableness of the winds, commonly protract it to two weeks.

SEPTEMBER THE 19TH

Grapes. Several people here in town have gotten the French grapevines and planted them in their gardens. They have two kinds of grapes, one of a pale green, or almost white; the other, of a reddish brown. From the white ones they say white wine is made, and from the red ones, red wine. The cold in winter obliges them to put dung round the roots of the vines, without which they would be killed by the frost. The grapes are beginning to ripen now; the white ones ripen a little sooner than the red ones. They make no wine of them here, because it is not worth while; but they are served for dessert like other berries. They say these grapes do not grow so big here as in France.

Watermelons (*Cucurbita citrullus* L.) are cultivated in great plenty in the English and French American colonies, and there is hardly a peasant here who has not a field planted with them. They are cultivated chiefly in the neighborhood of the town, and they are very rare in the north part of Canada. The Indians plant great quantities of watermelons at present, but whether they have done it of old is not easily determined, for an old Oneida Indian (of the six Iroquois nations) assured me that the Red Men did not know watermelons before the Europeans came into the country and showed them to the Indians. The French, on the other hand, asserted that the Illinois Indians had abundance of this fruit when the French first came to them, and that they declared they had planted them since time immemorial. However, I do not remember having read that the Europeans, who first came to North America, mention the watermelons, in speaking of the dishes of the Indians at that time. How great the summer heat is in those parts of America which I have passed through can easily be conceived, when one considers that in all those places they never sow watermelons in hot beds but in the open fields in spring, without so much as covering them, and they ripen in the season. Here are two species of them, *viz.* one with a red pulp, and one with a white one. The first is more common to the southward, with the Illinois Indians, and in the English colonies; the last is more abundant in Canada. The seeds are sown in the spring, after the cold has entirely left, in a good rich ground, at considerable distance from each other, because their stalks spread far and require much room if they are to be very productive. They were now ripe at Montreal, but in the English colonies they ripen in July and August. They usually require less time to ripen than the common melons. Those in the English colonies are usually sweeter and more agreeable than the Canadian ones. Does the greater heat contribute anything towards making them more palatable? Those in the province of New York are, however, reckoned the best. They contain a large percentage of water, and are cut into slices when eaten. They are always consumed raw, fresh.

The watermelons are very juicy; and the juice is mixed with the cooling pulp, which is very refreshing in the hot summer season. Nobody in Canada, Albany, or any other part of New York, could produce an example that the eating of watermelons in great quantities had hurt anybody; and there are examples of sick persons eating them without any danger. Further to the south, the frequent use of them, it is thought, brings on intermitting fevers and other bad distempers, especially in such people as are less used to them. Many Frenchmen assured me that when people born in Canada came to the Illinois Indians and ate several times of the watermelons there, they immediately got a fever; and therefore the Illinois advised the French not to eat of a fruit so dangerous to them. They themselves are subject to attack by fevers, if they cool their stomachs too often with watermelons. In Canada they keep them in a room which is a little heated, which means they keep fresh two months after they are ripe; but care must be taken that the frost does not spoil them. In the English plantations they keep them fresh in dry cellars during a part of the winter. They assured me that they keep better when they are carefully broken off from the stalk, and afterwards singed with a redhot iron in the place where the stalk was attached. In this manner they may be eaten at Christmas, and after. In Pennsylvania, where they have a dry sandy earth, they dig a hole in the ground, put the watermelons carefully into it with their stalks, by which means they keep very fresh during a great part of the winter. Few people, however, take this trouble with the watermelons; because they being very cooling, and the winter being very cold too, it seems to be less necessary to keep them for eating in that season, which is already very cold. They are of the opinion in these parts that cucumbers are more refreshing than watermelons. The latter are very strongly diuretic. The Iroquois call them *onóheserakáhti*.

Pumpkins of several kinds, oblong, round, flat or compressed, crook-necked, small, etc. are planted in all the English and French colonies. In Canada they fill the chief part of the farmers' kitchen gardens, though the onions are a close second. Each farmer in the English plantations has a large field planted with pumpkins, and the Germans, Swedes, Dutch and other Europeans settled in their colonies plant them. They constitute a considerable part of the Indian food; however, the natives plant more squashes than common pumpkins. They declare that they had the latter long before the Europeans discovered America, which seems to be confirmed by the accounts of the first Europeans that came into these parts, who mentioned pumpkins as common food among the Indians. The French here call them *citrouilles*, and the English in the colonies pumpkins. They are planted in spring, when they have nothing to fear from the frost, in an enclosed field, and in a good, rich soil. They are likewise frequently put into old hot beds. In Canada they ripen towards the beginning of September, but further south they are ripe at the end of July. As soon as the cold weather commences they remove all the pumpkins that remain on the stalk, whether ripe or not, and spread them on the floor in a part of the house, where the unripe ones grow perfectly ripe if they are not laid one upon the other. This is done round Montreal in the middle of September; but in Pennsylvania I have seen some in the fields on the nineteenth of October. They keep fresh for several months and even throughout the winter, if they be well secured in dry cellars (for in damp ones they rot very soon) where the cold cannot enter, or, which is still better, in dry rooms which are heated now and then to prevent the cold from damaging the fruit.

Pumpkins are prepared for eating in various ways. The Indians boil them whole, or roast them in ashes and eat them, or sell them thus prepared in the town; and they have, indeed, a very fine flavor when roasted. The French and English slice them and put the slices before the fire to roast; when are done they generally put sugar on the pulp. Another way of roasting them is to cut them through the middle, take out all the seeds, put the halves together again, and roast them in an oven. When they are quite done, some butter is put in while they are warm, which being imbibed into the pulp renders it very palatable. The settlers often boil pumpkins in water, and afterwards eat them either alone or with meat. Some make a thin kind of pottage of them, by boiling them in water and afterwards macerating the pulp. This is again boiled with a little of the water, and a good deal of milk, and stirred about while it is boiling. Sometimes the pulp is kneaded into a dough with maize or other flour; of this they make pancakes. Some make puddings and tarts of pumpkins. The Indians, in order to preserve the pumpkins for a very long time, cut them in long slices which they fasten or twist together and dry either in the sun or by the fire in a room. When they are thus dried, they will keep for years, and when boiled they taste very well. The Indians prepare them thus at home and on their journeys and from them the Europeans have adopted this method. Sometimes they do not take the time to boil the pumpkin, but eat it dry with dried beef or other meat; and I own they are eatable in that state, and very welcome to a hungry stomach. At Montreal they sometimes preserve them in the following manner: they cut a pumpkin in four pieces, peel them, and take the seeds out. The pulp is put into a pot with boiling water, in which it must boil from four to six minutes. It is then put into a strainer and left in it till the next day that the water may run off. Then it is mixed with cloves, cinnamon, and lemon peel and preserved in syrup, the quantity of the latter being the same as that of the pulp. After this operation it is boiled together till all the syrup is absorbed and the white color of the pulp is lost.

September the 20th

The crop this year in Canada was reckoned the finest they had ever had. In the province of New York, on the contrary, the crop was very poor. The autumn was very fine this year in Canada.

September the 22nd

Indian Trade. The French in Canada carry on a great trade with the Indians; and though it was formerly the only trade of this extensive country, its inhabitants were considerably enriched by it. At present they have besides the Indian goods, several other articles which are exported. The Indians in this neighborhood who go hunting in winter like the other Indian nations, commonly bring their furs and skins to sell in the neighboring French towns; however, this is not sufficient. The Red Men who live at a greater distance never come to Canada at all; and lest they should bring their goods to the English, or the English go to them, the French are obliged to undertake journeys and purchase the Indian goods in the country of the natives. This trade is carried on chiefly at Montreal, and a great number of young and old men every year undertake long and troublesome voyages for that purpose, carrying with them such goods as they know the Indians like and want. It is not necessary to take money on such a journey, as the Indians do not value it; and indeed I think the French, who go on these journeys, scarcely ever take a sol or penny with them.

Goods Sold to the Natives. I will now enumerate the chief goods which the French carry with them for this trade, and which have a good sale among the Indians:

1.[1] *Muskets, powder, shot,* and *balls.* The Europeans have taught the Indians in their neighborhood the use of firearms, and so they have laid aside their bows and arrows, which were formerly their only arms, and use muskets. If the Europeans should now refuse to supply the natives with muskets, they would starve to death, as almost all their food consists of the flesh of the animals which they hunt; or they would be irritated to such a degree as to attack the colonists. The savages have hitherto never tried to make muskets or similar firearms, and their great indolence does not even allow them to mend those muskets which they have. They leave this entirely to the settlers. When the Europeans came into North America they were very careful not to give the Indians any firearms. But in the wars between the French and English, each party gave their Indian allies firearms in order to weaken the force of the enemy. The French lay the blame upon the Dutch settlers in Albany, saying that the latter began in 1642 to give their Indians firearms, and taught the use of them in order to weaken the French. The inhabitants of Albany, on the contrary, assert that the French first introduced this custom, as they would have been too weak to resist the combined force of the Dutch and English in the colonies. Be this as it may, it is certain that the Indians buy muskets from the white men, and know at present better how to make use of them than some of their teachers. It is likewise certain that the colonists gain considerably by their trade in muskets and ammunition.

2, a. *Pieces of white cloth,* or of a coarse uncut material. The Indians constantly wear such cloth, wrapping it round their bodies. Sometimes they hang it over their shoulders; in warm weather they fasten the pieces round the middle; and in cold weather they put them over the head. Both their men and women wear these pieces of cloth, which have commonly several blue or red stripes on the edge.

---

[1] The goods are not numbered in the original.

b. *Blue or red cloth.* Of this the Indian women make their skirts, which reach only to their knees. They generally choose the blue color.

c. *Shirts and shifts of linen.* As soon as an Indian, either man or woman, has put on a shirt, he (or she) never washes it or strips it off till it is entirely worn out.

d. *Pieces of cloth,* which they wrap round their legs instead of stockings, like the Russians.

3. *Hatchets, knives, scissors, needles,* and *flint.* These articles are now common among the Indians. They all get these tools from the Europeans, and consider the hatchets and knives much better than those which they formerly made of stone and bone. The stone hatchets of the ancient Indians are very rare in Canada.

4. *Kettles of copper or brass,* sometimes tinned on the inside. In these the Indians now boil all their meat, and they produce a very large demand for this ware. They formerly made use of earthen or wooden pots, into which they poured water, or whatever else they wanted to boil, and threw in red hot stones to make it boil. They do not want iron boilers because they cannot be easily carried on their continual journeys, and would not bear such falls and knocks as their kettles are subject to.

5. *Earrings* of different sizes, commonly of brass, and sometimes of tin. They are worn by both men and women, though the use of them is not general.

6. *Cinnabar.* With this they paint their face, shirt and several parts of the body. They formerly made use of a reddish earth, which is to be found in the country; but, as the Europeans brought them vermilion, they thought nothing was comparable to it in color. Many persons told me that they had heard their fathers mention that the first Frenchmen who came over here got a heap of furs from the Indians for three times as much cinnabar as would lie on the tip of a knife.

7. *Verdigris,* to paint their faces green. For the black color they make use of the soot off the bottom of their kettles, and daub the whole face with it.

8. *Looking glasses.* The Indians like these very much and use them chiefly when they wish to paint themselves. The men constantly carry their looking glasses with them on all their journeys;

but the women do not. The men, upon the whole, are more fond of dressing than the women.

9. *Burning glasses.* These are excellent utensils in the opinion of the Indians because they serve to light the pipe without any trouble, which pleases an indolent Indian very much.

10. *Tobacco* is bought by the northern Indians, in whose country it will not grow. The southern Indians always plant as much of it as they want for their own consumption. Tobacco has a great sale among the northern Indians, and it has been observed that the further they live to the northward, the more tobacco they smoke.

11. *Wampum,* or as it is here called, *porcelaine.* It is made of a particular kind of shell and turned into little short cylindrical beads, and serves the Indians for money and ornament.[1]

12. *Glass beads,* of a small size, white or other colors. The Indian women know how to fasten them in their ribbons, bags and clothes.

13. *Brass* and *steel wire,* for several kinds of work.

14. *Brandy,* which the Indians value above all other goods that can be brought them; nor have they anything, though ever so dear to them, which they would not give away for this liquor. But on account of the many irregularities which are caused by the use of brandy, the sale of it has been prohibited under severe penalties; however, they do not always pay implicit obedience to this order.

These are the chief goods which the French carry to the Indians and they do a good business among them.

Furs Bought from the Natives. The goods which they bring back from the Indians consist almost entirely of furs. The French take them in exchange for their goods, together with the necessary food provisions which they may want on the return journey from the Indians. The furs are of two kinds; the best are the northern ones, and the poorer sort those from the south. In the northern parts of America there are chiefly the following skins of animals: bears, beavers, elks (*originacs*),[2] reindeer (*cariboux*), wolf-lynzes (*Loups cerviers*), and martens. They sometimes get marten skins from the south, but they are red and good for little. *Pichou du nord* is perhaps the animal which the English, near Hudson Bay, call

---

[1] An imitation wampum was made of porcelain, and sold to the Indians, hence the Canadian name *porcelaine,* presumably.

[2] Canadian-Basque form for *originaux* or *orignaux.*

the wolverene. To the northern furs belong that of the bear, which is rare, and of the fox, which is not very frequent, and generally black; and several other skins.

The skins of the southern parts are taken chiefly from: wild cattle, stags, roebucks, otters, *pichoux du sud*, of which P. Charlevoix makes mention,[1] and are probably a species of cat-lynx, or perhaps a kind of panther; foxes of various kinds, raccoons, cat-lynxes, and several others.

It is inconceivable what hardships the people in Canada must undergo on their hunting journeys. Sometimes they must carry their goods a great way by land. Frequently they are abused by the Indians, and sometimes they are killed by them. They often suffer hunger, thirst, heat, and cold, and are bitten by gnats, and exposed to the bites of poisonous snakes and other dangerous animals and insects. These [hunting expeditions] destroy a great part of the youth in Canada, and prevent the people from growing old. By this means, however, they become such brave soldiers, and so inured to fatigue that none of them fears danger or hardships. Many of them settle among the Indians far from Canada, marry Indian women, and never come back again.

The prices of the skins in Canada, in the year 1749, were communicated to me by M. de Couagne, a merchant at Montreal, with whom I lodged. They were as follows:

Great and middle sized bear skins, cost five livres; skins of young bears, 50 sols; lynxes, 25 sols; *pichoux du sud*, 35 sols; foxes from the southern parts, 35 sols; otters, 5 livres; raccoons, 5 livres; martens, 45 sols; wolf-lynxes (*Loups cerviers*), 4 livres; wolves, 40 sols; carcajous,[2] an animal which I do not know, 5 livres; skins of the visons, a kind of marten [the American mink], which live in the water, 25 sols; raw skins of elks (*originacs verts*), 10 livres; stags

---

[1] There seems to be no clear distinction in contemporaneous sources between the *pichou du nord* and the *pichou du sud*. The word is originally of Indian origin. Pierre F. X. de Charlevoix in his *Histoire et Description Générale de la Nouvelle France* (III, 407, in the three-volume edition) "uses the plural *Pichoux* for two species of wild cats: the one with a short tail, which is the common American wild cat; and the other, a larger animal that goes by the name of *cougar* or *puma* (*Felis concolor* L.). The latter is called also *catamount, mountain lion* and *American lion*." See William A. Read, *Louisiana-French* (1931), pp. 101-102.

[2] The wolverene. The term is somtimes applied to the Canadian lynx, cougar, or American badger.

(*cerfs verts*), 8 livres; bad skins of elks and stags (*originacs et cerfs passés*), 3 livres; skins of roebucks, 25, or 30 sols; red foxes, 3 livres; beavers, 3 livres.

I will now insert a list of all the different kinds of skins, which are to be gotten in Canada, and which are sent from there to Europe. I obtained it from one of the greatest merchants in Montreal. They are as follows:

Prepared roebuck skins, *chevreuils passés*.
Unprepared ditto, *chevreuils verts*.
Tanned ditto, *chevreuils tanés*.
Bears, *ours*.
Young bears, *oursons*.
Otters, *loutres*.
Pécans, [Woodshock, [or fisher] a species of Canadian marten (Marchand)].
Cats, *chats*.
Wolves, *loup de bois*.
Lynxes, *loups cerviers*.
North pichoux, *pichoux du nord*.
South pichoux, *pichoux du sud*.
Red foxes, *renards rouges*.
Cross foxes, *renards croisés*.
Black foxes, *renards noirs*.
Gray foxes, *renards argentes*.
Southern or Virginian foxes, *renards du sud ou de Virginie*.
White foxes, from Tadoussac, *renards blancs de Tadoussac*.
Martens, *martres*.
Visons, or *foutreaux*.
Black squirrels, *écureuils noirs*.
Raw stags skins, *cerfs verts*.
Prepared ditto, *cerfs passés*.
Raw elk skins, *originacs verts*.
Prepared ditto, *originacs passés*.
Reindeer skins, *cariboux*.
Raw hind skins, *biches vertes*.
Prepared ditto, *biches passées*.
Carcajoux. [Wolverene or Labrador badger].
Muskrats, *rats musqués*.

Fat winter beavers, *castors gras d'hiver.*
Ditto summer beavers, *castors gras d'été.*
Dry winter beavers, *castors secs d'hiver.*
Ditto summer beavers, *castors secs d'été.*
Old winter beavers, *castors vieux d'hiver.*
Ditto summer beavers, *castors vieux d'été.*

Native Copper. To-day I got a piece of native copper from Lake Superior. They find it there almost pure, so that it does not need melting over again, but is immediately fit for working. Father Charlevoix [1] speaks of it in his *History of New France.* One of the Jesuits at Montreal who had been at the place where this metal is native told me that it is generally found near the mouths of rivers and that there are pieces of pure copper too heavy for a single man to lift up. The Indians there say they formerly found a piece about seven feet long and nearly four feet thick, all pure copper. As it is always found in the ground near the mouths of rivers, it is probable that the ice or water carried it down from a mountain; but, notwithstanding the careful search that has been made, no place has been found where the metal lies in any great quantity but only in loose pieces.

Lead Ore. The head or superior of the priests of Montreal gave me a piece of lead ore to-day. He said it was taken from a place only a few French miles from Montreal, and it consisted of compact, shining cubes of lead ore. I was told by several persons here that further south in the country there is a place where they find a great quantity of this lead ore in the ground. The Indians nearby melt it and make balls and shot of it. I got some pieces of it consisting of a shining lead ore with narrow stripes through it and of a white hard earth or clay which effervesces with *aqua fortis.*

I likewise received some reddish brown earth to-day, found near the Lac des Deux Montagnes, or Lake of Two Mountains, a few French miles from Montreal. It may be easily crumbed into dust between the fingers. It is very heavy, and more so than the earth of that kind generally is. Outwardly it has a kind of glossy appearance, and when it is handled by the fingers for a time it looks as if it were covered with silver. It is, therefore, probably a kind of lead earth or an earth mixed with iron mica.

---

[1] See his *Hist. de la Nouv. Fr.,* Tom. VI., p. 415.—F.

The Women of Canada. The ladies in Canada are generally of two kinds: those who come over from France, and those who are natives. The former possess the politeness peculiar to the French nation; the latter may be divided into those of Quebec and Montreal. The first of these are equal to the French ladies in good breeding, having the advantage of frequently conversing with the French gentlemen and ladies, who come every summer with the king's ships, and stay several weeks at Quebec, but seldom go to Montreal. The ladies of this last place are accused by the French of being contaminated by the pride and conceit of the Indians, and of being much wanting in French good breeding. What I have mentioned above [1] about their dressing their head too profusely is the case with all the ladies throughout Canada. Their hair is always curled, even when they are at home in a dirty jacket and a short coarse skirt that does not reach to the middle of their legs. On Sundays and visiting days they dress so gayly that one is almost induced to think their parents in origin and social position to be among the best in the realm. The Frenchmen, who consider things in their true light, complain very much that a great number of the ladies in Canada have gotten into the pernicious custom of taking too much care of their dress, and squandering all their fortune and more upon it, instead of sparing something for future times. They are no less attentive to having the newest fashion; the best and most expensive dresses are discarded and cut to pieces; and they smile inwardly when their sisters are not dressed according to their fancy. But what they get as new fashions are often old and discarded in France by the time they are adopted in Canada, for the ships come but once every year from abroad, and the people in Canada consider that as the new fashion for the whole year which the people on board brought with them or which they imposed upon them as new.

The ladies in Canada, and especially at Montreal, are very ready to laugh at any blunders strangers make in speaking; but they are very excusable. People laugh at what appears uncommon and ridiculous.[2] In Canada nobody ever hears the French language spoken by any one but Frenchmen, for strangers seldom come

---

[1] See diary for July 25, 1749.
[2] Apparently the Canadian ladies laughed at Kalm's French.

there, and the Indians are naturally too proud to learn French, and compel the French to learn their language. Therefore it naturally follows that the sensitive Canadian ladies cannot hear anything uncommon without laughing at it. One of the first questions they put to a stranger is whether he is married; the next, how he likes the ladies in the country, and whether he thinks them handsomer than those of his own country; and the third, whether he will take one home with him. There are some differences between the ladies of Quebec and those of Montreal; those of the latter place seemed to be generally handsomer than those of the former. The women seemed to me to be somewhat too free at Quebec, and of a more becoming modesty at Montreal. The ladies at Quebec, especially the unmarried ones, are not very industrious. A girl of eighteen is reckoned very poorly off if she cannot enumerate at least twenty lovers. These young ladies, especially those of a higher rank, get up at seven, and dress till nine, drinking their coffee at the same time. When they are dressed they place themselves near a window that opens into the street, take up some needlework and sew a stitch now and then; but turn their eyes into the street most of the time. When a young fellow comes in, whether they are acquainted with him or not, they immediately lay aside their work, sit down by him, and begin to chat, laugh, joke, and invent "double-entendres" and make their tongues go like a lark's wings; this is considered *avoir beaucoup d'esprit*. In this manner they frequently pass the whole day, leaving their mothers to do all the work in the house. In Montreal the girls are not quite so flighty, and more industrious. It is not uncommon to find them with the maid in the kitchen. They are always at their needle work or doing some necessary business in the house. They are likewise cheerful and content; and nobody can say that they lack either wit or charms. Their fault is that they think too well of themselves. However, the daughters of all ranks, without exception, go to market, buy watermelons, pumpkins, and other food and carry it home themselves. They rise as soon and go to bed as late as any of the people in the house. I have been assured that in general their fortunes are not great and are rendered still more scarce by the number of children and the small revenues in a house. The girls at Montreal are very much displeased that those at Quebec get husbands sooner than they. The reason for this is that many young gentlemen who come over from France with the ships are captured by the ladies at Quebec, and marry them; but as these gentlemen seldom go up to Montreal the girls there are not often so happy as those of the former place.

September the 23rd

Sault au Recollet. This morning I went to Sault au Recollet, a place three French miles north of Montreal to describe the plants and minerals there and chiefly to collect seeds of various plants. Near the town there are farms on both sides of the road; but as one advances the country grows woody and varies in regard to height. It is generally very rough and there are pieces both of ordinary granite and a kind of gray limestone. The roads are bad and almost impassable for carriages. A little before I arrived at Sault au Recollet the woods came to an end, and the country was either cultivated or turned into meadows and pastures. Otherwise there was nothing especially pleasant on this journey. The parts visited could not be compared with those around Montreal.

Lime Kilns. About a French mile from the town are two lime kilns on the road. They are built in the ground of a gray infusible limestone, on the outside, and of pieces of granite rock nearest the fire. The height of the kiln from top to bottom is eighteen feet.

The limestone which they burn here is of two kinds. One is quite black and so compact that its constituent particles cannot be distinguished, some dispersed grains of white and pale gray spar excepted. Now and then there are thin cracks in it filled with a white small-grained spar.

I have never seen any fossils in this stone, though I looked very carefully for them. This stone is common on the Isle of Montreal, about ten or twenty inches below the upper soil. It lies in strata of five or ten inches thickness. This stone is said to give the best lime; for, though it is not so white as that of the following gray limestone, it makes better mortar, and almost turns into stone, growing harder and more compact every day. In repairing a house made partly of this mortar, it has happened that stones of the house crumbled sooner than the mortar itself.

The other kind is gray and sometimes a dark gray limestone,

consisting of a compact calcareous stone, mixed with grains of spar of the same color. When broken, it has a strong smell of stinkstone. It is full of petrified striated shells or pectinites. The greatest part of these petrifactions are, however, only impressions of the hollow side of the shells. Now and then I found also petrified pieces of the shell itself, though I could never find the same shells in their natural state on the shores; and it seems inconceivable how such a quantity of impressions could come together, as I shall presently mention.

I have had great pieces of this limestone, consisting of little else than pectinites lying close to one another. This limestone is found on several parts of the isle where it lies in horizontal strata of the thickness of five or ten inches. This stone yields a great quantity of white lime, but it is not so good as the former, because it grows damp in wet weather.

Fir wood is reckoned the best for the lime kilns and the thuya wood next to it. The wood of the sugar maple and other trees of a similar nature are not fit for it, because they leave a great quantity of coals.

Gray pieces of granite are to be seen in the woods and fields hereabouts.

The leaves of several trees and plants began now to get a pale hue; especially those of the red maple, the smooth sumach (*Rhus glabrum* L.), the *Polygonum sagittatum* L., and several of the ferns.

A great cross is erected on the road, and the boy who accompanied me told me that a person was buried there who had wrought great miracles. Those who went by touched their caps when they passed the cross.

At noon I arrived at Sault au Recollet, which is a little place situated on a branch of the St. Lawrence River that flows with a violent current between the Isle of Montreal and the Island of Jesus. It has gotten its name from an accident which happened to a Recollet friar called Nicolas Viel, in the year 1625. He went into a boat with a converted Indian and some native Hurons in order to go to Quebec; but on going over this place in the river, the boat upset, and both the friar and his proselyte were drowned. The Indians (who have been suspected of occasioning the upsetting of the boat) swam to the shore, saved what they could of the friar's effects and kept them.

The country hereabouts is full of stones, and settlers have but lately begun to cultivate it, for all the old people could remember the places covered with tall woods, which are now turned into grain fields, meadows, and pastures. The priests say that this place was formerly inhabited by some converted Hurons. These Indians lived on a high mountain, at a little distance from Montreal, when the French first arrived here, and the latter persuaded them to sell that land. They did so, and settled here at Sault au Recollet, and the church which still remains here was built for them, and they have attended divine service in it for many years. As the French began to increase on the Isle of Montreal, they wished to have it entirely to themselves, and persuaded the Indians again to sell them this spot and go to another. The French have since prevailed upon the Indians (whom they did not like to have with them because of their drunkenness and rambling idle life) to leave this place again and go to settle at the Lac des Deux Montagnes, where they are at present and have a fine church of stone. Their church at Sault au Recollet is of wood, looks very old and dilapidated, though its inside is tolerably good, and is used by the Frenchmen in this place. They have already brought a quantity of stones hither, and intend building a new church very soon. The botanical observations which I made during these days, I shall reserve for another publication.

Though there had been no rain for several days, the moisture in the air is so great that as I spread some papers on the ground this afternoon, in a shady place, intending to put the seeds I collected into them, they were so wet in a few minutes that they were useless. The whole sky was very clear and bright, and the heat as intolerable as in the middle of July.

Husbandry. One half of the grain fields are left fallow, alternately. The fallow grounds are never plowed in summer, so the cattle can feed upon the weeds that grow on them. All the seed used here is spring seed, as I have before observed. Some plow the fallow grounds late in autumn, others defer that business till spring; but the first way is said to give a much better crop. Wheat, barley, rye, and oats are harrowed, but peas are plowed into the ground. Farmers sow commonly about the fifteenth of April, and begin with the peas. Among the many kinds of peas which are to be gotten here, they prefer the green ones to all others for sowing. Peas require a high, dry, poor soil, mixed with coarse sand. (They did not know what it meant to stake the peas.) The harvest time commences about the end and sometimes in the middle of August. Wheat returns are generally fifteenfold and sometimes twentyfold; oats from fifteenfold to thirtyfold. The crop of peas is sometimes fortyfold, but at other times only tenfold, for it varies very much. The plow and harrow are the only implements of husbandry they have and those are not of the best sort. The manure is spread out in spring. The soil consists of a gray stony earth, mixed with clay and sand. They sow no more barley than is necessary for the cattle, for they make no malt here. They sow a good deal of oats, but merely for the horses and other cattle. Nobody knows here how to make use of the leaves of deciduous trees as a food for cattle, though the forests are furnished with no other than trees of that kind, and though the people are commonly forced to feed their cattle at home during five months of the year. No hobbled cattle were seen, and no ditches, unless there was enough water actually to drown the fields.

I have already repeatedly mentioned that almost all the wheat which is sown in Canada is spring wheat, that is, such as is sown in spring. Near Quebec it sometimes happens, when the summer is less warm, or the spring later than usual, that a great part of the wheat does not ripen entirely before the cold commences. I have been assured that some people who live on the Isle of Jesus sow wheat in autumn, which is better, finer, and gives a more plentiful crop than the spring wheat; but it does not ripen more than a week before the other wheat.

SEPTEMBER THE 25TH

In several places hereabouts, they enclose the fields with a stone fence instead of wooden pales. The large amount of stones which are to be gotten here renders the labor very trifling.

Here is an abundance of beech trees in the woods, and they now have ripe seeds. The people in Canada collect them in autumn, dry them and keep them till winter, when they eat them instead of walnuts and hazel nuts; and I am told they taste very good.

There is a salt spring, as the priest of this place informed me, seven French miles from here, near the river d'Assomption; they made a fine white salt from it during the war. The water is said to be very briny.

Fruit and Nut Trees. Some kinds of fruit trees succeed very well near Montreal, and I had here an opportunity of seeing some very fine pears and apples of various sorts. Near Quebec the pear trees will not grow because the winter is too severe for them, and sometimes they are killed by the frost in the neighborhood of Montreal. Plum trees of several sorts which were first brought over from France, succeed very well and withstand the rigors of winter. Three varieties of American walnut trees grow in the woods, but the walnut trees brought over from France died almost every year down to the very root, bringing forth new shoots in the spring. Peach trees cannot well thrive in this climate; a few bear the cold, but for greater safety they are obliged to put straw round them. Chestnut trees, mulberry trees, and the like, have never yet been planted in Canada.

Land Owned by Clergy and Noblemen. The whole cultivated part of Canada has been given away by the king to the clergy and some noblemen; but all the uncultivated parts belong to him, as likewise the place on which Quebec and Trois Rivières are built. The ground on which the town of Montreal is built, together with the whole isle of that name, belongs to the priests of the order of St. Sulpicius, who live at Montreal. They have given the land in tenure to farmers and others who are willing to settle on it, so much that they have no more upon their hands at present. The first settlers paid a trifling rent for their land; for frequently the whole lease for a piece of ground, three arpens broad and thirty long, consisted of a couple of chickens; and some pay twenty, thirty, or forty sols for a piece of land the same size. But those who came later had to pay near two écus (crowns) for such a piece of land, and thus the land rent became very unequal throughout the country. The revenues of the Bishop of Canada do not arise from any landed property. The churches are built at the expense of the congregations. The inhabitants of Canada do not yet pay any taxes to the king; and he has no other revenues from it than those which arise from the custom house.

A Mill. The priests of Montreal have a mill here where they take the fourth part of all that is ground. However, the miller receives a third part for his share. In other places he gets half of it. The priests sometimes lease the mill for a certain sum. Besides them, nobody is allowed to erect a mill on the Isle of Montreal, they having reserved that right to themselves. In the agreement drawn up between the priests and the inhabitants of the isle, the latter are obliged to get all their grain ground in the mills of the former. The mill is built of stone with three waterwheels and three pairs of stones. I noticed first that the wheels and axles were made of white oak; secondly that the cogs in the wheel and other parts were made either of the sugar maple or of *Bois dur* (*Carpinus ostrya*), because that was considered the hardest wood there; third, that the mill-stones had come from France and consisted of a conglomerate and quartz grains, both of the size of hazelnuts and ordinary sand, all bound together by white limestone. The stones were said to be sufficiently hard. The kernels were shaken down from the funnel in the manner described before.

They make a good deal of sugar in Canada of the juice running out of the incisions in the sugar maple, the red maple, and the sugar birch; but that of the first tree is most commonly used. The way of preparing it has been more minutely described by me in the *Memoirs of the Royal Swedish Academy of Sciences,* 1751.[1]

September the 26th

Autumn. Early this morning I returned to Montreal. Everything began now to look like autumn. The leaves of the trees were faded or reddish, and most of the plants had lost their flowers. Those which still preserved them were the following;[2]

Several sorts of asters, both blue and white.

Golden rods of various kinds.

Common milfoil.

---

[1] See Bibliography, item 16.

[2] *Asteres. Solidagines. Achillea millefolium. Prunella vulgaris. Carduus crispus. Oenothera biennis. Rudbeckia triloba. Viola Canadensis. Gentiana Saponaria.*

Common self-heal.

The crisped thistle.

The biennial oenothera.

The rough-leaved sun flower, with trifoliated leaves.

The Canada violet.

A species of gentian.

Wild grapevines are abundant in the woods hereabouts, climbing up very high trees.

Indian Food. I have made inquiry among the French, who travel far into the country, concerning the food of the Indians. Those who live far north I am told cannot plant anything on account of the great cold. They have, therefore, no bread, and do not live on vegetables; meat and fish are their only food, and chiefly the flesh of beavers, bears, reindeer, elks, hares, and several kinds of birds. Those Indians who live far southward, eat the following things. Of vegetables they plant corn, wild kidney beans (*Phafeoli*) of several kinds, pumpkins of different sorts, squashes, a kind of gourd, watermelons and melons (*Cucumis melo* L.). All these plants have been cultivated by the Indians long before the arrival of the Europeans. They likewise eat various fruits which grow in their woods. Fish and meat constitute a very large part of their food. And they like chiefly the flesh of wild cattle, roe-bucks, stags, bears, beavers and some other quadrupeds. Among their dainty dishes they reckon the water taregrass (*Zizania aquatica* L.), which the French call *folle avoine*, and which grows plentifully in their lakes, in stagnant waters, and sometimes in rivers which flow slowly. They gather its seeds in October, and prepare them in different ways, and chiefly as groats, which taste almost as good as rice. They make also many a delicious meal of the several kinds of walnuts, chestnuts, mulberries, acimine (*Annona muricata* L.), chinquapins (*Fagus pumila* L.), hazel nuts, peaches, wild prunes, grapes, whortleberries of several sorts, various kinds of medlars, blackberries and other fruit and roots. But the species of grain so common in what is called the Old World were entirely unknown here before the arrival of the Europeans; nor do the Indians at present ever attempt to cultivate them, though they see the use which the settlers make of the culture of them, and though they are fond of eating the dishes which are prepared from them.

September the 27th

Beavers. Beavers are abundant all over North America and they are one of the chief articles of trade in Canada. The Indians live upon their flesh during a great part of the year. It is certain that these animals multiply very fast; but it is also true that vast numbers of them are annually killed and that the Indians are obliged at present to undertake distant journeys in order to catch or shoot them. Their decreasing in number is very easily accounted for, because the Indians, before the arrival of the Europeans, only caught as many as they found necessary to clothe themselves with, there being then no trade with the skins. At present a number of ships go annually to Europe, laden chiefly with beavers' skins; the English and French endeavor to outdo each other by paying the Indians well for them, and this encourages the latter to extirpate these animals. All the people in Canada told me that when they were young all the rivers in the neighborhood of Montreal, the St. Lawrence River not excepted, were full of beavers and their dams; but at present they are so far destroyed that one is obliged to go several miles up the country before one can meet one. I have already remarked above that the beaver skins from the north are better than those from the south.

The Beaver a "Fish". Beaver meat is eaten not only by the Indians but likewise by the Europeans, and especially by the French, on their fasting days; for his Holiness the Pope has, like many of the old zoologists, classified the beaver among the fishes, since he spends most of his time in water. The meat is reckoned best if the beaver has lived upon vegetables, such as the aspen and the beaver tree (*Magnolia glauca* L.); but when he has eaten fish, it does not taste so well. To-day I tasted this meat boiled for the first time; and though everybody present besides myself thought it a delicious dish, yet I could not agree with them. I think it is eatable, but has nothing delicious about it. It looks black when boiled and has a peculiar taste. In order to prepare it well it must be boiled from morning till noon, that it may lose the strange taste which it has. The tail is likewise eaten, after it has been boiled in the same manner and roasted afterwards; but it consists of fat only, though they would not call it so, and cannot be swallowed by one who is not used to eating it.

Much has already been written concerning the dams or houses of the beavers; it is therefore unnecessary to repeat it. Sometimes, though but seldom, they catch beavers with white hair. In American cities one can now get as fine beaver hats made as one ever could in France or England.

The Fasting of Catholics. In connection with the eating of beavers the fasting of the Catholics appeared to me a bit strange. Those who first inaugurated the fast days did it undoubtedly with good and holy intentions to keep the people from eating too much meat, which is injurious to health, fattens the body too much, and makes it inadaptable for many things. But it seemed to be quite enough for them during the ordinary fast days to abstain from meat. If they could afford it they lived everywhere sumptuously and fed their body just as on the other days of the week; for they then had more courses prepared of eggs, of all kinds of fish, prepared with oils and fats, all kinds of milk dishes, and many especially sweet and good tasting fruits with a quantity of wine. So that for the most part wherever you ate on a fast day the table was better provided with varieties of food than on any other days, and still they called it fasting, *jours maigres* they named them.

Wine is almost the only liquor which people above the common class drink. They make a kind of spruce beer of the top of the white fir [1] which they drink in summer; but the use of it is not general and it is seldom drunk by people of quality. Great sums go annually out of the country for wine, as they have no grapevines here of which they could make a liquor that is fit to be drunk. The common people drink water, for it is not yet customary here to brew beer of malt; and there are no orchards large enough to supply the people with apples for making cider. Some of the people of rank who possess large orchards, sometimes out of curiosity get a small quantity of cider made. The people of quality here, who have been accustomed from their youth to drink nothing but wine, are greatly at a loss in time of war, when all the ships which bring wine are intercepted by the English privateers. Towards the end of the last war, they gave two hundred and fifty francs, and even one hundred écus, for a *barrique,* or hogshead, of wine.

---

[1] *Epinette blanche.* The way of brewing this beer is described at large in the *Memoirs of the Royal Swedish Academy of Sciences* for the year 1751, p. 190.—F.

Prices of Commodities. The present price of several things I have been told by some of the most prominent merchants here is as follows: an average horse cost forty francs and upwards; a good horse is valued at a hundred francs or more. A cow is now sold for fifty francs; but people can remember the time when they were sold for ten écus. A sheep costs five or six livres at present; but last year, when everything was dear, it cost eight or ten francs. A hog one year old, and of two hundred or a hundred and fifty pounds weight, is sold at fifteen francs. M. Couagne, the merchant, told me that he had seen a hog of four hundred pounds weight among the Indians. A chicken is sold for ten or twelve sols, and a turkey for twenty sols. A minot [1] of wheat sold for an écu last year; but at present it costs forty sols. Corn is always the same price as wheat because there is little of it here, and it is all used by those who trade with the Indians. A minot of oats costs sometimes from fifteen to twenty sols; but of late years it has been sold for twenty-six or thirty sols. Peas have always the same price as wheat. A pound of butter costs commonly about eight or ten sols; but last year it rose up to sixteen sols. A dozen eggs used to cost but three sols; however, now they are sold for five. They make no cheese at Montreal; nor is there any to be had, except what is gotten from abroad. A watermelon generally costs five or six sols, but if of a large size, from fifteen to twenty.

There are as yet no manufactures established in Canada, probably because France will not lose the advantage of selling off its own goods here. However, both the inhabitants of Canada and the Indians are very badly off for want of them in times of war.

Marriages. Those persons who wish to be married must have the consent of their parents. However, the judge may give them leave to marry if the parents oppose their union without any valid reason. Likewise if the man be thirty years of age, and the woman twenty-six, they may marry without waiting for their parents' consent. All they have to do is to go to the priest, who reads the banns three Sundays in succession in church, after which the ceremony may take place in the church in the presence of as few or as many people as they desire. Priests do not like to perform marriages in the homes.

---

[1] A French measure, about the same as two bushels in England.—F.

September the 29th

This afternoon I went out of town to the southwest part of the island in order to view the country and the husbandry of the people, and to collect several seeds. Just before the town are some fine fields, which were formerly cultivated but which now serve as pastures. To the northwest appears the high mountain which lies west of Montreal. It is very fertile and covered with fields and gardens from the bottom to the summit. On the southeast side is the St. Lawrence River, which is very broad here; and on its sides are extensive grain fields and meadows, and fine houses of stone which look white at a distance. At a great distance southeast appear the two high mountains near Fort Chambly, and some others near Lake Champlain, raising their tops above the woods. All the fields hereabouts are filled with stones of different sizes, and among them there is now and then some black limestone. About a French mile from the town the highroad goes along the river which is on the left; and on the right all the country is cultivated and inhabited. The farmhouses are three, four or five arpens distant from each other. The hills near the river are generally high and pretty steep; they consist of earth, and the fields below them are filled with pieces of granite and black limestone. About two French miles from Montreal the river runs very rapidly and is full of rocks; in some places there are waves. However, those who go by boats into the southern parts of Canada are obliged to work through such places.

Right outside of the city were a couple of windmills; they were built like others I have seen in this land, except that instead of having thin boards for wings they had linen, which was removed after a grinding.

Most of the farmhouses in this neighborhood are of stone, partly of the black limestone, and partly of other stones in the neighborhood. The roof is made of shingles or of straw. The gable is always very high and steep. Other buildings, such as barns and stables, are of wood.

Wild geese and ducks began now to migrate in great flocks to the southern countries.

OCTOBER THE 2ND

The two preceding days, and today, I employed chiefly in collecting seeds.

Last night's frost caused a great alteration in several trees. Walnut trees of all sorts were now shedding their leaves very fast. The flowers of a kind of nettle (*Urtica divaricata* L.) were entirely killed by the frost. The leaves of the American lime tree were likewise damaged. In the kitchen gardens the leaves of the pumpkins were all killed. However, the beech, oak, and birch, did not seem to have suffered at all. The fields were all covered with a hoarfrost. The ice in the pools of water was a geometrical line and a half in thickness.

The biennial oenothera (*Oenothera biennis* L.) grows in abundance on open woody hills, and fallow fields. An old Frenchman, who accompanied me as I was collecting its seeds, could not sufficiently praise its property of healing wounds. The leaves of the plant must be crushed and then laid on the wound.

The *Soeurs de Congregation* is an organization of religious women different from nuns. They do not live in a convent, but have houses both in the town and country. They go where they please, and are even allowed to marry if an opportunity offers; but this, I am told, happens very seldom. In many places in the country there are two or more of them: they have their house commonly near a church and generally the parsonage is on the other side of the church. Their business is to instruct young girls in the Christian religion, to teach them reading, writing, needlework and other feminine accomplishments. People of fortune board their daughters with them for some time. They have their boarding, lodging, beds, instruction, and whatever else they want, on very reasonable terms. The home where the whole community of these ladies live, and from which they are sent out into the country, is at Montreal. A lady that wants to join them must pay a large sum of money toward the common expenses, and some people believe it to be four thousand livres. If a person be once received, she is sure of a subsistence during her lifetime.

La Chine is a pretty village, three French miles to the southwest of Montreal, but on the same isle, close to the St. Lawrence River.

The farmhouses lie along the river side, about four or five arpens from each other. Here is a fine church of stone, with a small steeple; and the whole place has a very agreeable location. Its name is said to have had the following origin. When the unfortunate M. Salée was here, who was afterwards murdered by his own countrymen further up in the country, he was very intent upon discovering a shorter road to China by means of the St. Lawrence River. He talked of nothing at that time but this new short way to China. But as his project of undertaking the journey in order to make this discovery was stopped by an accident which happened to him here, and he did not at that time come any nearer China, this place got its name, as it were, by way of a joke.

This evening I returned to Montreal.

October the 5th

Government. The governor-general at Quebec, is, as I have already mentioned before, the chief magistrate in Canada. Next to him is the intendant at Quebec; then follows the governor of Trois Rivières. The intendant has the greatest power next to the governor-general; he pays all the money of the government, and is president of the board of finances and of the court of justice in this country. He is, however, under the governor-general; for if he refuses to do anything to which he seems obligated by his office, the governor-general can give him orders to do it, which he must obey. He is allowed, however, to appeal to the government in France. In each of the capital towns, the governor is the highest person, then the lieutenant-general, next to him a major, and after him the captains. The governor-general gives the first orders in all matters of consequence. When he comes to Trois Rivières and Montreal, the power of the governor ceases, because he always commands wherever he is. The governor-general commonly goes to Montreal once every year, and usually in winter, and during his absence from Quebec, the lieutenant-general commands there. When the governor-general dies, or goes to France before a new one has come in his stead, the governor of Montreal goes to Quebec to take command in the interim, leaving the major in command at Montreal.

Trade. One or two of the king's ships are annually sent from

France to Canada, carrying recruits to supply the places of those soldiers who have either died in service or have gotten leave to settle in the country and turn farmers, or to return to France. Almost every year France sends a hundred or a hundred and fifty people over in this manner. With these people it likewise sends over a great number of persons who have been found guilty of smuggling in France. They were formerly condemned to the galleys, but at present they send them to the colonies, where they are free as soon as they arrive, and may choose what manner of life they please, but are never allowed to go back to France without the king's special license. The king's ships likewise bring a great quantity of merchandise which the king has bought to be distributed among the Indians on certain occasions. The inhabitants pay very little to the king. In the year 1748 a beginning was made however, by laying a duty of three per cent on all the French goods imported by the merchants of Canada. A regulation was also made at the time that for all furs and skins exported to France from here one should pay a certain duty; but for what is carried to the colonies one pays nothing. The merchants of all parts of France and its colonies are allowed to send ships with goods to this place, and similarly the Quebec merchants are at liberty to send their goods to any place in France and its colonies. But the merchants at Quebec have but few ships, because the sailor's wages are very high. The towns in France which trade chiefly with Canada are Rochelle and Bourdeaux; next to them are Marseilles, Nantes, Hâvre-de-Grace, St. Malo, and others. The king's ships which bring goods to this country come either from Brest or from Rochefort. The merchants at Quebec send flour, wheat, peas, wooden utensils, etc. in their own bottoms, to the French possessions in the West Indies. The walls round Montreal were built in 1738 at the king's expense, on condition that the inhabitants should, little by little, pay off the cost to the king. The town at present pays annually 6000 livres for them to the government, of which 2000 are given by the seminary of priests. At Quebec the walls have likewise been built at the king's expense, but he did not redemand the expense of the inhabitants, because they had already the duty upon goods to pay, as above mentioned. The beaver trade belongs solely to the Indian Company in France, and nobody is allowed to carry it on here except the people appointed by that company. Every other fur trade is open to everybody. There are several places among the Indians far in the country where the French have stores of their goods; and these places they call *les postes*. The king has no other fortresses in Canada than Quebec, Fort Chambly, Fort St. Jean, Fort St. Frédéric, or Crown Point, Montreal, Frontenac, and Niagara. All other places belong to private persons. The king keeps the Niagara trade all to himself. Everyone who intends to trade with the Indians must have a license from the governor-general, for which he must pay a sum proportionate to the advantages for trade. A merchant who sends out a boat laden with all sorts of goods, and four or five persons with it, is obliged to give five or six hundred livres for the permission; and there are places for which they give a thousand livres. Sometimes one cannot buy the license to go to a certain trading place because the governor-general has granted or intends to grant it to some acquaintance or relation of his. The money arising from the granting of licenses belongs to the governor-general; but it is customary to give half of it to the poor: whether this is always strictly observed or not I shall not pretend to determine.

The Catholic Church Service. No other religion was tolerated here except the Catholic. It was said by all those who had been in France that people of both sexes in Canada were more devout than they were in France; nowhere could they go to church more regularly than here. Most of the service was in Latin. It seemed as if the whole service was too much of an external *opus operatum*. Most of it consisted in the reading of prayers with a rapidity which made it impossible to understand them, even for those who understood Latin. I could only get a word now and then and never a whole sentence, so that the common man could certainly get nothing of it nor derive any benefit from it. Even the best Latin scholar could not possibly keep his thoughts together and pray fervently at such break-neck speed. In fact, this must have been impossible for the priests themselves. The sermon was in French, and all quotations from the Scriptures were first given in the Latin Vulgate and then translated into French. Even the clericals, however, had difficulty in speaking Latin, since the words they needed did not appear in their missals. It was customary both in the city and country, both upon rising and retiring, to kneel in prayer; but whether this was in Latin or French I did not wish to be inquisitive enough to ask.

Although I paid particular attention to the matter, I never saw a Bible in any house, either in French or Latin, except at the residences of the clergy. But I saw a few French and Latin prayer books, though most of them were prayers to the Holy Virgin rather than to Almighty God.[1]

---

[1] This passage is omitted in Forster. Marchand condenses a portion of it from the Dutch translation, but in doing so makes Kalm declare that he had never seen a Bible in the hands of any priest or monk.

SUPPLEMENTARY DIARY

[HERE begins the recently discovered continuation of Kalm's Swedish journal of his American travels, which was published at Helsingfors, Finland, in 1929 by Fredr. Elfving. Since this part consists primarily of brief and more or less unpolished notes and accounts, unedited by the author himself, a certain freedom of form has necessarily been adopted in the following translation of it to make it a little more readable and more in conformity with the letter and spirit of the preceding Englished portion. However, in order to preserve a certain local color, Kalm's spelling of foreign words has been retained, wherever practical. The meteorological observations at the head of each daily entry have been assigned to the appendix, as were those of the original edited and published by Kalm].

OCTOBER 7, 1749

IN MONTREAL

Obstacles to the continuation of my journey, also the cause why I was not allowed to pass through Forts Frontenac and Niagara, but was forced to return by the same route as I had come, that is, through Fort St. Frédéric.—I was now almost ready to set out from here and was occupied yesterday with putting my seeds into small bags or cornets in order that I might to-day go from here up the river to Fort Frontenac; but just as I was about to start I received the Governor-general Marquis la Jonquière's letter, which frustrated this plan. On my departure from Quebec I received from him a passport to go through Fort St. Frédéric (now Crown Point), since I considered it impossible to get anyone in Oswego to take me from there to Albany. On my arrival in Montreal I was permitted not only to talk with the English who were bringing home the French prisoners (the latter of whom said that I could easily go through Oswego), but I also talked with the commander of Fort Frontenac, who informed me that on the plain was found an abundance of Indian rice (*Fol. avoine*), also red cedar and herbs, the medicinal value of which he praised most highly. Besides, he knew them well and promised to show them to me when I came there; thereupon he proceeded ahead to the place. As I was not able to get a tenth of the Indian rice in the region about Montreal which I ought to have had, and as red cedar and the other plants were not to be found here, I was compelled to write again to the governor-general and ask permission of him to plan my route through Fort Frontenac and Niagara to Albany. I gave as reasons that I had been sent by her Royal Highness [the Queen of Sweden]; that she had ordered me unconditionally to procure a generous supply of seeds of the Indian rice and other useful herbs and plants; that I was to take a route which neither I (nor any other botanist) had ever travelled before; and that consequently I hoped to discover much that was new and useful, at the same time fulfilling the request and hopes of the Swedish Academy of Sciences, not to mention other reasons. Monsieur Longueuil, governor of Montreal, placed no difficulties in the way of granting this request. But I considered it necessary that the governor-general should be consulted about the matter. Just as I was flattering myself the most over the great discoveries I was about to make on this journey and what [specimens] I should be able to collect, I received the following letter:

A Quebec le 26 S:bre 1749

Je suis bien fâché, Monsieur, de ne pouvoir pas changer l'arrangement que j'ai pris pour votre voyage, et d'être obligé de vous refuser la permission que vous me demandez de passer par les forts Frontenac et Niagara. En faisant votre route par le Fort S:t Frédéric, vous aurez tout l'agréement que pouvez desirer, j'ai donné des ordres exprés pour cela, et je vous souhaite beaucoup de plaisir et de satisfaction dans votre voyage.

J'ai l'honneur d'être parfaitement, Monsieur, votre très humble et très obéissant serviteur

La Jonquière.

[Quebec, September 26, 1749]
[I regret very much, sir, not to be able to change the arrangements which I have made for your journey, and to be compelled to refuse you the permission for traveling via Forts Frontenac and

Niagara. In taking your route via Fort St. Frédéric you will have all the comforts you desire—I have given definite orders for that—and I wish you much pleasure and satisfaction on your trip.

I have the honor, etc.

La Jonquière.[1]]

As soon as I received this letter I went at once to the governor in the city, M. Longueuil, and advanced many reasons why he should allow me to make the journey as I had planned it for myself and showed him how important it was for me. But he gave as a reply that he could not deviate a hair's breadth from the orders he had from Governor-general la Jonquière, which he showed to me, and wherein the words were even more rigid, namely: "Ne permettez pas le sieur Kalm de passer par Fort Frontenac et Niagara etc." I saw then that it was impossible for me to do otherwise than take the path of least resistance.[2] Yet in order that I might still fulfill my duty to the Royal Academy of Sciences I sought once more to prevail upon M. La Longueuil of this city to give me permission to take a route which would be far more productive for me. For that purpose I prepared the following memorial in French, as well as I was able, and had been busy with it the greater part of the day:

Monsieur.

M'excusez, Monsieur, si je suis forcé de Vous incommoder; m'excusez aussi, Monsieur, si je ne puis pas m'expliquer si bien dans la langue française, comme je bien souhaiterais; mais il suffit pour moi, si Vous, Monsieur, pouvez comprendre ce que je vais de dire.

J'avais l'honneur de recevoir hiere, quand j'étois sur le point de partir d'ici, la lettre très-gracieuse, que Monsieur le Gouverneur General Marquis la Jonquière m'a fait l'honneur de m'écrire, et dans cela la reponse sur ce que j'avois l'honneur de demander de lui, d'avoir la permission de retourner d'ici par Fort Frontenac à Nouvelle Angleterre; Monsieur Marquis la Jonquière m'écrit qu'il ne peut pas m'accorder cela, mais que je suis obligé de prendre la route par Fort S:t Frédéric.

---

[1] The part in brackets is a translation of the French.

[2] The reason for denying Kalm's request was technically a military one. Kalm had just come from the English colonies and was to return to them. No chances were taken with anyone. Kalm, though otherwise treated with the greatest hospitality, was undoubtedly closely watched, especially when in the neighborhood of a fortification.

Comme d'un côté le temps ne me permet pas ou d'aller moimême à Quebec pour demander de nouveau cette permission, ou d'envoyer un exprès pour cela, parceque je serais obligé de perdre le demi d'une semaine et davantage, qui beaucoup à perdre pour moi dans . . . tems d'année quand plusieurs grains sont meurs et prêts de tomber; et comme je d'un autre côté ay raison de croire, que Monsieur le Gouverneur General peut être, n'a bien compris le sujet de mon voyage, car je ne puis douter, que si M:r la Jonquière l'a bien compris, c'est tout impossible, qu'il a pû me refuser une demande, que je n'ay pas fait pour mes plaisirs, mais pour suivre les ordres et l'instruction, que par les ordres du Prince et Princesse hereditaires de la Suede l'Academie royale des sciences m'a donné; comça, Monsieur, je suis obligé à peu de mots de Vous dire le sujet de mon voyage, et en même tems a Vous très humblement, que vous me voulez accorder la permission de passer par Fort Frontenac.

Si tôt, Monsieur, que la Suede avoit cette joye inexprimable de voir la Princesse hereditaire, la soeur du Roi de Prusse, arriver à Stockholm, le premier soin de cette grande Princesse etoit de suivre les traces et l'example de son glorieux Pere pour rendre un royaume florissant et en état de pouvoir faire une veritable assistance aux ses allies contre leurs enemies; c'étoit pour cela, qu'elle parloit avec les senateurs du Royaume de Suede, qui étoient membres de l'Academie royale des Sciences de Suede, et depuis aussi avec tous les autres membres du dite Academie, de penser sur tous les moyens d'une affaire si importante.

L'Academie ne manqua pas de donner pour reponse à cette grande Princesse, qu'entre autres moyens pour reussir dans une proposition si utile et avantageuse pour la Suede, ce sera très important, si on envoyera qu'elqu'un de membres de la même Academie à l'Amerique Septentrionale; on savoit, que dans Canada et dans Nouvelle Angleterre ils se trouveroient plusieurs arbres, grains, herbes, froment et . . . , qui n'etoient pas dans l'Europe, et cependant, ou étoient très bonnes pour manger, ou pour la teinture, ou pour autres usages; on sçavoit aussi, que le froid étoit si dur dans l'Amerique Septentrionale comme dans la Suede, et par conséquent, que toutes les arbres et les herbes, qui peuvent croître dans Amerique Septentrionale, peuvent avec la même facilité croître et être plantés dans la Suede, sans être tué ou détruits par le froid,

comme ces plantes et arbres, qu'on a fait introduire de France et
d'autre pays chaudes d'Europe; si on pouvoit avoir des grains de ces
arbres et herbes, surtout de ceux, de lesquelles on sçavoit quelque
utilité, c'étoit un grand moyen pour rendre la Suede encore plus
florissante.

Si tôt que la Princesse a reçu cette reponse, Elle venoit Ellemême
dans l'Academie des Sciences, et ordonnoit, que la même Academie
choisiraient un de ses membres pour cet voyage le plûtôt que pouvoit se faire: Elle même voulait aller chez le Roy pour lui prier
de donner ordres à ses ambassadeurs à Paris et à Londres de pro-
curer pour celui, que l'Academie des Sc. jugeroit habile d'entreprendre cet voyage, tout l'agreement, toute soureté, toute liberté
de voyager par tout dans Amerique Septentrionale òu il voudra,
sans être empêché de suivre l'instruction que lui sera donné pour
satisfaire aux plaisirs de grande Princesse si utiles pour la Suede.
Le Roy donna tout à l'heure ces ordres avec plaisir, et on étoit per-
suadé, que le Roy de France n'aimeroit quelque chose tant que
satisfaire entierement les desirs d'une Princesse, pour laquelle il
avoit toujours eu une telle estime. Elle pressa Elle-même Monsieur
Lanmarie l'Ambassadeur de France d'aussi écrire à sa cour pour
cela; peu après j'étois choisi pour entreprendre cet voyage, et je
recevois les ordres de l'Academie de me preparer pour la même.
La Princesse même aussi bien que Monsieur Lanmarie m'assuroient,
que je pouvois être très assuré de cela, qu'ici dans Canada j'avois
liberté de voyager par tout comme j'étois dans ma patrie, tout
dans la même maniere, qu'on a permis dans Suede aux messieurs
Academiciens de Paris de fair par tout; de Nouvelle Angleterre Elle
n'oseroit dire le même, parcque la cour de Suede et Celui d'Angleterre n'étoient dans une telle alliance comme celui de France et
Suede. La Princesse me faisoit donner une instruction pour mon
voyage, de laquelle, Monsieur, voici quelques articles, et jugez de
la, si c'est permis à moi de suivre mes plaisirs.

Article 2. Il faut, que vous voyager dans ces endroits d'Amerique
Septentrionale, qui pour le froid ont le plus grand rapport avec
la Suede (N) sur tout dans Canada, parceque le froid là est si
grand, comme dans la Suede et les peu des herbes, que nous avons
ici dans la Suede de Canada peuvent resister au froid si beaucoup,
comme ceux de la Suede même.—Article 3. Quand vous trouvez
quelque arbre, ou quelque plante, qui est connue pour quelque

grand utilité, ou pour manager, ou pour la teinture, ou pour un excellent bois, etc. en prenez les grains toujours le plus au nord que vous pouvez trouver cette arbre ou cette plante.

Artic. 4. Les grains que nous specialement demandons exprés, que vous, quand vous retournez, aurez en grande quantité, sont les grains de l'arbre Meurier, Chataigne, Noix de toutes sortes, Bled d'Inde, Fol. Avoine, Myrtus de quoi on fait les chandelles, Cedar rouge et blanc, toutes les plantes que les sauvages mangent, Sassafras, Erable de quoi on fait le sucre, Chinkapins, Pommes de Terres, Taho, Taki, Raisins sauvages, etc.—Artic. 5. Nous avons de vous cette confiance, que vous comme un sçavant pouvez trouvei telles plantes, aussi, de quoi l'utilité sera aussi grande pour votre patrie, comme les predits.

Selon ces ordres, Monsieur, et selon cette instruction je suis parti de Suede. A Londres l'Ambassadeur de Suede ne pouvoit pas recevoir quelque passeport pour moi du Roy d'Angleterre pour quelque méfiances qui alors étoient.

[Sir:[1]

Pardon me, Sir, if I am obliged to bother you; excuse me also Sir, if I am not able to express myself as well in the French language as I should wish, but it will suffice if you, Sir, are able to understand what I am going to say.

I had the honor of receiving yesterday, when I was on the point of leaving here, the very gracious letter which M. the Governor-general Marquis la Jonquière had done me the honor of writing, and in it was the reply to the matter which I asked of him, namely, to have permission to return from here by way of Ft. Frontenac to New England. M. la Jonquière wrote me that he was not able to grant that, but that I must take the route via Ft. Frédéric.

Since, on the one hand, time does not permit me to go myself to Quebec to ask that permission again, or to send a messenger for it, because I should be obliged to lose half a week or more, which is too much for me to lose at this time of the year when all the seeds are ripe and ready to fall; and since I, on the other hand, have reason to believe that Monsieur the Governor-general, has not, perhaps, understood very well the purpose of my trip, because I

---

[1] The following in brackets is a translation of the French letter.

do not doubt that if Monsieur la Jonquière had understood it, it would have been quite impossible to refuse a request which I make not for my own pleasure but in order to follow the instructions which have been given me at the command of the Hereditary Prince and Princess by the Swedish Royal Academy of Sciences. Therefore, Monsieur, I am obliged to say a few words to you on the subject of my voyage and at the same time beg you to grant me permission to pass Ft. Frontenac.

As soon, Monsieur, as Sweden had that inexpressible joy of seeing the Hereditary Princess, sister of the King of Prussia, arrive in Stockholm, the first duty of that grand Princess was to follow the footsteps and example of her glorious father in making a flourishing Kingdom and a state able to give real assistance to her allies against her enemies. It was for that reason that she spoke with the senators of the Kingdom of Sweden who were members of the Royal Academy of Sciences, and later with all the other members of the Academy also [about a proposed scientific journey], thinking of all the means to carry out a plan so important.

The Academy did not hesitate to offer in response to that grand Princess other suggestions for succeeding in a proposition so useful and so advantageous to Sweden. It would be very important if they sent a member of that same Academy to North America. It was known that in Canada and New England there were many trees, grains, herbs, cereals, etc., which were not grown in Europe, and which nevertheless were very good for food, dyes or other uses. One knew also that the cold in [some parts of] North America was as severe as in Sweden, and that, as a result, all the trees and plants which could grow in North America could with the same success be planted and grown in Sweden, without being killed or destroyed by the frost, as the plants and trees had been which had been introduced from France and other warm countries of Europe. If they were able to get some seeds of American trees and plants, and especially those which they knew to be useful, this would be a potent means of making Sweden still more prosperous.

As soon as the Princess received that response, she came herself to the Academy of Sciences and commanded it to choose one of its members for that voyage, as soon as he could undertake it. She even wished to go to the King's palace to beg him to give orders to the ambassadors in Paris and London to procure for him whatever the Academy of Sciences considered necessary to undertake the voyage, such as permission and freedom to travel everywhere in North America that he wished, and without being hindered in following the instructions that would be given him, to the satisfaction of the Princess and the benefit of Sweden.

The King at once gave these orders and it was believed that the King of France desired nothing so much as the entire satisfaction of the Princess, for whom he had always had a great esteem.

She herself urged Monsieur Lanmarie, the Ambassador of France, also, to write to his court about this matter. A little afterward I was chosen to undertake that voyage, and I received orders from the Academy of Sciences to prepare myself for it. The Princess herself as well as M. Lanmarie informed me that I could be assured in this matter; that here in Canada I would have the liberty to travel everywhere, as if I were in my own country, just as they permitted gentlemen from the Academy of Paris to go about everywhere in Sweden. The Princess gave me instructions for my voyage; here are some of the articles from them, Monsieur; judge from them whether I am to be permitted to follow out my desires.

Article 2. It is necessary that you travel in those parts of North America which have the reputation in Sweden of being the coldest in all Canada, because the cold there is as great as that in Sweden, and the few plants which we have in Sweden from Canada are able to resist the cold as well as those native of Sweden.

Article 3. Whenever you find a tree or plant which is known for some particular use, as a food or dye, or for its excellent wood, etc., take the seeds always from the farthest north that you are able to find that tree or plant.

Article 4. The seeds that we wish you to bring in great quantities are seeds of the meurier [mulberry] tree, the chestnut, nuts of all kinds, Indian corn [maize], Indian rice, myrtle from which candles are made, bayberry, red and white cedar, all the plants which the Indians eat, sassafras, maples, from which they make sugar, dwarf chestnuts (chinkapins), potatoes, taho (*Peltandia Virginica*), taki (*Oroutium aquaticum*) wild grapes, etc.

Article 5. We have confidence in you that as a scientist you will be able to find plants whose usefulness to the fatherland will be as great as is predicted.

According to these orders, Monsieur, and conforming to these instructions I left Sweden. In London the Swedish Ambassador was unable to procure for me a passport from the King of England on account of some lack of confidence.]

Birch-bark is said to be quite scarce in Canada and birch-bark canoes daily more expensive.

Birch-bark Canoes. All the strips and ribs in them are made of white cedar (*Thuya*); the space between the latter varying in breadth between that of a palm and the width of three digits. The strips are placed so close to one another that one cannot see the birch-bark between them. All seams are held together by spruce roots or ropes made of the same material split. In all seams the birch-bark has been turned in double. The seams are made like a tailor's cross-stitch. In place of pitch they use melted resin on the outside seams. If there is a small hole in the birch-bark, resin is melted over it. The inner side of the bark or that nearest the tree always becomes the outer side of the boat. The whole canoe consists ordinarily of six pieces of birch-bark only, of which two are located underneath and two on either side. The bark strip directly underneath is sometimes so long that it covers three fourths of the canoe's length. I have not yet seen a boat whose bottom consisted of one piece only. Birch-bark canoes are dangerous to navigate, because if the sail is forced down in stormy weather, it may splinter the bottom of the boat. If one knocks against a sharp or rough stone, a large piece of the bottom of the canoe may be ripped out. It is therefore evident that these boats are continually subjected to adventures and must often be repaired. On that account no one should set out in them without bringing resin and even birch-bark along, though the latter can generally be procured wherever one goes. Likewise it is possible to procure the spruce roots nearly everywhere, and, lacking these, pine roots are said to be equally serviceable.

[After this the Diary reproduces a "Description of some Esquimaux words," taken from Mr. [Arthur] Dobbs' *An Account of the Countries Adjoining to Hudson Bay*, 1744. This list, which consists of 152 words and phrases, has not been reprinted].

OCTOBER THE 8TH

Monsieur Picquet, a missionary,[1] an odd man who has travelled much here in Canada, called on me to-day. He tried in every way to convince me that Father Charlevoix, who has described Canada,[2] was a big liar who had gone far astray from the truth.

All day I was occupied with putting my seeds into strong paper cornets, with labelling them, and with packing my other things. In the evening I wrote to Governor-general Marquis de la Jonquière, thanked him for all his favors, yet at the same time made it known how it grieved me that I was not permitted to direct my course via Fort Frontenac and Niagara, especially since it is not an easy matter to send someone hither again such a long distance from Sweden. I also wrote to Monsieur Gauthier (his name is pronounced as *Gō'-thié*).

OCTOBER THE 9TH

From Monsieur St. Lucas [I learned the following]: Algonquin is the mother, whose daughters are

Nepissin	Saki
Outayouis	Masguta
Saulteurs	Kikaps
Lutouatauani	Tête de boule
Renards	Gens de terre, on Hudson Bay

S(i)eoux is a language by itself.

Puants is an entirely different language, guttural in sound; [those who speak it are a] brave [people]; they have quite a fondness for the French.

Têtes de boule (quite stupid) are fond of the French but do not wish to have them among themselves for any length of time, as

---

[1] François Picquet (1708-1781) was noted for his influence among the Indians. He fought in the Seven Years' War; was wounded at Quebec, 1759; and after the battle of the Plains of Abraham escaped to New Orleans in Indian dress, the English having put a price on his head. He died in France in great poverty.

[2] *Histoire et description générale de la Nouvelle-France*, 1-3, Paris, 1744, by Pierre François Xavier de Charlevoix (1682-1761). Pére Charlevoix was noted as teacher, explorer and historian. His name is too well known to need elaboration.

they believe them capable of magic. If they see a Frenchman's gun they think that death is imminent. As soon as a Frenchman comes to them, they bring forth all their fur products in order to make the Frenchman depart soon.

OCTOBER THE 10TH

Length of Montreal, 723 toises (un tois-6 feet); width, where broadest, 190; where narrowest toward the north, only 90.

A rainy fall indicates a winter without snow and *vice versa*.

The current in the St. Lawrence is so strong that when one wants to go from Montreal up to Fort Frontenac, one requires from nine to ten days for the trip. But when one goes down stream it takes ordinarily only two days. Between Montreal and Fort Frontenac the distance is 60 French miles, or as far as from Montreal to Quebec. Common people ordinarily called Fort Frontenac Fort Cataracoui.

From Montreal to Prairie à Magdal is reckoned three French miles, but they must be very short ones; from Prairie to Fort S. Jean four miles, some said five; but Monsieur la Croix, who has travelled that distance more (often) than anyone else and one of his relatives who said he had measured it, said that it was not more than four lieues or French miles; but he is mistaken.

[I must] write down where I lived in Montreal, and also note with what love and kindness Monsieur de Couagne, Mademoiselle Charlotte and Mademoiselle Couagnette received me—just as if I had been a child of theirs, nothing less.

OCTOBER THE 11TH

I had indeed intended to continue my journey to-day, but since we did not get enough horses, because the road over which we were to travel was pretty bad, I had to postpone my journey to the following day. I shall wait and see if we can get any to-morrow. In every other way Baron Longueuil, Governor of Montreal, had done all that could be expected on his part, for he had given the sergeant who accompanied me strict orders to procure horses for me from the local residents. But those who lived nearest had already put their horses under shelter [evidently at some distance away] and the others had theirs out in the large forests.

French Language. All are of the opinion that in Canada the ordinary man speaks a purer French than in any province in France, yes, that in this respect it can vie with Paris itself. Those Frenchmen who were born in Paris must in this particular commend the inhabitants of Canada. The majority of them, men as well as women, could not only read whatever had been written, but also could write fairly well. I saw women who wrote as well as the best penman could have written. For my part I was ashamed that I could not write as well. That women write well here is largely due to the fact that in this country one has to learn to write one kind of letters only, namely the Latin or French. Besides, every girl is eager to write a message to her lover without having to ask the assistance of another. One thing I noticed especially in the French language was that in the style of writing it has not the advantage of the Swedish. I found that as far as the art of writing in all languages is concerned, that the Swedish is the most natural, because we write rarely more letters than we read, and we pronounce almost all the letters we write; we do not as a rule write any superfluous letters. The opposite holds in the French as well as in the English language, in which one writes many letters which are unnecessary and which one neither reads nor pronounces, e. g. *ils parlent* is written *ls, ent* more than is necessary, as one needs only *i parl,* and all the other characters are useless, and so mostly in all other cases. As a result one-third more characters are found in a French book than are necessary or, in other words, a French book could be one-third smaller than it is, if the unnecessary characters were removed. It was a pleasure to see how the women in Canada, who had not paid much attention to the French spelling and art of writing, followed in their writing the natural method so that they rarely wrote more characters than were essential and just as a Swede would write. As an instance I asked them to write *elles parlent* without telling them how it is spelled, only that it referred to more than one person. They wrote *ell parl* as a Swede would have written it, and likewise in nearly all words which I asked them to reproduce. This I observed to be so in Quebec, Montreal and here; if there were any words which they were familiar with through reading, they would write them correctly (written this A. M. in Prairie).

OCTOBER THE 12TH

The Journey. I had bethought myself of setting forth from Prairie during the morning before noon, but as the priest there, a rather civil and educated man, Monsieur Lignerie, together with the captain of said place, sent a message to me and urged me, if it were possible, to postpone my journey until after the mass so that the people could hear it, I found it my duty not to refuse such a just request. Monsieur Lignerie made the service shorter than was customary. After the mass at noon we set forth. I had as baggage four carts, two horses for each, in addition to the horse I rode, all of which was provided for me by the government in this country and paid for with the money of the French crown. The roads were unrivaled in wretchedness, wet and winding so that my horse sank in the mire up to his belly in most places. The weather was also rather mean and so rainy that one could scarcely lift up one's eyes. A large part of the trees had lost their leaves and the woods appeared rather barren.

The larch (*Larix*) grew in abundance on both sides of the road, in some places in sandy soil; the seeds were now ripe and I took a lot of them. The trees which stood nearest the road were not large, about twenty-four feet tall, but the persons who accompanied me said that one sometimes finds them as tall as the largest pines and proportionately broad. The tree is said to be good for lumber, likewise for fuel. They called them everywhere here the red American larch (*Epinette rouge*), quite distinct from that which bore that name in Quebec. Everyone knew that it lost its leaves during the winter and some of them had already begun to fade.

Indian Dances. In the evening of the fourth I arrived here at St. Jean. Several Indians were here who were out on a hunting expedition. When an Indian goes hunting, he does not go alone but takes his whole family with him, also his belongings; that is, his wife, children and dog. He then travels around in the forests shooting all kinds of animals. He eats the meat and preserves the skins to sell, for which he receives in return his clothes and ornaments, also his gunpowder, shot and other articles purchased from the Europeans. The Commander, Monsieur de Ganne (Gannes?), honored me by allowing some of the Indians to dance for me during the evening. The men danced first and then the women. (One of them, who was the leading dancer now stands beside me and is looking at what I am writing; he has a . . . over himself which he has daubed full of vermilion, smokes his pipe, etc.). Before they came out to dance, they painted their faces red and adorned themselves according to their custom. The women as well as the men painted their faces red. A drum was lent them which they struck regularly, one beat after the other, singing at the same time. One of them got up and began to dance; he pulled off his shirt and had on only enough to shield his nakedness, namely a blue cloth which was tied about his waistline and went from his back between his legs and up in front. Over this hung a cloth like a short apron or skirt. This cloth between the legs was of such a kind that it covered both the podices. When he danced he had an axe in his hand which he turned to and fro in the air. He indicated the time carefully with his feet. Sometimes the beats of the drum were further apart, sometimes quite close, and the Indian danced accordingly. Now and then he talked to the others who sang and beat upon the drum and they answered him. Sometimes they sang continuously, for the most part these words: Here I am, Here I am, etc. He turned now to this side, now to that, while he danced. He stood most of the time on one spot, but sometimes he hopped rapidly with both feet together over the whole floor or round about the floor. But the most amusing dance of all was the war dance which they danced when they were to go forth to war. In this they make known all the manners and motions which they use in warfare. He had an axe in his hand, danced a little while standing, then he threw himself down and began to creep about on his knees. Now he imitated with his hands the motion of paddling, as though to search out an enemy; now he looked backwards and with signs of the hand, nodding of the head, etc. he intimated that he saw the enemy and that the others should proceed slowly. Sometimes when on all fours he pressed his body near to the ground to make known that the enemy was close at hand. Again he sprang up hastily, ran away, struck with his axe and ran back to the others crying that he had now conquered. Sometimes when he crept on all fours he pressed his head hastily down upon the ground to hide himself. Now he proceeded slowly on hands and knees; he would motion with his hands to show how they remove the leaves as they go so that the enemy may not find their tracks. Sometimes when he was crawling about on all fours, he gathered together the sand on the floor, washed his hands with it, then heaped it up again and washed arms, waistline, etc. Then again when still crawling about, he turned suddenly backwards to make known that he thought the enemy had caught sight of him and that now it was time to withdraw. Sometimes while on all fours he suddenly sprang up, fell upon one of those who stood close at hand, threw this person under himself, gave him a good tug, ran away and thus intended to show that he had obtained the skull of his enemy, whereat he either sang or talked until he came to the others. While he was crawling on the floor he was rather quiet, but in the other dances he very frequently gave forth the tribal war cry, which indeed sounded horrible. Several others then danced, but all in the same manner. The women danced most simply; they all stood in a row, side by side, moved their feet forwards and back again, so that when they moved the right foot forward, they moved the left one back and then the opposite. When they had danced awhile thus, and turned their faces toward one side, they turned about, moving their feet in the same fashion as described. When they had danced for a time in this fashion they turned about in the same manner as they had danced before, and that constituted their whole dance. The hands were kept constantly hanging by their sides. When the men danced they often moved their hands back and forth. Once in awhile they all danced in a ring, the men foremost, the one after the other, and then the women often in the same manner, but they generally moved their hands back and forth at their sides, that is to say, one forward, when the other was back; that in brief was their dance.

OCTOBER THE 13TH

The Latitude of Various Places in Canada. Monsieur La Croix had a brass compass on which was engraved the latitude of various places in Canada, namely:

Paris	49	Cataracoui	45
Kebec (Quebec)	47	F. . .	44
Outouvas (Ottawa)	47	Michilimaquina	46
Tadoussac	48	Niagara	43

Les Sauteurs	48	Le Detroit	43
Trois Rivières	46	Les Illinois	40
L. Royalle	46	Les Miamis	40
Missisagé	46	Folles Avoines	44
Chambli	45	Pontaovas	41
Montréal	45	Fort Louis	41

The common man in Canada is more civilized and clever than in any other place of the world that I have visited. On entering one of the peasant's houses, no matter where, and on beginning to talk with the men or women, one is quite amazed at the good breeding and courteous answers which are received, no matter what the question is. One can scarcely find in a city in other parts, people who treat one with such politeness both in word and deed as is true everywhere in the homes of the peasants in Canada. I travelled in various places during my stay in this country. I frequently happened to take up my abode for several days at the homes of peasants where I had never been before, and who had never heard of nor seen me and to whom I had no letters of introduction. Nevertheless they showed me wherever I came a devotion paid ordinarily only to a native or relative. Often when I offered them money they would not accept it. Frenchmen who were born in Paris said themselves that one never finds in France among country people the courtesy and good breeding which one observes everywhere in this land. I heard many native Frenchmen assert this. The women in the country were usually a little better dressed than our [Swedish] women. They always had night-gowns, and the girls curled and powdered their hair on Sundays. During the week the men went about in their homes dressed much like the Indians, namely, in stockings and shoes like theirs, with garters, and a girdle about the waist; otherwise the clothing was like that of other Frenchmen. The women in the country frequently had such shoes too, except on Sundays. Everywhere the girls were alert and quick in speech and their manner rather impulsive; but according to my judgment and as far as I could observe, they were not as lustful and wanton as foreigners generally claim the French to be. They are somewhat free of speech, but indeed I believe them sufficiently restrained.

Thuya. This fortress, St. Jean, was constructed entirely of wood.

In place of masonry work they had put up heavy logs of arbor vitae of eighteen to twenty-four feet in length and height, one log being placed quite close to the other. They had chosen this tree because no other tree had been found in the whole of Canada which withstood the rot in the ground and was as durable as this one. I counted the annual rings on the largest and found them as follows: N. B. The annual rings were tolerably plain. One log had 92 annual rings, with a diameter of 12 inches; another had 139 annual rings, with a diameter of 15 inches; and still another had 136 annual rings with a diameter of 15 inches. One with 134 annual rings had a diameter of 16 inches was the largest I saw, and another with 142 annual rings had a diameter of 16 inches.

The sugar maples grew in great abundance in the woods here. They had already to a large extent lost their leaves, but the small trees still had fresh, green leaves. It was the largest only which had seeds; the others were without them, doubtless because they stood so close together in the forests and the small trees seldom had the opportunity of being exposed to the sun. I call small those which were from 48 to 60 feet high, as the sugar maple grows to a rather great height in Canada and is among the tallest trees in the country. The seeds had to some extent already fallen from the trees, and those which still had any hardly fell over when cut down, before all the seeds had fallen off and lay on the ground. I believe that I have written before that the sugar maple has like other kinds of maple two seeds side by side, but it differs in this respect from the others, that one of these seed pods is always empty, a hollow capsule only. This is invariable. Thus, for example, when I gathered a couple gallons of seeds from this sugar maple, half of them were useless. It was curious that almost everyone with whom I talked in Canada, although they had tapped the trees and made sugar from more than a thousand sugar maples, they did not know when I asked them whether the sugar maple had any seed. Many were of the opinion that they had already fallen at midsummer. The right time for these leaves to fall is at the end of the month of September, or more correctly, when the first frost has come in the fall; because shortly thereafter the leaves fall and with them the seed from the tree. Yet the seeds often remain after the leaves have fallen. The place where this tree grows is in level and low-lying forests in a rich and fertile soil.

Poisson armé [1] was said to be common in Lake Champlain, also in the brook and river which flow out of it and by Fort St. Jean. It was likewise plentiful at Niagara; yet no use had been found for it. This fish destroys and devours all other fishes.

Indians. A great number of the natives, i. e. the confederates of the French, had already begun to dress like the French: the same kind of jacket and vest, while on journeys they wore the same red cap or hat. But one could not persuade them to use trousers, for they thought that these were a great hindrance in walking. The women were not so quick to give up the customs of their forefathers and clothe themselves according to the new styles, but stuck to the old fashions in everything. But wait! Some had . . . caps of homespun or of coarse blue broad-cloth. When the French are travelling about in this country, they are generally dressed like the natives; they wear then no trousers, but do not carry Indian weapons. Monsieur Croix related that when the Indians go out during the summer to steal a march upon their enemies, they bind green grass about their heads, creep along the ground, pressing their bodies as close as possible to the earth, and move very stealthily to the place where their enemies are or those whom they wish to surprise. The enemy then cannot see them, but he thinks that it is the green grass only which is moving, and quick as a flash the adversary is upon his throat. When they . . . dance their war dance, they often bind their heads with green grass to depict this. The natives farther south among the Illinois have another way of sur-

---

[1] Through the efforts and courtesy of Professor Albert G. Feuillerat of Yale the editor has been able to identify this fish, an identification which, quite naturally, proved impossible to the Finnish editor of this part, in 1929. Professor Feuillerat quotes: "The *poisson armé* is the appropriate name for those species of gar [or garfish] that infest the waters of Louisiana [the French name having been brought from Canada to the South by the Acadians]. The gar has long, narrow jaws full of sharp teeth, and its body is protected by hard rhombic scales. It is highly destructive of other kinds of fish and its flesh is rank and tough. The alligator gar (*Lepisosteus tristoechus*, Bloch and Schneider) attain a length of eight to ten feet. Two other fresh water species are the long-nosed gar (*Lepisosteus osseus* L.) and the short-nosed gar (*Lepisosteus osseus* Raf.).

The Choctaw Indians knew the gar as the "strong fish"— *náni ḳállo* or *náni ḳamássa*. The Indians made use of the gar's sharp teeth to scratch or bleed themselves with, and their pointed scales to arm their arrows, says William Bartram (*Travels*, 176).

From *Louisiana French* by William A. Read, Ph. D. University Studies, No. 5, Louisiana State University Press, 1931, p. 61."

prising their enemy, namely by imitating the sounds of all kinds of birds and quadrupeds, a practice which they make use of when they run about in the woods at night. They lie in wait at a place where they know the Frenchmen or their enemies are, with their rifles cocked, and imitate the sound of some bird, etc. to entice the enemy to shoot at it. When the enemy comes close, he knows nothing before the others bear down upon him. They have a way of enticing the roe bucks to them. They tie the head of a roe to the back of their own head, crawl along the ground where they know the roe deer are, make sounds like one of these animals, which immediately comes to them. But as soon as the Indian gets the animal as close to himself as he wishes, he fires his gun, which he has had cocked and ready. When the French travel with their wares among those natives who live in the southernmost regions, they have to keep careful watch during the night and be alert in daytime, since they do not know what kind of Indians they come in contact with, and since a great many of them do not wish to let slip the opportunity to kill the Frenchmen in order to get their goods. The natives are tremendously rugged. I saw them going about these days with only a shirt on and a weapon hanging over it, often without shoes [moccasins], though they had on their . . . or stockings. The men wore no trousers, the women a short, thin skirt; neither of the sexes had anything on their heads. Thus they travelled at this time through the forests on their hunting trips, both in good and bad weather. They lay in this manner during cold and rainy nights in the damp and wet forests without having any other clothes to put under or on top of themselves at night than those they wore during the day. Consequently they carried their beds with them wherever they went. When they came in to Montreal to buy anything and when they left, the women had to carry heavy loads on their backs, but the men went as gentlemen without carrying anything except their guns, their pipes and their tobacco pouches.

October the 14th

At seven o'clock in the morning we set out in the name of the Lord in a canoe which was quite small and rather heavily loaded. Three natives accompanied us in a similar boat. There was a gentle wind, but it was directly against us.

We saw American natives frequently on both sides of the river on which we were paddling. They had their quarters for the night on the shore, as it was the season when they were out hunting deer.

Almost all the trees had faded leaves; some of them were bare. The red oak and also the aspen still had fairly green leaves.

The ordinary species of birch grew thickly on the sides of the river in the lowlands. The land on both sides of the river on which we were travelling was low and level. The birches had snowwhite, smooth bark or one just like ours.

I was shown one place on the western side of the river, about a mile and a half from St. Jean, where a small woodland had started up. It was said that in Count Frontenac's time 1800 natives who had come from Crown Point had camped on the place.

Yesterday we saw large numbers of wild geese in flight.

The red oak had nearly everywhere green leaves, as had the white oak here and there.

Expenses. At Fort St. Jean I received from the stores on the King of France's account two pounds of gunpowder, eight pounds of lead and shot and a lot of knives to be given to the natives whom I might encounter in the forests, in return for game which they have. Moreover the keg was filled with brandy for the same purpose. The fresh meat which I had brought along was cooked here and prepared for the journey by the commander's cook. I took along ½ quire of paper. They also asked me to state my needs and the mere mention of them meant fulfillment. The natives who had danced received considerable brandy in return, likewise the persons who acted as guides and drivers for me, not to mention the food and other provisions for them.

Bulrushes (*Scirpus pallidus altiss.*) grew profusely everywhere on the banks of the river.

Horsetails were also plentiful in some spots on the river banks.

Milium festucoides (millet grass) I shall call a grass which I found for the first time at Prairie à Magdal [1] near the shores of the St. Lawrence River and again to-day on the shores of this river which flows by Fort St. Jean. . . .

I took seeds from this grass. Some of the seeds had already fallen.

---

[1] La Prairie de la Madeleine, now called Laprairie.

Dactylus foliosus . . . I shall call a kind of grass which grew in lowlands near the banks of the river. . . . The seeds were now mature and the grass itself had begun to wither. I do not know what this grass can be used for, yet I took seeds from it and preserved a specimen of the plant.

Prinos. The inkberry grew as a rule here in the woods in level and somewhat low-lying places. The branches were now full of red berries which I tasted and found rather bitter. A man who accompanied me had told me the same; he called the shrub *bois de marque*, but did not know any use for it. I found this shrub quite common near the city of Montreal and learned that the shrub, the leaves of which I preserved at La Chine under the name of Andromeda, was absolutely the same.

The Canadians are generally good marksmen. I have seldom seen any people shoot with such dexterity as they. A bird that flew so close to them that they could reach it with their bullet or shot, had difficulty in escaping with its life. There was scarcely one of them who was not a clever marksman and who did not own a rifle.

The flowering fern (*Osmunda filix baccif.* Charlevoix) grows abundantly in low places in the woods. I gathered some of its seeds.

The buttonbush (*Cephalanthus*) flourished in the lowlands. The seeds seemed to be ripe.

We stopped for the night a little south of the windmill, still on the western side of the lake. They reckoned that it was about ten leagues (thirty English miles) from this place to St. Jean.

Three native women also came in their canoe and took shelter for the night next to us. They had no man with them, yet each of them had a gun, for they had set out to shoot ducks. One was married, the other two were said to be single. They were Abenaquis Indians. The native who accompanied us during the whole journey was an Iroquois Indian. It is singular that an Abenaquis and an Iroquois rarely take lodgings together, yet they now and then intermarry. The women who had come hither had their funnel-shaped caps, trimmed on the outside with white glass beads. They also had on the French women's waists and jackets which I had never before seen natives wearing. Their evening meal consisted of corn and native Iroquois beans boiled together.

The beach pea (*Lathyrus*) flourished among the rocks on the shore; it was very luxuriant as it hugged the rocks. The seeds had mostly fallen from the pods, although the stem itself and the leaves were green. It seems advantageous to sow these among rocks on the shore and thus make them useful.

The wild bean (*Feverolles*) also grew abundantly on the shore among the rocks. Its seeds were ripe, but had not yet to any extent dropped out of the pods. Some of the leaves and stalks had already begun to wither and fade.

Here the shores were full of cobble stones, most of them of the black variety of which Fort St. Frédéric is built. Parts of bed rock which consisted entirely of this variety of stone were also seen.

The French called the linden "bois blanc" (white wood). The Indian women used its bark in place of hemp for laces with which to sew up their shoes. They were busy during the evenings sewing up their footwear with this material and I could have sworn that it was a fine hemp cord they used. They take the bark, boil it in water for a long time, pound it with a wooden club until it becomes soft, fibrous and like swingled hemp. They sat twisting them on their thighs.

They had made mats of the rushes (*Scirpus pal. altiss.*) upon which they lay at night. These mats were very good looking.

The canoes were not allowed to remain in the water or near the river banks during the night; instead everything was taken out of them, they were carried up on the shore, turned upside down and left there until they were needed. The reason for this was that if a storm should come up during the night, they might be dashed to pieces against the rocks in the river; also this precaution prevented water from entering them and thus bring about rotting.

OCTOBER THE 15TH

At daybreak in the morning we continued our journey; the weather was a little cool. The shores were everywhere lined with black stones.

Arbor vitae grew nearly everywhere on the shore especially where crags and boulders or stones were plentiful. It preferred to grow in such places. I saw it nowhere of any great height, 24, 30 to 36 feet, seldom higher. In some spots it was rather thick. Most of the seeds were now ripe. Some of the cones had already opened and lost theirs, especially those growing in strong sunlight. The seeds of the others were about to drop. The arbor vitae grows readily in swamps and marshes.

The bed rock on the shore consisted of black limestone which was deposited in strata in the same way as that found at Fort St. Frédéric. I discovered in many places in it that petrified substance known as *Cornu ammonis,* which resembles somewhat a ram's horn because of its curves. The diameter of some was about a foot, yet they had three to four circles only. There was considerable space between the circles or twists so that they did not lie close upon one another.

The cranberry tree (*Femina*), a kind of *opulus,* flourished in some places on the shore. We consumed great quantities of the berries which were ripe. They had a pleasant, acid flavor and tasted right well. Even if we had had some other fruit here, we should not have scorned these.

Yesterday and to-day also we saw black squirrels in the woods. This squirrel is quite common about Fort St. Frédéric, but north of Montreal it is rather scarce and hard to find. They have these instead of the gray squirrel. Whenever either of these becomes aware of a human being in the woods, it begins to chatter and make considerable noise as it sits in trees, and one has difficulty in making it keep quiet. I have often been angry at it on that account.

The *brown partridges* which are described in the article on Raccoon were found in great numbers here in these forests. We shot several of them, which at this season were very fat birds. The Frenchmen called them *perdrix.*

Yesterday we shot a fat robin redbreast, also described in the account of Raccoon. He is called *Merula* by the Canadians.

The Indian dogs which had erect ears were said to be without equal in discovering wild cats and beavers. At this season the natives were busy hunting deer, but at the same time they took pains to see if they could discover any beaver dams, and if they found them they cut their mark into them. When a native comes to such a place and discovers that another has cut his mark into it before him, he does not touch it nor does he go there later to shoot the beavers, but considers it as a place that belongs to another which he is not supposed to touch.

Pike. The native who accompanied us to-day killed with his oar a pike which came close to his canoe when he was near the shore. It was one of our ordinary pikes, about two feet in length.

Bears are found in great numbers in this country, including the white bears in the Hudson Bay region. The ordinary bears found here are as a rule not as angry as ours, yet if they are harmed they seek to vent their anger upon the person who injured them and then sometimes do as great damage as ours. When one shoots a bear here, the meat is eaten and is considered almost as valuable as pork. The fat or suet is retained and melted; the oil made thereof is preserved. Not only the natives but also the French, especially on their journeys, use this oil in place of butter for stewing and preparing their food. Just recently when I was at Fort St. Jean, Madame la Croix had no oil with which to prepare the salad. She used bear oil therefore, and the salad tasted almost as good as with the usual cotton seed oil, though the flavor was a bit peculiar. When the French are travelling far up into the country their only food is corn. The latter is prepared thus: they put it in lye for one hour until the hull becomes loosened; then they wash it well so that the taste of the lye is removed. The kernels are then dried and carried along in bags on their journeys. They take these, add a little bear oil, some fat of the roe deer and hog's lard mixed, boil it and eat it. When the natives have lean and dried meat, they pour bear oil into a dish and dip strips of the meat before mentioned into it and eat it. The Indians, particularly the women, often oil their hair with it. It is said that the hair grows better because of it and that this oiling prevents the hair from becoming matted.

Polecat. The animal which the English call polecat, the French call *Bête puante* (stinking animal), also *Enfant du diable* (child of the devil), also *Pekan*. Monsieur Chaviodreuil said that he had eaten the meat of this animal a few times and found it as good as pork. But, one must be careful that the bladder is not injured and the contents mixed with the meat.

The *common juniper* (*Juniperus vulg.*) grew on a crag fifty-four miles (18 lieues) from Fort St. Jean on the side of a bay which the river made there and which was named for the roe deer. The juniper bushes were quite small and almost hugged the earth. I found no berries on them. Last summer when I was at Fort St. Frédéric the soldiers told me that juniper bushes were common in that neighborhood, although I do not remember at present whether I saw them. Old soldiers who now accompanied me and had formerly been stationed at Fort St. Frédéric said that they were found on many promontories round about that region and also believed that they were to be found at Lac St. Sacrement.

The French call those trees *Cedar rouge* which the English call red cedar and which the Swedes in New Sweden call *Röd En* (red juniper); whereas the French call arbor vitae *Cedar blanc* (white cedar), but according to the English, white cedar is a kind of cypress which grows in the swamps of New Sweden. Monsieur Lusignan, former commander at Fort St. Frédéric, assured me that there is another kind of *Cedar blanc* which grows in morasses near St. Joseph's River and according to all descriptions and comparisons was the cypress just mentioned. As far as the red juniper is concerned, I found it to-day in considerable numbers on the same point where I found the ordinary juniper. It did not grow there to any great size, the largest being about six inches in diameter. It grew in the clefts of the mountains and in other poor and barren places. Arbor vitae and the latter grew there together. Only a few had any berries. This is the place farthest north where I have found this red juniper. I saw that it liked the same habitat as the common juniper, because they grew there together. Monsieur Valsen, the commander at Fort Frontenac, told me that the red variety is unusually abundant about that fortress. The berries on the shrubs I discovered to-day were for the most part ripe. It receives its name from the fact that when it grows old, the wood inside or the pith (not the surface) becomes almost as red as blood. This is considered the most durable of all woods in the English provinces; in Canada where this shrub does not grow, they consider the arbor vitae as the most durable. The latter is, again, not found in the English provinces, so naturally I do not know which of the two is the more durable.

Gulls of the common variety flew in great numbers over the lake to-day.

The oars used in Canada were as a rule made of maple. This wood was preferred to all others. Lacking this they made the oars of Norway maple or ash. They did not propel the canoes in the same fashion as we do, turning our back in the direction we desired to go, and drawing the oars toward us, for they paddled forwards. In the same fashion they paddled the small canoes which were hollowed out of a part of a tree, but they rowed the boats in the same way as we.

The rocky ledges nearest the shore consisted entirely of the dense black limestone mentioned before. Sometimes small lighter-colored spar particles, frequently even fossil shells were discovered in them. They were arranged in strata of different thicknesses. Sometimes the stratum was three or less inches thick, again it was six, twelve or more inches. I saw in one place strata up to sixty inches in thickness. They were not horizontal, but lower toward the south and elevated toward the north. They were here and there cut off by perpendicular veins. There was petrified shell found in this rock, also the Cornu ammonis or likenesses of ram's horns. Frequently this rock crumbled into very small pieces or nearly dust; elsewhere it was as hard as granite. This made up the foundation of nearly all the islands in the lake, and if one happened upon those whose sides were steep, one had difficulty in getting up on them. They were overgrown with arbor vitae and other conifers.

The place or point where the common juniper and the red juniper or cedar grew is on that part of the map [1] which is designated as belonging to Monsieur Vincent, and just at the point where the red ink line runs out on the right of the same name, almost directly opposite Grand Isle.

In the evening we put up for the night just north of Rivière au Sable. This river is clearly indicated on the map, but has been given no name. It is on the western side of the lake, right opposite Isle Valcour. It is said to have received its name from the fact that along the shore lies a long sand bar of about six feet in height, with the lake on the eastern side and low-lying morasses and land on the western. A great many black ducks which the French called *outards* were swimming outside of this bar.

---

[1] The map is not reproduced in the original.

The beach pea (*Lathyrus maritimus*) grew in great quantities on the shore, mostly in the dry sand. It was green and luxuriant, but I could not discover any fruit on it or that it had had any this year.

Sweet-gale (*Myrica gale*) flourished here in low places. I took cones and catkins from this. The French called it *poivrié*, and I believe that it was said that one can dye yellow and other colors with its catkins. The leaves had begun to fade.

The buttonbush (*Cephalanthus*) grew here likewise abundantly. It was remarkable that although the greater number of them grew in the morass situated above here as their natural habitat, yet they were found in many places on the sand banks. Either they grew there when the place had been a swamp, and as the sand had been blown there they kept their earlier position and adapted themselves to the sand; or seeds had been carried thither and begun to germinate in the sand. The leaves of those growing on the sand banks had nearly all fallen, but those on the bushes in the morass were still left, though about to fall. The seeds were more nearly ripe on those growing in the sand than on those in the swamp.

Two kinds of pine (*Pinus*) flourished on the sand bars here and were fairly large.

The natives usually make a fire during the night, both summer and winter, when they camp in the woods. One would think that as a result there would be many forest fires during the summer, but I was given to understand that although at times fires do start during the summer, it seldom happens, for the natives themselves are very careful to put out the fires wherever they have made them, inasmuch as it serves their own interest. If a fire should break out and destroy the forest and the vegetation, the roe deer would flee from this region with the result that their hunting would be much less successful.

OCTOBER THE 16TH

We continued on our journey in the morning, but as the weather was against us, with a stiff wind blowing and our small boat heavily loaded, we neither dared nor could proceed. After rowing about a league we were compelled to seek the shore and wait until the weather changed or the wind spent itself. We found a haven at the mouth of a small stream, which it was said was a tributary of the same river on whose banks we had camped at night. The entire shore was sandy; some of the sand had the same characteristics as desert sand, namely that it was swirled about by the wind. The monkey flower (*Mimulus*) grew in moist places on the banks of the river. The seeds were ripe and had partly fallen from the pods.

Drift Sand. The sand mentioned above which was found everywhere on the shore, had almost the same qualities as drift sand, that is, it was whirled about by the wind. Whenever it had come in contact with a tree and come to a standstill, it formed a mound. It consisted of small, fine, clear, round and globular quartz, among which were to be found small particles of black mica. A few unusual plants which were found nowhere else grew in this sand.

The giant reed (*Arundo*) . . . , *or* the reed which the Dutch planted on their sandy banks, grew abundantly everywhere on the ridges here. The spikelets were already dry and the seeds either ripe or scattered. The leaves were still partially green, that is, the leaves of this year's growth, but it had also on the same stem or root some dry leaves, which were last year's growth. It was interesting to see how thick and profusely it grew on the sandy mounds with one small tuft close to the other.

The southernwood (*Artemisia abrotanum*). A species of this also grew in large quantities everywhere in the dry sand. A specimen was preserved and seeds gathered. The stems of most plants were dry and the seeds had fallen. Some had not yet advanced to this stage and some were still in bloom. It grew everywhere in the sand, one plant a short distance from the other. It is said to be a biennial because those plants which had borne fruit this year had dried up entirely, but everywhere green leaves without stems were visible. These I learned were those which get stems and bear fruit next year.

The beach pea flourished in many places in the sand. Nowhere did I find any pods upon them.

A kind of small willow (*Salix*), twenty-four to thirty inches tall, with laniferous leaves was found growing here and there. A specimen was preserved.

These [mentioned] were the plants which grew in the sand; otherwise there were several others that flourished here, some on the shore and some in the low-lying country, as, for example: the *water hemlock* (*Cicuta ramis bulbiferis*).

I recall an herb which grew here and which I called *Dentarioides umbellifera*. There was one plant only. I called it "Dentarioides" because its seeds resembled those of the toothwort, though attached to the end of the branches, and "umbellifera" because it is said to be in the shape of an umbel. The seeds were labelled with the former name and the herb was preserved.

The *bulrush* (*Scirpus culmo triquetro*) grew on the shore here and there in the damp sand. The spikelet was on the upper side; it was now full grown. Seeds were gathered and a few specimens of this, a perennial, were preserved.

Of the *bog rush* (*Juncus*) I took two varieties, one which resembled somewhat the *Juncus capit. Psyllii*, but had round leaves, the other had a cluster or clusters on the sides. Both were growing in wet sand on the shore. Seeds were taken from both.

A small *St. John's wort* came forth through the wet soil, (*Hypericum parvum in humidis proveniens*). Seeds were gathered and a specimen prepared. This wort always grows in very wet places and is about four inches high, perhaps slightly taller. Leaves are close to one another on the stem, and the creeping root here and there sends forth new shoots.

A larger St. John's wort came forth through the wet soil. (*Hypericum majus in humidis proveniens*). This wort grew in the same places as the former. It was not only very rare, but also eleven times larger, and among the largest Hyperica. I gathered some seed but could not save a specimen, as the leaves fell off as soon as I touched them.

I call by the name *Phlox* (?) a plant which flourished in sandy, dry places. The seeds had nearly all disappeared. At the root were green leaves. I got a few seeds and preserved a plant.

Buttonwood (*Platanus occidentalis*). One of the men who accompanied me went up the river near which we were stopping, and when he came back he brought with him some small balls of the fruit of the buttonwood which he called the cotton-tree (*Cotonier*). This is one of the places farthest north where it is found. The seed did not seem to be ripe. This tree is not found in the vicinity of Montreal.

In the afternoon the wind came up and we continued our journey. As the wind increased and our boat began to leak badly we did not dare to sail any farther, but were compelled to seek a haven at-a place opposite the Isle au Chapon, on the western side of the lake.

Description of Shore. The shore here as well as in many other places consisted entirely of rounded stones of varied composition. I do not know whether there had been more water in this lake formerly than now. I shall describe how the lake appears. Next to the water the shore is largely filled with rounded stones as just mentioned, either of the complex variety or the black limestone. The shore rises gradually up to six fathoms from the edge of the water, where there is ordinarily a rather steep bank of earth of about two to four fathoms height. The land above is usually level and free from rocks. This is how it appears at least in most places. Here and there are small steep cliffs of bed rock next to the water; in other places there are large rocks at the water's edge, then a high and steep bank of earth with round stones in it, and above it high mountains. Again there are sand banks which are next to the water, and often back of them is marshy land. Once in a while there are gentle, sloping sandy shores with a somewhat steep sand hill about thirty-six to sixty feet from the water's edge, above which the country is level. Occasionally the shore is so low and marshy that it is impossible to walk upon it without sinking down into the mud. This then, in brief, is the description of this lake's shore-line. Frequently there is a high bank next to the water, and back of it lowlands, level and large in area. On nearly all sides of the lake, a little distance away, are large and rather high mountains, nearly all overgrown with forests, and it is these mountains they hold responsible for the uncertain weather and the gales. I saw no visible indications that the water had decreased in the lake. It might have been possible, judging from the higher rocks which were three to five feet higher than the present water-line, but I am of the opinion that that condition might have obtained during the spring after a winter with much snow.

Hedysarum . . . flourished everywhere on the shore, even among the rocks, so that I wondered where it gathered any nourishment. It was entirely covered with seeds which clung to the clothes as one passed by. Many stems came from one root. It was now sending forth new shoots for next year. All this season's branches had died. It seems to be unusually well suited for sowing in stony places and on shores which are of little use to anyone. This plant makes splendid fodder for cattle. I gathered a sufficient quantity of its seeds.

Peas (*feverolles*) grew in the same spots as the preceding and may not be unsuited for fodder. They are good for planting in such places as I have mentioned above.

Tobacco Pouches. The native's way of protecting and carrying their tobacco with them was to cut it up and place it in the skin of an otter, a marten or of some other small animal. The feet, where they join the body, had been sewed up and an opening left under the chin. In other respects the skin was complete, as it had been taken from the animal, with head, legs and feet. The hairy side was on the outside. These skin pouches were adorned on the outside with red tassels, tin and brass trimmings. The Indian carried this tobacco pouch upon his arm, along with his tinder-box and pipe, wherever he went. As a rule the ordinary Frenchman made use of this same custom and the pouch of otter-skin was the most commonly used. A skin prepared thus cost thirty sous.

October the 17th

A Storm. At sunrise in the morning we continued our journey with the wind against us, though it was not yet very strong. About eight o'clock when we were almost opposite the Isle de Quatre Vents, the wind changed and became northeast, which was the best wind for us. We hoisted our sail and proceeded [on our journey], but our good fortune did not last long. The wind became steady, but it increased so that the waves began to be rather high. We found it wisest to seek the shore, and it was just in the nick of time, as before we reached land the wind had become so strong that we had to furl our sail, for fear that it might shatter our fragile boat. It looked rather bad for us to begin with, inasmuch as the shore nearest us was full of stones and so steep that we could not make a landing without endangering our lives. We were certain that even if we saved our lives, all that we had brought with us would be lost or spoiled. We were compelled, therefore, in spite of great danger, to row around a peninsula in order to get behind it into quiet waters. The waves were terribly high and the wind came in squalls, as it had done daily since we came to this lake. But God be praised! we got ashore safe and sound. May the Lord calm the winds so that we may soon get away from here! The place where we are now is in the inlet, which is formed by the promontory we rowed around and an extended cliff. From this cliff it is said to be six French miles to Fort St. Frédéric. It was 11 A. M. when we reached land.

Rocks embellished with peculiar markings were found about on the shore. I call them embellished from the fact that the water had eaten into them and had made figures just as if someone had carved a lot of foreign characters upon the whole rock. I have found here several such rocks. If one of these were carried far up into the forests to some little hillock and left there, some European, on coming to that place, would believe that it were a grave and that the symbols on the rock were a foreign and unknown script which the people who had come here in earlier times had made use of, and that it contained the life history of the one who was buried there. The rock was composed of a gray, closely grained sandstone mixed with a limestone of the same grain and color. Within were seen stripes of dark gray pyrites. Perhaps the same pyrites had covered the outside, and either the air, sun, water or rain had caused erosion and in some way had formed these characters.

We also found cockroaches in the place where we were resting. When we had made a fire by the side of an old decayed stump on the shore, a large number of these insects came creeping out of it driven out of their winter quarters by the fire and smoke. The French neither recognized them nor knew their name.

The *Shepherdia* (or *Lepargyræa Canadensis*) flourished everywhere on the shore of Lake Champlain, but I could never find any fruit on it. One of my followers called it *bois à perdrix* (a shrub for the partridge). I am not certain that he knew this shrub, yet he said that it had red berries and the partridges liked them.

The following were perennial plants, or at least they lasted for several years. They had begun to send forth new shoots at their roots: *Asclepias variegata* (*Periploca herbe à la puce*), a purplish weed; a species of *Hedysarum*; and the *Secale culmo multispiculo* [a variety of cereal grass].

A Traveler's Food. The French when making their journeys far up into the country to visit the Indians have as a rule during that long period of from one half to three years nothing else to live on than the hulled corn as described before, the fat of various animals which they mix with the corn and boil, and the game which they shoot in the forests. This is actually all the food they live on for such a long period, and still during this time they are making the most difficult and tiresome journeys of all. Yet they are in spite of it happy and merry, good-natured and healthy. It was on this account that the people of Canada considered corn a kind of grain which ought to be highly prized.

Oil Wells. A Frenchman in Montreal told me of an acquaintance of his who had travelled on the frontier of Carolina and there at the edge of a river had seen a well, out of which flowed an oil which floated on top of the water in the river.

Canadian Beverages. Canadians never used malt in brewing. The common man's drink was water. The better class, who had the means, used French wine, mostly red though sometimes white. Some made a drink from the spruce as described before. Cider was made occasionally by a person of rank just out of curiosity. Punch they thoroughly despised, and laughed at the English who made it. They said that they scarcely wished to taste it, although they had been in the English provinces several times. At mealtimes and whenever they ate, they drank frequently. Each one had his glass, which was filled. Between meals they seldom or never drank. The women drank much water, but also wine and wine mixed with water. The French drink much wine mixed with water at their meals. When one of them visited another between meals, the guest was never offered any refreshments except in the case of a close relative; the conversation furnished the only entertainment. Between the noonday and evening meals they sometimes would eat some fruit or sweetmeat, and they called that a collation.

In *Lake Champlain* I did not see anything that was new to me, neither plants nor anything else floating in it. The water was very clean, extremely clear and of excellent taste, with the exception of some places near the shore where some brook in the neighborhood emptied into the lake. Here the water tasted musty and somewhat swampy. Generally there were on both sides of this lake great stretches of level lowlands, which some day when they are populated and cultivated will make a glorious country, because the soil is so fertile. The greater part of the land about the lake has already been donated by the King to certain families of the gentry; consult my map.[1] The land about Fort St. Frédéric is said to belong to the King still, although it is to a great extent inhabited. On the eastern side of the lake are seen in the distance high mountains which separate Canada from New England.—The Abenaquis (Abnaki) Indians who wander about in these woods on the border are the Englishman's worst enemy.—They reckoned forty leagues (about 120 miles) from Fort St. Frédéric to Fort Chamblais. Fort St. Jean is four leagues this side of Fort Chamblais. The low-lying fields about the lake were said to be marshy.

Fishing. I have mentioned before various methods employed in fishing here. Now I wish to add another method to these. At Fort St. Jean I saw people fish by the light of torches in the same manner as we do. One of the men who was with me described how they sweep the bottom of the large lakes in Canada with seines, especially on Lake Superior, etc., and his account was just as if I had described the way in which we fish in the winter with a drag-net in Österbotten and Wöro [Finland].

Decrease in the Number of Animals, Birds, and Fishes. It is said that beavers and other animals, whose skins are sent to France, were formerly very numerous in the neighborhood of Montreal and the populated places in Canada. Now they have about disappeared there and it is necessary to travel far to shoot or bargain for them, and in the future it will be necessary to go still farther. Various kinds of birds have decreased in number, also the fish in some localities, though that is not yet so evident. Wolves are plentiful in this region.

Venomous animals that are a source of danger and injury were said not to exist in that part of Canada which is north of Fort Frontenac and Lake Ontario or north of Lake Champlain. But south of these places is found the rattlesnake, which is not to be trifled with. On this account Canada is considered to be very fortunate.

It was believed that the swallows migrated from here to a warmer region during the winter. The cuckoo was unknown in America.

---

[1] Whether or not Kalm prepared a special map to show the location of the various grants is not clear. It is not reproduced in the original. But see large map at end of this work.

Crickets were to be found in every house in Canada, but in the English provinces I have not yet seen them.

Fish of Peculiar Properties. Monsieur Valsen, Commander at Fort Frontenac, told me that a fish is caught in Lake Ontario which as yet has no name. If one eats this fish all the hairs on the body loosen so that one becomes bald. There are said to be several examples of this. Therefore the natives never eat this fish.

No *bees* are to be found in Canada, as they are said to perish during the winter. They cannot thrive there.

The *pearl oyster* has not yet been found in this country.

Oysters are found at the mouth of the St. Lawrence River, but seldom come the long distance to Quebec. The traveller occasionally brings to Montreal preserved oysters from New York. They are rather hard, but as they are scarce, one accepts them with the same appetite as one eats and accepts lobster in Stockholm, in spite of the fact that it is often half-decayed when it is received, and an inhabitant of Bohuslän [1] would never touch it.

The socalled tea bushes (*Chiogenes hispidula*) or creeping snowberries were found everywhere in the woods. My companions called them *grains de perdrix* because the partridges liked to eat their seeds.

A load of wood was said to cost 100 sous (4 shillings) at this time in Montreal. Formerly it was not priced so high. Monsieur Chavodreuil told me that when he first came to Montreal from France in 1716, he found in the city not more than three houses of stone; even some of the churches were built of wood. At present most of the houses are of stone and likewise all churches.

How the Natives Paint Designs on Their Bodies. I have related before that the Indians paint various designs on their bodies and that these are put on in such a way that they remain as long as the natives live. On their faces they paint figures of snakes, etc. Several of the French, especially the common people, who travel frequently about the country in order to buy skins, have in fun followed the example of the natives. However, they never paint their faces as the natives do, but another part of the body, as their chest, back, thighs and especially their legs. The designs they paint are

---

[1] A Swedish province on the southwestern coast of Sweden, where fresh lobster is plentiful.

made up of stripes, or they represent the sun, our Crucified Saviour, or something else which their fancy may dictate. As a rule the natives who are masters of the art adorn the Frenchmen. The color most used is black, and I do not recall seeing any other. Men who accompanied me, told me that they also use red paint and that black and red are the only colors used. The red dye comes from cinnabar which they here call vermilion. The black dye is made as follows: one takes a piece of alder, burns it completely and allows the charcoal to cool. Then the latter is pulverized. The natives do this by rubbing it between their hands. Then one puts this powder into a vessel, adds water to it, and allows it to stand until it is well saturated. When they wish to paint some figures on the body, they draw first with a piece of charcoal the design which they desire to have painted. Then they take a needle, made somewhat like a fleam, dip it into the prepared dye and with it prick or puncture the skin along the lines of the design previously made with the charcoal. They dip the needle into the dye between every puncture; thus the color is left between the skin and flesh.[1] When the wound has healed, the color remains and can never be obliterated. The men told me that in the beginning when the skin is pricked and punctured, it is rather painful, but the smart gradually diminishes and at the expiration of a day the smart and pain has almost ceased.

Every morning as we started to row one of the soldiers read the litany to the Virgin Mary which is found in the "Livre de Vie" for Saturday, if I remember correctly. This was never overlooked. It was always read in Latin and the other soldiers always answered either with the "misere nobis" or the "ora pro nobis" according to whether the prayer was directed to one of the saints or to the Virgin Mary. It was amusing to hear them read the Latin so zealously in spite of the fact that they did not understand the language. I also noticed that each and all of my companions hardly ever failed to kneel in prayer on arising and again at night on retiring.

OCTOBER THE 18TH

A storm continued throughout the day and prevented us from making any headway. We had to stay on that account in a place

---

[1] This is obviously a method of tattooing.

where there was no hunting because of the absence of animals.

Here and there in the woods was found the moosewood (or leatherwood) which the French generally called *bois de plomb*. It was used as cord for binding. One of the men said that if the root of this shrub was boiled in water and drunk, it acted as a strong purgative.

In the afternoon the storm abated somewhat and people began to venture forth on the lake. We saw first a few natives sail by us, and shortly thereafter the two English boats with the Englishmen on board who had escorted home the French prisoners captured in the last war with the English. This made me indignant, as I had left three days before they reached Montreal. But the treacherous weather made my journey to begin with a slow one, and now we had put our boat in such an inlet that the weather prevented our leaving it. We were compelled to remain until toward evening when the gale moderated and we eventually got away from there.

Wild ginger (*Asarum Canadense*) was found here and there in the woods where we stopped. The root had a strong aromatic odor and was said to be good to use in food. The French called it *gingembre*.

The *bearberry*[1] was also found here. The French gathered it and mixed it with the tobacco which they smoked.

Finally we left there in the evening and rowed far into the night until we came within six miles of Fort St. Frédéric. On some parts of the shore the rocks were perpendicular and in others they sloped slightly. In still other places where the rocks consisted of the black stone the water had worn away the rock at the base so that the upper part extended over the water. I paid particular attention this evening to the rocks and noticed that the height of the water in this lake at this time could not be more than three to four feet higher than at present, since the water had worn smooth a horizontal band at the latter height. From this it was possible to see how high the water rose when at its greatest height. Above this line the rocks were overgrown with lichens, and everywhere were trees, especially arbor vitae and pines.

In many places the rocks on the shore were composed of black limestone which generally lay like slate in strata. In some places

---

[1] *Arctostaphylos uva ursi.*

these did not lie horizontally but sloped, so that the southern part was lower and the northern higher, or they appeared as if they had been placed there during a south wind. In one place the black stone had separated into thin flag-stones, like a slate shingle, but as they had been exposed to the air for some time they could not be used for roof covering. Monsieur de la Galissonnière showed me in Quebec pieces which had been taken from this place.

In the evening we stopped for the night on the shore of the lake; the night was rather cold.

OCTOBER THE 19TH

We continued our journey from the place where we had passed the night. The sky was now almost entirely overcast and it looked as if snow might fall. It was quite cold. At 9:30 A. M. we arrived at Fort St. Frédéric. Before we reached the shore the soldiers accompanying me gave their customary salute with their muskets after which they called *vive le roi!* As soon as I reached the shore and stepped out of the boat, they gave a salute of five or six guns from the fortress in my honor, and all the officers together with the barefooted monk, Père Hippolite, came toward me on the shore and conducted me up to the commandant who was now Monsieur Herbin,[1] as Monsieur Lusignan, who was in command this summer when I first came here, had now been released from his post. I was received here with all possible good will and graciousness. As the Englishmen who had come hither yesterday afternoon and had left here this morning had taken away with them a large part of the bread supply, there was not enough left for my four men and myself for our whole journey. I was forced to postpone my departure to the following day, until fresh bread had been baked, which was indeed better for us. The officers here fared very well, for not only were all bodies of water full of ducks and game which now flew over in large numbers, preparing themselves for their fall migration, but it was also the season when the natives staged their hunt for roe deer in the large neighboring wastes and woodlands, whence they very frequently came to the

---

[1] Cf *Bulletin des Recherches Historiques*, XXII, 381, for Lieutenant Herbin's part in an exploit of November, 1747.

fortress with fresh meat to be exchanged for gunpowder, bullets, shot, bread and anything else they needed.

Natives. The officers told how the natives had learned to know almost every part of the forests. He (the native) goes out to hunt far from his home all day long, and at dusk he returns home in almost a straight line. Occasionally when he has killed more game than he is able to carry home, he leaves a part of it in the depths of the woods. On arriving home he tells his wife that he has left it in such and such a place and that she should go in such a direction to reach it. The wife does this and rarely fails to find it. Sometimes the Indians go through the woods when it is pitch dark, yet they find their way home safely.

OCTOBER THE 20TH

Collecting the Seeds of the Sugar Maple. In the morning when everything was ready we continued our journey. I went a little ahead down the country road in order to gather seeds from the sugar maple. Some of them had fallen from the trees. My men rowed along the bay which extends up to the Woodcreek. The commander, Monsieur Herbin, and the other officers followed me part of the way along the road and when I was a short distance outside of the fort I was honored with a salute of from five to six guns which were fired at the fort for my sake. When I had said farewell to the officers, I walked quickly to the last Canadian farm on this side. It was about three miles south of the fort. I stopped there until 2 P. M. to gather the seeds of the sugar maple which had fallen and were lying on the ground. There were indeed several trees which still retained their leaves, likewise a considerable numbers of their seeds, but they had such a weak hold that when I had several of these trees felled, they received on falling such a blow that all the seeds fell off; not a single one remained on the tree. It is worth noting that these sugar maples are among the highest trees found in Canada. I had all the trees cut down on the side where they were to fall, so that the seeds and leaves of the maples might not be damaged by falling against other trees. I had the maples cut only to the point where they gradually started to lean and sink downward, but still I could not prevent the trees from

falling heavily upon the ground and scattering all the seeds, so that
I was forced to hunt for them on the ground where they had been
strewn. But here I ran into a difficulty, because half the seed pods
of the sugar maple were empty. I had to squeeze every seed between my fingers to find out which were the solid ones and which
the soft ones; the former were good and the latter useless. I could
not burden myself with worthless seeds; I did not have room for
such. Then among those which had seeds, some were unripe,
others decayed or worm eaten. All of the latter had to be discarded.
In addition to the seeds from the sugar maple which had just fallen,
were also the newly fallen seeds of the American hornbeam (*Carpinus Caroliniana*), of the linden (*Tilia*) and of the ash (*Fraxinus*),
but not of any other trees.

I tried to get the ginseng, root and all, and had a soldier for that
purpose who knew where it grew, but the ground was so covered
with leaves that neither I nor he could procure a single plant.

We rowed from there at two o'clock and in the evening stopped
for the night on a promontory about twelve miles from Fort St.
Frédéric.

I shall call a certain long grass for the present "melic" grass
(*Melica*). It grew here in the rich soil in the less dense woodlands.
It had many soft leaves. I gathered seeds under this name and pre-
served several specimens.

I took some seeds from the brome grass (*Bromus*). A specimen
was preserved previously on my journey from Saratoga to Fort
St. Frédéric.

I called a certain grass here in Canada "Indian grass", of which I
also gathered seeds to-day. A specimen was prepared a few days
ago.

The woodlands which covered the stony and worthless promontories and islands ordinarily consisted of spruce, pine and arbor
vitae, and sometimes also birch.

OCTOBER THE 21ST

In the morning we continued our journey and after rowing one
and a half French miles we turned to the right and took the course
which led to Lac St. Sacrement (Lake George). On both sides we
were inclosed by rocky hills or mountains, steep on nearly all sides

and fairly high. They were valueless for purposes of cultivation, but no matter how rocky and useless they were, like our worst wooded hills, they were everywhere overgrown with the arbor vitae. Such hills must have been its native habitat. The channel from Lac Sacrement became indeed narrow, hardly a gunshot across, and also so shallow that the boat could scarcely proceed. After rowing three miles we came to the portage where we had to carry the canoe and our goods overland for a distance of a mile and a half. Here is a waterfall over a cliff of eighteen to twenty-four feet sloping height, and furthermore, above this same fall, it is so rocky and narrow that no boat can proceed.

Bustards and a few *ducks* lay in large flocks, swimming about at the entrance of the channel where it flows into Lac Sacrement. Hundreds of them rose into the air as we approached. At this time of the year the natives travel along the rivers and inlets killing large numbers of these birds.

Chestnuts with their burrs were lying here and there below the falls which Lac St. Sacrement made at the first portage. It was a sign that they either grew there near the shore or in the neighborhood of the lake. The soldiers at Fort St. Frédéric informed me that a chestnut tree is occasionally found between the afore-mentioned fort and Lac St. Sacrement.

We travelled overland a mile and a half carrying our belongings and the canoe. A native and his wife, both Iroquois Indians, followed us with their canoe, which the native without any effort carried on his head the whole distance. The region was slightly elevated, but yet fairly level and everywhere overgrown with woodland, which consisted largely of spruce and pine with no rocks of any considerable size. At 11 o'clock we came to the beginning of Lac St. Sacrement itself where we put our belongings aboard, after we had pushed our boat out into the water.

Lac St. Sacrement is a long, narrow lake which extends mostly from northeast to southwest, but with a few small bends. To begin with it was so shallow that we could scarcely proceed with our boat, which was somewhat heavily loaded. After rowing a distance of three or four gunshots we came to a place where the water flowed over a cliff and was only about twelve feet wide. We pulled the boat over this however without having to unload it, but we had to pull it six to twelve feet only before the water again became deep and the lake wider. On both sides of the lake are high, quite steep mountains covered with forests which send up into the air one high peak after another. The forests consist partly of pine and partly of leaf-bearing trees. There are a great many small islands scattered about in the lake. Bulrushes grow in many places in the middle of the lake, also near the shore. The islands are tops of small mountains or rocky formations. Some of the mountains are unusually high, especially on the northwestern side where they are more separated and not in a range as on the southeastern side. It was sometimes so shallow where the rushes grew that we could scarcely float along. The common reed also flourished here. The water was clear and had a pleasant taste. In some of the shallow places there were pebbles and sand. The width of the lake is half an English mile, now a little more, now a little less. In some places there were neither islands nor rushes to be found. The land on both sides between the mountains seemed to be of such a character that it would not be worth while cultivating, since it would not yield much of an income. He who settled here would doubtless have to live very frugally so far as grain was concerned, but he could have plenty of game, since here is where the natives start their hunt for the roe deer.

Juniper was found here and there on the stony islands and rocks of this lake. They were as large as our ordinary ones and grew in the same way.

The shore is filled with rocks, both large and small. The sides just above are rather steep. In some places are found pines, firs, and arbor vitae. The birch also is found now and then on the sides of this steep, rocky shore. The water is clear, so that it is possible to see the bottom even though the depth is great. The mountains on the edge of the lake are in some places very steep, extending out into the water where it is quite deep.

An Edible Lichen. Finally I came to know the *tripe-de-roche* which I had read about in Père Charlevoix's description of Canada. I had talked with several persons and asked them to show me this [lichen], but although they either were acquainted with it or at least said they knew it, they could not show it to me, since it was not to be found in the locality where they were. They who pointed out the lichen to me did not know it, because others and also the natives said afterwards that it was not the lichen, although it was somewhat similar to it. All the men who now accompanied me had travelled much in Canada and often eaten it. They had frequently talked of it on our journey and to-day when they discovered it, they called out in chorus: "Look, there is *tripe-de-roche*"! It grew upon the sides of the steep mountains mentioned above. It is a foliaceous lichen . . . and is somewhat like the crustaceous variety. . . . It is one of the largest of the foliaceous lichens, often twice the breadth of a palm in length and width and even more. It is turned up on all sides and is shaped almost like a skullcap or bowl. . . . The under side is pitch black, almost furry, the whole resembling black cordovan, or the black shoes which are like velvet on top. On top within the folds and creases it is grayish with a little olive coloring, even and smooth. When it is broken, the inside appears white and it is thin as a wafer. The upper side of some is a somewhat dirty black. This is the lichen which the natives sometimes eat and once in a while even the French travellers, though it happens only in the case of extreme necessity when they have no other food. This *tripe-de-roche* is boiled in water and is improved by the addition of some fat (suet). This moss swells so that it becomes almost half an inch thick. Most of the water [in which it is boiled] is poured off and it is eaten with a little of . . . It is a food which neither nourishes nor satisfies hunger, nor has it a pleasant taste. It serves only to sustain life. One of our men said that on one occasion he had subsisted for five days on this lichen and prunes (sour). If white fish is added thereto, it is fairly good. Salt should be added when it is boiled.

Toward evening quite a strong wind blew against us. It gained force from the high mountains situated on both sides of the lake and consequently became stronger and the waves grew higher. On much of our course we had many small rocky islands now overgrown with fir and sweet gale, which was especially plentiful on the lowlands. These islets or cliffs sloped gently on all sides. Pines were the trees most frequently found on them and they were not exceptionally tall. We sailed between and in back of the islands so that the wind and waves might not trouble us so much. The water was everywhere quite clear. The shore was sandy in places. We followed the northwest shore.

The *red juniper* was found now on the mountains and now in the clefts of the mountains. In some of the inlets where the mountains were farther apart, the land was level and it would not have been so hard to cultivate, for it was mostly overgrown by deciduous trees.

There were *chestnut trees* here and there near the water's edge.

At a certain place the lake was about two English miles in breadth, but it grew narrower again. On the northwest the land for a while remained level and not so high, but on the southeast it was just the opposite, high and steep, one high mountain after the other. The shore was rocky, yet not precipitous. Pines, firs (the *perusse*) and hardwood trees flourished on the plain. The width of the lake was one English mile.

The *reindeer moss* flourished on the rocks in the woods.

We encamped for the night back of a point northeast of the high mountains located on the southeast shore of the lake. These are almost the highest of the mountains. The length of the lake was said to be twenty-four miles and we estimated that we still had nine miles to its end. The neighboring forest consisted mostly of birches (?) which still retained their leaves. Next in number was the water-beech (*Carpinus*) and the mountain maple (*Acer spicatum*), but I could find no seeds of the latter. To be sure I found seeds here and there under the trees, but I am not certain that they were of this particular tree. Yet I gathered them.

Wild Geese. I inquired of my companions if they knew of any places in Canada where the wild geese remained throughout the summer. They answered that the geese laid their eggs and hatched the young in those places in Canada where the Illinois Indians lived. They also said that swans were found there; yet swans were said to be found on rivers in several localities in Canada.

The Saulteurs were said to be the worst of all natives in Canada. They live mostly near Lake Superior and Mishillmakira (?), but are also found scattered about the whole region. When they come across the French in the forests they treat them with kindness, bring game to them, etc., but when they [see an opportunity] . . . they kill them and carry off all that they have with them.

Linden or *basswood* (*Tilia*). The French called the kind of linden which grew abundantly in the woods *bois blanc* (white wood). They had with them bags in which they carried their food, which were made by the natives from the bark of this tree. The Indians take the bark, boil it in lye and pound it to make it soft. It becomes like a coarse hemp. They weave it in such a manner that the length-wise threads along the side are broad and scarcely twisted at all, while the crosswise threads which they have twisted on their thighs are about the size of small hemp cords and are not woven in with the lengthwise threads, but wound around as on a hamper.

OCTOBER THE 22ND

The wind was so strong that we could not proceed in our boat, which now was very heavily loaded with food which the men had taken along to have on their return journey. We had to stop for a considerable time. The wind came in gusts.

The *witch-hazel* grew here and there in the woods under the deciduous trees. Although the leaves had fallen from the majority of them, a few still remained, but so loosely attached that they fell off when one approached the bush. The size and shape of this shrub is much like that of the hazel, yet I believe that the hazel grows taller. It was now in full bloom, although the leaves on some of the bushes which had ceased blooming had already fallen. . . . The French who accompanied me did not know this shrub.

Seeds which had recently fallen from the trees were to be found upon the ground, namely, those of:

The sugar maple, which had lost its leaves, with the exception of the small new seedlings which had green leaves.

The mountain maple, which still had its leaves, though it had changed color.

The water-beech still had leaves, but had begun to fade.

The beech still had its leaves.

The leaves of the ash had fallen.

The walnut had no leaves.

The oak still had leaves.

The linden . . . had lost its leaves.

Red sand was found here and there on the shore of Lac St. Sacre-ment. It was made up of very small, round and shiny grains, just as if they had been either rubies or garnets. I never found it alone, but always mixed with the coarser grains of the light-colored quartz; yet I found it more often on top of the coarse sand than mixed with it. I discovered also the fine sand resembling iron on the shore.

A fog as thick as smoke arose to-day from the high mountains so that at a distance it appeared like the smoke from a burning charcoal kiln.

Bois de calumet (wood for pipes) is what the French called the *dogwood,* which had green branches and stem and not red like the [European] cornel. It sent forth each year long, narrow, even shoots without branches which had a fairly large pith. Both Indians and French took these shoots or branches, removed the outer bark so that they became smooth, then bored out the soft kernel and used the branches for their pipestems. The bowls of the pipes were made of a kind of limestone which was blackened so that they appeared pitch black. The process is described in my notes for my journey to and from Bay St. Paul.[1] The pith of this tree is so soft that a long narrow stick cut from the same tree will force out the kernel when it is pushed through a branch of it. This is not true of the red dogwood, the core of which is harder. On paring the bark from this branch people loosened it so that it remained attached at one end. Then they put the other end of the cutting in the ground a little distance from the fire and allowed the bark to dry a bit, whereupon they placed it in their tobacco pouch and smoked it mixed with their tobacco. They also used the bark of the red dogwood in the same manner. I inquired why they smoked this bark and what benefit they derived therefrom. They replied that the tobacco was too strong to smoke alone and therefore was mixed [with bark]. This is one of the customs which the French learned from the natives and it is remarkable that the Frenchmen's whole smoking etiquette here in Canada, namely, the preparation of the tobacco, the tobacco pouch, the pipe, the pipe-stem, etc. was derived from the natives, with the exception of the fire-steel and flint, which came from Europe and which the natives did not have before the French or Europeans came here. Red dogwood the French called *bois rouge* and knew of no other use for it than the one mentioned above.

The French call the water-beech (*Carpinus*) *bois dur*. They know of no other use for the wood than that it is good for fuel, and since it is hard and durable, it is used for cart axles.

I shall call that part of Canada a wilderness which lies between

---

[1] See page 498.

the French farms at Fort St. Frédéric and Fort Nicholson on the Hudson River, where Mr. Lydius and other Englishmen have their farms. Not a human being lives in these waste regions and no Indian villages are found here. It is a land still left to wild animals, birds, ect. At this time in the autumn Indians come hither from various localities; even natives who sometimes wage war against one another. They live here for several months by hunting alone, especially for roe deer which are plentiful in this vicinity. When this hunt is finished, they begin the hunt for beaver, which takes place toward spring. About the time when the beaver hunt begins, the meat of the roe deer loses its flavor, as it is the mating season. The skins are not as good as during other seasons, so that it seems as if nature itself commands them to cease the hunt. From the skins of these animals the natives as well as the French in Canada make their shoes which they use on their journeys and which somewhat resemble the Finnish (elastic) boots.

The so-called tea bushes (*Chiogenes hispidula* or moxie plum) were found where we were staying, growing along the ground in large numbers. I have also found them everywhere about the lake.

Fir (*Abies*). The hemlock spruce (*Epinette blanche*) which was used in brewing is found generally on the shore among the rocks.

Fir (*Abies*). The pine called *perusse* was also very common. I took a sufficient quantity of the seeds which were mature.

The pine with leaves in twos from the same sheath, with cones egg-shaped, were commonly found on the shore in a sandy and poor soil. The seeds were ripe, but had disappeared.

The pine with leaves in threes from the same sheath, with prickly cones, was also quite common and growing in the same kind of soil. The seeds had dropped from the cones.

Shepherdia grew here and there on the shore, but I could not find any fruit on it, except the small buds which are said to be the fruit. Might it be a kind of Myrica? I don't believe it.

Maple (*Acer spicatum*). The mountain maple was abundant here, but I do not know if it was the seed of this that I found under the tree. The French called it ordinarily *bois noir* (black wood) because the bark had black lines in it, especially when it was somewhat older. They also said that it was called by some *bois d'orignal* (wood of the elks) since these animals eat many of the twigs of this tree during the winter. When one passes through the forests at this season, one can tell from these trees the whereabouts of the elks, because the twigs are chewed off these trees. It is possible to see in what direction they have gone since the twigs have been broken off, hang and are bent in the direction in which they have gone.

Tales of Horror. During the evenings my companions were busy telling one another how they had gone forth in the last war to attack the English; how they had had Indians along and how they had beaten to death the enemy and scalped him. They also told how the natives often scalped the enemy while he was still alive; how they did the same thing with prisoners who were too weak to follow them, and of other gruesome deeds which it was horrible for me to listen to in these wildernesses, where the forests were now full of Indians who to-day might be at peace with one another and to-morrow at war, killing and beating to death whomsoever they could steal upon. A little while ago there was a crackling sound in the woods just as if something had walked or approached slowly in order to steal upon us. Almost everyone arose to see what was the matter, but we heard nothing more. It was said that we had just been talking about scalping and that we could suffer the same fate before we were aware of it. The long autumn nights are rather terrifying in these vast wildernsses. May God be with us!

OCTOBER THE 23RD

We continued our journey from this place at dawn, inasmuch as both the weather and wind were less severe.

The lake had at this point about the same appearance as before described; namely, on both sides high, fairly steep and wooded mountains. I do not know on which side they were the higher. There were small rocks here and there in the lake. The shore in some places was covered with a light sand, in other places with stones or bed rock. The water was clear and pleasant to the taste. Here it would have been impossible to paddle a canoe at night because of the many rocks along the shore. The lake extended from northeast to southwest with an occasional short curve. It seemed that the curves bent more toward the west than the east and left side, along which we were now proceeding.

Indians. One of the natives had put up his tent, if I may so call it, on the shore. He was one of those who had set out to hunt. The canoe lay upside down on the shore, as was the custom, and a short distance above in the woods, the Indian had made his hut which was constructed in the following way. He had placed pieces of birch bark and other bark on top of slender rods as a roof over himself where he lay and had hung an old [blanket?] .... [1] to protect himself on the sides from the wind and storms. His companion had done likewise on his side and their fire was between them. The wife and children were also sitting before the fire. The native had killed a great number of roe deer and hung up the flesh on all sides to dry. The skins were also stretched out to dry. At the fire's edge sticks were set into the ground perpendicularly and at the tip of these were pieces of meat for cooking. Indians discarded the horns, yet these were sometimes used for making knife handles and the like. The native men had pulled out the hair from the front part of their heads as far up as the part above their ears, so that the whole of this part of the head was bare, which gave them the appearance of having rather high foreheads. When they sat down they crossed their legs in front of them, one leg in front of the other. After we had bought a little of the meat of the roe deer we continued our journey.

Lake George. The lake had the same appearance as before mentioned with fairly high mountains on both sides, the one mountain piled upon the other so to speak. We followed along the northwest shore and in one place came upon a terribly high and steep mountain, which at the top, on the side toward the lake, was almost perpendicular. Below and next to the water there was a rather steep hill which was composed of large and small stones that had fallen down from the mountain. It was awe-inspiring when we rowed at the foot of the mountain and looked up, for it seemed as if the mountain hung right over our heads as we proceeded. The lake at the shore was very deep. The shore now for a while consisted of either stones or bed-rock and beyond that point the shore was precipitous. The mountains were everywhere overgrown with forests. The wind began to blow in strong gusts against us and the waves were fairly high, so that it was not especially pleasant to sit

---

[1] Word omitted in original, probably from defect or illegibility of manuscript.

here in the heavily loaded boat and write. I should not have wished
to have the boat founder here, since it was so fearfully deep that no
one of us could have saved his life. Fog was rising from the mountains in many places just like the smoke from a charcoal kiln. There
were many islands here and there in the lake.

The trees about this lake had not so generally lost their leaves
as those had in the neighborhood of Montreal and Lake Champlain. Many trees still retained their green leaves and the farther
south I proceeded the greater was the difference that I perceived.
These high mountains surrounding all sides prevented the cold
from being felt as early as around Lake Champlain, where the
mountains are not as high.

The *red juniper* grew here and there in the crevices of the rocks
and cliffs. Such are natural places for their growth. Some of them
doubtless had berries.

The shores for about six miles consisted of bed rock only or
large boulders, and was so steep that in case of a storm a birch-
bark canoe could not make a landing there. It is possible that one
might be able to save one's life with difficulty, but hardly the boat,
and one's belongings. In many places the rocks were so steep that
it would have been impossible to have escaped alive, if the waves
had forced one to seek land in this locality. Judging from the moss
on the rocks the water in this lake when at its full height is from
two to three feet higher than at present. Above this distance the
mountains and rocks were covered with lichens and mosses.

The pitch pine was generally found growing about the lake in
sandy and poor soil and in the crevices of the mountains. Also the
scrub pine flourished in similar places.

The arbor vitae was plentiful in some places and grew much under the same conditions as the red juniper, even among the rocks.

Birch likewise grew in the region about the lake and often in
the crevices in the bed-rock. The leaves were yellow, but had not
fallen from their branches. The opposite was true at Lake Champlain, where most of the trees had lost their leaves.

The mountains everywhere on both sides were fearfully high,
one close to the other and often quite steep, although covered with
forests of trees and pines. Here and there in the lake was a small
rocky island with a few trees upon it.

The wind was about the same as it had been the days before;

it blew in gusts, at times very strong, again more gentle and almost calm. Then it would change.

Juniper (the Swedish) also flourished here and there in the crevices of the rocks. The oak (the white, red and black) was found to grow on both sides of this lake.

There was an island in the lake of some considerable size in proportion to the other small islands. It consisted of a long low rock overgrown with shrubs.

The trees which grew among the stones that had tumbled down from the mountains were, among others, the following:

Birch in considerable numbers and even flourishing luxuriantly where the soil was the poorest: firs, both perusse and epinette; pines of all kinds; arbor vitae, in natural habitat; and a fair quantity of red juniper.

Note. Nearly all of these were trees which throve in the clefts of the mountains and grew quite rapidly. If anyone should wish to make use of such crevices, he should plant these trees and others which he finds will thrive there. Yet I did not find very many firs in the clefts of the bed-rock, but all varieties of pines, red and ordinary junipers, sometimes also the birch.

There were at this place in the lake several small islands covered with woodland and situated close to one another. All were of bed-rock covered with trees, mostly firs.

The lake was nearly everywhere about an English mile across, sometimes it grew a little wider and again it became narrower. Just about southwest of these islands it seemed to be quite broad. We rowed down toward the southeastern side and followed it. The lake now became about a couple of English miles broad with small islands here and there and surrounded by high mountains on all sides.

At noon the wind became so strong, augmented no doubt by the greater breadth of the lake here, that we could no longer continue our rowing. We were forced to seek the shore until it calmed down somewhat. We landed in back of a peninsula formed by a mountain. This was a barren place as far as herbs were concerned, as nothing much was to be found on a mountain. The rarer kinds found were these:

Bearberry plants covered the mountains in many places and flourished here in the same kind of places as ours [in Sweden].

The crevices in the mountains were full of them. Sweet fern (Myrica) grew everywhere in this locality. The so-called tea bush (moxie plum) was also common. The white pine (*Pinus alba*) throve in this region and grew to an unusual height. The Andromeda was found here also. The juniper was everywhere in the crevices. Indian grass, so called in New Sweden, was commonly found here. I do not recall that I have seen it farther north.

At 2:30 o'clock in the afternoon the weather became more calm and we set forth from here.

Flies now began to follow us, since we had received fresh venison which we were carrying along with us. They were the ordinary house flies.

The wind had nearly died down when we set out from the shore. It has not been very windy since. A squall accompanied with rain came, and suddenly there was such commotion in the lake that it looked almost like a boiling kettle. The waves went crisscross and were so large that we were in great danger. Yet it was almost calm. As soon as the squall and rain subsided, the strong agitation of the waters ceased. When the movement of the waters was at its height, we were in such a place that it would have been an impossibility for us to reach the shore, since it consisted of precipitous mountains reaching to the water's edge. The commotion of the waters was greatest about the promontories.

Arbor vitae flourished most where the shore consisted of fairly large rocks covered with a little soil or moss. The red American larch, which is used in making a beverage, grew here as did the perusse, its companion.

Lichen. Reindeer moss was everywhere abundant on the wooded mountains.

The shore on the southeastern side of the lake where we were rowing to-day consisted either of bed-rock, more or less steep, or it was covered with fairly large stones which could hardly be called cobble stones. The lake was very deep at the edge of the shore. We saw scarcely any other trees than varieties of pines and firs with an occasional birch among them.

The lake on which we had travelled this afternoon was nearly everywhere two English miles broad, if not a little more. My companions guessed that it was three miles or more in some places. Near the shore there was occasionally an island, but seldom any in the middle of the lake. There were mountains on both sides, yet they did not seem as high as those we had seen before, although some were as high if not higher: a mountain just now appeared on the southeastern side where we are rowing, and almost in front of it, in the center of the lake, an island is located. The rocks and mountains about the lake are of granite, and nowhere in this region have I found the black limestone. The sunshine about 3:30 P.M. was quite warm, yet the thermometer did not rise higher than fourteen degrees above 0°C. Toward the end of the lake were large islands covered with woodland. The forests were mostly pine or fir, a sign that the seeds had been carried there by the wind, but not so the seeds from hardwood trees. The seeds of the latter which had been carried by the water had either become decayed or had not reached land because of the rocky shore. There was a birch here and there. Perhaps the firs had come here first, grown up and when the leaf-bearing trees had come later, they did not thrive, since the firs had had the upper hand and had stifled them. This theory seems strengthened by the fact that oak and other trees are to be found here, but they are rare and few in number, small, miserable speci-mens, surrounded and crowded by the firs.

We had now approached the end of the lake. Now again I encoun-tered the same difficulty as last summer when I travelled to Canada through these wastes, namely, that the person who was to be our guide could find neither the way nor the portage a second time. We then had to begin by following along the shore, thus hunt-ing for the road again. If we do not find it, things will go wrong. Thank God, we have enough food, but if bad weather should set in, there is no pleasure in being in these vast wildernesses.

I have seen very few sugar maples in the vicinity of this lake, and in most places none at all.

The lake divides into two branches at its end, one toward the right and the other toward the left, or one toward the W.S.W. and the other S.S.W. We are now following the left, but after we had gone to the end of the bay and found only a small brook which ran out of a swamp or morass, but no trail nor sign of a portage, we had the pleasure of turning back to see if we were to be more successful along the other branch, which flows W.S.W. or toward the right when one comes from Canada. The land between these two branches or bays is a long peninsula of about a quarter of a

mile or so. It is a lowland, mostly overgrown with fir. Arbor vitae
is especially plentiful on the shore of this inlet and next to it in
number is the perusse. After four o'clock it was extremely calm
and the water in the lake very smooth. Once in a while there would
come a slight gust of wind.

The wolves were howling fearfully in the bay which we had just
left. It was said that they had just torn to bits a roe deer over which
they rejoiced, or they had killed one and were calling the others
to the feast.

A *black squirrel* was shot. It was the female who was not all
black. It was said that the male was entirely black. This one resembled our squirrels, and its size was about the same, but the
color varied. The greatest part of the hair on its body was black
with the outermost ends light gray. Its head between the ears was
almost entirely black. The cheeks . . . and stomach were almost
a brown, the whiskers long and black. From the nose and head
along its back there was a darker stripe. The tail and legs were the
same color as the rest of the body, namely, the end of the hair light
gray with the rest black. The tail was broad and fluffy just like that
of our squirrels.

We continued our journey farther in search of the right place
where one goes ashore and passes over to the English provinces.
We discovered smoke coming from a place on shore, toward which
we rowed in order to come in touch with people from whom we
could get all information [we wanted]. There were three boats of
Abenaquis Indians who had set up camp at this place. The greater part of them were out hunting, so that there were a couple of
men only and a few children left with the boats. The men were
almost drunk, since the Englishmen who had travelled through
here a couple of days before, had given them rum in payment for
the meat of the roe deer which they had given them. As soon as
we came ashore, they put on the pot to boil meat for us. According
to their wishes we must of necessity stay over night and eat with
them. But as the natives when they are intoxicated are often very
troublesome and even dangerous, we decided that it was wisest to
proceed from this place, especially since we learned that the portage
was at the end of the bay on which we were now rowing. We continued a little farther on and set up our quarters for the night, as
usual, on the shore.

October the 24th

We continued our journey from this place, rowing about a quarter of a Swedish mile before we came to the end of Lake St. Sacrement. Here the shore became sandy and sloping. My companions left their boat and the greater part of their food on the shore, since they had all they could do to take care of my belongings. They had to carry the latter fifteen miles if not more over land from the aforesaid lake to an arm of the Hudson River. The mountains at the end of the lake did not seem as high as those we had seen before on both sides of the lake. In the beginning, and almost along the whole way, we had mostly pine woods around us, though here and there was a clump of oaks. The pine woods consisted in part of red and white pine and cypress. Both white and red oak grew abundantly among the pines, but they were small.

Sweet fern flourished everywhere under and among the pines along our path. I have found the same variety in similar barren localities from Fort Nicholson way up to Albany. Most of the trees had now shed their leaves, but this shrub still had its leaves.

Tea bushes, so called in New Sweden, grew everywhere abundantly in the same barren places as just now mentioned.

The ground over which we were travelling was not exactly level, but slightly hilly, yet the hills were not especially high. Nearly everywhere there were numerous large pines which had fallen over and were left to decay. These had often fallen right across our path and hindered much those who were carrying the luggage. Now and again we saw a mountain in the distance. The region sloped uphill as a rule and was full of tea bushes. The greater part of the forest consisted of the pines and shrubs mentioned above.

We found *chestnut trees* in some spots in the forests. The soil in which they grew was in some places rather poor, hence it is easy to see that we could get this tree to grow in Sweden provided we could get some good nuts for seed.

Witch-hazel grew in many places in the forests, especially on level ground. Its flowers were now at their height and were very attractive from a distance since the leaves had fallen. The flowers were yellow.

We found that chestnut trees were among the common trees of the forests. The men who were with me said that the roe deer were very fond of the nuts after they had fallen from the trees. They eat acorns also.

Smuggling. When we had gone half the distance [to the English territory] we met a couple of Englishmen who came from Boston and had been sent to Quebec on the prisoners' behalf. They had several Iroquois with them as guides, some of whom carried their luggage, others their canoes. Merchants in Albany who carried on questionable business with those of Montreal, took advantage of this opportunity to send a lot of forbidden wares to Canada in exchange for which they were to receive the skins of beaver and other animals. In one spot we came upon an oak forest. We also saw now on the right and again on the left a high mountain some distance from us. We then passed over a mossy region. After that we travelled mostly through a sparse pine forest, the pine being mostly like ours [in Sweden].

I found several small larches growing in a swamp or morass. In Canada it grew more often in level and dry regions; yet also there in a swamp, e.g. between Fort St. Jean and Prairie de la Magdaleine.

The ground laurel (trailing arbutus) grew in some places in the same poor soil as the so called tea bushes, but I found no seeds on them.

On the journey from Lac St. Sacrement to Hudson River I did not see a single sugar maple. Wherever there were any, they were few in number, as the soil was too dry and poor.

Kalm Reaches the Hudson. About three o'clock in the afternoon we reached the Hudson River where on the downward journey one boards a boat, while on the upward journey one starts to carry it overland. We had no boat and still had eight English miles to the place where Fort Nicholson had stood and where was now the first English habitation. As it was late and the road poor and wooded, we had to postpone our journey to the following day. We took our quarters for the night under a cliff which projected out from the mountain and formed a room as it were beneath it. The natives had rested here before, otherwise it looked awesome when one looked up at the large cliff below which we were now lying, and contemplated that several rocks from the same steep mountain had but lately descended. We made a large fire under the rock. We stopped here so as to escape the rain if it were to come during the night. In summer this region is said to be infested with rattlesnakes, but the cold had long since driven them into their holes so that we did not need to have any fear of them.

The mountains of which the cliff mentioned above was a part, as well as the bottom and both the high steep sides of the river, were of the black limestone which lay here in strata. These were of varying thicknesses, some only six inches or less, others up to four feet or more, still others in between the two. Arbor vitae grew in the crevices.

OCTOBER THE 25TH

In the morning I went with three of the men from our night quarters to Fort Nicholson, where the first English colony was located. Our object was to procure a boat to take our things to Saratoga, since it would have been difficult for the men to carry the canoe through the woods from Lac St. Sacrement to this place. It was reckoned as eight short English miles from the place where we first reached the river to Fort Nicholson. We walked along the side of the river which was overgrown with woodland. The forest consisted mostly of tall white pines, with oaks here and there. The ground was generally level, sometimes slightly hilly. We rarely saw any rocks, let alone mountains. Now and then there were swamps or morasses, yet they were wooded. The most common tree growing here was the arbor vitæ, sometimes the fir called perusse.

To-day I saw *sassafras* for the first time since I turned south from Canada. I have not seen any native to Canada, only that which had been planted on the Isle de Magdaleine at Montreal. The plant which I found to-day was but eighteen inches tall and still retained its leaves. It grew in a barren place.

The *witch-hazel* (*Hamamelis*) grew in surprising numbers under and between the widely separated trees. It was in full bloom....

At 10 o'clock we arrived at *Fort Nicholson*. In the last war the house of the English here was burned by the French and the latters' native allies. Now they had built a new house. Colonel

Lydius who had lived here during the last war had not yet had his rebuilt, but intends to have it done next winter. The reason why the English, or rather the Dutch who live in Albany, build and live here is because of the trade with the natives. The English as well as the French also settle here. The natives bring their fur products to this place and in return receive almost all the wares from the English at a better price than from the French, with the [additional] advantage that they may buy and drink as much rum as they wish, a thing which is not permitted by the French traders. Here we see the reason why the native, even he who belongs to the French, prefers to sell his products to the Englishman rather than the Frenchman. There is nothing in the world which an American Indian prizes so highly as French brandy and rum, though he always prefers the French brandy. In Canada it was said that if an Indian saw brandy before him and the proposition were made to him to receive a good drink under the conditions that he was to be killed thereafter, he might well reflect a little thereupon before giving consent, provided he, somehow, before could still his longing for the brandy. The English know how to take advantage of this in a polite way, that is, to get from the Indian his products for next to nothing, under the pretext that they have only a little brandy or rum left, which they need themselves. A native who lives far to the west of Montreal might travel more than 200 to 300 Swedish miles, past the French colonies, to Oswego with his fur products, there to sell them at a low price, just for the satisfaction of once becoming drunk from rum.

OCTOBER THE 26TH

Letter writing consumed a good part of the day. I had four men with me from Montreal; namely, one sergeant and three civilians who had acted as guides the whole way and were therefore paid by the French government, as the original bill of my Canadian expenses which the French Crown had incurred on my behalf testifies. They had orders to go with me as far as Saratoga or four Dutch miles farther (N.B. A Dutch mile is equal to four English; a French lieue or mile, 3 English), but since head winds had delayed them and the season was so far gone, that the river might freeze over and they would encounter difficulties in returning home, I allowed them to turn back from here. I then wrote letters to some of my hosts and friends to thank them for their favors and friendship which they had shown me during my stay there. The letters I wrote were to the following:

Governor-general Marquis la Jonquière;

General de la Galissonnière, in France;

Commissary Bigat; [1]

The Governor of Montreal, Baron Longueuil;

Gauthier, the Royal Physician;

The Commander at Fort St. Frédéric, Mr. Herbin;

The Barrack-sergeant in Montreal, Mr. Martel;

Monsieur Renet de Couagne; [2]

Mademoiselle Charlotte de Couagne;

The Superior of the Jesuit [priests] in Quebec;

The Rev. Father Saint Pé, Society of Jesus, Montreal;

The Commander at Fort St. Jean, Monsieur de Ganne; [3] and

Barrack-sergeant, *ibid.*, Monsieur La Croix.

I continued my journey from Fort Nicholson in the afternoon. I had difficulty in starting out, as the natives who were there demanded so much [pay] that I could not afford to take them along. Finally the man who had charge of the trading post accompanied me with his boat. I thought I could make the descending journey easily, as he had another boat below the uppermost and largest waterfall. But when we arrived there and had begun to carry my luggage overland (about half a mile), we found that some one had carried off his other boat, and I was forced to leave all my goods in the unprotected forest during the night, since it was impossible for the two of us to get the first boat down the fall. Then we had to travel on foot along the country road a Dutch mile to Saratoga: a wretched road it was.

---

[1] This is possibly François Bigot (1703-c. 1777), who had been appointed commissary at Louisbourg in 1739 and intendant of New France, 1748. Because of gross fraud in Government funds he was thrown into the Bastile in 1759 and after a sensational trial was finally exiled from French territory.

[2] It has proved impossible to learn any details about the life of Renet de Couagne. The Couagnes in Canada were very numerous, all being the descendants of Charles de Couagne (1651-1706), who had emigrated to Canada in the seventeenth century.

[3] Probably either Michel de Gannes, sieur de Falaise (1702-52) or his brother, Charles-Thomas de Gannes (1714-65). Both were military officers in Canada in Kalm's day.

OCTOBER THE 27TH

Kalm in Saratoga. I occupied myself during the whole day with
getting my things home from the waterfall, where I had left them
the evening before. It became necessary to remove them, to prevent
the Indians who might happen that way from carrying off my
property, if they should take a fancy to it. Besides, the place where
my goods were left under the open skies demanded that they be removed from there, if they were not to be damaged. The road was
very poor for travelling with horses, but I finally succeeded in getting
my belongings away. Part of the time afterwards I was busy
wiping some of them dry, as the unusually heavy rain which had
fallen the night before had penetrated some of them, although
we had attempted to protect them well before we left them the
previous evening.

Dutch Food. The whole region about the Hudson River above
Albany is inhabited by the Dutch: this is true of Saratoga as well
as other places. During my stay with them I had an opportunity
of observing their way of living, so far as food is concerned, and
wherein they differ from other Europeans. Their breakfast here
in the country was as follows: they drank tea in the customary
way by putting brown sugar into the cup of tea. With the tea
they ate bread and butter and radishes; they would take a bite of
the bread and butter and would cut off a piece of the radish as they
ate. They spread the butter upon the bread and it was each one's
duty to do this for himself. They sometimes had small round
cheeses (not especially fine tasting) on the table, which they cut
into thin slices and spread upon the buttered bread. At noon
they had a regular meal and I observed nothing unusual about it. In
the evening they made a porridge of corn, poured it as customary
into a dish, made a large hole in the center into which they poured
fresh milk, but more often buttermilk. They ate it taking half a
spoonful of porridge and half of milk. As they ordinarily took
more milk than porridge, the milk in the dish was soon consumed.
Then more milk was poured in. This was their supper nearly
every evening. After that they would eat some meat left over from
the noonday meal, or bread and butter with cheese. If any of the
porridge remained from the evening, it was boiled with buttermilk
in the morning so that it became almost like a gruel. In order to

make the buttermilk more tasty, they added either syrup or sugar, after it had been poured into the dish. Then they stirred it so that all of it should be equally sweet. Pudding or pie, the Englishman's perpetual dish, one seldom saw among the Dutch, neither here nor in Albany. But they were indeed fond of meat.

OCTOBER THE 28TH

Leaving Saratoga. Early in the morning I resumed my journey from this place. It was indeed possible to go down the river in a boat, but as there were many places where there were cliffs and rocks in it, and as the water was very low, I chose the ordinary means of travel, by wagon along the country road. We crossed the river for the first time some distance above the Saratoga redoubt. The water was so deep here that it went over the front wheels of the wagon and the horses were just on the verge of swimming. The next time was just before we came to the place where Fort Saratoga had formerly been located. Here I was shown the spot where in the last war the French through an artful trick had taken a couple of hundred Englishmen prisoners right in view of the garrison of said fort. The story runs thus: Monsieur St. Luc, a Canadian officer with whom I have the pleasure of being well acquainted,[1] was ordered to make a sally. He stationed his men during the night in the woods not far from the fortress. In the morning, after daylight, he sent forth a few natives who were to shoot or take prisoners any who might leave the fort. The English shot at these, who pretended that they had been wounded and so could not run. Soon three hundred men rushed out of the fort to take them prisoners, ran along the field located on the northern side of the fort, and before they were aware of it, they were cut off from the fort and surrounded by the French. They saw no other way out than to give

---

[1] This is undoubtedly M. St. Luc de La Corne (Lacorne) (1712-1784), later legislative councillor of Quebec, who was an officer in the Canadian army about this time. He "took part in the defense of the province against Americans in 1775-76; and he commanded the Canadians and Indians in the campaign of 1778, under Burgoyne." The attack on Saratoga had taken place in May and June, 1747; but the attacking force seems to have been the larger, for St. Luc had two hundred men at his command, and only forty English prisoners were captured. Apparently the victory had been considerably exaggerated by the Frenchmen who told the story to Kalm. Cf. article on St. Luc de La Corne in the *Dictionaire Générale du Canada.*

themselves up as prisoners.[1] Shortly thereafter we drove by the fort, close to its gate, the fort being on our right and the river on our left. The fort, however, was now in ruins. Later places were pointed out to me where the French in the last war had killed the English or stolen upon them when they were out chopping wood or doing something else.

Triticum [a genus of cereal grass, wheat] was found at Saratoga and also in other places.

The poison ivy seen to-day had lost its leaves.

I saw a considerable amount of speedwell (*Veronica*) to-day in the woods and in the fields, but I do not remember seeing this plant farther north.

In the beginning of my journey there was an abundance of pine woods, mostly low pines. In other places there were many scrub oaks which thrive in barren spots.

The majority of the trees have now lost their leaves. I make an exception of the oaks whose leaves remain on the trees until spring, but they were either brown, reddish or brick colored.

We saw apple orchards, some of considerable size, on nearly every farm. It was a common custom in the English colonies, when anyone paid a visit to a house, to bring in a large dish of apples and invite the guests to partake of them. In the evening when we sat warming ourselves before the fire, a basket of apples was carried in and all in the house ate of them according to their desire. A large quantity of cider was made on every farm.

We took lodgings for the night about ten English miles from Albany. Ordinarily it is possible to drive from Saratoga to Albany in one day, but because so many of the bridges had fallen into decay on account of the war, we were of necessity greatly delayed on the road. New houses had been built this summer in many places to replace those burned by the enemy in the last war.

October the 29th

Approaching Albany. In the morning we continued our journey. The weather was not especially pleasant for travelling. It snowed fairly hard all day. The snow was the more disagreeable because it

---

[1] Substantially the same story was told by Kalm under the date of June 24. See p. 358.

was very wet. The greater part of the forest at the beginning of our journey consisted of low pines which grew in the barren lowlands. We now drove through all three branches of the river. There was here neither ferry nor bridge, so one drove through. The current in all three was indeed strong, but it did not come higher than the belly of the horse, except in a single place where it reached the side of his body. The river bottom was stony and uneven so that both the horses and the wagon swayed. The branches were a short distance apart. When the water is high as a result of heavy rains, it is difficult and even dangerous to get across. We saw now fields and now farms on the rest of the journey to Albany, where I arrived at 11 A.M.

Dutch Customs. I think that on the occasion of my former visit to this city I cited a few customs of the people in this locality. Inasmuch as I have not my notes with me, I shall record a few items here. With the tea [at breakfast] was eaten bread and butter or buttered bread toasted over the coals so that the butter penetrated the whole slice of bread. In the afternoon about three o'clock tea was drunk again in the same fashion, except that bread and butter was not served with it. At this time of the year since it was beginning to grow cold, it was customary for the women, all of them, even maidens, servants and little girls, to put live coals into small iron pans which were in turn placed in a small stool resembling somewhat a footstool, but with a bottom . . . upon which the pan was set. The top of the pan was full of holes through which the heat came. They placed this stool with the warming pan under their skirts so that the heat therefrom might go up to the *regiones superiores* and to all parts of the body which the skirts covered. As soon as the coals grew black they were thrown away and replaced by live coals and treated as above. It was almost painful to see all this changing and trouble in order that no part should freeze or fare badly. The women had however spoiled themselves, for they could not do without this heat.

OCTOBER THE 30TH

The snow now entirely covered the ground, but the sun to-day melted a considerable part of it on the southern sides of the hills. On the northern sides it remained untouched by the sun.

Wood. Hickory was considered the best wood of those trees which grew here, but it was said to have one disadvantage, namely, that when burning it caused great injury to the eyes. So far as excellence was concerned the sugar maple was said to be almost as good as the one just mentioned.

Dogwood. The wife of Colonel Lydius told me that when she had arrived in Canada she had suffered from pain in her legs as a result of the cold. It became so severe that for a period of three months she could not use one leg and had to go about with a crutch. She tried various remedies without avail. Finally, a native woman came to the house who cured her in the following manner. She went out into the forest, took twigs and cuttings of the dogwood, removed the bark, boiled them in water and rubbed the legs with this water. The pain disappeared within two or three days and she regained her former health.

Iris. Colonel Lydius related how the Indians make use of the iris root as a remedy for sores on the legs. This cure is prepared as follows. They take the root, wash it clean, boil it a little, then crush it between a couple of stones. They spread this crushed root as a poultice over the sores and at the same time rub the leg with the water in which the root is boiled. Mr. Lydius said that he had seen great cures brought about by the use of this remedy. It is the blue iris, which is extremely common here in Canada, that is used for this purpose.

Sassafras. He also told me that the natives consider the sassafras very valuable in the treatment of diseased eyes. They take the young slips, cut them into halves, scrape out the pith or the medulla, put it into water, and after it has been there for some time, wash the eyes with the same water. When the natives from Canada formerly came to his house at the . . . ying [1] place they were very anxious to hunt for sassafras. They cut the stems in two, took out the pith and preserved it and took it home with them to use as described above.

OCTOBER THE 31ST

Pumpkins. A certain kind of oblong and large gourds were called "pumpkins" in Dutch. They were much used by the Dutch, the

---

[1] The first part of the word is omitted in the original.

English, the Swedes and others here in America. The French in
Canada had also some use for them, but not as much as those mentioned before. They ripen in the middle of September here in
Albany and are able to stand a fair amount of cold. The natives
both in Canada and in the English provinces plant a considerable
amount of them, yet I think that they came here first from Europe.
They were delicious eating and they were prepared in various ways.
Sometimes they were cut in two or more parts, placed before the
fire and roasted. Frequently they were boiled in water and then
eaten. It was customary to eat them this way with meat. Here at
Albany the Dutch made a kind of porridge out of them, prepared
in the following way. They boiled them first in water, next mashed
them in about the same way as we do turnips, then boiled them
[again] in a little of the water they had first been boiled in, with
fresh milk added, and stirred them while they were boiling. What
a delicious dish it became! Another way of preparing these which
I observed, was to make a thick pancake of them. It was made by
taking the mashed pumpkin and mixing it with corn-meal after
which it was either boiled or fried. Both the gruel and the pancake
were pleasing to my taste, yet I preferred the former. The Indians
do not raise as many pumpkins as they do squashes. Some mix
flour with the pumpkins when making the porridge mentioned
above, others add nothing. They often make pudding and even pie
or a kind of tart out of them. Pumpkins can be preserved throughout
the winter until spring, when kept in a cellar where the cold cannot reach them. They are also cut into halves, the seeds removed,
the two halves replaced and the whole put into an oven to roast.
When they are roasted, butter is spread over the inside while it is
still hot so that the butter is drawn into the pumpkins after which
they are especially good eating.[1]

Discord had taken a firm hold among the inhabitants of Albany.
Although they were very closely related through marriage and
kinship, they had divided into two parties. Some members of these
bore such strong aversion to one another that they could scarcely
tolerate the presence of another member, nor could they even hear
his name mentioned. If a visit was made at the home of one of
them, you were then hated by his opponent, even though you had

---

[1] Cf. diary for September 19th, 1749.

visited him previously. It was interpreted that you were not satisfied with his friendship alone, but also wished to enjoy that of the other or that you liked him better.

November the 1st

Petroleum. Both Colonel Lydius and Mr. Rosbom and many others told me to-day about a kind of earth-oil which is to be had from the Indians allied with the English, in a place about three days' journey from the country of the Senecas. There is said to be there a small lake, about a musket's shot in length and breadth, upon the surface of which continually floats an oil. The natives go there in the spring when they find the lake almost covered with leaves which have fallen into it during the previous autumn and which are now saturated with oil. They set fire to them and allow them to burn up. The fire then continues to burn over the lake until all the leaves and all the oil on the water is consumed and only the water is left. They then take twelve-foot poles which they force down into the lake's bottom and an oil comes welling up from the bottom and the holes they have made therein. The bottom is not hard, but very soft. They skim with a spoon or scoop the oil which floats upon the water and preserve it in some vessel. Colonel Lydius accompanied me to the home of a man or gentleman who had some of this oil in a bottle. It was real petroleum and not at all different from what is sold at the apothecary's, except heavier; but I learned that this condition resulted from allowing it to stand several months in an open bottle, that is, without a stopper in the bottle. Its color is dark brown with a tendency toward red. The smell is absolutely that of petroleum or Jew's pitch (asphalt). The gentleman who had it in a bottle related that nothing was better for wounds than this. If a little were applied to the sore, it would heal within a short time, which he had experienced on his many journeys. He had a horse last year that injured itself badly by kicking one leg with the horseshoe on another foot. The wound was smeared with the oil and within a few days it was healed. Another horse had some time thereafter suffered a bad injury and was likewise cured within a short time. It is said to be this oil which Père Charlevoix tells about in his *Histoire de la Nouvelle France.*[1]

---

[1] Tome I, page 422.

The hawkbit (*Leontodon*) was found on the hills outside of the city in full bloom, although the ground had but recently been covered with snow.

Salt Springs. In Philadelphia Mr. Bartram related that he had seen several salt springs in the country occupied by the native allies of the English and that the Indians prepared salt from them by the boiling method. He also showed me a piece of it. In Albany several people told me the same story, and that when travelling from here to Oswego the place is on the left, and about a day's journey from the road. The brine is boiled by the natives in copper utensils. Salt springs are found there in more than one place. In one location is a salt spring from which they obtain salt, and on each side of it, hardly a musket's shot away, is a spring of fresh water in which it is impossible to detect a trace of salt. There is in Albany no one who is able to let me have a little of this salt. It is about these salt springs which P. Charlevoix tells in his *Histoire de la Nouvelle France*.[1]

Cabbage Salad. My landlady, Mrs. Visher, prepared to-day an unusual salad which I never remember having seen or eaten. She took the inner leaves of a head of cabbage, namely, the leaves which usually remain when the outermost leaves have been removed, and cut them in long, thin strips, about $\frac{1}{12}$ to $\frac{1}{6}$ of an inch wide, seldom more. When she had cut up as much as she thought necessary, she put them upon a platter, poured oil and vinegar upon them, added salt and some pepper while mixing the shredded cabbage, so that the oil etc. might be evenly distributed, as is the custom when making salads. Then it was ready. In place of oil, melted butter is frequently used. This is kept in a warm pot or crock and poured over the salad after it has been served. This dish has a very pleasing flavor and tastes better than one can imagine. She told me that many strangers who had eaten at her house had liked this so much that they not only had informed themselves of how to prepare it, but said that they were going to have it prepared for them when they reached their homes.

The gourds or melons raised by the Indians were generally called *squashes* by the English, *cascuta* by the Dutch and *citrouilles* by the French.

---

[1] Tome I, p. 421.

Melons d'eau so called by the French in Canada, were called "watermelons" by the English and "*watlemone*" (*watermeloen*) by the Dutch. The majority of those raised here were said to have red meat. As they had gone by, it was impossible now to see one, as they do not keep them but eat them only when they are fresh. It was said here that those obtained in New York are better and tastier than those in any other place in North America.

Mulberry trees are still found scattered through the forests about this town. I requested several people in this locality last summer to gather some of their berries when they became ripe so that I might have some seed. The majority had forgotten about it and I would have lost out entirely, if young Mr. Rosbom, a barber-surgeon, had not preserved a few for me. He said that he had gotten them with great difficulty and that they were the only ones he had seen during the summer, as they had been scarce. Formerly there had been more mulberry trees, but they had been wasted by cutting them down for the purpose of making dowels for boats and yachts. This wood withstands rot for an unusually long time. On a farm twenty English miles from this town on the way to Saratoga lived a farmer who told me that here and there in the woods in that region one might find a single mulberry tree. In Saratoga, which was twenty English miles further away, or forty miles from Albany, nearly due north, all the farmers said that they had never come across nor seen a mulberry in the woods. Colonel Lydius, however, assured me that he had seen such a tree in Saratoga and that Saratoga was the place farthest north where he had found these trees growing. He has lived for several years at the old Fort Nicholson, only twelve English miles north of Saratoga, and on a hundred journeys through the woods there he had never seen one of these trees. The fruit of the mulberry which grows here, he said, is much smaller than that of the trees planted in Holland, but as far as taste is concerned it is as pleasing and agreeable as the Dutch variety.

Oswego which is the trading center of the English on Lake Ontario is situated at 44° 47′ according to the observations made by a French engineer.

The Dutch began, they said, to settle New York in 1623.[1]

A kind of seed was called by the French *piment* (if I remember

---

[1] New Amsterdam was settled by the Dutch in 1614, and New Jersey colonized by them in 1617-1620.

rightly, by botanists called *Capsicum* or *Cor indum*) [*Cardiospermum halicacabum*]. This is used very commonly here to improve the flavor of food. It is used very much as pepper is. Nearly everyone here sowed the seeds, which were said to lie in the ground six weeks before they cracked open or germinated.

NOVEMBER THE 2ND

Walnuts. All the varieties of walnuts found in Canada flourish here with one additional kind. The oblong are called oil-nuts (butter nuts). *Noix amères* are called "bitter nuts"; *noix dures* are called "hickory nuts" (*kiskatom*, an American Indian name). In addition to these three there is a fourth kind which is smaller than the last mentioned, the shell being less hard yet the kernel sweet. When you paid a visit to any home a bowl of cracked nuts was also set before you, which you ate after drinking tea and even at times while partaking of the tea. A plate filled with large sweet apples of no particular name was also often added.

Ginseng (*Aralia Canad.*) was called . . . *wurtzel* by the Dutch. The old surgeon, Mr. Rosbom, said that the root not only is invaluable for wounds but taken internally it also affects the urine and is fine for stones in the bladder.

The *moccasin flower* (*Cypripedium*), which is found quite generally in the woods here, is said to be rather good for women in the throes of childbirth. I refer to the decoction made of the root.

Houses. Last summer on my journey up to Canada, on the day I arrived at Saratoga, I described how the houses belonging to the Dutch living in the country were built. They first put up the framework upon which the rafters and both roofs rested and then filled in the framework with unfired bricks. The inner side was brushed over with lime and whitewashed so that from the inside it looked like a stone house except where the perpendicular timbers which supported the rafters were visible. On the outside the houses were generally covered with clapboards so that the unfired brick might not be damaged by moisture, weather and wind. As a rule they did not have more perpendicular supports in the walls than they had cross beams, from three to five on each side or long wall. On the walls of the gables they had two or three upright beams, doubtless for the sake of firmness and strength when they put the bricks between them. The ceiling was horizontal and beamed. The roof was either of boards or shingles; there were several rooms under one roof. There was nearly always a cellar under the house and usually this was large enough to extend under the whole house. The walls of it were of masonry, yet the ceiling was not vaulted but made of plain boards or of planks with the floor of the house upon them. The fireplace in the houses in the country was built in an unusual way and it was nearly always placed in the wall on the gable end opposite the door. The fireplace for about six feet or more from the ground consisted of nothing more than the wall of the house which was six to seven feet wide and made of brick only. There were no projections on the sides of the fireplace, so it was possible to sit on all three sides of the fire and enjoy the warmth equally. Instead of forming this figure ⌐ or that one ⌐, namely of three sides, as our fireplaces do, the fireplaces here formed this one —, or a straight line without projections, corners or embellishments. Above, where the chimney began the bricks rested upon rafters and cross beams on three sides which had been arranged so as to support them. As the chimney was some distance above the floor they had put boards about these rafters, or as was more common, they had hung short curtains extending downward [and outward] to prevent the smoke from coming in. But in spite of this and because the fireplace had no sides, it frequently happened when the door was opened that the smoke was driven into the room. The hearth itself was always even with the floor. The fireplace was ordinarily six to eight feet in width. Occasionally a shelf had been made above it upon which teacups, etc. were placed. In Albany the fireplaces had small sides projecting out about six inches made of Dutch tiles with a white background and blue figures. As they knew nothing about dampers the fire burned during the winter all day, both in the houses in the towns as well as in the country. No tile stoves were used here. The houses in Albany were built in the same style as those in the country except that the gables which were always toward the street were constructed of fired brick so that, as previously noted, a stranger walking along the streets and not paying close attention might easily conclude that all houses here were built of fired brick. The floors of the houses were kept quite clean by the women who sometimes scrubbed them several times a week, and Saturday was the day especially set aside for that task. In many houses in the town they had partitioned off the part of the room where the beds stood by placing large doors before them, [like cupboards], and thus completely concealing the beds from view. Every house in the town as well as in the country had a bedroom or an attic where they kept miscellaneous household goods. This was generally reached by a ladder or staircase within the house. The ovens for baking which were made of brick were placed on a hill [apart from the house].

Decrease in the Amount of Water. Among the signs that the water in early times had been higher than at present, the following may be noted: Colonel Lydius told how just a little north of this town a whole tree had been found buried in the ground at a depth of twenty-five feet and not far distant from the river. A few English miles below this town there has been found on digging into the ground some very large teeth, almost like those of elephants, but shaped like human teeth. There has also been discovered here a fairly long tubular bone, so that it has been concluded that this skeleton was not that of an animal but of a human being. I inquired whether one had found any oyster shells or other shells in the ground. He answered that he did not know that any had been found in this locality, but farther down the river large quantities of them had been discovered in some places. It is not known whether there is a decrease of water in this river, but it is a fact that small banks are shifted from place to place, which generally occurs during the spring floods.

Rattlesnakes. Colonel Lydius has seen these in the clefts of the rocks on Lake Champlain. When the snake attacks a human being the blood in the beginning spurts high into the air from the wound. If it should strike certain parts of the body, as an artery, etc., the wound was believed to be fatal. It does not usually attack anyone if it is not trampled upon. On the other hand, it is possible to approach quite close to it without the snake minding it, though it looks sharply at a human being and, as it were, marvels at him. It cannot extend itself further than its own length; it floats like a bladder upon the water; it usually rattles before it strikes, yet not always. It is rare that one hears of anyone being attacked by it, despite the fact that the people travel so much about the woods; so it must be true what is said about it being stepped upon, etc. It is known that it has crawled over the stomach of a sleeping person without doing harm to him. The rattlesnakes first come forth from their dens into the sunshine of the early spring, sun themselves the whole day long and then in the evening they crawl back again. This they do from two to three weeks early in the spring until it becomes warmer. Colonel Lydius once found a great number of them in one place and he shot sixteen of them. His companion beat several to death, and the rest of them lay under a large stone and rattled vigorously. When he pointed toward them with his cane, they threw themselves toward him and then hastily drew back, just like a flash of lightning, and they repeated this at frequent intervals. The rattlesnake does not withdraw when it hears a human being approach, but lies still.

November the 3rd

I shall soon become tired of having to remain here and wait so long before I can get away. The morning after I arrived in this town two yachts departed, but it was impossible for me to get ready to leave on them, as I must procure various kinds of seeds of the walnut, chestnut, squash and other useful plants, a thing which I could not do in a hurry. People assured me that another boat was to leave as of Saturday last, but it is still tied up here, even though there is plenty of wind to-day.

The inhabitants of this town [*Albany*] are as a whole all Dutch or of Dutch extraction, descended from those who first came to settle this part of the country. Both sexes dress now very nearly like the English. In their homes and between themselves they always speak Dutch, so that rarely is an English word heard. They are so to speak permeated with a hatred toward the English, whom they ridicule and slander at every opportunity. This hatred is said to date back to the time when the English took this country away from the Dutch. Nearly all the books found in the homes are Dutch and it is seldom that an English book is seen. They are also more thrifty in their homes than the English. They are more frugal when preparing food, and seldom is more of it seen on the table than is consumed, and sometimes hardly that. They are careful not to load up the table with food as the English are accustomed to do. They are not so given to drink as the latter, and the punch bowl does not make a daily round in their households. When the men go out of doors, they frequently have only a white cap under the hat and no wig. Here are seen many men who make use of their own hair, cut short without a braid or knot, as both of the latter are considered a mark and characteristic of a Frenchman. The vast majority, in fact almost everyone here, carries on a business, though a great many have in addition their houses and farms in the country, close to or at some distance from the town. They have there good country estates, several sawmills and in many places even flour mills. The servants in this town are nearly all negroes. The children are instructed in both the English and Dutch languages. The English accuse the inhabitants here of being big cheats and worse than the Jews.

November the 4th

Corn. From Mrs. Vischer with whom I had lodgings, I learned one way of preparing corn so that it would be especially suitable for cooking and eating. They take the corn before it is ripe, boil it in a little water, allow it to dry in the sun and preserve it for future use. Corn prepared this way is then boiled with meat, etc. when it, as well as the soup in which it is boiled, is good to eat. The younger the corn is when picked, the better it is, provided, however, that it is not too young. When the corn is prepared this way it is not necessary to remove the hull as this is not yet hard. When it is boiled, it is done in the following manner: The whole cob is placed in the saucepan and when it is ready, the kernels are removed from it.

The departure from Albany took place at 2 P. M. on Capt. Wilj. Winov's boat. We had a favorable wind, but the captain had to stop for more freight at a couple of places, so we did not make as good headway as usual. Yet before sunset the freight was all on board.

The *Province of New York*, it was said, was not nearly as well populated and cultivated as the other English provinces here in America. The reason was said to be that for the most part the inhabitants were Dutch who in the past had acquired large stretches of land. Most of them are now very wealthy and their feeling of jealousy toward the English prevents their selling a piece of land unless they are able to get for it much more than it is worth. As the English as well as persons of other nations can buy land at a far more reasonable price in the other provinces here, they gladly allow the inhabitants of New York to keep their land. Moreover, in Pennsylvania there are greater advantages to be gained than here in New York. Because of this it so happens that people come from Germany every year to settle in this country. They come sometimes from London by boat to New York, but they hardly come ashore before they leave here for Pennsylvania. It is a fact that Pennsylvania alone has now almost more inhabitants than Virginia, Maryland and New York put together, due to the generous privileges which the sagacious Penn wrote into his wise laws and constitution for Pennsylvania. The inhabitants of New York console themselves with the thought that when once all the land in Pennsylvania has been filled up with inhabitants, the remainder will of necessity come hither, and then they can set whatever price they wish upon their land. But it is the general opinion that they will have to wait a long time.

The *red stone* which is found about New Brunswick, and of which I spoke last spring when I was there, is said to be useless for building purposes. Even though it seems firm and strong when in the ground, it happens that when it is taken up and exposed to the air for a time, it disintegrates and crumbles. One of the townsmen here had a house made of this stone, but it began so to crumble on the side which was exposed to the air that he was forced to have boards put on the outside to prevent its falling to pieces. There is, however, one use for this stone: when crushed and spread upon the fields it is a good fertilizer, and weeds do not thrive where this has been put upon the farm land and the kitchen gardens. Mr. Wiljams, a townsman, who had long lived in New Brunswick told me all this.

NOVEMBER THE 5TH

To-day we made very little headway on our journey, as there was hardly any wind, and in addition the captain was delayed in one place.

The *best cider* in America is said to be made in New Jersey and about New York, hence this cider is preferred to any other. I have scarcely ever tasted any better cider than that from New Jersey.

Squashes. The best squashes are said to be found in the region around Albany. Those from New York are said to be not so good and those from New Jersey, and Pennsylvania even less so. For this reason those living in New York procure their seed from Albany. The squashes are good then for three years, at the end of which period they must send for new seed. The squashes can be kept in good condition up to the month of March, if stored in cellars where the cold does not reach them. But the squashes keep longer than pumpkins. The latter are called *pumponen* by the Dutch. I was told about this at a large gathering of old men and women last evening.

Watermelons were also kept in Albany a part of the winter. They could even be eaten by the sick without harm, so far as the people of Albany knew.

Roofing slate, very beautiful and not affected by air, rain, etc., is said to be found in Highland which is a part of the Province of New York, situated about halfway between New York and Albany. The place where it is found is about five English miles from the Hudson River. They are thinking of roofing the houses in New York with it.

There is said to be a *sulphur spring* in the Mohawk country about seventy miles from Albany. The water comes down a mountain and when it reaches the gravel below it leaves so much sulphur that the deposit of it can be swept up from the ground. A man drank of the water with the result that his body swelled up therefrom. Mr. Wiljams from Brunswick has seen the place.

Iron ore is said to be found everywhere and in great quantities in this country, which everyone with whom I have conversed on this subject has confirmed as with one voice.

Ash. The baskets here in which food and the like was sold, were generally made of ash; the withes were thin and about a digit broad. The natives make quantities of these baskets and sell them to the Europeans.

Catskill Mountains. At two P. M. we reached the Catskill mountains which are also called the Blue Mountains. They form the long range in this part of America. They were north of us and their tops were all covered with snow, although on the ground not a bit of snow was to be seen. It had no doubt been washed away by the heavy and prolonged rain. These mountains tower above all the others, are visible as far as Albany, and as clear as if they were quite close, when in fact they are situated forty miles from the place just mentioned. There is found on these mountains many herbs and trees which are not to be found on the plains round about them and not any nearer than Quebec and the northern part of Canada. These mountains are said to be infested with rattlesnakes.

Limestone is found in abundance on both sides of the Hudson River. Formerly they burned the lime here and shipped it down to New York to be sold. Now, however, this is not done as much, for a considerable amount of lime is made in New York from oyster shells.

White pine is commonly used for roofs of houses in Albany, and such a roof is said to last forty years. This white pine is not found in New Jersey, so they say.

NOVEMBER THE 6TH

Our journey proceeded rather slowly as we had contrary winds and we did not advance faster than the tide carried us.

Charles XII. The passengers on the boat had the *History of Charles XII* [*King of Sweden*] with them to help while away the time.[1] It was the life written by Voltaire which had been translated into English. In London I saw it in nearly every bookseller's shop and bookstall. I had noticed during my travels that it was widely circulated and was one of the most read and best-known of books. They also had descriptions of his life by other authors. All marveled much at this great king and all had acquired a singular regard for him. But there was one thing of which nearly every one disapproved and that was his behavior at Bender when the Tartar Khan attacked him and he began to defend himself.

Commerce. We saw one boat after the other tied up at the shores of the river as we proceeded, all being loaded with wood, flour or

---

[1] Voltaire's *Histoire de Charles XII* appeared in 1731, and became immensely popular. It was immediately translated into English, and at least three editions were published in London during the year 1732. By 1740 seven English editions had been printed.

something else. The country was inhabited in most places, especially nearest the river. The captain told me that the soil on both sides of the river was said to be inferior, especially on the southern side, and on that account the inhabitants there were poor.

The *skulls of the American natives* are said to be much thicker and harder than those of the Europeans. This is a fact, according to what I learned from several Frenchmen with whom I talked in Canada and [from what I heard] also from Englishmen in this country.

The products which came from New Brunswick were said to be grain, flour in quite large quantities, bread, linseed in considerable amounts, and various utensils. All of these were sent on small sailing boats to New York, which is the only trading center, and to which wares are shipped. New York is situated forty English miles from New Brunswick.

The *Seneca Country* and the surrounding region is inhabited by Indians only. Formerly because of wars they lived in large villages surrounded by stockades, but now the natives have scattered, one living here, another there. In some places they do not allow unrestricted sale of rum, but only certain ones may buy it and sell it again in small quantities to prevent the harm occasioned by drunkenness. This was told by someone who had spent three years among them for purposes of trade.

There is said to be a street in New Brunswick which is inhabited by the Dutch only who have come from Albany and is therefore called Albany Street. These Dutchmen call upon one another, but seldom visit any other residents of the city; they keep themselves apart.

Prognosticon Tempestatum (Signs of the weather). In Albany it was considered a certain sign, especially in the fall when there had been two or three days of southeast or southwest wind with rainy weather, that a strong northwest wind would follow with a clear and cold atmosphere.

The boats from Albany were now making their last trip to New York for the season, since it was the general opinion that they would not be able to make another trip before the river was frozen over. Most of the boats are tied up for the winter at Albany, but a few remain at New York, since they continue their trips as long as possible, and then dare not go up the river, since it is frozen over earlier at Albany than at New York. The reasons for this are not only that Albany is situated farther north, but that the tide is not as strong there and the water is not salt but fresh.

November the 7th

The journey was to-day also a very slow one. We had to depend upon the tide, and when we came opposite the home of young Mr. Colden we had to come to anchor, as the tide was low, and besides the wind coming down from between the high mountains just below us was so strong against us that it was impossible for us to proceed. We were forced to remain here until half past twelve, when the tide turned and we again could continue our journey. The wind also changed and became favorable, but it was too gentle to be of any assistance to us. We floated along past the high mountains where we anchored late in the evening, as the skies did not look especially promising.

The chestnut oak grew near the home of Mr. Colden.

Wheat (*Triticum*) with waving spikes was also visible there.

That part of the country is called "Highlands" which is situated just below the high mountains, and the latter are right below Mr. Colden's house. The Highlands are located on both sides of the river.

Wild geese and ducks were flying about here and there in the neighborhood of the river, and large numbers were also swimming about in the water. Yet we were not permitted to shoot any of them.

November the 8th—In New York

The current prices on products in New York for this autumn were those given in the New York *Gazette* of November 13, 1749, as follows:

Wheat, per bush.	6 s.	Molasses,	1 s. 9 d. per Gal.
Flour, per C.	18 s.	Westindia Rum	3 s. 9 d.
Milk bread,	39 s.	New England d:o	2 s. 6 d.
White d:o	29 s.	Beef, per Bus.	36 s.
Middling,	24 s.	Pork	21.18 s.

Brown	18 s.	Flax-Seed	10 s.
Single refined			
Sugar	16 d.	Bohea Tea,	6 s. 6 d. per Box
Muscovado			
Sugar, per C.	50 s.	Indigo	7 s.
Salt, per Bush.	2 s. 6 d.	Chocolate	2 s. by the doz.

The *copper mine* which is located nine to twelve miles from New York on the side toward Philadelphia is as yet the only one known in this country. At least none of that ore is mined elsewhere. Nearly all the owners of it live in New York. The mine was worked previous to the last war. The ore from it is said to be of the best obtainable. Following an act of the Parliament in England the inhabitants here are no longer allowed to smelt and refine the silver and copper ores which they find, but are obliged to send them to England in their original state and have them smelted there. This ore has been so rich that they have sent it to England and sold it at a profit in spite of the high freight charges incurred. During the last war the work in the mine was abandoned, as it was not safe to send the ore over the seas to England for the privateers to seize. It was at this time that the mine became full of the water which is still in it. Now the owners intend to send to London for engines to pump the water out of it.

NOVEMBER THE 9TH

The Dutch Church, the Service, etc.—*Hunc diem perdidi.* (This day I have spent uselessly) I can in a certain way say about this day because I spent the morning in one of the Dutch churches and the afternoon in the other. I listened to two sermons, one of which lasted two hours, the other two and a half, and in neither instance could I understand much, because I was so far away from the pulpit; and even if I had understood all of it, I could not have remembered so long a sermon. But nevertheless, for the mere pleasure of it, I shall here record the ceremonies which are used here, a description of the church, etc. The morning service began exactly at ten A. M. and was to-day especially noteworthy, inasmuch as it was fifty years to-day since the minister preached his first sermon in that church. The fact was also stressed that the sum total of those who had heard his first (installation) sermon and were now present in the church, could not be very great. The church service was conducted as follows: first, the bell was sounded two or three times (they had here [in America] only one bell in each church tower) and this was rung by means of a rope which reached way down to the floor of the church, where it was customary to stand when ringing, or more correctly speaking, when tolling the bell. Then the cantor began to sing one of David's psalms rendered into verse. Only a few stanzas thereof were sung, and it seemed to me that nearly all of their hymns had similar tunes. While this hymn was being sung the minister mounted the pulpit, hung his hat on a peg, took off his robe and also hung that there, since the old fellow was eighty years of age and could not endure wearing his robe for two hours. He sat down on a chair and when the singing ceased, he stood up and preached for a while; then he read a lot of prayers, after which he began preaching again. At that time two men came forward, took the [longhandled] contribution bags and placed themselves before the pulpit, holding the handles upright, with the bags at the top. After the minister had mentioned a few facts about church work, they went about taking the offering. It was the custom here that everyone contributed, but no one was asked to contribute more than once during the same service. When the minister had finished the sermon, the clerk [or cantor] passed a note to him. He had put the note in a split at the end of a cane and thus passed it up to the minister from the place where he stood below the pulpit, saving himself the trouble of going up into it. It was the license to have the banns of marriage announced of those who were to be married. Finally he finished his sermon with the Lord's Prayer, announced the hymn to be sung, which was another [rhymed version] from the Psalms of King David, and when he had finished, the cantor began to sing it. They had also in the church several boards on which were indicated what hymns were to be sung before the sermon: e. g. to-day the following words were on the boards: "Psalm 18 pause"; but no post-sermon hymn had been indicated. The minister remained seated in the pulpit during the singing, and when this was finished he pronounced the benediction, after which everyone departed. No mass was said and no altar devotions with gospel, epistle, etc. were held. Even if they had had this chanting service at the altar, it could not have been performed since there were no altars in the Dutch churches here, but just a pulpit.

The church where I to-day attended morning worship was the so called New Dutch Church. I shall now describe the building. It is a large structure built of stone like the ordinary church, with a tower and bell. The church is located nearly in the direction of S. S. W. and N. N. E. with the tower at the latter end. Here I wish to note that the churches in this town are not constructed to face any particular direction, one standing east and west [for instance] in the customary way, as was almost true of the English church. The others were set north and south, the direction generally followed by the Lutheran and Reformed or Presbyterian churches, etc. Quite a large churchyard surrounds the temple, and about it are planted trees which give it the appearance of an enclosure. The church has several doors and large, high windows on all sides. Within there is not a sign of a painting or figure, only white walls and ceiling and an unpainted pulpit. The church is not vaulted, but has a ceiling built of boards in the form of a vault. There is no balcony in it. There are pews everywhere in the church, also several aisles; but both are rather narrow. The pews are made like ours except in the manner hereafter described. The backs of our pews extend perpendicularly from the top to the floor, while here the back extends from the top to the seat only. Then there is another perpendicular partition which is not in line with the back of the pew but placed a little ways under the seat of the pew so that a person can place his feet under the pew in front of him. The number of the pew is painted on its door. There are no chandeliers and no candlesticks in the church. There is no sacristy and no other room to take its place.

The ministers are dressed in black with a gown and collar like those of our ministers except that the gown is not quite as long. During the week when there is no service in the church, the shutters are closed. In the tower is a clock, which strikes the hour, the only one of its kind in the city, as far as I know.

The church was filled to-day with people who came out of curiosity to hear a man begin his fifty-first year as minister. For this reason more had flocked hither than usual. The men were dressed like the English; the majority wore wigs, but a few had their own hair, which was not very long, was not powdered, and had no more curls than nature had bestowed. A few elderly persons had worsted caps on and three or four wore hats. Perhaps they were Quakers whom curiosity had driven hither. The women as a rule had black velvet caps which they could fasten on by tying at the ears. Others wore the ordinary English gowns and short coats of broadcloth of various colors. Nearly everyone had her little container, with the glowing coals of which I have spoken before, under her skirt in order to keep warm. The negroes or their other servants accompanied them to church mornings carrying the warming pans. When the minister had finished his sermon and the last hymn had been sung, the same negroes, etc. came and removed the warming pans and carried them home.

In the afternoon I went to the so called Old Dutch Church. The service began there at two P. M. and was performed in exactly the same manner as the morning worship which I attended. There was singing before the sermon and the latter was preached in the same way. It ended with the Lord's Prayer, followed by a final hymn and the benediction. I observed one ceremony here which I did not see during the morning service, namely, the christening of a child. The service was as follows: as soon as the sermon was finished and the minister had read a few prayers, a woman carrying a child on her arm came forward to the pulpit. Then the minister began to read from there all the prayers [for the occasion] contained in the Dutch prayer book, and when he had finished she took the child to the rector, who sat in his pew, and let him christen it. This was performed, as we do in baptism, by putting water upon the child's head. After this the other minister in the pulpit talked briefly on the significance of the christening and then ended the service with the Lord's Prayer.

This church was also of stone with two pillars in the center which supported the wooden roof built in the form of a vault. Some of the windows had colored glass, but they were not old, since "City of New York" was inscribed upon them. There was a small organ, the gift of former Governor Burnet. The pulpit and vault were painted, but without any figures on them. There was not a single figure or painting in the church, only several coats-of-arms had been hung there. There were large balconies, and it was the custom here that the men sat there while the women occupied the ground floor, except the pews against the walls which were reserved for the men.

There were many chandeliers here. The examination in the Catechism was held late at night by candle-light.

Peter Kock. I learned very unexpectedly this afternoon of the death of Mr. Peter Kock, the merchant. I was telling those with whom I was staying that I visited him here last year just about this time, when they informed me that he was dead. He had first served with the admiralty and then for several years sailed to foreign ports. He married in Cusassov (?) a wealthy Dutch woman and in this marriage had several children, two of whom, daughters, are still living. He became a widower, then came to New York to build a ship in which to return to Sweden. He married here a woman of the Dutch family Van Horne and then took up his residence in Philadelphia. There he became one of the wealthiest of merchants. He was a pillar of the Swedish congregation there; he warded off the attempts of the United Brethren to get a foothold. It was to him that I took my bill of exchange. He received me with extraordinary kindness and performed unusual favors for me, so that it is not at all surprising that his death should affect me so deeply. He was quite well acquainted with the theological writings.

NOVEMBER THE 10TH

Letters. To-day I received letters from several of my friends; namely, two from Mr. [Abraham] Spalding in which he advises me not to leave here in the fall, inasmuch as I might experience difficulty on my arrival in London so late in the season to get an opportunity to continue [my return journey] to Sweden. Therefore he considered it wisest to postpone my return until next spring. In addition I received a letter from Mr. Collinson in London, also one from Jungström who described his whole journey from Albany, etc. From the gazettes which were printed here in town I discovered also that Mr. Franklin, the postmaster at Philadelphia, my very special friend, had had printed in the Philadelphia gazette an extract from the letter which I had written to him from Quebec, Canada. This extract had been reprinted in the gazette which was issued in this town. All of the French were especially pleased at this, as they had [in this article] been given considerable honor for their learning.

French Refinement vs. the Dutch. I have already told in my journal of the good breeding of the French in Canada. Now I must emphasize one item before I forget it: namely, that the inhabitant of Canada, even the ordinary man, surpasses in politeness by far those people who live in these English provinces, especially the Dutch. I just recently came from Canada and left behind me in the vicinity of Saratoga the French who had brought me to the English colonies. When I reached Saratoga and came in contact with the first English inhabitants who were of Dutch descent, I noticed a vast difference in the courtesy shown me in comparison with that shown me by the French; it was just as if I had come from the court to a crude peasant. Yet I must grant that although they [the Dutch] showed a lack of breeding in their speech, their intentions were of the best. I noticed that when they believed, or were persuaded to believe, that a person did not understand Dutch, they amused themselves by censuring the manners which differed [from their own]. The women, especially in the towns, had this habit, for they did not like the French mode of living at all. But never have I seen folks more ashamed than they when a person let them know that he understood every word they said. Some of them did not dare show themselves again.

The King of England's birthday was celebrated in town to-day, but the people did not make a great fuss over it. A cannon was fired at noon and the warships were decorated with many flags. In the evening there were candles in some windows and a ball at the governor's. Some drank until they became intoxicated, and that was all.

NOVEMBER THE 11TH

Dutch was generally the language which was spoken in Albany, as before mentioned. In this region and also in the places between Albany and New York the predominating language was Dutch. In New York were also many homes in which Dutch was commonly spoken, especially by the elderly people. The majority, however, who were of Dutch descent, were succumbing to the English language. The younger generation scarcely ever spoke anything but English, and there were many who became offended if they were taken for Dutch, because they preferred to pass for English.

Therefore it also happened that the majority of the young people attended the English church, although their parents remained loyal to the Dutch. For this same reason many deserted the Reformed and Presbyterian churches in favor of the English.

Lodgings, food, wood etc. were in this town much more expensive than in Philadelphia. The rooms were said to have grown more costly since so many people from Albany had lived here during the last war and thus brought about a shortage of living quarters. The prices paid then were still in force. Food was high-priced because so much flour, corn and other food stuffs had been exported in great quantities to the West Indies, and even to New England. The farming region round about here is not so well populated that it can supply the town in such quantity as is the case in Philadelphia, where the whole countryside is thickly inhabited by the Germans and others who have settled there. On the other hand, it is the general opinion that broadcloth and other merchandise can be had here at a more reasonable price than in Philadelphia.

November the 13th

My correspondence kept me occupied all day. Now that I had resolved to stay one more year in America (under H. E.'s [1] protection), since it would [otherwise] be impossible for me to accomplish all that the Academy of Sciences demanded of me and that I ought to accomplish, I wrote to several friends in Sweden about the matter. My letters were to be sent by a boat which was to leave New York for London within a fortnight. I wrote to the following:

1. His Excellency Count Tessin. I told him that I had written from Canada; also that I had resolved to stay another year, and about the receiving of my allowance.
2. The Royal Academy of Sciences. I wrote about my Canadian journey and what I had procured there. I related how I had been refused permission to go through Fort Frontenac, etc. I also told why it was necessary for me to remain another year; I told of . . . [2] from Mr. Spalding. I informed them

---

[1] H. E. = His Excellency (Count Tessin).
[2] The rest is omitted in the original.

that I would soon begin to send them seeds, and that I had received permission from the King of France.

3. The Vice-president, Baron Bielke, and the archiater, Dr. Linnæus, a letter together. I wrote them about the same things as I did the Academy, except at greater length. This was to be read by the Academy.

4. Mr. John Clason.[1] I told him that I had asked Mr. Spalding for a new draft.

5. The Secretary of the Academy of Sciences. I asked him to seal my letters and mail them.

6. Cousin Wilh. Ross. I requested him to countermand my house rent.

7. Mr. Abr. Spalding. I wrote him of my reasons for staying another year and asked for a draft of sixty pounds sterling. I told him of the letters sent, etc.

NOVEMBER THE 14TH

Dutch Customs. In New York I had lodgings with Mrs. van Wagenen, a woman of Dutch extraction whose dwelling was opposite the new Dutch church. She as well as everyone in her house was quite polite and kind. It is true that the Dutch both in speech and outward manners were not as polite and well-bred as the English, and still less so than the French, but their intentions were good and they showed their kindly spirit in all they did. When a Frenchman talks about a man in his presence and even in his absence he always uses "monsieur": e.g. "*donnez á monsieur* etc.". An Englishman says: "give the gentleman etc." while the Dutch always said: "*giw dese man*". The women were treated in the same way without ceremony, and yet the Dutchman always had the same good intentions as he who used more formality. If several persons of Dutch extraction should come into a house at this time of the year, as many as could be accommodated would sit down about the fire. Then if any others should happen in, they pretended not to see them. Even though they saw them and conversed with them, they did not consider it wise to move from the fire and give the others a little room, but they sat there like lifeless statues. The

---

[1] Johan Clason (1704-1790), wealthy merchant of Stockholm. He had become a member of the Swedish Academy of Sciences in 1745.

French and English always made room by moving a little.—When one spoke of refinement as the word is now used, and in applying it to the French and Dutch, it was just as if the one had lived a long time at the court while the other, a peasant, had scarcely ever visited the city. The difference between the English and the Dutch was like that of a refined merchant in the city and a rather crude farmer in the country. But it is well to remember that there are exceptions to every rule.

I have lived now for almost a week in a house with a good-sized family. There was the same perpetual evening meal of porridge made of corn meal. (The Dutch in Albany as well as those in New York called this porridge *Sappaan*.) It was put into a good-sized dish and a large hole made in its center into which the milk was poured, and then one proceeded to help oneself. When the milk was gone, more was added until all the porridge had been consumed. Care was usually taken that there should be no waste, so that when all had eaten, not a bit of porridge should remain. After the porridge one ate bread and butter to hold it down. I had observed from my previous contacts with people of Dutch extraction that their evening meal usually consisted of this "Sappaan". For dinner they rarely had more than one dish, meat with turnips or cabbage; occasionally there were two [dishes]. They never served more than was consumed before they left the table. Nearly all women who had passed their fortieth year smoked tobacco; even those who were considered as belonging to the foremost families. I frequently saw about a dozen old ladies sitting about the fire smoking. Once in a while I discovered newly-married wives of twenty and some years sitting there with pipes in their mouths. But nothing amused me more than to observe how occupied they were with the placing of the warming pans beneath their skirts. In a house where there were four women present it was well nigh impossible to glance in the direction of the fire without seeing at least one of them busily engaged in replacing the coals in her warming pan. Even their negro women had acquired this habit, and if time allowed, they also kept warming pans under their skirts.

My departure from New York for New Brunswick took place at 2:30 P. M. New Brunswick was said to be about forty English miles distant. The weather was fine, but the wind from the south was not very strong. There are two ways by water from New York to New Brunswick: namely, around the outside of Staten Island or on the inner side of the same. If the weather is good and there is some wind, the journey is much shorter by way of the outer passage. If, however, there are signs of a storm or bad weather, the inner passage is to be preferred. We now took the outer one. After we had sailed for a time, the wind died down so that we proceeded very slowly. Yet we crept along with the tide and the gentle wind which aided us until midnight, when the low tide forced us to drop anchor. This journey is made usually in rather small boats from New Brunswick. These have a small cabin, and as there was now a considerable number of passengers and the night was very cold, we had to crowd into the cabin. No one can expect to get a wink of sleep where there is scarcely room enough to sit. Larger yachts cannot be used, as the river which they navigate up to New Brunswick is in some places very shallow.

Signs of Future Weather Conditions. An elderly farmer who was with us on the boat prophesied that the winter was to be a very cold one this year. The reason for this he gathered from the fact that there were more squirrels this year than for many a year before and that they had been very busy gathering nuts and other things for preservation in their holes as food during the winter. As a result it happened that even though there had been plenty of nuts, it was difficult to obtain any quantity of them, because the squirrels had carried them off to their hiding places. He assured us that this was an old sign of a cold winter. He even intimated that when the winter was to begin, it would come suddenly and in a hurry, since the fall had been so beautiful.

To Prevent Cracking of Punch Bowls by Warm Water. Often when warm punch was poured into a porcelain bowl, even though the latter felt warm, it would crack. A Jew, who also was a passenger on the vessel from New York, said that if a new porcelain bowl is boiled for a time in water in which there are some husks ("brains" i.e. bran, hulls or chaff) the bowl will not be so apt to crack when a warm liquid is poured into it.

Jews. The Jew just mentioned was a rather good-natured and polite man and it would scarcely have been possible to take him for a Jew from his appearance. During the evening of this day which ushered in his Sabbath he was rather quiet, though he conversed with me about all kinds of things, and he himself often began the discourse. He told me that the Jews never cook any food for themselves on Saturday, but\that it is done on the day before. Yet, he said, they keep a fire in their houses on Saturdays during the winter. Furthermore, he said, that the majority of the Jews do not eat pork, but that this custom does not trouble the conscience of the young people when on their journeys, for then they eat whatever they can get, and that even together with the Christians.

NOVEMBER THE 15TH

The journey was continued in the morning after seven A.M. when the tide began to go up the river. The wind was right against us, but it was so gentle that it could neither hinder nor help us. We had to steer the boat as we rowed. I have described this river and the region about it in my notes for the month of May of this year [1749].[1] Now I just wish to add that the salt water runs almost up to New Brunswick, yet it never goes so far up that it actually reaches the town. In the river there are several oyster beds in which oysters are found and it is these which the boats run into when the water is low. The people living near the river have no evidence that it has grown less deep. It was shallow along its banks, but deep beyond them. We arrived at New Brunswick at ten A.M., where I met with an unusual piece of good luck. Just as I jumped off the boat and was about to run up into the town to procure lodgings until next week when the mail vans were to leave here, a man from Trenton who had just brought some travellers here and was about to return, came down to the boat and inquired whether there was anyone who desired to go to Trenton. I immediately took advantage of such a favorable opportunity, had my belongings put aboard his wagon and we set out from New Brunswick exactly at twelve o'clock noon. We had a distance of twenty-seven English miles to Trenton with good, dry roads. I have before described the whole of this road [2] so that I do not find anything of special interest to add at this time. We stopped at one place only, where we ate our dinner and fed our horses. Then we continued our journey. It became almost dark before we were halfway, because the roads are not as good in the vicinity of Brunswick

[1] See also pages 121, 122 and 323
[2] See pages 117 ff. and 322.

as they are nearer Trenton. The night was somewhat cloudy, yet whatever moonlight there was, assisted us. The gazettes had contained during the summer various accounts of how many travellers in the region about Philadelphia had been robbed on the roads of all their money. These facts made this journey by night particularly unpleasant for me, but thank God! we arrived safely in Trenton at a quarter of twelve.

Pennyroyal began to be found in large quantities at the sides of the roads about Brunswick and likewise along the road from there. It had entirely dried up.

The hawkbit (*Leontodon vulg.*) was everywhere along the road in full bloom, even though the ground here had already once been covered with snow. Compare also the [entry for the] 21st of October. This, i.e. the hawkbit, had short and somewhat broad leaves.

There was one plant of the chamomile (*Anthemis*) still in bloom. There were no signs of any other herbs in bloom at this season.

The products of the soil were said this year to be especially fine in New Jersey and Pennsylvania. There was a greater amount of wheat and corn than for several years past. Likewise there was an abundance of buckwheat, more than had been harvested for many years. It is often injured here by the frost in the autumn, before it has had a chance to ripen, but this year there had been no frost during this whole period, so that it had had an opportunity to mature. There had also been a rather plentiful crop of apples, but the spring sowing, such as of barley and oats, had failed, as had also the hay crop. The crop was small here, but still smaller in the vicinity of Boston. Boats loaded with hay had been sent there from Philadelphia. It was said that hay had been shipped there this year even from England.

The Bravery of the French Nation. I entered into conversation with an Englishman who had been a privateer for several years during the last war. While on these expeditions he had twice been taken prisoner by the Spanish and four times by the French. He had assisted in the capture of many ships, both French and Spanish. I asked him his opinion concerning the bravery of these two nations on the seas, their treatment of prisoners, etc. He answered me that if, for example, the French had a vessel with sixteen guns and a hundred men and the English likewise one with just as many men when they started to fight, the result would ordinarily be that they would have to leave one another, each having received equal injury. They cannot conquer one another, since the French are said to be merciless fighters on the seas. If a person is made a prisoner by the French, he is handled with extreme politeness and no prisoner can be shown greater kindness and consideration than that which the French nation allows its prisoners to experience. If an English vessel with sixteen guns and a hundred men should happen upon a Spanish man-of-war with the same number of guns and men, the English is usually able to conquer the Spanish within one half to an hour's time. If the English should remain at a distance and shoot at the Spanish, then the latter would be able to withstand the attack for a longer time and would return shot for shot. But if a sudden attack is made and an attempt made to board the ship, the Spanish cannot long withstand it. He said that an Englishman with such a ship and manned as described above much prefers to capture from two to four such Spanish ships than one of the French. If a man is taken prisoner by the Spanish, hardly a Turk or a heathen treats him worse than the Spanish do. The prisoner is treated as if he were a negro or a slave. They rob the prisoners, take off their clothes and leave them almost stark naked. Very frequently after they have captured an English vessel and made all on board prisoners, they attack them all with their swords and hew them down. If they obtain a prisoner whose life they wish to spare, they throw him into chains. They send the majority to their silver mines or to work as galley slaves. I remember when I was in England that they made the same distinction between the French and Spanish in their way of handling the prisoners, and they could not be particular enough in portraying the cruelty of the Spanish nation in this. Even neutral nations like the Swedish and Danish were quite rudely handled by the Spanish privateers during the last war between Spain and England, when they sometimes happened to come aboard some Swedish or Spanish (evidently an error in writing for *Danish*) ship. They robbed them of almost everything they owned, beat the sailors under their bare feet to make them say where they had concealed their money, etc. which tales I heard in London from those who had been forced to experience these things.

November the 16th

Farmers sow a considerable amount of buckwheat in this locality. It is used especially in preparing cakes similar to pancakes (griddle cakes). As these cakes come hot from the pan they are covered with butter which is allowed to soak into them. These cakes while they are still warm and prepared as above, are eaten in the morning with tea or coffee. The buckwheat straw is said to be useless; therefore it is left lying on the ground after it has been threshed. Neither the cattle nor other animals will eat it; only in cases of extreme necessity when the ground is covered with snow and they cannot obtain anything else will they occasionally chew a little of it.

Commerce. The products of this region are as a rule to be had at a lower price in Philadelphia than New York. The result is that many who live in the neighborhood of Trenton and also further down towards Philadelphia send the products of their farms, as butter, flour, etc., overland to Brunswick and then on to New York, there to be sold at a better price than they can obtain in Philadelphia. Throughout the summer the wagons travel between Trenton and New Brunswick.[1] They are loaded with flour and various kinds of merchandise which are being sent to New York. Many are carrying passengers only. On their return trips they have not much to transport except that they are frequently full of travellers of all classes and nationalities who during the whole summer do nothing but travel back and forth between Philadelphia and New York. The products mentioned above are brought in boats from Philadelphia here to Trenton and then overland from here to New Brunswick wherefrom they are carried on boats to New York. It is possible to travel this way from the Delaware River up to Trenton, but not further than here, as there is a fall or a cataract in this same river over which a small boat can scarcely pass, let alone a somewhat larger boat. The current above this fall or cataract was said to be quite strong. Yet it is possible to come down with boats from a distance of one hundred English miles from here, and therefore great quantities of the products of the region are transported here by boats and even to Philadelphia. But whereas it takes only one day to come down, it requires from ten to fifteen to go back against

---

[1] Cf. diary for October 28 and 29, 1748.

the current. The tide goes up the Delaware River to the falls here near the town of Trenton, but it can go no further on account of this cataract. At this time there were in here, in Trenton, four boats, each of which made a weekly trip to Philadelphia and back again. It often happens that on the return trip from New Brunswick, the wagons are loaded with all kinds of goods which are being sent to Philadelphia since sometimes, as I have mentioned before, certain articles are more expensive in Philadelphia; at other times in New York, and it is this fact that the merchants in the two towns take into account. Yet the man with whom I had lodgings, and who was a carter and continually carried these wares over the road, said that generally more goods were sent from Philadelphia to New York than *vice versa*. The country on both sides of the Delaware above this town is populated for about a distance of one hundred and fifty English miles. The land on this side of the river is more sandy and on the Pennsylvania side it consists of clay soil, especially a short distance from the river. About a hundred English miles further up in this New Jersey Government is a tract of land which is called Minisiek [Minisink or Meenesink].[1] It is lowland and rather thickly populated. The whole of this province, as well as Minisiek and other places on the western side of the road, is said to be nearly everywhere thickly inhabited, mostly by the Dutch or descendants of old Dutch families. From all of these places comes considerable corn, flour, hemp, linseed, lumber (since the country is full of forests and sawmills), butter, together with other products of the region, which are then sent either down the Delaware River to Philadelphia or down other rivers to New York. In various places in the country the merchants have stores to which they send, and where they sell, all kinds of merchandise. Besides these, there are to be found in every large village one or more stores.

The savin (*Juniperus sabina*) is said to grow as a common thing on the banks of the Delaware. It is found only as large shrubs and never grows to the size of a tree.

---

[1] Concerning this settlement see John W. Barber and Henry Howe, *Historical Collections of the State of New Jersey*, edition of 1855, p. 506. According to the map in this volume it seems to have been located in the southwestern part of what is now Delaware County, New York, near the Pennsylvania border, about where Cadosia and Hancock are now situated. It will be remembered that the Delaware River runs far up into New York State.

The red juniper, so called by the Swedes in New Sweden, or the Englishman's red cedar, grows abundantly further down New Jersey in sandy and dry places, according to what they told me.

Hvita cedern so called by the Swedes and *white cedar* by the Englishmen was to be found, as I was told, in quite a number of places in the deep swamps further down the province. In English such morasses were called *cedar swamps*. I described one of them last spring on my journey to Rapaapo in the month of May. The savin is also said to grow wild here, and I have likewise found it in the same state in the region about Albany.

Trenton. When I travelled through this little town last summer toward the end of May, I gave a short description of the same. Now I just wish to add one or two things. The soil upon which the town is located is mixed with sand, hence the location is considered very healthful. The old gentleman with whom I had lodgings, told me that when he came to this town some twenty years ago, there was not more than one house here, and since that time the town has grown so that there are now almost a hundred houses. These are built so that the street passes along on one side of the buildings, while on the other are kitchen gardens of varied dimensions, and in the furthermost corners of the gardens the privies are located. A little back of the houses in the gardens, also, are the wells, ordinarily with a draw bucket. The houses are two stories high with a cellar beneath them and a kitchen underground, next to the cellar. The houses are divided within into several rooms by thin board partitions.

November the 18th

Thermometrical Observation. I had my personal belongings, including the thermometer on board, so that I could not take a temperature reading before the men entered the boat, when I removed the thermometer and at half after nine exposed it for a long time in the open air to see how low the mercury might go. But no matter how long I kept it in the air, I could not get it lower than one degree above zero C., which appeared strange to me, since the wind was blowing so cold that it was impossible to stay outdoors very long before one became thoroughly chilled through; and when I held the instrument between my fingers, they became so stiff that I could scarcely move them. From this it may be seen that the body feeling of cold does not always correspond with the reading on the thermometer.[1]

The yachts [or boats] which were now en route from Trenton to Philadelphia were loaded with timber, wheat flour, several barrels of linseed, and many ditto of pork. The timber was of white and red or black oak.

Oakwood in Trenton cost seven shillings per cord; hickory, nine or ten.

At ten in the evening we left Trenton by boat.

November the 19th

Slow Sailing. The continuation of our journey proceeded very slowly to-day, for the wind was contrary, becoming a biting storm. The boat, too, was worn out and so loaded both above and below the deck that it sank down almost to the deck itself. When the tide was against us it was impossible to move, so that we had to cast anchor; and when the tide went out the storm was so violent and the waves so high that we dared not sail in the old, heavily-laden vessel much before evening, when the wind died down. We then tacked slowly forward and about seven o'clock in the evening reached Burlington, where we lay the following night until one in the morning, when the tide changed again and began to go out. The skippers confided that in the river between Trenton and Philadelphia are hidden rocks, but that one can easily dodge them with a boat.

Tobacco. The English chewed tobacco a great deal, especially if they had been sailors. Not an hour passed when they did not take as much cut tobacco as they could hold in the fingers of the right hand and stuff it in the mouth. Young fellows of from fifteen to eighteen years of age were often as bad as the older men.

The barrels that carried the flour (which constituted a large part of the cargo) were made of white oak, but the hoops were of hickory, and the ends of pine.

---

[1] A paragraph on watermelons which follows is but a repetition of earlier accounts and is therefore omitted here.

NOVEMBER THE 20TH—AT PHILADELPHIA

Oysters were sold in large quantities at this time in Philadelphia.[1]

NOVEMBER THE 21ST

Swedish Clergymen. Since I learned in my quarters this morning that two ministers had arrived from Sweden, I went out to find and talk with them. They were lodged at the home of Mr. Hesselius and were the Rev. Mr. [Israel] Acrelius, appointed dean of the Swedish congregations and pastor of the Christina parish, and Rev. Unander, who expected to be made pastor of the parishes of Penn's Neck and Raccoon. Both came from Roslagen [Uppland], had left Sweden in June, and had been on the way between Gravesend and Philadelphia more than six weeks.

In the forenoon, also, I was busy unpacking my goods from the boat, and in the afternoon in visiting some acquaintances.

The Library in this city was established a few years ago. Mr. [Benjamin] Franklin was one of the first to lay foundation for it, and was the real cause of its origin. A number of gentlemen then came together and each one gave forty shillings for the purchase of books, so that one hundred pounds were acquired for the purpose. Those who now are directors of the Library give annually so much that books may be bought for 50 pounds. Several others in the town have followed their example, and there are now several branches here, where books may be borrowed for a small fee. A rich gentleman from Rhode Island was here, and when he had the opportunity of examining this institution he liked it so well that when he had returned home he persuaded some gentlemen in that state to build a house for a library, to which he made a gift of 500 pounds sterling for books. Proprietor [Thomas] Penn presented the city with a piece of land for the building, with some telescopes, two globes, and a machine for [display of] electricity.[2]

---

[1] Here follows an article on oysters which is omitted in our translation, since the substance of it has already been given by Kalm in other entries of his diary. See especially entry for October 31, 1748.

[2] We must not forget that this is in 1749.

NOVEMBER THE 22ND

This morning we left Philadelphia for Raccoon, N. J. Jungström had gathered several seeds there, and at my departure last spring I had asked several of the Swedish people to collect seeds for me of various kinds. It was now time for me to find out whether they had heeded my request. At four in the afternoon I reached the Raccoon parsonage, where I found the wife [1] [now a widow] of the Rev. Mr. Sandin tolerably well. Jungström and the rest were also well.

Diseases. Last summer there was a great deal of sickness in New Sweden, and the reason is not so easy to understand. All asserted that the heat had been more severe than in many summers before, and according to Mr. Evans and others had been the worst on the Sunday I left Fort Anne.[2] They spoke of several who had fallen dead from the heat while walking. The epidemic began soon thereafter, and people over almost the whole country were ill, even those who otherwise had the strongest of health and were born in this land. They had to stay in bed a long period. Mr. [John] Bartram and most members of his household were sick; in Wilmington there was much illness too; also in Philadelphia. But nowhere was the epidemic so severe as in Raccoon and Penn's Neck. These two places and Salem had the reputation of being the most unhealthy localities in the whole country. There was scarcely a person who did not reiterate that these were the worst. Last summer there was hardly a house without several sick members, and it lasted a large part of the season. On some farms almost every person was ill, even the little children; and those who had been able before to keep their health there had this summer been obliged to take to the bed. Chills and fever [malaria probably] had been the predominating disease, but with several symptoms differentiating one person's illness from that of another. However, only a few died from it, and toward the end of October the disease had for the most part left. Mrs. Sandin had had a particularly difficult attack, from the beginning of June up to that time.

---

[1] This lady became, in the following year, Mrs. Kalm.
[2] June 29, 1749. Kalm in the entries for June 28 complains of the heat at Fort Anne. It must have been much hotter in New Jersey.

She had often been so weak that she thought her last hour had come.

November the 23rd—At Raccoon

There are many *black walnut trees* [1] in this locality, but not many further north. I never saw them north of New York. In planting them old van Neeman considered it best to bury the nuts where the trees were to stand permanently, because they would then better withstand the cold. Once he had planted some black walnuts, and the following summer transplanted some of the young trees, and allowed others to stand where they had come up. The next winter was unusually cold, and the majority of those transplanted froze off right near the roots, but those he had left untouched were not damaged.

Mulberry Trees. Old Kijhn (Keen) told me that sometimes during unusually severe winters one-year old mulberry trees freeze close to the ground. But this was not surprising, he claimed, for when a mulberry tree comes up it may grow as much as eight feet the first summer. In that case the pith is loose, juicy and large, and so is the tree itself. It can therefore not endure the frost so well. Nevertheless, this does not happen except in extreme cases, and only with young trees, and in the following spring new shoots will sprout from the old roots. But when a mulberry tree is two or three years old, no cold can damage it, no matter how severe the winter may be. These mulberry trees are of the black variety and grow wild here and there in the woods. None around here had ever seen any white mulberry trees.

November the 24th

Squashes are planted in large quantities everywhere in New Sweden, but they do not last after the end of October and consequently cannot be kept over winter. They become ripe in September, when

---

[1] By the Swedes of Raccoon called *svartnöttbom;* Ger., *Schwarznussbaum;* Dutch, *zwartnootboom,* translating the stems literally. The influence here is Dutch probably rather than German.

they are delicious; but later their shell hardens so that it becomes almost like wood, and the pulp and seeds decay. They have to be left out in the fields therefore, or thrown away, if not eaten in September or October.

November the 25th

In the morning I returned from Raccoon to Philadelphia, where I arrived at sunset. The roads were dry and the weather fine for the season. All hardwood trees here had already lost their leaves, and the forest looked everywhere wintry.

The Harvest. I have already, in my diary for the 15th of November, referred to the season's harvest in New Jersey and Pennsylvania, and all people I have met lately have confirmed what I said there. Farmers everywhere complained about the poor haycrop; it had not been so poor for years they said, and the reason was the terrible drought which had prevailed all summer in this locality. Many a husbandman did not know how he was to feed his cattle over winter. The crop of apples this autumn was bountiful, but many had the tendency to rot after they had been brought indoors. A large number of nut trees had borne no fruit this year, or at most very little. This was particularly the case with the various kinds of walnut and hickory trees, the chestnuts, and the many varieties of oak. There was so little fruit of the latter that the farmers of the place, who otherwise let their pigs feed and fatten on acorns, either by collecting them or allowing the pigs to run wild among the oaks, this year had to feed their swine on corn and other grain. This, again, despite the abundance of corn, has made the latter quite expensive. Nor has this season the crop of beechnuts been any larger than that of acorns; Jungström could not obtain any when he arrived here. On the other hand, there was an abundance of them in Canada this year: the trees were full of them. There was also a large quantity there of walnuts and hickory nuts and an appreciable crop of acorns. The hay crop, too, was heavy, excellent. The wheat harvest there was exceptional: they had not had one like it for many years. But in Albany and New York State in general the complaint was commonly heard that the wheat crop this season had been one of the poorest.

NOVEMBER THE 26TH—AT PHILADELPHIA

Remarks about Plants. To-day I met Mr. John Bartram for the first time since my return from Canada. He reiterated what he had often told me before, namely that all plants and trees have a special latitude where they thrive best, and that the further they grow from this region, whether to the north or south, the smaller and more delicate they become, until finally they disappear entirely. When I apprised him of the fact that I had found sassafras trees at Fort Anne, far to the north, but that these trees had been very small and without seeds, he answered that all such trees had grown there from seeds carried by birds, and that their slender stems showed it was not their natural habitat. In the Blue Mountains he had found the *Abies balsamifera* (balsam spruce), which generally grows in Canada. Near Oswego, he had come upon a mulberry tree, and one single specimen of *Arbor tulipifera* [*Liriodendron tulipifera*], which the local settlers called Old Woman's Smoke, and whose leaves were held to be a remedy for gout. He discovered an aloe plant in the northern part of Virginia, the farthest north that it grows.

Fol. Avoine. When I talked with him about Fol. avoine (*Zizania aquatica*, Indian rice), he told me he thought this to be that tall, thick grass which grows here in brooks and other bodies of water and has long, grain-bearing seeds. The Indians had formerly gathered these seeds for food. Now they are eaten by a bird which is described and pictured in [Mark] Catesby's Ornithology [1] and is called the ricebird (bobolink). This bird remains here until the seeds of the plant have fallen out, and then he moves on to Carolina, where the rice is ripe about that time. Mr. Bartram believed this wild Indian rice to be a good food, but encountered a difficulty in its gathering, since it ripens very unevenly and not all simultaneously. It begins to ripen in the beginning of August and continues to do so the rest of the month.

NOVEMBER THE 27TH

An Apple Beverage. Mr. Hesselius's daughter related how some

---

[1] Either *The Natural History of Carolina, Florida, and the Bahama Islands* (1731-1743, with appendix, 1748) or a paper on birds presented in 1747 before the Royal Society of London and printed in its *Philosophical Transactions* that same year. In this monograph Catesby "gave examples from among the South Carolina birds."

colonists in this vicinity made a pleasant beverage of apples, as follows: some apples—which need not be the best—and apple peelings are taken and dried. Half a peck of this dried fruit is then boiled in ten gallons of water and when removed from the fire the solid part taken out. Then yeast is added to the water, which is allowed to ferment, whereupon it is poured into vessels like any other drink. One who has not tasted it before would not believe that such a palatable beverage could be prepared from apples. It was said to be better than that made from persimmons, because it retains its quality longer and does not get sour.—I forgot one thing: when the apple ale is made, some bran should be added to the water.

Varieties of Stone. To-day Benjamin Franklin showed me several varieties of stone, which he had in part collected himself and in part received from others. All were formed in the English provinces of America and consisted of:

1. A rock crystal, the largest I had ever seen. It was four inches long and of a diameter of three fingers' breadth. I regretted it was not transparent but of a dingy, watery color and opaque texture. All six sides were smooth as if ground, and had been found in Pennsylvania.

2. *Asbestus stellatus*, with fibers radiating out from the center, as described in Wallerius's *Mineralogy,* page 145. Its color was a very dark gray, mostly blackish, and felt oily to the touch. It came from New Engand, where it is found in big stones that are utilized for fireplaces, because it does not change or crumble in the least from the action of fire.

3. *Stalactites.* These were discovered in a cave near Virginia and were of two kinds: the *stalactites conicus* which had depended from the roof of the cavern, and the [stalagmite] that had been deposited like a round, uneven, scraggy fungus on the floor of it, where the [calcareous] water had dripped from above. In color they resembled an unclean white.

November the 28th

Graphite. To-day I received in my quarters a piece of graphite that had been discovered some 20 English miles from here [Philadelphia]. It was dark blue, very soft so that it could be cut with a knife, and had an undulating surface. One could hardly touch it before the hands became black; it left a mark everywhere.

NOVEMBER THE 29TH

In the evening I accompanied Mr. John Bartram to his estate in the country.

The Falling Trees. I told Mr. Bartram about my observation of falling trees the summer before in the wilderness between Albany and Canada, describing how on calm nights I heard the trees crack to the ground. He offered the [highly fanciful] [1] explanation that it was due to a difference in atmospheric pressure, and that in his experience such a phenomenon was a sign of rain.

NOVEMBER THE 30TH

Mr. Bartram had found some *Salicornia* (glasswort) and *Potentilla anserina* L. (silver-weed) growing near salt springs. He had discovered some arbutus in the sand in western New Jersey, and found the taxus in several places near the Hudson, Delaware and Susquehanna Rivers, though always dwarfed specimens of it.

The persimmon is a very hard wood. Hammers and clubs are made of the sour gum tree. *Filipendula* (meadowsweet or dropwort) are also called tea bushes. There is a birch [sweet birch] which is chewed like sugar.

In the evening I returned with Mr. Bartram's son to the city. We met a man on the road who was complaining bitterly that two culprits had just attacked him and robbed him of 50 shillings and his overcoat, and had then beaten him and run away.

DECEMBER THE 1ST

Swedes. A man from Chester called on me last Friday morning. He was of old Dutch extraction whose ancestors had been in America since the Hollanders captured the land from the Swedes, but he was married to a Swedish woman. He gave me much information about the old Swedish settlers. They had been entirely satis-

---

[1] Words in brackets are not in the original text, of course.

fied to come under Dutch rule, he believed, because they had not heard anything for a long time from their mother country, nor received any aid from there. And more particularly because the Governor [Printz] of the Swedes had been rather severe, and treated them mostly as slaves.[1] A short time before the Dutch had captured New Sweden, a Dutch vessel appeared, loaded with all kinds of goods which they knew were needed by the Indians. At their arrival the Dutch asked the natives for permission to tie their ship with a rope to a tree on the shore, since otherwise the current might carry it away. When the Indians had permitted this, they were presented with various gifts and were promised more in return for a permanent permit to fasten their boat there and for a piece of land at the spot, where the Dutch might put up their tents and cook their food, it being so troublesome always to stay on the ship. The natives allowed this too: *hinc causa belli inter Suecos et Batavos.*[2] When Penn came he took much land away from the Swedes and the Dutch who lived near Philadelphia and gave it to the Quakers under the pretext that the Swedes had more land than they needed and that otherwise it would lie waste and uncultivated. Much land was taken away by fraud, and whenever [the higher authorities] got hold of the old deeds, describing how much land had been apportioned to each one, the Swedes and Dutch never had theirs returned.

DECEMBER THE 2ND

Violent Storms in North America (quotation). "The *American Weekly Mercury*, No. 231, Philadelphia May 31, 1724. On Saturday last, being the sixteenth of this instant, we had a violent storme of wind, wid very hard shower of rain and claps of thunder, which lasted about eight minutes, it has done very considerable damages to our orchards, and has killed several Cattle by the fall of the Trees, and in some parts of the country they had very large hail-stones, which they say has destroyed whole fields of corn.

N:o 243. Philadelphia aug. 13, 1724. On the third instant, about the hour of 12 (at New Garden in Chester) there began a most

---

[1] There is, possibly, some truth in this Dutch contention.
[2] "Hence a cause for war between the Swedes and the Dutch."

terrible and surprizing Whirl-wind; which took the roof of a barn and carried it into the air, and scattered it about 2 miles off, also a mill that had a large quantity of wheat in it, and has thrown it down and removed the millstones, and took a lath of the barn, and carried it into the air; it also carried a plough into the air." . . . Both were hurled down later; one into an oak lid and the other deep into the ground. A flock of geese and three or four hawks had to go along. From the *Philadelphia Country* I obtained the following: At Plymouth the whole roof was pulled off a big barn and carried out in the lot; a woman's skirt sailed seven or eight miles through the air, and grain stacks were strewn about the fields. "It took up almost all the apple trees in the orchard by the root and carried them some distance." People were in danger of being carried off right in their houses. From Bucks' county [came the report] of houses, fences and trees being torn down. In No. 245 it goes on to speak of this whirl-wind as follows: "Philad. aug. 27. We have a farther account of the Whirl-wind, or Hurrican, from the Great-Valley, where it took up very large trees by the roots; but particularly one about 3 foot over, and carried it up in the air a great height, so that when it fell, stuck very deep upright; it made clean work where it went, took up all as tho it had been pulled, and where it went across the roades, it laid trees so thick that it is very difficult to travel; it made a road of about 40 pole in breath, and in some places it parted, and then met again about two miles off.

N:o 244. Philad. Aug. 20. 1724 on th 17th, 18th and 19th of this instant, we had a violent storm of wind and Rain, which has caused such a fresh in our Creeks, and river, that it has broke several of our best mill dams, in this and our Neighbouring Provinces, and the fresh is so strong, that our Vessels in the road have not winded; Such a fresh has not been known this 20 years.

DECEMBER THE 4TH

To-day I went to Raccoon.

Cheese. In almost every place that I have been this summer, both in the English provinces and in the French ones in Canada, the residents have made cheese; but there has been an appreciable difference in the quality of it, which has varied with the locality. In Canada cheese was made only in certain places, such as on the

Isle d'Orleans below Quebec. The cheeses there were very small, round and thick, but not very good, about the poorest I have eaten in America. They tasted dry, lacked nutrition, and only common people for the most part were seen eating them; most of the better class left them alone. These bought cheese imported from France, which came in large, thick and round forms and tasted well, but smelled like sour herring. In New York State several kinds of cheese were made. Most of them appeared in an average, suitable-sized mold, were round, reasonably thick; and some tasted pretty well; but most varieties were poor and manufactured from sour milk. In Esopos [Kingston] good cheese was made in the shape of small globes that were flattened a little on top and bottom; one had added spices to it to improve its taste. Here in Pennsylvania we got cheese of both kinds, good and bad; but in general better cheese was made here than in any other place in America that I visited. The cheese made by the Swedes of Raccoon was especially good and looked very appetizing. It was molded in round, thick forms of from nine to twelve inches in diameter, and was the best made in this part of the world. Some of it could rival the English variety.

December the 5th

Centenarians in America. The *American Weekly Mercury*, No. 196, in a communication from Boston, September 2, 1723, mentions an old Indian by the name of John Quittamog, "living in Nipmug Countrey near Woodstock in New England. He is reckoned to be alone one hundred and twelve years old; he still confirms, that he was at Boston when the English first arrived; and when there was but one cellar in the place, and that near the common, and then brought down a bushel and a half of corn upon his back." He is still in good health and mind, has a retentive memory, "and is capable of travelling on foot ten miles a day."

The Pennsylvania *Gazette*, No. 138, in a note from New York of July 5, 1731: "We hear that about 26 miles above Albany there died lately one Johannes Legrange, aged 106."

"Dedham in New England June 5, 1732. This day died here the famous Sam Hide, Indian, in the 106th year of his age." He was well known for his "running jests and uncommon wit." He had slain 19 other Indians and made the same number of notches or dents on his gun.

No. 343. Boston, June 16, 1735. Thomas Curries died at Colchester, Connecticut in the 110th year of his age. He left 5 children, 39 grandchildren, and 28 great grandchildren; "some of the last are married." His hair was not gray, nor was he bald. He walked over six English miles a day during his last years.

The Germans [of the locality] were said to be very industrious and in good circumstances. Others lived mostly from hand to mouth.

December the 6th

Scurvy in the Mouth.—An Englishman by the name of Wood has offered the following remedy for scurvy in the mouth: one takes that sole of an old shoe which is nearest the foot in walking, burns it to a coal in a fire, and grinds it into a powder. Burn some alum, powder it, and add it to the shoe-sole powder. Take some of the mixture in a rag or cloth and rub the teeth and gums with it several mornings in succession.

Squirrels. "*The American Weekly Mercury*, No. 302. Boston, September 20, 1725. We have advice from Connecticut and several Towns in this province, that the Squirrels enter their fields, in droves, and destroy their corn; and that in some places they even run into the farmers' houses for food."

I left Racoon to-day at half after twelve, arriving in Philadelphia at exactly six o'clock.

December the 7th

Quakers. To-day I attended service in a Quaker meeting-house. I was once present at such a service in London, and once in this city, where there are two meeting-houses or churches. It is known that the Quakers are a religious sect that arose in England during the last century, of whom the majority is found in Pennsylvania, since this province was granted to Penn, who was a Quaker and who brought his religious followers to this territory. But I should like first to describe the service that took place in their church to-day, and then speak about their faith and customs.

The Quakers in this town attend meeting three times on Sunday—from ten to twelve in the morning, at two, and finally, at six in the evening. Besides, they attend service twice during the week, namely on Tuesdays from ten to twelve and on Thursdays at the same time. Then also a religious service is held in the church the last Friday of each month, not to mention their general gatherings, which I shall discuss presently.—To-day we appeared at ten, as the bells of the English church were ringing. We sat down on benches made like those in our academies on which the students sit. The front benches, however, were provided with a long, horizontal pole in the back, against which one could lean for support. Men and women sit apart. (In London they sat together). The early comers sit on the front seats, and so on down. Nearest the front by the walls are two benches, one on either side of the aisle, made of boards like our ordinary pews, and placed a little higher up than the other seats in the church. On one of them, on the men's side, sat to-day two old men; on the other, in the women's section, were four women. In these pews sit those of both sexes who either are already accustomed to preach or who expect on that particular day to be inspired by the Holy Ghost to expound the Word. All men and women are dressed in the usual English manner. When a man comes into the meeting-house he does not remove or raise his hat but goes and sits down with his hat on.—Here we sat and waited very quietly from ten o'clock to a quarter after eleven, during which the people gathered and then waited for inspiration of the Spirit to speak. Finally, one of the two old men in the front pew rose, removed his hat, turned hither and yon, and began to speak, but so softly that even in the middle of the church, which was not very large, it was impossible to hear anything except the confused murmur of the words. Later he began to talk a little louder, but so slowly that four or five minutes elapsed between the sentences; the words came both louder and faster. In their preaching the Quakers have a peculiar mode of expression, which is half singing, with a strange cadence and accent, and ending each cadence, as it were, with a full or partial sob. Each cadence consists of from two to four syllables, but sometimes more, according to the demand of the words and meaning; i. e. my friends//put in your mind//we can//do nothing//good of our self//without God's//help and assistance//etc. In the beginning the sobbing is not heard so plainly, but the deeper and farther the reader or preacher gets into his sermon the more violent is the sobbing between the cadences. The speaker to-day had no gestures, but turned in various directions; sometimes he placed a hand on his chin; and during most of the sermon kept buttoning and unbuttoning his vest. The gist of his sermon was that we can do nothing of ourselves without the help of our Savior. When he had stood for a while using his sing-song method he changed his manner of delivery and spoke in a more natural way, or as ministers do when they read a prayer. Shortly afterwards, however, he reverted to his former practice, and at the end, just as he seemed to have attained a certain momentum he stopped abruptly, sat down and put on his hat. After that we sat quietly for a while looking at each other until one of the old women in the front pew arose, when the whole congregation stood up and the men removed their hats. The woman turned from the people to the wall and began to read extemporaneously a few prayers with a loud but fearfully sobbing voice. When she was through she sat down, and the whole congregation with her, when the clock struck twelve, whereupon after a short pause each one got up and went home. The man's sermon lasted half an hour. During the sermon a man would get up now and then, but in order to show that he did not do so to speak he would turn his back to the front of the church—a sign that he did not arise from any spiritual inspiration. There were some present who kept their hats off; but these sheep were not of this flock, only strangers who had [like Kalm himself] come from curiosity and not because of any special prompting by the Spirit.

The meeting-house was whitewashed inside and had a gallery almost all the way around. The tin candle-holders on the pillars supporting the gallery constituted the only ornaments of the church. There was no pulpit, altar, baptismal font, or bridal pew, no prie-dieu or collection bag, no clergyman, cantor or church beadle, and no announcements were read after the sermon, nor were any prayers said for the sick.—This was the way the service was conducted to-day.

But otherwise there are often infinite variations. Many times after a long silence a man rises first, and when he gets through a woman rises and preaches; after her comes another man or woman;

occasionally only the women speak; then again a woman might start, [followed by a man], and so on alternately; sometimes only men rise to talk; now and then either a man or woman gets up, begins to puff and sigh, and endeavors to speak, but is unable to squeeze out a word and so sits down again. Then it happens, also that the whole congregation gathers in the meeting-house and sits there silently for two hours, waiting for someone to preach; but since none has prepared himself or feels moved by the Spirit, the whole audience rises again at the end of the period and goes home without the members having accomplished anything in the church except sitting and looking at each other. The women who hope to preach and therefore sit in a special pew generally keep their heads bowed, or hold a handkerchief with both hands over their eyes. The others, however, sit upright and look up, and do not cover their eyes. The men and women have separate doors, through which they enter and leave the church.

I shall now say something about their clothes and manners, in so far as they vary from those of others. The women have no clothing that differs from that of the other English [ladies], except that I do not remember having seen them wear cuffs, and although they censure all adornment I have seen them wear just as gaudy shoes as other English women. But the men's clothes differ somewhat from those of other gentlemen. For instance, they have no buttons on their hats, and these are neither turned up entirely nor turned down, but just a trifle folded up on the side and covered with black silk, so that they look like the headgear of our Swedish clergymen. They wear no cuffs; they never take off their hats, neither when they meet anyone nor when they enter a stranger's house or receive friends at their own home; they make no bows and hate all courtesies; and the only form of address is "My Friend" or "thee and thou". A son addresses his mother with "thee". In the plural for "they" one says "I and the Friends", naming them. Although they pretend not to have their clothes made after the latest fashion, or to wear cuffs and be dressed as gaily as others, they strangely enough have their garments made of the finest and costliest material that can be procured. So far as food is concerned, other Englishmen regard the Quakers as semi-Epicureans; for no people want such choice and well-prepared food as the Quakers. The staunchest Quaker families in the city are said to live the best. Yet in the matter of drinking they practice restraint. The majority of colonists did not look upon them as any *societas pia*, as they at first represented themselves, but as a political body. They cling together very close now, and the more well-to-do employ only Quaker artisans, if they can be found. If a skilled workman, laborer or someone else of their faith backslides and joins another church, they have no more to do with him. The Quakers have a general fund from which they lend money to their poor, at little or no interest, according to the circumstances of the borrower. Often when one of their tradesmen gets into financial difficulties, the Friends will collect a sum and present him with it. They have overseers who go about and see how the brothers live, which may be seen from their Code of Discipline.

When two members wish to marry they attend the monthly meeting together, rise and announce jointly and loudly that they expect to take each other as man and wife. This is repeated at three such meetings in succession, so that it requires three months before the banns are duly proclaimed. For further details, see the marriage code.

DECEMBER THE 8TH

The Newly Invented Pennsylvania "Fireplaces." Although this city [of Philadelphia] is situated at 40° north latitude and should be relatively warm in winter, when we compare it with European towns of the same latitude, experience has taught that sometimes it is as cold here as in old Sweden. It is necessary only to examine the meteorological observations in this travelogue for this year, the supplementary remarks here and there in the diary about weather conditions and temperature, and what Mr. Franklin records in his *Poor Richard's Almanac Improved.* It has therefore been imperative to make some provisions for heating the houses for a period of several months in winter, and for that purpose various types of stoves have been used. I shall not attempt to describe them all at this time, because they are not only described but their good features and serious faults elaborated in the book which Mr. Franklin published in Philadelphia, 1744, under the name of *An Account of the New Invented Pennsylvanian Fire-Places* etc. But since all these had their faults, and the Englishmen liked to see the fire burn instead of confining it in a stove, Benjamin Franklin invented a new type of stove, which not only provides plenty of heat, saves fuel and brings fresh air into the room, but is so constructed that the flame may be seen. An extensive description of it is found in the above-mentioned book by Mr. Franklin, so that I have but little to add, especially since Mr. [Lewis] Evans has in addition made copious marginal notes about it in his presentation copy of it to me [and which may be consulted]. Mr. Franklin invented the stove and Mr. Evans made the drawings and figures. The bottom should not rest flat on the floor or ground, but be elevated a little, and an opening or passage left to the air-box, so that the air which is heated beneath the bottom plate can get out into the room. There are several types of this stove: some have dampers, which in the description are called registers, and others have not; some have a front plate near the bottom which can be moved up and down, and when it is moved down the draft becomes stronger and the wood begins to burn quicker; others have a hole on the frontal plate with a small trap-door which when opened allows air to enter and fan the fire, for there is a narrow passage leading to this trap-door, either from the room below or from a space under the floor connected with the outside. Under the air-box is an opening through which fresh, cold air enters it, and here it is heated and sent into the room through side openings. If there is a cellar beneath the room with the stove, a hole is made to it, so that the air can come from that direction; but otherwise it must pass under the floor to the air-box from some side of the house, preferably an entrance hall, where a small opening may be made in the wall near the floor, through which fresh air may enter. Several who could not afford to purchase these stoves imitated them by making them either of brick or white Dutch tile [or brick], making only the top of iron. But these were not so warm.

December the 9th

More about the Franklin Stoves. I wrote yesterday about the newly invented fireplaces [stoves]. I shall now add a few more details. They are made or cast of iron, i. e. the iron which is obtained in this province. The size varies and so does the price.

Despite their usefulness they have been criticised. It was held that if the chimney could not be swept [as in big chimneys with open fireplaces] there was danger of fire; some thought the stoves gave too much heat, and since the Englishmen were not accustomed to this they liked open fires better; and the Germans preferred *their* small, oblong, square iron stoves, which are constructed in the same way as the stoves in Bohuslän and Norway, because they give more heat and cost less. Nevertheless, there were many who used these new stoves, both in Philadelphia, New York and elsewhere. In the country where they had plenty of fuel they used the large fireplaces, since many believed these new ones expensive, and did not reckon the cost of wood. Mr. Franklin loaned me one of the stoves for the winter. It kept the house quite warm, but then one had to use short wood in it. It proved often unnecessary to have a fire in the kitchen, and one could prepare chocolate and other food in the little stove. Also, it proved possible, by suspending a cord from above in front of the fire-box, to roast meat or fowl attached to it and turned. And curiously enough it was roasted better and quicker than in a regular kitchen fireplace, since the room was warm and heat came both from the fire and the iron of the stove. The chimney is seldom cleaned more than once a year, but Mr. Franklin was in the habit of setting fire to a sheet of paper every fortnight and let it pass through the flue leading to the stove and so burn off the soot there also. If the stove is narrow it is not so easy to sweep the chimney, after everything is closed up by masonry; but Mr. Franklin had a brick removed beside the stove, let a man pass down through the chimney, clean it, and when he reached the bottom near the stove had him force the soot through the hole made by the removal of the brick. When this was done the brick was replaced. Where the hearth is broad the stove is placed on one side of it, and a door made on the other through which the chimney sweep can enter and do his work. To get fresh air into the stove of a house with no cellar, and where no outside air is wanted, Mr. Franklin this year had had the stove in his own room set on a rim of masonry six inches from the floor, with an opening through the bricks on one side to let the cold air near the floor enter, pass through the air-box, where it was heated, and then pass through the holes on the iron sides of the stove into the room, etc. This brought about a constant circulation of air.

DECEMBER THE 10TH

Remarks about the Heat. After dinner I talked with Mr. Evans about the weather here in Philadelphia last summer, especially about the intensity of the heat. He said the severest heat had come on June 26th and 27th (N. S.), and more particularly the latter, when his thermometer showed 86° F., which he found to be the same as the temperature of the human body.[1] The heat on that day would have been unbearable, had it not been for a gentle breeze. According to a Fahrenheit thermometer owned by Mr. Franklin, and believed to be more accurate than Mr. Evans's the mercury on that day rose to 100°. No one remembered a hotter summer. [Then follows a brief comparison of the Fahrenheit and Centigrade (Celsius) thermometers, which is unnecessary here]. By referring to my meteorological observations for the 26th and 27th of June [1749] and the days immediately preceding and following one can notice the difference between the temperatures of Philadelphia and of Fort Anne, where I stayed on those dates. Strangely enough Mr. Evans reported that about the 14th of June, according to the newspapers, both England and France had had an unusually cold spell, and that there had been ice in England of the thickness of a silver dollar. In France the cold had damaged the grapevines and other fruit trees and vegetables.

DECEMBER THE 11TH

The Cold. It is now becoming quite cold. Yesterday especially it felt very penetrating, in particular when walking against the wind. All small streams and inlets are now covered with ice so that one can pass over on most of them without danger. The Schuylkill River was well frozen too. On the Delaware River there was so much ice that it was completely frozen this morning near the city. Yet the ice was very thin and disappeared as soon as the tide appeared. Otherwise there drifted so much ice in the river that at Gloucester's Ferry it was impossible to cross with a horse. One

---

[1] Mr. Evans's thermometer was not very accurate, as we can see. The temperature of the human body is about 98.4° F. It is possible, too, that Kalm's original manuscript showed, or meant to show, 96° instead of 86°, which still leaves an appreciable error.

could now cross only in small boats between the pieces of ice. People who had brought wood to town on boats expected to leave them and return by the highway. Ships that intended to sail hastened down the river before they would be frozen fast, and those only half loaded began to consider remaining where they were over winter, wherefore some took their boats into dock or to some safe place where it would be protected from the ice, both now and in the spring. The price of wood went up rapidly, because before that one had been able to buy a cord of hickory for 22 shillings, but now it had gone up to from 25 to 27 shillings per cord, and even then one had to hurry and take it lest it be snapped up by someone else. Oak wood rose from 16 to 19 and 20 shillings per cord, and one was glad to get it at that price. All wood bought now was green, cut this fall, for although the country people cut much wood in the spring and summer, which they let dry and cart into town in the early autumn, it is bought up at once by the wealthy and thoughtful, who then get their supply at a lower price. Later only green wood can be found on the market.—The following days the weather changed and became milder, so that by the 13th of December there was hardly a piece of ice to be seen near the Gloucester ferry; it had either melted or been carried away by the water.

DECEMBER THE 12TH

Newspapers in North America. There were said to be at present seventeen different newspapers published in the various English colonies in North America; for every week dailies or weeklies appear in Boston, New York, Philadelphia, Virginia and Carolina, and on the islands belonging to the English. Occasionally there are several printed in one city: in New York are both English and Dutch newspapers; in Philadelphia two English and one German. The first publication of this type in Philadelphia was *The American Weekly Mercury* which made its appearance on December 22, 1719. In the beginning it was only half a sheet, folio, in size, and printed by Andrew Bradford [1686-1742]. It has continued up to the present time.[1] In 1728 the second type of newspaper appeared, called

---

[1] Isaiah Thomas, *History of Printing*, II, 327, believes it was discontinued "soon after" 1746.

The Pennsylvanian Gazette. On October 1, 1728, the following announcement of it was circulated: "Whereas several gentlemen in this and the neighboring provinces, have given encouragement to the Printer hereof, to publish a paper of intelligence; and whereas the late[1] Mercury has been so wretchedly perform'd, that it has been not only a reproach to the province, but such a Scandal to the very name of printing, that it may, for its unparallel'd blunders and incorrectness, be truly stiled Nonsense in Folio, instead of a serviceable News-Paper." The first number by Samuel Keimer [1688-c. 1739],[2] was published on December 24, 1728. Mr. Keimer was a so-called Sabbatarian, who never designated the months as January, February, March, etc. but as the first, second, third month etc., and March was the first month in his system.[3] In place of Monday, Tuesday and so on, he put first day, second day etc. He was very queer and satirized everything so that it is not surprising that he had many enemies. On September 25, 1729, Mr. Franklin and Mr. [Hugh] Meredith began to publish this gazette, since Keimer left and gave up his position to the men just named. Newspapers then received a still better reputation, and as soon as the new editors started to print, they used selected paper and type. Januarius, Februarius, and Monday, Tuesday, etc. were recalled from their exile. The first two years the *Gazette* contained a weekly necrology, but later this section was given up.

DECEMBER THE 13TH

Prices Current in Philadelphia 1719, 1720 etc. From *The American Weekly Mercury* I have copied the following prices current in Philadelphia during the years 1719 and 1720, when newspapers here first appeared. The first list is printed on December 29, 1719 [in the second number].

---

[1] The *Mercury* had not in 1728 been discontinued of course, and was not, incidentally, by any means as black, relatively, as it was painted.

[2] See articles on Samuel Keimer and Andrew Bradford in the *Dictionary of American Biography*.

[3] This was common in the old Colonial style of reckoning.

Flour	9s. 6d. to 10s. per hundr.
Middling Bread	14 s. per Hundr.
Brown Bread	12 s. p. H.
Tobacco	14 s. p. H.
Muscovado sugar	40 to 45 sh. p. H.
Pork	45 sh. p. Barrel
Beef	30 sh. p. Barrel
Rum	3 sh. 9 d. p. Gallon
Molasses	1 sh. 6 d. p. Gallon
Wheat	3 sh. 3 d. to 3 sh. 5 d. p. Bushel
Indian corn	1 sh. 3 d. to 1 sh. 8 d. p. Bushel

The 26th of January 1720

Flour	9 to 10 sh. p. H.
White Bread	18 sh. p. H.
Middling Bread	14 sh. 6 per C.
Brown Bread	11 sh. 6 per C.
English salt	3 sh. per Bushel
Tobacco	14 sh. per H.
Muscovado sugar	30 to 45 C.
Pork	45 sh. per Barrel
Rum	3 sh. 8 d. per G.
Molasses	17 to 18 d. per G.
Wheat	3 sh. to 3 sh. 3 d. p. Bush.
Indian corn	1 sh. 6d. to 1sh. 8d.
Bohea Tea	24 sh. p. Pound.
Madera Wine	16 to 20 l. Pipe. [2 hogsheads]
Pitch	16 to 17 sh. p. Barrel
Tar	10 sh. p. Barrel
Turpentine	8 sh. p. H.
Rice	16 sh. p. H.
Pipe staves	3 l. per Thousand
Hoggshead stav.	45 sh. per Thousand
Barrel stav.	22 sh. 6 d. p. Thousand
Gunn Powder	7 l. 10 sh. p. Barrel
Brown oznabrigs	12 d. per Ell.
High coloured Malt	3 sh. 3 d. to 3 sh. 6 d. p. Bush.

December the 14th

Peas. I have previously mentioned in these *Travels* that but few peas are planted in New Sweden, because worms eat them up before they are ripe. But this is not all. I was shown a lot of peas that had been grown here during the summer of 1748 and had first been left untouched by worms, but which during the past winter, while lying dry in storage, had been so infested by vermin that hardly a pea had escaped. The kernels were so thoroughly consumed that there was hardly anything left but the shells. There was scarcely a pea that had not thus been hollowed out and spoiled.

December the 15th

Peach Trees. Old Kijhn (Keen) gave me the following observation and advice about planting peach trees in a cold country: they should be set out on high land or hills, where they are exposed to the wind. The reason for this idea came from his own experience on his farm in Raccoon, from which he had a view over a large part of the country thereabout. He noticed that one summer night, when frost had damaged almost all peach trees in the whole land so that little or no fruit was obtained that year, that the flowers on his own peach trees had not been injured at all. He received an excellent peach crop that autumn. People came from near and far to buy peaches from him. Those who lived in the valleys right below his own farm as well as others in the country, had their peach blossoms frozen. It is to be noted that Mr. Keen's estate lies on a very high elevation, which consists to a large extent of sand or loam, while in the vales here and there in the locality are marshes with gently flowing water. There is one kind of peach which is ripe two or three weeks before any other, and this is considered the best, both in smell and taste. The later varieties are not deemed so good.

December the 16th

To-day's prices of goods, from *The Pennsylvania Gazette*:

Flour	15 sh. 5 d.
Wheat	5 sh. 6 d.
Indian corn	2 sh. 7 d.
Shipbrood	17 sh.
Midling br.	26 sh.
White bisket	30 sh.
Beef	30 sh.
Pork	60 sh.

Pipe staves	9 l.
Hogshead staves	6 l.
Barrel staves	4 l.
Madera wine	27 l.
Westindia Rum	4 sh.
New England Rum	2 sh. 6 d.
Muscovado sugar	60 sh.
Molasses	20 d.
Coarse salt	2 sh.
Fine salt	3 sh.
Rice	20 sh.
Tobacco	18 sh.
Pennsylvania Loaf Sugar	1 sh. 5 d.
Indigo	7 sh.
Powder	10 l.
Hemp	3 d. halfpenny
Flaxseed	12 sh.
Barley	5 sh. 6 d.

Mortar. When some of the Swedes were building the fireplace in the Raccoon parsonage they mixed fresh horse manure with the mortar, maintaining that this made it better—it held together longer and did not crack. The mortar used here contained more sand than ordinary clay, so that it did not stick well together. In setting up the new Franklin stoves it was the custom at first to mix horsehair in the mortar to make it stick better, but now this has been given up, since it was found that the stoves smelled from it for a time. Perhaps horse manure in the mortar has the same effect, although it is not noticeable where there is no damper and most of the heat goes up through the chimney.

December the 17th

Prices Current in New York 1720. From *The American Weekly Mercury* of January 12.

Flour	14 to 15 sh. p. H.
White Bread	20 to 21 sh. p. H.
Midling Bread	18 to 19 sh. p. H.
Wheat	4 sh. to 4 sh. 6 d. p. B.
Indian corn	2 sh. p. B.
Pease	5 sh. p. B.

Beef	36 to 38 sh. per Barrel
Pork	56 sh. to 3 l. per Barrel
Logwood	12 l. per ton.
Rum	3 sh. 6 d. per G.
Molasses	1 sh. 6 d. per G.
Muscovado Sugar	40 to 45 sh. per H.
Madera wine	24 to 25 l. per Pipe
Pitch	16 to 17 sh. p. Barrel
Tar	13 sh. p. Barrel
Spanish plate	8 sh. 6 d. to 9 sh. per ounce
Pistoles	28 sh. p. piece
Addenda d. 28 Apr.	
Indigo	7 sh. p. P.

DECEMBER THE 18TH

Prices Current in Boston 1720. From *The American Weekly Mercury* of February 23.

Pitch	11 sh. p. H.
Tar	22 sh. p. Barrel
Turpentine	12 sh. p. H.
Train oil	36 pound p. Tun, and falling
Fish merchantable	23 sh. 6 p. per Quintal
D:o Jamaica	18 sh. per Q.
D:o Barbadoes	15 sh. per Q.
Barbadoes Rum	5 sh. per G.
Molasses	2 sh. 4 d. p. G.
Cocoa	7 pound p. H.
Beaver skins	3 sh. 10 d. p. pound
Buck and Doe Skins in oil	8 sh. 6 d. per P.
D:o indian dress	5 sh. per P.
D:o in the hair	1 sh. 8 d. per P.
Pine Boardes	55 sh. per Thousand
Flour	28 sh. p. H.
Bread coarse	25 sh. p. H.
Wheat	7 sh. 6 d. p. Bushel
Indian corn	4 sh. p. Bushel

Addenda of the 28th of April

Hops	4 d. halfpenny p. l.
Mackeril	35 sh. per B.
Isle of May salt	24 sh. p. Hog.
Whalebone	4 to 5 sh. p. l.

December the 19th

List of Births and Deaths in Philadelphia, 1722. I found in *The American Weekly Mercury* for the latter part of the year 1721 and the whole of 1722 a list of births and deaths of both sexes for each month during that period, not only of the members of the English Church, of the Quakers and of the Presbyterians, but also a list of the Negroes dead, the number of deaths from accidents, i. e. "casualties", and the number buried in the Strangers' Burying Ground. (See list on next page).

December the 20th

Weather Forecast. The weather to-day was glorious, and later in the evening we had starlight, with the moon appearing in its pure white radiance. Still there were signs of some change, because just before and at noon it was very warm, and not long after it the hens went to roost, as though it had been late in the afternoon. The following night it began to rain right after twelve o'cloock. See the meteorological observations for next day.

December the 21st

The Shortest Days. We were now passing through that time of the year when the northern hemisphere has its shortest days. But we learned that there was a large difference in this respect between the Old and the New Sweden. The sun rose here at 7:15 and set at 4:45. It was possible to read a book without artificial light at seven in the morning and five in the afternoon. But it was very dark at six, both morning and night, and before 7 A.M., and after 5 P.M. it was almost impossible to read by natural light. The frost had not yet had any serious results, for the ground was bare everywhere; but little snow had fallen so far, and what little had come had melted the same day it came, or the day after. The River was still open, and the smaller streams that had been covered with ice a little more than a week before had by the sunshine and recent rains been cleared of the ice again. This neighborhood now had about the appearance of a Swedish countryside in the beginning of October. The cattle went outdoors continually seeking their food, and but little was fed

	English Church				Quakers				Presbyterians				Buried in the Strangers' Burying Ground	Casualties	Negroes Dead
	Born		Died		Born		Died		Born		Died				
	Boys	Girls	Men	Women	Boys	Girls	Men	Women	Boys	Girls	Men	Women			
Year 1721 fr. the 21st of July to 25th of Dec.	16	18	26	22	—	—	—	—	—	—	—	—	—	1	—
Year 1722															
Ianuarius	2	1	2	—	—	—	4	3	2	—	1	—	2	—	—
Februarius	3	1	1	2	12	10	2	2	—	1	—	—	1	2	5
Martius	4	4	2	2	10	15	2	4	2	4	1	2	2	—	6
Aprilis	2	3	3	1	9	9	5	1	3	2	—	—	4	2	—
Maius	2	2	4	4	12	16	1	4	1	—	—	—	1	5	3
Iunius	—	1	—	—	13	12	6	2	—	2	—	—	1	1	1
Iulius	2	3	4	2	11	15	2	5	—	—	1	—	1	—	3
Augustius	4	4	9	6	20	18	7	4	4	2	2	3	1	1	3
September	4	4	5	0	13	16	1	4	2	1	1	1	4	3	3
October	4	2	5	2	4	9	3	1	2	1	1	2	—	—	2
November	2	3	2	2	9	11	1	3	1	3	3	2	—	—	3
December	2	1	—	—	18	12	5	1	—	1	1	—	2	—	2
Year 1723															
Ianuarius	4	3	2	1	—	—	5	1	—	1	1	—	—	—	—

them at home. It may be seen from the meteorological notes that, with few exceptions, the weather had for a long time been fair and pleasant.

DECEMBER THE 22ND

How to Prevent Candles from Dripping. I asked several people how to prevent candles from dripping. I was told [at first] that no remedy was known for it, but that a frequent cause of it was the adulteration of the tallow by lard. Mr. Franklin admitted that he had seen such candles, but that he had found no other remedy for the dripping than to wind a strip of paper round the candle. This would prevent the tallow from running. The paper will burn of course as fast as the tallow but not faster, since the tallow itself hinders it. We tested the suggestion and found it to be true. The candle will burn as brightly as otherwise, but it is necessary from time to time to remove the charred paper. Five sheets of paper suffice for twenty tolerably large candles. This remedy applies only when the candles are stationary; when they are being carried the hot tallow may easily, with an unsteady hand, run down over the paper and fingers and burn them. But a paper-wound candle is not consumed any more rapidly than a bare one of the same size.

Weather Forecast. To-day and for a couple of days following the sunsets colored the clouds in the west red. Mr. Turner, a merchant, said that in this city [Philadelphia] it meant fair weather for the next day, perhaps for several days, with relatively high temperatures. No winds followed as in the old countries.

DECEMBER THE 23RD

Candles from Spermaceti. This evening I had the opportunity to see some candles made of spermaceti. They looked like ordinary tallow or white wax candles, very white, translucent, with a cotton wick. They are made of the fat formed in the brain and head of the whale, a substance which in itself is said to be white and hard; but it needs some scientific knowledge to make these candles. There is a man in town who makes them, but he will not teach others the secrets of the process. The current price in the city for these candles is two shillings, nine pence per pound, while the tallow ones may be had for only eight pence per pound. The spermaceti candles burn as brightly as those of tallow, if not brighter. To-night we made an experiment to see which one of the two kinds would be consumed the faster, using candles of the same size, and learned that they burned about equally fast. The advantages which the new candles have, beside their rarity, are: (1) that they hardly ever need to be snuffed, but continue burning brightly, and the wick bends of itself and drops off as soon as it is burned; (2) they never drip, and though it may be warm in the room, such as when they are used in the summer time, they do not become soft, nor do they melt or soften in the hand when being carried, as tallow candles may do. Mr. Franklin and several others in the city made use of these candles for reading. It was said that there was plenty of them in London, and that they were not so expensive there as here, because more spermaceti was brought to England than to America. Consequently, candles are manufactured here (1) of spermaceti, (2) of various kinds of tallow, (3) of wax, and (4) of *myrica* or candleberries (bayberries), although it is claimed that now a lesser number of the latter are made than formerly. There is said to be a tree in the East Indies that yields wax or tallow from which candles are made.

DECEMBER THE 24TH

Raven quills are said to be the best of any to use as picks for plucking the strings of a spinette.

The following *current prices* of goods in the city have been taken from *The Pennsylvania Journal or Weekly Advertiser* for December 12, 1749. It will give a general idea of how to read and understand price lists that have previously appeared in this journal.

By the Hundred		By the Barrel	
Flour	15 sh.	Beef	30 sh.
White Bisket	28 sh.	Pork	60 sh.
Middling d:o	26 sh.	Pitch	14 sh.
Ship d:o	17 sh.	Tar	11 sh.
Muscovado sugar	60 sh.	Powder	10 l.
Rice	20 sh.		
Tobacco	18 sh.		
Turpentine	18 sh.		

By the Pound		By the Bushel	
London Loaf sugar	2 sh. 6 d.	Wheat	5 sh. 7 d to —
Pennsylvania sugar	1 sh. 4 d.	Indian corn	2 sh. 7 d.
Cotton	20 d.	Fine salt	3 sh.
Indigo	7 sh.	Coarse salt	2 sh. 6 d.
		Flax seed	12 sh.

By the Gallon		By the Thousand	
West India Rum	4 sh.	Pipe staves	10 l.
New England Rum	2 sh. 6 d.	Hogshead ditto	6 l.
Molasses	20 20 d.	Barrel ditto	4 l.

		By the Pipe.	
Butter one lb.	1 sh. à 10d.	Madeira Wine	27 l. to 30
Turnips ½ peck	4 d.		

DECEMBER THE 25TH

Extracts from *The American Weekly Mercury*. From the library in the city I brought home the first old newspapers of Philadelphia to make extracts about North American events that might have special interest.

Governor William Keith, Esq. of Pennsylvania issued on August 10, 1720, a proclamation concerning the administration of justice.[1]

Governor William Burnet, Esq. [of New York and New Jersey and later of Massachusetts], who entered office as chief magistrate on October 3, 1720, was on that date officially congratulated by the clergy and government offices of New York.

Under the date of February 7, 1721, the paper enters a complaint and protest over the number of criminals sent over from England to settle these colonies.

On February 20, 1721, Governor Burnet visited Philadelphia and was much eulogized everywhere.

New York, February 27. "Five or six weeks ago [came] the deepest snow fall at Albany that has been known for many years, and . . . all the fine Weather we have had, hath been winter-weather there."

---

[1] Unless given in quotation marks, the extracts are the editor's English translations of Kalm's summaries in Swedish. Only a few are literally copied from the *Mercury*.

Philadelphia, September 28, 1721. "Several Bears were seen yesterday near this place, and one killed at Germantown, and another near Derby." One was killed when he was eating acorns in an oak. Philadelphia, December 19, 1721. The river is so full of ice that no boats can enter the harbor. A similar item is repeated under the date of the 26th.

1722

Philadelphia, January 2. The river is frozen, so that no ships can move in or out. The same condition is announced under date of January 9th and 16th. In the middle of February boats began to leave.

All numbers for the years 1721 and 1722 are full of reports about pirates. The high seas were full of them, declares the weekly; ships were plundered, and mercantile trade almost stopped. The sea-robbers captured any ships they wanted.

On December 11th we are told that Monsieur Vaudreuil, the governor-general of Canada, had sent messages to the savages north of New England, inciting them to war against the English, and that the Englishmen had secured a copy of the message and sent it to the King of England.

The paper says nothing about the [Delaware River] being covered with ice in December of this year, nor is the matter mentioned during January of the following year, 1723.

1723

On the 30th of July a terrible storm raged over New York, coming first from the northeast and gradually shifting to the southeast. The rainfall was very heavy. "The water came up into the City higher than ever was known before." Many roofs were blown off and landingplaces damaged. "It . . . broke up all the Wharfs from one end of the city to the other, drove all the Vessels (except three) on shore", and spoiled wares stored in cellars. A severe storm visited Philadelphia on the same day, coming from the same direction. Many chimneys were hurled down. The storm "occasioned such a high tide here as has not been known there many years; the storm continued about 2 hours and a half; it has blown down a great many trees and very much damaged the fruit." In the number for August 22nd, it is reported that the storm did not reach Bermuda.

DECEMBER THE 26TH

A Continuation of the Extracts from the American Weekly Mercury.
February 1, 1726. No ships can enter the city because of ice. There is still much drifting ice on the fifteenth of the month. —On June 23, Patrick Gordon Esq. governor of Pennsylvania, arrived in Philadelphia. Two weeks previously the newspaper had printed the former Governor Keith's farewell address to the people of Pennsylvania, in which he sharply takes them to task for neglecting the interests of the King.[1]

Philadelphia, February 14, 1727. "We have had very cold weather for these four days past, which has filled our river full of ice." Ships that ventured out were obliged to return.

January 23, 1728. "We have had very hard weather here for near this two weeks past, so that it has froze our river up to such a degree that people go over daily, and they have set up two booths on the ice, about the middle of the said river." On the 30th the river was still frozen, but not by the 6th of February.

New York, January 29, 1728. No ships have passed in or out of the harbor for two weeks, because of the ice in the River. People have walked across on the ice from New York to Long Island.

N. 452. "At Boston upon the Lords day Aug. 11:th, 1728. p. m. a noble Rainbow was seen in the Cloud, after great thundering and darkness and rain; one foot . . . stood upon Dorchester neck, the Eastern end of it; and the other foot stood upon the Town; it was very bright, and the reflection of it, caused another faint rainbow to the westward of it. For the entire compleatness of it throughout the whole arch, and its duration, the like has been rarely seen. It lasted about a quarter of an hour. The middle part of it was dis-

---

[1] This was exactly half a century before the Declaration of Independence. Kalm was one of the first men of importance to note the budding neglect by the English government [of the colonists] and their growing discontent with it, though he did not always sympathize with the viewpoint of the colonists.

continued for a while; but the former integrity and splendor were quickly recovered."

DECEMBER THE 27TH

A Curious Phenomenon. The American Weekly Mercury N. 122, Newport, Rhode Island, March 30, 1722. There has lately a surprising appearance been seen at Narraganset, which is the occasion of much discourse here, and is variously represented; but for the substance of it, it is matter of fact beyond dispute, it having been seen by abundance of people, and one night about 20 persons at the same time, who came together for that purpose. The truth, as near as we can gather from the relations of several persons, is as follows. This last winter there was a woman died at Narraganset of the small pox, and since she was buried, there has appeared, upon her grave chiefly, and in various other places, a bright light as the appearance of fire. This appearance commonly begins about 9 or 10 of the clock at night, and sometimes as soon as it was dark. It appears variously as to time, place, shape and magnitude, but commonly on or about the grave, and sometimes about and upon the barn and trees adjacent; sometimes in several parts, but commonly in one entire body. The first appearance is commonly small, but increases to a great bigness and brightness, so that in a dark night they can see the grass and bark of the trees very plainly; and when it is at the height, they can see sparks fly from the appearance like sparks of fire, and the likeness of a person in the midst wrapt in a sheet with its arms folded. This appearance moves with incredible swiftness, sometimes the distance of half a mile from one place to another in the twinkling of an eye. It commonly appears every night, and continues till break of day. A woman in that neighbourhood says she has seen it every night for these six weeks past.

DECEMBER THE 28TH

The cold now became noticeably severe; the wind that blew was biting, and the River full of floating ice. A ship went down the River this morning, but encountered so much ice that it was compelled to turn back and either postpone the trip until a milder temperature should drive away the ice, or until spring.

DECEMBER THE 29TH

Price list of goods in New York according to the New York *Gazette* of December 22 (Dec. 11 O. S.), 1749.

Wheat per Bushel	6 sh.	Molasses	1 sh. 9 d. per G.
Flour, per H.	17 to 18 sh.	West India Rum	3 sh. 9 d.
Milk bread	40 sh.	New England "	2 sh. 6 d.
White d:o	30 sh.	Beef, per Bar,	34 sh.
Middling	24 sh.	Pork	2 l. 16 sh.
Brown	17 sh.	Flax see	10 sh.
Single refin'd sugar	16 d.	Bohea Tea	6 sh. 6 d. by the box
Muscovado sugar	50 sh. p. C.	Indigo	8 sh.
Salt, per Bushel	2 sh. 3 d.	Chocolate	24 sh. per Doz.

Season Forecast. In the above-mentioned *Gazette* I read the following: "Annapolis in Maryland, October 25. We are told by people from the back parts of this province, that they have had great numbers of Bears and other wild beasts, come down among them this fall; which they look upon as a certain token of an approaching hard winter."

DECEMBER THE 30TH

Pastor Dylander. Here I shall record a few facts about the late Johan Dylander, pastor of the Swedish church in Philadelphia, generally called Wicaco [now Southwark, Philadelphia], before the present incumbent of that office, Mr. Gabriel Näsman. When Mr. Dylander first arrived here there was no German minister, though a large number of Germans had settled in this locality. Consequently he was requested to preach to them occasionally, whenever the duties of his own parish would allow it. He travelled about among the Germans, therefore, preaching to them and suggesting, first of all, that they designate certain central localities where they might congregate for divine service. In Philadelphia he gave three sermons every Sunday: first one in the morning, in German; then at ten o'clock a second one in Swedish; and finally, after dinner, a third one in English. At every one of these sermons the church was full, but this was especially the case at the English service, when so many people appeared that some of them had to stand outside the doors and windows of the church. The Germans came to church in a body on Sunday morning with their psalmbooks under their arms. Some of the Swedes disapproved of having the Germans come to their temple, and did not like that Mr. Dylander gave them permission to come. Therefore Mr. Dylander [eventually] persuaded the Germans to hire a big old building in the city, where he attended to their spiritual needs. As a result he had no opportunity to accomplish anything else during the short time he was here from [1737 to 1741] than continually to travel around preaching, baptizing, administering the sacrament of the Holy Communion, etc. Once a month he went up to Lancaster to hold service, although this was far from Philadelphia. He had the German church in Germantown built, and preached in it at least once a month. He often said that in one week he gave sixteen sermons. He was incredibly beloved by all, high and low, rich and poor, by Englishmen, Swedes, Germans, and by members of almost all denominations. The reasons for this were not only his divine teaching and exemplary life but also his social affability, for in intercourse he had a special gift of being pleasant and entertaining with innocent and diverting sallies of wit. Mr. Dylander was so highly respected by the English that not only the most prominent in the city, like the proprietor, governor and others, visited him, but none felt well married unless the ceremony had been performed by him. His last illness came on the same afternoon that he had attended [officiated at?] a funeral. He rode from that to Mr. Kock's but soon felt so ill that he was compelled to go home that night and take to his bed. This was on a Monday. When Mr. Kock visited him a few days later he did not seem to be fatally ill, but he himself was certain that he would never get well, and requested that Rev. Mr. Petrus Tranberg administer the Holy Communion of the following Sunday to those of his church who were entitled to it, for, he said, he would never again perform that service. When he was thus seriously ill a bridal couple appeared to be married by him, and his wife sought to prevent it because of his weakness, but when he saw the wishes of the party concerned he bade them come forward, for he desired in this manner to end his official work on earth. So he had pillows placed behind his back, took the missal in his hands, which he held open, and read the ceremony by heart and gave them his blessing. His sickness was yellow fever, which he had had three months before when an epidemic of it killed so many people in this country, and had almost recovered from it when he again fell a victim to it, and this time he was forced to succumb. His funeral was one of the best attended in the land, and the services were conducted by Mr. Tranberg. There was such a crushing throng in the Swedish Church that many of the foremost ladies of the province lost large pieces of their skirts when they tried to get in; in fact some lost half of their skirts, an apron, or some other garment. Many of the men of rank had to climb in through the windows, and over the ladies, to get to their pews. He had the reputation of having in two years learned to speak better English than Mr. Tranberg, who had been in the country for several years. He spoke so well that it was difficult to notice that he was not a born Englishman or at least born among the English. If he had lived a couple of years longer, he would have returned to Sweden, for he maintained that he did not have the right conscience to remain in this country, where people cared so little for true godliness, especially his Swedish congregation, which cared least of all. A little before his death he lost a little prestige among a few people because of the following incidents. Mr. [George] Whitefield came over [from England] about that time, began to preach, and with his eloquence won many followers. He had accepted the Calvinistic dogma on election of grace, whereupon Mr. Dylander translated the work of Gerardius [Gerhard] [on the same subject] into English and had it published, with his own notes and commentaries.[1] Since this appealed considerably to all right-thinking minds among the Lutheran and English Churches, the Whitefieldians believed themselves offended. Otherwise Dylander was so popular among the English that he sometimes preached in the English Church in the city, and on those occasions the church was always crowded if the congregation knew in time that he was to appear.

Here follows a [brief] notice that was published in *The Penn-*

---

[1] *Free Grace in Truth: The XXIVth Meditation of Dr. John Gerhard Translated from Latin into English, with Notes for the Better Understanding of the Author's Meaning.* Printed by Benjamin Franklin, 1741.

sylvania Gazette for November 5, 1741 [relative to Mr. Dylander].
—"Monday last died of a lingering indisposition, the Reverend Mr.
John Dylander, Pastor of the Swedish Church, at Wicaco, near
Philadelphia. His sincerity, affability, and other good qualities
make his death lamented by many."[1]

DECEMBER THE 31ST

Pestilence. I have asked several people whether or not there ever
had been a pestilence in this country. All agreed that they had never
heard of any. It is claimed that certain insects are the cause of pests.
This is maintained in a volume published in London in 1738 under
the name of *A National Account of the Weather*[2] etc. and where
the last article entitled "The Cause of the Plague" contains the following words: "It has been observed that Plagues and the most contagious distempers, have commonly happened in those years, when
the Easterly Winds have more than ordinary prevail'd in the spring
and summer seasons; then the air comes to be infected, and rarely or
never at other times." If it is true that there are but few easterly
winds in this region, and the few that exist come over the ocean
where such insects necessarily must die—because the sea air is so different from all other,—it would not be surprising if no plague came
here. And all other winds in this vicinity are salubrious. . . . The
author asserts also that India, China, and the southern parts of Africa
and America have never had a real pest, for the same reason. But
he is wrong when he claims that none of the plants and insects found
in Europe are to be seen in America. These *Travels* can demonstrate
the contrary. He is of the opinion that these pestilential insects first
came from Tatary, where they are born in the swamps and the stagnant waters of the wilderness.

JANUARY THE 2ND, 1750

The cold was now quite severe. The Delaware River had yesterday
been covered by ice near Philadelphia, and to-day the whole river

---

[1] It should be remembered perhaps in reading this rather long account of Mr.
Dylander that Kalm himself was later ordained in the Lutheran Church and eventually became a doctor of divinity.

[2] The title should read *A Rational Account of the Weather*, by John Pointer, 1723;
2nd ed. 1738.

674 PETER KALM'S TRAVELS

was full of boys, girls and older people moving about upon it, the majority of them skating. But at Gloucester the River was still open, because it is narrower there and the current therefore swifter. On the third of the month several shoppers walked over the River, but on the fifth a large part of it was so open again that horses were being carried across in boats.

JANUARY THE 3RD

Inundations in North America, From *The American Weekly Mercury,* Philadelphia, June 29, 1721.—"It seems the rains of late have been very violent up in the country, especially to the Westward of the Skullkill river, by which there came down such a sudden fresh, that in some places the water rose 20 feet perpendicular from its usual bounds in a few hours time." Many cattle were drowned, mills and bridges carried away, and dams torn down. "A large stone bridge near pennypack mill is wholly destroyed. It is esteemed to be the greatest fresh we have known here these twenty years, & the most suddain & unaccountable, because we have not had such unusual rains."

November 8, 1722. "We have news from South Carolina, that a storm began there the 9th of September last, which continued in all 5 days. The rain was more violent than the wind, doing considerable damage to the corn & rice & carried away some houses and cattle in the country. The water rose upwards of 30 feet more than usual."

JANUARY THE 4TH

Cold Temperature in North America. From *The American Weekly Mercury* of January 15, 1722, a communication from New York. "It is excessive cold, and the river full of ice from the narrows to New York. Yesterday a great many people went upon the ice from New York to the Ferry on Long Island."—Mr. Edw. Holley's thoughts of the reason for the severe cold in North America may be seen in *The Philosophical Transactions,* No. 363.

JANUARY THE 5TH, 1750

(CHRISTMAS DAY, 1749, OLD STYLE)

To-day Christmas Day was celebrated in the city, but not with such reverence as it is in old Sweden. On the evening before, the bells of the English Church rang for a long time to announce the approaching Yuletide. In the morning guns were fired off in various parts of the town. People went to church, much in the same manner as on ordinary Sundays, both before and after dinner.[1] This took place only in the English, Swedish, and German churches. The Quakers did not regard this day any more remarkable than other days.[2] Stores were open, and anyone might sell or purchase what he wanted. But servants had a three-day vacation period. Nowhere was Christmas Day celebrated with more solemnity than in the Roman Church. Three sermons were preached there, and that which contributed most to the splendor of the ceremony was the beautiful music heard to-day. It was this music which attracted so many people. It must be emphasized that of all the churches in Philadelphia only the Swedish and the Catholic possessed organs. There had formerly been one in the English temple, but it had later become useless, and there had not yet been any measures taken to procure a new one. The organ in the Swedish church had also through improper care become worthless. Consequently an organ was to be heard only in the papal place of worship. The officiating priest was a Jesuit, who also played the violin, and he had collected a few others who played the same instrument. So there was good instrumental music, with singing from the [back] organ-gallery besides. People of all faiths gathered here, not only for the high mass but particularly for the vespers. Pews and altar were decorated with branches of mountain laurel, whose leaves are green in winter time and resemble the lauro-cerasus (cherry laurel). At the morning service the clergyman stood in front of the altar; but in the afternoon he was in the gallery, playing and singing.

---

[1] In Sweden Christmas was, until recently at least, regarded as a very sacred religious festival. Service is still held in all Swedish churches in Sweden and elsewhere at five or six in the morning on Christmas Day, and no unnecessary work or visiting is done on that day.

[2] Cf. diary for Jan. 5, 1749.

There was no more baking of bread for the Christmas festival than for other days; and no Christmas porridge on Christmas Eve.[1] One did not seem to know what it meant to wish anyone a merry Christmas. However, [after I had written this] I heard several members of the English Church wish one another a happy Christmas holiday. In the English church a sermon was preached in the morning; but after dinner only a prayer meeting was held, and on the day after Christmas again, only a prayer meeting.[2] But, as I have already noted, the Quakers paid not the slightest attention to Christmas; carpentry work, blacksmithing and other trades were plied on this day just as on other days. If Christmas Day falls on a Wednesday or Saturday, which are market days, the Quakers will bring all kinds of food into the market as usual; but no others will, and only Quakers will buy anything of them on such a day. Others make provisions so that purchases will prove unnecessary until the first market day after Christmas. The same custom is observed at New Year's. At first the Presbyterians did not care much for celebrating Christmas, but when they saw most of their members going to the English church on that day, they also started to have services.

JANUARY THE 7TH

Animal Diseases in America. The American Weekly Mercury, in a communication from New York, September 4, 1721, says: "A mortal distemper is got among the horses this way; many hundreds are dead and dying daily. There are 200 and odd dead in the town of Hackinsack, and as many in several other towns. There are 250 dead at Elisabeth town, and there abouts."

New York, September 11, 1721. The distemper among the horses continues, and spreads upon Long Island and Westchester County; & not only horses, but many neat cattle and hogs are dead, and continue to die with the same distemper.

---

[1] In Sweden, where the Christmas season lasted for three weeks, there was enormous gastronomic preparations made of both solid and liquid material for these holidays. Special drinks and special dishes were an imperative part of the menu and general celebration.

[2] In the Swedish church calendar the day after Christmas is also a religious holiday with regular morning service. The same is true of the Easter holidays.

JANUARY THE 8TH

Customs. It was a custom here among the English to express good wishes to a newly-married couple by paying them a personal visit during the first week or first month of their married life. The nearest relatives and friends would come and say "I wish you joy." Men generally made these visits before noon; the women, in the afternoon. The men received each a glass of wine; the women, wine and tea. Upon leaving, each guest was given a piece of wedding-cake [1] done up in clean paper, which he brought home with him. If the cake was not provided, the bride was either considered stingy or ignorant of *savoir-vivre*. The bridal couple was then in duty bound to return all calls by the well-wishers. Failure to return a visit was held as a sign that further visits from the party in question were undesired. In order to receive these visits and felicitations the bride and groom, and especially the bride, were obliged to remain at home for two weeks after the wedding, and to be always dressed after lunch. If she did not stay at home but went out, she was considered afraid of visits and unwilling to receive them. The wedding-cake was made of eggs, flour, butter and sugar mixed and thoroughly beaten, with some sweetmeats added. At baptisms, also, the guests were regaled with tea and wine, and the same kind of cake was served, in the same way, at the exit. The tea and the cake were given out by the woman who attended the young mother; [2] and politeness required that a gratuity of four or five shillings or even a dollar, [3] be given to her. At weddings, however, no tips were given. An average wedding cake cost 30 shillings [or about four dollars].

JANUARY THE 9TH (1750)

The Delaware River was now frozen in most places at Philadelphia. For the last three days there had been a large number of young men

---

[1] This is very free translation of the Swedish *puderkaka*, made of powdered brown sugar, etc.

[2] The christening took place probably while the young mother was still in bed, since early baptisms were not uncommon. Kalm's Swedish version is not absolutely clear on this point.

[3] The Pennsylvania shilling at this time was worth about 13 1/2 cents, so that it took between seven and eight to make a dollar.

and boys on the ice, some walking but most of them skating. There was still an open place here and there in the middle of the River; nevertheless, to-day at eleven I saw a man successfully driving a horse and sleigh on the ice directly in front of the city. The next day was mild and beautiful, when a section of the ice before the town suddenly broke up and began to move downstream. There were a good many people on this piece of ice: booths had been set up to sell brandy and such things to the skaters, and now they all found something else to do besides enjoying themselves. People rushed away precipitously, and fortunately all reached *terra firma* safely. The ice remained, but for a few days no one dared go out on it. There had been some people on the other side of the river starting to cross when the ice began to loosen, but these were obliged to turn back. On the 13th of this month the river was wholly open again so that ships could move in and out. The English youth is very fond of skating, and so are men of thirty years or over. Men of all classes have a passion for this sport. They would sometimes go three or four miles to reach a place where the ice was safe. Sheltered spots were flooded with men skaters, but I saw no women on the ice here.

JANUARY THE 10TH AND 11TH

Much Snow in America. The American Weekly Mercury N. 272. Philad. March 3, 1725. We hade such abundance of snow fall here yesterday, and last night, that it's near two foot deep. which has not been known here for some years past.

Earthquakes in North America. The American Weekly Mercury N. 237, Boston, June 15, 1724. On Thursday morning last some shocks of an Earthquake were felt here, by a considerable number of persons in different parts of the Town.

Ibid. N. 242. Philadelphia, Aug. 6, 1724. We had a small convulsion of the earth (or Earthquake) last night about the hours of 9 or 10 of the clock, which lasted about half a minute and was felt by many people in this city.

N. 409. Philad. Nov. 2, 1727. We have advice from New York, that on sunday night (:d. 29 Octob.) about ten a-clock they had a shock of an Earthquake, and about Two they felt a second shock which shook the pewter from off the shelves, and the China from off the cupboardsheads & chimney pieces, & set all the clocks a running down.

N. 412. Boston, Nov. 6, 1727. On the 20:th past, about 30 minutes past 10 a night, which was very calm & serene, and the sky full of stars, the town was on a suddain exceedinly surprised with the most violent shock of an Earthquake that ever was known. It began with a loud noise like thunder, the very earth reel'd and trembled to such a prodigious degree, that the houses rock'd and shook in so much that every body expected they should be buried in the ruins. Abundance of the inhabitants were wakened out of their sleep, with the utmost astonishment, & others so terribly afraighted, that they ran into the streets thinking themselves more safe there, but through the infinite goodness and mercy of God, the shock continued but about 2 minutes, and tho' some small damage was done in a few houses, yet by God's great blessing, we don't hear that any body received any hurt thereby. There was several times till the next morning heard some distant rumblings of it, but since then the earth has been quiet, tho' the minds of the people have still a great & just terror & dread upon them. On the next day prayers were offered in almost all the churches and the day was set apart as a public fast-day.

N. 420. Boston, Dec. 7, 1727. We hear from Newbury, that last week, *viz.* on wednesday and friday, they had there the repeated shocks of an Earthquake.

N. 422. Boston, Dec. 28, 1727. By Capt. Cooper late from Barbados, we have advice, that the Earthquake we had here October the 29:th, about half an hour past 10 in the evening was felt there the day before about noon; which is nigh 2000 miles from this place. The houses were in great convulsion, and the streets arose and fell like the waves of the sea, so that they were afraid the earth would sink under them, and they ran down to the wharves, to get into boats and vessels for their safety.

N. 428. Marblehead, Jan. 31, 1728. Yesterday between 1 & 2 a clock p. m. we had a terrible shock of an Earthquake, which began with a rumbling noise like the rolling a log over an hollow floor, & increased until it seemed like the discharging of several cannon at a distance; at which time the earth trembled so as to jar the pewter on the shelves in many houses; the whole shock lasted about 50 seconds. It's thought that had this shock been in the night in still weather, it would have appeared the greatest since the great shock on the 29 of October. This is the third shock we have had within these six days last past; and about the 30th since the 30th of October last.

EXTRACTS FROM THE MANUSCRIPT CONTINUATION OF KALM'S TRAVELS IN AMERICA

European News. A few moments ago I received the newspapers that were published to-day. In these I learned, from a dispatch of September 30, 1749, that the minister of the Russian Czarina [1] in Stockholm had delivered a memorial to the Royal Swedish Court requesting the permission to station troops in Finland, since she was a guarantee of Swedish freedom. [2] As soon as I had read but half of the Czarina's insolent, damnable and super-immoral demand, I became so angry, I must confess, and my blood circulation so violent, that every limb in my body shook for an hour as if I had had the ague, and I could not read or write a word. It would be better if the devil removed the whole gang of sympathisers with the Russian Czarina in Sweden, where she commands a few demons, or that they were all hanged in a gallows, rather than let them work their own will. I venture to believe that many a satan has heartily rejoiced at this damned shamelessness on the part of the Czarina. But there must be some honorable Swedish men left. God bless and keep His Royal Highness [Frederick av Hesse—Cassel], the apple and delight of all righteous Swedish eyes, Her Royal Highness [Ulrica Eleonora], and the whole Royal House! May our Lord also further His Excellency Count [C. G.] Tessin's useful projects for Sweden! [3] May God induce many others to follow his footsteps! (February 3, 1750, N. S.)

The language of the Indians is difficult to speak, as you can see from the following *Nummatchekodtantamoonganunnonash,* which

---

[1] Elizabeth.

[2] The last war between Sweden and Russia had ended in 1743 with the treaty of Åbo. Kalm represents the turbulent feelings of a Swedish Finn toward Russia in 1750.

[3] Count Tessin was the contemporary champion of Swedish national independence and of the economic, scientific and artistic development of his country. He sponsored especially the friendship of France and Sweden at the expense of Russia, a policy which had a strong appeal to Kalm.

signifies *our lusts*. *Kummogdadonattoottummooc iteaon gannun-
nonash* means *our question*. Mr. [John] Eliot [apostle to the In-
dians], the first English clergyman in New England, learned this
language from an Indian man-servant and later wrote a grammar
of the dialect. He began his mission in October, 1646. (February
14, 1750).

Swedish seamen came often to Philadelphia on the English ships,
either as boatswains, ordinary seamen, carpenters, gunners, or second mates. For the most part they performed the duties of a gun-
ner. I was told that during the war between France and England,
which terminated with the peace of 1748 [Aix-la-Chapelle], there
were seventeen English privateers in North American waters and
that the gunners on all of them were Swedes. All the English sea
captains assured me that there was a large number of Swedes on
the English merchant vessels, but that the number on the Dutch
ships was still larger. They could earn quite a good sum here in a
year, but many of them understood but little of economizing, and
as soon as they landed directed their steps to saloons and gambling
dens, where most of their pay was spent. Those who served on
board the English ships especially did not become rich; for the
British sailors were accustomed to live well and have no care for
the morrow, and the Swedish seamen imitated them. On the other
hand, those in the Dutch employ usually became well-to-do; because Holland is a parsimonious nation that knows the value of
money, and gradually the Swedes working for it acquired that same
quality. A large proportion of the Swedish sailors on foreign vessels were such as belonged to the Swedish navy or admiralty, and
who had either obtained leave to enter service elsewhere for outside training in order to serve their own country skillfully at some
later date, or had simply left without official permission. Almost
all the Swedish sailors whom I met in America, and particularly
those that belonged to the Swedish naval department, had neither
thought nor desire of ever returning to Sweden. When it was
pointed out to them that it was their duty to serve their fatherland
in time of need, they always answered, as if with one voice, that
they felt no obligation to do so, since their native land had been
and continued to be ingrateful for their services. No matter how
faithfully they had served or would serve they were never promoted

to any higher position, and wholly ignorant and untrained youths were always given the preference instead of them. In fact, [the officials] had generally promoted such as hardly knew the name of a single rope or anything else that belonged to a ship, and when these commanded they had been obliged to keep a paper fastened on their arm on which were written the words of command. Nor did these green officers dare to take their eyes away from the paper. Yet they had treated their subordinates severely, not much better than if they had been dogs, and given them only spoiled food, although these subordinates could easily have been the teachers of the others so far as seamanship was concerned. (April 28 and May 17, 1750).

In one place we passed a house built of clay. It had been constructed according to a half-timber method with a cross-work of wood on which the roof had been laid. Between the wood the spaces had been filled with clay, which then had been allowed to dry. (May 5, 1750).

Swedish vs. English. In the morning we continued our journey from near Maurice River down to Cape May. We had a Swedish guide along who was probably born of Swedish parents, and was married to a Swedish woman but who could not, himself, speak Swedish. There are many such here of both sexes; for since English is the principal language in the land all people gradually get to speak that, and they become ashamed to talk in their own tongue, because they fear they may not in such a case be real English. Consequently many Swedish women are married to Englishmen, and although they can speak Swedish very well it is impossible to make them do so, and when they are spoken to in Swedish they always answer in English. The same condition obtains among the men; so that it is easy to see that the Swedish language is doomed to extinction in America; and in fifty or sixty years' time there will not be many left who can understand Swedish, and still less of those who can converse in it (May 7, 1750).

"Irish Bull." An Irishman had recently written a letter to a fellow countryman, in which he had also enclosed a copy of the same letter, with the postscript that he was sending both, for fear that in these troublous times one might be lost, and if the original didn't arrive the copy might. (May 17, 1750).

The Indian Catechism. The Rev. Mr. Lars Nyberg[4] told me that he had taken a copy of the Indian Catechism, which had been sent over from Sweden, and tested it on the savage natives of Virginia to see whether or not they could understand it, since it was said to be translated into their dialect.[5] But they could not comprehend a word of it, whereas it was found upon examining the Indians on the Delaware River that they understood a good deal. The late Mr. Peter Kock asserted several times that it was stupid to have spent so much money in the printing of the Red Man's Catechism, since it had not been translated into the right Indian dialect (May 21, 1750).

Smilax herbacea. The flower of this plant has the most disagreeable odor in the world, for it smells like a dead snake, and when you have once smelled of it the terrible odor will never leave your nose (June 7, 1750).

Honorary Titles. The following resolution by his Royal [Swedish] Majesty in regard to a petition by Bishop [Jesper] Svedberg [father of Emanuel Swedenborg] was graciously granted in Stockholm on April 1, 1721: the title of Magister,[6] in accordance with the request in the petition, was bestowed upon Dean Eric Björk of Falun and Dean Anders Sandel in Hedemora [both formerly pastors of New Sweden], with the further provision that they henceforth be excused from praesidium duties at ministerial conferences. After due consideration, too, of all the reasons set forth, His Royal Majesty likewise conferred the same honorary title on the preachers Abr. Lidenius and Samuel Hesselius, who now have charge of Swedish congregations in America (June 10, 1750).

Skin Gifts. It appears from Bishop Svedberg's letter that some people in America had sent skins both to the King of Sweden and to the Bishop, for which the Bishop at first expressed his thanks. But when the colonists continued sending the skin gifts he took offence, and wrote that he was attending to his duties from love for

---

[4] Later a member of the Moravian Church.

[5] Johan Companius's "Vocabularium Barbaro . . . Virgineorum" had appeared 1696 in Stockholm at Royal expense, in an exquisite monographed binding, together with a translation of Luther's Catechism into the Lenape dialect.

[6] The nearest American equivalent here is Master of Arts, but occasionally it corresponded more to our honorary doctor's title.

the Swedes and not for love of the skins, which he did not want (*ibid.*).

More about the Indian Catechism. In Christina, or the Swedish Church in Wilmington, there are something over a hundred copies of those catechisms which had been published in Sweden in the Swedish and Indian tongues and later sent over here. They are bound in a French binding and printed on excellent paper, which shows what an act of vanity it was. I was told that Dean Anders Hesselius had written home to Bishop Svedberg informing him that he had converted several heathen savages, when as a matter of fact, only one single Indian had through this means (*taliter qualiter*) been converted (*ibid.*).

The doors in the Ephrata Protestant Convent, about thirteen or fourteen miles from Lancaster [Pa.], are so narrow that only one person can pass through at a time, and if he is fat he cannot get in at all. Our Royal Councillor Cedercreutz [7] would therefore have to stay out. The doors are made of a single board of the *Liriodendron tulipifera* or tulip tree (June 13, 1750).

This same convent had its own printing press. Among the books that had been printed there was a German translation of the Gospel of Nicodemus (June 14, 1750).

In the city of Lancaster the Town Hall is located almost in the center of the town and round about is the marketplace, about the same as in Fredrikshamn, [Finland] (June 16, 1750).

King Charles XII Sends Books to New Sweden. In 1707 King Charles XII sent over [to New Sweden] a large number of books, Bibles, psalmbooks (hymnal, missal and prayerbook combined), A B C-books, and works of religious meditation. Half of these were sent to the Christina Congregation in Wilmington. Payments for them were to be made to a widow of a former clergyman (Anders Rudman?) (June 20, 1750).

Skins. The following year something over thirty skins, most of them mink, were dispatched to the Royal Secretary Peringer Liljeblad for his trouble with the packing and sending of the books, as decided in the parish meeting of May 13 (*ibid.*).

In a letter dated at Brunnsbo the 24th of November, 1714, Bishop Svedberg again acknowledges gratefully the receipt of the skins

---

[7] Herman Cedercreutz (1684-1754), count and diplomat, was noted for his obesity.

sent by the congregations, and specifies that the skins sent to the King are still in his custody (*ibid*).[8]

The same year, at the departure of Mr. Eric Björk [for Sweden], the Swedes sent 115 skins with him including pelts of the bear, mink, cat, fox, muskrat, wolf, otter, panther and wild cat. Some were sent to the King in Sweden, and others to Bishop Svedberg. Those belonging to the King were forwarded to Pomerania, where they were confiscated by the [German] enemy at the surrender of Stralsund [1715] (*ibid*).[9]

On the fifth of December, 1719, Bishop Svedberg again expresses his thanks for skins sent (*ibid*.).

About ten years later, on November 19, 1729, in his letter of recall to Rev. Jonas Lidman, Bishop Svedberg writes: "Bring back something nice with you, especially pelts, as Magister Björck did." Later, in a letter to the congregation of October 15, 1731, he thanks them for the skin gifts which they had sent him, but which he had not yet received (*ibid*.).

Phlox maculata had the peculiarity that when its seed vessels were thoroughly dried they would burst with a crack, leave their paper enclosure and scatter all over the room (July 3, 1750).

Origin of the Indians. Concerning their origin the Indians had the following legend: A large turtle floated on the water. Around it gathered more and more slime and other material that fastened itself to it, so that it finally became all America. The first savage was sent down from heaven, and rested on the turtle. When he encountered a log he kicked it, and behold, people were formed from it. In every city (of the Red Men) there is ordinarily one family which takes the name of "Turtle" (July 30, 1750).

Impatiens (*noli tangere*) was called "the crowing cock" by the Indians, because of the form of the flower (August 8, 1750).

The Dog. An old Indian said that when God had created the world and its people, he took a stick, cast it on the ground, and spoke unto man, saying, "Here thou shalt have an animal which will be of great service to thee, and which will follow thee wherever

---

[8] Charles XII was at this time, as always, busy with his numerous wars, though he returned to Sweden from Turkey in 1714.

[9] Frederick William I of Prussia declared war against Sweden in 1715 and in December captured Stralsund in Swedish Pomerania.

thou goest," and in that same moment the stick turned into a dog (August 11, 1750).

In number 1136 of the *Pennsylvania Gazette*, for September 20, 1750, Mr. [Benjamin] Franklin published under my name my whole article on Niagara Falls (October 3, 1750).[10] Herr Adolph Benzel (later Benzelstjerna), the son of the late Archbishop Dr. Eric Benzelius, arrived yesterday in Philadelphia on a ship from Holland. He had been absent from Sweden about six years, and had spent most of that time in the French service. He had now come hither to examine this land, and to settle here if he could find suitable employment and living conditions (September 29, 1750).[11]

The Catechism which the followers of Count Zinzendorf had translated into Swedish and published is of historical consequence, because it is the first and only Swedish book printed in America (October 23, 1750).

A brief list of expressions is given below illustrating the present (1748-1750) language and method of speaking in New Sweden, which shows how far one has already deviated from the Swedish that is used in the Old Country. Soon it will be a new tongue and we can see that Swedish will in a short time die out.[12] [The illustrations follow. The English words or stems mixed in with the Swedish are reproduced in *italics*. For the benefit of those who may not read Swedish, a translation is given in parenthesis. Note the custom of placing Swedish endings on English roots, a practice well known of Swedish-Americans of to-day].

A. Denna hästen *amblar* braf. (This horse is a good stepper).

C. Jag vill *considerera* det innan jag *concluderar* något. (I shall consider it before I make a decision).

---

[10] This was the first description in English of Niagara Falls based on first-hand information. See bibliography of Kalm's writings on America at end of this volume, item 12. Cf. also J. L. Odhelius, *Åminnelsetal öfver Kalm*, p. 24.

[11] Benzelstjerna stayed in America and became a fortification officer and an intendant of forests in the English service. He married the daughter of a Swedish pastor at Raccoon.

[12] Kalm did not dream that a century later there would commence an immigration tide from Scandinavia that was ultimately to bring hundreds of thousands of Swedes to America, and perpetuate the Swedish language in North America for several generations at least.

D. Ni är *desperat* flitig. (You are "desperately" diligent).

G. Jag vill gå och öppna *gäten*. (I shall go and open the gate).

H. Om det skulla *happna* så. (If it should happen thus).

I. Var *isi* och *trubla* er intet. (Be at ease and don't trouble yourself).

Han tog in något, som *isade* honom mycket. (He took something [medicine] which eased [relieved] him very much).

K. Jag *kärar* det intet. (I have no care about it, i. e. pay no attention to it).

Rum voro *kipare* fordom än nu. (Formerly rooms were cheaper than now).

Det är en *klöfver* karl. (That is a clever fellow).

En å (a river) was always called a *kil* by the Swedes, as in *Schuylkill*, and corresponded to *creek* among the English.

L. Jag *lusa* min hustru på sjön. (I lost my wife on the ocean), said the Rev. Sam Hesselius when he came back home to Sweden. The Swedes in this locality used the word *lusa* for *mista* (lose) a great deal.[13]

P. Vill ni *plisa* sitt ner och äta sådant som vi ha? (Will you please sit down and eat of what we have)?

Jag har varit mycket *pårli* denna vinter. (I have been very poorly this winter).

Det är denna tiden ej så mycket vatten *pannorna* som i min barndom. (There is not so much water in the ponds nowadays as there was in my youth).

S. Ni kommer körandes så *smärtli*. (You are out driving in a dashing style, i. e. smartly).—Vill ni *småka* tobak? (You wish to smoke)?—Han *simar* vara en braf karl. (He seems to be a fine fellow).—Jag har legat och *slipat* så länge. (I have been lying down sleeping for such a long time). De begynna *skeda ut*[14] att tala svenska. (They are beginning to give up talking Swedish).—*Stenhäst* (stone horse) was always used for a stallion.

---

[13] The modern Swedish *lusa* means to louse, to free from lice, or delouse, hence the tragic-comic element in Hesselius's statement as written. Kalm evidently represented the sound of *o* in *lose* with a *u*. The modern careless Swedish-American would spell and pronounce it *losa*, reproducing the vowel sound of the English infinitive.

[14] Probably from the Dutch *uitscheiden*, to cease, to stop.

T. Hästen vill *trötta*. (The horse wants to trot).—*Trubla* er intet därom. (Don't trouble yourself about it).—All Swedes called the kitchen garden a *tina*, which is said to have come from the Dutch [*tuin*, garden].

V. Jag *vantar* intet. (I don't want, or need, anything).—Jag vill *väta på* er. (I shall wait upon, or come to see, you). Madame Tranberg used this expression in speaking to the Rev. Mr. Acrelius.—Vill ni komma och ta en liten *vak* med mig? (Will you take a walk with me)?—Mäster Professor, han rider efter och *väter på* frun och magistern. (He will ride behind and wait on your wife and yourself, Professor).— Hästen *väcker*. (The horse shies).[15]

Anglicism. Vill ni *komma våra vägar*. (Do you want, or intend to come our way, i. e. to visit us)?—Så göra de *våra vägar*. (That's what they do up our way, i. e. among our people, at our homes).— Jag *fattas veta* om han kommer, hvad det kostar. (I should like to know what it costs, if he comes).—Han bor just *öfver vägen*. (He lives right across the street, lit. over the way).—The word *karl*(*en*)[16] is always used for "this man", no matter how distinguished he may be, or whether he be present or not, for instance, Hvarifrån är denna *karlen*? (Where does this man come from)?

Swedish Songs. In Wicaco I received some Swedish songs that had been printed there in the year 1701. There were eight of them, all composed by the Rev. [Andreas] Rudman. They are of two kinds, or on two types of paper, and maybe of two different years, for on one part the title is given before the (two) songs, and the date printed, while the other six are undated. All are printed in octavo, with Latin script, and are said to be not only the first items printed in Swedish in this vicinity (i. e. in Philadelphia), but one of the first writings in any language to be printed here, since before that time there was no printing establishment in Philadelphia and everything was published in New York. (January 18, 1751).

The [Royal Scientific] Society in London has degenerated con-

---

[15] *Väcker* is possibly from the stem in the Dutch word *ontwijken*, to shun; Swedish *undvika*, to avoid; German, *weichen*; or simply the Swedish *viker*, sing., pres. ind., of *vika*, to turn or move suddenly, to turn to one side.

[16] German *Kerl*, English *churl*. To-day there is nothing really derogatory about the word in Swedish, but it is not used in polite society about and in the presence of a man from the upper classes. It is used in the sense of "fellow." *Herrn* should be used instead of *karlen*.

siderably since it was founded, and does not have the standing it
enjoyed for several years thereafter. In the beginning there were
members thoroughly expert in all sciences, and everything published at that time in the *Transactions of the Royal Society* was
learned, useful, informative, and quite well prepared. That period
may be said to extend from 1670 to 1720, or a little longer, and
embraced England's great savants, such as Boyle, Rajus, Tyson,
Whiston, Wallis, Halley, Flamstead, Newton, Lister, Sloane, and
many others. It is true that many of them lived on after 1720, but
very little by them is found in the *Transactions* subsequent to that
date. Almost everything printed there during the period noted
will be read with pleasure, because it is so well prepared; but later
the society seemed to weaken more and more, to tire, to decline,
so that that it became difficult to find anything of value in its publication. One had to hunt for it, and occasionally one perused almost
a whole quarter of the journal without finding anything worth
reading. Among the present members of the Society there are very
few of outstanding merit. Sloane is still living, but is so old that
he has hardly been able to talk for several years, to say nothing
of writing.[17] Much the same condition obtains in the case of Dr.
Mead Collinson, who has a scholar's inquisitiveness and is an ar-
dent promoter of the natural sciences, but does not publish anything of his own. Miller [author of the Botanical Dictionary]
publishes whatever is of value in his work in separate books, and
only sends in to the *Transactions* of the Society his minor articles.
The same is true of most of the other members that amount to
anything. For this reason the Society has fallen into disrepute,
even here in England,[18] so that even the more sensible people
speak of it with mockery. Recently, too, the well-known Dr.
[John] Hill (c. 1716-1775) published a small pamphlet which is
specifically directed against the [Royal] Society, and in which he
critically reviews a number of their treatises, showing by numerous examples how wretchedly their articles had been written, and
how carelessly they had been printed. Dr. Hill's particular brochure, however, appears to have been written with too great heat
of passion.[19] (April 16, 1751).

---

[17] Hans Sloane was at this time 91 years old. He died in 1753.
[18] Kalm was now in England on his return journey.
[19] Hill apparently attacked the Royal Society in more than one article. The *Dic-*

In Gothenburg they are said to have discovered an efficient exterminator of bedbugs, namely fresh coriander seeds stuffed into the holes and cracks where the vermin is found. Others claimed that the leaves of the *Lepidium rudevale* were best for that purpose. (May 5, 1751).

The [Swedish] commercial adviser [Magnus L.] Lagerström (1691-1759) asserted that he had discovered a method of planting and transplanting spruce trees, without running the risk of having them die in the process. It consisted in covering the bottom of the hole where the spruce is to be planted with small stones, placing the tree directly upon the stones, and then pressing the dirt around the roots in the usual way. If the spruce is placed right on the soil without any stones it usually perishes from the transplanting. (May 11, 1751).

The Beginning of Swedish Manufacturing. The councillor of commerce, [the noted Jonas] Alström (later Alströmer) told me that he first went to England in 1707, when he was 22 years old; that he was born in Alingsås; that he about the year 1714 first entertained the plan of introducing factories in his native land, since, travelling about in England a great deal, he had been employed by Swedish merchants in sending a large amount of English-made goods to Sweden. He therefore visited the British manufacturing establishments, to which he had free access, and studied them diligently. In those days he was allowed to inspect everything because they [the factory directors] were not then jealously apprehensive [of competition]. Thereupon he returned to Sweden when King Charles XII came back from Turkey (1714), thinking that there would immediately be peace, and that thoughts would be turned to industry. But when he came home he found that the war was waged more violently than ever, and that the subject least considered was factories. Consequently, a little while afterwards he

---

tionary of National Biography says: "Failing to obtain the requisite number of names for his nomination to the Royal Society, he attacked the society in several satirical pamphlets, specially vituperating Folkes and Baker, his former patrons, and in 1751 published "A Review of the Works of the Royal Society", holding up to ridicule the "Philosophical Transactions," to which he had himself contributed two papers a few years previously."

Hill was a noted author, extremely prolific, and conducted the *British Magazine,* 1746-1750. In 1759-1775 he published *The Vegetable System,* for which he obtained the Royal Order of Vasa of Sweden.

started off again; was taken prisoner by the Danes; but was set free again after seven weeks, under the pretext that he was an Englishman. After peace had been proclaimed he returned, but travelled first through a part of France and Holland, hired there a number of Frenchmen who were specialists in their respective manufacturing lines, whom he sent to Sweden at his own expense. At the Riksdag of 1723 he obtained several privileges for the industries. At the session of 1727 he caused the middle estate to send a deputation to the nobles, demanding the erection of factories, the total prohibition of certain imported goods, and a tariff of 5 per cent on others. It did not look very encouraging to him at first in this Riksdag, because various merchants of the middle class sought to persuade their fellow tradesmen to overthrow the proposal and destroy the recently constructed shops. And the latter, from all appearances, would have succeeded in their schemes if the burgomaster of Malmö, Stobée,[20] had not championed the cause. The latter was a very enterprising, aggressive man. (May 15, 1751).

[20] Possibly Lorentz Christoffer Stobée (1676-1756), army officer and civil administrator, who later became governor of "Göteborgs och Bohus Län."

A DESCRIPTION OF NIAGARA WATER-
FALL IN NORTH AMERICA

[THE following description of Niagara Falls is translated from a contemporary letter by Kalm to the librarian Carl Christoffer Gjörwell
of Stockholm. It was originally the author's intention that this
account should conclude the promised fourth part of his *Travels*.
It is more detailed than the description in English sent to Benjamin
Franklin. Cf. the *Förord* in Elfving's *Tilläggsband*].

A DESCRIPTION OF NIAGARA WATERFALL IN
NORTH AMERICA

Waterfalls are found in many places in the world, where water
in some river, creek or brook hurls itself down a rocky precipice
from a considerable height. There are few countries which cannot boast of some such phenomenon. In Sweden, for instance, we
have Trollhättan. The Ammä waterfall at Kajana [Finland] is
not to be despised either. The well-known fall at Woxen [Voxna,
Sweden] is large. But there are probably few waterfalls in the
world that can be compared to Niagara in North America, when
one considers its height and the amount of water which passes over
it. Therefore we must without contradiction reckon it as among
the largest on the globe.

In order that the truth of this statement may appear more clearly,
I must mention the following before I describe the Falls.

In North America are five fresh-water lakes, each one large
enough in size to be more like an ocean than a lake. These are
Lake Superior, Lake Huron, Lake Michigan, Lake Erie and Lake
Ontario. The Upper Lake (Lake Superior) is calculated to be 200
French miles long from east to west, and in several places 80 French
miles broad, from north to south.[1] It is pretty big, the largest of

---

[1] Lake Superior is 400 miles long, hence about 140 French miles.

them all. Lake Huron is about 100 French miles long and between 30 and 40 miles broad. Lakes Michigan and Erie are about of the same size, namely 100 miles in length and 30 in width. I shall speak of Lake Ontario later. Into each of these bodies of water a large number of brooks and rivers empty. The nature of the outlet from these inland lakes is this: the water from Lake Superior flows through a narrow sound into Lake Huron. A little south of this sound Lake Michigan sends its water also into the same lake. The water of Lake Huron together with what it has received from the two above-mentioned lakes seeks an outlet in a long, narrow strait which is called Le Detroit and runs into Lake Erie. All the water thus collected in the four inland seas mentioned and in the numerous streams running into them makes its way first through a short, narrow sound, and then over Niagara's lofty falls, whereupon it forms a broad river of about six French miles in length and empties into Lake Ontario, and from there flows with the St. Lawrence River through the most thickly settled parts of Canada and to the ocean.

In order to learn more of the nature of the interior of America, its natural products, the customs of the inhabitants, etc. I undertook during the summer of 1750 a journey through the land of the Iroquois. These Indians or savages have for a long time had the reputation of being cannibals, because they were occasionally wont to roast their prisoners of war and eat their flesh, so as to inspire a greater terror among their enemies. This custom has now for the most part been given up. A traveller who visits their villages or cities must not be terrified if he sometimes should find the outside of the gable walls of their houses covered with human skulls. These are trophies of war and serve as victory proofs of the number of enemies slain. Nevertheless, however cruel they may be in warfare, when it is in no way advisable to meet them, they are, on the other hand, very friendly and hospitable when at home in times of peace, especially if they are sober, for then they exhibit greater hospitality than most of the Christians. A stranger has scarcely time to enter their dwelling before the Indian mistress offers food, while her husband tries in his way to entertain the visitor. Since Niagara is located near their land I felt it worth while to see it, as one of the most remarkable sights in nature. I knew that no

Swede before that time had ever had the opportunity to behold it.[2] Well, after a very difficult and quite adventurous journey on horse-back through the territory of the Iroquois, I finally arrived on the 13th of August, 1750, new style, at Fort Oswego belonging to the English and located on the great Lake Ontario, which more resembles a sea than a lake. This lake is situated between the 42nd and 44th degrees north latitude; its length from east to west is about 80 French miles; and its breadth is about half of that distance. In this lake there are only a few small islands to be found, and these near the shores; there are none further out. The water is as fresh as spring water, clear, and in some places over 60 fathoms deep. It never freezes over in winter, and ice forms only near the shores. From Oswego I made my way along the coasts of this lake in a flatbottomed boat or battoe as far as Fort Niagara, then belonging to the French, and where I arrived after six days of rowing on August 23. Fort Niagara is situated on the west shore of Lake Ontario, right near the place where Niagara River flows into it. The river and the lake have already washed away a part of the land nearest the fort, so that to prevent further damage it has become necessary to construct retaining walls there. The location of the place is pleasant: one has a view over Lake Ontario towards the north, and of Niagara River in the west and south. Woods of beautiful hardwood trees appear on the mainland. Here I was received with much courtesy by the French officers.

Early in the morning of the 24th of August, accompanied by the French officers and three soldiers, I started off for the famous Niagara waterfall, the distance from Fort Niagara to the Falls being six French miles, of which half is travelled on water and half on land. We first ascended Niagara River in a birch boat. The width of the stream here was said to be about twelve arpents, but this varied. The banks of the river consisted of high precipitous hills of red sandstone, in layers. The water in the river flowed very slowly at first; but the further up we came, the faster the current, so that after rowing a French mile we had great difficulty in work-

---

[2] This may be true, but it is not certain. Several Scandinavians had come to New York State in the seventeenth century, and among them may well have been a Swede who penetrated as far as Niagara Falls. But if he did, he did not write up his observations.

ing our way upstream, although we rowed close to the shore. Shortly after we had left Fort Niagara we saw the vapor of the Falls rising high toward the sky like a thick cloud, and this cloud could be seen during our whole journey, gradually increasing in size as we approached the falls.

After three French (about nine English) miles of hard rowing we stepped ashore to continue our trip on foot. It is difficult to come nearer with a boat, because the number of steep rapids encountered. First we had to climb up the high, steep river banks, then proceed three French miles by land, which has two high and tolerably steep hills to be crossed. On this road we met a great number of Indians of both sexes, who were engaged in carrying their skins and other goods to Quebec. These goods had either been purchased originally from the Indians by the French and were being sent on, or the Indians were taking them to Quebec on their own initiative. Several natives had their own horses which carried such wares in return for pay. In as much as one cannot row a boat from Lake Erie to Lake Ontario because of Niagara Falls, boats have to be carried over land this distance (of nine English miles). Of course only birch canoes as boats can thus be conveniently transported. To-day I saw four men carrying a birch canoe that was five and a half fathoms long and about five and a half feet wide in the middle. Finally about half past ten in the forenoon we reached the Falls ourselves. The air was clear and the wind southwest, which was the best we could have wished for, for it drove away the vapors, rising like dense smoke, from the side where we sat, and we could see the falls much more distinctly. Had the wind been contrary we could not have seen many feet through the fog. The temperature was rather high, too; at three in the afternoon the thermometer registered 26½ degrees C. I also tried to determine the temperature of the water in Niagara River. In that part of the River over which we had travelled in the morning the thermometer showed a uniform temperature of 22°. But the water close to the awe-inspiring falls showed a constant temperature of 24½°.

Now I shall, without any alterations, reproduce that description of the Falls which I made with pen and ink as I sat on the utmost brink of them, at hardly a fathom's distance from the place where

an enormous mass of water hurls itself perpendicularly down from a height of 135 French feet, i. e. $147^{69}\!/_{1000}$ Swedish feet, or about $34\frac{2}{3}$ fathoms.[3]
The river at the falls runs from S.S.E. to N.N.W., and the falls themselves from southwest to northeast, not in a straight line but in the shape of a horseshoe or semicircle, since the island in the middle of the falls lies higher up than the ends of the semicircle. Above the falls and about in the middle of the river is an island, extending from S.S.E. to N.N.W., which is said to be about eight arpents long [about 1450 feet]. (The length of a French arpent here was $\frac{1}{84}$ of a French mile [and the latter is 2.9 of an English mile]). The island tapered in width toward its ends, being in the middle about a quarter of its length, with its lower extremity extending right up to the falls, so that no water tumbles down where the island touches the falls. The width of the latter, in its curved line, was said to be six arpents [about 1094 English feet]. The island lies exactly in the middle, so that the falls are split in two. The breadth of the end of the island in the falls is about $\frac{1}{3}$ of an arpent [61 ft.], and the length of the island is, as I have just said, estimated by all to be eight arpents.

On the two sides of this island flows all the water collected in the great and small sea-like lakes of Superior, Michigan, Huron and Erie and the many streams, brooks and rivers emptying into them. Before arriving at the island the water does not flow so rapidly, but as soon as it reaches the former, and divides into two branches, it begins to rush with such a fearful velocity that in many places it turns as white as the strongest rapids, and shoots up into the air. If a boat should ever reach that spot, no matter how good it might be, or how brave and strong the occupants were, it would still be impossible for them to reach land. They would be compelled to follow along down the terrific falls. In fact, they would not be able to keep their boat right side up for even the shortest time before it would be capsized by the violent current and the rocks in it. On the west side of the island the current is stronger, swifter and more snow-white, and there flows a greater quantity of water.

When a person stands down near the spot where the water begins to cast itself perpendicularly down the rock and looks back

---
[3] 135 French feet = 24 1/4 fathoms. Note by Elfving.

up the river, it is easy to see that the stream runs at a considerable incline: it seems as sloping as the side of a fairly steep hill. It is so inclined that the forest a short distance above and beyond it cannot be seen. From this it is easily perceived what forceful speed the water must have, even before it reaches the angle of the falls proper. And when it does reach that point, it immediately [of course] hurls itself downward. It is enough to make the hair stand on end on any observer who may be sitting or standing close by, and who attentively watches such a large amount of water falling vertically over a ledge from such a height. The effect is awful, tremendous!

During my stay in Montreal, 1749, I was anxious to learn the exact vertical height of the Falls from M. [Etienne Rocbert] de La Morandière [1701-1762] who was royal [army officer and] engineer and who a few years before had been commissioned by the governor-general of Canada, Monseigneur Beauharnois [4] to make a careful measurement of the same. He told me that he had measured them very accurately at three different occasions, almost every time he had happened to pass by, and found them at every measurement to be 135 French feet.[5] When I arrived at Niagara I asked all the gentlemen who were there, and who had often seen the Falls, how high they believed them to be. They answered almost unanimously that they had found their height to be exactly the same as that given by M. de La Morandière. When I to-day beheld the Falls themselves, where I met M. [Daniel] Joncaire [1716-71] who had spent ten full years here as an officer and had had charge of the goods that had almost daily been shipped between the Lakes of Erie and Ontario and therefore was better acquainted with the location than anyone else, I inquired of him how high he thought they were. He answered that he himself had never measured them, but that a few years ago a Jesuit had been there who had measured the Falls with a line or cord and that he [Joncaire] had helped him in his undertaking. He had there found the falls to be 150 French

---

[4] See page 440, note. He was governor-general from 1726-1747.

[5] The actual height of Niagara Falls on the Canadian side is 158 feet and on the American side 167 feet. The French estimate of 135 of their feet would be the equivalent of about 144 English feet, which is too short, if the length of the French foot as given in reference works is correct, viz. 1.066 English feet.

feet.[6] M. Joncaire assured me that they were certainly not less than 135 feet, and surely not over 150, but somewhere in between. He felt that M. de La Morandière's measurements must be nearer the true distance since it was a difficult matter to measure the height with a line; the line would be crooked because of the ledge and water. Since others had so accurately determined the height, I did not feel it was necessary to measure it again with a cord, especially since my time did not allow it, and some places were so slippery that the attempt would have been quite hazardous.

When the water rushes down the ledge and strikes the bottom it jumps up again in some places to a considerable distance. Elsewhere it boils and seethes with a snow-white foam, behaving just like the water does in a glass into which more is poured from above. The falling waters cause a loud roar, as can easily be imagined from the amount of water running over from such a great height. The noise of the falls can sometimes be heard at Rivière à la Boeuf, which is said to be located about fifteen French miles to the south; and at Fort Niagara, which is six French miles away, one can hear it very plainly in calm weather. One would be able to hear it there at almost any time, were it not for Lake Ontario, on which the water is seldom still, [drowning out the noise from Niagara] through its own din and roar in dashing against its shores.

M. Joncaire and all others declared that the falls occasionally made a louder racket than at other times. When its roar was unusually strong in any direction, it was said infallibly to indicate rain and bad weather. This was the reason, it was claimed, why the Indians of the neighborhood were able to prophecy the weather so accurately. All residents of Fort Niagara asserted that when the falls were heard clearly, i. e. as far as the fort, it signified invariably a northeast wind, and they thought this peculiar because the falls were southwest from the fort and one would have rather expected a southwest wind to hear it stronger.

Several who have spoken of these falls have declared that the

---

[6] This is about 159.7 English feet, which is reasonably accurate. Obviously the Jesuit with his string came much nearer to the actual height of the falls than did the supposedly more scientific calculations of the French engineer officer. Or, were the Falls lower in Kalm's day?

roaring noise is so deafening that people standing near them cannot hear each other speak unless they yell loudly close to the ears; but I did not find it so. We were close to the falls on all sides, but one could well hear what another spoke, provided he talked a little louder than usual; it was not necessary at all to shout. When the locusts or grasshoppers, which are described in the *Transactions* of the Royal Swedish Academy of Sciences for the year 1756,[7] made their piercing shrieks in the neighboring trees they easily drowned out the noise of the falls.

A large mass of vapors rise from the bottom of the falls that resemble a thick smoke rising high toward the sky. Caused by the violence of water contact these vapors, if the weather is calm, rise straight up to a great height and look like the heaviest cloud. But if there is a wind they are blown about as in a driving storm, and anybody enveloped by them will get as wet as if he had been dragged out of the sea. A couple of the Frenchmen who accompanied me climbed down a short distance below the falls, to examine the spot. The wind drove the mist at them so turbulently that they stood as in an impenetrable fog and thought they would suffocate. They left at once, and when they came up they were so drenched they were forced to take off almost all their clothes and dry them in the sun, in the interim walking about half naked.

If a spectator takes up his position anywhere near the falls whether it be in the woods or elsewhere, and the direction of the wind is from the falls, he will soon be so wet from the mist, even in the most glaring sunshine, that he will be glad to make his escape as soon as possible.

Everyone contended that from the other, east side of Lake Ontario, where only a small portion of this section can be seen, one could on a still morning behold something like a heavy mist rising above the woods in the vicinity of the falls, and that it looked like the smoke from a forest fire, although it was but the vapor from Niagara. The same phenomenon is noticed several miles from the falls, on Lake Erie, if one looks in the right direction. The morning seems to be particularly favorable to the rising of a formidable-looking fog from the falls, so much so that it is under those cir-

---

[7] This clause was of course inserted when Kalm later prepared his notes for the press.

cumstances hard to see the falls at all; but if a wind comes up the mist is dissipated considerably.

Almost daily, and especially in the spring and autumn at the migration time of the birds, a large number of sea fowl are found dead beneath the falls. There was said to be two reasons for this, 1. when the birds swim in the water a little above the island located in the middle of the falls, they like to let the current carry them slowly onward. This current approaching the precipice becomes swifter and swifter, and finally near the beginning of the falls becomes so rapid that when the birds hope to rise from the water they are no longer able to do so because of this swiftness, and are obliged to follow the current over the falls and perish. Long-time settlers of this vicinity declared unanimously that they had often seen a whole flock of sea birds sail over the falls. 2. When the mist in calm weather rises high and is very thick it happens that birds fly into it and never return. Either their wings get too soaked from the fog—because it drenches faster and more thoroughly than rain or other water—or the thundering noise from the falls frightens them so when they pass over into the heavy vapors that they try to alight, [are lost and killed]. All present agreed that they had often seen acquatic birds fly into the heavy mist and that they had found them dead beneath the falls a short time thereafter. As a further corroboration of the second reason it was pointed out that not only water fowl but land birds had been found killed at the bottom of the falls. The death of these cannot be ascribed to swimming too long on the waters above until they are borne to their destruction; they must have been lost in the fog. M. Joncaire, who has lived here for many years and visited the falls almost daily, assured me that he had an innumerable number of times seen whole flocks of swimming sea birds go over the falls, and that he had also perceived repeatedly how birds flew into the thick smoke and were later found killed below the falls. He could not remember, however, having seen any fauna from the land or forest suffer this fate.

In the autumn of 1749, among other birds, a swan sailed over the falls; but it was not seriously hurt, as it happened, and remained alive. It swam about beneath the falls for a month, but was not able to fly. Many tried to shoot it, but it was clever enough to move either to the other side or nearer the falls when a hunter appeared, so that it was impossible to get near it. Finally the swan disappeared, nobody knows whereto. Either it was eventually able to use its wings and fly off or some Indian killed it.

It is especially in the autumn, when birds fly south in large flocks, that a large number of them are lost in the falls, and that is the season therefore when they are collected. Both Indians and resident French soldiers from Fort Niagara are said to appear here daily then to gather the [killed] supply of sea fowls. The commandant of Fort Niagara, M. Beaujeu [8] assured me that the soldiers there lived in the autumn a long time principally from the birds found dead beneath the falls, which they prepared for food in various ways. The birds were said to make good food if they had not been dead too long.

During my sojourn in the English colonies the Englishmen claimed they had been told that in the autumn one could gather bags full of down near these falls from the birds killed and rotted. I asked those present if this story was true, and they answered that while it was not literally so, it was true that much down was obtained by plucking the birds found killed beneath the falls.

Besides birds a number of animals like deer are frequently found killed at the bottom of the cataract. They attempt to swim across the river above the falls to the island in the middle, are pulled into the current, and pass over the brink. They are usually dashed to pieces in the descent. Bears, seeing deer on the island, occasionally try to visit them, but are with much growling compelled to change their course and go over the falls. Later they have been discovered crushed to death at the bottom. Sometimes, in fact almost daily, fishes suffer the same fate. These are said to be good to eat.

Often it happened, too, according to a unanimous testimony, that drunken savages, who had propelled their birch canoes a considerable distance above the cataract, had gradually been drawn into the current and disappeared over the watery precipice. Only an arm was found of some of them.—Several told me that they had purposely thrown large trees into the stream above in order to

---

[8] Either Daniel-Hyacinthe-Marie Beaujeu (1711-1755) or his brother Louis Liénard de Beaujeu (1716-1802), both military men noted in Canadian history. See *Dictionnaire Générale du Canada.*

see how they would behave when they passed over the falls, but that they were completely lost in the fall; wherefore it was concluded that there existed a bottomless pit below. But although the trees might have stuck to the bottom at first, they might well have come loose afterwards unnoticed.

One cannot help being amazed and awed when standing above the cataract gazing down. It felt as if one were looking down from the highest church steeple. The falls that I had seen previously— Cohoes and Montmorency in North America, and Trollhättan in Sweden seemed but child's play in comparison. One could not gaze and contemplate without feelings of wonder and astonishment. The water seemed to flow gently and well-nigh lazily over the cliff at first; but the farther down it went the greater the speed. Even now there was an immense amount of water hurling itself out into space; but in the spring there is of course more, the most of any season. In its descent, near the bottom, the water strikes the rock in some places and there it foams and froths and roars in gigantic leaps, so that it gives the impression of a continuous firing of cannon, with a heavy, persistent and hastily appearing mist dashing forth; for the liquid is creamy white and looks like a thick vapor. When falling, the water at the top first appears green; but further down it takes on a snow-white tinge as in the strongest waterfall in the world, which this in fact really is. When the water reaches the bottom it sets all in motion beneath it, so that it swirls about in circles and whirls like the liquid in a seething cauldron. About two musketshots below the falls the current is not stronger but that one could row a good boat across there without danger; but a little below we meet again some strong and steep rapids, where the water jumps forward with such great violence that it turns white with froth. It would not be advisable to run these cascades in a boat, for it would be capsized and dashed to pieces.

Between the island right above the falls and the land on the east side was a distance of about two short arpents [365 ft.], it was said. To me it seemed but two good stone's throws. This islet is covered with tall trees, and is sometimes full of deer. These in attempting to swim across the river above the falls have been caught by the current and pulled along, a few being fortunate enough to strike this holm and thus escape the falls.

This islet had always been considered entirely inaccessible and no human being was thought ever to have visited it; but about 1739 a method of reaching it was discovered by the following adventure. Two natives of the Seneca Nation, who lived north of Lake Erie, were travelling on business down to Fort Niagara. Here they procured a quantity of French brandy, whereupon they started back over land, past the Niagara Falls, to Lake Erie. Here they stepped into their birch canoe, intending to visit an island in the lake a good distance above the falls to hunt deer. They had already imbibed so much of the French liquor that they knew but little of what they were doing, and during their paddling they continued to refresh themselves from time to time. Finally the brandy made them so heavy and sleepy that they lay down to sleep, letting the boat drift on the lake. The water which here moves gradually toward the falls carried the boat further and further down, until it finally reached the rapids a short distance above the upper end of the islet that lies in the middle of the river near the falls. Here one of the Indians, hearing the roar of the waterfall, woke up. He was horribly frightened, became sober immediately, and shouted with all his might to his companion that they were lost and in the twinkling of an eye would be right in the precipitous cataract. The companion woke up at once, both seized their oars (paddles?) with all the energy at their command, and since the boat had drifted quite close to the island, they sought desperately to reach it, and were lucky enough to land there. Now they were happy at first, to be sure, that they had so far saved their lives; but when they had had time to consider the situation, they felt that they had not gained much, and that it might have been just as well to have let the current carry them over the falls once and for all, for then the agony would have been brief. Here they would either have to starve to death, gradually, or else eventually hurl themselves into the waterfall anyway to end their days quickly. For how they should ever be able to return to their families again, alive, became a new problem, which both to themselves and others had always been deemed impossible to solve. But necessity urged them to try everything. On the holm grew the American linden tree; they peeled off its bark, and of its bast made a ladder as long as the height of the falls. One end of this ladder was fastened

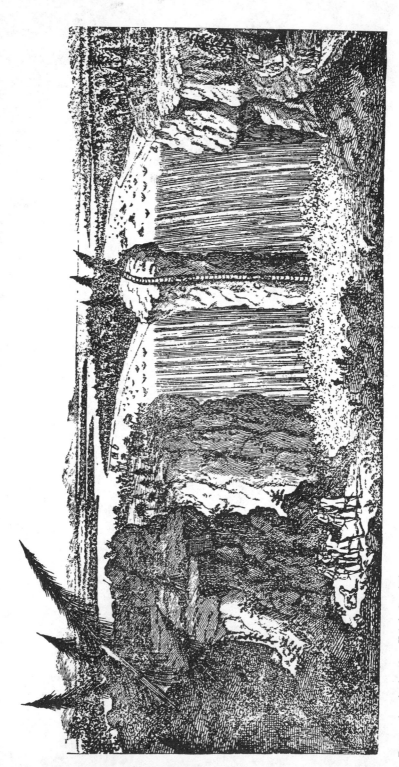

Niagara Falls

From early print based on Kalm's description

to large trees that grew on the brink of the perpendicular ledge which constituted the nether extremity of the islet between the falls, and the other was let down to the water below the cataract. Then they climbed down the ladder to the rock at the bottom, where they threw themselves into the water, thinking that they might be able to swim to land, especially if they could reach the more quiet water below the falls before the place where the abovementioned rapids commence. As I have noted previously, one could safely row a boat on this intervening space. But the descending waters on the two sides of the island make strong surging waves at the bottom, which with great force were hurled back against the rock directly below the islet. Consequently as soon as the Indians had jumped into the water, these huge waves drove them so forcibly back on the rock that when they had tried a few times to swim they had been so bruised and buffeted about that most of their skin had been torn off their bodies. They realized then that it was impossible for them to reach the shore, and were therefore compelled to climb back up the ladder to the island. Here they were obliged to remain for nine days in all, without any food except what they had brought with them in the boat and the few deer that they had been able to catch. But of water to slack their thirst they had plenty.

Fortunately for them, some savages who happened to pass by on the mainland, caught sight of the prisoners and heard them call. The former hurried down to Fort Niagara, told the commandant of the plight of their brothers, and exhorted him to bring about their rescue, if possible. Immediately he called his officers and men into consultation and questioned the Indians, who were well acquainted with the current on both sides of the holm where the two others were marooned. They remembered that the water at the upper end of said holm was not very deep, nor was the current there very strong. They had occasionally observed six Indian boys wading far out into the water there. So they adopted the following plan: the commander had a few poles shod at one end with a ferrule-like spike, so that they became quite sharp. Thereupon he urged the visiting Indians to do their best to save their brothers, and showed them how they should go about it. He and his officers went along [with them] to see the plan put into execution. A couple of young

Indians, each one provided with four iron-shod poles and some food in a bag with which to refresh the wretches on the island, set out on the venture. They made their way along the shore at the east side of the river to a point a little beyond the extreme end of the island, and from there began to wade diagonally across to that spot, driving their iron-shod poles into the bottom at every step in order to support themselves against the violence of the current. At that time the water in the deepest places went just up above the knees. Consequently they reached the island safely, fed the wretches a little, and giving each one of the latter a pair of shod poles conducted them back to land. After the trail in this way had been blazed, the Indians continued at later times to visit the island without danger. That which attracted the natives to the place was, as noted above, the number of deer, elks and other animals that tried to cross the stream but which were carried to this islet, if they escaped passing over the falls. When the natives from the mainland saw a sufficiently large number of animals on the island they waded across and killed them. The two first Indians, who were so luckily rescued from death, were said to be still alive, when I visited the falls.

The bottom of the river above the falls, between the island and the mainland on both sides, consists of ledges or rock of a gray, compact limestone. The cliff which forms the falls, is of the same material. Now and then a stone of considerable size is found in the rapids above the falls, and also below where the water strikes the bottom. These stones are granite. The limestone of the falls lies stratum super stratum. The banks of the river below the falls are also of the same gray, massive lime rock, and are exceedingly high and steep, almost vertical. It is however, possible to climb up in some places; but it is certainly dangerous and the attempt should be made with great caution. On the west side of the river the bank is less precipitous.

The land about the falls is stony, and here and there a large bit of gneiss or granite is found.

Several people told me that one could pass below the falls right between the rocky wall and the mass of falling water, since the latter does not fall perpendicularly close to the rock but describes a curve. I asked M. Joncaire if this were so, and he answered in a decided negative, for, he said, the water in its descent hugged the wall very closely.[9] On the west side of the island there had formerly been a cliff projecting out farther than the rest of the wall, and when the water ran over this protruding part it had been possible to pass beneath the falls below that point; but this jutting rock had fallen down some time before, so that it was now impossible to go between the wall and the water.

The Indians make small bark boats which they use in fishing below the falls. Among other fish they catch a large number of small eels of nine or twelve inches in length, and all the dexterity needed for their capture is to go below the cataract and feel around with the fingers in the cracks, holes and crevices of the wet rock, find them and grab them. A large quantity is gotten this way. Some small native boys tried the method to-day for my benefit, and returned very soon with a large heap of them. The Indian lads were real daredevils; they walked right out to the very edge of the cataract or river and looked down, where it was not only vertical but where the rock besides had been worn away by the water. They waded a long distance into the water right above the cataract, and then proceeded to approach the falls themselves so close that there was not more than a foot to the outer ledge of the cliff where the water spills over from its terrifying height. There stood those rascals, gazing down! I was chilled inside, when I saw it, and called to them; but they only smiled, and still stood a while on the outermost brink. A single false step would have cost them their lives, had they had a thousand of them.—I suspect that the officers who accompanied me had commanded these Indian boys to show me their skill and daring.

When the wind blows the most furious hurricane seems to reign below the cataract, because of the mist being blown in all directions.

On a certain time of day when the sun is shining, we can always see a rainbow against the mist. I arrived a little after eleven in the forenoon, and we had clear, sunshiny weather all day, so that we could see almost constantly a beautiful rainbow on the east side of the river, near the bottom of the cataract, about six fathoms below it. When we stood on the edge or bank of the river the rainbow appeared beneath us. I saw it at eleven, twelve, one, and half

---

[9] M. Joncaire's reply was wrong, of course, as every visitor to Niagara knows. Obviously, after this denial, Kalm did not proceed to test the facts.

past one o'clock, but after I had been away to dinner and returned at a quarter past three no more rainbow was to be observed, presumably because the sun was lower in the sky. While it lasted the following phenomena about it were noticed. If the vapors were thick, two rainbows were seen, one outside the other; but the latter was then very faint. When the fog was thin, only one rainbow appeared. It had quite a large arc, and was on the outside of a flaming red color, the tinge of a flame mixed with smoke; but the inner part of the arc was greenish-yellow, and the central portion of a heavenly blue, though very light, the flame-colored and greenish-yellow parts made up the arc or bow, and the light blue filled the remainder. Because the mist was driven over the river with the wind, the rainbow looked like a floating stream, for although it stood still the vapors moved forward. The thicker the mist, the plainer the rainbow; too much fog made a double one. Occasionally when the mist was driven away, the rainbow disappeared almost entirely, and only pieces of it were visible. As one moved from one place to another the rainbow seemed to move also, so that one could constantly see a new one. Some have claimed that a rainbow may be seen here only at eleven in the morning; others that it can be seen all day. My own observations are recorded as above. The external rainbow was almost always of a green color, narrow, thin and faint.

The basin or the space which the water immediately below the falls occupies is near them as broad as the falls themselves, but a short distance below it becomes gradually narrower.

The boat landing above Niagara, when one wished to sail up Lake Erie was said to be about 15 arpents from the falls [about a third of a mile]. I thought it was but a little less than ⅛ of a Swedish mile [¾ of an English mile]. It is not considered safe to approach the falls nearer with a boat, lest one suffers the same fate as the two Indians, or something worse.

In the rocky sides of the river below the falls are a large number of holes, since it is stratified limestone. It was claimed that rattlesnakes hibernated there. In summer time they are scattered about in the woods and on the hills. To-day we saw three of them near the road, which were killed at once. In the beginning of autumn when it begins to grow colder, they gradually gather here from all directions and creep into the aforementioned holes; and when in spring it gets warmer they come out again from their hiding places, when one can sometimes see them by the hundreds. It was asserted that about six hundred of these hideous beasts had been killed this spring (1750) in this neighborhood, when they first came out of these winter quarters and were still torpid. It is well known that sometimes [—Kalm might have said frequently or generally—] there is no cure for the bite of this dangerous serpent; it is so poisonous or deadly.

ADDENDA TO THE DIARY

[THE following material, on New Sweden, is not all new of course. In fact, a large proportion of it, in its original form, has been used before, and notably by Israel Acrelius in his well-known work on New Sweden (1759).[1] But Kalm's account in its simplicity gives us many new slants or viewpoints, much interesting and illuminating historical gossip, and some details omitted by Acrelius. Besides, it serves as a useful corroboration of many facts of history connected with New Sweden. It represents first-hand sources in any event, and as such is some contribution to knowledge. However, a number of pages, which Kalm copied from the Wicaco Church Records, have been omitted here as being primarily of local interest.]

THE ARRIVAL OF THE FIRST SWEDES

I have talked with several of the oldest Swedes and sought to learn in what year the first Swedes arrived; but there was none of them who could give me definite information about it.[2] Their sentiments and stories I have already introduced [in these volumes].

SEPTEMBER THE 18, 1748

From Mr. Kock and Jacob Bengtsson I learned that after people had been sent over from Sweden to settle and cultivate this land, several years passed before the mother country cared to make any inquiries about the countrymen it had sent hither, or to learn how they were, and if they were living or dead. The Swedes who then came over settled near bays and rivers, where they had good opportunities for fishing. Most of them had no horses or beasts of burden, so that when they needed salt, which could not be procured

---

[1] This has been translated into English in part by Nicholas Collin (1841) and in full by William M. Reynolds (1874). The latter is called *A History of New Sweden or the Settlements on the River Delaware*, printed in the *Memoirs of the Historical Society of Pennsylvania*.

[2] The first Swedish settlers landed on the Delaware in 1638.

elsewhere than in New York, then belonging to the Dutch and called New Amsterdam, they went thither after it, riding in part on oxen and in part on cows, bought salt there and whatever they needed and brought it back home on the backs of the just-mentioned animals. And since they had no other people to associate with than the native Indians, they soon began to differ more and more in their actions and manners from the Europeans and old Swedes and began to resemble the Indians. At the arrival of the English, therefore, the Swedes to a large extent were not much better than savages. One of the first reasons for sending Swedes over here was connected with a man by the name of Printz.[1] He, together with some others, had been in prison in Sweden for some misdeed, and suggested to the Swedish government that if he were released he would leave the land, go to America and seek out such places as could be settled by Swedes and be of such advantage to Sweden as other colonies had been to other countries. His request was granted; a ship [*Fama*] was fitted out; and Printz, who was an experienced sailor, made captain of it. Several criminals who had been serving sentences in prison with him were released and made his companions on condition that they would settle here, start to till the land, and hunt for gold, silver and similar things. He sailed hither on the ship just mentioned, let the newcomers settle wherever they pleased, stayed here a while to govern them, and then after a time returned to Sweden. When the Swedes saw that they could not get what they needed from Sweden they surrendered to the Dutch [2] and became their subjects, enjoying their former privileges.

The nonagenarian, *Nils Göstafson* [Gustafson], told me that when the Dutch came and took the land the Swedes did not accept any favors from them; nor did the Dutch harm them in any way. Then when the English came and seized this territory there were a great many Dutchmen living among the Swedes; but a large proportion of them left for Surinam, which the Hollanders had received from the English in return for the American settlement.

---

[1] Johan Printz (1592-1663), governor of New Sweden. His misdemeanor, mentioned in the next sentence, consisted in a breach of discipline while serving in the Thirty Years' War. It was not in Printz's case regarded very seriously.

[2] The Swedes surrendered to the Dutch involuntarily in 1655 because of the sheer superiority in numbers of the latter, but undoubtedly there were some who did not care who the ruling power was.

Zachris Peterson told about his father, who died seven years ago at the age of ninety-five. He had emigrated from Uppland in Sweden, and related how the Swedes had formerly gone to New York to purchase some cows, but that on the return journey, when the English fleet had been at anchor in the Delaware River, the English had taken away a number of those same cows and left only one cow to two families.

Governor Printz is said to have been a big, very tall man, who inspired such a terror among the Indians that although they had conspired to annihilate the Swedes they became so frightened that they dared not do anything but agree to everything he proposed. In the beginning the danger from the Indians was so great that when the Swedes were plowing, someone had to walk behind the plowman with a gun in his hand to defend him if the savages should appear.[1]

Mr. Peter Rambo, who lived down in Raccoon, N. J., related to me on the 30th of January, 1749, the following concerning the first coming of the Swedes, which he had heard from his father who had died six years ago and had at the time of his demise been very old. His grandfather's name was Peter Rambo [also]; he had been born in Stockholm, had with others been hired to come here, and had gotten his freedom after three years to return to Sweden if he had so desired. Several of these so hired had returned to their homeland, but he had remained. But those who had been sent here because of some misdeed had not been allowed to return. The first Peter Rambo landed here when the original settlers had been here four years (1642). He was then unmarried, and when he had been here for a short while he married and had several sons of whom this Peter Rambo's father was the youngest. He was born 1661 and was 12 years younger than his oldest brother. The original Peter Rambo, when he emigrated, had brought apple seeds and several other tree and garden seeds with him in a box. He had also taken some rye and barley along. Later when the Englishmen came he had often told them that his hands had been the very first to sow seed in the settlement, thereby announcing that the first Swedes had not brought these seeds with them, and that conse-

---

[1] The relations between the settlers of New Sweden and the Indians are generally supposed to have been exceptionally friendly.

quently no European seed had been sown here before he upon his arrival had made a beginning. His grandfather had prospered, so that Governor Penn had often lodged at his house; and when the English first came here it had been rather difficult for some of them, so that Rambo not only helped them as much as he could but for ten years gave to everyone that came to him free food and lodging. The old man was very kind, but liked to drink a bit at times.—All the Swedes who first came had built cow barns, as the living Peter Rambo affirmed, and they still had them when he was a boy; but afterwards when they saw that the English did not use them they had abolished them.[1]

The Finnish Ship. Among the ships sent here from Sweden was one loaded almost entirely with Finnish passengers, who had been sent here to settle the land; but when they came near the American continent the vessel sprung a leak so that they were unable to pump out all the water that came in. They kept pumping however for three days, though the water finally got the upper hand and the damage was irreparable. Besides the crew, there were three hundred people on board. When the sailors saw that all hope of saving the ship was gone, they jumped into the [life]-boat under the pretext of investigating the leak, but in reality to save their own lives. But when one of the Finns, by the name of Lickoven (otherwise known as Jacob Eit) noticed it, he jumped into the boat also. The ship sank with all its passengers, but some of those in the lifeboat reached the shores of New England; yet no one came here, and information was received only through rumors. This ship was called *Det Finska Skeppet* (the Finnish ship). The ship doctor had silently been exhorted by the captain to board the lifeboat but he had not wished to leave the people, had stayed on the vessel and perished with the rest.

There seems to be no evidence that the Swedes here ever used dampers in their rooms or huts [as Kalm has mentioned before], either when they first came or later, because in addition to the fact that the winters were not so severe or long here as in Sweden they had a vast amount of forests of big trees right near their home [so that no saving of fuel was needed], and, besides, they had not brought any dampers with them from Sweden, and none were ob-

---

[1] Kalm has mentioned this fact before.

tainable here. I asked many old settlers if they had not heard them say that their ancestors had some at first, but they all declared they had never heard of any.

Some cows, a horse and a mare were brought over on the ship on which the first Rambo sailed to America, and the same evening they sighted land the mare foaled. It also snowed quite hard that night.—When the Swedes had been in this country for a while the English of New England learned that the former did not have all the cattle they needed. Therefore they drove down a large herd of cattle from the north and sold them to the Swedes, receiving good pay in skins and other goods that the Swedes had previously bought from the Indians.

The First Cause of Enmity between the Dutch and the Swedes. The Dutch and Swedes had always lived on good, friendly terms with one another until a number of Dutchmen came and settled in New-Castle, and until a ship arrived from Sweden that carried a captain by the name of [Sven] Scute, who had begun to shoot at the Hollanders to drive them away: this was the origin of the dissension between the two.

When the Swedes first arrived they lacked all kinds of tools, both for agriculture and other purposes.

Swedes and Indians. All the old Swedes told me with one voice that in former times the Indians had on several occasions banded together to kill the Swedish colonists, but through God's providence some old Indian man or woman had always secretly run to the Swedes and warned them about what their fellow-Indians had in mind. Sometimes the Swedes wanted to pay these secret messengers, but they would not accept anything, and hastily returned to their people. The Swedes then collected, and when the natives saw them prepared, they dared not attack, and a new peace treaty was drawn up between the two parties. The Indians always liked the Swedes better than the English, and the English better than the Dutch, whom they still hate a good deal.

UNDATED

Although the Swedes constituted only a handful of people in comparison with the Red Men, these were quite afraid of the former, and never ventured to attack them. Mr. Jacob Bengtsson told me, from assertions by his father and grandfather, that sometimes when the Indians stole a pig from the Swedes, the latter not only went and brought it back but gave the former a sound beating, without the Indians daring to strike back, to say nothing of a more serious revenge, and would return to the Swedes soon thereafter and be very humble and friendly.

NOVEMBER THE 22ND, 1748

Erich Rännilson had an old paper about his family and its immigration which read as follows: "Erich Mulleen aged 46 year, from Helsingland in Swedland, arrived in Dellaware River in the Ship Örn the 26th day of May, 1652. His wife Ingeri Philips aged 36 year from Wermland in Swedland, arrived in Delaware River in the Ship Mercurius March 1654. This family as follow" listed a daughter Anna, 16 years; a son, Anders, 14; Olle, 11; Erich, 9; Johan, 7; a daughter, Eleonora, 4; and one Catharina, 1 year old. N. B. These ships were the last that brought people from Sweden.

Church Records on Early Swedish Settlements. I could obtain no information from the church records about [the time of the first settlement], for there were no such records before the Rev. Mr. Rudman and Rev. Björk came here. Everything that happened before is enveloped in the darkness of forgetfulness. However, according to their story, the first Swedes arrived about the time given by [Thomas] Campanius [Holm] in his description of New Sweden,[1] namely 1630 or soon thereafter [1638]. I shall not here repeat what I have already said on the matter, but only add what others have later related. There have formerly been several records extant which have dealt with the coming of the earliest Swedish colonists, their number and names, and the locations in Sweden from which they came, and an account of their first activities, not to mention several deeds concerning their properties; but after the English took possession, William Penn in particular has sought to get at all these documents, and they are now said to be among his collections.

The old man Gustafson, 91, whom I have mentioned before, told me that several ships with people, had come from Sweden; that

---

[1] *Kort Beskrifning om Provincien Nya Swerige uti Amerika,* 1702.

Printz was one of the first who arrived; and that he returned after he had been here a while, nobody after that hearing anything more from him; that the name of one of the ships which had brought settlers was *Örnen* (The Eagle), and another *Mercurius*.[1] Otherwise he could not give the year of the first settlement; only when he was a boy [about 1665] he had heard it said that the Swedes had then been a long while in the country, so that he felt they must have been here forty or fifty years before his birth.[2] In his youth the Swedes lived mostly near the kills, rivers, and the bays found in them, and only seldom in the interior where dwelt thousands of Indians, who had daily relations with the Swedes.

Cooperative Farming. The wife of the old man Måns Keen said she had heard from her father and grandfather that when the first Swedes arrived, every family, besides the grain brought for food, had carried a keg, pail or stoup of every kind of seed to sow when they came. They had also arranged it between themselves for one family to bring a cow, another an ox, etc. So that after arrival, when they were to plow for instance, it happened often that they [pooled their resources and] hitched a cow and an ox together to form a team.

Kidnaping and Scalping. Formerly the Indians used to steal a Swedish child now and then, which was never returned.—Once they killed a few Swedes and scalped them. They also scalped completely a little girl, and would have slain her if they had not seen a Swedish boat appear, when they fled. The girl was later cured, though she never got any more hair on her head. She was married, had several children, and lived long thereafter.

Of old *Åke Helm* I learned the following facts: His father came to this country with Governor Printz, who brought him along [as a servant or companion?] for his son. He was then twelve years old. He lived to be 75. This year, 1749, it is forty-five or forty-six years since he passed away, which agrees entirely with the record given by Campanius that Gov. Printz arrived in 1642 [1643].

---

[1] These are now well known facts among historians and need no elaboration. What we learn from Kalm here is the incredible ignorance of the Swedish colonists of 1750 about the details of their ancestors' settlements.

[2] Since this conversation took place in 1748, and Gustafson was 91 years old at the time, he must have been born in 1657, when the first Swedes had been in America only nineteen years.

The First Swedish church was located on Penako (Tinicum) Island, between Philadelphia and Chester, where Governor Printz had his residence, and where the Swedes also possessed a fort or redoubt.[1] Now the island is inhabited mostly by Quakers, though there is said to be one Swede left there. There is no sign left of the Swedish temple, and the cemetery is washed away by water; because the Delaware [River] has eroded the soil there so that the graves are now under water.

Old Mr. Helm reiterated that the Swedes had in part brought cattle with them, but that the largest number had been procured from New York. The cattle of his younger days had been larger than now, so that there was no comparison—they were getting smaller and smaller. They also gave more milk formerly than now. But in his youth settlers had not possessed very many of them; he who owned a cow was held to be rich and prosperous.

Finnish Settlers. Finns have also settled here. They have never had clergymen of their own, but have always had themselves served by the Swedish. They have always spoken Finnish among themselves. Most of them settled in Penn's Neck, where people have been found who until very recently spoke Finnish. But now most of them are dead, and their descendants changed into Englishmen. Helm believed that the copy of the Finnish Psalmbook, which had been presented to me by Zachris Peterson, was not only the oldest of all Finnish and Swedish books available here at this time, and which the Finns had brought with them, but that it was the only Finnish book procurable here; for Mr. Helm said he had often seen the copy, and had also had it as a loan, without ever seeing any more in that language. Most of the Finns came over on the ship *Örnen.*

Mr. Jacob Bengtson in Philadelphia imparted the information obtained from both his grandfathers, who were among the first Swedish settlers, that the majority of the early Swedes came via Gothenburg, and that they hailed from Västergötland. His own paternal grandparents came from that province. It was also possible to tell from the manner of speech of the oldest colonists that

---

[1] This may have been the second church of the Swedish Colony, being erected in 1646. The Rev. Reorus Torkillus "apparently" built the first edifice in 1641. See Amandus Johnson, "History of the Swedes in the Eastern States from the Earliest Times until 1782" in *The Swedish Element in America,* II, 21.

they had originated there. But a number came from Uppland and a large number from Finland. He knew of no other Swedish provinces from which early settlers had come.

EXTRACTS FROM THE RECORDS OF WICACO CHURCH

On the eighth of June, 1750, I borrowed from the Rev. Gabriel Näsman the Wicaco Church records to see whether I might find anything that could throw light on the history of this land, especially as the late Anders Rudman had exerted himself more to obtain and search the old transactions than anyone of the other congregations had done. The title of the book in translation is "A καὶ Ω. Church Records of the Parish Wicaco from the First Coming of the Swedes to America and its Part New Sweden, afterwards called New Netherlands and now Pennsylvania, so far as They have been Compiled from Friends and a few Writings, by Anders Rudman, Pastor."

Ever since Providence through the wise Christopher Columbus in the year 1492 discovered this new part of the world (although in itself as old as other parts), and Americus Vespucius had the good fortune to have the whole country named after him, the European nations have from time to time sought to colonize this land and become masters of its wealth. Among other nations our old Goths and Swedes have not lagged behind. I shall not dwell on their ancient and many commendable deeds, but limit myself to [an account of] their settlement in this country, and relate the events connected with it, such as I have learned them from many oral sources, and particularly from Capt. Israel Helm.

Before the Swedes' arrival on the [Delaware] River a few Dutchmen had a fort on the other side, on a place which the Indians called Hermaomissing, now Glo[uce]ster; but the Dutch named it Fort Nassau. The commander's name was Menewe [Peter Minuit or Minnewit], who could not get along well with the people and therefore returned to Holland, where after certain charges had been advanced he was removed.

Knowing the land and its condition he went to Sweden and informed the distinguished gentlemen there that the Dutch had settled on the east side of the [Delaware] River, but that the whole continent was still a wilderness, and that although there were so many heathens there that it was a bit adventurous to settle there, if they, as gentlemen of means and quality, would like to venture a settlement he would be willing to be its leader. The chancellor, Count Axel Oxenstjerna, the chief of the others, spoke to Queen Christina, who was so pleased by the proposal that through her order and permission a vessel was fitted out and sent from Gothenburg, called *Calmar Nyckel* (The Key of Kalmar), with a number of people who safely arrived in [America] and negotiated with the Indians for land extending all the way from the mouth of the [Delaware] to the [Trenton] falls. There they drove a few stakes into the ground, which the old folks' of my early days in America told me they had often seen; and if one investigated the matter it was believed that the stakes could still be located, on both sides of the River, all the land of which was purchased of the heathens and the sale thereof recorded in deed and letter with the signature or marks of the Indians attached. The document was sent back to Sweden, and when I was in Stockholm it lay in the archives there. It is to be regretted that there is no copy here. Such a one was really made, to be sure, but it disappeared during the many troublous vicissitudes of the day. When [Governor] Rising and [Per] Lindeström [1] came to New Sweden they brought with them the names of the signers of the agreement, which all the Indians who were still living recognized, but they did not wish to hear the names again, since it is a custom of theirs not to speak of or name a dead person. The people settled at a place called Christina, erected a fort there which they named Fort Christina, after the Queen, and the locality has that name to this day. Måns Kling was the surveyor who mapped the whole settlement and reproduced the whole River with all its kills and streams on a chart that was sent to Sweden and deposited in its archives, which I saw the day before my departure and which Mr. Aurén copied in miniature and brought here. Minnewit returned; [his] ship came back laden with goods; but he died and Peter Holländare, a native Swede, took his place, stayed a year and a half and returned. Capt. Israel Helm said he had seen him in 1655 on Skeppsholmen, and that he was a major there.

---

[1] An engineer who in 1691 completed his *Geographia Americæ*, which was not published, however, in Swedish until 1923, and in English, in a translation by Amandus Johnson, 1925.

In the year [1643] Johan Printz arrived with two war vessels, the *Swan* and the *Charitas*, with full powers to become Governor over the settlement, which he really was, for ten years, during which time only two ships arrived namely *Svarta Kattan* (The Black Cat), on which there were no settlers or passengers, and later the *Swan*, which brought the Rev. Johan Campanius, Capt. Fisk, etc. Since the time passed but slowly for Mr. Printz, and no more ships came over because of the war between Sweden and the Emperor [Ferdinand II in the Thirty Years' War], and because Governor Printz had made himself detested among his own people by his ultra-severity, and especially after the Dutch had begun to come too close, having established themselves five miles below Christina—for those and other reasons he returned to Sweden.[1] His influence and command of man-power had not been sufficient to prevent the Dutch from making that settlement. In the meantime the ship *Örnen* had left Sweden with Director Johan Rising, the chief surveyor and engineer Peter [Per] Lindeström, and several officers and other men, who arrived safe to be sure, but unfortunately got into a quarrel with the Dutch on Sandhuken (New Castle), capturing their fort on their first arrival, Trinity Sunday, 1654, which caused all the later trouble. To understand this rightly we must go back to the very beginning: before the coming of the Swedes to the [Delaware] River, the Dutch had settled on the Hudson River, where they had begun to build a town, calling it New Amsterdam. Then when they started to prosper, some came across to this River, and as a protection erected a fort on the New Jersey side, where Gloucester now is located, calling it Fort Nassau. Their Governor did not get along well with his people, and therefore Minnewit returned to Holland, and from there went up to Sweden with a history of the settlement. Immediately thereupon the influential Swedish gentlemen with the sanction of the higher authorities came together and decided to establish a colony here, and so sent him back with a number of settlers, who upon their arrival negotiated

---

[1] Kalm's source is not wholly accurate here. The *Charitas* had been in New Sweden in 1641. Printz and Rev. Campanius sailed from Sweden together in the fall of 1642, on the *Fama* and the *Swan*, arriving in New Sweden, 1643. In 1646 the *Gyllene Haj* arrived, and in 1648 the *Swan*, which in the interim had returned to Sweden, arrived again in New Sweden with a large cargo for the Colonists.

Governor Printz was probably not as black, i.e. as severe, as Kalm's source has painted him. Cf. Amandus Johnson, *The Swedes on the Delaware*, 1927, pp. 142 ff.

with the Indians for land on the River, and built Fort Christina, allowing the Dutch to dwell peacefully on the other side. Later Printz appeared with full powers as governor. During his time the Dutch commenced to interfere with the trade between the Indians and the Swedes, for which reason Printz held them in subjection, and also had Fort Elseborg [Elfsborg] constructed, where all their ships had to pay toll. Then they moved from Nassau— since there was no longer any great danger from the natives—down to this side, on Sandhuken, and erected a fortification there. Mr. Printz forbade it, but since his power was too weak to prevent it, the Dutch went on with their plan. This irritated some good Swedes, who would not suffer the intrusion of strangers, who against the wishes of the real owners flauntingly settled and encroached upon their purchased territory. Consequently some crossed over to Sweden to tell the tale, and among them Mr. Scute, whereupon the ship *Örnen* sailed hither with the command to set about with gentle means to recover Sandhuken and place the Dutch in obedience. But the Germans [Dutch] would have none of that, and after Sandhuken had been taken, Governor Hugwesand [Peter Stuyvesant] of New Amsterdam arrived with a troop of soldiers and captured the whole River on behalf of the West India Company in Amsterdam. Stuyvesant regretted it, for which reason he became ready at once to return the Fort, but Mr. Rising, expecting the means of a greater revenge from Sweden, would not accept the offer without orders from abroad. One can judge the condition in the land from the following pass made out by Mr. Rising to old Nils Matson:

Be it known through the Governor of New Sweden and the most humble servant of His Royal Majesty and King that since the bearer, the honest and intelligent Nils Matson, a freeman from Silleson, at this inopportune time in New Sweden, when we Swedes have so unexpectedly been attacked by the Dutch with hostile intent, is unable to move away so hastily, and must for the sake of his belongings remain here until a more convenient time, and since he has on that account requested a testimonial letter from me, which I have not for the sake of justice been able to refuse him, I hereby testify that he, during the whole time that he has stayed in the country has acted like a reliable and faithful servant of the Crown,

that he has willingly aided in the repair of the Fort and buildings
and in the performance of other public service, that he volunteered
in the last struggle by repairing to Fort Trefaldighet and there par-
ticipating in the defense of the land, and that he on the way was
made a prisoner and taken on a ship where he for about three weeks
had to endure much mockery. In the meantime his home was robbed by the enemy, and his wife divested of everything, having constantly maintained her loyalty as a good subject during these troublous times. In testimony of which, I attach my seal and signature.
Actum Fort Christina. September 24, 1655.

L. S.   Johan Rising.

Now the settlement was under Dutch rule; Sweden was involved
in war; the gentlemen who formerly had constitued a "company"
were incapable of action, and cared but little about any colony,
since the venture gave but small or no profit. The king and government, who had received title to the settlement lands from the
[New Sweden] Company were busy with other big affairs, and so
the situation had to remain as it was, and the land was now under
Dutch and now under English rule, since these two peoples were
at war. But the fact that this River belonged to the Swedish crown
is attested among other proofs by a letter in the hand of Queen
Christina, dated at Stockholm, August 20, 1653, and given Captain
Hans Amundsson Bask. So that the Dutch try in vain to veil their
actions by declaring that they took the land from a managing company and not from the Swedish government. It is true that in the
beginning it was a company which under the protection of Queen
Christina carried on its commerce here, but the real power lay vested
in the queen, all the ships were sent by her, and all costs were defrayed by her generous hand in that prosperous time, all of which
is duly attested by this document literally translated into English
by M. Carl Springer.

Wee Christina by the Grace of God, Queen of Swedland, Gothen
& Wenden, Great Princess of Finland, Dutchess of Estland etc.

Be it known, that we of favour & because of the true & trusty service, which is done unto us & the Crown of our true & trusty servant Capt. Hans Amundsson Bask, for which service he has done

& further is obliged to do so long as he yet shall live, so have we granted and given unto him freely, as the virtue of this our open Letter is, & doth Shew & Specify, that is, we have freely given & granted unto him, his wife and heirs, that is heirs after heires, one certain piece & tract of land, being & lying in New Sweadland, Marcus Hook by name, which doth reach up to and upwards to Upland Creek & that, whit all the priviledges, appartenances and conveniences thereunto belonging, both in wett and dry, what soever name or names they have and may be called none excepted of them, that is, which hath belonged unto the afores:d tract of land of age, & also by law and judgement may be claimed unto it, & he & his heirs to have & to hold it unmolested for ever for their Lawfull possession & inheritance, so that all which will unlawfully lay any claim there unto, they may regulate themselves hereafter, so that they may not lay any further claim or pretence unto the aforsaid tract of land for ever hereafter. Now for the true confirmation here of have we this with our own hand underwritten and also manifested with our seale in Stockholm the 20 of august in the year of our Lord 1653.

Christina

L. S.   Niels Tungell.

Secretary.

After the settlement had so innocently declined, the Swedes underwent many sufferings, which if all were to be enumerated would make a reader resentful. I shall only mention two, from which an opinion may be formed. 1. In the year 1656 the *ship Mercurius* arrived with several Swedes, among them Anders Bengtsson, who is still living and to-day, April 5, 1703, gave a vivid narrative of the event. He told how the Dutch forbade the vessel to pass up the River, and how they ingloriously would have sent it back, had it not been for the heathens who liked the Swedes and who collected, boarded the ship and in defiance brought it up past the fort. 2. The daughter of Governor Printz, Madame Armgott, sold Tennakong and neglected to make a reservation for the church and the bell. The buyer of the land laid claim to both, and the Swedes were compelled to repurchase their church bell for two days . . . since Madame Armgott had gone away, who had promised them their bell, but who had not kept her promise. [Then follows a copy of her written obligation, in Dutch].[1]

Just as New Amsterdam, now New York, a few times passed under Dutch, now English, rule, so this [Delaware] River which by the Germans is called South River to distinguish it from East and North River at New York—passed in its entirety under English jurisdiction, the Hollanders getting possession of Surinam again. Afterwards Penn received it as a property reward for the large services rendered the English government by his father. . . .

Mr. *William Penn* arrived with a group of his followers, namely Quakers, whom England was glad to get rid of, and began settling the country. He founded a city at Wicaco, called Philadelphia, to which the landowners objected for a long time, but who were finally won over by kind words and other means. Philadelphia had a modest beginning, but could within twenty years take pride in its own splendor and strength. Penn, as proprietor and governor absolute, returned home, and put Mr. William Markham in his place. The Swedes were offered the opportunity of banding together and living after their own laws, but they submitted to the English, which they and their children bitterly but uselessly regret.

Thereupon it happened that an impostor appeared, who claimed the name of Königsmarck.[2] He started a revolt; got many followers, especially among the Finns; was captured, branded and exiled. His confederates lost their land and suffered great harm; and besides, many innocent colonists were brought into ill repute with W. Penn, who otherwise was very kind to them, presenting them with a small box of catechisms and books as well as a copy of a Bible in folio for their church.

[Then follows several pages (193-221) of local church history up

---

[1]   Laus Deo den 24 Maius A:o 1673.
Iik ondergeschrewen Armegot Printz bekenne gelewert te habben aen de gement hier Oowen von de Augsburgsche Confessie, de Klock di aen Tennakom geweesen sijnde, dat sij dar meede daen wat haer belifwen sar en belowe haer te bevrijden voon alle naemanige dit gedaen sijnde, voor ondergeschrewen getuygen. Datun als oowen.
Het merk van
    P.K.   Armegast Prins.
Peter Kock
Het merk van
    X.
Jonas Nilson
[2] His real name was Marcus Jacobson. See below, pp. 732 ff.

to about the year 1735, giving the plight of the colonists because of the lack of pastors, their urgent correspondence with the Swedish authorities to procure clergymen, and their final success in these negotiations which resulted in the arrival in New Sweden (1697) of Anders Rudman, Eric Björk and Jonas Aurén. A list of parishioners and statistical details of births, burials and weddings are given, as are also the wages of the laborers who helped build the new church. A description of the unique administration and troubles of Jacob Fabritius, German pastor of a Swedish congregation, is briefly sketched, and an enumeration made of the religious tracts, catechisms and other books which were sent to New Sweden with Rudman [1] *et. al.* by the Swedish king and government. Since most of these local details, however, are already reasonably well known to the readers most interested, and have been pretty thoroughly exploited by previous writers they are here omitted].

Remarks on [*William*] *Penn.* Mr. Lewis Evans told me the following about Penn: there is good reason to believe that W. Penn was not a real, good Quaker at heart, for he was in reasonable favor with King James II of England, who as we know, was a strong Catholic. It is true that W. Penn in a letter to the archbishop in London assured him that he was not a Catholic; but as a politician he might well have been clever enough to make that assertion; it would not have been wise in those days to let anyone know that he was of the Catholic faith. However, all Quakers wish to believe that he was a member of their Society; but it is known that often Catholics conceal themselves under that name among the English.

It is believed by most people that the form of government which Penn furnished to his province, and in accordance with which it was to be ruled, was drawn up by Penn himself, since it bears his name; but there is strong evidence that it had first been outlined and formulated by Sir William Jones [1631-1682] in England, who was one of the best men that England at that time possessed in the field of law. He is said to have been attorney-general of England [from 1675-1679].

Thomas Penn [1702-1775], who is now [hereditary] proprietor, is one of the meanest and stingiest of men. [For example] he tries

---

[1] Rudman and his party spent ten weeks on the ocean between England and America. The MS copy of his diary of the trip is preserved in the Yale University Library.

to obtain a personal examination of all former deeds on the land which his father gave away, and when he under some pretext succeeds in procuring them he often does not return them at all but has the property in question resurveyed, and then he lops off a piece, maintaining that the party has more than he is entitled to. Many residents in Philadelphia have lost land in this way, and it makes no difference of what nationality they may be, if he only can find some excuse to get the land. It is known that the land on which Philadelphia is built formerly belonged to a Swede,[1] and that William Penn made a trade with him, giving him twice the amount of land in another locality, four English miles from Philadelphia. But the present proprietor thought it was too large a grant, so he had the land surveyed three or four times, and each time he cut off a slice from it.

A New Academy Established. When this year, 1749, a new academy was established for the education of youth, the trustees of the same wrote to the proprietor Thomas Penn, in London, thinking the news would give him a singular pleasure. He answered he wished them luck in their undertaking, but he felt that the country was still too young to found academies;[2] he feared the public would not be able to stand the expense. The residents of the city, who heard this, thought he would have done better if he had donated a few acres of land for said academy, for he himself would in the future have profited the most from the gift. Whenever they had wanted to build lodging houses (dormitories) for the children (students), he would have received a good income in rent if these had been erected on his land.

Houses and Buildings in New Sweden (Addenda). Since it has frequently happened that a disastrous fire has broken out because of having the [large] oven in the cabin or dwelling house, the cus-

---

[1] It belonged to the Svensson (or Swensson) brothers, as Kalm has previously pointed out.

[2] The University of Pennsylvania had been founded in 1740. The article on Thomas Penn in *The National Cyclopædia of American Biography*, lists him nevertheless as a "benefactor if not a founder of the College of Philadelphia (afterward merged in the University of Pennsylvania), as well as of the hospital and library there," but adds that "these gifts were well within his power, as he was the chief proprietor of one of the largest feudal estates in the world." The proprietary estates were exempt from taxation, which later led to serious complaints and disputes. Benjamin Franklin was in 1757 sent to England to present the Colonial views to Thomas Penn, who had returned to London in 1747.

tom has now been abandoned entirely. So I have not seen an oven in a cottage anywhere. It is now built separately in the yard, a short distance from the houses, and is generally covered by a little roof of boards, to protect it from rain, snow and storms. Usually it is elevated a few inches above the ground so that chickens and other small animals can stand under it when it rains. The reason why some colonists still construct both fireplaces and ovens of nothing but clay is that in many places here it is impossible to find a stone as big as a fist, not to mention larger ones; that but little experience has been had in brick-making over a fire; and that, besides, a first settler who erects a house on a plantation cannot afford to buy brick.

From his maternal grandfather Rambo, who had been one of the earliest settlers, Mr. Jacob Bengtson had learned that the board ceilings in the first colonial houses had been covered with earth to prevent the heat from escaping through the top. No dirt is used now, only thin boards, and not any too many of these. Probably there was no real damper in those days, and when it was very cold and the fuel in the fireplace had all been burned, someone in the house very likely crawled up on the roof and shut in the heat by placing a board wrapped in cloth over the chimney. There were no stone houses before the English came. The Swedes knew nothing of brick-making or burning lime; the first settlers of New Sweden were ignorant folk.

The Swedes. From an old man, Mårtin Gäret [Martin Garret], seventy-five, I learned the following observations: he said he had never seen or heard of any Swedes who used dampers here, but he had heard them say that they used them in Sweden.[1] However, he himself had made a cover which he placed over the chimney on cold nights, thereby retaining much more heat than usual. But it was a lot of trouble to climb up on the roof of the house every night and morning.

The fireplaces made at the time were built in a corner of the dwelling room. He still had such a hearth in his room, which resembled those we build in our guest chambers or bedrooms. They are now called Swedish fireplaces here, and are said to be quite

---

[1] Kalm was extremly interested in the economy of heat, as was natural from one who came from the Scandinavian North. Time and again he comes back to the subject of dampers.

rare. The most common ones now are English, which are as large as our kitchen hearth, though the bottom of them is not higher than the floor of the room.

The Swedes raised many peach trees in his youth, and a sufficient number of apple trees, but no cider was made then. They made a malt beverage at that time.

Formerly the houses of the Swedes were all of wood, with clay smeared between the logs, like those now built here by the Irish. They have no glass in their windows, only small loopholes with sliding shutters before them, just like our Finnish cabin windows.

The weather was much the same as it had been in years past.

He did not know whether the people formerly had better teeth than now; he still had most of his own.[1] He had never liked tea;[2] according to Rev. Mr. Jonas Lidman,[3] the eating of hot bread was responsible for the bad teeth.

Fever and ague had formerly been rare, and when it did come it was not so severe as it is now; that virulent form of fever was not known.

Jacob Bengtson was now a man of about sixty years of age. His father was Anders Bengtson, who had emigrated on the ship *Mercurius* that arrived in America right after the Dutch had taken possession of it. This Anders Bengtson is the same man mentioned so often [in the records] as being among those who wrote to Sweden after clergymen immediately before Mr. Rudman's day (before 1697). He was also the one who [on Sundays when there was no preacher] read to the Swedish congregation out of a book of religious reflections and sermons [postilla], after the days of Fabritius, and before Rudman and Björk had arrived. From this Mr. Bengtson I learned the following:

Rev. Jacob Fabritius was blind for several [9] years. He was of German nationality but born in Poland. He could not preach in Swedish—he did this in German—which most of the Swedes of that day understood. When he was to preach he had someone read the sermon to him and then he would expound it.

---

[1] This is opinion or information supplementary to what Kalm had said about the subject before.

[2] It will be remembered that Kalm thought the excessive tea-drinking in the Colonies might have something to do with the prevalence of poor teeth.

[3] Pastor of the Wicaco Church from 1719-1730.

Photograph by Philip B. Wallace

Old Swedes' Church, Wilmington, Delaware

The bell which now hangs in the Wicaco Church [Gloria Dei] belonged formerly to the church on Tinicum, and was brought hither from Sweden when the settlement belonged to the Swedish government. When the Tinicum Church was abandoned the Swedes of Wicaco had it brought hither; but when Mr. Rudman and Mr. Björk came, and two separate congregations were formed, Wicaco and Christina [Old Swedes' at Wilmington], a dispute arose as to who should have it. The members of Christina felt it belonged to them as much as it did to Wicaco, and wanted to claim it, while those of Wicaco refused to give it up. The quarrel lasted until a Swedish captain by the name of Jacob Trent arrived and settled it. He pointed out that while the bell really belonged to both parties, it could not very well be cut in two, and that therefore it had better remain where it was. He had on his vessel two ship-bells, he said, and was willing to present one of them to Christina Church. This he did, and the quarrel was settled.

This Captain Jacob Trent was said to be a Scotchman by birth, though born of a Swedish mother. He arrived in New Sweden the year after Mr. Rudman (i. e. in 1698) with a large ship, well-manned, for he had more than ninety men on board, and armed with big cannon. Most of the officers were Swedish, and among them was also a nobleman by the name of Oxhufvud. Mr. Bengtson thought it was a Swedish ship. It brought a large number of Scotchmen who came to settle here, and whom the captain sold to the Englishmen, who payed for their passage. He spent a whole winter in this country, near Chester. The English did not know what to think of him. At times they thought of making a prisoner of him, but there were too few people in the locality, and his ship too formidable, so that they did not dare attack him. His crew were for the most part Swedes and it was believed that he had been sent by the King of Sweden to inspect the country. Many Swedes who were carried here on this vessel remained in America. From here Trent went to the Barbados, where he was deprived of most of his men, because he seemed too strong and dangerous on the high seas. He was allowed only as big a crew as was absolutely necessary to the navigation of his ship. Some time afterwards he died, and only a few of his men returned to Sweden, among whom, however, was the above-mentioned Oxhufvud. Captain Trent had a brother in town who was a wealthy merchant, and from whom Trenton is said to have received its name, because he first settled in that locality.[1]

Mr. Jacob Bengtson's maternal grandfather was Peter Gunnarson Rambo. He is the man mentioned in Postmaster [Johan] Thelin's letter to these [Swedish] congregations, in which it is mentioned that he has a sister in Stockholm, etc.[2] His name first was Peter Gunnarson Ramberg, but changed the latter to Rambo. He came to this country from Västergötland with the Swedish Governor [Printz], and was then a bachelor, and was still living when Mr. Rudman arrived, though he died soon thereafter. He is the ancestor of all the many Rambos who now inhabit this land or have ever dwelt here. His oldest son's name was Gunnar.

Mr. Isaac Bengtson's father was Anders Bengtson, also from Västergötland.—I asked the former whether he knew of any Swedish clergymen here from Lock (enius)'s time.[3] He answered that he could not name any, but that he heard that two or three lay buried in Traanhuken, a place just below Christina-kill, where one of the first Swedish churches was built, but that is all he knew.

The Swedes whose names are Johnson are also from Västergötland.

Those who bear the name of Keen were said to be of Dutch extraction.[4] The ancestor's name was Kusk (?) which was later changed to Keen, pronounced "Kijn."

Kock, or as it is now written, Cock, represents a large family here. The ancestor of that family, who came here from Sweden, was Pehr Johnson, but since he was cook on the ship he was given the name Cock, and kept it ever since.[5] He was born in Uppland or in that vicinity, and died before Rudman's days.

---

[1] Trenton was named in 1719 after William Trent, Speaker of the House of Assembly, probably the brother of Jacob, as Kalm relates.

[2] Johan Thelin, postmaster at Gothenburg, served as effectual intermediary between the Swedes of America and his own government, when they needed more clergymen. The letter in question is dated Nov. 16, 1692, and is printed in an atrocious English translation by Carl Springer in the omitted part of the Kalm-Elfving *Tilläggsband*, pp. 192-201.

[3] Rev. Lars Carlsson Lockenius arrived in New Sweden in January, 1648.

[4] This may be true, but the original (Swedish) name was Kyn. See Gregory B. Keen, *The Descendants of Jöran Kyn of New Sweden*. Jöran Keen was a Swedish-born soldier, who sailed from Stockholm with Johan Printz in 1642.

[5] Cock arrived 1641 on the ship *Charitas*. Amandus Johnson gives his original name as Pehr *Larsson* Cock. See *Swedish Settlements*, 712.

Most (?) of the people who settled at Christina or Traanhuken were from Finland, such as Mr. Björk's mother-in-law, and others. The ancestor of the Holsten (Holstein) family came from Germany, according to the testimony transmitted by the first Holsten settler. Tolsa, Mullika, and Likonen may be recognized as Finnish names. Toy was said to have come from Holland, but old Garret claimed the family of that name came from Finland.

[As mentioned before], the name of the family who owned the land where Philadelphia is located is Svensson (Swensson). The original settler was Sven Gunnarson. He had three sons, Sven, Olof and Anders; of these the first was born in Sweden, Olof on the way hither, and Anders in this country, one of the first Swedes born in America. There had only been a couple of children born here before him. From these three brothers Penn purchased the land on which the city was built. . . . North of Wicaco Church there stands yet an ancient wooden structure, which belongs to this family, and which is now rented to some Germans. This house is older than Philadelphia, but now it is rapidly approaching its final doom.

The wife of Lars Lock was the first female child born of Swedish parents in the colonies.

Isaac Bengtson's mother had often told him that when she was a girl she had seen a man carrying the mail between Boston and Virginia on an ox which he rode. He had visited her father both on the down as well as on the return trip, and had lodged there over night. He went away in August and returned the following May. There were no horses here at that time.

Usually one or two merchant boats appeared in the Delaware every year, laden with salt, linen, etc. The Swedish bought goods from them; but sometimes it happened that no "yacht" came during the entire summer. In that predicament the colonists had to obtain their materials in New York, and to get there either had to walk or ride on oxen or cows. If they went on foot there were generally several together in a party, and then they carried their purchases on their backs. Now and then Indians were hired to do the lugging. They had no money and paid for their goods either with skins or wampum, which they got from the Indians.

The Swedes whose name is Iockum (Jokum or Yokum) came originally from Germany.—Anders Whiler (Wheeler) who is counted among the Swedes, was an Englishman on his father's side but a Swede on his mother's.—The first Stille to appear in America was Olof Stille.—Bure is believed to be Finnish,[1]—Wallrave is a common name in Christina, but the Swedish ancestral place of the family is unknown.

From the above it may be seen that not all who here pass for Swedes have come from Sweden; but some have come from Germany, Holland, England, and other places, either because one of the parents was Swedish or they were married to Swedes.[2]

Among the living old Swedes here is a man by the name of Garret. His older sister is still living and is over seventy years old. Her husband, Johan Scute, is dead, and was the son of the [Sven] Scute who was vice governor of New Sweden when that belonged to the Swedish crown. There are no Scutes left on the male side.

In Mr. Rudman's observations quoted above I mentioned an impostor by the name of Königsmarck. Mr. Isaac Bengtson told the following tale about him.—This rascal, of Swedish birth, had been transplanted from England to Maryland to serve a number of years as a servant or slave. He was duly sold, but escaped and joined his fellow-countrymen in New Sweden, pretending to be of a distinguished family in Sweden whose name was Königsmarck. He said the Swedish fleet lay outside in the Bay and would at the first opportunity capture the land from the English; that he had been sent to encourage the Swedes who dwelt here to cast off the foreign yoke; and exhorted them to join in slaying the English as soon as they heard the Swedish fleet approach. A large proportion of the Swedish colonists let themselves be persuaded, and concealed the alleged Königsmarck in the Colony for a long time, that no one might learn about his presence. They carried the best food and drink they had to him, so that he lived exceedingly well, and what is more, they went to Philadelphia and bought powder, bullets, lead, etc. to be ready at the first signal. He had the Swedes called together to a supper, and after the drinks had been passed he ex-

---

[1] That is, Swedish-Finnish, in this case. Probably a large proportion of the so-called Finns who came to America at the time were of Swedish origin, though there were some, as we have seen with pure Finnish names, in which case the father at least was of the original Finnish blood.

[2] It should be remembered, too, that up to the eighteenth century, Sweden had possessions in Northern Germany, whose inhabitants, like the Finns, were (politically) Swedish nationals.

horted them to throw off the old rule, reminded them of what they had suffered, and finally asked them whether they sympathized with the King of Sweden or the King of England. A few declared themselves at once for the Swedish ruler, but Peter Kock pointed out that since the land was English and the settlement had been duly ceded to the English crown he ought to support the English sovereign. Thereupon he ran out, slammed the door, and braced himself in front of it so that the alleged Königsmarck could not get away, and called for help to arrest him. The impostor tried to force open the door, and Kock stabbed his hand with a knife; though the swindler got away [temporarily]. But Kock reported the matter to the English, who set out and made the alleged Königsmarck a prisoner. Captain Kock then demanded his real name, for, he said, "We can see that you are not of noble blood." He then admitted that his name was Marcus Jacobson. He was so ignorant that he could neither read nor write. After being branded he was sold in the Barbados as a slave.[1] The Swedes who had sided with him lost half of what they owned—land, cattle, clothes and other goods.

Mr. Bengtson could not remember the Swedish church on Tinicum, for it had fallen into decay before his day, but he recalled clearly the old wooden church at Wicaco which had remained standing many years after the present stone [brick] structure had been erected. It was located immediately south of the present temple, and so close to it that one could barely squeeze into the church between the two. Rev. Fabritius is buried in the choir of the old church, or now right in front of the southern church door.—Iongh [Jongh or Young] who lived during the days of Fabritius is said to have been a schoolmaster in this place.

Mr. Jacob Bengtson related the following story about Governor Printz: when he arrived in his ship the Indians came by the thousands to see him, as soon as they learned that he had come. He then entertained a few of them and had such trifles distributed among them as were considered very precious by them. The shores were full of their boats. They had heard the roar of the cannon and marveled mightily over it, calling them Manito or God. The Governor asked them if they would like to have such a Manito, since they

---

[1] James Kirke Paulding, The American author, wrote a hasty novel of New Sweden. He called it *Koningsmarcke, the Long Finne,* the only striking name he could remember, apparently.

had none before. At first they did not know what to answer, but
finally they approved of it and were happy at the prospect of possessing such a gift. The Governor had the cannon loaded heavily
with balls and brought on land; a long thick cable was attached to
it, and the natives were told to pull it away. All took hold of the
cable and pulled; the cannon was heavy and sank down in the
mud so that they could scarcely jerk it from the spot. The Governor told them they must all walk in a straight line in front of it,
otherwise it would not follow them. They did so, but without
much success. On the shore lay some embers from the many fires
they had first built on the shores; so the Governor suggested to
them to take a firebrand and exhort the Manito to action by burning its tail, when it would move faster. An Indian took a burning
ember, smiled and laughed, and applied it to the fuse, which he
took to be the tail. The cannon went off and killed all the Indians
in the cable-team in front. The others were so terrified at this, believing Manito had become wroth over their action that they dashed
headlong for their boats and began to row with all their might
toward the other side of the river. Governor Printz then had the
cannon loaded with chain shot and fired at the fleeing Indians. This
not only mowed down many more natives but the trees where they
were trying to land. This increased the terror of the Indians, who
felt they must be safe on the opposite shore, when Manito was on
the one behind. They abandoned their boats and anything else
they had, fled to the woods, and did not stop until they had arrived
at the sea near Cherbour (?), which was over six or seven Swedish
miles distant, and dared not return for a long while.[1]

Otherwise Governor Printz exercised rather good control over
the natives, who were afraid of him. One of their chiefs had the
honor of being admitted to his table whenever he visited, and the
Governor favored him in several other respects. Once when this
sachem had been intoxicated among his own people he had ventured to assert that next time he dined with the Governor he would

---

[1] So far as the editor can learn there is no foundation whatsoever for this stupid
tale. Governor Printz did not like or trust the Indians, but did not go out of his
way to provoke them; he was too prudent to do otherwise, and had been instructed
to keep on friendly terms with the natives. This was not always possible, but the
killings on either side were relatively few. The above legend shows what proportions
a story may assume during the course of a century.

cut his throat. The Governor learned about it. A short time later the chief came on his visit to the Governor and expected as usual to be entertained; but the latter kicked him out, and reminding him of the former hospitality told him that thereafter he would be forced to eat with his dogs since he [the Governor] had no more respect for him.

When the Swedes had dwelt in this country for a while the Indians on the New Jersey side conspired to kill all the Swedes on a certain night. In these days the colonists were not so scattered as they are to-day, and had concentrated in just three or four locations, with several living together in each place in a common dwelling place like a blockhouse. The Indians had planned to start the massacre at one end [of the settlement] down by the River, and then to proceed inland during the night up to the other sections. The plan was to slay before the victims could learn of the fates of those already killed. But during the night or the day before [the massacre was to take place] the sachem's squaw or queen appeared and warned the Swedes. They made preparations for defense, and also warned their neighbors so that they might be prepared. When the savages arrived and perceived that the Swedes stood ready to receive them, they realized that they had been betrayed, returned hastily, and sought to ferret out the traitor. After a thorough investigation they learned that it had been their own chief's wife. Thereupon they transfixed a stake lengthwise through her body, and thus let her miserably perish. The next day they came and made a new treaty with the Swedes.—Thus Jacob Bengtson.

Extracts of Pennsylvania Laws. Law concerning liberty of conscience. Whosoever believeth in God the Father, His Son and Holy Ghost, and acknowledges the Holy Scriptures and wishes to live in peace, shall have religious freedom.

The governor alone has the right to purchase land from the Indians. [N. B. One observes through the whole book of statutes] that a large number of laws had formerly been enacted, but that the majority of them had subsequently been repealed.[1]

Swearing is punishable with a fine of five shillings for the first offense, or five days in jail; second offense, six shillings; third, 10

---

[1] This was 186 years ago. It is clear that we have not made much progress in the method of lawmaking since 1750.

shillings or as many days in prison.—The blasphemer has to give 10 £ to the poor and spend three months in the penitentiary at hard labor.

Justices have the right to lay out roads. If anyone fills them in [i. e. blocks them by obstructions of any kind] he is fined 5 £.

Fences must be five feet high.

A ship carrying sick people may not approach within a mile of Philadelphia or any other city. The breaking of this ordinance is punished by a fine of 100 £.

If someone cashes a check on an English bank and it is returned with a protest, he must pay a fine of 20 £ and restore the money.

The produce of the land, such as wheat, rye, corn, barley, oats, pork, meat and tobacco shall in trade be valid as money, unless the contract has definitely stipulated payment in silver.

Fire in a chimney entails a fine of 40 shillings. Every house must have a water barrel and a leather bucket; failure to comply with this provision brings a fine of 10 shillings. Smoking on the street costs 1 shilling and the fine is used for purchasing leather buckets, pumps, etc. for the public welfare.

There is a bounty of three pence per dozen on blackbirds, and threepence apiece on crows. The heads should be shown to a town official.

Anyone selling rum to the Indians, secretly or otherwise, must pay a fine of 10 £.

He who works on Sunday is fined 20 shillings.

An adulterer must suffer twenty-one lashes on his bare back and spend a year in prison at hard labor or pay a fine of 50 £. Besides, the wife who has suffered shall have the right to divorce the guilty from bed and board. If he be guilty a second time, he shall receive the same number of lashes and seven years of hard labor in prison, or pay a fine of 100 £, and a like amount for every subsequent offense. Fornication is punishable with a fine of 10 £ or twenty-one lashes on his bare body at a whipping-post. If a woman gives birth to an illegitimate child and at the birth or in court accuses someone, he shall be father to the child. Anyone proven father shall support the child.

Bigamy is punished by thirty-nine lashes on the bare back, life imprisonment at hard labor, and declaration of invalidity of the second marriage.

Pigs must have a yoke, and rings in their snout. Swine without these may be killed with impunity within fourteen miles of the Delaware or any other navigable river. (Afterwards this law was extended to include all Pennsylvania.) No pigs may run loose in Philadelphia, Chester or Bristol. Offenders against this ordinance lose the pigs thus running wild.

All saloons and taverns must be licensed.

To prevent degeneration of horses small stallions, eighteen months old or more are forbidden to run loose. They must have a height of thirteen hands, counting four inches to a hand. Such a stallion may be gelded by anyone at the risk of the owner, who should besides pay a fine of 10 shillings.

A bounty of fifteen shillings was first paid for wolves, but this was later raised to 20 shillings. For a young wolf cub the bounty at first was 7 shillings sixpence, which was raised later to 10 shillings. An old red fox brought 2 shillings' reward, a young red fox cub, 1 shilling.[1]

---

[1] In addition to the material of the diary notes, which ends here, readers of Swedish will find some supplementary observations on America in Kalm's private correspondence with his teacher Linné and others, and especially in his letters from America, of course. This correspondence, to which we have already referred, has been edited and published by J. M. Hulth in series of *Bref och skirfvelser af och till Carl von Linné* (under the general editorship of Th. M. Fries and J. M. Hulth), Första Afdelningen, Del VIII, Uppsala, 1922, pp. 1-118. A few other references to Kalm's travels in America are found in other volumes of this large collection of Linné letters, the publication of which began in 1907, but all of the essential material of general interest to America is found in his diary as here published. Yet we refer to the above correspondence for some personal, intimate matters.

METEOROLOGICAL OBSERVATIONS

In the first column of these tables, the Reader will find the days of the month; in the second, the time or hour of the day when the observations were made: in the third, the rising and falling of the thermometer; in the fourth, the wind: and in the fifth, the weather in general, such as rainy, fair, cloudy, etc.

The thermometer which I have made use of is that of Mr. Celsius, or the Swedish thermometer so called [the Centigrade], as I have already pointed out in the preface.[1] To distinguish the degrees above freezing point from those below it, I have expressed the freezing point itself by oo, and prefixed o to every degree below it. The numbers therefore which have no o before them, signify the upper degrees. Some examples will make this still more intelligible. On the 17th of December it is remarked, that the thermometer, at eight o'clock in the morning was at 02.5. It was therefore at 2 degrees and 5/10 below the freezing point: but at two in the afternoon it was at 00.0, or exactly upon the freezing point. If it had been 00.3, it would have signified that the thermometer had fallen 3/10 of a degree below the freezing point: but 0.3 would signify that it had risen 3/10 of a degree above the freezing point. Thus likewise 03.0 is three degrees below the freezing point; and 4.0, four degrees above it.

The numbers in the columns of the winds signify as follows: o, is a calm; 1, a gentle breeze; 2, a fresh gale; 3, a strong gale; and 4, a violent storm or hurricane. When, in some of the last tables, the winds are only marked once a day, it signifies that they have not changed that day. Thus, on the 21st of December, stands N.o fair. This shows that the weathercocks have turned to the north all day; but that no wind has been felt, and the sky has been clear all the day long.

Before I went to Canada in summer 1749, I desired Mr. John Bartram to make some meteorological observations in Γennsyl-

---

[1] This preface has not been translated literally in this work but the substance of it has been incorporated in the Introduction.

vania, during my absence, in order to ascertain the summer heat
of that province. For that purpose, I left him a thermometer, and
instructed him in the proper use of it; and he was so kind as to
write down his observations at his farm, about four English miles
to the south of Philadelphia. He is very excusable for not putting
down the hour, the degree of wind, etc., for being employed in
business of greater consequence, that of cultivating his grounds, he
could not allow much time for this. What he has done, is however
sufficient to give an idea of the Pennsylvanian summer.

AUGUST 1748

D.	H.		THER.	WIND	THE WEATHER IN GENERAL
1	5	m	20.0	E S E 2	Fair
	2	a	24.5	E 2	
2	5	m	22.0	E 2	
	2	a	24.5	E 2	
3	5	m	22.0	E 1	
	2	a	25.5	S S W 1	Cloudy with some rain.
4	5	m	22.0	S 1	Alternately fair, cloudy and rainy all day.
	1	a	21.0	S 1	
5	5	m	17.0	S S W 1	Chiefly rainy.
6	7	m	17.0	S 2	Cloudy.
	2	a	19.0	S 2	Somewhat cloudy, but chiefly fair.
7	5	m	15.5	S S W 2	Alternately fair and cloudy.
8	5	m	18.0	S S W 0	Fair all day.
	3	a	19.0	S S W 0	
9	6	m	17.5	W N W 0	
	4	a	21.0	W N W 1	
10	6	m	18.5	E 1	Fair.
	3	a	20.5	E 1	
11	6	m	47.0	E N E 1	Somewhat cloudy.
	12:30	a	18.5	S W 1	Fair.
	4		22.0	S W 1	
	6		22.00	W 3	
12	6	m	16.0	N W 1	Cloudy with some drizzl. rain at ten.
	4	a	19.0	N W 1	Cloudy, fair, some drizzl. rain altern.
13	6	m	17.0	W N W 2	Cloudy with some rain; foggy; sometimes
	2	a	18.5	W N W 2	fair.
14	5	m	18.0	W S W 0	Somewhat cloudy; fair from 11m. to 3a.
	4	a	20.0	W S W 0	Cloudy.
15	5	m	18.0	W S W 0	Cloudy; sometimes fair; at ten o'clock fell
	2	a	19.5	N E 2	a thin fog.

D.	H.	THER.	WIND	THE WEATHER IN GENERAL
16	6	m 18.3	N N E 2	Somewhat cloudy; sometimes fair.
	2	a 18.5		Dark; rainy at night.
17	6	m 18.5	E N E 2	Dark; with some drizzling rain.
	2	a 19.5		Drizzling rain all the afternoon.
18	6	m 19.0	E 2	Drizzling rain all the day.
	2	a 20.5		
19	6	m 19.5		Cloudy.
	2	a 20.0		Scattered clouds.
20	6	m 19.5		Fair.
	2	a 21.5		Scattered clouds: sometimes rain.
21	6	m 20.8	E 1	Somewhat cloudy, fair at nine.
	2	a 21.3		Thin clouds.
22	5	m 21.0		Fair: about twelve it became cloudy.
	1	a 23.5	E S E 1	Cloudy.
23	5	m 22.2		Scattered clouds.
	7		S E 2	
	2	a 24.2		Scattered clouds, dark towards eve.
24	5	m 23.5	W S W 2	Violent rain.
	6		W 2	
	7		W N W 1	About seven it cleared up.
	9		N W 1	
	2	a		Scattered clouds.
25	6	m 24.5	W 1	Scattered clouds.
	10		W N W 3	
	2	a 23.5		
26	6	m 24.0	W 2	Fair. At night a great halo appeared round the sun.
	2	a 24.5	S W 2	Dark. A strong redness at sunset.
			W S W 1	Cloudy. At ten it began to rain, and it rained all day.
27	6	m 24.5	S E 2	
	11		E 3	
	1	a	N E 4	Rain.
	4	21.5	N 1	Scattered clouds.
28	7	m 23.0		
	2	a 23.5	S W 1	
29	6	m	S W 3	Towards evening drizzling rain and light-
	2	a 25.5	N W 2	Scattered clouds; air very cool. [ning.
30	6	m 23.5		
	2	a 21.5	S W 1	Fair: in the morning it began to grow cloudy; at night lightning, hard rain, and some thunder.
31	6	m 22.2		

September 1748

D.	H.	Ther.		Wind	The Weather in General
1	7	m	20.0	N W 2	Scattered clouds.
	2	a	21.5		Clouds passing by. Rain and strong winds all the afternoon.
2	6	m	19.0	N W 1	Scattered clouds all day.
	2	a	20.5	N W 0	At night a great halo round the moon.
3	6	m	21.5	W S W 0	Scattered clouds.
	2	a	23.0	S 1	It became more cloudy. In the evening appeared a great halo round the sun.
4	6	m	23.3	E 1	Scattered clouds.
	12	n	27.5	E S E 1	
	2	a	24.0		
5	6	m	24.5	S E 3	Scattered clouds.
	12	n	26.5		
6	6	m	27.0	S E 2	Scattered clouds.
	1	a	28.5		At night a great halo round the moon, and the sky very red.
7	6	m	27.5	E 3	Dark sometimes. The sun shone through the clouds.
8	12	n	28.5	N E 2	Scattered clouds.
8	6	m	26.0	N N E 2	Scattered clouds all day.
	1	a	26.5		
9	6	m	24.5	N 1	Scattered clouds all day.
	1	a	24.5		
10	5	m	24.0	N N W 1	Fair.
	1	a	24.5		
11	6	m	23.2	W N W 1	Fair.
	2	a	25.0		At night a halo round the moon.
12	6	m	24.0	A Calm	Fair, and very hot.
	12:30	a	26.0		
13	5	m	25.5	S E 1	Fair.
	1	a	26.5		
14	6	m	25.5	S E 1	Fair; but a cool wind all the morning.
	1	a	26.5		
15	5	m	23.0	S E 1	Scattered clouds.
	1	a	27.5		It grew more cloudy. In the evening and ensuing night, violent rain and winds.
16	5	m	21.5	N N E 1	It rained hard all day.
	2	a	21.5		
17	5	m	25.5	N W 1	Cloudy.
	1	a	21.0		Scattered clouds.

D.	H.	THER.	WIND	THE WEATHER IN GENERAL
18	6 m	13.0	Calm	Fair.
19	1 a	24.5	N N E 1	Fair all day.
20	6 m	14.0	N E 1	Scattered clouds.
21	6 m	11.0	N E 0	Scattered clouds.
	1 a	23.0		
22	7 m	10.5	N E 1	Fair.
	1 a	25.0		
23	6 m	11.0	N N E 1	Fair.
	2 a	28.0		
24	6 m	14.0	N E 1	Fair
	2 a	28.0		It grew dark. At night came rain, which continued late.
25	6 m	18.0	N W 1	Dark. At 8, scattered clouds.
	2 a	28.0	N E 1	Scattered clouds.
26	6 m	15.5	N N E 1	Fair.
	2 a	27.5		
27	6 m	17.0	N E 1	Cloudy. Fair at 8, and all the morning.
	2 a	27.0		Cloudy.
28	6 m	14.0	N E 1	Fair and cloudy alternately.
	2 a	20.0		
29	7 m	15.5	N E 1	Cloudy.
	2 a	20.5		Fine, drizzling rain.
30	7 m	16.0	N E 0	Alternately fair and cloudy.

OCTOBER 1748

D.	H.	THER.	WIND	THE WEATHER IN GENERAL
1	6 m	19.0	W 1	Fair. Scattered clouds at 8.
	2 a	18.5		Scattered clouds. Dark towards night.
2	6 m	18.5	S W 0	Cloudy.
3	6 m	15.0	N W 1	Cloudy.
	1 a	18.0		Scattered clouds. Late at night a great halo round the moon.
4	7 m	6.0	N W 1	Fair.
	1 a	16.0		
5	7 m	2.0	N 1	Fair.
6	7 m	2.0	N E 1	Fair.
	1 a	18.0		At night a great halo round the moon.
7	6 m	7.0	E N E 1	Cloudy. Fair at 9, and all day.
8	6 m	14.0	E N E 1	Cloudy. Scattered clouds at 8.
9	6 m	18.0	S S E 1	Rain all the morning.
	3 a	23.0		Cloudy.
10	6 m	20.0	S W 0	Fog, and a drizzling rain.
	2 a	23.0		Fair.

D.	H.	Ther.	Wind	The Weather in General
11	7	m 20.0	S W 1	Fog, which fell down. Fair at 8.
	2	a 26.0		Fair.
12	6	m 8.0	W N W 1	Fair all day.
	8		W 1	
	2	a 20.0	W S W 1	
13	6	m 2.0	W N W 1	In the morning, hoary frost on the plants.
	2	a 17.0	W S W o	Fair all day.
14	6	m 5.0	S S W o	Fair.
	2	a 21.0		
15	6	m 4.5	S S E o	Fair.
	2	a 24.0		
16	6	m 11.0	E N E o	Cloudy.
17	6	m 8.0	N E 1	Cloudy.
	2	a 18.0		Cloudy. Violent rain all night.
18	6	m 12.0	N W o	Cloudy.
	5	a 4.0	S W o	
19	6	m 00.0	W S W 1	Scattered clouds.
	2	a 9.0		
20	5	m 01.0	W N W 1	Fair.
	2	a 9.0		
21	7	m 00.0	W o	In the morning ice on standing water, white hoary frost on the ground; fair all day.
	1	a 15.0		
22	6	m 00.0	W o	Fair.
23	6	m 4.5	N N E 1	Fair.
	1	a 16.0		
24	6	m 4.5	N o	Fair.
	2	a 18.0		
25	6	m 4.5	S W 1	Fair. Air very much condensed in the afternoon.
26	6	m 4.0	S W o	Fair.
	3	a 19.0		
27	6	m 1.0	S W o	Fair.
	3	a 17.0		
28	6	m 9.0	E 2	Heavy rain all day.
29	6	m 14.0	W 1	Fair.
	1	a 20.0		At night I saw a meteor, commonly called the shooting of a star, going far from N. W. to S. E.
30	6	m 3.0	N W 1	Fair.
31	7	m 4.0	W 1	Fair.
	1	a		

November 1748

D.	H.	Ther.	Wind	The Weather in General
1	7	m	3.0	S 1	Fair.
2	6	m	4.0	N 0	Fair.
	3	a	18.0		
3	7	m	7.0	N W 1	Fair.
	1	a	14.0	S E 0	
4	7	m	1.0	S W 0	In the morning the fields were covered with white frost.
	12	n	19.0		A fair day.
5	7	m	4.0	S W 1	Fair.
	1	a	17.0		
6	7	m	4.5	N E 1	Fair.
	1	a	12.0		Towards evening somewhat cloudy.
7	7	m	7.0	E N E 1	Cloudy.
	4	a	11.5		
8	7	m	11.5	E N E 2	Drizzling rain.
	12:30	a	18.0	E S E 3	Heavy rain.
9	7	m	17.0	S E 1	Drizzling rain.
	9		15.0	S S W 1	At eight it cleared up.
	1	a	17.0		Scattered clouds.
10	7	m	6.0	S S W 2	Fair.
	12:30	a	13.0	W N W 2	
11	7	m	4.0	W S W 1	Cloudy.
	12:30	a	12.0		Scattered clouds.
12	6	m	03.0	S W 1	Fair.
	2	a	11.5	N W 2	Cloudy.
	4		5.0		
13	7	m	00.0	N N E 1	This morning ice on the water.
	2	a	5.5		Fair.
14	7	m	0.5	N 3	Fair.
	1	a	8.0	N 2	
15	7	m	3.0	S 2	A strong red aurora.
	1	a	8.0		Cloudy and continual drizzling rain.
16	7	m	4.5	W 1	Fair.
17	7	m	01.0	W 1	Fair and cloudy alternately.
	1	a	8.0		Sometimes drizzling rain.
18	7	m	4.0	S 1	Fair.
	3	a	6.5	N W 2	
19	7	m	03.0	W 0	Fair.
	2	a	11.5		
20	7	m	01.0	N N E 1	Fair.
	2	a		S 1	

D.	H.		THER.	WIND	THE WEATHER IN GENERAL
21	7	m	15.0	S W 2	Fair.
	1	a	19.0		
22	7	m	20.0	E 1	Rain all day.
	2	a	10.0		
23	8	m	16.0	S 1	Cloudy, foggy, and rain now and then.
	8	a		S W 4	
24	7	m	00.0	W N W 3	Fair.
25	7	m		N W 0	It was very cold last night, and fair today.
26				N W 0	Alternately fair and somewhat cloudy, and always pretty cold.
27					Fair; scattered clouds: pretty warm in the air.
28					Cloudy, foggy, and quite calm.
29					Somewhat cloudy.
30				N 1	Fair, and a little cold.

DECEMBER 1748

D.	H.		THER.	WIND	THE WEATHER IN GENERAL
1				N 1	Fair.
2				W S W 1	Fair, and cold; a great halo round the moon at night.
3				W S W 1	A pretty red aurora, however a fair day.
4	7	m	6.0	S S W 0	Fair.
	3	a	18.0		
5	7	m	5.5	N N E 1	
	4	a	9.5		
6	7	m	6.5	S S W 1	Cloudy.
	3	a	14.0		Somewhat fairer: hard rain in the next night.
7	7	m	13.5	S W 1	Cloudy.
	2	a	19.0		Fair.
8	7	m	5.0	S 1	Cloudy.
	2	a	13.5		Rain and wind next night; thick, but
9	7	m	12.0	S W 2	scattered clouds.
	2	a	10.0	W N W 2	
10				W N W 2	Scattered clouds.
11	7	m	2.0	S S W 1	Fair.
	2	a	12.5		
12	7	m	0.5	N E 1	Cloudy, rain, and fog all day from nine o'clock.
	2	a	10.5		

D.	H.		THER.	WIND	THE WEATHER IN GENERAL
13	8	m	7.5	S W o	Foggy, and cloudy.
	2	a	10.0		Next night a strong N. W. wind.
14	8	m	1.0	N W 2	Scattered clouds.
	2	a	2.0		
15	8	m	07.0	W N W 1	Fair and cloudy alternately.
	2	a	01.0		
16	8	m	01.0	W 1	Fair.
	2	a	1.5		
17	8	m	02.5	N W 1	Cloudy, some snow, the first this winter.
	2	a	00.0		
18	8	m	03.0	W 1	Fair.
	2	a	4.0		
19	8	m	1.0	W 1	Cloudy.
	2	a	8.0		Fair.
20	8	m	01.5	W S W 2	Scattered clouds: about six at night were
	2	a	7.5	W S W 1	quite red stripes in the sky, to the north.
21	8	m	07.0	N o	Fair.
	2	a	2.0		
22	8	m	04.5	S E o	Fair.
	2	a	13.0		It grew cloudy in the afternoon.
23	8	m	13.0	S S W o	Heavy rain.
	2	a	18.0		Foggy and cloudy.
24	8	m	13.0	W S W o	Thick fog.
	2	a	17.0	S W 1	Fair; but late in the evening a hard shower of rain.
25	8	m	18.0	S 3	Last night was a storm, rain, thunder, and lightning.
	2	a	18.5	S S E 2	Heavy rain all day.
26	8	m	3.0	W 3	Last night a violent storm from W. and S. and heavy rain. The morning was cloudy, and some snow fell.
	2	a	3.5	W N W 3	Clears up.
27	8	m	04.0	W N W 3	Fair.
28	8	m	07.0	W o	Fair.
	2	a	8.0		
29	8	m	3.0	N N E 1	Somewhat cloudy, and intermittent showers.
	2	a	13.0	— — — o	
30	8	m	8.0	N N E 1	Cloudy and foggy all day.
	2	a	10.0	— — — o	
31	8	m	6.0	W 3	Fair.
	2	a	4.0	N W 1	At night a halo round the moon.

JANUARY 1749

D.	H.	THER.	WIND	THE WEATHER IN GENERAL
1	7:30 m	07.0	N W o	Fair.
	2 a	4.0	— — o	
2	7:30 m	04.5	W N W 1	Alternately fair and cloudy.
	2 a	5.5	— — — 1	
3	7:30 m	2.0	N W 1	Cloudy.
	2 a	2.0	— — 1	
4	7:30 m	02.0	W 1	Fair.
	2 a	11.0	— 1	
5	7:30 m	03.0	W o	Fair.
6	7:30 m	03.0	W o	Fair, but darkened towards night, with
	2 a	14.5	— o	some snow.
	5 a	14.5	N W 3	
7	7:30 m	01.0	W N W 1	Somewhat cloudy.
	2 a	3.0	— — — 1	
8	7:30 m	04.0	W N W 1	Fair.
	2 a	8.0	— — — 1	
9	7:30 m	03.0	W N W 1	Aurora, cloudy, heavy rains at night.
	2 a	8.0	— — — 1	
10	7:30 m	15.0	S 2	Cloudy, and showers, some snow at night;
	2 a	2.0	W 4	at 9 m. W.S.W. 3; at 11 m. S.W. 4; at
	4 a			2 a W. 4.
11	7:30 m	03.0	W N W 3	Cloudy.
	2 a	04.0	— — — 3	
12	7:15 m	04.0	W N W 3	Fair.
	2 a	01.5	N N W 2	
13	7:15 m	07.5	W N W 2	Fair.
	1 a	03.0	— — — 2	Cloudy.
14	7:15 m	05.5	W N W 1	Cloudy, and snowed all day; it lay more
	1 a	02.0	— — — 1	than two inches deep.
15	7 m	07.0	W N W o	Fair.
	2 a	3.0	— — — o	
16	7 m	08.9	N W 3	All last night W N W 4.
	8 m	09.0		Fair all day.
	2 a	08.0	— — 1	
17	7 m	011.0	N N E o	Cloudy, snowed all day, and ensuing
	7 a	09.0	— — — o	night.
18	7 m	012.0	N W 1	Cloudy, and snowed in the morning, fair
	10 m	011.0	— — 1	all the afternoon, and the ther. at 011.0:
				snow lay 5 inches deep.
19	7 m	015.5	W 1	Fair.
	1 m	010.5	— 1	

D.	H.		THER.	WIND	THE WEATHER IN GENERAL
20	7	m	012.5	W 1	Fair.
	2	a	07.0		
21	7	m	022.0	W N W o	Fair.
	2	a	03.0	W 1	
22	7	m	05.0	W 1	Fair.
	2	a	01.0	W 1	Cloudy.
23	7	m	010.0	W N W 1	Fair; a great halo round the moon at
	7	a	3.0		night.
24	7	m	01.0	N N E o	Cloudy, snowed all day.
	2	a	4.0	N E o	
25	7	m	00.0	W N W o	Fair.
	2	a	4.0	W o	
26	7	m	013.0	W N W 1	Fair.
	2	a	1.0	— — — 1	Cloudy; at three in the afternoon it began
					to snow.
27	7	m	07.0	W 1	Fair; halo round the moon at night.
	2	a	00.0	— 1	
28	7	m	01.0	W N W 1	Cloudy; snowed almost all day.
	3	a	4.0	— — — 1	
29	7	m	05.0	N N E 1	Fair.
	3	a	03.0	— — — 1	
30	7	m	013.0	W N W 1	Fair; halo round the moon at night.
	3	a	4.0	— — — 1	
31	7	m	04.0	W N W 1	Fair; halo round the moon at night.
	3	a	8.0	— — — 1	

FEBRUARY 1749

D.	H.		THER.	WIND	THE WEATHER IN GENERAL
1	7	m	03.0	W N W 1	Fair; a halo round the moon at night.
	1	a	11.0	W 1	
2	7	m	5.0	W N W o	Fair.
	2	a	6.0	W o	
3	7	m	00.0	W o	Fair.
	2	a	19.5	— o	
4	7	m	5.5	W o	Cloudy; at ten at night wind N N E 3,
	2	a	11.0	—	snow.
	4	a		N N E 2	
5	7	m	06.0	N N W 2	Fair.
	1	a	03.0	N W 2	

D.	H.		THER.	WIND	THE WEATHER IN GENERAL
6	7	m	010.5	N W 0	A cracking noise was heard in all houses
	2	a	3.0	W S W 1	the night before. Aurora.—Fair all day,
					— at 7 in morn. N.W. 0 — at 9,
					W.N.W. 1 — at 11, W 1 — at 2 in the
					afternoon, W.S.W. 1.
7	7	m	01.0	N N E 1	Cloudy — fair — at 7 in the morn. N. N.
	2	a	1.0	N W 1	E. 1 — at 9, N 1 — at 10, W.N.W. 1 —
					at 12, N W 1.
8	7	m	09.0	N W 0	Fair.
	2	a	7.0	W 1	
9	7	m	03.0	W 1	Fair.
	3	a	16.0	— 1	
10	7	m	7.0	W 1	Pretty clear; a violent storm with rain all
	1	a	11.0	S S W 4	the ensuing night.
11	7	m	9.0	S S W 2	Fair; rain towards night; at night a light
	1	a	11.0		similar to an Aurora Borealis in S. W.
12	7	m	4.0	S S W 3	Fair; about nine at night a faint Aurora
	1	a	10.0		Borealis in S.W.
13	7	m	2.0	W N W 2	Cloudy.
	3	a	5.0	N W 2	Fair.
14	7	m	06.0	N W 1	Fair.
	3	a	02.5	W N W 2	Flying clouds.
15	6:45	m	010.5	N W 1	Fair; at eight in the evening an Aurora
	2	a	03.0	W N W 0	Borealis.
16	6:45	m	013.0	W N W 0	Fair.
	2	a	00.0	N W 1	
17	6:30	m	02.0	W N W 1	Cloudy and snow; wind all the afternoon
	2	a	00.0	W 1	long.
18	6:30	m	2.0	W N W 1	Cloudy.
	2	a	00.0		
19	6:30	m	03.0	N N E 2	Cloudy; rain all day, mixed with snow and
	2	a	01.0		hail.
20	6:30	m	1.5	N W 1	Cloudy.
	2	a	4.5		
21	6:30	m	00.8	N W 0	Cloudy; at 5 in the morning we heard a
	4	a	4.0	N N E 1	waterfall near a mill, about a mile S S of
					us making a stronger noise than com-
					mon, though the air was very calm—at
					10 began a rain which continued the
					whole day.
22	6:30	m	3.0	W N W 2	Fair.
	2	a	3.5		

D.	H.		THER.	WIND	THE WEATHER IN GENERAL
23	6:30	m	06.0	W 2	Fair.
	4	a	4.0		Some clouds gathered round the sun.
24	6:30	m	4.0	S S W 1	Cloudy.
	3	a	10.0	W 1	
25	6	m	3.0	W N W 0	Alternately fair and cloudy.
	2	a			
26	6	m	012.0	N N W 1	Fair; cloudy at night; at eight in the even-
	3	a	02.0		ing was a halo round the moon and the clouds in S. quite red.
27	6	m	04.0	N 2	Cloudy, and snow in the morning; but
	3	a	01.0		fair at 4 in the afternoon.
28	6	m	04.5	N W 4	Flying clouds.
	3	a	03.5	W N W 4	

MARCH 1749

D.	H.		THER.	WIND	THE WEATHER IN GENERAL
1	6	m	09.0	W N W 2	Fair. A great halo round the moon at
	3	a	01.5		night.
2	6	m	06.0	N W 2	Fair. A faint halo round the moon at
	4	a	2.5		night.
3	6	m	04.0	N W 1	Fair. Cloudy afternoon. About 8 at
	2	m	6.5	S 1	night the clouds in S W were quite red. At 9 it began to snow.
4	6	m	0.5	E S E 1	Cloudy. Heavy rain at night.
	2	a	7.0	S 1	
5	6	m	4.0	W 1	Alternately fair and cloudy. The next
	2	a	11.0	W 3	night calm.
6	6	m	4.0	W 2	Fair.
7	6	m	00.0	W S W 1	Alternately fair and cloudy in the morn- ing. In the afternoon cloudy, with intermittent rain and thunder.
8	6	m	2.0	W N W 0	Fair. About 8 at night we saw what is
	3	a	20.0	W S W 2	called a snowfire to the S.W. — See Vol. I, p. 252.
9	6	m	5.0	N 1	Fair.
	3	a	13.5		Cloudy. Snowfire in S.W. about 8 at night.
10	6:30	m	5.0	S S E 1	Cloudy. Snow and rain all day, and next
	2	a	6.5	S E 1	night.
11	6	m	9.0	S S E 1	Cloudy and heavy rain in the morning.
	3	a	14.0	W 1	Cleared up in the afternoon.

D.	H.		THER.	WIND	THE WEATHER IN GENERAL
12	6	m	9.0	N N W o	Cloudy in the morning. Cleared up at 10.
	3	a	15.0	E N E o	Towards night cloudy, with rain.
13	6	m	9.5	N N E 2	Cloudy, with heavy rain. Fair at 4 in the
	2	a	8.0	10 m. N 3	afternoon.
14	6	m	4.0	W N W 2	Fair.
	2	a	10.0		
15	3	m	00.0	W S W o	Fair. Cloudy towards night.
	3	a	13.0	W 2	
16	6	m	2.5	N N E 3	Snow violently blown about all day.
	3	a	01.0		
17	6	m	01.0	N W 2	Cloudy. Cleared up at 8 in the morning.
	3	a	5.0		
18	6	m	02.0	W S W o	Fair. The fields were now covered w'th
	3	a	4.0	W 2	snow.
19	6	m	02.0	W N W 1	Fair.
	3	a	6.0	N W 2	
20	6	m	05.5	W o	Fair. Cloudy towards night.
	3	a	11.5	S W 1	Cloudy.
21	6:30	m	2.0	S S E o	Cloudy. Intermittent showers.
	3	a	14.5		
22	6	m	10.0	S S E o	Cloudy.
	3	a	19.5		
23	6	m	15.0	S S E 1	Heavy rain.
	3	a	19.0		
24	6	m	8.0	S W 1	Fair.
	3	a	15.0		
25	6:30	m	6.5	W N W 3	Fair.
	3	a	11.0		Flying clouds.
26	6	m	00.0	W N W 2	Fair.
	5	a	11.0	S W 2	Flying clouds. About 8 at night a snow-fire on the horizon in S.W.
27	6	m	3.0	W N W 1	Fair.
	3	a	9.0		
28	6:30	m	3.0	S 1	Rain all the day, and the next night.
	3	a	12.0		
	11	a		N N W 3	
29	6	m	1.0	N N W 2	Fair.
	2	a	6.0		
30	6	m	03.0	E 1	Fair. Cloudy at noon: began to snow,
	2	a	4.0	S E 1	which continued till night, when it turned into rain.
31	6:15	m	5.0	N 1	Cloudy.
	3	a	14.0		

April 1749

D.	H.	Ther.		Wind	The Weather in General
1	6	m	5.5	N N E 1	Rain in the morning,—aftern,—and in the night.
	3	a	3.5	E 1	Snow, with much thunder and lightning.
2	6	m	0.5	N N E 1	Snow almost the whole day.
	3	a	0.5		
3	6	m	02.0	N W 1	Fair.
	3	a	9.0		
4	6	m	02.0	W 1	
	3	a	16.0		
5	6	m	00.5	N 1	Fair.
	3	a	19.0	S W 1	Sun very red at setting.
6	6	m	4.0	S W 1	Fair.
	3	a	23.0		
7	6	m	13.0	S 2	Fair. Cloudy afternoon.
	3	a	24.0		About 7 in the evening it began to rain, and continued till late at night.
8	7	m	9.0	N W 3	Flying clouds.
	3	a	13.0		
9	6	m	1.0	N 1	Alternately fair and cloudy. Snowed in the evening, and at night.
	3	a	7.0		
10	7	m	2.5	N E 1	Cloudy. Began to rain at ten, and continued all day till night.
	3	a	6.5		
11	6	m	5.0	N E 1	Rain almost the whole day.
	3	a	9.0		
12	6	m	2.0	W N W 2	Fair. Afternoon cloudy, with hail and rain.
	2	a	13.0		
13	6	m		N W 2	Fair.
	2	a		S W 1	Cloudy.
14	6	m		E 1	Cloudy; fair at eight. Cloudy towards night.
	2	a			
15	6	m		E 1	Almost quite fair.
	2	a			
16	6	m	6.5	W N W 2	Fair.
	2	a	13.5	— — — 1	
17	6	m	7.0	S 1	Alternately fair and cloudy.
	3	a	16.0	S W 1	Rain.
18	7	m	6.0	N 0	Fair.
	3	a	18.0	N W 3	
19	5:30	m	2.0	N N W 0	Fair.
	3	a	20.0	W 2	

D.	H.		THER.	WIND	THE WEATHER IN GENERAL
20	6	m	2.0	S W o	A hoar frost this morning. Fair and very
	3	a			hot all day.
21				S W 1	Fair; with hot vapors raised by the sun.
22	5	m	13.0	S o	Almost fair.
	3	a	23.0		
23	5:30	m	11.0	W 1	Fair.
	3		25.5		
24	6	m	12.0	S 1	Cloudy, intermittent drizzling showers.
	3	a	22.0		
25	6	m	18.0	S o	Rain the preceding night, and now and
	3	a	24.0		then this day. At night thunder and
					lightning.
26	6	m	28.0	W 1	Fair.
	3	a	30.0		
27	6	m	17.0	W 2	Fair.
	3	a	25.0		
28	6	m	7.0	W o	Fair.
	3	a	24.0		
29	6	m	7.0	N 2	Fair.
	3	a	17.0	E 2	
30	5	m	3.0	E 1	Flying clouds.
	3	a	15.5	S 1	

MAY 1749

D.	H.		THER.	WIND	THE WEATHER IN GENERAL
1	4	m	01.5	S o	Hoar frost this morning,—fair.
	3	a	18.5	S W 1	
2	5	m	1.0	W 1	Fair.
	3	a	23.0		
3	5:30	m	4.0	W 1	Fair.
	3	a	27.5		
4	5	m	16.0	W 1	Fair.
5	5	m	13.0	S 3	Flying clouds.
	3	a	27.0		
6	5	m	14.5	N o	Fair.
7	5	m	13.0	N o	Somewhat cloudy.
8	5	m	4.0	N o	Fair.
9	6	m	14.0	S 1	Rain almost the whole day.
	3	a	14.0		
10	6	m	13.0	S S W o	Intermittent showers.
	3	a	16.0		

D.	H.		Ther.	Wind	The Weather in General
11	6	m	12.0	W S W o	Fair.
	3	a	28.0		
12	6	m	13.0	W N W 2	Fair.
	3	a	20.0		
13	5	m	9.0	N W 1	Fair.
	3	a	18.5		
14	5	m	∞.5	N W o	Fair.
15	5	m	9.0	S S W 2	Cloudy.
	3	a	20.0		Rain.
16	5	m	17.0		Cloudy.
	4	a	23.0		
17	5	m	20.0	S 1	Rain intermittently all day; and lightning
	3	a	24.0		very frequent at night.
18	5	m	13.0		Fair.
19	5	m	17.0	W 2	Fair.
20	5	m	19.0	W 1	Fair.
	3	m	24.0		
21	6	m	20.0		Fair.
22				S W 1	Fair. Very hot.
23	5	m	17.0	S W 1	Fair.
	3	a	33.5		
24	12	m	32.0	S W 1	Fair.
25	8	m	23.0	S W 1	Fair, and very warm.
	2	a	28.0		
26	8	m	21.0	W N W 2	Flying clouds; at night thick clouds, with
	3	a	25.0		storm and rain.
27	7	m	17.0	W 2	Thick, scattered clouds.
	2	a	25.0		Pretty cool.
28	7	m	15.0	W 1	Flying clouds.
	2	a	25.0		
29	7	m	16.0	W 2	Flying clouds.
	2	a	25.0		
30	5	m	13.0	W N W 1	Fair.
	—	a	25.0	W 1	Cloudy.
31	5	m	13.0	S W 1	Somewhat cloudy.
	1	a	27.0		Fair.

June 1749

D.	H.		Ther.	Wind	The Weather in General
1	5	m	23.0	S W 1	Rain the preceding night.
2				S E 1	Morning cloudy,—cleared up at ten,— flying clouds.

D.	H.		THER.	WIND	THE WEATHER IN GENERAL
3	7	m	24.0	S W 1	Flying clouds; afternoon, thunder clouds with rain from the N. W.
4	3	a	26.0	N W 1	Flying clouds.
5	5:30	m	15.5	S 1	Fair.
	3	a	22.0		
6	5	m	18.5	S W 1	Alternately fair and cloudy.
	3	a	23.0		
7	All d.		20.0		Cloudy and rainy.
8	6	m	15.5	N W 0	Cloudy.
	3	a	23.0	— — 1	Flying clouds.
9	5	m	13.0		Fair.
10	5	m	11.0	S W 1	Fair.
	3	a	22.5		
11	7	m	20.0	N 1	Flying clouds.
	2	a	33.0	S W 1	Thunder storm, with rain.
12	6	m	23.0	N 0	Fair.
	3	a	32.0	S 2	Somewhat cloudy.
13	5	m	19.0	S E 2	Almost fair.
	3	a	27.0		
14	6	m	26.0	S 1	Fair.
	3	a	25.0		Thunder clouds, with rain.
15	6	m	18.0	N 0	Fair.
	3	a	26.5		
16	6	m	20.0	N N E 1	Fair.
	2	a	28.0		
17	5:30	m	18.0	N 0	Fair.
	3	a	27.5		
18	5	m	21.0	E S E 1	Fair.
	3	a	32.0	N E 1	Thunder, with heavy showers.
19	6	m	20.0	N N W 1	Fair.
	3	a	27.0		
20	5	m	18.0	S 1	Fair.
	3	a	26.0		Cloudy.
21	5	m	23.0	S W 0	Cloudy, with some showers.
22	5	m	9.0	W 1	Fair.
23	6	m	17.0	S 1	Fair.
	—	a		N W 1	Cloudy.
24	6	m	20.5	S 1	Cloudy, afterwards fair.
	—	a		S W 1	Thunder and rain.
25	5	m	23.0	S 1	Fair.
	2	a	32.0		
26	5	m	14.0	N 1	Fair.

D.	H.		THER.	WIND	THE WEATHER IN GENERAL
27	6	m	15.0		Fair.
28	6	m	18.0	S 1	Fair.
	1	a	35.0		
29	7	m	6.0		Fair.
30	5	m	11.0	S 1	Fair.
	3	a	31.0	W 1	

JULY 1749

D.	H.		THER.	WIND	THE WEATHER IN GENERAL
1				N 3	Flying clouds.
2	5	m	7.5	N 2	Fair.
3	8	m	26.0	N 1	Fair.
	2	a	28.0	— 1	Thunder storm, and rain at night.
4	6	m	20.0	S 1	Cloudy; intermittent showers in the
	—	a		N 2	afternoon.
5				W 1	Fair.
	4	a	26.0	— 1	Cloudy; rain at night.
6	5:30	m	18.0	S W 1	Rain all the preceding night; fair in daytime.
7	4:30	m	17.0	N W o	Fair.
8	6	m	16.0	N o	Alternately fair and cloudy. A halo round the sun in the forenoon.
9	7	m	21.0	S W o	Rain the preceding night. In daytime,
	3	a	22.0		cloudy with some showers.
10	4:45	m	18.0	S W o	Fair; sometimes flying clouds and showers.
	3	a	24.5	— — 1	
11	5	m	17.0	S S E 1	Fair.
	2	a	26.0	— — — 1	
12	5	m	22.0	W 1	Fair.
13	6	m	20.0	S S W 1	Fair.
	3	a	33.0	— — — 1	
14	5	m	21.0	W S W 1	Fair.
	2	a	28.0	— — — 1	
15	5	m	26.0	N N E 1	Fair.
	3	a	28.0	— — — 1	
16	5	m	14.0	S o	Fair; sometimes cloudy.
	10	m		S S E 1	
17	5	m	19.0	S 1	Fair.
	3	a	24.0	— 1	Cloudy.
18	5	m	15.0	N N E o	Fair.
	2	a	25.0	— — — o	

D.	H.		THER.	WIND	THE WEATHER IN GENERAL
19	5	m	19.0	S S W 1	Cloudy; rain.
—		a			Pretty fair.
20	5	m	19.0	S 1	Fair.
	3	a	24.0	— 1	Cloudy; some rain.
21				S 0	Fair.
	3	a	27.0	— 0	Flying clouds.
22	5	m	16.0	S W 2	Fair.
	3	a	27.0	S W 2	
23	6	m	19.0	S S W 1	Alternately fair and cloudy.
	3	a	28.5	— — — 1	
24	6	m	20.0	S W 1	Fair.
	3	a	29.0	— — 1	
25	5	m	20.0	W S W 0	Fair.
	3	a	29.5	— — — 0	
26	5	m	21.0	S 0	Fair.
	3	a	30.0	— 1	
27	5	m	22.0	W 1	Cloudy; intermittent showers.
	3	a	21.5	— 1	
28	6	m	17.0	W 1	Fair.
	3	a	27.0	— 1	
29	6	m	16.0	N W ½	Fair; flying clouds at night, and showers.
	2	a	24.0	— — 1	
30	6	m	14.0	W N W 1	Fair.
	2	a	26.0	— — — 1	
31	6	m	16.0	E 1	Cloudy; rain almost all day.
	3	a	22.0	— 1	

AUGUST 1749

D.	H.		THER.	WIND	THE WEATHER IN GENERAL
1	6	m	22.0	N E 1	Cloudy. Some showers.
	3	a	28.0	— — 1	
2	4:30	m	16.0	N E 1	Fair.
		a		S E 1	Cloudy. Fair towards night.
3	5	m	13.0	S W 2	Fair.
4		m		N E 2	Cloudy. Some showers.
	2	a	21.0	— — 2	
5		m		N E 1	Fair.
		a		S W 1	
6	5	m	16.0	N E 3	Heavy rain all day.
	3	a	16.0	— — 3	Some thunder.

D.	H.		THER.	WIND	THE WEATHER IN GENERAL
7	6	m	13.0	E S E 1	Cloudy. Frequent showers.
	3	a	16.0	— — — 1	
8	6	m	16.0	S W 1	Cloudy. Some showers.
	3	a	27.0	— — 1	
9	6	m	14.0	S W 1	Flying clouds.
	1	a	20.0	— — 1	Rain at night.
10	6	m	14.0	S W 1	Flying clouds.
	3	a	24.0	— — 1	
11	6	m	15.5	W 1	Cloudy.
12	6	m	14.0	W 1	Flying clouds.
	2	a	25.0	— 1	
13	7	m	15.5	N W 1	Fair.
	2	a	30.0	— — 1	
14	6	m	16.0	N E 2	Fair.
	2	a	26.0	— — 2	
15	6	m	14.0	N E 1	Fair.
	2	a	28.0	— — 1	
16	5	m	14.0	S E 1	Fair. At night thunder and rain.
	3	a	26.0	— — 1	
17	5	m	14.5	S o	Flying clouds.
	3	a	27.0	— o	
18	5	m	16.0	W 1	Thunder and rain in the morning. At ten
	3	a	29.0	— 1	in the morning flying clouds.
19	6	m	17.0	W 1	Fair.
	3	a	30.0	— 1	
20	5	m	16.5	S W o	
	3	a	28.0	— — o	
21	5	m	17.0	S W 1	Fair.
	2	a	29.0	— — 1	
	5	a	27.0	— — 1	
22	5	m	19.0	N E 2	Rain all day.
	3	a	17.5		
23	5	m	16.5	S W 3	Rain early in the morning. At 10 m.
	2	a	22.5	— — 3	flying clouds.
24	6	m	13.5	S W 2	Flying clouds.
	2	a	22.0	— — 2	
25	5	m	7.0	S W 2	Fair.
	4	a	20.5	— — 2	
26	5	m	13.0	N E 1	Alternately fair and cloudy.
	3	a	18.0	— — 1	Much rain this afternoon.
27	5	m	10.5	S W 1	Flying clouds.
	2	a	23.0	— — 1	

D.	H.		THER.	WIND	THE WEATHER IN GENERAL
28	5	m	10.0	S W 1	Fair.
	2	a	20.0	— — 1	
29	5	m	13.0	N E 2	Fair.
30	5:30	m	11.0	N E 2	Fair.
31	6	m	13.6	S 1	Fair and cloudy alternately.
	3	a	18.5	— 1	Intermittent showers.

SEPTEMBER 1749

D.	H.		THER.	WIND	THE WEATHER IN GENERAL
1	5:30	m	14.5	N N W 1	Fair.
	3	a	30.0	— — — 1	
2	5:30	m	9.0	N 1	Fair.
	2	a	18.0	S S W 1	
3	5:30	m	7.5	S 1	Somewhat cloudy. Fair now and then.
	2	a	20.0	— 1	
4	6	m	14.0	S 1	Now and then a shower; and in the inter-
	2	a	17.5	— 1	vals fair.
5	6	m	14.0	N E 2	Fog. Rain all day. Now and then thunder.
6	10:30	m	15.0	N E 2	Fog, and drizzling rain all day.
	10:30	a	15.0	— — 2	
7	7	m	17.0	S W 1	Fog and rain.
	3	a	22.0	— — 1	Fair.
8	5:30	m	15.0	S S W 1	Fair.
	4	a	28.0	— — — 1	
9	5	m	17.5	E N E 2	Fair.
	3	a	25.0	— — — 2	
10	5:30	m	16.0	N E 2	Fair.
	3	a	26.0	— — 2	
11	5:30	m	15.0	E N E 0	Fair.
	3	a	25.0	— — — 0	
12	7	m	14.5	N N E 1	Fair.
		a		S W 1	
13	5:30	m	14.0	N E 1	Fair.
	1:30	a	24.5	— — 1	
14	5	m	15.0	N E 2	Fair.
	1	a	22.5	— — 2	
15	5:30	m	16.0	N N E 3	Fair. Forenoon, a halo round the sun.
	2	a	19.0	— — — 3	
16	5:30	m	8.5	N N E 1	Fair.
	3	a	20.5	— — — 1	
17	5	m	12.0	S W 0	Fair.

D.	H.		THER.	WIND	THE WEATHER IN GENERAL
18	6	m	17.0	S W 1	Fair.
	3	a	27.0	— — 1	
19	6	m	14.0	S W 1	Fair.
	3	a	26.0	— — 1	
20	6	m	19.0	S W 1	Fair.
	3	a	26.0	— — 1	Cloudy. Rain towards night.
21	6	m	15.0		Fair.
	3	a	19.5		
22	6	m	13.0	E 0	Somewhat cloudy.
	3	a	22.0	— 0	
23	6	m	14.0	S W 0	Fair.
24	6	m	18.0	S W 2	Fair. Rain at noon.
	2	a	26.0	— — 2	Flying clouds in the afternoon.
25	7	m	16.0	W 1	Alternately clear and cloudy.
	2	a	17.0	— 1	
26	8	m	12.5	N E 1	Fair.
	3	a	11.5	— — 1	Cloudy and rainy.
27	6	m	9.3	N 1	Rain all day.
	3	a	14.0	— 1	
28	6	m	8.0	S W 1	Heavy rain all day.
	3	a	14.0	— — 1	
29	6	m	8.0	S 1	Fog.
	1	a	13.0	— 1	Flying clouds.
30	8	m	14.0	S W 2	Drizzling rain.
	2	a	18.0	— — 2	Somewhat clear.

OCTOBER 1749

D.	H.		THER.	WIND	THE WEATHER IN GENERAL
1	7:30	m	9.0	N W 1	Rain.
		n		— — 1	Somewhat fairer.
2	7	m	2.0	W 1	Hoarfrost this morning. Fair all day.
3	6	m	3.5	S W 1	Fair.
	1	a	12.0	— — 1	
4	6	m	11.0	S 1	Rain.
5	6	m	10.5	N E 1	Cloudy.
		a	11.0	— — 1	
6	6:30	m	10.0	E N E 1	Rain all day.
	3	a	12.0	— — — 1	
7	6:30	m	10.0	E N E 1	Flying clouds.
	2	a	14.0		
8	6:30	m	7.0	S 1	Fair.
	3	a	18.0	S 1	

METEOROLOGICAL OBSERVATIONS,

Made by Mr. John Bartram, near Philadelphia,
During my absence, in the Summer of the year 1749.

JUNE 1749

D.	THER. MORN.	THER. AFT.	WIND	THE WEATHER IN GENERAL
1	22	25	W	Cloudy.
2	20	27	W	Cloudy.
3	23	28	W	Showers.
4	22	28	W	Fair.
5	18	25	W	Fair.
6	18	25	W	Cloudy.
7	22	22	N E	Cloudy.
8		21	N E	
9		21	N	
10	14	22	E	
11	22	23	E	
12	25	25	E	
13	23	25	E	
14	25	27	E 3	
15	24	28	E	Fair.
16	22	26	E	
17	23	27	E	
18	25	27	E	
19	23	24	N W	
20	17	26	W	
21	24	26	W	
22	18	27	W	
23	15	29	W	
24	22	30	W	
25	22	31	W	
26	23	30	N	
27	19	32	W	
28	24	36	W	
29	25	37	W	
30	25	36	N	

July 1749

D.	Ther. Morn.	Ther. Aft.	Wind	The Weather in General
1	21	30	W	
2	18	27	N W	
3	26	28	S W	Heavy showers.
4	24	36	N W	
5	22	32	W	
6	22	34	N W	Rain.
7	20	35	W	Hard showers.
8	20	35	N E	Rain.
9	20	29	N	Fair.
10	16	29	N	Fair.
11	17	33	N W	Fair.
12	20	35	W	Fair. Rain at night.
13	22	33	W	Fair.
14	26	30	W	Hard showers.
15	20	29	N	Fair.
16	21	30	E	Rain.
17	29	29	N E	Cloudy.
18	18	19	N E	Rain.
19	18	33	W	Fair.
20	19	33	W	Fair.
21	22	31	W	Fair.
22	23	23	W	Heavy showers.
23	23	25	W	Heavy showers.
24	20	36	W	Fair.
25	27	36	W	
26	28	32	W	
27	24	30	W	Fair.
28	19	27	W	Fair.
29	23	30	W	Rain.
30	30	34		
31	21	34		

August 1749

D.	Ther. Morn.	Ther. Aft.	Wind	The Weather in General
1				
2	18	32		
3	17	30		
4	18	33		
5	22	39	W	
6	18	37	N 2	
7	17	27	W	
8	14	25	N W	
9	12	24	N W	
10	13	24	N W	
11	11	25	N W	
12	14.5	30	N W	
13	18	31	N W	
14	18	30	W	
15	15	30	W	Rain.
16	23	33	N	
17	14	34	N W	
18	18	37	W	
19	18	25	S W	
20	20	26	N E	Rain.
21	20	25	N W	
22	23	34	N W	
23	17	34	W	
24	18	30	W	
25	20	32	N W by W	
26	10	24	N W	Fair.
27	12	20	N W	Fair.
28	13	23	N W	Fair.
29	22	24	W	Fair.
30	17	25	E	
31	20	29	E	

SEPTEMBER 1749

D.	THER. MORN.	THER. AFT.	WIND	THE WEATHER IN GENERAL
1	19	30	E	Hard showers.
2	18	20	E	Rain.
3	19	25	E	Rain.
4	22	25	E	Foggy.
5	23	21	N E	Cloudy.
6	23	37	N E	Cloudy.
7	24	34	N E	Cloudy.
8	24	32	N E	Cloudy.
9	23	33	N E	Rain.
10	23	32	W	Rain.
11	19	25	N E	
12	13	25	N E	
13	12	20	N E	
14	12	33	N E	
15	13	27	N E	
16	20	26	N E	
17	17	27	E	
18	16	34	S E	
19	12	30	S W	
20	17	26		
21	17	25	W	
22	15	30	E	
23	20	29	E	
24	21	29	W	
25	23	28	W 3	
26	20	15	E by N	Thunderstorm.
27	15	19	N W	
28	10	20	N W	
29				
30	6	26	W	

OCTOBER 1749

D.	THER. MORN.	THER. AFT.	WIND	D.	THER. MORN.	THER. AFT.	WIND
1	13	25	W	5	17	30	E
2	14	29	N W	6	18	30	E
3	8	15	N	7	16	21	N W
4	13	29	W	8	11	22	N W

October 1749

D.	H.	THER.	WIND	THE WEATHER IN GENERAL
6	6:30 m	8.0	E N E 1	Cloudy and rain.
	3 a	10.0		
7	6:30 m	8.0	E N E 1	Cloudy, nine o'clock clear; then clear and
	2 a	12.0		scattered clouds.
8	6:30 m	5.0	W 1	Fair.
	3 a	10.0		
9	6:30 m	5.0	S W 1	Fair and scattered clouds.
	3 a	15.0		
10	6:30 m	5.0	S W 2	Fair and scattered clouds.
11	6:30 m	13.5	S 1	Cloudy, occasional rain.
	2 a	16.5		
12	8 m	13.0	S S E 1	Cloudy and rainy; afternoon heavy rain.
13	11 m	4.0	W N W 1	Cloudy and drizzling.
	2 a	7.0		Rain stopped, toward evening clear.
14	6 m	02.5	W	Fair. Ground white with hoar frost.
15	6 m	1.0	W N W 2	Fair, later partly cloudy.
	2 a	11.0		Wind at 10 P. M., N 2.
16	6 m	6.0	S W 2	Fair and scattered clouds.
	2 a	12.0	E N E 2	
17	6 m	6.0	S 1	Cloudy.
	2 a	—	N E 3	
18	6 m	02.0	N E 3	Strong gale during night. Day, fair.
	2 a	4.0		
19	6:30 m	04.0	S W 2	Cloudy during morning. Ice one half inch thick on standing water.
20			N E 2	Cloudy with occasional sunshine. Air cold.
21	6:30 m	1.0	N E 1	Fair, cloudless.
	2 a	8.5	S W 1	Sunset without a cloud.
22	6:30 m	4.5	S W 3	Cloudy, rain occasionally.
	2 a	4.0		Mostly fair.
23	6:30 m	5.5	S W 2	At sunrise clear, otherwise at intervals cloudy with an occasional light shower.
	2 a	11.0		
	3:15 a	14.0		Mostly fair.
24	6 m	4.5	S W 1	Fair.
25	6:30 m	3.0	S W 1	Fair and scattered clouds.
	a			Somewhat cloudy.
26	6 m	05.0	S W 1	Last night N W 2 and cold; Morning, fair. Afternoon partly cloudy.
27	1 a	10.0		During night heavy rain and thunder; rain continued until 9 A. M., then fair.

D.	H.		THER.	WIND	THE WEATHER IN GENERAL
28	6:30	m	06.0	N E 1	Cold at night. Generally fair, though partly cloudy in afternoon, Air chilly. Wind during night, N W 2.
29	1	a	0.5	S S E 1	Snow during preceding night and during most of day.
30	7	m	00.0	N N W 1	Fair.
	2	a	2.0		
31	7	m	09.5	N N E 1	Fair.
	2	a	0.5		

NOVEMBER 1749

D.	H.		THER.	WIND	THE WEATHER IN GENERAL
1	7	m	05.0	N N E 0	Morning, fair; afternoon somewhat cloudy
	2	a	10.0	S 2	cloudy. Change of wind about noon.
2	7	m	6.0	S 1	Cloudy, rain, heavy atmosphere.
	3	a	8.3		
3	7	m	6.0	W N W 1	Cloudy and torrential rain all day.
	2	a	7.0		
4	7	m	5.0	N N W 2	Morning, heavy rain. Afternoon, cloudy, without rain.
5	7	m	3.0	W S W 1	Morning, somewhat cloudy. Afternoon,
	3:30	a	6.5		fair.
6	6:30	m	04.0	S 1	Fair, cloudless. Glorious weather for time
	2:30	a	11.5		of year.
7	7	m	2.5	S 1	Morning, fair until 9, increasing cloudiness. Quick shifting of wind as follows: At 12 noon, N 1; 6 P. M. N W 2; 10 P. M. N 4; 12 midnight, N E 4.
	2	a	11.0		
8					
9	8	m	02.0	S W 1	Cloudy.
	2	a	1.5	N 2	
10	7	m	0.5	N E	Fair. Glorious autumn weather.
	2	a	8.0		
11	7	m	4.0	W S W 1	Morning fair but atmosphere seemed
	2	a	6.0		heavy.
12	7	m	02.0	N W 1	Fair.
	3	a	5.0		
13	7	m	03.5	N W 1	Fair.
	4	a	5.0		
14	7	m	02.0	N E	Fair. In afternoon, patches of filmy
	3	a	6.0	S 1	clouds.

D.	H.		THER.	WIND	THE WEATHER IN GENERAL
15	10	m	4.5	N E	Heavy mist until about 11 A.M. Fair for
		a		S 1	remainder of day. Evening, cloudy with a little rain.
16	8	m	7.5	S 1	Cloudy.
	2	a	8.5		
17	7	m	5.0	S W 1	Cloudy.
18	9:30	m	1.0	N W 2	Fair.
		a	6.0		
19	7:30	m	3.5	S 3	Fair until 10 A. M. Increasingly cloudiness until sunset, when it cleared.
20	8	m	4.3	S 2	Cloudy. Sun shone all day, but most of the time through clouds, like a red disk.
	3:15	a	11.5		
21				N N E 2	Cloudy, but sun shone through clouds all day like a red disk.
22	7	m	4.5	S W o	Cloudy, with some sunshine.
23			14.0	S W 1	Thin clouds, sultry weather. In the evening, rain. Temperature given is estimated only.
24	12		n8–10.0	W 1	Fair.
25	5	a	5.0	W 2	Severe frost during preceding night; Day, fair; beautiful for this time of year.
26	7	m	03.0	N W 1	Early morning, cloudy. After 9 A. M.,
	2	a	5.0		Fair.
27	7:30	m	05.0	S S W o	During the night, rain; day, cloudy.
	2	a	5.0		
28	7:30	m	00.0		
	2	a	2.5	N W 1	Fair, with scattered clouds in the morning; small flurries of snow.
29	7	m	03.0	N W 1	Cloudy with flurries of snow, especially toward evening.
	3	a	3.5		
30	8	m	03.0	W 1	Fair, later cloudy.
	3	a	1.0		

DECEMBER 1749

D.	H.		THER.	WIND	THE WEATHER IN GENERAL
1	7	m	1.5	W S S 2	Early morning, cloudy. After 10 A. M.
	2	a	4.0		Fair. Wind in the evening, N W 1.
2	7:15	m	04.0	W S W 1	Fair in the morning; in the afternoon,
	3	a	3.0	N W 2	cloudy and snow, occasionally becoming a storm. Late evening, clear.

D.	H.		Ther.	Wind	The Weather in General
3	7:15	m	010.0	N W 2	Fair.
	9	m	010.5		
	3	a	9.0		
4	7	m	013.0	S S W 2	Early morning, fair; 9 A.M. cloudy, with
	3	a	6.0	N W 1	sun appearing at intervals.
5	8	m	5.0	S W 2	
	3	a	3.0	W N W 3	Fair. At 10 A.M. wind shifted to W 2.
6	8	m	08.0	W 1	Fair. Glowing sunset; later, cloudy.
	2	a	5.0		
7	7:30	m	04.0	S W 0	Cloudy, quite foggy, thick atmosphere.
	2	a	2.0		
8	7	m	03.5	S W 0	Cloudy, with heavy, misty atmosphere.
	2	a	4.0		Rain at 4 P. M. continuing into night.
9	7	m	02.0	S W 1	Early morning, fair. 10 A.M., cloudy;
	2	a	3.0	N W 1	Early afternoon, fair. At 4 P. M. a cloud from N W with snowstorm. Clear again at 6 P.M.
10	7	m	010.0	N W 1	Fair.
	3	a	04.0		
11	7:30	m	011.0	N W 0	Fair, later cloudy; at 8 P. M., snow.
	3	a	3.0	E 0	At 1 P. M. the wind was S W 0.
12	7	m	1.0	S W	Cloudy.
13	7	m	1.0	W 2	Early morning, cloudy; at 9 A.M., fair.
14	8	m	02.0	W 1	Fair; in evening, cloudy.
	2	a	6.0		
15	8	m	5.0	E 2	Rain all day and far into night. Thunder
	2	a	4.0		and lightning late at night.
16	8	m	00.0	W N W 2	Cloudy.
	2	a	5.0		
17	8	m	5.0	W N W 2	Early morning, fair; after 9 A.M. scat-
		a	4–5.0		tered clouds with a little hail.
18	7	m	3.25	N W 2	Fair.
	2:30	a	1.5		
19	7	m	05.0	N W 2	Fair.
	2	a	01.0		
20	7	m	06.0	W N W 1	Fair.
21	7:15	m	01.25	N E	Early morning, rain; showers all day.
	2	a	00.0		
22	7:15	m	00.0	E	Cloudy at first, at 10 A. M. fair.
	2	a	4.0		
23	7:30	m	04.0	S W 1	Fair.

D.	H.		THER.	WIND	THE WEATHER IN GENERAL
24	7	m	06.5	S W 1	Fair; at 4 P.M., cloudy. 7 P.M., fair.
		a		W N W 1	
25	7:15	m	08.0	S W	Fair.
	2	a	01.5	W N W	
26	7:30	m	08.0	E o	Cloudy in morning. Afternoon, snow.
27	7:15	m	03.5	W N W	Fair, scattered clouds.
	2	a	01.5	S W o	
28	7:30	m	08.0	N W 3	Fair, with large scattered clouds.
	3:30	a	04.5		Afternoon, fair.
29	7:15	m	012.0	W N W	Fair, moderating in evening.
	2		02.0	W S W 1	
30	7	m	04.0	S W 1	Cloudy.
	2	a	0.5		
31	7:30	m	09.0	N W	Scattered clouds; at 5 P. M. snow flurries.
	2	a	05.5	W 1	At 7 P.M., fair.
	10:30	a	011.0		

JANUARY 1750

D.	H.		THER.	WIND	THE WEATHER IN GENERAL
1	7	m	014.5	N W	Fair.
	2	a	010.0		
2	7:15	m	017.0	N W 1	Fair.
	2	a	06.0		
3	7	m	08.0	N W o	Cloudy. At 7 A.M. rain, turning into ice,
	2	a	03.0		making the ground slippery.
4	7	m	01.0	W N W 1	Cloudy in early morning; at 8 A.M., fair;
	2	a	2.0		at 12 noon, cloudy; at 4 P.M., fair.
5	7	m	04.0	W N W	Cloudy with flurries of snow.
	2	a	1.0		
6	7	m	06.5	N W 1	Fair.
	3	a	02.0		
7	7	m	010.5	W N W 1	Fair with scattered snow clouds.
	2	a	04.0		
8	7	m	011.5	N N W	Fair.
	2	a	02.0		
9	7	m	05.0	W N W 1	Fair, bright sunshine.
	2	a	2.0		
10	7	m	08.0	N W 1	Beautiful sunshine, with scattered clouds
	3	a	5.0		in the evening. Wind at 10 A.M., S W.
11	7	m	05.5	S W	Fair.
	2	a	7.0		

A BIBLIOGRAPHY OF

PETER KALM'S WRITINGS ON AMERICA

1. *En Resa til Norra America.* På Kong. Swenska Vetenskaps-academiens befallning, och Publici kostnad, I-III, Stockholm, 1753-1761.

   The second and third volumes deal with North America. Now rare.

2. *Pehr Kalms Resa till Norra Amerika,* utgiven av Fredr. Elfving och Georg Schauman. Skrifter utgivna av Svenska Litteratursällskapet i Finland. Första delen, 1904; andra delen, 1910; tredje delen, 1915; fjärde delen (Tilläggsband) sammanställt av Fredr. Elfving, 1929.

   A reprint published in Helsingfors. It has valuable prefaces in vols. I and IV, and the last volume contains the previously unpublished diary notes by Kalm, which were found by Georg Schauman in the university library at Helsingfors. The Preface to vol. I contains also a record of the more important reviews of Kalm's *Resa,* both in Sweden and abroad. I refer to this preface.

Translations of Kalm's Resa

3. *Reise nach der nordlichen America,* welche auf Befehl der Königlichen Schwedischen Akademie der Wissenschaften und auf allgemeine Kosten, von Peter Kalm, . . . ist verrichtet worden. Leipzig, G. Kiesewetter, 1754–1764. 3 Th.[1]

---

[1] Part I of the Leipzig edition was translated by Carl Ernst Klein, a Pomeranian Legation-preacher, who had settled in Stockholm; Parts I–II of the Göttingen version by J. P. Murray; and Part III by his brother J. A. Murray. Parts II and III are identical in the two German versions. The two Murrays were Swedish-Germans of Scotch ancestry—Johan Philip Murray (1726–1776), professor of philosophy at Göttingen, and the more famous Johan Andreas Murray (1740–1792), botanist, Linné pupil, professor of medicine and author of *Apparatus medicaminum,* who in *Biografiskt Lexicon* is credited with the German translation of Part III of Kalm's work. J. Andreas Murray was born in Stockholm, whither his father had moved in 1736 and become pastor of the German church. It should be noted, incidentally, that the German versions, the only ones to be made directly from the original, include that portion of the work relating to England, which is omitted in the English, Dutch, and French translations. (See next item).

PETER KALM'S TRAVELS   771

4. Des Herren Peter Kalm . . . mitgliedes der Königlichen Schwedischen Akademie der wissenschaften, *beschreibung der reise* die er
nach dem Nördlichen Amerika auf den befehl gedachter akademie und öffentliche kosten unternommen hat . . . Eine ueber-
setzung, Göttingen, Wittwe A. Vanderhoek, 1754-1764. 3 Th.

Appeared in the series of *Sammlung neuer und merkwürdiger
Reisen zu Wasser und zu Lande.* ix–xiter Theil. Both German
editions are quite rare, in America at least.

5. *Travels into North America;* containing its natural history, and a
circumstantial Account of its Plantations and Agriculture in
general, etc. By Peter Kalm, Professor of Economy in the University of Aobo in Swedish Finland, etc. Translated into English
by John Reinhold Forster, F.A.S. [With map, cuts "for the
illustration of Natural History," not found in the original, and
notes.] Vols. i–iii. Warrington, 1770–1771. Part dealing with
Norway and England omitted.

It is obvious from a statement in the preface, xv, that the translation
is made from the German version.[2] But it was a decided success
and a second edition was issued almost immediately.

6. Same, second edition, abridged, London, T. Downdes, 1772. 2 vols.

---

[2] This English translation was the result of temporary financial distress on the
part of the translator, who was invited to London from Germany for another plan
that did not materialize, found himself stranded there, and so took up the Englishing
of certain "Reisebeschreibungen" to get a living. So far as Kalm's work goes, most
of the translating was in reality done by Forster's talented sixteen-year old son,
Johann George Adam Forster (1754–1794). We may assume, however, that his
father superintended the job and supplied the prefaces and notes. See the *Allgemeine
Deutsche Biographie*, vii, 168, 173.
A preface by Forster to the third volume accuses Kalm of prejudice against the
English in favor of the French. Maybe there is a little extra kindliness toward the
French of Canada, who were exceedingly polite to Kalm during his visit there, but
if so, it is an unconscious favoritism and in no way serious. Kalm is in the main
a just, objective observer, scientific, and sometimes honest to the point of a childish
naiveté. He constantly cites his authorities for phenomenon, real or alleged, that he
has not observed with his own eyes, and leaves the final judgment to his reader. The
usual criticism of Kalm is that he is hardly at all *subjective* and certainly not at all
literary or sensational. Style, for example, has but little meaning to this Swede, but
fact everything, and to record this in a simple, practical form is his great objective.
In comparing English colonial women with those of French Canada Kalm gives the
palm of glory to the Canadians, but in the budding troubles between England and her
American colonies, for instance,—troubles which Kalm was one of the first to observe—Kalm's sympathy is with England. He had no prejudice against Englishmen.
On the other hand, Forster, either because of conviction or circumstances, was or had
to be favorable to the English, of course.

7. Same, reprinted in *A General Collection of the best and most interesting voyages and travels*, edited by J. Pinkerton, 1808–1814, vol. 13, 4to. London, 1812.[3]

8. *Kalm's Account of his Visit to England on his way to America in 1748*. Translated by Joseph Lucas. With two maps and several illustrations. London and New York, 1892.

   This is a separate translation of the part of Kalm's *Resa* which deals with England only, and before omitted in any foreign version. It contains a life of the author, a translator's preface and a facsimile of the title-page of the original opposite its own title-page. It is a careful work, octavo, of 458 pages plus an index. Many words and sectional titles are given in both Swedish and English, viz., "Snake-Oil, Orm-olja." The character â is used instead of å.

9. *Reis door Noord Amerika*, gedaen door den Heer Pieter Kalm . . . Vercierd met kopern platen . . . Te Utrecht, 1772. 1–2.

   This Dutch translation is based on Forster's English and Murray's German version. It is a handsome work in quarto, and, as noted before, the part relating to England is omitted. The name of the translator is unknown.

10. Jacques Philibert Rousselot de Surgy, *Histoire naturelle et politique de la Pensylvanie et de l'établissement des Quakers dans cette contrée*. Tr. de l'allemand. P.M. d.s. censeur royal. Précédée d'une carte géographique. Paris, Garneau, 1768.

    The "compiler's chief sources were a German translation of P. Kalm's Resa till Norra America" and Gottlieb Mittelberger's *Reise nach Pensylvanien im Jahr 1750*.

11. *Voyage de Kalm en Amérique*, analysé et traduit par L. W. Marchand. Montréal, par T. Berthiaume, 1880. Two vols. Published in *Mémoires de la societé historique de Montréal*. Provided with notes and index.

    The part dealing with the United States is condensed into an "analyse" of 151 pages with occasional citations of literal trans-

---

[3] In addition, according to the British Museum Catalogue, vol. 2 of John Hamilton Moore's *A new and complete Collection of Voyages and Travels*, London, 1778 (2nd ed. 1785?) contains some material based on Pehr Kalm.

lations. The emphasis here is, of course, on Canada, and the portion treating of it is carefully translated.[4]

Kalm did pioneer work in describing Niagara Falls, a task which was intended to conclude a finished fourth part of his *Resa,* but which never appeared in the original edition. A contemporaneous letter to the librarian Gjörwell on Niagara was recently published, however, in Elfving's abovementioned *Tilläggsband* in 1929, pp. 162–180. But Kalm did more than that. On September 2, 1750, he addressed a letter on Niagara Falls "to his friend in Philadelphia," Benjamin Franklin. It was composed in English and was the earliest account of Niagara in that language.[5] It was printed as follows:

12. (a) In No. 1136 of the *Pennsylvania Gazette* for September 20, 1750.
(See Kalm's own testimony in Elfving's *Tilläggsband,* p. 157.)
I have not seen this first printing of the article. Dow does not mention it, but Professor A. J. Uppvall of the University of Pennsylvania has verified this printing.
(b) A letter from Mr. Kalm, a gentleman of Sweden, now on his travels in America, to his friend in Philadelphia, a particular account of the Great Fall of Niagara. *Gentleman's Magazine,* Jan., 1751, 21:15–19.
An engraved picture based in all probability on Kalm's description appeared the following month in the same magazine. It was the first view after Hennepin's (1697) to be founded on actual sight of the Falls. Various printed and pictorial reproductions of the Falls soon appeared which were based on Kalm's published letter.
(c) Same, reprinted in John Bartram's *Observations,* etc., London, 1751, where it is termed "a curious Account of the Cataract of Niagara." Bartram's *Observations* with Kalm's description of Niagara was reprinted at Rochester, New York, in 1895. Kalm's letter is found on pages 79–94.
This description is much less scientific and detailed than the one given in his letter to Gjörwell.
(d) Same, in Dodley's *Annual Register,* 4th ed., London, 1765, 2:388–394.

---

[4] In 1900, in the city of Lévis, Canada, near Quebec, there appeared a pamphlet, *Voyage de Kalm au Canada* by J. Edmond Roy, a member of the Royal Society of Canada. Though the brochure contains some new material it is of course based chiefly on Kalm's *Travels.*

[5] Charles Mason Dow, *Anthology and Bibliography of Niagara Falls,* 1–11, Albany, 1921. 1, 62–63. Dow refers to Kalm as an "eminent Swedish Botanist" and reproduces his account of Niagara, pp. 53–63 of vol. 1.

(e) The Falls of Niagara, 1764. From a newspaper of the day. In *Mass. Mag.*, 1790, 2:592. This is Kalm's account almost word for word. . . . It "reads like a careful revision of the earlier description." [6]

(f) Same, in Dow's *Anthology*. See note 5.

The *Enciclopedia Universal Ilustrada* [7] states in its article on "Pedro Kalm" that his letter to Franklin on Niagara Falls "fue traducida á muchos idiomas," and the *Biografiskt Lexicon*, its probable source, assures us that six editions of it appeared in one year in England and America, and that it was translated into both German and French. I have been unable to verify these statements, but assume them to be true. A condensation by an anonymous writer of the account of Niagara Falls appeared in *Uppfostringssälsk. Tidning*, 1782, nos. 45 and 46.

13. The passenger Pigeon . . . Accounts by Pehr Kalm (1759) [8] and John James Audubon (1831). Smithsonian Institution, *Annual Report* for 1911. Washington, 1912, 407–424.

Between the years 1749 and 1778 Kalm contributed seventeen articles on American subjects to Kong. Vetenskaps Academiens *Handlingar*, two of them running through two or three continuations. They deal with the climate, trees, insects, animals, plants, and other agricultural topics. We shall reproduce the titles of the whole list.

14. 1749. Anmerkningar om historia naturalis och Climatet af Pensylvanien. Pp. 70–79.—An abstract from a letter of October 14, 1748.

15. 1750. Lobelia, såsom et specificum mot Lues Venerea, 280–290.

Five species of this herb, discovered by Kalm, described as a cure for venereal diseases. Kalm possibly learned this from the Indians, who made a medicine from the lobelia plant. There are hundreds of species of it. See Webster's *Unabridged Dictionary*.

16. 1751. Huru socker göres i America af lönnens saft, 143–159. A treatise on maple sugar.

17. 1751. Huru dricka göres i America af gran-ris, 190–196. An account of how to prepare spruce beer.

---

[6] Dow, *Ibid.*, 1, 63.
[7] It is interesting to note in this connection that neither the *Encyclopedia Britannica* (14th ed.) nor *Der Grosse Brockhaus* devotes any special article to Pehr Kalm.
[8] Obviously a translation of Kalm's Swedish article on "Vilda Dufvor i Norra America" which has been printed in Svenska Vetenskaps Academiens *Handlingar* for 1759. See below, item no. 23.

18. 1751. (a) Om Amerikanska Maysen,[9] 305–319.

    1752. (b) Om maisens [9] skötsel och nytta, 24–43.

19. 1752. Några Nord-sken observerade i America, 145–155.

20. 1752. (a) Om Skaller-Ormen, 308–319.[10]

    1753. (b) Fortsättning af berättelsen om Skaller-Ormen, 52–67.[10]

    Kalm was immensely interested in rattlesnakes.

    1753. (c) Om Botemedlet emot Skaller-Ormens bett, 185–194.
    Kalm gives a bibliography on the subject.

21. 1754. Om de Amerikanske skogs-lössen [ticks], 19–31.

22. 1756. Beskrifning om et slags gräshopper i Norra America, 100–116.

23. 1759. Beskrifning på de vilda Dufvor i Norra America, 275-295.

    As always, Kalm furnishes a bibliography. See note 8.

24. 1764. Om maskar, som fördärfva skogarna i America, 124–139.

25. 1767. Om Norr-Americanska Svarta Valnöts-Trädets egenskaper, nytta och Plantering, 51–64.

26. 1769. Om Norr-Americanska hvita Valnöts-Trädets egenskaper och nytta, 119–127.

27. 1771. Thermometrika Rön på Hafs och Sjöars vattens värma, 52–59.

    Based largely on observations made in America.

28. 1773. Om Tuppsporre-Hagtorns nytta till lefvande Häckar, 343–349.

    On cockspur-hawthorn.

29. 1776. Beskrifning på Norr-Americanska Mulbärs-trädet, 143–163.

30. 1778. Om Americanska Valnöts-Trädet Hiccory, 262–283.

    Based in part on information received from Benjamin Franklin.

31. In 1751 there appeared, also, in Stockholm a separate anonymous pamphlet by Kalm with the title "Berättelse om naturliga stället, nyttan samt skötseln af några växter ifrån N. America."

---

[9] Note the slight difference in spelling.
[10] "Vide, Medical, & cases and experiments, translated from the Swedish, London, 1758, p. 282."—A propos of rattlesnakes, J. R. Forster in his English translation of Kalm's *Resa* gives this reference, 1, 116. Evidently Kalm's articles on the subject were already mentioned, and perhaps translated, in this London publication of 1758. I have not seen the book.

Besides the writings above, Kalm as frequent praeses at the University of Åbo was naturally the chief source of information or inspiration, or both, for a large number (146) of theses by his student respondents. Among these there are at least six items of *Americana* for which we can unhesitatingly give the presiding officer main credit of authorship. The six theses, none of them very long, have in the Yale library been assigned to the Rare Book Room. They are:

32. Anders Chydenius, Americanska näfwerbåtar. . . Åbo, J. Merckell. 1753.
A master's thesis on Indian birch canoes.

33. Daniel Backman, Med Guds wälsignande nåd och wederbörandes tilstånd yttrade tankar om nyttan, som kunnat tillfalla wårt kjära fädernesland, af des nybygge i America, fordom Nya Swerige kalladt. Åbo, 1754.

34. Andreas Abraham Indrenius, Specimen Academicum de Esquimaux, gente americana, quod in regio Fennorum lycaeo . . . Åbo, 1756.

35. Georg A. Westman, Itinera priscorum Scandianorum in Americam. Dissertatione graduli.

Publicly read in Åbo, 1747, but not printed there, it appears, until 1757.

36. Sven Gowinius, Enfaldiga tankar om nyttan som England kan hafva af sina nybyggen i Norra America. Åbo, 1763.

An important pamphlet of 22 pages.

37. Esaias Hollberg, Norra americanska färge-örter. Åbo, 1763.[11]

---

[11] This bibliography was first published in *Scandinavian Studies and Notes*, May, 1933, pp. 89–98. It is here reproduced with the permission of the editor of that journal, Professor A. M. Sturtevant.

INDEX

This comprehensive index covers both volumes of the work. Volume I contains pages 1 through 401 and Volume II contains 402 through 776.

lina, see *Picus Carolinus*, 254; crested, see *Picus pileatus*, 79, 254; gold winged, see *Cuculus auratus*, 254; king of the, see *Picus principalis*, 254; red headed, see *Picus erythrocephalus*, *pileatus*, 254; spotted, hairy, see *Picus villosus*, 254; least spotted, see *Picus pubescens*, 255; yellow bellied, see *Picus varius*, 255

Worms, grass-worms, 213

Wormseed, see *Chenopodium anthelminticum*, 90

Yams, see *Dioscorea alala*, 320

Yellow bedstraw, see *Galium luteum*, 380

Yellow hammer, see *Emberiza citrinella*, 12, 70

Yew (berry-bearing), see *Taxus baccata*, 461

Yokes and hobbles, 115

Yokum (Iockum, Jokum), family, 731

Zea Mays, Indian corn, 74

Zizania aquatica, wild rice, Indian rice, 250, 389, 401, 533, 642

Zinzendorf, Count 32; catechism of, 687 ff.

A CATALOG OF SELECTED
DOVER BOOKS
IN ALL FIELDS OF INTEREST

A CATALOG OF SELECTED DOVER
BOOKS IN ALL FIELDS OF INTEREST

DRAWINGS OF REMBRANDT, edited by Seymour Slive. Updated Lippmann, Hofstede de Groot edition, with definitive scholarly apparatus. All portraits, biblical sketches, landscapes, nudes. Oriental figures, classical studies, together with selection of work by followers. 550 illustrations. Total of 630pp. 9⅛ × 12¼.
21485-0, 21486-9 Pa., Two-vol. set $25.00

GHOST AND HORROR STORIES OF AMBROSE BIERCE, Ambrose Bierce. 24 tales vividly imagined, strangely prophetic, and decades ahead of their time in technical skill: "The Damned Thing," "An Inhabitant of Carcosa," "The Eyes of the Panther," "Moxon's Master," and 20 more. 199pp. 5⅜ × 8½. 20767-6 Pa. $3.95

ETHICAL WRITINGS OF MAIMONIDES, Maimonides. Most significant ethical works of great medieval sage, newly translated for utmost precision, readability. Laws Concerning Character Traits, Eight Chapters, more. 192pp. 5⅜ × 8½.
24522-5 Pa. $4.50

THE EXPLORATION OF THE COLORADO RIVER AND ITS CANYONS, J. W. Powell. Full text of Powell's 1,000-mile expedition down the fabled Colorado in 1869. Superb account of terrain, geology, vegetation, Indians, famine, mutiny, treacherous rapids, mighty canyons, during exploration of last unknown part of continental U.S. 400pp. 5⅜ × 8½. 20094-9 Pa. $6.95

HISTORY OF PHILOSOPHY, Julián Marías. Clearest one-volume history on the market. Every major philosopher and dozens of others, to Existentialism and later. 505pp. 5⅜ × 8½. 21739-6 Pa. $8.50

ALL ABOUT LIGHTNING, Martin A. Uman. Highly readable non-technical survey of nature and causes of lightning, thunderstorms, ball lightning, St. Elmo's Fire, much more. Illustrated. 192pp. 5⅜ × 8½. 25237-X Pa. $5.95

SAILING ALONE AROUND THE WORLD, Captain Joshua Slocum. First man to sail around the world, alone, in small boat. One of great feats of seamanship told in delightful manner. 67 illustrations. 294pp. 5⅜ × 8½. 20326-3 Pa. $4.50

LETTERS AND NOTES ON THE MANNERS, CUSTOMS AND CONDITIONS OF THE NORTH AMERICAN INDIANS, George Catlin. Classic account of life among Plains Indians: ceremonies, hunt, warfare, etc. 312 plates. 572pp. of text. 6⅛ × 9¼. 22118-0, 22119-9 Pa. Two-vol. set $15.90

ALASKA: The Harriman Expedition, 1899, John Burroughs, John Muir, et al. Informative, engrossing accounts of two-month, 9,000-mile expedition. Native peoples, wildlife, forests, geography, salmon industry, glaciers, more. Profusely illustrated. 240 black-and-white line drawings. 124 black-and-white photographs. 3 maps. Index. 576pp. 5⅜ × 8½. 25109-8 Pa. $11.95

CATALOG OF DOVER BOOKS

THE BOOK OF BEASTS: Being a Translation from a Latin Bestiary of the Twelfth Century, T. H. White. Wonderful catalog real and fanciful beasts: manticore, griffin, phoenix, amphivius, jaculus, many more. White's witty erudite commentary on scientific, historical aspects. Fascinating glimpse of medieval mind. Illustrated. 296pp. 5⅜ × 8¼. (Available in U.S. only)   24609-4 Pa. $5.95

FRANK LLOYD WRIGHT: ARCHITECTURE AND NATURE With 160 Illustrations, Donald Hoffmann. Profusely illustrated study of influence of nature—especially prairie—on Wright's designs for Fallingwater, Robie House, Guggenheim Museum, other masterpieces. 96pp. 9¼ × 10¾.   25098-9 Pa. $7.95

FRANK LLOYD WRIGHT'S FALLINGWATER, Donald Hoffmann. Wright's famous waterfall house: planning and construction of organic idea. History of site, owners, Wright's personal involvement. Photographs of various stages of building. Preface by Edgar Kaufmann, Jr. 100 illustrations. 112pp. 9¼ × 10.
23671-4 Pa. $7.95

YEARS WITH FRANK LLOYD WRIGHT: Apprentice to Genius, Edgar Tafel. Insightful memoir by a former apprentice presents a revealing portrait of Wright the man, the inspired teacher, the greatest American architect. 372 black-and-white illustrations. Preface. Index. vi + 228pp. 8¼ × 11.   24801-1 Pa. $9.95

THE STORY OF KING ARTHUR AND HIS KNIGHTS, Howard Pyle. Enchanting version of King Arthur fable has delighted generations with imaginative narratives of exciting adventures and unforgettable illustrations by the author. 41 illustrations. xviii + 313pp. 6⅛ × 9¼.   21445-1 Pa. $5.95

THE GODS OF THE EGYPTIANS, E. A. Wallis Budge. Thorough coverage of numerous gods of ancient Egypt by foremost Egyptologist. Information on evolution of cults, rites and gods; the cult of Osiris; the Book of the Dead and its rites; the sacred animals and birds; Heaven and Hell; and more. 956pp. 6⅛ × 9¼.
22055-9, 22056-7 Pa., Two-vol. set $20.00

A THEOLOGICO-POLITICAL TREATISE, Benedict Spinoza. Also contains unfinished *Political Treatise*. Great classic on religious liberty, theory of government on common consent. R. Elwes translation. Total of 421pp. 5⅜ × 8½.
20249-6 Pa. $6.95

INCIDENTS OF TRAVEL IN CENTRAL AMERICA, CHIAPAS, AND YUCATAN, John L. Stephens. Almost single-handed discovery of Maya culture; exploration of ruined cities, monuments, temples; customs of Indians. 115 drawings. 892pp. 5⅜ × 8½.   22404-X, 22405-8 Pa., Two-vol. set $15.90

LOS CAPRICHOS, Francisco Goya. 80 plates of wild, grotesque monsters and caricatures. Prado manuscript included. 183pp. 6⅜ × 9⅜.   22384-1 Pa. $4.95

AUTOBIOGRAPHY: The Story of My Experiments with Truth, Mohandas K. Gandhi. Not hagiography, but Gandhi in his own words. Boyhood, legal studies, purification, the growth of the Satyagraha (nonviolent protest) movement. Critical, inspiring work of the man who freed India. 480pp. 5⅜ × 8½. (Available in U.S. only)
24593-4 Pa. $6.95

ILLUSTRATED DICTIONARY OF HISTORIC ARCHITECTURE, edited by Cyril M. Harris. Extraordinary compendium of clear, concise definitions for over 5,000 important architectural terms complemented by over 2,000 line drawings. Covers full spectrum of architecture from ancient ruins to 20th-century Modernism. Preface. 592pp. 7½ × 9⅝. 24444-X Pa. $14.95

THE NIGHT BEFORE CHRISTMAS, Clement Moore. Full text, and woodcuts from original 1848 book. Also critical, historical material. 19 illustrations. 40pp. 4⅝ × 6. 22797-9 Pa. $2.25

THE LESSON OF JAPANESE ARCHITECTURE: 165 Photographs, Jiro Harada. Memorable gallery of 165 photographs taken in the 1930's of exquisite Japanese homes of the well-to-do and historic buildings. 13 line diagrams. 192pp. 8⅜ × 11¼. 24778-3 Pa. $8.95

THE AUTOBIOGRAPHY OF CHARLES DARWIN AND SELECTED LETTERS, edited by Francis Darwin. The fascinating life of eccentric genius composed of an intimate memoir by Darwin (intended for his children); commentary by his son, Francis; hundreds of fragments from notebooks, journals, papers; and letters to and from Lyell, Hooker, Huxley, Wallace and Henslow. xi + 365pp. 5⅜ × 8. 20479-0 Pa. $5.95

WONDERS OF THE SKY: Observing Rainbows, Comets, Eclipses, the Stars and Other Phenomena, Fred Schaaf. Charming, easy-to-read poetic guide to all manner of celestial events visible to the naked eye. Mock suns, glories, Belt of Venus, more. Illustrated. 299pp. 5¼ × 8¼. 24402-4 Pa. $7.95

BURNHAM'S CELESTIAL HANDBOOK, Robert Burnham, Jr. Thorough guide to the stars beyond our solar system. Exhaustive treatment. Alphabetical by constellation: Andromeda to Cetus in Vol. 1; Chamaeleon to Orion in Vol. 2; and Pavo to Vulpecula in Vol. 3. Hundreds of illustrations. Index in Vol. 3. 2,000pp. 6¼ × 9¼. 23567-X, 23568-8, 23673-0 Pa., Three-vol. set $36.85

STAR NAMES: Their Lore and Meaning, Richard Hinckley Allen. Fascinating history of names various cultures have given to constellations and literary and folkloristic uses that have been made of stars. Indexes to subjects. Arabic and Greek names. Biblical references. Bibliography. 563pp. 5⅜ × 8½. 21079-0 Pa. $7.95

THIRTY YEARS THAT SHOOK PHYSICS: The Story of Quantum Theory, George Gamow. Lucid, accessible introduction to influential theory of energy and matter. Careful explanations of Dirac's anti-particles, Bohr's model of the atom, much more. 12 plates. Numerous drawings. 240pp. 5⅜ × 8½. 24895-X Pa. $4.95

CHINESE DOMESTIC FURNITURE IN PHOTOGRAPHS AND MEASURED DRAWINGS, Gustav Ecke. A rare volume, now affordably priced for antique collectors, furniture buffs and art historians. Detailed review of styles ranging from early Shang to late Ming. Unabridged republication. 161 black-and-white drawings, photos. Total of 224pp. 8⅜ × 11¼. (Available in U.S. only) 25171-3 Pa. $12.95

VINCENT VAN GOGH: A Biography, Julius Meier-Graefe. Dynamic, penetrating study of artist's life, relationship with brother, Theo, painting techniques, travels, more. Readable, engrossing. 160pp. 5⅜ × 8½. (Available in U.S. only) 25253-1 Pa. $3.95

CATALOG OF DOVER BOOKS

HOW TO WRITE, Gertrude Stein. Gertrude Stein claimed anyone could understand her unconventional writing—here are clues to help. Fascinating improvisations, language experiments, explanations illuminate Stein's craft and the art of writing. Total of 414pp. 4⅝ × 6⅜. 23144-5 Pa. $5.95

ADVENTURES AT SEA IN THE GREAT AGE OF SAIL: Five Firsthand Narratives, edited by Elliot Snow. Rare true accounts of exploration, whaling, shipwreck, fierce natives, trade, shipboard life, more. 33 illustrations. Introduction. 353pp. 5⅜ × 8½. 25177-2 Pa. $7.95

THE HERBAL OR GENERAL HISTORY OF PLANTS, John Gerard. Classic descriptions of about 2,850 plants—with over 2,700 illustrations—includes Latin and English names, physical descriptions, varieties, time and place of growth, more. 2,706 illustrations. xlv + 1,678pp. 8½ × 12¼. 23147-X Cloth. $75.00

DOROTHY AND THE WIZARD IN OZ, L. Frank Baum. Dorothy and the Wizard visit the center of the Earth, where people are vegetables, glass houses grow and Oz characters reappear. Classic sequel to *Wizard of Oz*. 256pp. 5⅜ × 8. 24714-7 Pa. $4.95

SONGS OF EXPERIENCE: Facsimile Reproduction with 26 Plates in Full Color, William Blake. This facsimile of Blake's original "Illuminated Book" reproduces 26 full-color plates from a rare 1826 edition. Includes "The Tyger," "London," "Holy Thursday," and other immortal poems. 26 color plates. Printed text of poems. 48pp. 5¼ × 7. 24636-1 Pa. $3.50

SONGS OF INNOCENCE, William Blake. The first and most popular of Blake's famous "Illuminated Books," in a facsimile edition reproducing all 31 brightly colored plates. Additional printed text of each poem. 64pp. 5¼ × 7. 22764-2 Pa. $3.50

PRECIOUS STONES, Max Bauer. Classic, thorough study of diamonds, rubies, emeralds, garnets, etc.: physical character, occurrence, properties, use, similar topics. 20 plates, 8 in color. 94 figures. 659pp. 6⅛ × 9¼. 21910-0, 21911-9 Pa., Two-vol. set $14.90

ENCYCLOPEDIA OF VICTORIAN NEEDLEWORK, S. F. A. Caulfeild and Blanche Saward. Full, precise descriptions of stitches, techniques for dozens of needlecrafts—most exhaustive reference of its kind. Over 800 figures. Total of 679pp. 8⅛ × 11. Two volumes. Vol. 1 22800-2 Pa. $10.95
Vol. 2 22801-0 Pa. $10.95

THE MARVELOUS LAND OF OZ, L. Frank Baum. Second Oz book, the Scarecrow and Tin Woodman are back with hero named Tip, Oz magic. 136 illustrations. 287pp. 5⅜ × 8½. 20692-0 Pa. $5.95

WILD FOWL DECOYS, Joel Barber. Basic book on the subject, by foremost authority and collector. Reveals history of decoy making and rigging, place in American culture, different kinds of decoys, how to make them, and how to use them. 140 plates. 156pp. 7⅞ × 10¾. 20011-6 Pa. $7.95

HISTORY OF LACE, Mrs. Bury Palliser. Definitive, profusely illustrated chronicle of lace from earliest times to late 19th century. Laces of Italy, Greece, England, France, Belgium, etc. Landmark of needlework scholarship. 266 illustrations. 672pp. 6⅛ × 9¼. 24742-2 Pa. $14.95

CATALOG OF DOVER BOOKS

ILLUSTRATED GUIDE TO SHAKER FURNITURE, Robert Meader. All furniture and appurtenances, with much on unknown local styles. 235 photos. 146pp. 9 × 12. 22819-3 Pa. $7.95

WHALE SHIPS AND WHALING: A Pictorial Survey, George Francis Dow. Over 200 vintage engravings, drawings, photographs of barks, brigs, cutters, other vessels. Also harpoons, lances, whaling guns, many other artifacts. Comprehensive text by foremost authority. 207 black-and-white illustrations. 288pp. 6 × 9.
24808-9 Pa. $8.95

THE BERTRAMS, Anthony Trollope. Powerful portrayal of blind self-will and thwarted ambition includes one of Trollope's most heartrending love stories. 497pp. 5⅜ × 8½. 25119-5 Pa. $8.95

ADVENTURES WITH A HAND LENS, Richard Headstrom. Clearly written guide to observing and studying flowers and grasses, fish scales, moth and insect wings, egg cases, buds, feathers, seeds, leaf scars, moss, molds, ferns, common crystals, etc.—all with an ordinary, inexpensive magnifying glass. 209 exact line drawings aid in your discoveries. 220pp. 5⅜ × 8½. 23330-8 Pa. $3.95

RODIN ON ART AND ARTISTS, Auguste Rodin. Great sculptor's candid, wide-ranging comments on meaning of art; great artists; relation of sculpture to poetry, painting, music; philosophy of life, more. 76 superb black-and-white illustrations of Rodin's sculpture, drawings and prints. 119pp. 8⅜ × 11¼. 24487-3 Pa. $6.95

FIFTY CLASSIC FRENCH FILMS, 1912-1982: A Pictorial Record, Anthony Slide. Memorable stills from Grand Illusion, Beauty and the Beast, Hiroshima, Mon Amour, many more. Credits, plot synopses, reviews, etc. 160pp. 8¼ × 11.
25256-6 Pa. $11.95

THE PRINCIPLES OF PSYCHOLOGY, William James. Famous long course complete, unabridged. Stream of thought, time perception, memory, experimental methods; great work decades ahead of its time. 94 figures. 1,391pp. 5⅜ × 8½.
20381-6, 20382-4 Pa., Two-vol. set $19.90

BODIES IN A BOOKSHOP, R. T. Campbell. Challenging mystery of blackmail and murder with ingenious plot and superbly drawn characters. In the best tradition of British suspense fiction. 192pp. 5⅜ × 8½. 24720-1 Pa. $3.95

CALLAS: PORTRAIT OF A PRIMA DONNA, George Jellinek. Renowned commentator on the musical scene chronicles incredible career and life of the most controversial, fascinating, influential operatic personality of our time. 64 black-and-white photographs. 416pp. 5⅜ × 8¼. 25047-4 Pa. $7.95

GEOMETRY, RELATIVITY AND THE FOURTH DIMENSION, Rudolph Rucker. Exposition of fourth dimension, concepts of relativity as Flatland characters continue adventures. Popular, easily followed yet accurate, profound. 141 illustrations. 133pp. 5⅜ × 8½. 23400-2 Pa. $3.50

HOUSEHOLD STORIES BY THE BROTHERS GRIMM, with pictures by Walter Crane. 53 classic stories—Rumpelstiltskin, Rapunzel, Hansel and Gretel, the Fisherman and his Wife, Snow White, Tom Thumb, Sleeping Beauty, Cinderella, and so much more—lavishly illustrated with original 19th century drawings. 114 illustrations. x + 269pp. 5⅜ × 8½. 21080-4 Pa. $4.50

SUNDIALS, Albert Waugh. Far and away the best, most thorough coverage of ideas, mathematics concerned, types, construction, adjusting anywhere. Over 100 illustrations. 230pp. 5⅜ × 8½. 22947-5 Pa. $4.00

PICTURE HISTORY OF THE NORMANDIE: With 190 Illustrations, Frank O. Braynard. Full story of legendary French ocean liner: Art Deco interiors, design innovations, furnishings, celebrities, maiden voyage, tragic fire, much more. Extensive text. 144pp. 8⅜ × 11¾. 25257-4 Pa. $9.95

THE FIRST AMERICAN COOKBOOK: A Facsimile of "American Cookery," 1796, Amelia Simmons. Facsimile of the first American-written cookbook published in the United States contains authentic recipes for colonial favorites—pumpkin pudding, winter squash pudding, spruce beer, Indian slapjacks, and more. Introductory Essay and Glossary of colonial cooking terms. 80pp. 5⅜ × 8½. 24710-4 Pa. $3.50

101 PUZZLES IN THOUGHT AND LOGIC, C. R. Wylie, Jr. Solve murders and robberies, find out which fishermen are liars, how a blind man could possibly identify a color—purely by your own reasoning! 107pp. 5⅜ × 8½. 20367-0 Pa. $2.00

THE BOOK OF WORLD-FAMOUS MUSIC—CLASSICAL, POPULAR AND FOLK, James J. Fuld. Revised and enlarged republication of landmark work in musico-bibliography. Full information about nearly 1,000 songs and compositions including first lines of music and lyrics. New supplement. Index. 800pp. 5⅜ × 8¼. 24857-7 Pa. $14.95

ANTHROPOLOGY AND MODERN LIFE, Franz Boas. Great anthropologist's classic treatise on race and culture. Introduction by Ruth Bunzel. Only inexpensive paperback edition. 255pp. 5⅜ × 8½. 25245-0 Pa. $5.95

THE TALE OF PETER RABBIT, Beatrix Potter. The inimitable Peter's terrifying adventure in Mr. McGregor's garden, with all 27 wonderful, full-color Potter illustrations. 55pp. 4¼ × 5½. (Available in U.S. only) 22827-4 Pa. $1.75

THREE PROPHETIC SCIENCE FICTION NOVELS, H. G. Wells. *When the Sleeper Wakes, A Story of the Days to Come* and *The Time Machine* (full version). 335pp. 5⅜ × 8½. (Available in U.S. only) 20605-X Pa. $5.95

APICIUS COOKERY AND DINING IN IMPERIAL ROME, edited and translated by Joseph Dommers Vehling. Oldest known cookbook in existence offers readers a clear picture of what foods Romans ate, how they prepared them, etc. 49 illustrations. 301pp. 6⅛ × 9¼. 23563-7 Pa. $6.00

SHAKESPEARE LEXICON AND QUOTATION DICTIONARY, Alexander Schmidt. Full definitions, locations, shades of meaning of every word in plays and poems. More than 50,000 exact quotations. 1,485pp. 6½ × 9¼. 22726-X, 22727-8 Pa., Two-vol. set $27.90

THE WORLD'S GREAT SPEECHES, edited by Lewis Copeland and Lawrence W. Lamm. Vast collection of 278 speeches from Greeks to 1970. Powerful and effective models; unique look at history. 842pp. 5⅜ × 8½. 20468-5 Pa. $10.95

CATALOG OF DOVER BOOKS

THE BLUE FAIRY BOOK, Andrew Lang. The first, most famous collection, with many familiar tales: Little Red Riding Hood, Aladdin and the Wonderful Lamp, Puss in Boots, Sleeping Beauty, Hansel and Gretel, Rumpelstiltskin; 37 in all. 138 illustrations. 390pp. 5⅜ × 8½.   21437-0 Pa. $5.95

THE STORY OF THE CHAMPIONS OF THE ROUND TABLE, Howard Pyle. Sir Launcelot, Sir Tristram and Sir Percival in spirited adventures of love and triumph retold in Pyle's inimitable style. 50 drawings, 31 full-page. xviii + 329pp. 6½ × 9¼.   21883-X Pa. $6.95

AUDUBON AND HIS JOURNALS, Maria Audubon. Unmatched two-volume portrait of the great artist, naturalist and author contains his journals, an excellent biography by his granddaughter, expert annotations by the noted ornithologist, Dr. Elliott Coues, and 37 superb illustrations. Total of 1,200pp. 5⅜ × 8.
Vol. I 25143-8 Pa. $8.95
Vol. II 25144-6 Pa. $8.95

GREAT DINOSAUR HUNTERS AND THEIR DISCOVERIES, Edwin H. Colbert. Fascinating, lavishly illustrated chronicle of dinosaur research, 1820's to 1960. Achievements of Cope, Marsh, Brown, Buckland, Mantell, Huxley, many others. 384pp. 5¼ × 8¼.   24701-5 Pa. $6.95

THE TASTEMAKERS, Russell Lynes. Informal, illustrated social history of American taste 1850's–1950's. First popularized categories Highbrow, Lowbrow, Middlebrow. 129 illustrations. New (1979) afterword. 384pp. 6 × 9.
23993-4 Pa. $6.95

DOUBLE CROSS PURPOSES, Ronald A. Knox. A treasure hunt in the Scottish Highlands, an old map, unidentified corpse, surprise discoveries keep reader guessing in this cleverly intricate tale of financial skullduggery. 2 black-and-white maps. 320pp. 5⅜ × 8½. (Available in U.S. only)   25032-6 Pa. $5.95

AUTHENTIC VICTORIAN DECORATION AND ORNAMENTATION IN FULL COLOR: 46 Plates from "Studies in Design," Christopher Dresser. Superb full-color lithographs reproduced from rare original portfolio of a major Victorian designer. 48pp. 9¼ × 12¼.   25083-0 Pa. $7.95

PRIMITIVE ART, Franz Boas. Remains the best text ever prepared on subject, thoroughly discussing Indian, African, Asian, Australian, and, especially, Northern American primitive art. Over 950 illustrations show ceramics, masks, totem poles, weapons, textiles, paintings, much more. 376pp. 5⅜ × 8. 20025-6 Pa. $6.95

SIDELIGHTS ON RELATIVITY, Albert Einstein. Unabridged republication of two lectures delivered by the great physicist in 1920–21. *Ether and Relativity* and *Geometry and Experience*. Elegant ideas in non-mathematical form, accessible to intelligent layman. vi + 56pp. 5⅜ × 8½.   24511-X Pa. $2.95

THE WIT AND HUMOR OF OSCAR WILDE, edited by Alvin Redman. More than 1,000 ripostes, paradoxes, wisecracks: Work is the curse of the drinking classes, I can resist everything except temptation, etc. 258pp. 5⅜ × 8½.   20602-5 Pa. $3.95

ADVENTURES WITH A MICROSCOPE, Richard Headstrom. 59 adventures with clothing fibers, protozoa, ferns and lichens, roots and leaves, much more. 142 illustrations. 232pp. 5⅜ × 8½.   23471-1 Pa. $3.95

PLANTS OF THE BIBLE, Harold N. Moldenke and Alma L. Moldenke. Standard reference to all 230 plants mentioned in Scriptures. Latin name, biblical reference, uses, modern identity, much more. Unsurpassed encyclopedic resource for scholars, botanists, nature lovers, students of Bible. Bibliography. Indexes. 123 black-and-white illustrations. 384pp. 6 × 9.
25069-5 Pa. $8.95

FAMOUS AMERICAN WOMEN: A Biographical Dictionary from Colonial Times to the Present, Robert McHenry, ed. From Pocahontas to Rosa Parks, 1,035 distinguished American women documented in separate biographical entries. Accurate, up-to-date data, numerous categories, spans 400 years. Indices. 493pp. 6½ × 9¼.
24523-3 Pa. $9.95

THE FABULOUS INTERIORS OF THE GREAT OCEAN LINERS IN HISTORIC PHOTOGRAPHS, William H. Miller, Jr. Some 200 superb photographs capture exquisite interiors of world's great "floating palaces"—1890's to 1980's: Titanic, Ile de France, Queen Elizabeth, United States, Europa, more. Approx. 200 black-and-white photographs. Captions. Text. Introduction. 160pp. 8⅜ × 11¾.
24756-2 Pa. $9.95

THE GREAT LUXURY LINERS, 1927–1954: A Photographic Record, William H. Miller, Jr. Nostalgic tribute to heyday of ocean liners. 186 photos of Ile de France, Normandie, Leviathan, Queen Elizabeth, United States, many others. Interior and exterior views. Introduction. Captions. 160pp. 9 × 12.
24056-8 Pa. $9.95

A NATURAL HISTORY OF THE DUCKS, John Charles Phillips. Great landmark of ornithology offers complete detailed coverage of nearly 200 species and subspecies of ducks: gadwall, sheldrake, merganser, pintail, many more. 74 full-color plates, 102 black-and-white. Bibliography. Total of 1,920pp. 8⅜ × 11¼.
25141-1, 25142-X Cloth. Two-vol. set $100.00

THE SEAWEED HANDBOOK: An Illustrated Guide to Seaweeds from North Carolina to Canada, Thomas F. Lee. Concise reference covers 78 species. Scientific and common names, habitat, distribution, more. Finding keys for easy identification. 224pp. 5⅜ × 8½.
25215-9 Pa. $5.95

THE TEN BOOKS OF ARCHITECTURE: The 1755 Leoni Edition, Leon Battista Alberti. Rare classic helped introduce the glories of ancient architecture to the Renaissance. 68 black-and-white plates. 336pp. 8⅜ × 11¼.
25239-6 Pa. $14.95

MISS MACKENZIE, Anthony Trollope. Minor masterpieces by Victorian master unmasks many truths about life in 19th-century England. First inexpensive edition in years. 392pp. 5⅜ × 8½.
25201-9 Pa. $7.95

THE RIME OF THE ANCIENT MARINER, Gustave Doré, Samuel Taylor Coleridge. Dramatic engravings considered by many to be his greatest work. The terrifying space of the open sea, the storms and whirlpools of an unknown ocean, the ice of Antarctica, more—all rendered in a powerful, chilling manner. Full text. 38 plates. 77pp. 9¼ × 12.
22305-1 Pa. $4.95

THE EXPEDITIONS OF ZEBULON MONTGOMERY PIKE, Zebulon Montgomery Pike. Fascinating first-hand accounts (1805–6) of exploration of Mississippi River, Indian wars, capture by Spanish dragoons, much more. 1,088pp. 5⅜ × 8½.
25254-X, 25255-8 Pa. Two-vol. set $23.90

A CONCISE HISTORY OF PHOTOGRAPHY: Third Revised Edition, Helmut Gernsheim. Best one-volume history—camera obscura, photochemistry, daguerreotypes, evolution of cameras, film, more. Also artistic aspects—landscape, portraits, fine art, etc. 281 black-and-white photographs. 26 in color. 176pp. 8⅜ × 11¼. 25128-4 Pa. $12.95

THE DORÉ BIBLE ILLUSTRATIONS, Gustave Doré. 241 detailed plates from the Bible: the Creation scenes, Adam and Eve, Flood, Babylon, battle sequences, life of Jesus, etc. Each plate is accompanied by the verses from the King James version of the Bible. 241pp. 9 × 12. 23004-X Pa. $8.95

HUGGER-MUGGER IN THE LOUVRE, Elliot Paul. Second Homer Evans mystery-comedy. Theft at the Louvre involves sleuth in hilarious, madcap caper. "A knockout."—Books. 336pp. 5⅜ × 8½. 25185-3 Pa. $5.95

FLATLAND, E. A. Abbott. Intriguing and enormously popular science-fiction classic explores the complexities of trying to survive as a two-dimensional being in a three-dimensional world. Amusingly illustrated by the author. 16 illustrations. 103pp. 5⅜ × 8½. 20001-9 Pa. $2.00

THE HISTORY OF THE LEWIS AND CLARK EXPEDITION, Meriwether Lewis and William Clark, edited by Elliott Coues. Classic edition of Lewis and Clark's day-by-day journals that later became the basis for U.S. claims to Oregon and the West. Accurate and invaluable geographical, botanical, biological, meteorological and anthropological material. Total of 1,508pp. 5⅜ × 8½.
21268-8, 21269-6, 21270-X Pa. Three-vol. set $25.50

LANGUAGE, TRUTH AND LOGIC, Alfred J. Ayer. Famous, clear introduction to Vienna, Cambridge schools of Logical Positivism. Role of philosophy, elimination of metaphysics, nature of analysis, etc. 160pp. 5⅜ × 8½. (Available in U.S. and Canada only) 20010-8 Pa. $2.95

MATHEMATICS FOR THE NONMATHEMATICIAN, Morris Kline. Detailed, college-level treatment of mathematics in cultural and historical context, with numerous exercises. For liberal arts students. Preface. Recommended Reading Lists. Tables. Index. Numerous black-and-white figures. xvi + 641pp. 5⅜ × 8½.
24823-2 Pa. $11.95

28 SCIENCE FICTION STORIES, H. G. Wells. Novels, *Star Begotten* and *Men Like Gods,* plus 26 short stories: "Empire of the Ants," "A Story of the Stone Age," "The Stolen Bacillus," "In the Abyss," etc. 915pp. 5⅜ × 8½. (Available in U.S. only)
20265-8 Cloth. $10.95

HANDBOOK OF PICTORIAL SYMBOLS, Rudolph Modley. 3,250 signs and symbols, many systems in full; official or heavy commercial use. Arranged by subject. Most in Pictorial Archive series. 143pp. 8⅛ × 11. 23357-X Pa. $5.95

INCIDENTS OF TRAVEL IN YUCATAN, John L. Stephens. Classic (1843) exploration of jungles of Yucatan, looking for evidences of Maya civilization. Travel adventures, Mexican and Indian culture, etc. Total of 669pp. 5⅜ × 8½.
20926-1, 20927-X Pa., Two-vol. set $9.90

DEGAS: An Intimate Portrait, Ambroise Vollard. Charming, anecdotal memoir by famous art dealer of one of the greatest 19th-century French painters. 14 black-and-white illustrations. Introduction by Harold L. Van Doren. 96pp. 5⅜ × 8½.
25131-4 Pa. $3.95

PERSONAL NARRATIVE OF A PILGRIMAGE TO ALMANDINAH AND MECCAH, Richard Burton. Great travel classic by remarkably colorful personality. Burton, disguised as a Moroccan, visited sacred shrines of Islam, narrowly escaping death. 47 illustrations. 959pp. 5⅜ × 8½.   21217-3, 21218-1 Pa., Two-vol. set $17.90

PHRASE AND WORD ORIGINS, A. H. Holt. Entertaining, reliable, modern study of more than 1,200 colorful words, phrases, origins and histories. Much unexpected information. 254pp. 5⅜ × 8½.   20758-7 Pa. $4.95

THE RED THUMB MARK, R. Austin Freeman. In this first Dr. Thorndyke case, the great scientific detective draws fascinating conclusions from the nature of a single fingerprint. Exciting story, authentic science. 320pp. 5⅜ × 8½. (Available in U.S. only)   25210-8 Pa. $5.95

AN EGYPTIAN HIEROGLYPHIC DICTIONARY, E. A. Wallis Budge. Monumental work containing about 25,000 words or terms that occur in texts ranging from 3000 B.C. to 600 A.D. Each entry consists of a transliteration of the word, the word in hieroglyphs, and the meaning in English. 1,314pp. 6⅜ × 10.
23615-3, 23616-1 Pa., Two-vol. set $27.90

THE COMPLEAT STRATEGYST: Being a Primer on the Theory of Games of Strategy, J. D. Williams. Highly entertaining classic describes, with many illustrated examples, how to select best strategies in conflict situations. Prefaces. Appendices. xvi + 268pp. 5⅜ × 8½.   25101-2 Pa. $5.95

THE ROAD TO OZ, L. Frank Baum. Dorothy meets the Shaggy Man, little Button-Bright and the Rainbow's beautiful daughter in this delightful trip to the magical Land of Oz. 272pp. 5⅜ × 8.   25208-6 Pa. $4.95

POINT AND LINE TO PLANE, Wassily Kandinsky. Seminal exposition of role of point, line, other elements in non-objective painting. Essential to understanding 20th-century art. 127 illustrations. 192pp. 6½ × 9¼.   23808-3 Pa. $4.50

LADY ANNA, Anthony Trollope. Moving chronicle of Countess Lovel's bitter struggle to win for herself and daughter Anna their rightful rank and fortune— perhaps at cost of sanity itself. 384pp. 5⅜ × 8½.   24669-8 Pa. $6.95

EGYPTIAN MAGIC, E. A. Wallis Budge. Sums up all that is known about magic in Ancient Egypt: the role of magic in controlling the gods, powerful amulets that warded off evil spirits, scarabs of immortality, use of wax images, formulas and spells, the secret name, much more. 253pp. 5⅜ × 8½.   22681-6 Pa. $4.00

THE DANCE OF SIVA, Ananda Coomaraswamy. Preeminent authority unfolds the vast metaphysic of India: the revelation of her art, conception of the universe, social organization, etc. 27 reproductions of art masterpieces. 192pp. 5⅜ × 8½.
24817-8 Pa. $5.95

CHRISTMAS CUSTOMS AND TRADITIONS, Clement A. Miles. Origin, evolution, significance of religious, secular practices. Caroling, gifts, yule logs, much more. Full, scholarly yet fascinating; non-sectarian. 400pp. 5⅜ × 8½.
23354-5 Pa. $6.50

THE HUMAN FIGURE IN MOTION, Eadweard Muybridge. More than 4,500 stopped-action photos, in action series, showing undraped men, women, children jumping, lying down, throwing, sitting, wrestling, carrying, etc. 390pp. 7⅞ × 10⅝.
20204-6 Cloth. $19.95

THE MAN WHO WAS THURSDAY, Gilbert Keith Chesterton. Witty, fast-paced novel about a club of anarchists in turn-of-the-century London. Brilliant social, religious, philosophical speculations. 128pp. 5⅜ × 8½.
25121-7 Pa. $3.95

A CEZANNE SKETCHBOOK: Figures, Portraits, Landscapes and Still Lifes, Paul Cezanne. Great artist experiments with tonal effects, light, mass, other qualities in over 100 drawings. A revealing view of developing master painter, precursor of Cubism. 102 black-and-white illustrations. 144pp. 8¾ × 6⅝.
24790-2 Pa. $5.95

AN ENCYCLOPEDIA OF BATTLES: Accounts of Over 1,560 Battles from 1479 B.C. to the Present, David Eggenberger. Presents essential details of every major battle in recorded history, from the first battle of Megiddo in 1479 B.C. to Grenada in 1984. List of Battle Maps. New Appendix covering the years 1967–1984. Index. 99 illustrations. 544pp. 6½ × 9¼.
24913-1 Pa. $14.95

AN ETYMOLOGICAL DICTIONARY OF MODERN ENGLISH, Ernest Weekley. Richest, fullest work, by foremost British lexicographer. Detailed word histories. Inexhaustible. Total of 856pp. 6½ × 9¼.
21873-2, 21874-0 Pa., Two-vol. set $17.00

WEBSTER'S AMERICAN MILITARY BIOGRAPHIES, edited by Robert McHenry. Over 1,000 figures who shaped 3 centuries of American military history. Detailed biographies of Nathan Hale, Douglas MacArthur, Mary Hallaren, others. Chronologies of engagements, more. Introduction. Addenda. 1,033 entries in alphabetical order. xi + 548pp. 6½ × 9¼. (Available in U.S. only)
24758-9 Pa. $11.95

LIFE IN ANCIENT EGYPT, Adolf Erman. Detailed older account, with much not in more recent books: domestic life, religion, magic, medicine, commerce, and whatever else needed for complete picture. Many illustrations. 597pp. 5⅜ × 8½.
22632-8 Pa. $8.50

HISTORIC COSTUME IN PICTURES, Braun & Schneider. Over 1,450 costumed figures shown, covering a wide variety of peoples: kings, emperors, nobles, priests, servants, soldiers, scholars, townsfolk, peasants, merchants, courtiers, cavaliers, and more. 256pp. 8⅜ × 11¼.
23150-X Pa. $7.95

THE NOTEBOOKS OF LEONARDO DA VINCI, edited by J. P. Richter. Extracts from manuscripts reveal great genius; on painting, sculpture, anatomy, sciences, geography, etc. Both Italian and English. 186 ms. pages reproduced, plus 500 additional drawings, including studies for Last Supper, Sforza monument, etc. 860pp. 7⅞ × 10⅝. (Available in U.S. only) 22572-0, 22573-9 Pa., Two-vol. set $25.90

CATALOG OF DOVER BOOKS

AMERICAN CLIPPER SHIPS: 1833–1858, Octavius T. Howe & Frederick C. Matthews. Fully-illustrated, encyclopedic review of 352 clipper ships from the period of America's greatest maritime supremacy. Introduction. 109 halftones. 5 black-and-white line illustrations. Index. Total of 928pp. 5⅜ × 8½.
25115-2, 25116-0 Pa., Two-vol. set $17.90

TOWARDS A NEW ARCHITECTURE, Le Corbusier. Pioneering manifesto by great architect, near legendary founder of "International School." Technical and aesthetic theories, views on industry, economics, relation of form to function, "mass-production spirit," much more. Profusely illustrated. Unabridged translation of 13th French edition. Introduction by Frederick Etchells. 320pp. 6⅛ × 9¼. (Available in U.S. only)
25023-7 Pa. $8.95

THE BOOK OF KELLS, edited by Blanche Cirker. Inexpensive collection of 32 full-color, full-page plates from the greatest illuminated manuscript of the Middle Ages, painstakingly reproduced from rare facsimile edition. Publisher's Note. Captions. 32pp. 9⅜ × 12¼.
24345-1 Pa. $4.50

BEST SCIENCE FICTION STORIES OF H. G. WELLS, H. G. Wells. Full novel *The Invisible Man,* plus 17 short stories: "The Crystal Egg," "Aepyornis Island," "The Strange Orchid," etc. 303pp. 5⅜ × 8½. (Available in U.S. only)
21531-8 Pa. $4.95

AMERICAN SAILING SHIPS: Their Plans and History, Charles G. Davis. Photos, construction details of schooners, frigates, clippers, other sailcraft of 18th to early 20th centuries—plus entertaining discourse on design, rigging, nautical lore, much more. 137 black-and-white illustrations. 240pp. 6⅛ × 9¼.
24658-2 Pa. $5.95

ENTERTAINING MATHEMATICAL PUZZLES, Martin Gardner. Selection of author's favorite conundrums involving arithmetic, money, speed, etc., with lively commentary. Complete solutions. 112pp. 5⅜ × 8½. 25211-6 Pa. $2.95

THE WILL TO BELIEVE, HUMAN IMMORTALITY, William James. Two books bound together. Effect of irrational on logical, and arguments for human immortality. 402pp. 5⅜ × 8½. 20291-7 Pa. $7.50

THE HAUNTED MONASTERY and THE CHINESE MAZE MURDERS, Robert Van Gulik. 2 full novels by Van Gulik continue adventures of Judge Dee and his companions. An evil Taoist monastery, seemingly supernatural events; overgrown topiary maze that hides strange crimes. Set in 7th-century China. 27 illustrations. 328pp. 5⅜ × 8½. 23502-5 Pa. $5.00

CELEBRATED CASES OF JUDGE DEE (DEE GOONG AN), translated by Robert Van Gulik. Authentic 18th-century Chinese detective novel; Dee and associates solve three interlocked cases. Led to Van Gulik's own stories with same characters. Extensive introduction. 9 illustrations. 237pp. 5⅜ × 8½.
23337-5 Pa. $4.95

Prices subject to change without notice.
Available at your book dealer or write for free catalog to Dept. GI, Dover Publications, Inc., 31 East 2nd St., Mineola, N.Y. 11501. Dover publishes more than 175 books each year on science, elementary and advanced mathematics, biology, music, art, literary history, social sciences and other areas.